Unlimited Practice!

MyStatLab offers a wide variety of problems that let you practise to improve your understanding of the course material.

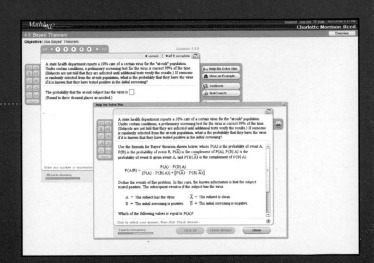

Help Me Solve This

Stuck on an exercise and don't know where to begin? Click the Help Me Solve This button to see a walkthrough that shows you how to set and solve the exercise.

Videos

Many of the exercises include a short video that provides an explanation of how to solve the type of exercise you are working on.

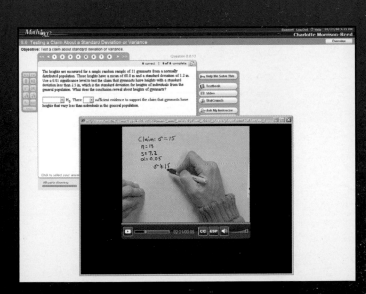

Similar Exercise

Once you have solved an exercise, *MyStatLab* allows you to practice similar exercises to reinforce concepts and prepare for mid-terms/finals.

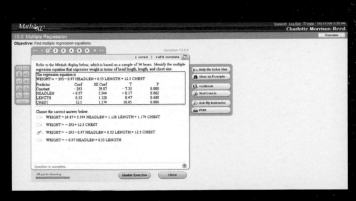

Personalized Learning!

The *MyStatLab* Study Plan is based on your specific learning needs.

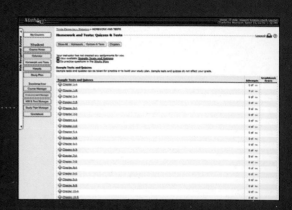

Auto-Graded Tests and Assignments

MyStatLab comes with two pre-loaded Sample Tests for each chapter so you can self-assess your understanding of the material.

Personalized Study Plan

A Study Plan is generated based on your results on Sample Tests and instructor assignments. You can clearly see which topics you have mastered and, more importantly, which topics you need to work on!

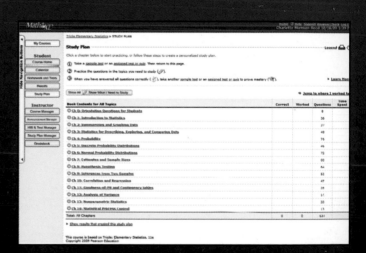

"I'm a visual learner so having the video lectures was helpful, and the homework allowed me to see which areas I was struggling in. The helpful hints for solving the homework questions were very helpful."

Practice Problems

Use the Study Plan exercises to get practice where you need it. To check how you're doing, click Results to get an overview of all your scores.

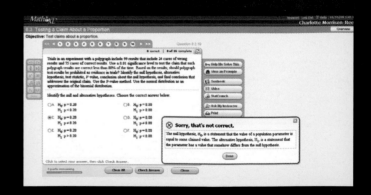

Save Time. Improve Results. www.mystatlab.com

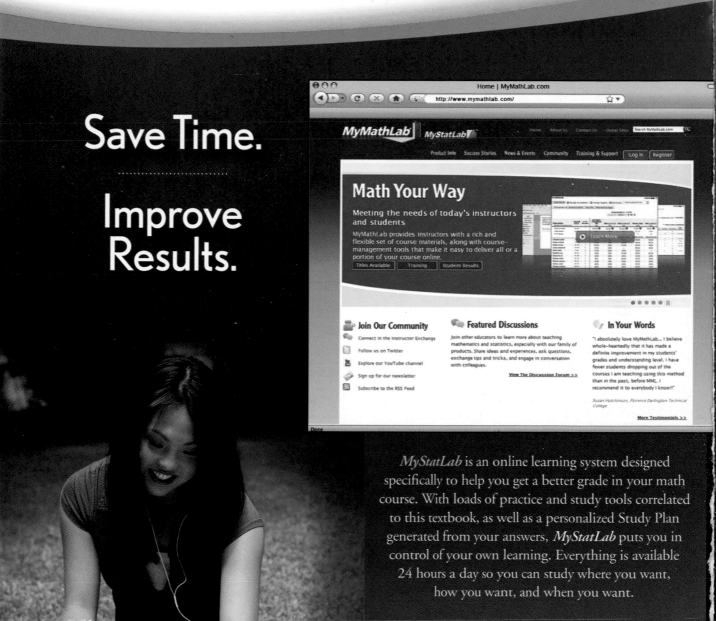

ELEMENTARY
statistics

THIRD CANADIAN EDITION

MARIO F. TRIOLA
Dutchess Community College

WILLIAM M. GOODMAN
Durham College

RICHARD LAW
Humber College

GERRY LABUTE
Mount Royal University
University of Calgary

Pearson Canada
Toronto

Library and Archives Canada Cataloguing in Publication

Elementary statistics / Mario F. Triola ... [et al.]. — 3rd Canadian ed.

Includes index.
Second Canadian ed. written by: Mario F. Triola, William M. Goodman, Richard Law.
ISBN 978-0-321-22597-9

1. Statistics—Textbooks. I. Triola, Mario F. II. Triola, Mario F. Elementary statistics.

QA276.12.E44 2010   519.5   C2009-902121-8

ISBN 978-0-321-22597-9

Vice-President, Editorial Director: Gary Bennett
Editor-in-Chief: Nicole Lukach
Acquisitions Editor: Cathleen Sullivan
Marketing Manager: Colleen Gauthier
Senior Developmental Editor: Paul Donnelly
Production Editor: Imee Salumbides
Copy Editor: Tom Gamblin
Proofreader: Lu Cormier
Production Coordinator: Lynn O'Rourke
Compositor: International Typesetting & Composition (ITC)
Photo and Permissions Researcher: Amanda Campbell
Art Director: Julia Hall
Cover and Interior Designer: Opus House Inc. / Sonya V. Thursby
Cover Image: Veer

1 2 3 4 5   13 12 11 10 09

Printed and bound in the United States of America.

CONTENTS

Preface ... vii

1 Introduction to Statistics ... 2

1-1   Overview ... 4

1-2   The Nature of Data ... 5

1-3   Uses and Abuses of Statistics ... 9

1-4   Design of Experiments ... 15

2 Describing, Exploring, and Comparing Data ... 30

2-1   Overview ... 32

2-2   Summarizing Data with Frequency Tables ... 33

2-3   Graphs of Data ... 41

2-4   Measures of Central Tendency ... 55

2-5   Measures of Variation ... 71

2-6   Measures of Position ... 88

2-7   Exploratory Data Analysis (EDA) ... 97

3 Probability ... 114

3-1   Overview ... 116

3-2   Fundamentals ... 116

3-3   Addition Rule ... 128

3-4   Conditional Probabilities ... 135

3-5   Multiplication Rule and Bayes' Theorem ... 145

3-6   Counting ... 151

4 Discrete Probability Distributions ... 172

4-1   Overview ... 174

4-2   Random Variables ... 175

4-3   Binomial Probability Distributions ... 186

4-4   Mean, Variance, and Standard Deviation
      for the Binomial Distribution ... 201

4-5   The Poisson Distribution ... 207

4-6   The Hypergeometric Distribution ... 214

5 Continuous Probability Distributions ... 224

5-1 Overview ... 226

5-2 The Standard Normal Distribution ... 229

5-3 Normal Distributions: Finding Probabilities ... 243

5-4 Normal Distributions: Finding Values ... 252

5-5 The Central Limit Theorem ... 258

5-6 Normal Distribution as Approximation
to Binomial Distribution ... 270

5-7 Exponential Distribution ... 280

6 Estimates and Sample Sizes ... 292

6-1 Overview ... 294

6-2 Estimating a Population Mean: Large Samples ... 294

6-3 Estimating a Population Mean: Small Samples ... 315

6-4 Estimating a Population Proportion ... 326

6-5 Estimating a Population Variance ... 338

7 Hypothesis Testing ... 358

7-1 Overview ... 360

7-2 Fundamentals of Hypothesis Testing ... 361

7-3 Testing a Claim About a Mean: Large Samples ... 374

7-4 Testing a Claim About a Mean: Small Samples ... 398

7-5 Testing a Claim About a Proportion ... 410

7-6 Testing a Claim About a Standard
Deviation or Variance ... 420

8 Inferences from Two Samples ... 438

8-1 Overview ... 440

8-2 Inferences About Two Means: Dependent
Samples (Matched Pairs) ... 441

8-3 Inferences About Two Means: Independent
and Large Samples ... 452

8-4 Comparing Two Variances ... 464

8-5 Inferences About Two Means: Independent
and Small Samples ... 473

8-6 Inferences About Two Proportions ... 488

9 Analysis of Variance ... 506

    9-1   Overview ... 508

    9-2   One-Way ANOVA ... 510

    9-3   Two-Way ANOVA ... 524

10 Goodness-of-Fit and Contingency Tables ... 546

    10-1   Overview ... 548

    10-2   Goodness-of-Fit ... 549

    10-3   Contingency Tables: Independence
and Homogeneity ... 566

    10-4   Tests of Normality ... 582

11 Correlation and Regression ... 598

    11-1   Overview ... 600

    11-2   Correlation ... 600

    11-3   Regression ... 619

    11-4   Variation and Prediction Intervals ... 637

    11-5   Multiple Regression ... 652

12 Nonparametric Statistics ... 672

    12-1   Overview ... 674

    12-2   Sign Test ... 677

    12-3   Wilcoxon Signed-Ranks Test ... 688

    12-4   Wilcoxon Rank-Sum Test for Two Independent Samples ... 698

    12-5   Tests for Multiple Samples ... 709

    12-6   Rank Correlation ... 722

    12-7   Runs Test for Randomness ... 733

13E Statistical Process Control (on Pearson eText and CD only) ... 1

    13-1   Overview ... 3

    13-2   Control Charts for Variation and Mean ... 3

    13-3   Control Charts for Attributes ... 19

14E Time Series (on Pearson eText and CD only) ... 34

    14-1   Overview ... 36

    14-2   Index Numbers ... 36

    14-3   Time-Series Components ... 46

14-4 Seasonal Components and Seasonal Adjustments ... 53

14-5 The Trend Component and Forecasting with Trend and Seasonal Factors ... 60

14-6 Smoothing Data with Moving Averages ... 67

14-7 Other Forecasting Terms ... 70

15E Project and Epilogue (on Pearson eText and CD only) ... 79

15-1 A Statistics Group Project ... 79

15-2 Which Procedure Applies? ... 82

15-3 A Perspective ... 83

Glossary (on Pearson eText and CD only) ... 85

Appendix A: Tables ... 756

Appendix B: Data Sets ... 779

Appendix C: Bibliography ... 809

Appendix D: Answers to Odd-Numbered Exercises (and ALL Review Exercises and Cumulative Review Exercises) ... 810

Index ... 852

About This Book

Statistics is used everywhere. From opinion polls to clinical trials in medicine, from professional sports to the development of wind power projects, statistics influences, enriches, and shapes the world around us. *Elementary Statistics*, Third Canadian Edition, illustrates the relationship between statistics and our world with a variety of real applications, bringing life to theory in a clear and engaging way.

Here are some of the questions we pose:

- Are commercial vehicles more likely to be involved in accidents than noncommercial vehicles?
- Do hockey players weigh more than they used to?
- Is the mean body temperature really 37°C (98.6°F), as is commonly believed?
- Does "El Nino" really affect the weather (as reported by the media)?
- Are people spending more time on the Internet?
- How can we forecast next year's Canadian retail sales?

Audience and Prerequisites

Elementary Statistics, Third Canadian Edition, is written for students majoring in any field except mathematics. A strong mathematics background is not necessary, but students should have completed a high school or college elementary algebra course. Although underlying theory is included, this book does not stress the mathematical rigour more suitable for mathematics majors. Because the many examples and exercises cover a wide variety of different and interesting statistical applications, *Elementary Statistics* is appropriate for students pursuing majors in a wide range of disciplines, ranging from business and economics to the social sciences, physical and biological sciences, and humanities.

Like its predecessor, this edition has been prepared for the Canadian market and makes extensive use of Canadian data. Where possible, the text presents data using metric measurement; however, because some industries use imperial measurements, this is also conveyed in selected data.

Professors are given great flexibility to choose how much detail to include in the course: Subsections on the "rationale" for statistical procedures, and on specifics of related computer software, can easily be included or omitted in the course content.

Content and Organization Changes in the Third Canadian Edition

- To streamline the text, the chapters on Statistical Process Control (13E), Time Series (14E), and Project and Epilogue (15E) now appear exclusively in the Pearson eText and on the accompanying CD-ROM. These chapters are designated with an "E" to indicate that they are electronic only.

- We have added instructions for Minitab 15 and Excel 2007 in the Using Technology sections.

- In Chapter 3, we have expanded the treatment of Bayes' Theorem and added material on the construction of 2×2 contingency tables.

- Chapter 4 now includes a section on hypergeometric distribution.

- In Chapter 5, we have expanded the discussion of uniform distribution and added a new section on exponential distribution.

- In Chapter 6, we have added more details regarding the factors affecting confidence intervals and sample sizes for both means and proportions.

- In Chapter 7, we have reorganized the discussion of hypothesis testing, including the power of the test for μ.

- Chapter 9 now features coverage of construction of one-way and two way ANOVA tables by hand, and Tukey simultaneous confidence intervals.

- Chapter 10 now covers chi-square goodness-of-fit tests for binomial and Poisson distributions and expansion of tests for normality, including the Lilliefors test.

- In Chapter 11, we have expanded the discussion of regression.

- In Chapter 12, we have added discussions of the Wilcoxon signed-ranks test for one median, small sample procedure for the Wilcoxon rank-sum test for two means, and the Friedman test.

Additional Features

INTERNET PROJECT

Probability Distributions and Simulation

Probability distributions are used to predict the outcomes of the events they model. For example, if we toss a fair coin, the distribution for the outcome is a probability of 1/2 for heads and 1/2 for tails. If we toss the coin ten consecutive times, we expect 5 heads and 5 tails. We might not get this exact result, but in the long run, over hundreds or thousands of tosses, we expect the split between heads and tails to be very close to "50–50". Go to either of the following websites

http://www.mathxl.com   or   http://www.mystatlab.com

and click on Internet Project, and then on Chapter 4, where you will find two explorations. In the first exploration you are asked to develop a probability distribution for a simple experiment, and use that distribution to predict the outcome of repeated trial runs of the experiment. In the second exploration, we will analyze a more complicated situation: the paths of rolling marbles as they move in pinball-like fashion through a set of obstacles. In each case, a dynamic visual simulation will allow you to compare the predicted results with a set of experimental outcomes.

- Internet Projects included at the end of each chapter involve the student with applications using data found on the Internet. The projects can be found at www.mathxl.com or www.mystatlab.com.

- ActivStats icons throughout the text encourage students to go to the ActivStats CD when encountering difficult concepts.
- The CD-ROM, packaged free with the text, contains:
  - STATDISK statistical software
  - Data Desk/XL statistical software (an Excel "add in")
  - Data Sets (Appendix B) in Excel and text file formats
  - PDFs of Chapters 13E, 14E, 15E, and the Glossary

EXAMPLE

At the 0.05 significance level, use the data in Table 10-1 to test whether a married couple has children or not significantly depends on the region of residence.

SOLUTION

The null hypothesis and alternative hypothesis are as follows:

H_0: A married couple having children is independent of the region of residence.

H_1: A married couple having children depends on the region of residence.

The significance level is $\alpha = 0.05$.

Because the data are in the form of a contingency table, we use the χ^2 distribution with this test statistic, together with the expected values from Table 10-12:

$$\chi^2 = \sum \frac{(O - E)^2}{E}$$

$$= \frac{(457 - 476.688)^2}{476.688} + \frac{(604 - 608.1297)^2}{608.1297} + \frac{(1522 - 1427.8074)^2}{1427.8074}$$

Hallmark Features

Beyond an interesting and accessible (and sometimes humorous) writing style, great care has been taken to ensure that each chapter of *Elementary Statistics* will help students understand the concepts presented. The following features are designed to help meet that objective:

Chapter Opener: A list of chapter sections with a brief description of their contents previews the chapter for the student; a chapter-opening problem, using real data, then motivates the chapter material; and a chapter overview provides a statement of the chapter's objectives.

EXAMPLE

Use the sample data from the preceding example to construct a 95% confidence interval estimate of μ_d.

SOLUTION

Using the values of $\bar{d} = -42.5$, $s_d = 33.2$, $n = 14$, and $t_{\alpha/2} = 2.160$, we first find the value of the margin of error E:

$$E = t_{\alpha/2} \frac{s_d}{\sqrt{n}} = 2.160 \frac{33.2}{\sqrt{14}} = 19.2$$

The confidence interval can now be found:

$$\bar{d} - E < \mu_d < \bar{d} + E$$
$$-42.5 - 19.2 < \mu_d < -42.5 + 19.2$$
$$-61.7 < \mu_d < -23.3$$

The result is sometimes expressed as $\mu_d = -42.5 \pm 19.2$ or as $(-61.7, -23.3)$.

Interpretation

In the long run, 95% of such samples will lead to confidence interval limits that actually do contain the true population mean of the differences. Note that the confidence interval limits do not contain 0, indicating that the true value of μ_d is significantly different from 0. That is, the mean value of the "right − left" differences is different from 0. On the basis of the confidence interval, we conclude that there is sufficient evidence to support the claim that there is a difference between the right- and left-hand reaction times. This conclusion agrees with the conclusion in the preceding example.

Solved Examples: Each section contains one or more solved examples (based on real data) that demonstrate the concept or method under discussion.

Exercises: Exercises are arranged in order of increasing difficulty by dividing them into two groups: (a) Basic Skills and Concepts and (b) Beyond the Basics. The Beyond the Basics Exercises address more difficult concepts or require a somewhat stronger mathematical background.

 5-1 Exercises A: Basic Skills and Concepts

In Exercises 1–5, suppose that the temperature readings for a certain gauge are uniformly distributed between 0°C and 5°C. Find the probability of a randomly selected temperature reading falling in the following range:

1. Greater than 2°C
2. Less than 3°C
3. Between 2°C and 4°C
4. Between 0.8°C and 4.7°C
5. Find the mean and standard deviation for the gauge readings.

In Exercises 6–8, suppose that the amount of paint that goes into a 4-L can is uniformly distributed between 3.85 L and 4.15 L.

6. What is the probability a can has less than 3.9 L?
7. What is the probability a can has more than 4.05 L?
8. Suppose the manufacturer wants the amount of paint in a can to be within 0.5 standard deviations of the mean. Based on the probability of this happening, are these realistic expectations?

 5-1 Exercises B: Beyond the Basics

9. For a uniform distribution, show why 100% of the distribution lies within 2 standard deviations of the mean, regardless of the values for a and b with $a < b$.

Statistical Planet Sidebars: These sidebars explore the uses and abuses of statistics in real, practical, and interesting applications that span many topics.

STATISTICAL PLANET

How Likely Is an Asteroid Strike? NASA astronomer David Morrison says that there are about 2000 asteroids with orbits that cross Earth's orbit, yet we have found only 100 of them. It's therefore possible for an undetected asteroid to crash into our planet and cause a global catastrophe that could destroy most life. Morrison says that there is a 1/10,000 probability that within a human lifetime,

Flowcharts: Flowcharts appear throughout the text to simplify and clarify complex concepts and procedures.

End-of-Chapter Features

- Vocabulary list of important terms
- Chapter Review, which summarizes the key concepts and topics of the chapter
- Review Exercises for practice on the chapter concepts and procedures
- Cumulative Review Exercises to reinforce earlier material
- Technology Projects for use with Excel and STATDISK applications
- From Data to Decision, a final review problem requiring critical thinking and a writing component
- Cooperative Group Activities to encourage active learning in groups
- Internet Project, which applies important chapter concepts

Appendices

- Appendix A provides the statistical tables required for the procedures covered in the text.
- Appendix B lists 22 data sets, 12 of which are Canadian. These are used extensively throughout the entire book, in chapter problems, examples, and exercises. These data sets are provided in printed form in Appendix B and in electronic form on the CD packaged with the text.
- Appendix C is a bibliography of recommended texts and reference books.
- Appendix D contains answers to all the odd-numbered section exercises, as well as answers to Review Exercises and Cumulative Review Exercises.

CANADIAN DATA SETS

Data Set 2:	Body Temperatures of Healthy Adults in Varied Postures
Data Set 4:	Temperature and Precipitation for Canadian Cities
Data Set 5:	Sitting Days and Durations of Sessions of Parliament
Data Set 6:	The *Financial Post* Top 100 Companies for 1997
Data Set 7:	Temperature and Precipitation Departures from Seasonal Means
Data Set 10:	Annual Rate of Return of 40 Mutual Funds
Data Set 12:	Lotto 6/49 Results
Data Set 14:	Labour Force 15 Years and Over by Occupation, by Province
Data Set 17:	Sample Dates of Last Spring Frost at Two Groups of Alberta Locations
Data Set 18:	Body Temperatures of Adults Accepted for Voluntary Surgical Procedures
Data Set 20:	Person-Days Lost in Labour Stoppages in the Transportation Industry
Data Set 22:	Canadian Exports and Imports by World Area

Quick-Reference Tables

A symbol table is included for quick reference on the front and back inside cover pages.

Technology

Elementary Statistics, Third Canadian Edition, can be used without reference to any specific technology. For those who choose to supplement the course with technology, where appropriate, sections called Using Technology provide written descriptions and screen grabs to help students access the statistical capabilities of Minitab, Excel (including the DDXL add-in), and STATDISK.

STATDISK is an easy-to-use statistical software package developed specifically for use with *Elementary Statistics.* The latest version of the application is included on the CD-ROM accompanying the text.

For students using other statistical software, the accompanying CD-ROM has plain text files for the data sets in Appendix B. Graphing calculators may also be used for many of the exercises.

EXCEL:	In Excel it is possible—but usually not necessary—to reproduce a full binomial probability distribution table for all values of x, for a given n and p.
EXCEL (prior to 2007):	Simply click on *fx* on the toolbar, and select the function category **Statistical** and then the function name **BINOMDIST**. In the dialog box, enter the number of successes x, as well as the values of n (number of trials) and p (the probability of success each trial), plus 0 for the binomial distribution (instead of 1 for the cumulative binomial distribution). The probability value for *exactly* x successes will be displayed. If, instead, you enter a 1 for the cumulative binomial distribution, the value returned is the probability of *up to or including* x successes. (*Hint:* For the probability of "at least x successes," interpret as: $1 -$ [the probability of "*up to or including* $(x - 1)$ successes"].)
EXCEL 2007:	Click on **Menus** on the main menu, then the *fx* button, then more functions and select the function category **Statistical** and then the function name **BINOMDIST**. Proceed as above.
MINITAB 15:	From the **Calc** menu, choose **Probability Distributions** and then **Binomial**. Suppose $n = 10$ and $p = 0.24$. If you want to calculate $P(x = 2)$, click the **Probability** radio button and enter 10 for the number of trials and 0.24 for the probability of success.

Supplements

Instructor's Guide and Solutions Manual contains solutions to all in-text exercises, quizzes (with answers), and other instructor material.

Electronic Transparencies in PowerPoint provide lecture material that follows the text's sequence and theory.

Active Learning Questions allow instructors to quiz students on key concepts using PowerPoint slides.

Test Generator enables instructors to build, edit, print, and administer tests using an electronic bank of questions developed to cover all the key concepts in the text.

CourseSmart is a new way for instructors and students to access textbooks online anytime from anywhere. With thousands of titles across hundreds of courses, CourseSmart helps instructors choose the best textbook for their class and give their students a new option for buying the assigned textbook as a lower cost eTextbook. For more information, visit **www.coursesmart.com**.

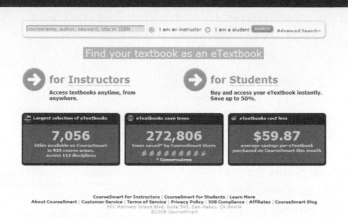

MyStatLab™ is a text-specific, easily customizable online course that integrates interactive multimedia instruction with textbook content. Powered by CourseCompass™ and MathXL, MyStatLab gives you the tools you need to deliver all or a portion of your course online, whether your students are in a lab setting or working from home. MyStatLab provides a rich and flexible set of course materials, featuring free-response tutorial exercises for unlimited practice. Students can also use online tools, such as video lectures, animations, and a multimedia textbook, to independently improve their understanding and performance. Instructors can use MyStatLab's homework and test managers to select and assign online exercises correlated directly to the textbook, and they can also create and assign their online exercises and import TestGen tests for added flexibility. MyStatLab's online gradebook automatically tracks students' homework and test results and gives the instructor control over how to calculate final grades. For more information, visit our website at **www.mystatlab.com** or contact your sales representative.

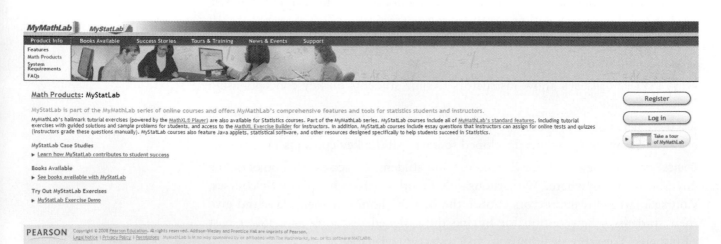

ActivStats, developed by Paul Velleman and Data Description, Inc., is an award-winning multimedia introduction to statistics and a comprehensive learning tool that works in conjunction with the book. It complements this text with interactive features such as videos of real-world stories, teaching applets, and animated expositions of major statistical topics. Contact your sales representative for details.

StatCrunch is an online statistical software website that allows users to perform complex analyses, share data sets, and generate compelling reports of their data. Developed by Webster West, Texas A&M University, StatCrunch already has more than 10,000 data sets available for students to analyze, covering almost any topic of interest. Interactive graphics help users understand statistical concepts and are available for export to enrich reports with visual representations of data. Additional features include:

- A full range of numerical and graphical methods that allow users to analyze and gain insights from any data set

- Flexible upload options that allow users to work with their .txt or Excel® files, both online and offline

- Reporting options that help users create a wide variety of visually-appealing representations of their data

StatCrunch access is available to qualified adopters and can be packaged with Pearson statistics textbooks. For more information, visit **www.statcrunch.com**, or contact your Pearson sales representative.

Acknowledgments

The Third Canadian Edition would not have been possible without the help of some truly talented and dedicated people: the librarians at Mount Royal University in Calgary, my copy editor, Tom Gamblin, and the production team at Pearson, consisting of production editor, Imee Salumbides, and production coordinator, Lynn O'Rourke.

A very special thanks to my right-hand man, my developmental editor, Paul Donnelly, who taught me that it is possible to do enough work to resink the Titanic and still maintain a measure of sanity.

I also gratefully acknowledge the following reviewers whose comments and feedback helped inform this edition:

Alan Chan, Atlantic Baptist University

Michelle Boué, Trent University

Natasha Davidson, Douglas College

Suzanne Fredericks, Ryerson University

Andreas J. Guelzow, Concordia University College of Alberta

David Holloway, British Columbia Institute of Technology

Wayne Horn, Carleton University

Ngok Yeung Lai, Nova Scotia Community College

Patricia Lake, College of the North Atlantic

Manon Lemonde, University of Ontario Institute of Technology

Dorothy Levay, Brock University

Deli Li, Lakehead University

Shawn Liu, Mount Royal University

Nancy Luckai, Lakehead University

William J. Montelpare, Lakehead University

Julia Morton, Nipissing University

Joy Rose, Herzing University

Elaine Santa Mina, Ryerson University

Catherine Stanley, Acadia University

Steven Webber, Ryerson University

Dedications

To Ginny, Marc, Dushana, and Marisa, Scott, Anna, Siena, and Kaia.
—M.F.T.

To Kathryn, my wife, and to my children, Rachel and Aleksey. Even through the chaos, you have added a rich balance to my life.
—W.M.G.

I thank my intelligent and resourceful wife, Beth. She kept me focused and inspired when the need arose, and was a helpful source of original ideas.
—R.L.

To my wife, Susan, for her love and support while I slaved away in the basement dungeon (a.k.a. my office).
—G.L.

1 Introduction to Statistics

1-1 Overview

The term *statistics* is defined, along with the terms *population*, *sample*, *parameter*, and *statistic*.

1-2 The Nature of Data

Quantitative data, *qualitative data*, *discrete data*, and *continuous data* are defined, along with the four *levels of measurement* (*nominal*, *ordinal*, *interval*, *ratio*).

1-3 Uses and Abuses of Statistics

Examples of beneficial uses of statistics are presented, along with some of the common ways in which statistics are used to deceive. Deceptive uses include small samples, precise numbers, distorted percentages, loaded questions, misleading graphs, and bad samples.

1-4 Design of Experiments

Observational studies and experiments are described, along with good statistical methodology. The importance of good sampling is emphasized. Different sampling methods are defined and described, including random sampling, stratified sampling, systematic sampling, cluster sampling, and convenience sampling.

CHAPTER PROBLEM

As a student, are you in the most dangerous "profession"?

On his fourth day of work, a young worker is fatally burned by a highly flammable chemical. While guiding a truck into a loading bay, another worker is crushed to death because the truck driver loses sight of him. A forklift driver is crushed when his forklift topples over. Sadly, these are among the hundreds of Canadian workers (about three a day, according to North American Occupational Safety and Health (NAOSH) statistics) who are killed as a result of occupational injuries.

Some professions are inherently more dangerous than others. Fatal injuries are sometimes suffered by police officers in car crashes or firefighters in collapsed buildings. Taxicab drivers are sometimes killed by passengers determined to steal fares. Some miners' lives have been shortened by dangerous levels of coal dust.

For obvious reasons, governments and health workers around the world take a serious interest in collecting and analyzing statistics relating to these hazards. For example, the Swiss physician H.C. Lombard once compiled longevity data for different professions. He used death certificates that included name, age at death, and profession. He then proceeded to compute the average (mean) length of life for the different professions, and he found that students were lowest with a mean age at death of only 20.7 years! (See "A Selection of Selection Anomolies" by Wainer, Palmer, and Bradlow in *Chance*, Volume 11, No. 2.) Similar results would be obtained if the same data were collected today in Canada. Could being a student really be more dangerous than working in a chemical plant, upholding the law, working in a mine, or driving a taxicab? Not likely, as you will see in Section 1-3 of this chapter, *Uses and Abuses of Statistics*.

1-1 Overview

We begin our study of *statistics* by noting that the word has two basic meanings. In the first sense, the term is used in reference to actual and specific numbers derived from data, such as the results of a Canadian Automobile Association survey, according to which 58% of the respondents paid cash for their used vehicles.

A second meaning refers to statistics as a method of analysis.

DEFINITION

Statistics is a collection of methods for planning experiments, obtaining data, and then organizing, summarizing, presenting, analyzing, interpreting, and drawing conclusions based on the data.

Statistics involves much more than simply drawing graphs and calculating averages. In this book, we will learn how to develop generalized and meaningful conclusions that go beyond the original data. In statistics, we commonly use the terms *population* and *sample*. These terms are at the very core of statistics and we define them now.

DEFINITIONS

A **population** is the complete collection of all elements (scores, people, measurements, and so on) to be studied.
A **census** is the collection of data from *every* element in a population.
A **sample** is a subcollection of elements drawn from a population.

For example, a typical Nielsen television survey uses a *sample* of 1700 households, and the results are used to form conclusions about the *population* of all 10,820,050 households in Canada. Every five years, Statistics Canada tries to obtain a complete census, with information on every Canadian; but in practice, it is impossible to reach everyone.

Closely related to the concepts of population and sample are the concepts of *parameter* and *statistic*. The following definitions are easy to remember if we recognize the alliteration in "population parameter" and "sample statistic."

DEFINITIONS

A **parameter** is a numerical measurement describing some characteristic of a *population*.
A **statistic** is a numerical measurement describing some characteristic of a *sample*.

1. **Parameter:** According to the 1881 Canada Census, 12.4% of the population of Yale District, British Columbia, belonged to the Buddhist religion. Assuming that the list of 8951 residents for that region did not overlook anyone, then the 12.4% is a *parameter*, not a statistic.

2. **Statistic:** In a survey of 1031 tournament-level golfers, 44% had the career-threatening condition known as the "yips" (*The Globe and Mail*, January 8, 2001). The figure 44% is a *statistic* because it is based on a sample, not the entire population of all professional golfers.

STATISTICAL PLANET

Misleading Statistics in Journalism
New York Times reporter Daniel Okrant wrote that although every sentence in his newspaper is copy-edited for clarity and good writing, "numbers, so alien to so many, don't get nearly this respect. The paper requires no specific training to enhance numeracy, and no specialists whose sole job is to foster it." He cites an example of the *New York Times* reporting an estimate of more than $23 billion that New Yorkers spend for counterfeit goods each year. Okrant writes that "quick arithmetic would have demonstrated that $23 billion would work out to roughly $8000 per city household, a number ludicrous on its face."

An important consideration when evaluating sample statistics is whether the data were collected in an appropriate way, such as through a process of *random* selection. If sample data are not collected appropriately, then no amount of statistical torturing can salvage the data.

We are convinced that if you apply yourself diligently, and with an open mind, to the materials in this book, you will discover that statistics is an interesting and rich subject—with many real and meaningful applications within your grasp.

1-2 The Nature of Data

Data are observations (such as measurements, genders, and survey responses) that have been collected. Some data sets consist of numbers (such as heights), and others are nonnumeric (such as gender). The terms *quantitative data* and *qualitative data* are often used to distinguish between these two types.

DEFINITIONS

Quantitative data consist of numbers representing counts or measurements.
Qualitative (or **categorical** or **attribute**) **data** can be separated into different categories that are distinguished by some nonnumeric characteristic.

Data Set 4 in Appendix B includes amounts of precipitation in different Canadian cities. Those amounts are quantitative data, but the names of the cities are qualitative data.

We can further describe quantitative data by distinguishing between the discrete and continuous types.

DEFINITIONS

Discrete data result from either a finite number of possible values or a countable number of possible values. (That is, the number of possible values is 0, or 1, or 2, and so on.)
Continuous (numerical) data result from infinitely many possible values that can be associated with points on a continuous scale in such a way that there are no gaps or interruptions.

When data represent counts, they are discrete; when they represent measurements, they are continuous. The numbers of eggs that hens lay are *discrete* data because they represent counts; the amounts of milk that cows produce are *continuous* data because they are measurements that can assume any value over a continuous span. (For convenience, however, continuous quantitative data are often scaled into seemingly discrete units, such as "to the nearest litre," or "in 1000s of dollars.")

Another common way to classify data is to use four levels of measurement: nominal, ordinal, interval, ratio. In applying statistics to real problems, the level of measurement of the data is an important factor in determining which procedure it is appropriate to use.

DEFINITION

The **nominal level of measurement** is characterized by data that consist of names, labels, or categories only. The data cannot be arranged in an ordering scheme (such as low to high).

EXAMPLES

The following are examples of sample data at the nominal level of measurement:
1. Survey responses of yes, no, and undecided
2. The "kingdoms" into which life forms can be classified

Because nominal data lack any ordering or numerical significance, they cannot be used for calculations. Numbers are sometimes assigned to categories (especially when data are computerized), but these numbers lack any real computational significance. However, you *can* count how often data fall into different nominal categories.

DEFINITION

The **ordinal level of measurement** involves data that may be arranged in some order, but differences between data values either cannot be determined or are meaningless.

EXAMPLES

The following are examples of data at the ordinal level of measurement:
1. An editor rates some manuscripts as "excellent," some as "good," and some as "bad." (We can't find a specific quantitative difference between "good" and "bad.")
2. An Olympic screening committee ranks Gail 3rd, Diana 7th, and Kim 10th. (We can find a difference between ranks of 3 and 7, but the difference of 4 doesn't mean anything.)

Ordinal level data provide information about relative comparisons, but not about the degrees of differences. They are restricted in the types of calculations they can be used for.

DEFINITION

The **interval level of measurement** is like the ordinal level, with the additional property that we can determine meaningful amounts of differences between data. However, there is no inherent (natural) zero starting point (where *none* of the quantity is present).

EXAMPLES

The following are examples of data at the interval level of measurement.

1. Body temperatures of 37.0°C and 36.39°C. Such values are ordered, and we can determine their difference (often called the *distance* between the two values). However, there is no natural starting point. The value of 0°C might seem like a starting point, but it is arbitrary and does not represent "no heat." It is wrong to say that 50°C is twice as hot as 25°C.

2. The years 1000, 2002, 1867, and 1944. (Time did not begin in the year 0, so the year 0 is arbitrary instead of being a natural zero starting point.)

DEFINITION

The **ratio level of measurement** is the interval level modified to include the inherent zero starting point (where zero indicates that *none* of the quantity is present). For values at this level, differences and ratios are both meaningful.

EXAMPLES

The following are examples of data at the ratio level of measurement:

1. Weights of plastic discarded by households (0 kg does represent no plastic discarded, and 10 kg does weigh twice as much as 5 kg).

2. Distances (in kilometres) travelled by cars in a test of fuel consumption. (A distance of 0 km is possible, and it would be meaningful to calculate that one car drove three times as far as another.)

3. Temperature readings on the Kelvin scale are at the ratio level of measurement; that scale has an absolute zero.

This level is called the ratio level because the starting point makes ratios meaningful. Because a 200-kg weight is *twice* as heavy as a 100-kg weight, but 50°C is *not* twice as hot as 25°C, weights are at the ratio level while Celsius temperatures are at the interval level. For a concise comparison and review, study Table 1-1 to see the differences among the four levels of measurement.

Table 1-1 Levels of Measurement of Data

Level	Summary	Example	
Nominal	Categories only. Data cannot be arranged in an ordering scheme.	Student cars: Corvettes Ferraris Porsches	Categories or names only.
Ordinal	Categories are ordered, but differences cannot be determined or they are meaningless.	Student cars: compact mid-size full-size	An order is determined by "compact, mid-size, full-size."
Interval	Differences between values can be found, but there is no inherent starting point. Ratios are meaningless.	Campus temperatures: 15°C 20°C 30°C	90°C is not twice as hot as 45°C.
Ratio	Like interval, but with an inherent starting point. Ratios are meaningful.	Weights of university football players: 150 lb 195 lb 300 lb	300 lb is twice 150 lb.

Exercises A: Basic Skills and Concepts

In Exercises 1–4, identify each number as discrete or continuous.

1. Each regular strength Dristan tablet has 325 mg of acetaminophen.

2. An Air Canada altimeter reports a height of 10,000 m.

3. A poll of 1015 people shows that 40 of them now subscribe to an on-line computer service.

4. Radar indicated that Paul Tracy reached a top speed in the straight-away of 275 km/h.

In Exercises 5–8, determine whether the given value is a statistic or a parameter.

5. A sample of students is selected, and the average (mean) age is 20.7 years.

6. 93% of those questioned in a consumer survey recognized the Campbell's Soup brand name.

7. According to the latest census, 4.83% of families in Calgary have an annual income less than $10,000.

8. The average (mean) of the atomic numbers for the elements that are listed on a student's chemistry data tables is 52.0.

In Exercises 9–18, determine which of the four levels of measurement (nominal, ordinal, interval, ratio) is most appropriate.

9. Ratings of superior, above average, average, below average, or poor for blind dates

10. Phenylephrin contents (in milligrams) of Dristan tablets

11. Social insurance numbers (SIN)

12. Temperatures (in degrees Celsius) of a sample of angry taxpayers who are being audited

13. Years in which the Winnipeg Blue Bombers won the Grey Cup

14. Final course grades (A, B, C, D, F) for statistics students

15. Telephone area codes

16. Annual incomes of nurses

17. Cars described as subcompact, compact, intermediate, or full-size

18. Colours in a handful of M&M candies

1-2 Exercises B: Beyond the Basics

19. A pollster in Saskatchewan surveys 200 people and asks them their preference of political party. She codes the responses as 0 (for Green), 1 (for Liberal), 2 (for New Democratic Party), and 3 (for Conservative). She then calculates the average (mean of the numbers) and gets 1.78. How can that value be interpreted?

20. In "The Born Loser" cartoon strip, Brutus expresses joy over an increase in temperature from 1° to 2°. When asked what is so good about 2°, he answers, "It's twice as warm as this morning." Why is Brutus wrong once again?

1-3 Uses and Abuses of Statistics

Uses of Statistics

The applications of statistics have grown to the point that practically every field of study now benefits in some way from the use of statistical methods. Manufacturers now provide better products at lower costs through the use of statistical quality control techniques. Diseases are controlled through analyses designed to anticipate epidemics. Endangered species of fish and other wildlife are protected through regulations and laws that react to statistical estimates of changing population sizes. By pointing to lower fatality rates, legislators can better justify laws such as those governing air pollution, auto inspections, seat belt and air bag use, and drunk driving. We will cite only these few examples, because a complete compilation of the uses of statistics would easily fill the remainder of this book (a prospect not totally unpleasant to some readers).

As you proceed with this course, you will encounter many different applications of statistics.

AS

Abuses of Statistics

Abuses of statistics have occurred for some time. For example, about a century ago, statesman Benjamin Disraeli famously said, "There are three kinds of lies: lies, damned lies, and statistics." It has also been said that "figures don't lie; liars figure," and that "if you torture the data long enough, they'll admit to anything." Historian Andrew Lang said that some people use statistics "as a drunken man uses lampposts—for support rather than illumination." These statements refer to abuses of statistics in which data are presented in ways that may be misleading. Some abusers of statistics are simply ignorant or careless, whereas others have personal objectives and are willing to suppress unfavourable data while emphasizing supportive data. We will now present a few examples of the many ways in which data can be distorted.

Bad Samples   A major source of deceptive statistics is inappropriate methods of collecting data. One very common sampling method allows the sample subjects to decide for themselves whether to be included.

DEFINITION

> A **self-selected sample** is one in which the respondents themselves decide whether to be included.

In such surveys, people with strong opinions are more likely to participate, so the responses obtained do not necessarily represent the whole population. Phone-in polls are good examples of self-selected samples because the callers themselves decide whether to be included in the survey.

A self-selected sample is only one way in which the method of collecting data can be seriously flawed. As another example, consider our chapter problem, which described how Swiss physician H.C. Lombard once compiled longevity data for different professions and found that students ranked lowest, with a mean of only 20.7 years. The problem is that the sample is not appropriate, because most people are students when they are relatively young; they are no longer students when they become a little older and are employed in some other profession. Also, all of the students in the Lombard study *died*, so the 20.7-year longevity figure merely suggests that if students die, they are likely to be young. A similar error would be made if you computed the average age of teenagers who died. That figure could not possibly be more than 19 years, but this does not really mean that being a teenager is more dangerous than being a test pilot, a profession in which members rarely join at ages younger than 19 years.

Small Samples   In Chapter 6 we will see that small samples are not necessarily bad, but small sample results are sometimes used as a form of statistical "lying." The toothpaste preferences of only 10 dentists should not be used as a basis for a generalized claim such as "Covariant toothpaste is recommended by 7 out of 10 dentists." Even if a sample is large, it must be unbiased and representative of the population from which it comes. Sometimes a sample might seem relatively large (as in a survey of "2000 randomly selected adult Canadians"), but if conclusions are made about subgroups, such as the Catholic male Conservatives in the survey, such conclusions might be based on very small samples.

Precise Numbers   Sometimes the numbers themselves can be deceptive. A very precise figure, such as an annual salary of $37,735.29, might be used to sound precise, so that people assume incorrectly that it is also *accurate*. In this case, the number is an estimate, so it would be better to state that the number is about $37,700.

Guesstimates   Another source of statistical deception involves estimates that are really guesses and can therefore be in error by substantial amounts. We should consider the source of the estimate and how it was developed. At a recent demonstration in Ottawa, organizers estimated the crowd size to be 50,000, but the media used aerial photographs and grids to come up with a more accurate estimate that was much smaller.

Distorted Percentages   Misleading or unclear percentages are sometimes used. An airline ran full-page ads boasting better service. In referring to lost baggage, these ads claimed that this was "an area where we've already improved 100% in the last six months." An editorial criticizing this statistic interpreted the 100% improvement figure to mean that no baggage is now being lost—an accomplishment not yet achieved.

Partial Pictures   "Ninety percent of all our cars sold in this country in the last ten years are still on the road." Millions of consumers heard that impressive commercial message. What the manufacturer failed to mention was that 90% of the cars had been sold within the last three years. The claim was technically correct, but it was very misleading through not presenting the complete story.

Deliberate Distortions   In her book *Tainted Truth*, Cynthia Crossen cites as an example the publication by the magazine *Corporate Travel* of results showing that among car rental companies, Avis was the winner in a survey of people who rent cars. When Hertz requested detailed information about the survey, the actual survey responses disappeared and the magazine's survey coordinator resigned. Hertz sued Avis (for false advertising based on the survey) and *Corporate Travel*; a settlement was reached.

Loaded Questions   Survey questions can be worded to elicit a desired response. For example, in 1999 a *Toronto Star* poll asked, "Should the provincial government change the law to make it easier to get squeegee kids off our streets?" The results were 69% in favour of changes to the law. The so-called squeegee kids had just been at the centre of some very negative press. The wording of the question appealed to those *Star* readers who had very strong negative feelings toward these individuals. It might have been more appropriate to word the question, "Should the government change the law to discourage street vendors from approaching motor vehicles?"

Figure 1-1
Earnings of Full-Time
Professional Workers

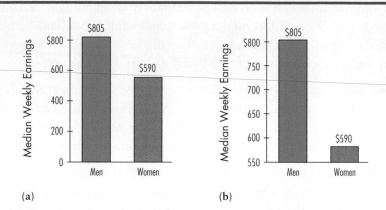

(a)   (b)

Sometimes questions are unintentionally loaded by such factors as the order of the items being considered. For example, one German poll asked these two questions:

- Would you say that traffic contributes more or less to air pollution than industry?
- Would you say that industry contributes more or less to air pollution than traffic?

When traffic was presented first, 45% blamed traffic and 27% blamed industry; when industry was presented first, those percentages changed dramatically to 24% and 57%, respectively.

Misleading Graphs   Many visual devices—such as bar graphs and pie charts—can be used to exaggerate or de-emphasize the true nature of data. (Such devices will be discussed in Chapter 2.) The two graphs in Figure 1-1 depict the *same data* from Statistics Canada, but graph (b) is designed to exaggerate the difference between the earnings of men and women. By not starting the vertical axis at zero, graph (b) tends to produce a misleading subjective impression. Figure 1-1 carries an important lesson: We should analyze the *numerical* information given in the graph, so that we won't be misled by its general shape.

Pictographs   Drawings of objects, called *pictographs*, may also be misleading. Some objects commonly used to depict data include moneybags, stacks of coins, army tanks (for military expenditures), cows (for dairy production), barrels (for oil production), and houses (for home construction). When drawing such objects, artists can create false impressions that distort differences. If you double each side of a square, the area doesn't merely double; it increases by a factor of four. If you double each side of a cube, the volume doesn't merely double; it increases by a factor of eight. If taxes double over a decade, an artist may depict tax amounts with one moneybag for the first year and a second moneybag that is twice as deep, twice as tall, and twice as wide. Instead of appearing to double, taxes will appear to increase by a factor of eight, so the truth will be distorted by the drawing.

Entire books have been devoted to such abuses of statistics, including Darrell Huff's classic *How to Lie with Statistics*, Robert Reichard's *The Figure Finaglers*, and Cynthia Crossen's *Tainted Truth*. Understanding these practices will be extremely helpful in evaluating the statistical data found in everyday situations.

1-3 Exercises A: Basic Skills and Concepts

1. You've been hired to research recognition of the Roots brand name, and you must conduct a telephone survey of 1500 consumers in Canada. What is wrong with using telephone directories as the population from which the sample is drawn?

2. The *Calgary Herald* held an online poll asking people if they twitter. Of the 791 respondents, 7% said yes. What is wrong with this survey?

3. A report sponsored by British Columbia fruit farmers concluded that cholesterol levels can be lowered by eating fruit products. Why might the conclusion be suspect?

4. An employee has an annual salary of $40,000 but is told that she will be given a 10% cut in pay because of declining company profits. She is also told that next year, she will be given a 10% raise. This doesn't seem too bad because the 10% cut seems to be offset by the 10% raise.

   a. What is the annual income after the 10% cut?
   b. Use the annual income from part (a) to find the annual income after the 10% raise. Did the 10% cut followed by the 10% raise get the employee back to an annual salary of $40,000?

5. Sixty-nine percent of those who responded to a *Toronto Star* poll question said that the provincial government should change the law to make it easier to get squeegee kids off the streets. How valid is the 69% result?

6. McLaren NightVision, a manufacturer of "night-vision" lenses for drivers, claims in an advertisment that darkness is the cause of most highway mayhem. Their advertisement stated that "according to Ministry of Transportation statistics, over 25% of highway fatalities occur between midnight and 6 a.m." What is misleading about this statement?

7. In a study on campus crimes committed by students high on alcohol or drugs, a mail survey of 1875 students was conducted. A report on the survey noted, "Eight percent of the students responding anonymously say they've committed a campus crime. And 62% of that group say they did so under the influence of alcohol or drugs." Assuming that the number of students responding anonymously is 1875, how many actually committed a campus crime while under the influence of alcohol or drugs?

8. A study conducted by the U.S. Insurance Institute for Highway Safety found that the Chevrolet Corvette had the highest fatality rate—5.2 deaths for every 10,000. The car with the lowest fatality rate was the Volvo, with only 0.6 deaths per 10,000 vehicles. Does this mean that the Corvette is not as safe as the Volvo?

9. An East Coast newspaper claims that pregnant mothers can increase their chances of having healthy babies by eating lobster. That claim is based on a study showing that babies born to lobster-eating mothers have fewer health problems than babies born to mothers who don't eat lobsters. What is wrong with this claim?

10. A survey includes this item: "Enter your height in inches." It is expected that actual heights of respondents can be obtained and analyzed. Identify the two major problems with this item.

11. "According to a nationwide survey of 250 hiring professionals, scuffed shoes was the most common reason for a male job seeker's failure to make a good first impression." Newspapers carried this statement based on a poll commissioned by Kiwi Brands, producers of shoe polish. Comment on why the results of the survey might be questionable.

12. The increases in expenditures made by the government of Newfoundland and Labrador for pollution abatement are shown in the graph below (based on data from Statistics Canada). What is wrong with the figure?

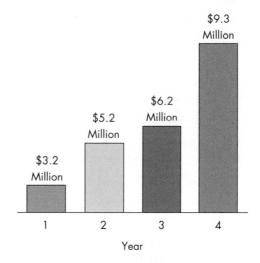

1-3 Exercises B: Beyond the Basics

13. A study noted that the mean life span for 35 male symphony conductors was 73.4 years, in contrast to the mean of 69.5 years for males in the general population. The longer life span was attributed to such factors as fulfillment and motivation. There is a fundamental flaw in concluding that male symphony conductors live longer. What is it?

14. A researcher at the Sloan-Kettering Cancer Research Center was once criticized for falsifying data. Among his data were figures obtained from 6 groups of mice, with 20 individual mice in each group. These values were given for

the percentage of successes in each group: 53%, 58%, 63%, 46%, 48%, 67%. What's wrong?

15. Try to identify each of the four major flaws in the following. A daily newspaper ran a survey by asking readers to call in their response to this question: "Do you support the development of atomic weapons that could kill millions of innocent people?" It was reported that 20 readers responded and 87% said "no," while 13% said "yes."

16. A chart caption in an advertisement described a dental rinse as one that "reduces plaque on teeth by over 300%."
    a. If you remove 100% of some quantity, how much is left?
    b. What does it mean to reduce plaque by over 300%?

1-4 Design of Experiments

Sometimes we have an interesting and important data set, but we do not have a particular objective in mind, so we want to *explore* the data to see what insights we might acquire. (The basic tools for exploring and describing data sets are presented in Chapter 2.) More often, though, we do have a specific objective, and we want to collect the data and do the analysis that will help us to meet that objective. We typically get our data from two common sources: *observational studies* (such as polls) and *experiments* (such as using irradiation to improve the shelf life of food products).

DEFINITIONS

In an **observational study**, we observe and measure specific characteristics, but we don't attempt to manipulate or modify the subjects being studied.

In an **experiment**, we apply some *treatment* and then proceed to observe its effects on the subjects.

For example, an observational study might involve a survey of citizens to determine what percentage of the population favours the registration of handguns. An experiment might involve a drug treatment given to a group of patients in order to determine its effectiveness as a cure. With the handgun survey, we collect data without modifying the people being polled, but the drug treatment consists of modification of the subjects.

When designing an experiment, there are a few basic steps that should be followed:

1. *Identify your objective.* Identify the exact question to be answered, and clearly identify the relevant population. (For example, "Is the Salk vaccine effective in reducing the incidence of polio in the population of children?")

2. *Collect sample data.* The way in which sample data are collected is absolutely critical to the success of the experiment. The sample data must be representative

STATISTICAL PLANET

Political Polls Grow
The publication of political polls during the period leading up to an election has been a controversial issue; some people believe that such polls have too much influence on voters' decisions. The use of polls, however, has grown over the last three decades. In the lead-up to the 1980 federal election, only 4 polls were published; that number grew to 24 in the period before the 1988 election. Polling is complicated by answering machines and people who decline to take part, but good polls include repeated attempts to get responses from those who are not at home and those who refuse to answer. Ignoring those who don't answer or respond could result in a sample that is not representative of the population.

of the population in question, the sample must be large enough so that the effects of the treatment can be known, and the question should be addressed without interference from extraneous factors.

3. *Use a random procedure that avoids bias.* (In the Salk vaccine experiment, for example, children were assigned to the two groups through a process of random selection.)

4. *Analyze the data and form conclusions.*

Controlling the Effects of Variables

When conducting an experiment, it is all too easy to receive interference from variable factors that are not relevant to the issue being studied. These effects can be controlled through good experimental design. Careful experiments often involve a **treatment group** that is given a particular treatment and a second, **control group** that is not given the treatment. For example, a 1954 polio experiment involved a treatment group of children who were injected with the Salk vaccine and a control group of children who were injected with a placebo that contained no medicine or drug. In experiments of this type, a **placebo effect** occurs when an untreated subject incorrectly believes that he or she is receiving a treatment and reports an improvement in symptoms. The placebo effect can be countered by using **blinding**, a technique in which the subject doesn't know whether he or she is receiving a treatment or a placebo. The polio experiment was **double-blind**, meaning that the children being injected didn't know whether they were given the Salk vaccine or a placebo, and the doctors who gave the injections and evaluated the results didn't know either.

When designing an experiment to test the effectiveness of one or more treatments, you should be careful to assign the *experimental units* (or subjects) to the different groups in such a way that those groups are very similar. (Such similar groups of experimental units are called **blocks**.) One effective approach is to use a **completely randomized design**, which requires that the experimental units be divided into different groups through a process of *random* selection. For example, such a design might involve random assignment of people to a group treated with aspirin and a control group that is not treated. Another approach is to use a **rigorously controlled design**, with experimental units carefully chosen so that the different groups (or blocks) are carefully arranged to be similar. With a rigorously controlled design, you might try to design treatment and control groups to include people similar in age, weight, blood pressure, and so on. The polio experiment was a completely randomized experimental design because the subjects in both the treatment group and the control group were selected randomly. It incorporated replication by including very large (200,000) numbers of subjects in each group to reduce the effects of chance sample variation.

When conducting experiments, results are sometimes ruined because of *confounding.*

Confounding occurs when the effects from two or more variables cannot be distinguished from each other.

For example, if you're conducting an experiment to test the effectiveness of a new fire retardant on a brush fire but it begins to rain, then confounding occurs because it's impossible to distinguish between the effect from the retardant and the effect from the rain.

One of the worst mistakes is to collect data in a way that is inappropriate. We cannot overstress this very important point:

Data carelessly collected may be so completely useless that no amount of statistical torturing can salvage them.

As well as being chosen appropriately, your sample must be sufficiently large to avoid misleading results, which can arise from the erratic behaviour of very small samples. Repetition of an experiment to increase sample size is called **replication**.

Randomization

We have just stressed the importance of selecting samples by appropriate techniques. In five of the more common methods, described below, *randomization* plays a crucial role in how the samples are chosen.

In a **random sample**, members of the population are selected in such a way that each individual has an *equal chance* of being selected.

A **simple random sample** of n subjects is selected in such a way that every possible sample of size n has the same chance of being chosen.

Random samples have historically been selected by a variety of methods, including using computers to generate random numbers and using tables of random numbers. With random sampling, we expect all groups of the population to be (approximately) proportionately represented in the sample. Careless or haphazard sampling can easily result in a biased sample with characteristics very unlike the population from which the sample came. Random sampling, in contrast, is carefully designed to avoid any bias. For example, using telephone directories automatically eliminates anyone with an unlisted number or a cell phone, and ignoring those segments of the population could easily yield misleading results. Pollsters commonly circumvent this problem by using computers to generate phone numbers so that all numbers are possible. They must also be careful to include those who are initially unavailable or initially refuse to comment. One polling company has found that the refusal rate for telephone interviews is generally at least 20%. If you ignore those people who initially refuse, you run a real risk of having a biased sample.

> With **stratified sampling**, we subdivide the population into at least two different subpopulations (or strata) that share the same characteristics (such as gender), then we draw a sample from each stratum.

In surveying views on pay equity, we might use gender as a basis for creating two strata. After obtaining a list of men and a list of women, we use some suitable method (such as random sampling) to select a certain number of people from each list. When the various strata have sample sizes that reflect the general population, we say that we have *proportionate* sampling. If it should happen that some strata are not represented in the proper proportion, then the results can be adjusted or weighted accordingly.

Given a fixed sample size, if you randomly select subjects from different strata, you are likely to get more consistent (and less variable) results than by simply selecting a random sample from the general population. For that reason, stratified sampling is often used to reduce the variation in the results.

> In **systematic sampling**, we select some starting point and then select every kth (such as every 50th) element in the population.

For example, if Noranda wanted to conduct a survey of its 33,000 employees, it could begin with a complete roster, then select every 100th employee to obtain a sample of size 330. This method is simple and is often used.

> In **cluster sampling**, we first divide the population area into sections (or clusters), then randomly select a few of those sections, and then choose *all* the members from those selected sections.

One important way that cluster sampling differs from stratified sampling is that cluster sampling uses *all* members from selected clusters, whereas stratified sampling uses a *sample* of members from each strata. An example of cluster sampling can be found in a survey of Tim Hortons customers whereby we randomly select 30 locations (from the more than 3000 locations across Canada) and then survey all the customers over, say, a 12-hour period from each of those chosen locations. This would be much faster and much less expensive than selecting a few customers from each of the many locations in Canada. The results can be adjusted or weighted to correct for any disproportionate representations of groups. Cluster sampling is used extensively by government and private research organizations.

> With **convenience sampling**, we simply use results that are readily available.

In some cases results from convenience sampling may be quite good, but in other cases they may be seriously biased. In investigating the proportion of left-handed people, it would be convenient for a student to survey his or her classmates, because they are readily available. Even though such a sample is not random, the results should be quite good. In contrast, it might be convenient for CBC Radio to ask people to call a toll-free number to register their opinions, but this would be a self-selected survey and the results would likely be biased.

Figure 1-2 illustrates the five common methods of sampling just described. The descriptions are intended to be brief and general. Thoroughly understanding these different methods so that you can successfully use them requires much more extensive study than is practical in a single introductory course. To keep this section in perspective, we should note that this text will make frequent reference to "randomly selected" data, and you should understand that such data are carefully selected so that all members of the population have the same chance of being chosen. Although we will not make frequent reference to the other methods of sampling, you should understand that they exist and that the method of sampling requires careful planning and execution. Note that if you are obtaining measurements of a characteristic (such as height) from people, you will get more accurate results if you do the measuring yourself instead of asking the subjects for the value. Asking tends to yield a disproportionate number of rounded results, as well as many results that reflect *desired* values instead of *actual* values.

No matter how well you plan and execute the sample collection process, there is likely to be some error in the results. For example, randomly select 1000 adults, ask them if they graduated from high school, and record the sample percentage of "yes" responses. If you randomly select another sample of 1000 adults, it is likely that you will obtain a *different* sample percentage. This unavoidable error is called *sampling error* and you should take it into account when interpreting sample results.

DEFINITIONS

> A **sampling error** is the difference between a sample result and the true population result; such an error results from chance sample fluctuations.
> A **nonsampling error** occurs when the sample data are incorrectly collected, recorded, or analyzed (such as by selecting a nonrandom and biased sample, using a defective measuring instrument, or copying the sample data incorrectly).

If we carefully collect a sample so that it is representative of the population, we can use the methods in this book to analyze the sampling error, but we must exercise extreme care so that nonsampling error is minimized.

Figure 1-2
Common Sampling Methods

Random Sampling

Each member of the population has an equal chance of being selected. Computers are often used to generate random telephone numbers.

Stratified Sampling

Classify the population into at least two strata, then draw a sample from each.

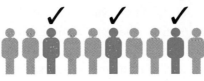

Systematic Sampling

Select every kth member.

Cluster Sampling

Divide the population area into sections, randomly select a few of those sections, and then choose all members in them.

Convenience Sampling

Use results that are readily available.

Exercises A: Basic Skills and Concepts

In Exercises 1–4, determine whether the given description corresponds to an observational study or an experiment.

1. Temperatures and amounts of precipitation and snowfall are recorded in different locations across the country (as in Data Set 4 in Appendix B).

2. People who smoke are asked to halve the number of cigarettes consumed each day so that any effect on pulse rate can be measured.

3. In a physical education class, the effect of exercise on blood pressure is studied by requiring that half of the students walk a mile each day while the other students run a mile each day.

4. The relationship between weights of bears and their lengths is studied by measuring bears that have been anaesthetized.

In Exercises 5–16, identify which of these types of sampling is used: random, stratified, systematic, cluster, or convenience.

5. When she wrote *Women and Love: A Cultural Revolution,* author Shere Hite based her conclusions on 4500 responses from 100,000 questionnaires distributed to women.

6. A psychologist at the University of Saskatchewan surveys all students from each of 20 randomly selected classes.

7. A sociologist at Grant MacEwan University selects 12 men and 12 women from each of 4 English classes.

8. Sony selects every 200th CD from an assembly line and conducts a thorough test of quality.

9. A tobacco lobbyist writes the name of each member of Parliament on a separate card, shuffles the cards, and then draws 10 names.

10. The marketing manager for Sympatico tests a new sales strategy by randomly selecting 250 consumers with less than $50,000 in gross income and 250 consumers with gross income of at least $50,000.

11. Planned Parenthood polls 500 men and 500 women about their views concerning the use of contraceptives.

12. A market researcher for Air Canada interviews all passengers on each of 10 randomly selected flights.

13. A medical researcher from Acadia University interviews all leukemia patients in each of 20 randomly selected hospitals.

14. A reporter for the *Financial Post* interviews every 25th chief executive officer identified in that magazine's listing of the 500 companies with the highest stock market values.

STATISTICAL PLANET

Hawthorne and Experimenter Effects
The well-known *placebo effect* occurs when an untreated subject incorrectly believes that he or she is receiving a real treatment and reports an alleviation of symptoms. The *Hawthorne effect* occurs when treated subjects somehow respond differently, simply because they are part of an experiment. (This phenomenon is called the "Hawthorne effect" because it was first observed in a study of factory workers at Western Electric's Hawthorne plant.) An *experimenter effect* (sometimes called a Rosenthal effect) occurs when the researcher or experimenter unintentionally influences subjects through such factors as facial expression, tone of voice, or attitude.

15. A reporter for *Report on Business* obtains a numbered listing of the 1000 companies with the highest stock market values, uses a computer to generate 20 random numbers between 1 and 1000, and then interviews the chief executive officers of companies corresponding to these numbers.

16. In conducting research for the Ottawa evening news, a reporter for CBC interviews 15 people as they leave Revenue Canada audits.

17. Describe a procedure for obtaining a simple random sample of 200 students from the population of full-time students at your college.

18. Describe a procedure for obtaining a simple random sample of 100 eligible voters from your local riding.

1-4 Exercises B: Beyond the Basics

19. Two categories of survey questions are *open* and *closed*. An open question allows a free response, while a closed question allows only a fixed response. IIerc are examples:

    Open question: What do you think can be done to reduce crime?
    Closed question: Which of the following approaches would be most effective in reducing crime?

    - Hire more police officers.
    - Get parents to discipline children more.
    - Correct social and economic conditions in slums.
    - Improve rehabilitation efforts in jails.
    - Give convicted criminals tougher sentences.
    - Reform the courts.

    a. What are the advantages and disadvantages of open questions?
    b. What are the advantages and disadvantages of closed questions?
    c. Which type is easier to analyze with formal statistical procedures, and why is that type easier?

20. Among the ten provinces, one province is randomly selected. Then, a province-wide voter registration list is obtained and one name is randomly selected. Does this procedure result in a randomly selected voter?

21. We noted that systematic sampling could result in a random sample, but not a simple random sample.

    a. Does stratified sampling result in a random sample? A simple random sample? Explain.
    b. Does cluster sampling result in a random sample? A simple random sample? Explain.

We will make frequent reference to three particular software packages: Microsoft Excel, Minitab, and STATDISK. STATDISK has two advantages: (1) it is easy-to-use software designed specifically as a supplement to this textbook, and (2) it's *free* to colleges that adopt this textbook. Excel and Minitab are higher-level statistical software packages, but they are also relatively easy to use.

With Excel (prior to 2007) and STATDISK, programs are selected from a main menu bar at the top of the screen, as shown:

EXCEL (prior to 2007): File Edit View Insert Format Tools Data Window Help

STATDISK: File Edit Analysis Data Help

To enter a *new* data set:

EXCEL (prior to 2007): Select **File** from the main menu bar and then select **New**, or click the **New File** icon on the toolbar.

EXCEL 2007: Click the round Microsoft Office button in the upper left corner for the pull-down menu. Proceed as above.

STATDISK: Select **Data** from the main menu bar, then select **Sample Editor**.

To *save* and name a data set:

EXCEL (prior to 2007): Select **File** from the main menu bar, then select **Save As**, or click the **Disk** icon on the toolbar, which brings up a **Save As** dialog box.

EXCEL 2007: Click the round Office button in the upper left corner for the pull-down menu. Proceed as above.

STATDISK: Select **File** from the main menu bar, then select the option of **Save As**.

To retrieve (or *open*) a previously stored data set:

EXCEL (prior to 2007): Select **File** from the main menu bar, then select **Open**, or click the **Open File** icon on the toolbar.

EXCEL 2007: Click the round Office button in the upper left corner for the pull-down menu. Proceed as above.

STATDISK: Select **File** from the main menu bar, then select **Open**. In procedures where STATDISK is expecting input, you can also import data that have been stored in a column of an Excel spreadsheet. Open the relevant Excel file, and highlight the column containing the data; click on **Edit** then on **Copy**. Then open STATDISK, and place the cell pointer at the top of the column that requires the data; click on **Edit** then on **Paste**.

To *print* results:

EXCEL (prior to 2007): Select **File** from the main menu bar, then select the option of **Print,** or select the **Printer** icon on the toolbar.

EXCEL 2007: Click the round Office button in the upper left corner for the pull-down menu. Proceed as above.

STATDISK: Select **File** from the main menu bar, then select **Print.**

To *quit* the program:

EXCEL (prior to 2007): Select **File** from the main menu bar, then select **Exit.**

EXCEL 2007: Click the round Office button in the upper left corner for the pull-down menu. Click the Close icon.

STATDISK: Select **File** from the main menu bar, then select **Quit.**

Excel and STATDISK are each capable of performing almost all of the important operations discussed in this book.

Some statistics professors prefer to use other software packages, such as SPSS, SAS, BMDP, Execustat, Systat, Mystat, or Statgraphics. Whichever software package is chosen, students will generally benefit by improving the computer skills that have become so important in today's world.

VOCABULARY LIST

attribute data **05**
blinding **16**
categorical data **05**
census **04**
cluster sampling **18**
completely randomized
  design **16**
confounding **17**
continuous data **05**
control group **16**
convenience sampling **19**
data **05**
discrete data **05**
double-blind **16**
experiment **15**

interval level of
  measurement **07**
nominal level of
  measurement **06**
nonsampling error **19**
numerical data **05**
observational study **15**
ordinal level of
  measurement **06**
parameter **04**
placebo effect **16**
population **04**
qualitative data **05**
quantitative data **05**
random sample **17**

ratio level of
  measurement **07**
replication **17**
rigorously controlled
  design **16**
sample **04**
sampling error **19**
self-selected sample **10**
simple random
  sample **17**
statistic **04**
statistics **04**
stratified sampling **18**
systematic sampling **18**
treatment group **16**

REVIEW

This chapter began with a general description of the nature of statistics, then discussed different aspects of the nature of data. Uses and abuses of statistics were illustrated with examples. We then discussed the design of experiments, emphasizing the importance of good sampling methods. On completing this chapter, you should be able to do the following:

- Distinguish between a population and a sample
- Distinguish between a parameter and a statistic
- Identify the level of measurement (nominal, ordinal, interval, ratio) of a set of data
- Recognize the importance of good sampling methods, as well as the serious deficiency of poor sampling methods
- Understand the importance of good experimental design, including the control of variable effects, sample size, and randomization

REVIEW EXERCISES

1. The Consumer Product Testing Laboratory selects a dozen batteries (labelled 9 volts) from each company that makes them. Each battery is tested for its actual voltage level.
   a. Are the values obtained discrete or continuous?
   b. Identify the level of measurement (nominal, ordinal, interval, ratio) for the voltages.
   c. Which type of sampling (random, stratified, systematic, cluster, convenience) is being used?
   d. Is this an observational study or an experiment?
   e. What is an important effect of consumers using batteries that are labelled 9 volts when, in reality, the voltage level is very different?

2. Researchers at the Consumer Product Testing Laboratory test samples of electronic surge protectors to find the voltage levels at which computers can be damaged. For each of the following, determine which of the four levels of measurement (nominal, ordinal, interval, ratio) is most appropriate.
   a. The measured voltage levels that cause damage
   b. Rankings (first, second, third, and so on) in order of quality for a sample of surge protectors
   c. Ratings of surge protectors as "recommended, acceptable, not acceptable"
   d. The room temperatures of the rooms in which the surge protectors are tested
   e. The countries in which the surge protectors were manufactured

3. *Report on Business* magazine conducts a survey by mailing a questionnaire to 5000 people known to invest in securities. On the basis of the results, the

magazine editors conclude that most investors in Canada are pessimistic about the economy. What is wrong with that conclusion?

4. You plan to conduct an experiment to test the effectiveness of Statistiszene, a new drug that allegedly improves performance in learning statistics. You will use a sample of subjects who are treated with the drug and another sample of subjects who are given a placebo.
   a. What is blinding, and how might it be used in this experiment?
   b. Why is it important to use blinding in this experiment?
   c. What is a completely randomized design?
   d. What is a rigorously controlled block design?
   e. What is replication, and why is it important?

5. Identify the type of sampling (random, stratified, systematic, cluster, convenience) used in each of the following.
   a. A sample of products is obtained by selecting every 100th item on the assembly line.
   b. Random numbers generated by a computer are used to select serial numbers of cars to be chosen for sample testing.
   c. An auto parts supplier obtains a sample of all items from each of 12 different randomly selected retail stores.
   d. A car maker conducts a marketing study involving test drives performed by a sample of 10 men and 10 women in each of 4 different age brackets.
   e. A car maker conducts a marketing study by interviewing potential customers who request test drives at a local dealership.

6. Census takers have found that in obtaining people's ages, they get more people of age 50 than of age 49 or 51. Can you explain how this might occur?

7. You plan to conduct a survey on your campus. What is wrong with selecting every 50th student leaving the cafeteria?

8. The *Ottawa Citzen* reports that a pro-choice rally was attended by 8725 people. Comment.

CUMULATIVE REVIEW EXERCISES

This book's cumulative review exercises are designed to incorporate some material from preceding chapters, a feature that will be implemented in each chapter. The exercises in this section use concepts learned before beginning the study of this book.

1. The following survey question raised concerns when responses suggested that about 22% of respondents thought that the Holocaust might not have occurred:

   "Does it seem possible or does it seem impossible to you that the Nazi extermination of the Jews never happened?"

A subsequent poll showed that respondents were probably confused by the double negative in the wording of the question. Here is the wording used in a subsequent Roper poll:

"Does it seem possible to you that the Nazi extermination of the Jews never happened, or do you feel certain that it happened?"

Is this second version substantially less confusing? Can you write the question in a way that is clearer than both of these versions?

2. Refer to the graph that follows. It is similar to one that Edwin Tufte, author of *The Visual Display of Quantitative Data*, refers to in observing: "This may well be the worst graphic ever to find its way into print." He notes that the graph reports "almost by happenstance, only five pieces of data (since the division within each year adds to 100 percent)." First examine this graph and identify the information it attempts to relate. Then design a new graph that relates the same information.

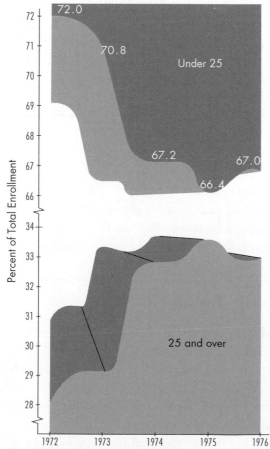

AGE STRUCTURE OF
COLLEGE ENROLLMENT

The objective of this project is to introduce the technology resources that can assist in collecting simple random samples. Suppose that the 108 temperatures in Data Set 18 in Appendix B were, instead of a sample, the complete data listing from a population. Use your statistics software to select a simple random sample of size 6 from this population.

EXCEL: In an empty cell, type in the Excel formula =randbetween (**1108**). This results in an integer in that range. Copy/paste this formula down a column. (*Hint:* Paste to a larger number of cells than the intended size *n* of the sample.)

STATDISK: Click on **Data,** then on **Uniform Generator.** In the dialog box, input a **Sample Size (n)** that is larger than the actual *n* intended for your sample, and input a **Minimum** of 1, a **Maximum** of 108, and 0 (zero) for **Number of Decimals.** Click on **Generate.**

In either of the two preceding methods, you have generated a list of random integers in the range 1 to 108. Each *distinct number* in the output represents one selected member of the sample; simply pick out the patient whose patient number corresponds. If a patient number has already been used, skip to the next number in the output, until all members of the sample have been chosen.

Adapt the method just shown to select a sample of size 5 from the "population" of Bonus numbers listed in Data Set 12 in Appendix B.

■ FROM DATA TO DECISION

Misrepresented Data

Collect an example from a current newspaper or magazine in which data have been presented in a deceptive manner. Identify the source (including the publication date) from which the example was taken. Explain the way in which the presentation is deceptive, and suggest how the data might be presented more fairly.

■ COOPERATIVE GROUP ACTIVITIES

1. **Out-of-class activity:** Divide into groups of 5, then collect 50 data values by using random sampling as described in Section 1–4. Then repeat the collection of 50 data values for each of the other four methods of sampling: stratified, systematic, cluster, convenience. In each case, calculate the "mean." (The mean is defined in Chapter 2 as the average obtained when the scores are added, and the total is then divided by the number of scores.) First describe in detail the procedure used for each method of sampling, then list

the scores, then compare the five means. Does it appear that the different methods of sampling produce the same results? The sample data should be selected from a population such as the ages of books in your college library or the ages of cars in the student parking lot.

2. **In-class activity:** Divide into groups of three or four and use the data given below to construct a graph that exaggerates the increases in the high points of the TSE 300 Composite Index (now the S&P/TSX Composite Index). Also construct a graph that de-emphasizes those increases, and construct a third graph that represents the data fairly.

Time Period	TSE 300 High (S&P/TSX High)
1980–84	2,598.3
1985–89	4,112.9
1990–94	4,609.9
1995–99	8,498.8
2000–04	11,402.0
2005–08	15,154.8

3. **In-class activity:** Divide into groups of three or four. Assume that you must conduct a survey of full-time students at your college. Design and describe in detail a procedure for obtaining a *random* sample of 100 students.

▋ INTERNET PROJECT

In this section of each chapter, you will usually be directed to the following websites

http://www.mathxl.com   or   http://www.mystatlab.com

where a great variety of chapter-related activities, simulations, and examples (and links to other, related sites) await you. These activities will help you to explore and understand the rich nature of statistics and its importance in our world. So visit these websites often and enjoy the activities!

2

Describing, Exploring, and Comparing Data

2-1 Overview

This chapter presents important tables, graphs, and measurements that can be used to describe a data set, explore a data set, or compare two or more data sets. Later chapters will use many of the important concepts introduced in this chapter.

2-2 Summarizing Data with Frequency Tables

The construction of frequency tables, relative frequency tables, and cumulative frequency tables is described. These tables are useful for condensing a large set of data to a smaller and more manageable summary.

2-3 Graphs of Data

Methods for constructing histograms, relative frequency histograms, stem-and-leaf plots, pie charts, Pareto charts, and scatterplots are presented. These graphs are extremely helpful in visually displaying characteristics of data that cannot be seen otherwise.

2-4 Measures of Central Tendency

Measures of central tendency are attempts to find values that are representative of data sets. The following measures of central tendency are defined: *mean*, *median*, *mode*, *midrange*, and *weighted mean*. The concept of skewness is also considered.

2-5 Measures of Variation

Measures of variation are numbers that reflect the amount of scattering among the values in a data set. The following

measures of variation are defined: *range*, *standard deviation*, *mean deviation*, and *variance*. Such measures are extremely important in statistical analyses.

2-6 Measures of Position

The *standard score* (or *z* score) is defined and used to illustrate how unusual values can be identified. Also defined are *percentiles*, *quartiles*, and *deciles* that are used to compare values within the same data set.

2-7 Exploratory Data Analysis (EDA)

Techniques for exploring data with five-number summaries and boxplots are described. Boxplots are especially useful for comparing different data sets.

CHAPTER PROBLEM

Can 12-oz aluminum cans be made thinner to save money?

Data Set 15 in Appendix B includes these two samples:

1. 12-oz aluminum cans that are 0.0109 in. thick (reproduced as Table 2-1)
2. 12-oz aluminum cans that are 0.0111 in. thick

We will explore the values in Table 2-1, which lists the axial loads (in pounds) of the sample of aluminum cans that are 0.0109 in. thick. This data set was supplied by a student who used a previous edition of this book. She is an employee of the company that manufactures the cans considered here, and she uses methods she learned in her introductory statistics course. The authors are grateful for her contributions.

The axial load of a can is the maximum weight supported by its sides, and it is measured by using a plate to apply increasing pressure to the top of the can until it collapses. It is important to have an axial load high enough so that the can isn't crushed when the top lid is pressed into place. In this particular manufacturing process, the top lids are pressed into place with pressures that vary between 158 lb and 165 lb.

Thinner cans have the obvious advantage of using less material so that costs are lower, but thinner cans are probably not as strong as the thicker ones. The company manufacturing these cans is currently using a 0.0111-in. thickness, but is testing to determine if the thinner cans could be used. Using the methods of this chapter, we will explore the data set (reproduced in Table 2-1) for these thinner cans (0.0109 in. thick). Ultimately, we will determine whether these thinner cans could be used.

Table 2-1 Axial Loads of 0.0109-in. Cans

270	273	258	204	254	228	282
278	201	264	265	223	274	230
250	275	281	271	263	277	275
278	260	262	273	274	286	236
290	286	278	283	262	277	295
274	272	265	275	263	251	289
242	284	241	276	200	278	283
269	282	267	282	272	277	261
257	278	295	270	268	286	262
272	268	283	256	206	277	252
265	263	281	268	280	289	283
263	273	209	259	287	269	277
234	282	276	272	257	267	204
270	285	273	269	284	276	286
273	289	263	270	279	206	270
270	268	218	251	252	284	278
277	208	271	208	280	269	270
294	292	289	290	215	284	283
279	275	223	220	281	268	272
268	279	217	259	291	291	281
230	276	225	282	276	289	288
268	242	283	277	285	293	248
278	285	292	282	287	277	266
268	273	270	256	297	280	256
262	268	262	293	290	274	292

2-1 Overview

We sometimes collect data in order to address a specific issue. For example, a safety study of the elevators in the CN Tower would require data concerning the average weight of the people who ride the elevators. In other cases, we collect or obtain data without having a specific objective, but because we wish to explore the data to see what might be revealed. A geologist may wonder about the time intervals between eruptions of the Old Faithful geyser—are they equally distributed over the range of times or do some time intervals occur more often than others? In both circumstances, we need a variety of tools that will help us *understand* the data set. This chapter presents those tools.

When analyzing a data set, we should first determine whether we have a *sample* or a complete *population*. That determination will affect both the methods we use and the conclusions we form. We use methods of **descriptive statistics** to summarize or describe the important characteristics of an available set of data, and we use methods of **inferential statistics** when we specifically use sample data to make inferences (or generalizations) about a population. When your professor calculates the final exam average for your statistics class, that result is an example of a descriptive statistic. However, if we state that the result is an estimate of the final exam average for all statistics classes, we are making an inference that goes beyond the known data.

Descriptive statistics and inferential statistics are the two general divisions of the subject of statistics. This chapter deals with the basic concepts of descriptive statistics.

Important Characteristics of Data

When describing, exploring, and comparing data sets, the following characteristics of data are usually most important.

1. *Centre:* A representative or average value that indicates where the middle of the data set is located

2. *Variation:* A measure of the amount that the values vary among themselves

3. *Distribution:* The nature or shape of the distribution of the data (such as bell-shaped, uniform, or skewed)

4. *Outliers:* Sample values that lie very far away from the vast majority of the other sample values

5. *Time:* Changing characteristics of the data over time

In this chapter we begin to show how the tools of statistics can be applied to these characteristics of data to provide instructive insights. The chapter includes detailed steps for important procedures, although it is recognized that, today, a computer or calculator can automate many of these. We recommend that in each case you perform a few of the calculations manually, to enhance your understanding of the underlying techniques. (However, using Excel to help with repetitive multiplying and so on is a handy compromise—if your computer is nearby.)

2-2 Summarizing Data with Frequency Tables

When investigating large data sets, it is generally helpful to organize and summarize the data by constructing a frequency table.

DEFINITION

A **frequency table** lists classes (or categories) of values, along with frequencies (or counts) of the number of values that fall into each class.

Table 2-2 is a frequency table with 10 classes. The **frequency** for a particular class is the number of original values that fall into that class. For example, the first class in Table 2-2 has a frequency of 9, indicating that there are 9 values between 200 and 209 inclusive.

We will first present some standard terms used in discussing frequency tables, and then describe a procedure for constructing them.

DEFINITIONS

Lower class limits are the smallest numbers that can actually belong to the different classes. (Table 2-2 has lower class limits of 200, 210, . . . , 290.)

Upper class limits are the largest numbers that can actually belong to the different classes. (Table 2-2 has upper class limits of 209, 219, . . . , 299.) For continuous data, these upper class limits actually represent the largest values in the classes—*when rounded to the indicated number of decimal places*. For example, the class 200–209 may, in fact, include a can with an axial load of 209.49, which has been rounded to 209; but a can with an axial load of 209.50 will have been rounded to 210 and included in the next class.

Class boundaries are the numbers used to separate classes, but without the gaps created by class limits. They are obtained as follows: Find the size of the gap between the upper class limit of one class and the lower class limit of the next class. Add half of that amount to each upper class limit to find the upper class boundaries; subtract half of that amount from each lower class limit to find the lower class boundaries. (Table 2-2 has class boundaries of 199.5, 209.5, 219.5, . . . , 299.5.)

Class midpoints are the midpoints of the classes. (Table 2-2 has class midpoints of 204.5, 214.5, . . . , 294.5.) Each class midpoint can be found by adding the lower class limit to the corresponding upper class limit and dividing the sum by 2.

Class width is the difference between two consecutive lower class limits or two consecutive lower class boundaries. (Table 2-2 uses a class width of 10.)

Table 2-2 Frequency Table of Axial Loads of Aluminum Cans

Axial Load	Frequency
200–209	9
210–219	3
220–229	5
230–239	4
240–249	4
250–259	14
260–269	32
270–279	52
280–289	38
290–299	14

The definitions of class midpoints and class boundaries are tricky. Be careful to avoid the easy mistake of making the class width the difference between the lower class limit and the corresponding upper class limit. See Table 2-2 and note that the class width is 10, not 9. You can simplify the process of finding class boundaries by understanding that they basically fill the gaps between classes by splitting the difference between the end of one class and the beginning of the next class. Carefully examine the definition of class boundaries, and spend some time until you understand it.

It should be noted that if the data are continuous, the end of one class may be the beginning of the next class. For example, one class may be 5 to under 10 and the next class 10 to under 15. In this situation, the class boundary between the two classes would be 10, not 9.5.

Constructing Frequency Tables

The main reason for constructing a frequency table is to use it for constructing a graph that effectively shows the distribution of the data (for example, see *histograms*, introduced in the next section.) It is important to observe the following *guidelines* when constructing a frequency table—whether manually or by using technology.

1. *Be sure that the classes are mutually exclusive.* That is, each of the original values must belong to exactly one class.
2. *Include all classes*, even those for which the frequency is zero.
3. *Try to use the same width for all classes*, although sometimes open-ended intervals such as "65 years or older" are impossible to avoid.
4. *Select convenient numbers for class limits.* Round up to use fewer decimal places or use numbers relevant to the situation.
5. *Use between 5 and 20 classes.*
6. *The sum of the class frequencies must equal the number of original data values.*

A common convention is for a frequency table to have four columns: class, frequency, relative frequency, and cumulative relative frequency. This allows the construction of a range of graphs from one table.

If you are constructing a frequency table manually, follow the procedure given below. (While some instructors include these detailed steps, others may skip them, perhaps preferring a computer-based approach.)

STEP 1: *Decide on the number of classes your frequency table will contain.* As a guideline, the number of classes should be between 5 and 20. The actual number of classes may be affected by the convenience of using round numbers or other subjective factors.

STEP 2: *Determine the class width by dividing the range by the number of classes.* (The range is the difference between the highest and the lowest scores.) Round the result *up* to a convenient number so as to guarantee that all of the data will be included in the frequency table.

$$\text{class width} = \text{round } up \text{ of } \frac{\text{range}}{\text{number of classes}}$$

STEP 3: *Select as the lower limit of the first class either the lowest score or a convenient value slightly less than the lowest score.* This value serves as the starting point.

STEP 4: *Add the class width to the starting point to get the second lower class limit.* Add the class width to the second lower class limit to get the third, and so on. If you need to add one more class in order to include all the data, that is generally not a problem.

STEP 5: *List the lower class limits in a vertical column, and enter the upper class limits, which can be easily identified at this stage.*

STEP 6: *Represent each data value by a tally in the appropriate class, then use those tally marks to find the total frequency for each class.*

Determination of the number of classes is not yet part of federal law, so it is generally okay to use a different number of classes—which will result in a different frequency table—providing the table still has convenient and understandable values. (Exercise 25 in this section is about a possible exception.)

EXAMPLE

Construct a frequency table for the 175 axial loads of aluminum cans given in Table 2-1.

SOLUTION

We will list the steps that lead to the development of the frequency table shown in Table 2-2.

Step 1: We begin by selecting 10 as the number of desired classes. (Within the guideline of 5 to 20 classes, choose a small number of classes with smaller data sets and a larger number of classes with larger data sets.)

Step 2: With a minimum of 200 and a maximum of 297, the range is $297 - 200 = 97$, so

$$\begin{aligned} \text{class width} &= \text{round up of } \frac{97}{10} \\ &= \text{round up of } 9.7 \approx 10 \end{aligned}$$

Step 3: The lowest value is 200. Because it is a convenient value, it becomes the starting point and we use it for the lower limit of the first class.

Step 4: Add the class width of 10 to the lower limit of 200 to get the next lower limit of 210. Continuing, we get the other lower class limits of 220, 230, and so on.

Step 5: The lower class limits suggest these upper class limits:

200	209
210	219
etc.	

(The raw data are all rounded to the nearest whole unit.)

Step 6: Count up the data values in each class and enter the frequencies as shown in the right column of Table 2-2.

Table 2-2 provides useful information by making the list of axial loads more intelligible, but we cannot reconstruct the original 175 axial loads from the frequency table. We have traded the exactness of the original data for a better understanding of the data.

Relative Frequency Table

An important variation of the basic frequency table is to include **relative frequencies,** which are easily found by dividing each class frequency by the total of all the frequencies. The relative frequencies are often expressed as percentages.

$$\text{relative frequency} = \frac{\text{class frequency}}{\text{sum of all frequencies}}$$

Table 2-3 is an example of a **relative frequency table** and includes the relative frequencies for the 175 axial loads summarized in Table 2-2. The first class has a relative frequency of $9/175 = 0.051$, or 5.1%. The second class has a relative frequency of $3/175 = 0.017$, or 1.7%, and so on. If constructed correctly, the sum of the relative frequencies should total 1 (or 100%), with small discrepancies allowed for rounding errors.

Table 2-3 Relative Frequency Table of Axial Loads of Aluminum Cans

Axial Load	Frequency	Relative Frequency
200–209	9	0.051
210–219	3	0.017
220–229	5	0.029
230–239	4	0.023
240–249	4	0.023
250–259	14	0.080
260–269	32	0.183
270–279	52	0.297
280–289	38	0.217
290–299	14	0.080

Table 2-4 Cumulative Frequency Table of Axial Loads

Axial Load	Frequency	Relative Frequency	Cumulative Relative Frequency
200–209	9	0.051	0.051
210–219	3	0.017	0.069
220–229	5	0.029	0.097
230–239	4	0.023	0.120
240–249	4	0.023	0.143
250–259	14	0.080	0.223
260–269	32	0.183	0.406
270–279	52	0.297	0.703
280–289	38	0.217	0.920
290–299	14	0.080	1.000

Relative frequency tables make it easier for us to understand the distribution of the data and to compare different sets of data.

Cumulative Relative Frequency Table

Another variation of the standard frequency table is used when cumulative totals are desired. The **cumulative relative frequency** for a class is the sum of the relative frequencies for that class and all previous classes. Table 2-4 is an example of a **cumulative relative frequency table**, complete with frequencies, relative frequencies, and cumulative relative frequencies. For example, in the first row there are 9 values less than 210 for a cumulative relative frequency of $9/175 = 0.051 = 5.1\%$, in the second row there are $9 + 3 = 12$ values less than 220 for a cumulative relative frequency of $12/175 = 0.069 = 6.9\%$, and so on. If constructed correctly, the last cumulative relative frequency must be equal to 100%.

Once the table is complete, we can ask questions about it.

EXAMPLE

    a. What percentage of the axial loads are less than 240?

    b. What percentage of the axial loads are at least 260?

    c. What percentage of the axial loads are at least 220 but less than 260?

SOLUTION

    a. From the cumulative relative frequency column, we observe the value of 0.120, meaning that 12.0% of the axial loads are less than 240.

    b. From the cumulative relative frequency column, 0.223 or 22.3% are less than 260. From this, we can calculate that the percentage that are at least 260 is $100\% - 22.3\% = 77.7\%$.

c. The best approach is to add the relevant frequencies of 5, 4, 4, and 14 for a sum of 27 and divide 27 by 175 to obtain 0.154 = 15.4%. If we were to add the relative frequencies of 0.029, 0.023, 0.023 and 0.080, we would get 0.155 instead. This approach would be incorrect since we would be introducing unnecessary rounding errors.

USING TECHNOLOGY

Bin	Frequency	Cumulative %
J	K	L
209	9	5.14%
219	3	6.86%
229	5	9.71%
239	4	12.00%
249	4	14.29%
259	14	22.29%
269	32	40.57%
279	52	70.29%
289	38	92.00%
299	14	100.00%
More	0	100.00%

EXCEL DISPLAY

EXCEL: Although Excel cannot generate a frequency distribution table from scratch, it can automate the job of tallying the frequencies within each class. Follow the guidelines of this section to decide on the best set of class limits. Input the *upper class limits* into an Excel column (for example, J5:J14). Input the raw data into a column or block of cells (for example, B4:H28).

EXCEL (Prior to 2007): On Excel's menu bar, click **Tools**, then select **Data Analysis**, then **Histogram**. For *Input Range*, type in the address range for the raw data; for *Bin Range*, type in the address range for the column of upper class limits. For *Output Range*, type in the desired cell address for the *upper left corner* of the output. For cumulative relative frequencies, there is an optional checkbox. The output that corresponds to the chapter problem is shown in here.

EXCEL 2007: Click the **Data** tab and then **Data Analysis** in the **Analysis** group. Proceed as above.

MINITAB 15: Put the data in column C1, starting at row 1. From the **Stat** menu, choose **Tables**, then choose **Tally Individual Variables** from the submenu. Click on C1 in the left window and then click **Select**. C1 will now appear in the **Variables** window to the right. By default, the **Counts** box is checked. You can also check the **Percents, Cumulative Counts**, and **Cumulative Percents** boxes.

MINITAB DISPLAY

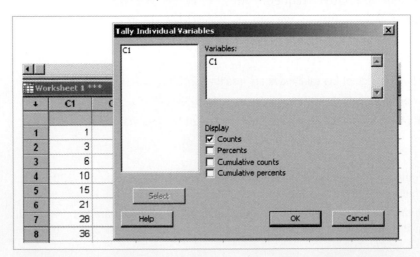

2-2 Exercises A: Basic Skills and Concepts

In Exercises 1–4, identify the class width, class midpoints, and class boundaries for the given frequency table.

1.
Absences	Frequency
0–5	39
6–11	41
12–17	38
18–23	40
24–29	42

2.
Absences	Frequency
0–9	22
10–19	40
20–29	71
30–39	44
40–49	23

3.
Mass (kg)	Frequency
0.00–1.99	20
2.00–3.99	32
4.00–5.99	49
6.00–7.99	31
8.00–9.99	18

4.
Mass (kg)	Frequency
0.0–4.9	60
5.0–9.9	58
10.0–14.9	61
15.0–19.9	62
20.0–24.9	59

In Exercises 5–8, complete the frequency table that corresponds to the frequency table in the exercise indicated by adding the relative frequency and cumulative relative frequency columns, accurate to 3 decimals.

5. Exercise 1   6. Exercise 2   7. Exercise 3   8. Exercise 4

In Exercises 9–12, compute the probabilities requested for the specified tables.

9. Exercise 1
   a. What percentage of the absences are less than 12?
   b. What percentage of the absences are more than 17?
   c. What percentage of the absences are at least 12 but less than 24?

10. Exercise 2
   a. What percentage of the absences are no greater than 19?
   b. What percentage of the absences are at least 30?
   c. What percentage of the absences are from 10 through 29 inclusive?

11. Exercise 3
   a. What percentage of the weights are less than 6 kg?
   b. What percentage of the weights are at least 2 kg?
   c. What percentage of the weights are at least 4 kg but less than 8 kg?

12. Exercise 4
   a. What percentage of the weights are less than 15 kg?
   b. What percentage of the weights are at least 10 kg?
   c. What percentage of the weights are at least 5 kg but less than 20 kg?

13. Compare the *distribution* of the data in Exercise 1 to the *distribution* of the data in Exercise 2. What is the basic difference?

14. Compare the *distribution* of the data in Exercise 3 to the *distribution* of the data in Exercise 4. What is the basic difference?

In Exercises 15–16, use the given information to find upper and lower limits of the first class. (The data are in Appendix B, but there is no need to refer to the appendix for these exercises.)

15. A data set consists of weights of metal collected from households for one week. Those weights range from 0.26 lb to 4.95 lb. You wish to construct a frequency table with 10 classes.

16. A sample of M&M candies has weights that vary between 0.838 g and 1.033 g. You wish to construct a frequency table with 12 classes.

In Exercises 17–20, construct the frequency table and answer the associated questions.

17. Refer to the labour force data in Data Set 14 in Appendix B, using the column for male Newfoundlanders. Construct a frequency table of the sizes of occupational groups—interpreting the "size" of an occupational group as the number of male Newfoundlanders having a given SOC (Standard Occupational Classification) code. Use 12 classes, beginning with a lower class limit of 0. Describe two different notable features of the result.

18. Refer to the TEMPAGE column in Data Set 18 in Appendix B, and construct a frequency table of the ages of patients admitted for voluntary surgery. Start at age 13 and use a class width of 17.5. Does the result violate any of the guidelines for constructing frequency tables?

19. Refer to Data Set 20 in Appendix B, and construct a frequency table with 10 classes of the annual unemployment rates for Canada. The table you construct will have no reference to time or year. Compare the table with the original data, and evaluate whether the unemployment numbers appear random with respect to time, or whether time is associated in some way with the results of the frequency distribution.

20. Refer to Data Set 21 in Appendix B. A set of high-price diamonds is sold at auction. Construct a frequency table for the variable colour of the diamonds. If the colour of diamonds that fetch over $50,000 is never greater than 3, are these diamonds likely to be bought at prices over $50,000?

21. Below are two frequency tables for 32 cities, showing annual precipitation and annual snowfall. First complete the corresponding relative frequency columns, and then use those results to compare the two samples.

Precipitation (mm)	Frequency	Snowfall (cm)	Frequency
266 up to 492	8	32 up to 68	2
492 up to 718	4	68 up to 104	1
718 up to 944	7	104 up to 140	5
944 up to 1170	6	140 up to 176	3
1170 up to 1396	2	176 up to 212	4
1396 up to 1622	4	212 up to 248	4
1622 up to 1848	0	248 up to 284	2
1848 up to 2074	0	284 up to 320	4
2074 up to 2300	0	320 up to 356	5
2300 up to 2526	1	356 up to 392	2

22. Listed below are two sets of scores that are supposed to be heights (in inches) of randomly selected adult males. One of the sets consists of heights actually obtained from randomly selected adult males, but the other set consists of numbers that were fabricated. Construct a frequency table for each set of heights. By examining the two frequency tables, identify the set of data that you believe to be false, and state your reason.

    a. 70 73 70 72 71 73 71 67 68 72 67 72 71 73
       72 70 72 68 71 71 71 73 69 73 71 66 77 67
    b. 70 73 70 72 71 66 74 76 68 75 67 68 71 77
       66 69 72 67 77 75 66 76 76 77 73 74 69 67

23. Refer to Data Set 15 in Appendix B. The axial loads are the weights that were applied before the cans collapsed. Construct two separate frequency tables for the cans that are 0.0111 in. thick. In the first table, do not include the value of 504 lb. Start the first class at 200 lb and use a class width of 20 lb. The load of 504 lb is called an outlier because it is far away from all the other values. Construct a second frequency table which does include that value. Compare the results, and interpret the effect of the outlier on the results.

24. In constructing a frequency table, Sturges' guideline suggests that the ideal number of classes can be approximated by $1 + (\log n)/(\log 2)$, where n is the number of values. Use this guideline to suggest a tentative number of classes (rounded off) for constructing a frequency table from a data set with each number of values.

    a. 50   b. 100   c. 150
    d. 500   e. 1000   f. 50,000

25. Is it possible that the number of classes in a frequency table can have a dramatic effect on the apparent distribution of the data? If so, construct a data set for which a change from five classes to six results in a dramatic change in the apparent distribution of the frequencies.

2-3 Graphs of Data

In Section 2–2 we showed how frequency tables can be used to describe, explore, or compare distributions of data sets. In this section we continue the study of distributions by introducing graphs that show the distributions of data in pictorial form. As you read through this section, keep in mind that the objective is not simply to construct a graph, but rather to learn something about a data set—that is, to understand the nature of the distribution.

Histograms

A common and important graphic device for presenting frequency of information is the *histogram*.

DEFINITION

> A **histogram** is a bar graph in which the horizontal scale represents classes and the vertical scale represents frequencies. The heights of the bars correspond to the frequency values, and the bars are drawn adjacent to each other (usually without gaps—at least for manually drawn versions).

We generally construct a histogram after we have first completed a frequency table representing a data set. Each bar of a histogram is marked with its lower class boundary at the left and its upper class boundary at the right. Improved readability, however, is often achieved by using class midpoints instead of class boundaries. The histogram in Figure 2-1 corresponds directly to the frequency table (Table 2-2) in the previous section.

Before constructing a histogram from a completed frequency table, we must give some consideration to the scales used on the vertical and horizontal axes. The maximum frequency (or the next highest convenient number) should suggest a value for the top of the vertical scale; 0 should be at the bottom. In Figure 2-1 we designed the vertical scale to run from 0 to 60. The horizontal scale should be

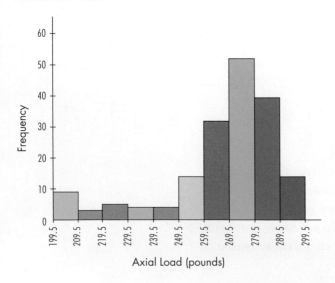

Figure 2-1 Histogram of Axial Loads of Aluminum Cans

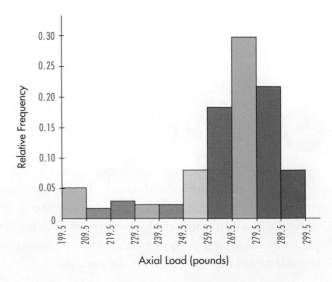

Figure 2-2 Relative Frequency Histogram of Axial Loads of Aluminum Cans

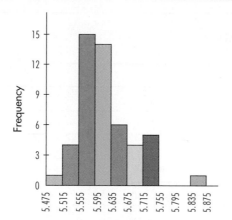

Figure 2-3
Histogram of Weights of
Quarters

designed to accommodate all the classes of the frequency table. Ideally, we should try to follow the rule of thumb that the vertical height of the histogram should be about three-fourths of the total width. Both axes should be clearly labelled.

Relative Frequency Histogram

A **relative frequency histogram** has the same shape and horizontal scale as a histogram, but the vertical scale is marked with *relative frequencies* instead of actual frequencies, as in Figure 2-2.

AS

Frequency tables and graphs such as histograms enable us to see how our data are distributed, and the distribution of data is an extremely important characteristic. Figure 2-3 is roughly bell-shaped in the sense that it resembles the shape shown in the margin. Many procedures in statistics require that a distribution have a bell-shaped distribution similar to the one shown in Figure 2-3, and one way to check that requirement is to construct a histogram.

Bell Shape

Compare the ideal bell shape with Figure 2-1, the axial loads of aluminum cans. The fact that the left tail of the distribution extends far to the left, showing that a number of cans had apparently low values for their axial loads, might be an indicator of a quality problem.

Frequency Polygon

A **frequency polygon** uses line segments connected to points located directly above the class midpoint values for a frequency distribution. Figure 2-4 shows the frequency polygon corresponding to Table 2-2. The heights of the points correspond to the class frequencies, and the line segments are extended to the left and right

Figure 2-4
Frequency Polygon for Axial
Loads of Aluminum Cans

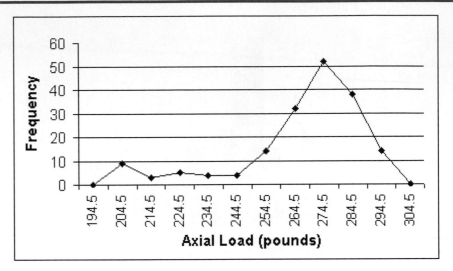

so that the graph meets the *x*-axis at one class width below the lowest class mid-point, and at one class width above the largest class midpoint.

Ogive

An **ogive** (pronounced "oh-jive") is a line graph that depicts cumulative frequencies, just as the cumulative frequency table (see Table 2-4 in the preceding section) lists cumulative frequencies. Figure 2-5 is an ogive corresponding to Table 2-4. Note that the ogive uses class boundaries along the horizontal scale and that the graph begins with the lower boundary of the first class and ends with the upper boundary of the last class. Ogives are useful for determining the number of values *less than* some particular class boundary. For example, Figure 2-5 shows that about 160 of the values are less than 289.5.

Figure 2-5
Ogive of Axial Loads
of Aluminum Cans

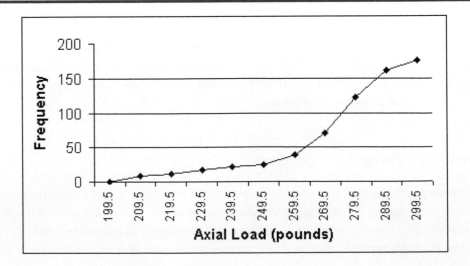

Stem-and-Leaf Plots

In a **stem-and-leaf plot** we sort data according to a pattern that reveals the underlying distribution. The pattern involves separating a number (such as 257) into two parts, usually the first one or two digits (25) and the remaining digits (7). The stem consists of the leftmost digits (in this case, 25) and the leaves consist of the rightmost digits (in this case, 7). The method is illustrated in the following example.

EXAMPLE

Use the axial loads of aluminum cans listed in Table 2-1 to construct a stem-and-leaf plot.

SOLUTION

If we use the two leftmost digits for the stem, the stems consist of 20, 21, . . . , 29. We then draw a vertical line and list the leaves as shown below. The first value in Table 2-1 is 270, and we include that value by entering a 0 in the stem row for 27. We continue to enter all 175 values, and then we arrange the leaves (the digits positioned to the right) so that the numbers are arranged in increasing order. The first row represents the numbers 200, 201, 204, 204, 206, and so on.

Stem	Leaves
20	014466889
21	578
22	03358
23	0046
24	1228
25	01122466677899
26	012222233333455567788888888889999
27	00000000112222233333344445555566666777777778888888999
28	000111122222233333344445556666677899999
29	00011222334557

By turning the page on its side, we can see a distribution of these data. Here's the great advantage of the stem-and-leaf plot: We can see the distribution of the data and yet retain all the information in the original list; if necessary, we could reconstruct the original list of values.

You might notice that the rows of digits in a stem-and-leaf plot are similar in nature to the bars in a histogram. One of the guidelines for constructing histograms is that the number of classes should be between 5 and 20, and that same guideline applies to stem-and-leaf plots for the same reasons. Stem-and-leaf plots can be *expanded* to include more rows and can also be *condensed* to include fewer rows. The stem-and-leaf plot of the preceding example can be expanded by subdividing rows into those with the digits 0 through 4 and those with digits 5 through 9. This is commonly called a split stem-and-leaf plot since the stem is split into two or more rows. This expanded stem-and-leaf plot is shown here.

Stem	Leaves
20	0144
20	66889
21	
21	578
22	033
22	58
23	004
23	6
24	122
24	8
25	011224
25	66677899
26	0122222333334
26	5556778888888889999
27	000000001122222333333344444
27	5555666667777777788888888999
28	000111122222233333344444
28	555666667789999
29	00011222334
29	557

When it becomes necessary to *reduce* the number of rows, we can *condense* a stem-and-leaf plot by combining adjacent rows, as in the following illustration. Note that we separate digits in the leaves associated with the numbers in each stem by an asterisk. Every row in the condensed plot must include exactly one asterisk so that the shape of the plot is not distorted.

```
78-79  07*4      ←This row represents 780, 787, 794.
80-81  *55       ←This row represents 815, 815.
82-83  9*        ←This row represents 829.
84-85  *         ←This row has no data.
86-87  79*0      ←This row represents 867, 869, 870.
```

Another advantage of stem-and-leaf plots is that their construction provides a fast and easy procedure for *sorting* data (arranging data in order). Data must be sorted for a variety of statistical procedures, such as finding the median (discussed in Section 2–4) and finding percentiles or quartiles (discussed in Section 2–6).

Pareto Charts

Consider this statement: Among 186,291 deaths in Canada in 1995, 79,107 were attributable to heart disease, 58,817 to cancer, 18,886 to respiratory diseases, 15,537 to stroke, 7723 to accidents, 3968 to suicide, 1764 to AIDS, and 489 to murder (based on data from Statistics Canada). The information contained in this statement concerning the relationships among the data would be much more effectively conveyed in a Pareto chart. A **Pareto chart** is a bar graph for qualitative data, with the bars arranged in order according to frequencies. As in histograms, vertical scales in Pareto charts can represent frequencies or relative frequencies. The tallest bar is at the left, and the smaller bars are at the right, as in Figure 2-6. By arranging the bars in the order of frequency, the Pareto chart focuses attention on the more important categories. From Figure 2-6 we can see that heart disease is a problem that is much more prevalent than the other categories.

Pie Charts

Like Pareto charts, pie charts are used to depict qualitative data in a way that makes them more understandable. Figure 2-7 shows an example of a **pie chart**, which graphically depicts qualitative data as slices of a pie. Construction of such a pie chart involves slicing up the pie into the proper proportions. If the category of heart disease represents 42.5% of the total, then the wedge representing heart disease should be 42.5% of the total. (The central angle should be $0.425 \times 360° = 153°$.)

The Pareto chart of Figure 2-6 and the pie chart of Figure 2-7 depict the same data in different ways, but a comparison will probably show that the Pareto chart does a better job of showing the relative sizes of the different components.

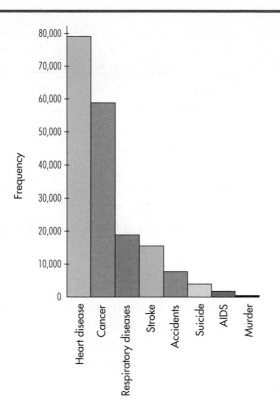

Figure 2-6
**Pareto Chart of Deaths
by Type**

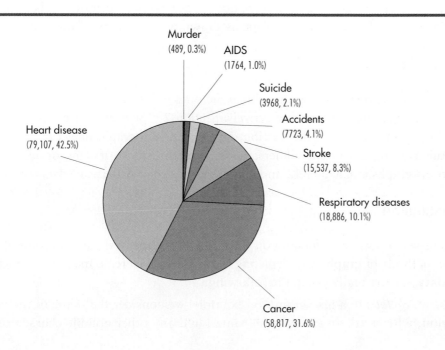

Figure 2-7
**Pie Chart of Deaths by
Type**

Figure 2-8
Scatter Diagram
Comparing Canadian
Imports and Exports

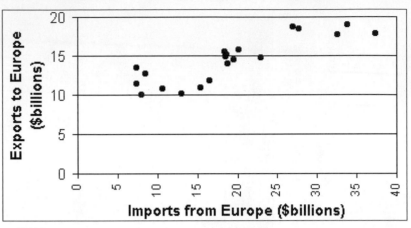

Scatter Diagrams

A **scatter diagram** is a plot of paired (x, y) data with a horizontal x-axis and a vertical y-axis. The data are paired in a way that matches each value from one data set with a corresponding value from a second data set. To manually construct a scatter diagram, construct a horizontal axis for the values of the first variable, construct a vertical axis for the values of the second variable, and then plot the points. The pattern of the plotted points is often helpful in determining whether there is some relationship between the two variables. (This issue is discussed at length when the topic of correlation is considered in Section 11–2.)

Using the data on Canadian exports and imports to and from Europe in Data Set 22 in Appendix B, we used Excel to generate the scatter diagram shown in Figure 2-8. On the basis of that graph, there does appear to be a relationship between the dollar values of imports and exports, as shown by the pattern of the points.

Other Graphs

Numerous pictorial displays other than the ones just described can be used to represent data dramatically and effectively. Section 2–7 will present boxplots, which are very useful for seeing the distribution of data. Pictographs depict data by using pictures of objects, such as soldiers, tanks, airplanes, stacks of coins, or moneybags. Various graphs in Chapter 12 and Chapter 14E depict patterns of data over time.

Conclusion

In this section we have focused on the nature or shape of the distribution of data and methods of graphically depicting data. When we create meaningful pictures of data, we are really doing the following:

Describing data: In a histogram, for example, we consider the shape of the distribution, whether there are extreme values, and any other notable characteristics.

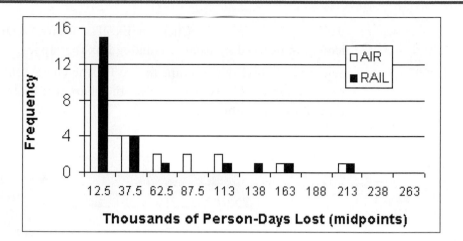

We note the overall shape of the distribution. Are the values evenly distributed? Is the distribution skewed (lopsided) to the right or left? Does the distribution peak in the middle? We observe any outliers (values located far away from most of the other values) that might hide the true nature of the distribution.

Exploring data: We look for any features of the graph that reveal useful and/or interesting characteristics of the data set. In Figure 2-9, for example, we see a common distribution pattern among graphs of *severities*—such as the severities of loss or damage caused by different storms or accidents. Based on Data Set 20 in Appendix B, it appears that for both the air and rail transportation industries, most stoppages result in relatively few person-days lost; but there is an extended tail of work stoppages that involve many more people.

Comparing data: Figure 2-9 allows us to compare the person-days-lost distributions for work stoppages in the air and rail transportation industries. It appears that the air transportation industry has been more prone to work stoppages that involve 62,500 or more person-days lost.

USING TECHNOLOGY

EXCEL (Prior to 2007): Excel can generate many of the graphs discussed in this section. In fact, Figure 2-4 (frequency polygon), Figure 2-5 (ogive), Figure 2-8 (scatter diagram), and Figure 2-9 (histogram comparisons) were all created with Excel. (Pie charts can also be generated in Excel.) For any of these graph types, begin by entering the relevant data in separate Excel columns. (For a histogram, for example, enter the class [or category] labels in one column, and the corresponding data values in another.) Click on Excel's "Chart Wizard" icon, and proceed by using the dialog box that appears.

EXCEL 2007:
Select the cells that you want to use for the chart. On the **Insert** tab, in the **Charts** group, select the chart type and sub-type. Proceed as above.

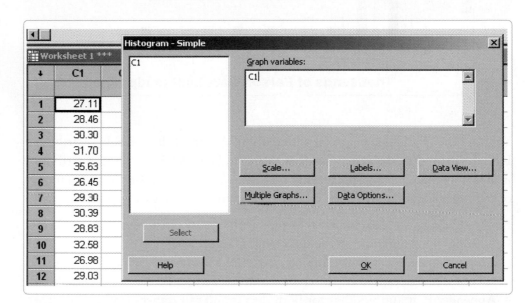

MINITAB 15:
From the **Graph** menu you can choose various types of graphs. Put the data in separate columns for each group of data. For example, for a simple histogram of one data set, put the data in column C1, choose **Histogram** from the **Graph** menu, then **Simple** from the sublist. Click C1 in the left window and then **Select**. C1 will now appear in the right window. For the default options, click OK. However, you can adjust different features such as **Labels**, which enables you to put a title to your graph.

STATDISK:
An advantage of STATDISK is that you can generate histograms directly from the raw data—without needing to first construct a frequency table. Click to select **Data**, then **Histogram**. As displayed in the figure, you have the choice of specifying the class width and class start (i.e., the lowest lower class limit), or you can click on **Auto Fit** and let

STATDISK decide on these values for you. You can also click **Data,** then **Scatter Diagram,** to generate a scatter diagram from the raw data for two paired variables.

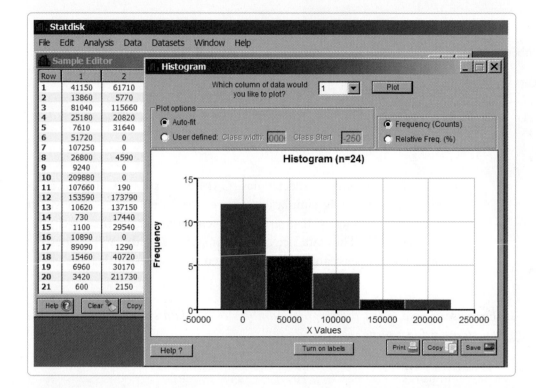

2-3 Exercises A: Basic Skills and Concepts

1. Visitors to Yellowstone National Park consider an eruption of the Old Faithful geyser to be a major attraction that should not be missed. The given frequency table summarizes a sample of times (in minutes) between eruptions. Construct a histogram corresponding to the given frequency table. If you're scheduling a bus tour of Yellowstone, what is the minimum time you should allocate to Old Faithful if you want to be reasonably sure that your tourists will see an eruption?

Time	Frequency
40–49	8
50–59	44
60–69	23
70–79	6
80–89	107
90–99	11
100–109	1

Age	Students	Faculty/Staff
0–2	23	30
3–5	33	47
6–8	63	36
9–11	68	30
12–14	19	8
15–17	10	0
18–20	1	0
21–23	0	1

2. Samples of student cars and faculty/staff cars were obtained at the author's college, and their ages (in years) are summarized in the accompanying frequency table. Construct a relative frequency histogram for student cars and another relative frequency histogram for faculty cars. Based on the results, what are the noticeable differences between the two samples?

Income Tax Payable	Number of Clients
$0 to less than $1000	15
$1000 to less than $2000	21
$2000 to less than $3000	35
$3000 to less than $4000	12

3. The given group data were collected on the amount of income tax payable for the clients of an accounting firm. Construct a histogram corresponding to this frequency table.

Minutes	Frequency
100 up to 200	1
200 up to 300	8
300 up to 400	5
400 up to 500	1
500 up to 600	0
600 up to 700	1
700 up to 800	4
800 up to 900	5

4. Canadian Automated Electronics (CAE) did a study of time to complete tasks entailing simulation control for 25 employees. Construct a relative frequency histogram that corresponds to the given frequency table. The data are a summary of the study. What does the histogram suggest about the time taken to complete a task?

In Exercises 5 and 6, list the original numbers in the data set represented by the given stem-and-leaf plots.

5.
Stem	Leaves
57	017
58	13349
59	456678
60	23

6.
Stem	Leaves
10	21 45 91
11	11 32 77 83
12	04 22 49
13	69

In Exercises 7 and 8, construct the stem-and-leaf plots for the given data sets found in Appendix B.

7. The lengths (in inches) of the bears in Data Set 3. (*Hint:* First round the lengths to the nearest inch.)

8. Weights (in pounds) of plastic discarded by 62 households as listed in Data Set 1. Start by rounding the listed weights to the nearest tenth of a pound (or one decimal place). (Use an expanded stem-and-leaf plot with about 11 rows.)

Job Sources of Survey Respondents	Frequency
Help-wanted ads	56
Executive search firms	44
Networking	280
Mass mailing	20

9. A study was conducted to determine how people get jobs. The table below lists data from 400 randomly selected subjects. Construct a Pareto chart that corresponds to the given data. If someone would like to get a job, what seems to be the most effective approach?

10. Refer to the data given in Exercise 9, and construct a pie chart. Compare the pie chart to the Pareto chart, and determine which graph is more effective in showing the relative importance of job sources.

11. A local power commission sent a survey to homeowners to determine household power efficiency. Following is a list of major electrical appliances and their total kilowatt-hour (kWh) usage for one year. Construct a pie chart representing the given data.

Appliance	Total Use (kWh)
Refrigerator	1525
Freezer	1206
All other	1190
Lights	1109
Heating fan/pump	1082
Laundry	750
Air conditioning	725
Cooking	547
Television	189
Stereo	186
Dishwasher	176
Microwave oven	94

12. Refer to the data given in Exercise 11 and construct a Pareto chart. Compare the Pareto chart to the pie chart, and determine which graph is more effective in showing the relative power consumption of the appliances.

13. Refer to the labour force data in Data Set 14 in Appendix B, using the two columns for Male and Female Newfoundlanders. Compare the sizes of the occupational groups for each of the two genders by constructing two frequency polygons on the same axis. (Interpret the "size" of an occupational group as the number of individuals having a given SOC code.) (Compare Exercise 17 in Section 2–2.) Does the distribution of job-group sizes appear to differ substantially for male and female members of the Newfoundland labour force? Can we tell from the results if it is the same job groups for both genders that have the higher frequencies?

14. Construct an ogive based on the sizes of occupational groups for Male Newfoundlanders in the labour force. (See Exercise 13.) As above, interpret the "size" of an occupational group as the number of individuals having a given SOC code. Use the ogive to estimate the median size of the occupational groups (that is, the size below which half of the sizes of occupational groups fall).

In Exercises 15–16, use the given paired data from Appendix B to construct a scatter diagram.

15. In Data Set 4, use average annual temperature for the horizontal scale and use precipitation for the vertical scale. Based on the result, does there appear to be a relationship between average annual temperature and precipitation?

16. In Data Set 3, use the distances around bear necks for the horizontal scale and use the bear weights for the vertical scale. Based on the result, what is the relationship between a bear's neck size and its weight?

In Exercises 17–20, refer to the data sets in Appendix B.
   a. *Construct a histogram.*
   b. *Describe the general shape of the distribution, such as bell-shaped, uniform, or skewed (lopsided).*

17. Data Set 3 in Appendix B: weights of bears (use 11 classes with a class width of 50 and begin with a lower class boundary of 20.5).

18. Data Set 11 in Appendix B: weights of 100 M&Ms (use 12 classes with a class width of 0.017 and begin with a lower class boundary of 0.8375).

19. Data Set 1 in Appendix B: weights of paper discarded by 62 households in one week (use 10 classes).

20. Data Set 12 in Appendix B: the 300 numbers selected in the Lotto 6/49 (not the bonus numbers).

2-3 Exercises B: Beyond the Basics

	π		22/7
x	f	x	f
0	8	0	0
1	8	1	17
2	12	2	17
3	11	3	1
4	10	4	17
5	8	5	16
6	9	6	0
7	8	7	16
8	12	8	16
9	14	9	0

21. Frequency tables are given for the first 100 digits in the decimal representation of π and the first 100 digits in the decimal representation of 22/7.
   a. Construct histograms representing the frequency tables, and note any differences.
   b. The numbers π and 22/7 are both real numbers, but how are they fundamentally different?

22. In an insurance study of motor vehicle accidents, fatal crashes are categorized according to time of day, with the results given in the accompanying table.

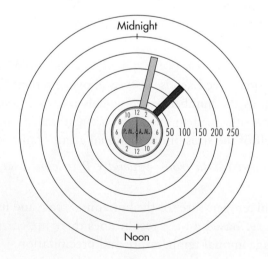

Time of Day	Number of Fatal Crashes
A.M. 12–2	194
2–4	149
4–6	100
6–8	131
8–10	119
10–12	160
P.M. 12–2	152
2–4	221
4–6	230
6–8	211
8–10	223
10–12	178

a. Complete the circular bar chart and construct a histogram.

b. Which is more effective in depicting the data? Why?

c. Because the time period 4:00 a.m. to 6:00 a.m. has the lowest number of fatal crashes, is that time period the safest time to drive? Why or why not?

23. Using a collection of sample data, we construct a frequency table with 10 classes and then construct the corresponding histogram. How is the histogram affected if the number of classes is doubled but the same vertical scale is used?

24. In "Ages of Oscar-Winning Best Actors and Actresses" by Richard Brown and Gretchen Davis (*Mathematics Teacher* magazine), stem-and-leaf plots are used to compare the ages of actors and actresses at the time they won Oscars. Here are the results for 34 recent winners from each category:

Actors:	32	37	36	32	51	53	33	61	35	45	55	39
	76	37	42	40	32	60	38	56	48	48	40	
	43	62	43	42	44	41	56	39	46	31	47	
Actresses:	50	44	35	80	26	28	41	21	61	38	49	33
	74	30	33	41	31	35	41	42	37	26	34	
	34	35	26	61	60	34	24	30	37	31	27	

a. Construct a back-to-back stem-and-leaf plot for the above data. The first two scores from each group have been entered below.

Actors' Ages	Stem	Actresses' Ages
	2	
	3	
72	4	4
	5	0
	6	
	7	
	8	

b. Using the results from part (a), compare the two different sets of data, and explain any differences.

2-4 Measures of Central Tendency

The main objective of this section is to present the important measures of central tendency, and to show how to compute them.

DEFINITION

A **measure of central tendency** is a value at the centre or middle of a data set.

Figure 2-10
Mean as a Balance Point
If a fulcrum is placed at the
position of the mean, it will
balance the histogram.

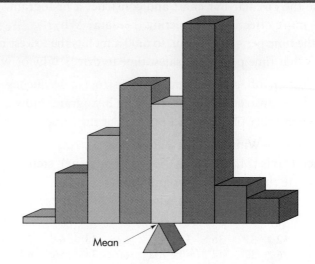

Mean

Whereas Sections 2–2 and 2–3 considered frequency tables and graphs that reveal the nature or shape of the *distribution* of a data set, this section focuses on finding values that are typical or at the *centre* of a data set. There are different criteria for determining the centre, and so there are different definitions of measures of central tendency, including the mean, median, mode, and midrange. We begin with the mean.

Mean

The (arithmetic) mean is generally the most important of all numerical descriptive measurements, and it is what most people call an average. In Figure 2-10 we illustrate that the mean is at the centre of the data set in the sense that it is a balance point for the data.

DEFINITION

> The **arithmetic mean** of a set of values is the number obtained by adding the values and dividing the total by the number of values. This particular measure of central tendency will be used frequently throughout the remainder of this text, and it will be referred to simply as the **mean**.

This definition can be expressed as Formula 2-1, where the Greek letter Σ (uppercase Greek sigma) indicates *summation* of values, so that Σx represents the sum of all data values. Also, the symbol n denotes the **sample size**, which is the number of values in the data set.

Formula 2-1
$$\text{mean} = \frac{\Sigma x}{n}$$

The mean is denoted by $\bar{x}$ (pronounced "x-bar") if the available values are a sample from a larger population; if all values of the population are available, then we can denote the computed mean by μ (lowercase Greek mu).

NOTATION

Σ denotes the *addition* (called the *summation*) of a set of values.

x is the *variable* usually used to represent the individual data values.

n represents the *number of values in a sample.*

N represents the *number of values in a population.*

$$\bar{x} = \frac{\Sigma x}{n}$$

$$\mu = \frac{\Sigma x}{N}$$

EXAMPLE

A potato chip packaging plant selects 10 bags for a quality control check. The weights in grams are listed below. Find the mean for this sample.

454.1   455.0   454.2   454.9   454.7   454.4   455.1   454.6   454.4   455.2

SOLUTION

The mean is computed by using Formula 2-1. First add the values:

$$\Sigma x = 454.1 + 455.0 + 454.2 + 454.9 + 454.7 + 454.4$$
$$+ 455.1 + 454.6 + 454.4 + 455.2 = 4546.6$$

Now divide the total by the number of values present. Because there are 10 values, we have $n = 10$ and get

$$\bar{x} = \frac{4546.6}{10} = 454.66$$

The mean value is therefore 454.66 g.

For the 10 values in the above example, 454.66 is at the centre, according to the definition of the mean. One disadvantage of the mean is that it is sensitive to every data value, so even one unusually large or small value can affect the mean dramatically. The median largely overcomes that disadvantage.

STATISTICAL PLANET

Six Degrees of Separation
Social psychologists, historians, political scientists, and communications specialists are among those interested in the "Small World Problem": Given any two people in the world, how many intermediate links are required in order to connect the two original people? Social psychologist Stanley Milgram conducted an experiment using the U.S. mail system. Subjects were instructed to try to contact other target people by mailing an information folder to an acquaintance whom they thought would be closer to the target. Among 160 such chains that were initiated, only 44 were completed. The number of intermediate acquaintances varied from 2 to 10, with a median of 5. A mathematical model was used to show that if those missing chains were completed, the median would be slightly greater than 5. (See "The Small World Problem," by Stanley Milgram, *Psychology Today,* May 1967.)

Median

The **median** of a data set is the middle value when the values are arranged in order of increasing (or decreasing) magnitude. The median is often denoted by $\tilde{x}$ ("x-tilde," pronounced "x-til-duh").

To find the median, first sort the values (arrange them in order), then use one of these two procedures:

1. If the number of values is odd, the median is the number that is located in the exact middle of the list.

2. If the number of values is even, the median is found by computing the mean of the two middle numbers.

EXAMPLE

Find the median of the weights of the first 5 bags of potato chips in the previous example.

454.1   455.0   454.2   454.9   454.7

SOLUTION

Begin by rearranging the scores so that they are in order from lowest to highest:

454.1   454.2   454.7   454.9   455.0

Because the number of values is an odd number (5), we find the number that is at the exact middle. The number 454.7 is in the middle, so the median for this data set is 454.7 g.

EXAMPLE

The following values are the incomes (in dollars) that performers received for one rock concert. The mean is $8900. Find the median.

500   600   800   50,000   1000   500

SOLUTION

Begin by rearranging the values so that they are in order:

500   500   600   800   1000   50,000

Because the number of values is an even number (6), we find the two values that are at the middle and then find their mean. The two values at the middle are 600 and 800, so the median is found by adding these two values and dividing by 2. The median is $700.

For this data set, the mean of $8900 was strongly affected by the extreme value of $50,000, but that extreme value did not affect the median of $700.

Mode

The **mode** of a data set is the value that occurs most frequently. When two values occur with the same greatest frequency, each one is a mode and the data set is **bimodal**. When more than two values occur with the same greatest frequency, each is a mode and the data set is said to be **multimodal**. When no value is repeated, we say that there is no mode. The mode is often denoted by M.

EXAMPLE

Find the modes of the following data sets.
a. 5   5   5   3   1   5   1   4   3   5
b. 1   2   2   2   3   4   5   6   6   6   7   9
c. 1   2   3   6   7   8   9   10

SOLUTION

a. The number 5 is the mode because it is the value that occurs most often.
b. The numbers 2 and 6 are both modes because they occur with the same greatest frequency. The data set is bimodal.
c. There is no mode because no value is repeated.

Among the different measures of central tendency we are considering, the mode is the only one that can be used with data at the nominal level of measurement, as illustrated in the next example.

EXAMPLE

A study of reaction times involved 30 left-handed subjects, 50 right-handed subjects, and 20 ambidextrous subjects. Although we cannot numerically average these characteristics, we can report that the mode is right-handed, because that is the characteristic with the greatest frequency.

Midrange

DEFINITION

The **midrange** is the value midway between the highest and lowest values in the data set. It is found by adding the highest value to the lowest value and then dividing the sum by 2, as in the following formula:

$$\text{midrange} = \frac{\text{highest value} + \text{lowest value}}{2}$$

Find the midrange of the weights of the 10 bags of potato chips in the sample.

454.1   455.0   454.2   454.9   454.7   454.4   455.1   454.6   454.4   455.2

SOLUTION

The midrange is found as follows:

$$\frac{\text{highest value} + \text{lowest value}}{2} = \frac{455.2 + 454.1}{2} = 454.65 \text{ g}$$

The midrange isn't used much (it is too sensitive to a single extreme value), but we include it mainly to emphasize the point that there are several different ways to define the centre of a set of data. (For two other methods, see Exercises 20–22 below.) Figure 2-11 illustrates the differences among the mean, median, mode, and midrange. (Assume the boxed numbers are six original data values.)

Unfortunately, the term *average* is sometimes used for any measure of central tendency and is sometimes used specifically for the mean. Because of this ambiguity, we should avoid using the term *average* when referring to a particular measure of central tendency. Instead, we should use the specific term, such as mean, median, mode, or midrange. Also, be cautious when you encounter a value reported by someone as an average; unless we are told specifically, the value could have been computed by any one of several different approaches.

**Figure 2-11
Measures of Central Tendency**

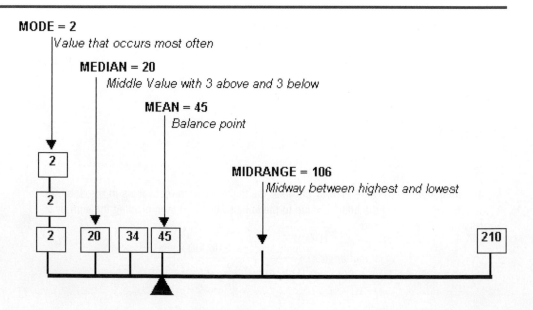

MODE = 2
Value that occurs most often

MEDIAN = 20
Middle Value with 3 above and 3 below

MEAN = 45
Balance point

MIDRANGE = 106
Midway between highest and lowest

2   2   2   20   34   45   210

Refer back to Table 2-1, where we list 175 axial loads of aluminum cans. Find the (a) mean, (b) median, (c) mode, and (d) midrange.

SOLUTION

a. Mean: The sum of the 175 values is 46,745, so

$$\bar{x} = \frac{46,745}{175} = 267.1 \text{ lb}$$

b. Median: After arranging the values in increasing order, we find that the 88th value of 273 is in the exact middle, so the median is 273.0 lb. We express the result with an extra decimal place by using the round-off rule that follows this example.

c. Mode: The most frequent axial load is 268 lb, which occurs 9 times. It is therefore the mode.

d. Midrange: We use the following formula to find the midrange:

$$\text{midrange} = \frac{\text{highest value} + \text{lowest value}}{2} = \frac{297 + 200}{2} = 248.5 \text{ lb}$$

We now summarize these results:

mean:	267.1 lb
median:	273.0 lb
mode:	268 lb
midrange:	248.5 lb

Previously, we constructed a frequency table and histogram for the data in Table 2-1 and we saw the distribution of the data. We now have important information about the centre of the data.

ROUND-OFF RULE

A simple rule for rounding calculations of measures of central tendency is this:

Carry one more decimal place than is present in the original set of data.

When implementing this rule, we round only the final answer and not intermediate values. For example, the mean of 2, 3, and 5 is 3.33333333 . . . , and it can be rounded as 3.3. Because the original data were whole numbers, we rounded the answer to the nearest tenth. As another example, the mean of 2.1, 3.4, and 5.7 is rounded to 3.73 with two decimal places (one more decimal place than was used for the original values).

Mean from a Frequency Table

When data are summarized in a frequency table, we can approximate the mean by replacing class limits with class midpoints and assuming that each class midpoint

STATISTICAL PLANET

The Earned Run Average
A weighted mean is used in baseball's earned run average (ERA), which is a measure of a pitcher's effectiveness. It is computed as follows: Multiply the number of earned runs scored against the pitcher by 9, then divide by the total number of innings pitched. (Earned runs do not include runs scored by players who got on base because of errors.) The ERA represents the total number of runs a pitcher would yield in nine innings. It has the advantage of being weighted to correspond to a full game, so different pitchers can be compared. However, it is criticized because it does not take into account the very different circumstances faced by starting pitchers and relief pitchers.

is repeated a number of times equal to the class frequency. In Table 2-2, for example, the first class of 200–209 contains 9 values that fall somewhere between those class limits, but we don't know the specific values of those 9 values. We make calculations possible by assuming that all 9 values are the class midpoint of 204.5. With 9 values of 204.5, we have a total of $9 \times 204.5 = 1840.5$ to contribute toward the grand total of all values combined. The number of values is equal to the sum of the frequencies, so Formula 2-2 can be used to find the mean from a frequency table. Formula 2-2 is not really a new concept; it is simply a variation of Formula 2-1.

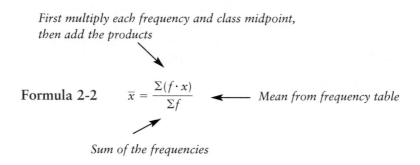

First multiply each frequency and class midpoint, then add the products

Formula 2-2 $\bar{x} = \dfrac{\Sigma(f \cdot x)}{\Sigma f}$ ← *Mean from frequency table*

Sum of the frequencies

The aluminum can axial load data from Frequency Table 2-2 have been entered in Table 2-5, where we apply Formula 2-2. When we used the original collection of values to calculate the mean directly, we obtained a mean of 267.1, so the value of the estimated mean obtained from the frequency table is just a little off.

Table 2-5 Finding Σf and $\Sigma(f \cdot x)$

Axial Load (lb)	Frequency f	Class Midpoint x	$f \cdot x$
200–209	9	204.5	1,840.5
210–219	3	214.5	643.5
220–229	5	224.5	1,122.5
230–239	4	234.5	938.0
240–249	4	244.5	978.0
250–259	14	254.5	3,563.0
260–269	32	264.5	8,464.0
270–279	52	274.5	14,274.0
280–289	38	284.5	10,811.0
290–299	14	294.5	4,123.0
Total	$\Sigma f = 175$		$\Sigma(f \cdot x) = 46{,}757.5$

$$\bar{x} = \frac{\Sigma(f \cdot x)}{\Sigma f} = \frac{46{,}757.5}{175} = 267.2$$

Weighted Mean

In some situations the values vary in their degree of importance, so we may want to compute a **weighted mean**, which is a mean computed with the different values assigned different weights. In such cases, we can calculate the weighted mean by assigning different weights to different values, as shown in Formula 2-3:

Formula 2-3 $$\text{weighted mean: } \overline{x} = \frac{\Sigma(w \cdot x)}{\Sigma w}$$

For example, suppose we need a mean of 5 test scores (85, 90, 75, 80, 95), but the first 4 tests count for 15% each, while the last score counts for 40%. We can simply assign a weight of 15 to each of the first 4 tests, and a weight of 40 to the last test, then proceed to calculate the mean by using Formula 2-3 as follows:

$$\overline{x} = \frac{\Sigma(w \cdot x)}{\Sigma w}$$

$$= \frac{(15 \times 85) + (15 \times 90) + (15 \times 75) + (15 \times 80) + (40 \times 95)}{15 + 15 + 15 + 15 + 40}$$

$$= \frac{8750}{100} = 87.5$$

As another example, college grade-point averages can be computed by assigning each letter grade the appropriate number of points (A = 4, B = 3, etc.), then assigning to each number a weight equal to the number of credit hours. Again, Formula 2-3 can be used to compute the grade-point average.

The Best Measure of Central Tendency

So far, we have considered the mean, median, mode, and midrange as measures of central tendency. Which one of these is best? Unfortunately, there is no single best answer to that question because there are no objective criteria for determining the most representative measure for all data sets. The different measures of central tendency have different advantages and disadvantages, some of which are summarized in Table 2-6. An important advantage of the mean is that it takes every value into account, but an important disadvantage is that it is sometimes dramatically affected by a few extreme values. This disadvantage can be overcome by using a trimmed mean, as described in Exercise 25.

Skewness

A comparison of the mean, median, and mode can reveal information about the characteristic of skewness, defined below and illustrated in Figure 2-12.

Table 2-6 Comparison of Mean, Median, Mode, and Midrange

Average	Definition	How Common?	Existence	Takes Every Score into Account?	Affected by Extreme Scores?	Advantages and Disadvantages
Mean	$\bar{x} = \dfrac{\Sigma x}{n}$	most familiar "average"	always exists	yes	yes	uses all the data; works well with many statistical methods; sensitive to extreme values
Median	middle value	commonly used	always exists	no	no	often a good choice if there are some extreme scores
Mode	most frequent value	sometimes used	might not exist; may be more than one mode	no	no	appropriate for data at the nominal level
Midrange	$\dfrac{\text{high + low}}{2}$	rarely used	always exists	no	yes	very sensitive to extreme values

General comments:
- For a data collection that is approximately symmetric with one mode, the mean, median, mode, and midrange tend to be about the same.
- For a data collection that is obviously asymmetric, it is a good idea to report both the mean and the median.
- The mean is relatively *reliable*. That is, when samples are drawn from the same population, the sample means tend to be more consistent than the other averages (consistent in the sense that the means of samples drawn from the same population don't vary as much as the other averages).

DEFINITION

A distribution of data is **skewed** if it is not symmetric and extends more to one side than the other. (A distribution of data is **symmetric** if the left half of its histogram is roughly a mirror image of its right half.)

Figure 2-12
Skewness

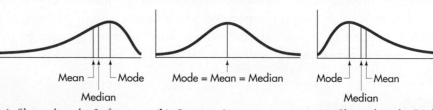

(a) Skewed to the Left (Negatively Skewed): The mean and median are to the *left* of the mode.

(b) Symmetric (Zero Skewness): The mean, median, and mode are the same.

(c) Skewed to the Right (Positively Skewed): The mean and median are to the *right* of the mode.

Data skewed to the *left* are said to be **negatively skewed;** the mean and median are to the left of the mode. Although not always predictable, negatively skewed data generally have the mean to the left of the median. (See Figure 2-12(a).) Data skewed to the *right* are said to be **positively skewed;** the mean and median are to the right of the mode. Again, although not always predictable, positively skewed data generally have the mean to the right of the median. (See Figure 2-12(c).)

If we examine the histogram in Figure 2-1 for the axial loads of aluminum cans, we see a graph that appears to be skewed to the left. In practice, many distributions of data are symmetric and without skewness. Distributions skewed to the right are more common than those skewed to the left because it's often easier to get exceptionally large values than values that are exceptionally small. With annual incomes, for example, it's impossible to get values below the lower limit of zero, but there are a few people who earn millions of dollars in a year. Annual incomes therefore tend to be skewed to the right, as in Figure 2-12(c).

Many computer programs allow you to enter a data set and use one operation to get several different sample statistics, referred to as descriptive statistics. (See Section 2–6 for sample Excel and STATDISK displays.)

EXCEL: First enter the relevant data in a spreadsheet range (such as A1:A20).

EXCEL (Prior to 2007):   Select **Tools,** then **Data Analysis,** then select **Descriptive Statistics** and click **OK.** In the dialog box, enter the data input range, click on **Summary Statistics,** then click **OK.** (If Data Analysis does not appear in the Tools menu, it must be installed using Add-Ins.) Excel also offers a set of useful statistical functions that can be used individually as needed. Suppose the raw data are in range A1:A20; to find the mean, median, mode, count (n), or largest or smallest value, respectively, type the corresponding function into an empty cell: =**average**(A1:A20), =**median**(A1:A20),   =**mode**(A1:A20),   =**count**(A1:A20), =**max**(A1:A20), or **min**(A1:A20). (*Caution:* =**mode**() returns the first mode according to the order in which the data is listed, even if there is actually more than one mode.)

EXCEL 2007:   Click the **Data** tab and then **Data Analysis** in the **Analysis** group. Proceed as above.

MINITAB 15:   Put the data in column C1. From the **Stat** menu, choose **Basic Statistics** and then **Display Descriptive Statistics.** Click on C1 in the left window and then click on **Select.** C1 will now appear in the Variables window to the right. Clicking **OK** will give you the default statistics, such as mean and median. Clicking the **Statistics** button allows you to specify which statistics you want by checking the desired boxes. Minitab rounds the values to 3 decimals.

MINITAB DISPLAY

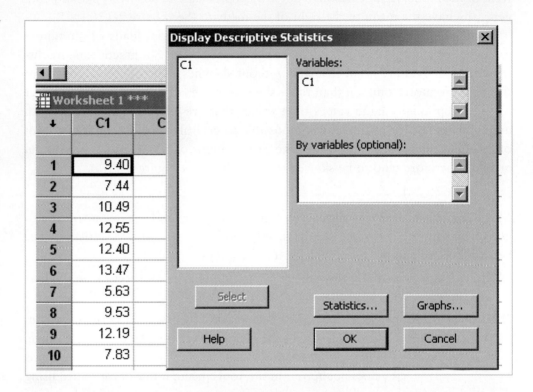

STATDISK: Click to select **Data**, then **Descriptive Statistics**. Choose which column you want. Then click the **Evaluate** button to get the mean, median, midrange, and other descriptive statistics.

 Exercises A: Basic Skills and Concepts

In Exercises 1–4, find the (a) mean, (b) median, (c) mode, and (d) midrange for the given sample data.

1. A sample of 20 chocolate chip cookies was taken from a box. The cookies were examined for the number of chocolate chips in each cookie. The numbers recorded were

    18 15 17 17 16 18 16 15 16 14
    16 17 18 15 14 16 15 18 17 16

2. A survey of families in a particular neighbourhood showed the number of children in each household:

0 3 4 7 2 2 3 0 1 5 2 4 1 2 3
3 2 1 6 1 3 2 9 2 3 2 1 0 6 2

3. Prices of a set of windshield wiper blades vary from store to store. Listed are the prices of blades for a particular make and model of car (in dollars):

7.45   5.51   4.96   5.62   6.26   5.10   8.29   6.82   5.34   7.00   5.19

4. Based on data from Statistics Canada, the average annual snowfall (in centimetres) for 15 Canadian cities is

359   271   208   290   271   293   387   320
227   131   170   122   195   138   60

In Exercises 5–8, find the mean, median, mode, and midrange for each of the two samples, then compare the two sets of results.

5. Masses (in kg) of samples of the contents in cans of regular Coke and diet Coke:

Regular: 0.3724   0.3705   0.371   0.3732   0.3719   0.3749

Diet: 0.3533   0.3526   0.3589   0.3576   0.3565   0.3573

Does there appear to be a significant difference between the two data sets? How might such a difference be explained?

6. Masses (in kg) of samples of the contents in cans of regular Coke and regular Pepsi:

Coke: 0.3724   0.3705   0.371   0.3732   0.3719   0.3749

Pepsi: 0.3754   0.3707   0.3732   0.3714   0.3735   0.3774

Does there appear to be a significant difference in the weights of the two different brands?

7. Maximum breadth of samples of male Egyptian skulls from 4000 BCE and 150 CE (in mm, based on data from *Ancient Races of the Thebaid* by Thomson and Randall-Maciver):

4000 BCE: 131 119 138 125 129 126 131 132 126 128 128 131

150 CE: 136 130 126 126 139 141 137 138 133 131 134 129

Changes in head sizes over time suggest interbreeding with people from the other regions. Do the head sizes appear to have changed from 4000 BCE to 150 CE?

8. Annual values of Canadian exports to Europe and to the areas designated Far East/Oceania (in millions of dollars) were collected in independent random samples:

Far East/Oceania: 19,407 16,889 16,688 15,887 13,262 8288

Europe: 17,834 15,794 14,686 14,490 13,928 9946

Does there appear to be a significant difference between the export amounts to these two regions? Do they differ by another measure, besides a measure of centre?

In Exercises 9–12, refer to the data set in Appendix B, and find the (a) mean, (b) median, (c) mode, and (d) midrange.

9. Data Set 2: 22 body temperatures for supine position

10. Data Set 4: the average annual temperature of all cities listed

11. Data Set 3: the weights of the bears

12. Data Set 11: the weights of the red M&M plain candies

In Exercises 13–16, find the mean of the data summarized in the given frequency table.

13. Visitors to Yellowstone National Park consider an eruption of the Old Faithful geyser to be a major attraction that should not be missed. The given frequency table summarizes a sample of times (in minutes) between eruptions.

Time	Frequency
40–49	8
50–59	44
60–69	23
70–79	6
80–89	107
90–99	11
100–109	1

14. Samples of student cars and faculty/staff cars were obtained at the author's college, and their ages (in years) are summarized in the frequency table below. Find the mean age of student cars and find the mean age of faculty/staff cars. Based on the results, are there any noticeable differences between the two samples? If so, what are they?

Age	Students	Faculty/Staff
0–2	23	30
3–5	33	47
6–8	63	36
9–11	68	30
12–14	19	8
15–17	10	0
18–20	1	0
21–23	0	1

15. The given frequency table shows the distribution in megatonnes (Mt) of carbon dioxide emissions from fossil fuel in Canada over the 30-year period from 1965 to 1994. If the level this year is 492 Mt, how does that compare to the average of the 30-year period?

Carbon Dioxide Emissions (Mt)	Frequency
250 to less than 300	3
300 to less than 350	4
350 to less than 400	12
400 to less than 450	9
450 to less than 500	2

16. Canadian Automated Electronics (CAE) did a study of time to complete tasks entailing simulation control for 25 employees. The data shown are a summary of the study. What do you conclude from the result?

Minutes	Frequency
100–200	1
200–300	8
300–400	5
400–500	1
500–600	0
600–700	1
700–800	4
800–900	5

17. A student receives quiz grades of 60, 84, and 90. The final exam grade is 88. Find the weighted mean if the quizzes each count for 20% and the final exam counts for 40% of the final grade.

18. A student's transcript shows an A in a 4-credit course, an A in a 3-credit course, a C in a 3-credit course, and a D in a 2-credit course. Grade points are assigned as follows: A = 4, B = 3, C = 2, D = 1, F = 0. If grade points are weighted according to the number of credit hours, find the weighted mean (grade-point average) rounded to three decimal places.

2-4 Exercises B: Beyond the Basics

19. a. Find the mean, median, mode, and midrange for the following labour force populations (based on data from Statistics Canada):

    Halifax: 186,600   Quebec City: 356,800   Oshawa: 148,100

    Winnipeg: 381,300   Saskatoon: 121,600   Victoria: 164,600

    b. If a constant value k is added to each population, how are the results from part (a) affected?

    c. If each population given in part (a) is multiplied by a constant k, how are the results from part (a) affected?

    d. Data are sometimes transformed by replacing each score x with $\log x$. For the given x values, determine whether the mean of the $\log x$ values is equal to $\log \overline{x}$.

20. The **harmonic mean** is often used as a measure of central tendency for data sets consisting of rates of change, such as speeds. It is found by dividing the number of scores n by the sum of the *reciprocals* of all scores, expressed as

$$\frac{n}{\sum \dfrac{1}{x}}$$

(No score can be zero.) For example, the harmonic mean of 2, 4, and 10 is

$$\frac{n}{\sum \dfrac{1}{x}} = \frac{3}{\dfrac{1}{2} + \dfrac{1}{4} + \dfrac{1}{10}} = \frac{3}{0.85} = 3.5$$

a. Four students drive from Toronto to Saint John (1440 km) at a speed of 60 km/h and return at a speed of 100 km/h. What is their average speed for the round trip? (The harmonic mean is used in averaging speeds.)

b. Jacques Villeneuve posted the average speeds (in kilometres per hour) shown below for his last 10 laps of practice, and in doing so qualified for the Canadian Grand Prix at Montreal. What was Villeneuve's average speed for these 10 laps?

$$192.8 \quad 191.6 \quad 197.5 \quad 198.2 \quad 199.1 \quad 197.6 \quad 196.5 \quad 197.5 \quad 193.5 \quad 190.0$$

21. The **geometric mean** is often used in business and economics for finding average rates of change, average rates of growth, or average ratios. Given n scores (all of which are positive), the geometric mean is the nth root of their product. For example, the geometric mean of 2, 4, and 10 is found by first multiplying the scores to get 80, then taking the cube root (because there are 3 scores) of the product to get 4.3. The **average growth factor** for money compounded at annual interest rates of 10%, 8%, 9%, 12%, and 7% can be found by computing the geometric mean of the growth factors 1.10, 1.08, 1.09, 1.12, and 1.07. Find that average growth factor.

22. The **quadratic mean** (or **root mean square**, or **R.M.S.**) is usually used in physical applications. In power distribution systems, for example, voltages and currents are usually referred to in terms of their R.M.S. values.

    The quadratic mean of a set of scores is obtained by squaring each score, adding the results, dividing by the number of scores n, and then taking the square root of that result. For example, the quadratic mean of 2, 4, and 10 is

$$\sqrt{\frac{\Sigma x^2}{n}} = \sqrt{\frac{4 + 16 + 100}{3}} = \sqrt{\frac{120}{3}} = \sqrt{40} = 6.3$$

Find the R.M.S. of these power supplies (in volts): 151, 162, 0, 81, 268.

23. Frequency tables often have open-ended classes, such as the accompanying table (based on data for Ontario from Statistics Canada), which summarizes annual household incomes for a sample of 5197 families of 3–4 persons. Formula 2-2 cannot be directly applied because we can't determine a class midpoint for the class "more than 80,000." Calculate the mean by assuming this class is really (a) 80,000 up to 100,000, (b) 80,000 up to 110,000, and (c) 80,000 up to 120,000. What can you conclude?

Annual Household Income ($)	Frequency
0 up to 20,000	404
20,000 up to 40,000	1081
40,000 up to 60,000	982
60,000 up to 80,000	1310
More than 80,000	1420

24. When data are summarized in a frequency table, the median can be found by first identifying the *median class* (the class that contains the median). We then assume that the scores in that class are evenly distributed and we can interpolate. This process can be described by

$$(\text{lower limit of median class}) + (\text{class width}) \left(\frac{\left(\dfrac{n + 1}{2} \right) - (m + 1)}{\text{frequency of median class}} \right)$$

where n is the sum of all class frequencies and m is the sum of the class frequencies that *precede* the median class. Use this procedure and the frequency data in Table 2-2 to find the median axial load.

25. Because the mean is very sensitive to extreme scores, it can be criticized as not being a *resistant* measure of central tendency. The *trimmed mean* is more resistant. To find the 10% trimmed mean for a data set, first arrange the data in order, then delete the bottom 10% of the scores and the top 10% of the scores, and calculate the mean of the remaining scores. For the weights of the bears in Data Set 3 from Appendix B, find (a) the mean, (b) the 10% trimmed mean, and (c) the 20% trimmed mean. How do the results compare?

26. Using an almanac, a researcher finds the average teacher's salary for each province and territory. He adds those 13 values, then divides by 13 to obtain their mean. Is the result equal to the national average teacher's salary? Why or why not?

2-5 Measures of Variation

This section deals with the characteristic of variation, which is one of the most important topics in the entire book. The following key concepts should be mastered: (1) Variation refers to the amount that values vary among themselves, and it can be measured with specific numbers, collectively called **measures of variation**; (2) values that are relatively close together have low measures of variation, whereas values that are spread farther apart have measures of variation that are larger; (3) the standard deviation is a particularly important measure of variation, which can be computed; (4) the values of standard deviations must be *interpreted* correctly.

Many banks once required that customers wait in separate lines at each teller's window, but most have now changed to one single main waiting line. Why did they make that change? The mean waiting time didn't change, because the waiting-line configuration doesn't affect the efficiency of the tellers. They changed to the single line because customers prefer waiting times that are more *consistent* with less variation. Thus thousands of banks made a change that resulted in lower variation (and happier customers), even though the mean was not affected. The listed values are waiting times (in minutes) of customers:

Humber Valley Credit Union (Single line)	6.5	6.6	6.7	6.8	7.1	7.3	7.4	7.7	7.7	7.7
Durham Credit Union (Multiple lines)	4.2	5.4	5.8	6.2	6.7	7.7	7.7	8.5	9.3	10.0

In Figure 2-13 we compare frequency polygons for both data sets, and we include the measures of central tendency. A comparison of the measures of central tendency does not reveal any differences between the two data sets, but their compared frequency polygons provide clear and strong visual evidence that the Humber Valley Credit Union has waiting times with substantially less variation than the times for the Durham Credit Union.

Let us now proceed to develop some specific ways of actually *measuring* variation. We begin with the range.

Range

The **range** of a set of data is the difference between the highest value and the lowest value. To compute it, simply subtract the lowest value from the highest value. For

Figure 2-13
Frequency Polygons of Waiting Times

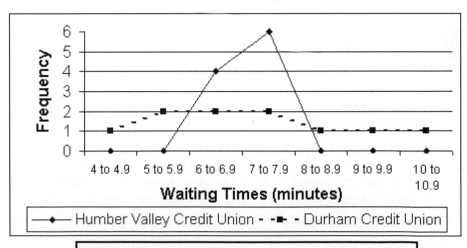

Humber Valley Credit Union (single line)	Durham Credit Union (multiple lines)
Mean = 7.15	Mean = 7.15
Median = 7.20	Median = 7.20
Mode = 7.7	Mode = 7.7
Midrange = 7.10	Midrange = 7.10

the Humber Valley customers, the range is $7.7 - 6.5 = 1.2$ min. Durham has waiting times with a range of 5.8 min, and this larger value suggests greater variation.

The range is very easy to compute, but because it depends on only the highest and the lowest values, it is not as useful as the other measures of variation that take account of every value. (See Exercise 25 for an example in which the range is misleading.)

Standard Deviation of a Sample

The standard deviation is the measure of variation that is generally the most important and useful. We define the standard deviation in the following box, but to understand this concept fully you will need to read the remainder of this section very carefully.

DEFINITION

The **standard deviation** of a set of sample values is a measure of variation of values about the mean. It is calculated by using Formula 2-4 or Formula 2-5.

Formula 2-4
$$s = \sqrt{\frac{\Sigma(x - \bar{x})^2}{n - 1}}$$
sample standard deviation

Formula 2-5
$$s = \sqrt{\frac{n\Sigma(x^2) - (\Sigma x)^2}{n(n - 1)}}$$
shortcut formula for sample standard deviation

Formulas 2-4 and 2-5 are equivalent in the sense that they will always yield the same result. Formula 2-4 has the advantage of reinforcing the concept that the standard deviation is a type of average deviation. Formula 2-5 is a "shortcut" (i.e., easier to calculate) if you must calculate standard deviations on your own. Formula 2-5 also has the advantages of eliminating rounding errors when the exact value of the mean is not used, and (for calculators and statistical software) it requires only three memory locations to calculate (for Σn, Σx, and Σx^2), instead of one location per data value. So which formula should you use? We advise the following: Use Formula 2-4 for a few examples, then learn how to find standard deviations on your calculator or by using statistical software; you might want to consult your calculator manual now to find the procedure that yields the value of the standard deviation.

Why define a measure of variation in the way described by Formula 2-4? In measuring variation in a set of sample data, it's reasonable to begin with the individual amounts by which values deviate from the mean. For a particular value x, the amount of **deviation** is $x - \bar{x}$, which is the difference between the value and the mean. However, the sum of all such deviations is always zero, which really doesn't

AS

do anything for us. To get a statistic that measures variation (instead of always being zero), we could take absolute values, as in $\Sigma |x - \overline{x}|$. If we find the mean of that sum, we get the **mean absolute deviation** described by the following expression:

$$\text{Mean absolute deviation} = \frac{\Sigma |x - \overline{x}|}{n}$$

Instead of using absolute values, we can obtain an alternative, and more com-monly used, measure of variation by making all deviations $(x - \overline{x})$ nonnegative by squaring them. Finally, we take the square root to compensate for that squaring. As a result, the standard deviation has the same units of measurement as the original scores. For example, if customer waiting times are in minutes, the standard deviation of those times will also be in minutes. On the basis of the format of Formula 2-4, we can describe the procedure for calculating the standard deviation as follows.

Procedure for Finding the Standard Deviation with Formula 2-4

STEP 1: Find the mean of the values $(\overline{x})$.

STEP 2: Subtract the mean from each individual value $(x - \overline{x})$.

STEP 3: Square each of the differences obtained from Step 2. [This produces numbers of the form $(x - \overline{x})^2$.]

STEP 4: Add all of the squares obtained from Step 3 to get $\Sigma(x - \overline{x})^2$.

STEP 5: Divide the total from Step 4 by the number $(n - 1)$; that is, 1 less than the total number of values present.

STEP 6: Find the square root of the result of Step 5.

EXAMPLE

Find the standard deviation of the Humber Valley Credit Union customer waiting times. Those times (in minutes) are reproduced below:

    6.5   6.6   6.7   6.8   7.1   7.3   7.4   7.7   7.7   7.7

SOLUTION

We will follow the six steps just given. (See Table 2-7, where the steps are executed.)

Step 1: Obtain the mean of 7.15 by adding the values and then dividing by the number of values:

$$\overline{x} = \frac{\Sigma x}{n} = \frac{71.5}{10} = 7.15 \text{ min}$$

Step 2: Subtract the mean of 7.15 from each value to get these values of $(x - \overline{x})$: −0.65, −0.55, . . . , 0.55.

Table 2-7 Calculating Standard Deviation for Humber Valley Credit Union Customers

x	$x - \bar{x}$	$(x - \bar{x})^2$
6.5	−0.65	0.4225
6.6	−0.55	0.3025
6.7	−0.45	0.2025
6.8	−0.35	0.1225
7.1	−0.05	0.0025
7.3	0.15	0.0225
7.4	0.25	0.0625
7.7	0.55	0.3025
7.7	0.55	0.3025
7.7	0.55	0.3025
Totals: 71.5		2.0450

$$\bar{x} = \frac{71.5}{10} = 7.15 \text{ min} \qquad s = \sqrt{\frac{2.0450}{10 - 1}} = \sqrt{0.2272} = 0.48 \text{ min}$$

Step 3: Square each value obtained in Step 2 to get these values of $(x - \bar{x})^2$: 0.4225, 0.3025, . . . , 0.3025.

Step 4: Sum all of the preceding values to get the value of

$$\Sigma(x - \bar{x})^2 = 2.0450$$

Step 5: There are $n = 10$ values, so divide by 1 less than 10:

$$2.0450 \div 9 = 0.2272$$

Step 6: Find the square root of 0.2272. The standard deviation is

$$\sqrt{0.2272} = 0.48 \text{ min}$$

Ideally, we would now interpret the meaning of the resulting standard deviation of 0.48 min, but such interpretations will be discussed a little later in this section.

EXAMPLE

This time use Formula 2-5 to find the standard deviation of the Humber Valley Credit Union customer waiting times (in minutes) (assuming the same data as before).

SOLUTION

Formula 2-5 requires that we find the values of n, Σx, and Σx^2. Because there are 10 scores, we have $n = 10$. The sum of the 10 scores is 71.5, so $\Sigma x = 71.5$. The third required component is calculated as follows:

$$\begin{aligned}
\Sigma x^2 &= 6.5^2 + 6.6^2 + 6.7^2 + \cdots + 7.7^2 \\
&= 42.25 + 43.56 + 44.89 + \cdots + 59.29 \\
&= 513.27
\end{aligned}$$

Formula 2-5 can now be used to find the value of the standard deviation.

$$s = \sqrt{\frac{n(\Sigma x^2) - (\Sigma x)^2}{n(n-1)}} = \sqrt{\frac{10(513.27) - (71.5)^2}{10(10-1)}}$$

$$= \sqrt{\frac{20.45}{90}} = 0.4766783 = 0.48 \text{ min (rounded)}$$

A great self-test is to stop here and calculate the standard deviation of the waiting times for the Durham Credit Union. Follow the same procedures used in the preceding two examples and verify that the standard deviation is 1.82 min. Although the interpretations of those standard deviations will be discussed later, we can now compare them and note that the standard deviation of the times for Humber Valley (0.48 min) is much lower than the standard deviation for Durham (1.82 min). This supports our subjective conclusion that the waiting times at the Humber Valley Credit Union have much less variation than those at the Durham Credit Union. It should be noted for Formula 2-4 that if $\bar{x}$ does not terminate (such as when the sample size is a multiple of 3, 7, etc.), you should use Formula 2-5 instead in order to avoid unnecessary rounding errors.

Standard Deviation of a Population

In our definition of standard deviation, we referred to the standard deviation of *sample* data. A slightly different formula is used to calculate the standard deviation σ (lowercase Greek sigma) of a *population*: Instead of dividing by $n-1$, divide by the population size N, as in the following expression.

$$\sigma = \sqrt{\frac{\Sigma(x-\mu)^2}{N}} \qquad \text{population standard deviation}$$

For example, if the 10 scores in Table 2-7 constitute a *population,* the standard deviation is as follows:

$$\sigma = \sqrt{\frac{\Sigma(x-\mu)^2}{N}} = \sqrt{\frac{2.0450}{10}} = 0.45 \text{ min}$$

Because we generally deal with sample data, we will usually use Formula 2-4, in which we divide by $n-1$. Many calculators do both the sample standard deviation and the population standard deviation; but for some creative but strange reason, calculators use a variety of different notations. Be sure to identify the notation used by your calculator.

Variance of a Sample and Population

If we omit Step 6 (taking the square root) in the procedure for calculating the standard deviation, we get the *variance*, defined in Formula 2-6:

Formula 2-6 $$s^2 = \frac{\Sigma(x - \bar{x})^2}{n - 1} \quad \text{sample variance}$$

Similarly, we can express the population variance as

$$\sigma^2 = \frac{\Sigma(x - \mu)^2}{N} \quad \text{population variance}$$

DEFINITION

> The **variance** of a set of values is a measure of variation equal to the square of the standard deviation.
>
> Sample variance: Square of the sample standard deviation s.
>
> Population variance: Square of the population standard deviation σ.

The variance is an important statistic used in some important statistical calculations, such as analysis of variance (discussed in Chapter 9). The sample variance s^2 is an **unbiased estimator** of the population variance σ^2, which means that values of s^2 tend to target the value of σ^2 without systematically tending to overestimate or underestimate σ^2. As useful as variance is, we should concentrate first on the concept of standard deviation as we try to get some sense of this statistic. A major difficulty with the variance is that it is not in the same units as the original data. For example, a data set might have a standard deviation of $3.00 and a variance of 9.00 square dollars. Because a square dollar is an abstract concept that we can't relate to directly, we find variance difficult to understand.

Here is the notation and round-off rule we are using.

NOTATION

s denotes the standard deviation of a set of *sample* data

σ denotes the standard deviation of a set of *population* data

s^2 denotes the variance of a set of *sample* data

σ^2 denotes the variance of a set of *population* data

Note: Articles in professional journals and reports often use SD for standard deviation and Var for variance.

ROUND-OFF RULE

As in Section 2–4, we use this rule for rounding final results:

Carry one more decimal place than was present in the original data.

We should round only the final answer and not intermediate values. (If it's absolutely necessary to round intermediate results, we should carry at least twice as many decimal places as will be used in the final answer.)

Finding Standard Deviation from a Frequency Table

We sometimes need to compute the standard deviation of a data set that is summarized in the form of a frequency table, such as Table 2-2 in Section 2–2. If the original list of sample values is available, use those values with Formula 2-4 or 2-5 (or with the help of your calculator or statistical software) so that the result will be more exact. If the original data are not available, we can use the following formula (where the x's are not the individual data values—which are not available to us—but the class midpoints in the frequency table that *are* available).

$$s = \sqrt{\frac{\Sigma f \cdot (x - \bar{x})^2}{n - 1}}$$

We will express this formula in an equivalent expression that usually simplifies the actual calculations.

Formula 2-7 $\quad s = \sqrt{\dfrac{n[\Sigma(f \cdot x^2)] - [\Sigma(f \cdot x)]^2}{n(n - 1)}}$ $\quad$ standard deviation for frequency table

where $\quad x =$ class midpoint
$\qquad\quad f =$ class frequency
$\qquad\quad n =$ sample size (or $\Sigma f =$ sum of the frequencies)

EXAMPLE

Estimate the standard deviation of the 175 axial loads of aluminum cans by using Formula 2-7 with the frequency table (Table 2-2).

SOLUTION

Application of Formula 2-7 requires that we find the values of n, $\Sigma(f \cdot x^2)$, and $\Sigma(f \cdot x)^2$. After finding those values from Table 2-8, we apply Formula 2-7 as follows:

$$s = \sqrt{\frac{n[\Sigma(f \cdot x^2)] - [\Sigma(f \cdot x)]^2}{n(n - 1)}} = \sqrt{\frac{175(12,579,173.75) - (46,757.5)^2}{175(175 - 1)}}$$

$$= \sqrt{\frac{15,091,600}{30,450}} = \sqrt{495.6190476} = 22.3 \text{ lb}$$

The 175 axial loads have a standard deviation estimated to be 22.3 lb. (The exact value calculated from the original set of data is 22.1 lb, so the result obtained here is quite good.)

Table 2-8 Calculating Standard Deviation from a Frequency Table

Axial Load	Frequency f	Class Midpoint x	$f \cdot x$	$f \cdot x^2$
200–209	9	204.5	1,840.5	376,382.25
210–219	3	214.5	643.5	138,030.75
220–229	5	224.5	1,122.5	252,001.25
230–239	4	234.5	938.0	219,961.00
240–249	4	244.5	978.0	239,121.00
250–259	14	254.5	3,563.0	906,783.50
260–269	32	264.5	8,464.0	2,238,728.00
270–279	52	274.5	14,274.0	3,918,213.00
280–289	38	284.5	10,811.0	3,075,729.50
290–299	14	294.5	4,123.0	1,214,223.50
Total	$\Sigma f = 175$		$\Sigma(f \cdot x) = 46,757.5$	$\Sigma(f \cdot x^2) = 12,579,173.75$

Interpreting and Understanding Standard Deviation

We will now attempt to make some intuitive sense of the standard deviation. First, we should clearly understand that the standard deviation measures the variation among values. Values close together will yield a small standard deviation, whereas values spread farther apart will yield a larger standard deviation. Refer again to Figure 2-13, which illustrates samples with clearly different amounts of variation.

Because variation is such an important concept and because the standard deviation is such an important tool in measuring variation, we will consider three different ways of developing an interpretation of standard deviations. The first is the **range rule of thumb,** a very rough estimate. (We could improve the accuracy of this rule by taking into account such factors as the size of the sample and the nature of the distribution, but we prefer to sacrifice accuracy for the sake of simplicity. We want a simple rule that will help us interpret values of standard deviations; later methods will produce more accurate results.)

RANGE RULE OF THUMB

For estimation: For typical data sets, the range of a set of data is approximately 4 standard deviations (4s) wide, so the standard deviation can be approximated as follows:

$$\text{standard deviation} \approx \frac{\text{range}}{4} \quad \text{range rule of thumb}$$

where range = (highest* value) − (lowest* value)

but if the actual highest or lowest value clearly appears to be an outlier, then use the next highest or lowest value—because we are looking here for the range that holds most of the data.

For interpretation: If the standard deviation s is known, we can use it to find rough estimates of the minimum and maximum "usual" sample values as follows:

$$\text{minimum "usual" value} \approx (\text{mean}) - 2 \times (\text{standard deviation})$$
$$\text{maximum "usual" value} \approx (\text{mean}) + 2 \times (\text{standard deviation})$$

When calculating a standard deviation using Formula 2-4 or 2-5, you can use the range rule of thumb as a check on your result, but realize that although the approximation will get you in the general vicinity of the answer, it can be off by a fairly large amount.

EXAMPLE

Use the range rule of thumb to find a rough estimate of the standard deviation of the sample of 175 axial loads of aluminum cans listed in Table 2-1.

SOLUTION

In using the range rule of thumb to estimate the standard deviation of sample data, we find the range and divide by 4. By scanning the list of scores, we find that the lowest is 200 and the highest is 297, so the range is $297 - 200 = 97$. The standard deviation s is estimated as follows:

$$s \approx \frac{\text{range}}{4} = \frac{97}{4} = 24.3 \text{ lb}$$

Interpretation
This result is in the ballpark of the correct value of 22.1 lb that is obtained by calculating the exact value of the standard deviation with Formula 2-4 or 2-5.

The above example illustrated how we can use known information about the range to estimate the standard deviation. The following example is particularly important as an illustration of one way to *interpret* the value of a standard deviation.

EXAMPLE

Suppose that the average weekly earnings for manufacturing industries in Canada in 2008 was $825.16 and the standard deviation was $146.82. Using the range rule of thumb, estimate the highest and lowest "usual" weekly earnings of these industries.

SOLUTION

The lowest and highest "usual" earnings are estimated from the range rule of thumb as follows:

$$\text{minimum} \approx (\text{mean}) - 2 \times (\text{standard deviation})$$
$$= \$825.16 - 2(\$146.82)$$
$$= \$531.52$$
$$\text{maximum} \approx (\text{mean}) + 2 \times (\text{standard deviation})$$
$$= \$825.16 + 2(\$146.82)$$
$$= \$1118.80$$

We expect that most workers in these industries will earn between $531.52 and $1,118.80 in weekly wages, although there may be a few workers with unusually low or high wages. We now have a much better understanding of how the wages vary.

Another rule that is helpful in interpreting a standard deviation is the **empirical (or 68-95-99.7) rule,** *which applies only to a data set having a distribution that is approximately bell-shaped*, as in Figure 2-14.

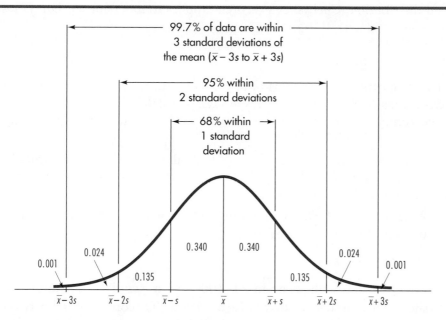

**Figure 2-14
The Empirical Rule**

EMPIRICAL (OR 68-95-99.7) RULE FOR DATA WITH A BELL-SHAPED DISTRIBUTION

- About 68% of all scores fall within 1 standard deviation of the mean.
- About 95% of all scores fall within 2 standard deviations of the mean.
- About 99.7% of all scores fall within 3 standard deviations of the mean.

EXAMPLE

Adult IQ scores have a bell-shaped distribution with a mean of 100 and a standard deviation of 15. Use the empirical rule to find the percentage of adults with IQ scores between 55 and 145.

SOLUTION

The key to solving this problem is to recognize that 55 and 145 are each exactly 3 standard deviations away from the mean of 100. (Because the standard deviation is $s = 15$, it follows that $3s = 45$, so 3 standard deviations below the mean is $100 - 45 = 55$, and 3 standard deviations above the mean is $100 + 45 = 145$.) The empirical rule states that 99.7% of all scores are within 3 standard deviations of the mean, so it follows that 99.7% of adults should have IQ scores between 55 and 145. Because values outside of that range are so rare, someone with an IQ above 145 or below 55 is considered exceptional.

A third concept helpful in understanding or interpreting a standard deviation is **Chebyshev's theorem**. The empirical rule applies only to data sets with a bell-shaped distribution. Chebyshev's theorem applies to *any* data set, but its results are very approximate.

CHEBYSHEV'S THEOREM

The proportion (or fraction) of *any* set of data lying within K standard deviations of the mean is always *at least* $1 - 1/K^2$, where K is any positive number greater than 1. For $K = 2$ and $K = 3$, we get the following two specific results:

- At least 3/4 (or 75%) of all scores will fall within the interval from 2 standard deviations below the mean to 2 standard deviations above the mean ($\bar{x} - 2s$ to $\bar{x} + 2s$).
- At least 8/9 (or 89%) of all scores will fall within 3 standard deviations of the mean ($\bar{x} - 3s$ to $\bar{x} + 3s$).

Using IQ scores with a mean of 100 and a standard deviation of 15, Chebyshev's theorem tells us that at least 75% of IQ scores will fall between 70 and 130, and at least 89% of IQ scores will fall between 55 and 145.

After studying this section, you should understand that the standard deviation is a measure of variation among values. Given sample data, you should be able to compute and interpret the standard deviation. You should recognize that for typical data sets, it is unusual for a score to differ from the mean by more than 2 or 3 standard deviations.

USING TECHNOLOGY

Computer programs that can output descriptive statistics generally include some measures of variation. (See Section 2–6 for sample Excel and STATDISK displays.)

EXCEL: First enter the relevant data in a spreadsheet range (such as A1:A20).

EXCEL (Prior to 2007): Select **Tools**, then **Data Analysis**, then select **Descriptive Statistics** and click **OK**. In the dialog box, enter the data input range, click on **Summary Statistics,** then click **OK.** (Excel also offers some related statistical functions that can be used individually as needed. Suppose the raw data are in range A1:A20. To find the sample standard deviation, population standard deviation, sample variance, or population variance, respectively, type the corresponding function into an empty cell: =**stdev**(A1:A20), =**stdevp**(A1:A20), =**var**(A1:A20), =**varp**(A1:A20).)

EXCEL 2007:   Click the **Data** tab and then **Data Analysis** in the **Analysis** group. Proceed as above.

MINITAB 15:   Put the data in column C1. From the **Stat** menu, choose **Basic Statistics** and then **Display Descriptive Statistics.** Click on C1 in the left window and then click **Select.** C1 will now appear in the **Variables** window to the right. Clicking **OK** will give you the default statistics, such as mean and median. Clicking the **Statistics** button allows you to specify which statistics you want by checking the desired boxes. Minitab rounds the values to 3 decimals.

STATDISK: Click to select **Data,** then **Descriptive Statistics.** Choose which column you want, then click the **Evaluate** button.

2-5 Exercises A: Basic Skills and Concepts

In Exercises 1–4, find the range, variance, and standard deviation for the given data. (The same data were used in Section 2–4 where we found measures of central tendency. Here we find measures of variation.)

1. A sample of 20 chocolate chip cookies was taken from a box. The cookies were examined for the number of chocolate chips in each cookie. The numbers recorded were

    18   15   17   17   16   18   16   15   16   14
    16   17   18   15   14   16   15   18   17   16

2. A survey of families in a particular neighbourhood showed the number of children in each household:

    0   3   4   7   2   2   3   0   1   5   2   4   1   2   3
    3   2   1   6   1   3   2   9   2   3   2   1   0   6   2

3. Prices of a set of windshield wiper blades vary from store to store. Listed are the prices of blades for a particular make and model of car (in dollars):

    7.45 5.51 4.96 5.62 6.26 5.10 8.29 6.82 5.34 7.00 5.19

4. Based on data from Statistics Canada, the average annual snowfall (in centimetres) for 15 Canadian cities is

    359 271 208 290 271 293 387 320
    227 131 170 122 195 138 60

In Exercises 5–8, find the range, variance, and standard deviation for each of the two samples, then compare the two sets of results.

5. Masses (in kg) of samples of the contents in cans of regular Coke and diet Coke:

    Regular: 0.3724   0.3705   0.371   0.3732   0.3719   0.3749
    Diet: 0.3533   0.3526   0.3589   0.3576   0.3565   0.3573

6. Masses (in kg) of samples of the contents in cans of regular Coke and regular Pepsi:

    Coke: 0.3724   0.3705   0.371   0.3732   0.3719   0.3749
    Pepsi: 0.3754   0.3707   0.3732   0.3714   0.3735   0.3774

7. Maximum breadth of samples of male Egyptian skulls from 4000 BCE and 150 CE (in mm, based on data from *Ancient Races of the Thebaid* by Thomson and Randall-Maciver):

    4000 BCE: 131 119 138 125 129 126 131 132 126 128 128 131
    150 CE: 136 130 126 126 139 141 137 138 133 131 134 129

8. The annual values of Canadian exports to Europe and to the areas designated Far East/Oceania (in $millions) were collected in independent random samples:

    Far East/Oceania: 19,407 16,889 16,688 15,887 13,262 8288
    Europe: 17,834 15,794 14,686 14,490 13,928 9946

In Exercises 9–12, refer to the data set in Appendix B and find the standard deviation.

9. Data Set 2 in Appendix B: 22 body temperatures for supine position

10. Data Set 4 in Appendix B: the average annual temperature of all cities listed

11. Data Set 3 in Appendix B: the weights of the bears

12. Data Set 11 in Appendix B: the weights of the red M&M plain candies

In Exercises 13–16, find the standard deviation of the data summarized in the given frequency table.

13. Visitors to Yellowstone National Park consider an eruption of the Old Faithful geyser to be a major attraction that should not be missed. The given frequency table summarizes a sample of times (in minutes) between eruptions.

Time	Frequency
40–49	8
50–59	44
60–69	23
70–79	6
80–89	107
90–99	11
100–109	1

14. Samples of student cars and faculty/staff cars were obtained at the author's college, and their ages (in years) are summarized in the accompanying frequency table. Find the standard deviation in age of student cars and of faculty/staff cars. Based on the results, are there any noticeable differences between the two samples? If so, what are they?

Age	Students	Faculty/Staff
0–2	23	30
3–5	33	47
6–8	63	36
9–11	68	30
12–14	19	8
15–17	10	0
18–20	1	0
21–23	0	1

15. The given frequency table shows the distribution in megatonnes (Mt) of carbon dioxide emissions from fossil fuel in Canada over the 30-year period from 1965 to 1994.

Carbon Dioxide Emissions (Mt)	Frequency
250 to less than 300	3
300 to less than 350	4
350 to less than 400	12
400 to less than 450	9
450 to less than 500	2

16. The frequency table opposite indicates the weekly earnings of 46 students.

Weekly Wages ($)	Number of Students
80 to less than 90	6
90 to less than 100	10
100 to less than 110	14
110 to less than 120	6
120 to less than 130	10

17. If you must purchase a replacement battery for your car, would you prefer one that comes from a population with $\sigma = 1$ month, for time to failure, or one that comes from a population with $\sigma = 1$ year? (Assume that both populations have the same mean and price.) Explain your choice.

18. As a manager, you must purchase lightbulbs to be used in a hospital. Should you choose halogen bulbs that have lives with $\mu = 3000$ h and $\sigma = 200$ h, or should you choose compact fluorescent bulbs with $\mu = 3000$ h and $\sigma = 250$ h? Explain.

19. If a data set consists of the prices of textbooks expressed in dollars, what are the units used for standard deviation? What are the units used for variance?

20. A data set consists of 20 values that are fairly close together. Another value is included, but this new value is an outlier (very far away from the other values). How is the standard deviation affected by the outlier? No effect? A small effect? A large effect?

21. A typing-competency test yields scores with $\bar{x} = 80.0$ and $s = 10.0$, and a histogram shows that the distribution of the scores is roughly bell-shaped. Use the empirical rule to answer the following:
a. What percentage of the scores should fall between 70 and 90?

b. What percentage of the scores should fall within 20 points of the mean?

c. About 99.7% of the scores should fall between what two values? (The mean of 80.0 should be midway between those two values.)

22. Heights of adult women have a mean of 63.6 in. and a standard deviation of 2.5 in. What does Chebyshev's theorem say about the percentage of women with heights between 58.6 in. and 68.6 in.? Between 56.1 in. and 71.1 in.?

2-5 Exercises B: Beyond the Basics

23. a. Find the range and standard deviation s for the following labour force populations (based on data from Statistics Canada):

Halifax: 186,600	Quebec City: 356,800	Oshawa: 148,100
Winnipeg: 381,300	Saskatoon: 121,600	Victoria: 164,600

b. If a constant value k is added to each population, how are the results from part (a) affected?

c. If each population given in part (a) is multiplied by a constant k, how are the results from part (a) affected?

d. Data are sometimes transformed by replacing each score x with log x. For the given x values, determine whether the standard deviation of the log x values is equal to log s.

e. The following exercise summarizes parts (b) and (c): For the body temperature data listed in Data Set 18 of Appendix B (first 10 scores), $\bar{x} = 36.36°C$ and $s = 0.35°C$. Find the values of $\bar{x}$ and s for the data after each temperature has been converted to the Fahrenheit scale. [*Hint:* F = (C × 9) ÷ 5 + 32.]

24. If we consider the values 1, 2, 3, . . . , n to be a population, the standard deviation can be calculated by the formula

$$\sigma = \sqrt{\frac{n^2 - 1}{12}}$$

This formula is equivalent to Formula 2-4 modified for division by n instead of $n - 1$, where the data set consists of the values 1, 2, 3, . . . , n.

a. Find the standard deviation of the population 1, 2, 3, . . . , 100.

b. Find an expression for calculating the sample standard deviation s for the *sample* values 1, 2, 3, . . . , n.

c. Computers and calculators commonly use a random-number generator that produces values between 0.00000000 and 0.99999999. In the long run, all values occur with the same relative frequency. Find the mean and standard deviation for the *population* of those values.

25. Two different sections of a statistics class take the same surprise quiz and the scores are recorded below. Find the range and standard deviation for each section. What do the range values lead you to conclude about the variation in the two sections? Why is the range misleading in this case? What do the standard deviation values lead you to conclude about the variation in the two sections?

   Section 1:   1   20   20   20   20   20   20   20   20   20   20
   Section 2:   2   3   4   5   6   14   15   16   17   18   19

26. a. The **coefficient of variation**, expressed as a percent, is used to describe the standard deviation relative to the mean. It allows us to compare variability of data sets with different measurement units (such as centimetres versus kilograms), and it is calculated as follows:

$$\frac{s}{\overline{x}} \cdot 100 \quad \text{or} \quad \frac{\sigma}{\mu} \cdot 100$$

Find the coefficient of variation (C.V.) for the following two samples:

Sample A (Service in years of 10 employees):
   10   35   40   12   15   11   17   16   15   25

Sample B (Number of new cars sold at a dealership over 7 days):
   8   5   12   3   0   6   3

   b. Genichi Taguchi developed a method of improving quality and reducing manufacturing costs through a combination of engineering and statistics. A key tool in the Taguchi method is the **signal-to-noise ratio**. The simplest way to calculate this ratio is to divide the mean by the standard deviation. Find the signal-to-noise ratio for the sample data given in part (a).

27. In Section 2–4 we introduced the general concept of skewness. Skewness can be measured by **Pearson's index of skewness:**

$$I = \frac{3(\overline{x} - \text{median})}{s}$$

If $I \geq 1.00$ or $I \leq -1.00$, the data can be considered to be *significantly skewed*. Find Pearson's index of skewness for the axial loads of the aluminum cans listed in Table 2-1, and then determine whether there is significant skewness.

28. a. A sample consists of 6 scores that fall between 1 and 9 inclusive. What is the largest possible standard deviation?
   b. For any data set of n scores with standard deviation s, every score must be within $s\sqrt{n-1}$ of the mean. A statistics teacher reports that the test scores in her class of 17 students had a mean of 75.0 and a standard deviation of 5.0. Kelly, the class's self-proclaimed best student, claims that she received a grade of 97. Could Kelly be telling the truth?

2-6 Measures of Position

In this section we introduce *z* scores, which enable us to standardize values so that they can be compared more easily. We also introduce quartiles, percentiles, and deciles, which help us better understand data by showing their positions relative to the whole data set.

z Scores

Most of us are reasonably familiar with IQ scores, and we recognize that an IQ of 102 is fairly common, whereas an IQ of 170 is rare. The IQ of 102 is fairly common because it is very close to the mean of 100, but the IQ of 170 is rare because it is so far above 100. This might suggest that we can differentiate between typical scores and rare scores on the basis of the difference between the score and the mean $(x - \bar{x})$. However, the size of such differences is relative to the scale being used. With IQ scores, a 2-point difference is insignificant, but for college grade-point averages, the 2-point difference between 2.00 and 4.00 is very significant, especially to parents. It would be much better if we could use a standard that doesn't require an understanding of the scale being used. With the standard score, we get such a result.

DEFINITION

The **standard score**, or **z score**, is the number of standard deviations that a given value *x* is above or below the mean. It is found by using

$$\text{Sample} \qquad\qquad \text{Population}$$
$$z = \frac{x - \bar{x}}{s} \quad \text{or} \quad z = \frac{x - \mu}{\sigma}$$

(Round *z* to two decimal places.)

EXAMPLE

Heights of all adult males have a mean of $\mu = 69.0$ in., a standard deviation of $\sigma = 2.8$ in., and a distribution that is bell-shaped. Basketball player Michael Jordan earned a giant reputation for his skills, but at a height of 78 in., is he exceptionally tall when compared to the general population of adult males? Find the *z* score for his 78-in. height.

SOLUTION

Because we are dealing with population parameters, the *z* score is calculated as follows:

$$z = \frac{x - \mu}{\sigma} = \frac{78 - 69.0}{2.8} = 3.21$$

Interpretation
Michael Jordan's height of 78 in. is 3.21 standard deviations above the mean; that is, he is unusually tall.

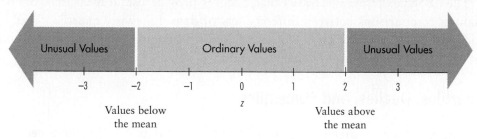

Unusual Values   Ordinary Values   Unusual Values

−3   −2   −1   0   1   2   3

z

Values below   Values above
the mean   the mean

Figure 2-15
Interpreting z Scores
Unusual values are those with z scores less than $z = -2.00$ or greater than $z = 2.00$.

The role of z scores in statistics is extremely important because they can be used to differentiate between ordinary values and unusual values. Values with standard scores between -2.00 and 2.00 are ordinary, and values with z scores less than -2.00 or greater than 2.00 are unusual. (See Figure 2-15.) Michael Jordan's height converts to a z score of 3.21, so we consider it unusual because it is greater than 2.00. In comparison to the general population, Michael Jordan is exceptionally tall.

Our criterion for unusual z scores follows from the empirical rule and Chebyshev's theorem. Recall that for data with a bell-shaped distribution, about 95% of the values are within 2 standard deviations of the mean. Also, Chebyshev's theorem states that for any data set, at least 75% of the values are within 2 standard deviations of the mean.

We noted earlier that z scores are also useful for comparing values from dissimilar populations. The following illustrates this use of z scores.

EXAMPLE

A statistics professor gives two different tests to two sections of her course. Which score is relatively better: an 82 on the Section 1 test, or a 46 on the Section 2 test?

$$\text{Section 1: } \bar{x} = 75 \text{ and } s = 14$$
$$\text{Section 2: } \bar{x} = 40 \text{ and } s = 8$$

SOLUTION

We can't directly compare the scores of 82 and 46 because they come from different scales. Instead, we convert them both to z scores. For the score of 82 on the Section 1 test, we get a z score of 0.50, because

$$z = \frac{x - \bar{x}}{s} = \frac{82 - 75}{14} = 0.50$$

For the score of 46 on the Section 2 test, we get a z score of 0.75, because

$$z = \frac{x - \bar{x}}{s} = \frac{46 - 40}{8} = 0.75$$

Interpretation

A score of 82 on the Section 1 test is 0.50 standard deviation above the mean, whereas a score of 46 on the Section 2 test is 0.75 standard deviation above the mean. This implies that the 46 on the Section 2 test is the better relative score, when considered in the context of the other test results.

AS

The above example illustrated that z scores provide useful measurements for making comparisons between different sets of data. Likewise, quartiles, deciles, and percentiles are measures of position useful for comparing scores within one set of data or between different sets of data.

Quartiles, Deciles, and Percentiles

Just as the median divides the data into two equal parts, the three **quartiles**, denoted by Q_1, Q_2, and Q_3, divide the *sorted* values into four equal parts. (Values are sorted when they are arranged in order.) Roughly speaking, Q_1 separates the bottom 25% of the sorted values from the top 75%, Q_2 is the median, and Q_3 separates the top 25% from the bottom 75%. To be more precise, at least 25% of the data will be less than or equal to Q_1, and no more than 75% will be greater than Q_1. At least 75% of the data will be less than or equal to Q_3, while no more than 25% will be greater than Q_3.

Quartiles are useful for exploring and describing the distribution of a data set. We describe a procedure for finding quartiles after we discuss percentiles. No procedure for calculating quartiles is universally approved (not even ours), so be aware that different techniques and computer programs can yield different results (and may not agree with our answers at the back of the book). For example, if you use the data set of 1, 3, 6, 10, 15, 21, 28, and 36, Excel would calculate Q_1 at 5.25, STATDISK would return an answer of 4.5 and Minitab the lowest of the three at 3.75.

Just as there are three quartiles separating a data set into four parts, there are nine **deciles**, denoted by D_1, D_2, D_3, . . . , D_9, which partition the data into 10 groups with about 10% of the data in each group. There are also 99 **percentiles**, which partition the data into 100 groups with about 1% of the scores in each group. (Quartiles, deciles, and percentiles are examples of **quantiles**—or **fractiles**—which partition data into parts that are approximately equal.)

The process of finding the percentile that corresponds to a particular value x is fairly simple, as indicated in the following expression:

$$\text{percentile of value } x = \frac{\text{number of values less than } x}{\text{total number of values}} \cdot 100$$

EXAMPLE

Table 2-9 lists the 175 axial loads of aluminum cans, sorted from lowest to highest. Find the percentile corresponding to 241.

SOLUTION

From Table 2-9 we see that there are 21 values less than 241, so

$$\text{percentile of 241} = \frac{21}{175} \cdot 100 = 12$$

The axial load of 241 is the 12th percentile.

Table 2-9 *Ranked* Axial Loads of Aluminum Cans

200	201	204	204	206	206	208	208	209	215	217	218	220	223	223
225	228	230	230	234	236	241	242	242	248	250	251	251	252	252
254	256	256	256	257	257	258	259	259	260	261	262	262	262	262
262	263	263	263	263	263	264	265	265	265	266	267	267	268	268
268	268	268	268	268	268	268	269	269	269	269	270	270	270	270
270	270	270	270	271	271	272	272	272	272	272	273	273	273	273
273	273	274	274	274	274	275	275	275	275	276	276	276	276	276
277	277	277	277	277	277	277	277	278	278	278	278	278	278	278
279	279	279	280	280	280	281	281	281	281	282	282	282	282	282
282	283	283	283	283	283	283	284	284	284	284	285	285	285	286
286	286	286	287	287	288	289	289	289	289	289	290	290	290	291
291	292	292	292	293	293	294	295	295	297					

The preceding example illustrated the procedure for finding the percentile corresponding to a given value. There are several different methods for the reverse procedure of finding the value corresponding to a particular percentile, but the one we will use is summarized in Figure 2-16, which uses the following notation.

NOTATION

n number of values in the data set

k percentile being used (Example: For the 25th percentile, $k = 25$.)

L locator that gives the *position* of a value (Example: For the 25th value in a sorted list, $L = 25$.)

P_k kth percentile (Example: P_{25} is the 25th percentile.)

EXAMPLE

Refer to the 175 axial loads of aluminum cans in Table 2-9, and find the value corresponding to the 25th percentile. That is, find the value of P_{25}.

SOLUTION

We refer to Figure 2-16 and observe that the data are already sorted from lowest to highest. We now compute the locator L as follows:

$$L = \left(\frac{k}{100} \right) n = \left(\frac{25}{100} \right) \cdot 175 = 43.75$$

We answer no when asked in Figure 2-16 if 43.75 is a whole number, so we are directed to round L up (not off) to 44. The 25th percentile, denoted by P_{25}, is the 44th score, counting from the lowest. Beginning with the lowest score of 200, we count through the list to find the 44th score of 262, so $P_{25} = 262$.

Figure 2-16
Finding the Value of the
***k*th Percentile**

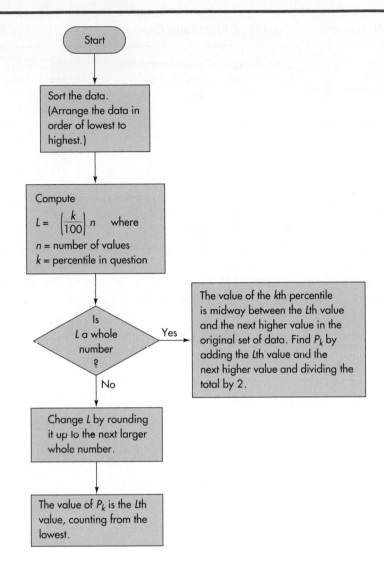

Suppose we want to find the percentile corresponding to a score of 262. Verify that there are 41 scores below 262; be sure to count each individual score, including duplicates. Finding the percentile for 262 therefore yields $(41/175) \cdot 100 = 23$ (rounded). There is a small discrepancy: In the preceding example we found the 25th percentile to be 262, but when we reverse the process we find that 262 is the 23rd percentile. As the amount of data increases, such discrepancies become smaller. We could eliminate the discrepancy by using a more complicated procedure that includes interpolations instead of rounding.

Because of the sample size in the preceding example, the locator L first became 43.75, which was rounded to 44 because L was not originally a whole number. In the next example we illustrate a case in which L does begin as a whole number. This condition will cause us to branch to the right in Figure 2-16.

EXAMPLE

Refer to the axial load data for aluminum cans, as listed in Table 2-9. Find P_{40}, which denotes the 40th percentile.

SOLUTION

Following the procedure outlined in Figure 2-16 and noting that the data are already sorted from lowest to highest, we compute

$$L = \left(\frac{k}{100}\right)n = \left(\frac{40}{100}\right) \cdot 175 = 70 \quad \text{(exactly)}$$

We note that 70 is a whole number, and Figure 2-16 indicates that P_{40} is midway between the 70th and 71st values. Because the 70th and 71st values are both 269, we conclude that the 40th percentile is 269.

Once you have mastered these calculations with percentiles, similar calculations for quartiles and deciles can be performed with the same procedures by noting the relationships given in the margin.

Using these relationships, we can see that finding Q_1 is equivalent to finding P_{25}. In an earlier example we found that $P_{25} = 262$, so it follows that the first quartile can be described by $Q_1 = 262$. If we need to find the third quartile, Q_3, we can restate the problem as that of finding P_{75}, and we can then proceed to use Figure 2-16.

In addition to the measures of central tendency and the measures of variation already introduced, other statistics are sometimes defined using quartiles, deciles, or percentiles, as in the following:

Quartiles	Deciles
$Q_1 = P_{25}$	$D_1 = P_{10}$
$Q_2 = P_{50}$	$D_2 = P_{20}$
$Q_3 = P_{75}$	$\vdots$
	$D_9 = P_{90}$

$$\text{interquartile range} = Q_3 - Q_1$$

$$\text{semi-interquartile range} = \frac{Q_3 - Q_1}{2}$$

$$\text{midquartile} = \frac{Q_1 + Q_3}{2}$$

$$10-90 \text{ percentile range} = P_{90} - P_{10}$$

USING TECHNOLOGY

Procedures to generate *descriptive statistics* in Excel, Minitab 15 or STATDISK have been described in Sections 2–4 and 2–5, in the Using Technology sections. STATDISK's display also includes some measures of position. (In Excel, percentile values can be generated by the function **=percentile**(*range,percentile*), where *range* is the Excel range containing the raw data, and *percentile* indicates the percentile number (for example 0.28, for the 28th percentile). The corresponding function to find quartiles is **=quartile**(*range,quartile*) where quartile takes the values

1, 2, 3, or 4, depending on the quartile wanted. Minitab 15 reports Q_1, Q_2, and Q_3 by default when displaying descriptive statistics.) The Excel and STATDISK outputs, based on the 175 axial loads listed in Table 2-1, are shown here. It should be noted that computers generally use more complicated formulas than those we have presented. Thus, the values they compute may not necessarily equal the values computed by hand, particularly Q_1 and Q_3.

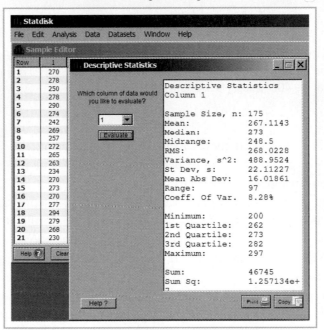

Mean	267.1143
Standard Error	1.67153
Median	273
Mode	268
Standard Deviation	22.11227
Sample Variance	488.9524
Kurtosis	1.679651
Skewness	-1.4777
Range	97
Minimum	200
Maximum	297
Sum	46745
Count	175
Confidence Level(95.0%)	3.299083
Minimum	200
Maximum	297
Sum	46745
Count	175
Confidence Level(95.0%)	3.299083

2-6 Exercises A: Basic Skills and Concepts

In Exercises 1–4, express all z scores with two decimal places.

1. The Goodstone Tire Company claims that the useful life of its tires is an average of 68,000 km with a standard deviation of 10,000 km. Find the z scores corresponding to the following:
   a. A tire that lasts 50,000 km
   b. A tire that lasts 78,000 km
   c. A tire that lasts 70,000 km

2. Student cars at the author's college have ages with a mean of 7.90 years and a standard deviation of 3.67 years. Find the z scores for cars with the given ages:
   a. A 12-year-old Corvette
   b. A 2-year-old Ferrari
   c. A brand-new Porsche

3. A large airline found that, on average, each passenger's luggage weighed 50.6 kg with a standard deviation of 8.8 kg. Find the z score for a passenger whose luggage weighed 75 kg.

4. The mean score on a statistics test was 70 and the standard deviation was 5.5. Find the z score for students scoring 80 or better on the test.

In Exercises 5–8, express all z scores with two decimal places. Consider a score to be unusual if its z score is less than −2.00 or greater than 2.00.

5. The Beanstalk Club is limited to women and men who are very tall. The minimum height requirement for women is 70 in. Women's heights have a mean of 63.6 in. and a standard deviation of 2.5 in. Find the z score corresponding to a woman with a height of 70 in. and determine whether that height is unusual.

6. According to Statistics Canada, the monthly earnings of workers in the mining industry was $3840, with a standard deviation of $240. A mine worker claims to earn $4325 each month. Find the z score corresponding to this worker's wage. Is the amount unusual?

7. One of the few working vending machines is located and is found to accept quarters with weights that are not unusual. Weights of quarters have a mean of 5.67 g and a standard deviation of 0.070 g. Find the z score for a quarter weighing 5.50 g. Will it be accepted by this vending machine?

8. For men aged between 18 and 24 years, serum cholesterol levels (in mg/100 mL) have a mean of 178.1 and a standard deviation of 40.7. Find the z score corresponding to a male, aged 18–24 years, who has a serum cholesterol level of 275.2 mg/100 mL. Is this level unusually high?

9. Which of the following two scores has the better relative position?
   a. A score of 60 on a test for which $\bar{x} = 50$ and $s = 5$
   b. A score of 250 on a test for which $\bar{x} = 200$ and $s = 20$

10. Two similar groups of students took equivalent language facility tests. Which of the following results indicates the higher relative level of language facility?
   a. A score of 65 on a test for which $\bar{x} = 70$ and $s = 10$
   b. A score of 455 on a test for which $\bar{x} = 500$ and $s = 80$

11. Three prospective employees take equivalent tests of critical thinking. Which of the following scores corresponds to the highest relative position?
   a. A score of 37 on a test for which $\bar{x} = 28$ and $s = 6$
   b. A score of 398 on a test for which $\bar{x} = 312$ and $s = 56$
   c. A score of 4.10 on a test for which $\bar{x} = 2.75$ and $s = 0.92$

12. Three students take equivalent tests of a sense of humour and, after the laughter dies down, their scores are calculated. Which is the highest relative score?
   a. A score of 2.7 on a test for which $\bar{x} = 3.2$ and $s = 1.1$
   b. A score of 27 on a test for which $\bar{x} = 35$ and $s = 12$
   c. A score of 850 on a test for which $\bar{x} = 921$ and $s = 87$

In Exercises 13–16, use the 175 ranked axial loads of aluminum cans listed in Table 2-9. Find the percentile corresponding to the given value.

13. 254   14. 265   15. 277   16. 288

In Exercises 17–24, use the 175 ranked axial loads of aluminum cans listed in Table 2-9. Find the indicated percentile, quartile, or decile.

17. P_{70} 18. P_{20} 19. D_6 20. D_3
21. Q_3 22. Q_1 23. D_1 24. P_1

In Exercises 25–28, use the annual values for person-days lost in the truck-based transportation industry due to work stoppages (listed in Data Set 20 in Appendix B). Find the percentile corresponding to the given number of person-days lost.

25. 1000   26. 54,000   27. 12,000   28. 40,000

In Exercises 29–36, use the annual values for person-days lost in the truck-based transportation industry due to work stoppages (listed in Data Set 20 in Appendix B). Find the indicated percentile, quartile, or decile.

29. P_{85} 30. P_{35} 31. Q_1 32. Q_3
33. D_9 34. D_3 35. P_{50} 36. P_{95}

2-6 Exercises B: Beyond the Basics

37. Use the ranked axial loads of aluminum cans listed in Table 2-9.
   a. Find the interquartile range.
   b. Find the midquartile.
   c. Find the 10–90 percentile range.

d. Does $P_{50} = Q_2$? If so, does P_{50} *always* equal Q_2?

e. Does $Q_2 = (Q_1 + Q_3)/2$? If so, does Q_2 *always* equal $(Q_1 + Q_3)/2$?

38. When finding percentiles using Figure 2-16, if the locator L is not a whole number, we round it up to the next larger whole number. An alternative to this procedure is to interpolate so that a locator of 23.75 leads to a value that is 0.75 (or 3/4) of the way between the 23rd and 24th scores. Use this method of interpolation to find P_{35}, Q_1, and D_3 for the weights of bears listed in Data Set 3 of Appendix B.

39. Distribution of z Scores:

a. A data set has a distribution that is uniform. If all of the values are converted to z scores, what is the shape of the distribution of the z scores?

b. A data set (whose units are cm) has a distribution that is bell-shaped. If all the values are converted to z scores, what is the shape of the distribution of z scores?

c. In general, how is the shape of a distribution affected if all values are converted to z scores?

d. If all the values in the data set mentioned in (b) are converted from centimetres into inches, what is the effect on the distribution of z scores?

40. Using the scores 2, 5, 8, 9, and 16, first find $\bar{x}$ and s, then replace each score by its corresponding z score. (Don't round the z scores; carry as many decimal places as your calculator can handle.) Now find the mean and standard deviation of the five z scores. Will these new values of the mean and standard deviation result from every set of z scores?

2-7 Exploratory Data Analysis (EDA)

The theme of this chapter is describing, exploring, and comparing data; the focus of this section is exploration. We will first define exploratory data analysis, then introduce some new tools—outliers, five-number summaries, and boxplots. We can then add these to the techniques presented earlier in this chapter.

DEFINITION

> **Exploratory data analysis** is the process of using statistical tools (such as graphs, measures of central tendency, and measures of variation) to investigate data sets in order to understand their important characteristics.

Recall that in Section 2–1 we introduced these three important characteristics of data: (1) *central tendency*, a representative or average value; (2) *variation*, a measure of the amount of spread among the sample values; and (3) *distribution*, the nature or shape of the distribution of the data. When exploring a data set, we usually want to calculate the mean and the standard deviation and to generate a histogram, but we should not stop there. Are there any other notable features—especially features that could have a strong effect on results and conclusions? One such feature is the presence of outliers.

Outliers

An **outlier** is a value that is located very far away from almost all the other values. Relative to the other data, an outlier is an *extreme* value. The following example illustrates the possible effects if an outlier is included in a data set.

EXAMPLE

When using computer software it is easy to make typing errors. Referring to the axial loads listed in Table 2-1, suppose that the first entry of 270 was incorrectly entered as 2700. How does that affect your data set?

SOLUTION

After the first entry of 270 is mistakenly replaced with the value of 2700, the data in Table 2-1 will result in a mean of 281.0 (instead of the correct value of 267.1), and a standard deviation of 185.2 (instead of the correct value of 22.1). The accompanying STATDISK display shows the histogram for the mistakenly modified data set. Compare this histogram with the correct version, shown in Figure 2-2 of Section 2-3, and you can clearly see the difference.

Our example demonstrates these important principles:

1. An outlier can have a dramatic effect on the mean.

2. An outlier can have a dramatic effect on the standard deviation.

3. An outlier can have a dramatic effect on the scale of the histogram so that the true nature of the distribution is totally obscured.

STATDISK DISPLAY
Incorrect Data

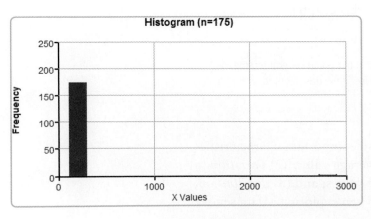

An easy way to find outliers is to examine a *sorted* list of the data. In particular, look at the minimum and maximum sample values and determine whether they are very far away from the other typical values. In the preceding example, the outlier of 2700 is an error. When such an error is identified, we should either correct it or delete it. Remember, however, that some data sets include outliers that are correct values, not errors. (It is often useful to "troubleshoot" exactly where and how the specific extreme value was collected—in order to judge if an error was made.) When exploring data, we might study the effects of outliers by constructing graphs and calculating statistics with and without the outliers. (See Exercise 12 for a way to depict outliers on boxplots.)

Boxplots

Boxplots are graphs that are useful for revealing central tendency, the spread of the data, the distribution of the data, and the presence of outliers (extreme values). The construction of a boxplot requires that we obtain the minimum value, the first quartile Q_1, the median (or second quartile Q_2), the third quartile Q_3, and the maximum value (together these are called the *five-number summary*.) Because medians are used to reveal central tendency and quartiles are used to reveal the spread of data, boxplots have the advantage of not being as sensitive to extreme values as other devices based on the mean and standard deviation. Boxplots don't show as much detailed information as histograms or stem-and-leaf plots, so they might not be the best choice when dealing with a single data set. However, boxplots are often more useful when comparing two or more data sets. When using two or more boxplots for comparing different data sets, it is important to use the same scale so that meaningful comparisons can be made.

> DEFINITIONS
>
> For a set of data, the **five-number summary** consists of the minimum value, the first quartile Q_1, the median, the third quartile Q_3, and the maximum value.
>
> A **boxplot** (or **box-and-whisker diagram**) is a graph of data that consists of a line extending from the lowest score to the highest, and a box with lines drawn at the first quartile Q_1, the median, and the third quartile Q_3.

EXAMPLE

Refer to Data Set 9 of Appendix B and use the pulse rates of smokers.

a. Find the values constituting the five-number summary.

b. Construct a boxplot for the pulse rates of smokers.

SOLUTION

a. The five-number summary consists of the minimum, Q_1, median, Q_3, and maximum. To find those values, we should first arrange the pulse rates of smokers in order from low to high. Here is the *sorted* list of the 22 pulse rates of smokers from Data Set 9:

52 52 60 60 60 60 63 63 66 67 68
69 71 72 73 75 78 80 82 83 88 90

From this sorted list, it is easy to identify the minimum of 52 and the maximum of 90. Using the flowchart of Figure 2-16, we find that the first quartile Q_1 (or P_{25}) is 60, which is located by calculating $L = (25/100)22 = 5.5$ and rounding up the result to 6. Q_1 is the 6th value in the sorted list, namely 60. The median is 68.5, which is the number midway between the 11th and 12th values. We also find that $Q_3 = 78$ by using Figure 2-16 for the 75th percentile. The five-number summary is therefore 52, 60, 68.5, 78, and 90.

b. In Figure 2-17 we graph the boxplot for the data. We use the minimum (52) and the maximum (90) to determine a scale of values, then we plot the values from the five-number summary as shown.

In Figure 2-18 we show some generic boxplots along with common distribution shapes.

Exploring

Including the options from this section, we now have the following arsenal of tools for data exploration:

- *Measures of central tendency:* mean, median, and mode
- *Measures of variation:* standard deviation and range
- *Measures of spread and relative location:* minimum and maximum values, and quartiles
- *Unusual values:* outliers
- *Distribution:* histograms, stem-and-leaf plots, and boxplots

Now that you have all these tools in hand, begin to *think creatively!* Rather than simply cranking out statistics and graphs, try to identify those that are particularly interesting or important. As a first step, investigate outliers and consider their effects by finding measures and graphs with and without the outliers.

Figure 2-17
Boxplot of Pulse Rates (Beats per Minute) of Smokers

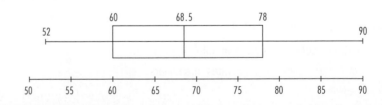

Figure 2-18
Boxplots Corresponding to Bell-Shaped, Uniform, and Skewed Distributions

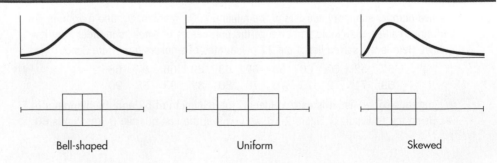

Bell-shaped   Uniform   Skewed

If you have spent time in the four Atlantic provinces, it is easy to conclude that they have (on average) more annual rainfall than do other provinces. Later in this book we will describe statistical methods that can be used to formally address such a claim, but for now, let us explore the related data in Data Set 4 in Appendix B, to see what can be learned. (Even if we know the more formal statistical tests, it is useful to explore the data first.) We will use the Average Precipitation column, and assume that the selected cities are representative of areas within Atlantic or non-Atlantic Canada.

SOLUTION

In the following table and STATDISK display we apply our arsenal of data exploration tools, listed above this example. Note that the table (based on Excel) and the display disagree about the first and third quartiles for the two groups of provinces; but since we are just exploring the data, this is not a great concern.

	Atlantic Provinces	Other Provinces
Mean	1313.250	772.583
Standard Deviation	171.873	453.035
Minimum	1097.000	266.000
Q_1	1154.000	463.750
Median	1341.000	737.000
Q_3	1455.750	908.250
Maximum	1514.000	2523.000

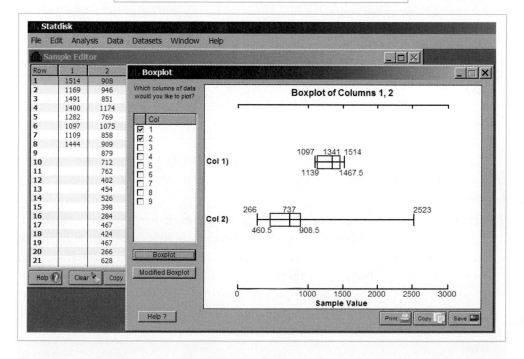

Interpretation

Examining and comparing the statistics and graphs, we make the following important observations.

- *Outliers:* The maximum value of 2523 for precipitation in non-Atlantic Canada stands out as a possible outlier. A further check by sorting the data for non-Atlantic Canada shows that this maximum value is more than double the next largest value (of 1174). Because the data source (Environment Canada) is generally reliable, it is probable that this outlier is not a mistaken value, but indicates that at least one city (Prince Rupert, B.C.) may have an atypical value, compared to the average precipitations in other non-Atlantic cities.

- *Means:* Comparison of means (1313 for Atlantic Canada, and 773 for other provinces) may support the claim of higher precipitation in Atlantic Canada, but this may be misleading, given the large standard deviation (453) for non-Atlantic provinces. However, if the mean and standard deviation for non-Atlantic cities are recalculated, omitting the apparent outlier (2523), they become 696 and 263, respectively—further strengthening the claim of difference from Atlantic Canada.

- *Variation:* Data from the non-Atlantic provinces have a larger standard deviation (453 *with* the outlier, 263 *without*) than data from the Atlantic provinces (172). This is logical, since the former are drawn from a greater diversity of geographical and meteorological contexts.

- *Distributions and five-number summaries:* In comparison with the non-Atlantic provinces, data from the Atlantic provinces appear to have a more symmetrical distribution, with less variation; and the measures of location and position clearly suggest (except for the outlier) greater precipitation values for cities from the Atlantic region.

We have now gained considerable insight into the comparative values for "average precipitation" from cities in two parts of Canada (Atlantic provinces, the other provinces). Based on our exploration we can conclude that, in general, the Atlantic cities experience the greater levels of precipitation, but in the broad expanse of "the other provinces" there are some notable exceptions to the rule.

USING TECHNOLOGY

We have already seen how Excel, Minitab 15, and STATDISK can generate measures of location, variation, and position. These are their procedures for generating boxplots.

EXCEL: To generate boxplots with Excel, use the Data Desk XL add-in that is a supplement to this book. In a block of cells (such as M1:N24) enter the data for each desired boxplot (if there is more than one column, Excel will generate one boxplot per column). Data columns should start on the same row, but do *not* need to contain the same number of values.

EXCEL (Prior to 2007):	Click on **DDXL** and select **Charts** and **Plots**. Under **Function Type**, select the option **Boxplot**. In the dialog box, click on the pencil icon and enter the range of data (the range should be large enough to include all the data values in the input block; it does not matter if some cells are empty). Click on **OK**. The result is actually a modified boxplot, as described in Exercise 12.
EXCEL 2007:	Click on the **Add-in** tab on the main menu. You will see **DDXL** to the left. Click on it and proceed as above.
MINITAB 15:	Put the data in column C1. For a simple boxplot, from the **Graph** menu, choose **Boxplot** and then **Simple**. Click on C1 in the left window and then **Select**. C1 will now appear the right window. For the default boxplot, click **OK**. If you want to identify the values in the boxplot such as Q1, Q3, etc., click the **Labels** box on the **Labels** dialog box, click the **Data Labels** tab and select **Individual data** from the pull-down menu and the **Use y-value labels** radio button.
STATDISK:	Choose the main menu item of **Data**, then select **Boxplot**. Choose which columns you want, then click the **Evaluate** button.

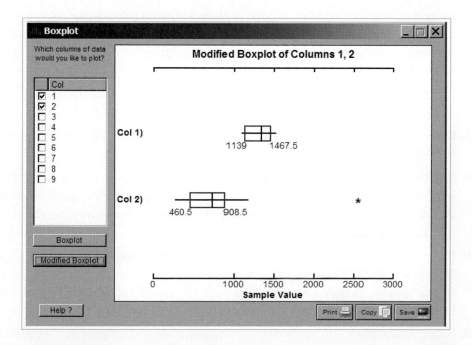

2-7 Exercises A: Basic Skills and Concepts

Include values of the five-number summary in all boxplots.

1. Refer to Data Set 4 in Appendix B and construct a boxplot for the average annual temperatures of the cities listed.

2. Refer to Data Set 4 in Appendix B and construct a boxplot for the annual precipitation of the cities listed.

3. In "Ages of Oscar-Winning Best Actors and Actresses" by Richard Brown and Gretchen Davis (*Mathematics Teacher* magazine), boxplots are used to compare the ages of actors and actresses at the time they won Oscars. The results for 34 recent winners from each category are listed below. Use boxplots to compare the two data sets.

Actors:	32	37	36	32	51	53	33	61	35	45	55	39
	76	37	42	40	32	60	38	56	48	48	40	
	43	62	43	42	44	41	56	39	46	31	47	
Actresses:	50	44	35	80	26	28	41	21	61	38	49	33
	74	30	33	41	31	35	41	42	37	26	34	
	34	35	26	61	60	34	24	30	37	31	27	

4. Refer to Data Set 9 in Appendix B for these two data sets: pulse rates of those who smoke and pulse rates of those who don't smoke. Construct a boxplot for each data set. Based on the results, do pulse rates of the two groups appear to be different? If so, how? Is this the result you would expect? (Exclude the 8 and 15, which must be errors.)

5. Refer to Data Set 9 in Appendix B for these two data sets: pulse rates of males and pulse rates of females. Construct a boxplot for each data set. Based on the results, do pulse rates of the two groups appear to be different? If so, how? (Exclude the 8 and 15, which must be errors.)

6. Refer to Data Set 10 in Appendix B. Use boxplots to compare the one-year return of U.S. equity funds to that of the funds other than U.S. equity.

7. Refer to Data Set 11 in Appendix B. Use boxplots to compare the weights of red M&M candies to the weights of yellow M&M candies.

8. Refer to Data Set 13 in Appendix B. Construct a boxplot for the weights of quarters. Compare the shape of the resulting boxplot to the generic shapes shown in Figure 2-18. Based on the boxplot, what do you conclude about the nature of the distribution?

9. Refer to Data Set 12 in Appendix B. Construct a boxplot for the 50 digits from the Lotto 6/49 bonus draw. Compare the shape of the resulting boxplot to the generic shapes shown in Figure 2-18. Based on the boxplot, do the Lotto 6/49 results appear to be occurring as expected?

10. Refer to Data Set 1 in Appendix B. Use boxplots to compare the weights of discarded paper to the weights of discarded plastic.

2-7 Exercises B: Beyond the Basics

11. Compare the two boxplots from Exercises 1 and 2. How would you describe the shape of the distribution in the two sets of data? Are they uniform, skewed, or bell-shaped?

12. The boxplots discussed in this section are often called *skeletal (or regular)* boxplots. **Modified boxplots** are constructed as follows:

  a. Calculate the difference between the quartiles Q_3 and Q_1 and denote it as *IQR (interquartile range)*, so that $IQR = Q_3 - Q_1$.

  b. Draw the box with the median and quartiles as usual, but when extending the lines that branch out from the box, go only as far as the scores that are within 1.5 *IQR* of the box.

  c. *Mild outliers* are scores above Q_3 by an amount of 1.5 *IQR* to 3 *IQR*, or below Q_1 by an amount of 1.5 *IQR* to 3 *IQR*. Plot mild outliers as solid dots.

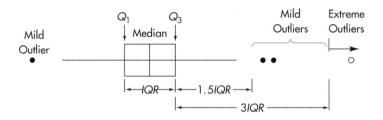

  d. *Extreme outliers* are scores that exceed Q_3 by more than 3 *IQR* or are below Q_1 by an amount more than 3 *IQR*. Plot extreme outliers as small hollow circles.

The accompanying figure is an example of the boxplot described here. Use this procedure to construct the boxplot for the given scores, and identify any mild outliers or extreme outliers.

    3   15   17   18   21   21   22   25   27   30 38 49 68

VOCABULARY LIST

arithmetic mean **56**
average growth factor **70**
bimodal **59**
box-and-whisker diagram **99**
boxplot **99**
Chebyshev's theorem **82**
class boundaries **33**
class midpoints **33**

class width **33**
coefficient of
  variation **87**
cumulative relative
  frequency **37**
cumulative relative
  frequency table **37**
deciles **90**

descriptive statistics **32**
deviation **73**
empirical (or 68–95–99.7)
  rule **81**
exploratory data analysis
  (EDA) **97**
five-number summary **99**
fractiles **90**

frequency   33
frequency polygon   43
frequency table   33
geometric mean   70
harmonic mean   69
histogram   42
inferential statistics   32
lower class limits   33
mean   56
mean absolute
   deviation   74
measure of central
   tendency   55
measures of variation   71
median   58
midrange   59
mode   59
modified boxplots   105

multimodal   59
negatively skewed   65
ogive   44
outlier   98
Pareto chart   46
Pearson's index of
   skewness   87
percentiles   90
pie chart   46
positively skewed   65
quadratic mean   70
quantiles   90
quartiles   90
range   72
range rule of thumb   79
relative frequency   36
relative frequency
   histogram   43

relative frequency
   table   36
root mean square
   (R.M.S.)   70
sample size   56
scatter diagram   48
signal-to-noise ratio   87
skewed   64
standard deviation   73
standard score   88
stem-and-leaf plot   45
symmetric   64
unbiased estimator   77
upper class limits   33
variance   77
weighted mean   63
z score   88

REVIEW

Chapter 2 considered methods and techniques of describing, exploring, and comparing data. We noted these important characteristics of data:

1. **Central Tendency:** A representative or average value

2. **Variation:** A measure of the amount that the values vary

3. **Distribution:** The nature or shape of the distribution of the data (such as bell-shaped, uniform, or skewed)

4. **Outliers:** Sample values that lie very far away from the vast majority of the other sample values

5. **Time:** Changing characteristics of the data over time

After completing this chapter, you should be able to do the following:

- Summarize the data by constructing a frequency table or relative frequency table (Section 2–2)

- Visually display the nature of the distribution by constructing a histogram, stem-and-leaf plot, pie chart, or Pareto chart (Section 2–3)

- Calculate measures of central tendency by finding the mean, median, mode, and midrange (Section 2–4)

- Calculate measures of variation by finding the standard deviation, variance, and range (Section 2–5)

- Compare individual scores by using z scores, quartiles, deciles, or percentiles (Section 2–6)

- Investigate and explore the spread of data, the centre of the data, and the range of values by constructing a boxplot (Section 2–7)

In addition to obtaining the above tables, graphs, and measures, we should *understand* and *interpret* those results. For example, we should clearly understand that the standard deviation is a measure of how much the data vary, and we should be able to use the standard deviation to distinguish between scores that are usual and those that are unusual.

REVIEW EXERCISES

1. A family living in Ontario collected the data from their natural gas bills to monitor their consumption of gas and to try to save money. The following values are the cubic feet of gas used each month for five years. Using the lowest volume as the lower limit of the first class, construct a frequency table with 10 classes.

Month	Year				
	1	2	3	4	5
Jan	303	263	260	104	219
Feb	192	212	254	170	176
Mar	126	125	156	142	102
Apr	57	83	83	38	70
May	34	42	31	40	27
June	24	18	15	19	20
July	18	15	17	19	18
Aug	17	14	17	19	18
Sept	58	75	60	48	62
Oct	116	118	91	102	104
Nov	181	203	200	205	223
Dec	254	292	206	200	223

2. Construct a relative frequency table (with 10 classes) for the data in Exercise 1.

3. Construct a histogram that corresponds to the frequency table from Exercise 1.

4. For the data in Exercise 1, find (a) Q_1, (b) P_{45}, and (c) the percentile corresponding to the volume of 125 cu. ft.

5. Use the range rule of thumb to estimate the standard deviation of the data in Exercise 1.

6. Use the frequency table from Exercise 1 to find the mean and standard deviation for the volumes.

7. Use the data from Exercise 1 for the first two years to construct a stem-and-leaf plot with 7 rows.

8. Use the data from Exercise 1 to construct a boxplot.

9. Given below are times (in seconds) between an order being placed and the food being received at a McDonald's drive-through window. Find the (a) mean, (b) median, (c) mode, (d) midrange, (e) range, (f) standard deviation, and (g) variance.

$$135 \quad 90 \quad 85 \quad 121 \quad 83 \quad 69 \quad 87 \quad 159 \quad 177 \quad 135 \quad 227$$

10. Given below are numbers taken from the hydro bills of the family mentioned in Exercise 1. The numbers represent the usage of hydro (in kilowatt-hours) every two months for six years.

$$\begin{array}{cccccccccccc}
22 & 21 & 20 & 22 & 27 & 25 & 19 & 18 & 18 & 18 & 23 & 21 \\
26 & 24 & 16 & 15 & 23 & 24 & 35 & 32 & 18 & 28 & 31 & 45 \\
24 & 23 & 20 & 23 & 23 & 25 & 19 & 19 & 18 & 19 & 23 & 24
\end{array}$$

Find the (a) mean, (b) median, (c) mode, (d) midrange, (e) range, (f) standard deviation, (g) variance, (h) Q_1, (i) P_{30}, and (j) D_7.

11. Scores on a test of depth perception have a mean of 200 and a standard deviation of 40.
    a. Is a score of 260 unusually high? Explain.
    b. What is the z score corresponding to 185?
    c. Assuming that the scores have a bell-shaped distribution, what does the empirical rule say about the percentage of scores between 120 and 280?
    d. What is the mean after 20 points have been added to every score?
    e. What is the standard deviation after 20 points have been added to every score?

Time (years)	Number
4	147
5	81
6	27
7	15
7.5–11.5	30

12. The accompanying table lists times (in years) required to earn an honours bachelor's degree for a sample of undergraduate students. Use the table to find the mean and standard deviation. Based on the results, is it unusual for an undergraduate to require eight years to earn a bachelor's degree? Explain.

13. Using the frequency table given in Exercise 12, construct the corresponding relative frequency histogram.

14. An industrial psychologist gave a subject two different tests designed to measure employee satisfaction. Which score is better: a score of 57 on the first test, which has a mean of 72 and a standard deviation of 20, or a score of 450 on the second test, which has a mean of 500 and a standard deviation of 80? Explain.

15. Refer to the two boxplots below. The first boxplot represents a sample of skulls of male Egyptians from about 4000 BCE, whereas the second boxplot represents a sample of male Egyptian skulls from about 150 CE (in mm, based on data from *Ancient Races of the Thebaid* by Thomson and Randall-Maciver). A shift in head sizes would suggest some societal changes, such as

interbreeding with other cultures. By comparing the two boxplots, is there a shift in the maximum skull breadth? Explain.

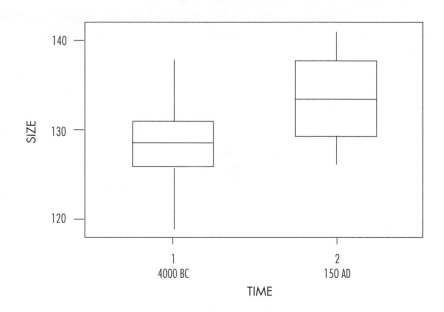

16. Statistics Canada did a study of environmental impacts on health in five provinces. Listed below are five age-standardized mortality rates selected by cause (per 100,000 population). Construct a Pareto chart summarizing the given data.

Lung cancer: 296   Heart disease: 959
Respiratory disease: 297   Chronic liver disease: 45
All accidents: 303

CUMULATIVE REVIEW EXERCISES

1. The amounts of time (in hours) spent on paperwork in one day were obtained from a sample of office managers with the results given below:

   3.7 2.9 3.4 0.0 1.5 1.8 2.3 2.4 1.0 2.0
   4.4 2.0 4.5 0.0 1.7 4.4 3.3 2.4 2.1 2.1

a. Find the mean, median, mode, and midrange.
b. Find the standard deviation, variance, and range.
c. Are the given scores from a population that is discrete or continuous?
d. What is the level of measurement of these scores? (nominal, ordinal, interval, ratio)

2. a. A set of data is at the nominal level of measurement and you want to obtain a representative data value. Which of the following is most appropriate: mean, median, mode, or midrange? Why?

   b. A sample is obtained by telephoning the first 250 people listed in the local telephone directory. What type of sampling (random, stratified, systematic, cluster, convenience) is being used?

   c. An exit poll is conducted by surveying everyone who leaves the polling booth at 30 randomly selected election ridings. What type of sampling (random, stratified, systematic, cluster, convenience) is being used?

3. Auto dealer A sold 5 Volkswagen Beetles for $28,000 each; auto dealer B sold 9 Beetles for $31,000 each; and dealer C sold 3 Beetles for $27,000 each. Can the dealers state that the mean selling price of Beetles is $28,666.67? If not, how should they calculate the mean price of this vehicle?

TECHNOLOGY PROJECT

It is commonly believed that the mean body temperature of healthy adults is 37.0°C. Refer to Data Set 18 in Appendix B and consider the body temperatures there. Open the data file that you created for the Chapter 1 Technology Project in Excel or STATDISK. Proceed to obtain a histogram, measures of central tendency, measures of variation, Q_1, Q_3, the minimum, and the maximum values. Use the results to describe important characteristics of the data set. Based on this sample, what do you conclude about the common belief that the mean body temperature is 37.0°C? Is this the result you would have expected?

FROM DATA TO DECISION

Garbage In, Insight Out

Refer to Data Set 1 in Appendix B. That data set consists of weights of different categories of garbage discarded by a sample of 62 households. With such data sets, there are often several different issues that can be addressed. In Chapter 11 we will consider the issue of whether there is some relationship between household size and amount of waste discarded so that we might be able to predict the size of the population of a region by analyzing the disposed-of garbage. For now, we will work with descriptive statistics based on the data.

   a. Construct a Pareto chart and a pie chart depicting the relative amounts of the total weights of metal, paper, plastic, glass, food, yard waste, textile waste, and other waste. (Instead of frequencies, use the total weights.) Based on the results, which categories appear to be the largest components of the total amount of waste? Is there any single category that stands out as being the largest component?

b. A pie chart depicted metal, paper, plastic, glass, food, yard waste, and other waste with the percents of 14%, 38%, 18%, 2%, 4%, 11%, and 13%, respectively. Do these percentages appear to be consistent with Data Set 1 in Appendix B?

c. For each category, find the mean and standard deviation, and construct a histogram of the 62 weights. Enter the results in the table below.

d. The amounts of garbage discarded are listed by weight. Many regions have household waste collected by commercial trucks that compress it, and charges at the destination are based on weight. Under these conditions, is the volume of the garbage relevant to the problem of community waste disposal? What other factors are relevant?

e. Based on the preceding results, if you had to institute conservation or recycling efforts because your region's waste facility was almost at full capacity, what would you do?

	Metal	Paper	Plastic	Glass	Food	Yard	Textile	Other
Mean								
Standard deviation								
Shape of distribution								

COOPERATIVE GROUP ACTIVITIES

1. Out-of-class activity: Are estimates influenced by anchoring numbers? In the article "Weighing Anchors" in *Omni* magazine, author John Rubin observed that when people estimate a value, their estimate is often "anchored" to (or influenced by) a preceding number, even if that preceding number is totally unrelated to the quantity being estimated. To demonstrate this, he asked people to give a quick estimate of the value of $8 \times 7 \times 6 \times 5 \times 4 \times 3 \times 2 \times 1$. The average answer given was 2250, but when the order of the numbers was reversed, the average became 512. Rubin explained that when we begin calculations with larger numbers (as in $8 \times 7 \times 6$), our estimates tend to be larger. He noted that both 2250 and 512 are far below the correct product, 40,320. The article suggests that irrelevant numbers can play a role in influencing real estate appraisals, estimates of car values, and estimates of the likelihood of nuclear war.

Conduct an experiment to test this theory. Select some subjects and ask them to quickly estimate the value of

$$8 \times 7 \times 6 \times 5 \times 4 \times 3 \times 2 \times 1$$

Then select other subjects and ask them to quickly estimate the value of

$$1 \times 2 \times 3 \times 4 \times 5 \times 6 \times 7 \times 8$$

Record the estimates along with the particular order used. Carefully design the experiment so that conditions are uniform and the two sample groups are selected in a way that minimizes any bias. Don't describe the theory to subjects until after they have provided their estimates. Compare the two sets of sample results by using the methods of this chapter. Provide a written report that includes the data collected, the detailed methods used, the method of analysis, any relevant graphs and/or statistics, and a statement of conclusions. Include a critique of reasons why the results might not be correct and describe ways in which the experiment could be improved.

A variation of the preceding experiment is to survey people about their knowledge of the population of Kenya. First ask half of the subjects whether they think the population is above 5 million or below 5 million, then ask them to estimate the population with an actual number. Ask the other half of the subjects whether they think the population is above 80 million or below 80 million, then ask them to estimate the population. (Kenya's population is 29 million.) Compare the two sets of results and identify the "anchoring" effect of the initial number that the survey subjects are given.

2. **In-class activity:** In each group of three or four students, find the total value of the coins possessed by each individual member. Find the group mean and standard deviation, then exchange those statistics with the other groups. Using the group means as a separate data set, find the mean, standard deviation, and shape of the distribution. How do these results compare to the mean and standard deviation originally found in the group?

3. **In-class activity:** Given below are the ages of motorcyclists at the time they were fatally injured in traffic accidents. If your objective is to dramatize the dangers of motorcycles for young people, which would be most effective: histogram, Pareto chart, pie chart, mean, median, . . . ? Construct the graph and find the statistic that best meets that objective. Is it okay to deliberately distort data if the objective is one such as saving lives of motorcyclists?

17	38	27	14	18	34	16	42	28	24	40	20	23	31
37	21	30	25	17	28	33	25	23	19	51	18	29	

INTERNET PROJECT

Data on the Internet

The Internet is a host to a wealth of information. Much of that information comes in the form of raw data, which can be studied and summarized using the statistics introduced in this chapter. The Internet Project for this chapter, located at either of the following websites

http://www.mathxl.com   or   http://www.mystatlab.com

will point you to data sets in such areas as sports, finance, and the weather. (When you reach the site, click on Internet Project, then on Chapter 2.) Once you have assembled a data set, you can use the methods in this chapter to summarize and classify the data.

3 Probability

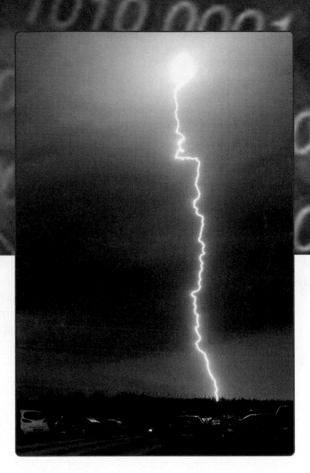

3-1 Overview

Chapter objectives are identified. The importance of probability is discussed, along with its role in basic statistical methods.

3-2 Fundamentals

The relative frequency and classical definitions of probability are both presented and illustrated. Methods are given for finding probabilities of simple events. The law of large numbers is described, and the complement of an event is defined and illustrated. Odds are also considered.

3-3 Addition Rule

The addition rule is described as a method for finding the probability that either one event *or* another event (or both events) occurs when an experiment is conducted. Mutually exclusive events are defined. The rule of complementary events is also introduced.

3-4 Conditional Probabilities

Construction of 2×2 contingency tables is illustrated. Conditional probability is defined and illustrated. Independent events are defined.

3-5 Multiplication Rule and Bayes' Theorem

The multiplication rule is introduced as a method for finding weighted averages. Bayes' Theorem is introduced and illustrated.

3-6 Counting

The following important counting techniques are described: the fundamental counting rule, factorial rule, permutations rule (when all items are different), permutations rule (when some items are identical to others), and combinations rule. Such counting devices are used to find the total number of outcomes.

Do you have a better chance of being hit by lightning or winning the lottery?

Many of us take action based on the likelihood of events occurring. Some of us fly in airplanes, recognizing that while there is a chance of a crash, the likelihood of this happening is really quite small. Some of us sprint to our cars during a thunderstorm knowing that we could be struck by lightning, but, again, the likelihood of that event is actually quite small. Many of us buy lottery tickets knowing that we could win, although the likelihood of that event is very small, but do we really understand just how small? Is the chance of winning the lottery actually smaller than the chance of being struck by lightning? In this chapter we discuss *probability* as we determine specific ways of measuring the chance of various events. Among other things, we will find probabilities for being struck by lightning and for winning a lottery. We will then know which event is more likely.

3-1 Overview

The primary objective of this chapter is to develop a sound understanding of probability values, which we will build upon in subsequent chapters. A secondary objective is to develop the basic skills necessary to solve simple probability problems.

3-2 Fundamentals

In considering probability problems, we deal with procedures (sometimes called experiments) that produce outcomes.

DEFINITIONS

An **event** is any collection of results or outcomes of a procedure.

A **simple event** is an outcome or an event that cannot be broken down into simpler components.

The **sample space** for a procedure consists of all possible *simple* events. That is, the sample space consists of all outcomes that cannot be broken down any further.

EXAMPLES

Procedure	Event	Sample Space
Roll one die	A 3 is rolled (simple event)	{1, 2, 3, 4, 5, 6}
Roll two dice	Dice add to 7 (not a simple event)	{1-1, 1-2, . . . , 6-6}

When rolling one die, rolling a 3 is a *simple event* because it cannot be broken down any further, and the *sample space* consists of these simple events: 1, 2, 3, 4, 5, 6. When rolling a pair of dice, the result of 7 is not itself a simple event because it can be broken down into simple events, such as 3-4 and 6-1. In rolling a pair of dice, the sample space consists of 36 simple events: 1-1, 1-2, . . . , 6-6.

There are different ways to define the *probability* of an event, and we will present three approaches. We begin by listing some basic notation.

NOTATION FOR PROBABILITIES

P denotes a probability. A, B, and C denote specific events. $P(A)$ denotes the probability of event A occurring.

Conduct (or observe) a procedure a large number of times, and count the number of times that event A actually occurs. Then the **probability** of A, $P(A)$, is *estimated* as follows:

$$P(A) = \frac{\text{number of times } A \text{ occurred}}{\text{number of times trial was repeated}}$$

CLASSICAL APPROACH TO PROBABILITY

Assume that a given experiment has n different simple events, each of which has an *equal chance* of occurring. If event A can occur in s of these n ways, then the **probability** of A is given by

$$P(A) = \frac{\text{number of ways } A \text{ can occur}}{\text{number of different simple events}} = \frac{s}{n}$$

SUBJECTIVE PROBABILITIES

$P(A)$, the **subjective probability** of event A, is found by simply guessing or estimating its value based on knowledge of the relevant circumstances.

It is very important to note that *the classical approach requires equally likely outcomes*. If the outcomes are not equally likely, we must use the relative frequency estimate or rely on our knowledge of the circumstances to make an *educated* guess. Figure 3-1 illustrates this important distinction.

When finding probabilities with the relative frequency approach, we obtain an *approximation* instead of an exact value. As the total number of observations increases, the corresponding approximations tend to get closer to the actual probability. This property is stated as a theorem commonly referred to as the **law of large numbers**.

LAW OF LARGE NUMBERS

As a procedure is repeated again and again, the relative frequency probability of an event tends to approach the actual probability.

The law of large numbers tells us that the relative frequency approximations tend to get better with more observations. This law reflects a simple notion supported by

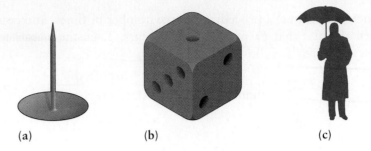

Figure 3-1

Comparison of Relative Frequency and Classical Approaches

(a) Relative Frequency Approach: When trying to determine P(the tack lands point up), we must repeat the experiment of tossing the tack many times and then find the ratio of the number of times the tack lands with the point up to the number of tosses. That ratio is our estimate of the probability.

(b) Classical Approach: When trying to determine $P(2)$ with a balanced and fair die, each of the six faces has an equal chance of occurring.

$$P(2) = \frac{\text{number of ways 2 can occur}}{\text{total number of simple events}}$$

$$= \frac{1}{6}$$

(c) Subjective Probability: When trying to estimate the probability of rain tomorrow, meteorologists use their expert knowledge of weather conditions.

common sense: A probability estimate based on only a few trials can be off by substantial amounts, but with a very large number of trials, the estimate tends to be much more accurate. For example, if you conduct an opinion poll of only a dozen people, your results could easily be in error by large amounts, but if you poll thousands of *randomly selected* people your sample results will be much closer to the true population values.

Figure 3-2 illustrates the law of large numbers by showing computer-simulated results. Note that as the number of births increases, the proportion of girls approaches the 0.5 value.

After examining the relative frequency and classical approaches to probability, it might seem that we should always use the classical approach when a procedure has equally likely outcomes. It frequently happens, though, that many procedures are so complicated that the classical approach is impractical to use. Instead, we can more easily get estimates of the desired probabilities by using the relative frequency approach. In such cases simulations are often helpful. (A **simulation** of a procedure is a process that behaves in the same ways as the procedure itself, thus producing similar results.) For example, it's much easier to use the relative frequency approach for estimating the probability of winning at solitaire—that is, to play the game many times (or to run a computer simulation)—

Figure 3-2

Illustration of the Law of Large Numbers

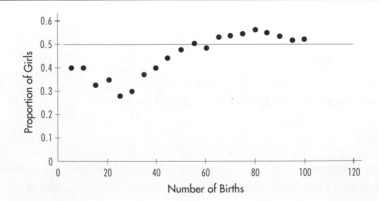

than to perform the extremely complex calculations required with the classical approach.

The examples that follow are intended to illustrate the use of the three approaches. In some of these examples we use the term *random*. Recall these definitions from Section 1–4: In a **random sample** of one element from a population, all elements available for selection have the same chance of being chosen; a sample of *n* items is a (**simple**) **random sample** if it is selected in such a way that every possible sample of *n* items from the population has the same chance of being chosen. The general concept of randomness is extremely important in statistics. When making inferences based on samples, we must have a sampling process that is representative, impartial, and unbiased. If a sample is not carefully selected, it may be totally worthless.

STATISTICAL PLANET

Shakespeare's Vocabulary
According to Bradley Efron and Ronald Thisted, Shakespeare's writings included 31,534 different words. They used probability theory to conclude that Shakespeare probably knew at least another 35,000 words that he didn't use in his writings. The problem of estimating the size of a population is an important problem often encountered in ecology studies, but the result given here is another interesting application. (See "Estimating the Number of Unseen Species: How Many Words Did Shakespeare Know?" in *Biometrika*, Vol. 63, No. 3.)

EXAMPLE

Find the probability that a person will achieve a hole-in-one when playing a round of golf.

SOLUTION

It is extremely unlikely that a person will get a hole-in-one when playing a round of golf. We can, though, find the approximate probability of this event. Recently, a Canadian golf course did a hole-in-one survey. They found 8 holes-in-one scored in 64,000 rounds of golf played at the course. The probability of a hole-in-one during a round of golf at that course is estimated to be

$$\frac{8}{64,000} = 0.000125$$

EXAMPLE

A typical multiple-choice question on a test has 5 possible answers. If you make a random guess on one such question, what is the probability that your response is wrong?

SOLUTION

There are 5 possible outcomes or answers, and there are 4 ways to answer incorrectly.

$$P(\text{wrong answer}) = \frac{4}{5} = 0.8$$

The preceding example included the information that the total number of outcomes is 5, but the following examples require us to calculate the total number of possible outcomes.

EXAMPLE

A survey of songs played by three Vancouver music stations shows the following breakdown of the type of music played:

Rock	Country	Blues
64	43	21

If one of the songs played is randomly selected, find the probability that it is a blues song.

SOLUTION

The total number of songs surveyed is found by computing $64 + 43 + 21 = 128$. With random selection, the 128 songs are equally likely:

$$P(\text{blues}) = \frac{\text{number of blues songs}}{\text{total number of songs}} = \frac{21}{128} = 0.164$$

There is a 0.164 probability that a song selected at random will be a blues song.

EXAMPLE

Find the probability that a couple with 3 children will have exactly 2 boys. Assume that boys and girls are equally likely and that the gender of any child is not influenced by the gender of any other child.

SOLUTION

We first list the sample space that identifies the 8 outcomes. Of those 8 different possible outcomes, 3 correspond to exactly 2 boys, so

$$P(\text{2 boys in 3 births}) = \frac{3}{8} = 0.375$$

There is a 0.375 probability that if a couple has 3 children, exactly 2 will be boys.

<u>1st</u> <u>2nd</u> <u>3rd</u>

boy-boy-boy

exactly 2 boys →

boy-boy-girl
boy-girl-boy
boy-girl-girl
girl-boy-boy
girl-boy-girl
girl-girl-boy
girl-girl-girl

EXAMPLE

In choosing among several computer suppliers, a purchasing agent wants to know the probability of a personal computer breaking down during the first two years. What is that probability?

SOLUTION

There are only two outcomes: A personal computer either breaks down during the first two years or it does not. Because those two outcomes are not equally likely, the relative frequency approximation must be used. This requires that we somehow observe a large number of personal computers. A *PC World* survey of 4000 personal computer owners showed that 992 of them broke down during the first two years. (The computers broke down, not the owners.) On the basis of that result, we *estimate* that the probability is 992/4000, or 0.248.

Figure 3-3
Possible Values
for Probabilities

If we select an odd-numbered year at random—for example, 1999—what is the probability that it is a leap year?

SOLUTION

All leap years (for example, 1996) are divisible by 4, and so a leap year must be even. It is impossible for an odd-numbered year to be a leap year. When an event is impossible, we say the probability is 0. When an event is certain (for example, it will rain at least one day of the year in St. John's, Newfoundland), we say the probability is 1.

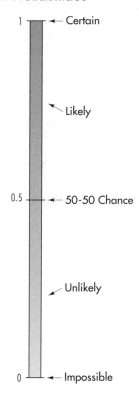

Because any event imaginable is impossible, certain, or somewhere in between, it is reasonable to conclude that the mathematical probability of any event is 0, 1, or a number between 0 and 1 (see Figure 3-3).

PROBABILITY VALUES

- The probability of an impossible event is 0.
- The probability of an event that is certain to occur is 1.
- $0 \leq P(A) \leq 1$ for any event A.

In Figure 3-3, the scale of 0 through 1 is shown on the left, whereas the more familiar and common expressions of likelihood are shown on the right.

Complementary Events

Sometimes we need to find the probability that an event A does *not* occur.

DEFINITION

The **complement of an event** A, denoted by $\overline{A}$, consists of all outcomes in which event A does *not* occur.

EXAMPLE

A survey of some mutual funds is conducted. A sample group of 20 Canadian equity funds and 30 international equity funds is assembled. If one fund is selected at random from the sample group, find the probability that a Canadian equity fund is *not* selected.

SOLUTION

First, observe that the total sample space consists of 50 equity funds. Second, because 20 of the 50 funds are Canadian, it follows that 30 of the 50 funds are international, so

$$P(\text{not selecting a Canadian fund}) = P(\overline{\text{Canadian}})$$
$$= P(\text{international})$$
$$= \frac{30}{50} = 0.6$$

Although it is difficult to develop a universal rule for rounding off probabilities, the following guide will apply to most problems in this text.

ROUNDING OFF PROBABILITIES

When expressing the value of a probability, either give the *exact* fraction or decimal or round off final decimal results to three significant digits. (*Suggestion:* When a probability is not a simple fraction, such as 2/3 or 5/9, express it as a decimal, so it can be better understood.)

All the digits in a number are significant except for the zeros that are included for proper placement of the decimal point.

EXAMPLES

- The probability 0.0000128506 has six significant digits (128506), and it can be rounded to three significant digits as 0.0000129.
- The probability of 1/3 can be left as a fraction or rounded in decimal form to 0.333, but not 0.3.
- The probability of heads in a coin toss can be expressed as 1/2 or 0.5; because 0.5 is exact, there's no need to express it as 0.500.
- The fraction 7659/32785 is exact, but its value isn't obvious, so express it as the decimal 0.234.

An important concept of this section is the mathematical expression of probability as a number between 0 and 1. This type of expression is fundamental and common in statistical procedures, and we will use it throughout the remainder of this text. A typical computer output, for example, may include a "*P*-value" expression such as "significance less than 0.001." We will discuss the meaning of *P*-values later, but they are essentially probabilities of the type discussed in this section. For now, you should recognize that a probability of 0.001 (equivalent to 1/1000) corresponds to an event so rare that it occurs an average of only once in a thousand trials.

Odds

Expressions of likelihood are often given as *odds*, such as 50:1 (or "50 to 1"). A serious disadvantage of odds is that they make many calculations much more difficult. As a result, statisticians, mathematicians, and scientists prefer to use probabilities. The advantage of odds is that they make it easier to deal with money transfers associated with gambling, so they tend to be used in casinos, in lotteries, and at racetracks. There are three definitions that apply.

DEFINITIONS

The **actual odds against** event A occurring are the ratio $P(\overline{A})/P(A)$, usually expressed in the form of $a : b$ (or "a to b"), where a and b are integers having no common factors.

The **actual odds in favour** of event A are the reciprocal of the odds against that event. If the odds against A are $a : b$, then the odds in favour are $b : a$.

The **payoff odds** against event A represent the ratio of net profit (if you win) to the amount bet.

$$\text{payoff odds against event } A = (\text{net profit}):(\text{amount bet})$$

EXAMPLE

Find the actual odds against rolling a 5 when a single die is rolled once.

SOLUTION

$P(\text{roll a 5}) = 1/6$ because there is only one way to roll a 5, but there are 6 simple events in the sample space. With $P(\text{roll a 5}) = 1/6$, we calculate:

$$\text{Actual odds against rolling a 5} = \frac{P(\overline{\text{roll a 5}})}{P(\text{roll a 5})} = \frac{5/6}{1/6} = \frac{5}{1}$$

We express this as 5:1, or "5 to 1." Because the actual odds against rolling a 5 are 5:1, the actual odds in favour are 1:5.

See Exercise 31 for conversions from odds into probabilities.

EXAMPLE

If you bet $4 on the number 13 in roulette, your probability of winning is 1/38, and the payoff odds are given by the casino as 35:1.

a. Find the actual odds against the outcome of 13.

b. How much net profit would you make if you win by betting on 13?

c. If the casino were operating just for the fun of it, and the payoff odds were changed to match the actual odds against 13, how much would you win if the outcome were 13?

STATISTICAL PLANET

Subjective Probabilities at the Racetrack
Researchers studied the ability of racetrack bettors to develop realistic subjective probabilities. (See "Racetrack Betting: Do Bettors Understand the Odds?" by Brown, D'Amato, and Gertner, *Chance* magazine, Vol. 7, No. 3.) After analyzing results for 4400 races, they concluded that although bettors slightly overestimate the winning probabilities of "longshots" and slightly underestimate the winning probabilities of "favourites," their general performance is quite good. The subjective probabilities were calculated from the payoffs, which are based on the amounts bet, and the actual probabilities were calculated from the actual race results.

SOLUTION

a. With $P(13) = 1/38$ and $P(\text{not } 13) = 37/38$, we get

$$\text{Actual odds against } 13 = \frac{P(\text{not } 13)}{P(13)} = \frac{37/38}{1/38} = \frac{37}{1} \text{ or } 37:1$$

b. Because the payoff odds against 13 are 35:1, we have

$$35:1 = (\text{net profit}):(\text{amount bet})$$

so that there is a $35 profit for each $1 bet. For a $5 bet, the net profit is $35 \times 5 = $175. The bettor would collect $175 plus the original $5 bet.

c. If the casino were not operating for profit, the payoff odds would be equal to the actual odds against the outcome. If the payoff odds were changed from 35:1 to 37:1, you would obtain a net profit of $37 for each $1 bet, which is a profit of $37 \times 5 = $185 for a $5 bet. (The difference between the answers for parts (b) and (c) indicates the profit normally received by the casino.)

3-2 Exercises A: Basic Skills and Concepts

1. Which of the following values *cannot* be probabilities?

$$0, \ 0.0001, \ -0.2, \ 3/2, \ 2/3, \ \sqrt{2}$$

2. a. What is $P(A)$ if event A is that February has 30 days this year?
   b. What is $P(A)$ if event A is that November has 30 days this year?
   c. A sample space consists of 500 separate events that are equally likely. What is the probability of each?
   d. On a mid-term exam, each question has 5 possible answers. If you make a random guess on the first question, what is the probability that you are correct?

3. Find the probability that when a coin is tossed, the result is heads.

4. We noted in this section that for the experiment of rolling a pair of dice, there are 36 simple events that form the sample space: 1-1, 1-2, . . . , 6-6. Find the probability of rolling a pair of dice and getting a total of 4.

5. Refer to Data Set 11 in Appendix B. Based on those sample results, estimate the probability that when a plain M&M candy is randomly selected, it will be red.

6. Refer to Data Set 8 in Appendix B. Based on those sample results, estimate the probability that a randomly selected statistics student has at least one credit card.

7. According to a survey sent out by Deloitte Touche Tohamatsu to British Columbia's top CEOs, 20 out of the 80 respondents possess an MBA. What is the

estimated probability of a British Columbian top CEO having an MBA? Does it appear that an MBA is a requirement for the position?

8. The Kelly-Lynne Advertising Company is considering a computer campaign that targets teenagers. In a survey of 1066 teens, 181 had a computer on-line service in their household. If a teen is randomly selected, estimate the probability that he or she will have access to an on-line service in his or her household. Would you advise this company to use a computer advertising campaign?

9. In a survey of college students, 1162 stated that they cheated on an exam and 2468 stated that they did not (based on data from the Josephson Institute of Ethics). If one of these college students is randomly selected, find the probability that he or she cheated on an exam.

*In each of Exercises 10–22, you will be asked to calculate the probability of a specific event occurring. Based on your solutions, identify also which of these events would be **unusual** if they occurred—considering an event to be "unusual" if its probability is less than or equal to 0.05.*

10. A survey was done across five provinces to determine the mortality rate of lung cancer among male patients. Of 1000 patients who were tracked during this survey, 295 eventually succumbed to the disease (based on data from Statistics Canada). If a male lung cancer patient is selected at random, what is the probability that he will die of the disease?

11. In a study of blood donors, 225 were classified as group O and 275 had a classification other than group O. What is the approximate probability that a person will have group O blood?

12. A recent census by Statistics Canada showed there are 54,946 police officers in Canada. Of this number, 188 serve in Prince Edward Island. If a police officer is randomly selected from across Canada, what is the probability that she or he is serving in Prince Edward Island?

13. a. If a person is randomly selected, find the probability that his or her birthday is October 18, which is National Statistics Day in Japan. Ignore leap years.
    b. If a person is randomly selected, find the probability that his or her birthday is in November. Ignore leap years.

14. In a study of brand recognition, 831 consumers knew of Campbell's Soup, and 18 did not (based on data from Total Research Corporation). Use these results to estimate the probability that a randomly selected consumer will recognize Campbell's Soup. How do you think this probability value compares to typical values for other brand names?

15. In a Bruskin/Goldring Research poll, respondents were asked how a fruitcake should be used. One hundred thirty-two respondents indicated that it should be used for a doorstop, and 880 other respondents cited other uses, including

birdfeed, landfill, and a gift. If one of these respondents is randomly selected, what is the probability of getting someone who would use a fruitcake for a doorstop?

16. In a survey of Canadian workers who worked more than 50 hours per week, it was found that 334 workers surveyed were under 35 years of age and 401 workers were 35 to 55 years old. If one worker from this survey is selected at random, what is the probability that she or he is 35 years or older? Based on this result, are workers slowing down as they get older?

17. Among 400 randomly selected drivers in the 20–24 age bracket, 136 were in a car accident during the last year (based on data from the National Safety Council). If a driver in that age bracket is randomly selected, what is the approximate probability that he or she will be in a car accident during the next year? Is the resulting value high enough to be of concern to those in the 20–24 age bracket?

18. In a survey of Canadians who had just purchased new vehicles, 1521 bought pick-up trucks, 1849 bought sport utility vehicles (SUVs), and 1671 bought passenger vehicles. Use these results to estimate the probability that a new vehicle purchaser will buy an SUV. What would this result indicate to auto manufacturers about the SUV market?

19. When the allergy drug Seldane was clinically tested, 70 people experienced drowsiness and 711 did not (based on data from Merrell Dow Pharmaceuticals, Inc.). Use this sample to estimate the probability of a Seldane user becoming drowsy. Based on the result, is drowsiness a factor that should be considered by Seldane users?

20. A Statistics Canada study of deaths by cancer by age group showed 579 deaths in the 0–29 age group, 10,456 deaths in the 30–59 age group, and 42,928 deaths in the 60-and-over age group. Find the probability that a randomly selected cancer mortality will be a person in the first age group.

Method of Fraud	Number
Stolen card	243
Counterfeit card	85
Mail/phone order	52
Other	46

21. A study of credit-card fraud was conducted by MasterCard International, and the accompanying table is based on the results. If one case of credit-card fraud is randomly selected from the cases summarized in the table, find the probability that the fraud resulted from a counterfeit card.

22. A Gallup survey about tooth brushing resulted in the sample data in the given table. If one of the respondents is randomly selected, find the probability of getting someone who brushes three times per day, as dentists recommend.

Tooth-Brushings per Day	Number
1	228
2	672
3	240

23. A couple plans to have 2 children.
    a. List the different outcomes according to the gender of each child. Assume that these outcomes are equally likely.
    b. Find the probability of getting 2 girls.
    c. Find the probability of getting exactly 1 child of each gender.

24. A couple plans to have 4 children.
    a. List the 16 different possible outcomes according to the gender of each child. Assume that these outcomes are equally likely.
    b. Find the probability of getting all girls.
    c. Find the probability of getting at least 1 child of each gender.
    d. Find the probability of getting exactly 2 children of each gender.

25. On a quiz consisting of 3 true/false questions, an unprepared student must guess at each one. The guesses will be random.
    a. List the different possible solutions.
    b. What is the probability of answering all 3 questions correctly?
    c. What is the probability of guessing incorrectly for all questions?
    d. What is the probability of passing the quiz by guessing correctly for at least 2 questions?

26. Both parents have the brown/blue pair of eye-colour genes, and each parent contributes one gene (for one of those colours) to a child. Assume that if the child has at least one brown gene, that colour will dominate and the eyes will be brown. (Actually, the determination of eye colour is somewhat more complex.)
    a. List the different possible eye-colour gene combinations that the child could have. Assume that these outcomes are equally likely.
    b. What is the probability that a child of these parents will have the blue/blue pair of genes?
    c. What is the probability that the child will have brown eyes?

27. Find the actual odds against correctly guessing the answer to a multiple-choice question with 5 possible answers.

28. Find the actual odds against randomly selecting someone who is left-handed, given that 10% of us are left-handed.

29. a. The probability of a 7 in roulette is 1/38. Find the actual odds against 7.
    b. If you bet $2 on the number 7 in roulette and you win, the casino gives you $72, which includes the $2 bet. First identify the net profit, then find the payoff odds.
    c. How do you explain the discrepancy between the odds in parts (a) and (b)?

30. a. In the casino game craps, you can bet that the next roll of the two dice will result in a total of 2. The probability of rolling 2 is 1/36. Find the odds against rolling 2.
    b. If you bet $5 that the next roll of the dice will be 2, you will collect $155 (including your $5 bet) if you win. First identify the net profit, then find the payoff odds.
    c. How do you explain the discrepancy between the odds in parts (a) and (b)?

3-2

31. If the actual odds against event *A* are *a*:*b*, then $P(A) = b/(a + b)$. Find the probability of Horse Cents winning his next race, given that the actual odds against his winning are 10:3.

32. A double-blind experiment is designed to test the effectiveness of the drug Statistiszene as a treatment for number blindness. When treated with Statistiszene, subjects seem to show improvement. Researchers calculate that there is a 0.04 probability that the treatment group would show improvement even if the drug has no effect. What should you conclude about the effectiveness of Statistiszene?

33. The stem-and-leaf plot summarizes the time (in hours) that managers spend on paperwork in one day (based on data from Adia Personnel Services). Use this sample to estimate the probability that a randomly selected manager spends more than 2.0 hours per day on paperwork.

```
0. | 00
1. | 0578
2. | 00113449
3. | 347
4. | 445
```

34. After collecting IQ scores from hundreds of subjects, a boxplot is constructed with this five-number summary: 82, 91, 100, 109, 118. If one of the subjects is randomly selected, find the probability that his or her IQ score is greater than 109.

35. In part (a) of Exercise 13, leap years were ignored in finding the probability that a randomly selected person will have a birthday on October 18.
   a. Recalculate this probability, assuming that a leap year occurs every 4 years. (Express your answer as an exact fraction.)
   b. Leap years occur in years evenly divisible by 4, except they are skipped in 3 of every 4 centesimal years (years ending in 00). The years 1700, 1800, and 1900 were not leap years, but 2000 was a leap year. Find the exact probability for this case, and express it as a fraction.

36. a. If two flies land on an orange, find the probability that they are on points that are within the same hemisphere.
   b. Two points along a straight stick are randomly selected. The stick is then broken at those two points. Find the probability that the three resulting pieces can be arranged to form a triangle. (This is a difficult problem.)

3-3 Addition Rule

The main objective of this section is to introduce the **addition rule** as a device for finding $P(A \text{ or } B)$, the probability that either event *A* occurs or event *B* occurs (or they both occur), as the single outcome of a procedure. The key word to remember is *or*. (Throughout this text we use the *inclusive or*, which means either one or the other or both. Except for Exercise 27, we will not consider the *exclusive or*, which means either one or the other but not both.)

In the previous section we presented the fundamentals of probability, and considered events that could be categorized as *simple*. In this and the following section we consider *compound events*.

NOTATION FOR ADDITION RULE

$P(A$ or $B) = P($event A occurs or event B occurs or they both occur$)$

Let us begin with a simple example. In Figure 3-4 we have a Venn diagram with 10 dots.

Note that $P(A$ or $B) = 8/10 = 0.8$. We then calculate $P(A) = 4/10 = 0.4$ and $P(B) = 5/10 = 0.5$. From this, $P(A) + P(B) = 9/10 = 0.9$. However, this does not equal $P(A$ or $B) = 0.8$ since the dot in the intersection of A and B was counted twice. To adjust, we subtract $P(A$ and $B) = 1/10 = 0.1$. So $P(A) + P(B) - P(A$ and $B) = 0.4 + 0.5 - 0.1 = 0.8$, which now equals $P(A$ or $B)$. We can now formalize the result:

FORMAL ADDITION RULE

$$P(A \text{ or } B) = P(A) + P(B) - P(A \text{ and } B)$$

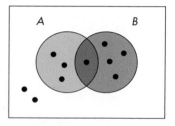

Figure 3-4
Venn Diagram to Illustrate Addition Rule

Because of the relationship between the addition rule and the Venn diagram shown in Figure 3-4, the notation $P(A \cup B)$ is often used in place of $P(A$ or $B)$. Similarly, the notation $P(A \cap B)$ is often used in place of $P(A$ and $B)$, so the formal addition rule can be expressed as

$$P(A \cup B) = P(A) + P(B) - P(A \cap B)$$

The addition rule is simplified whenever A and B cannot occur simultaneously, so $P(A$ and $B)$ becomes zero. Figure 3-5 illustrates that with no overlapping of A and B, we have $P(A$ or $B) = P(A) + P(B)$. The following definition formalizes the lack of overlapping shown in Figure 3-5.

Total Area = 1

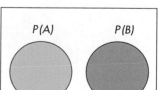

Figure 3-5
Venn Diagram Showing Nonoverlapping Events

Figure 3-6
Applying the Addition Rule

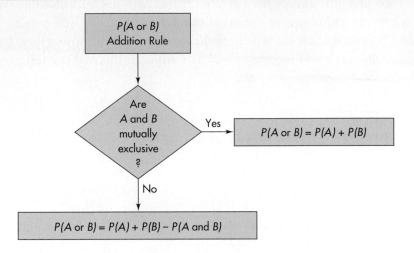

The flowchart of Figure 3-6 shows how mutually exclusive events affect the addition rule.

EXAMPLE

The probability that an executive owns a cell phone is 95%, the probability that an executive owns a PDA is 68%, and the probability that an executive owns both a cell phone and a PDA is 65%. What is the probability that an executive owns either a cell phone or a PDA?

SOLUTION

The probability we seek can be denoted by P(cell phone or PDA) $= P$(cell phone) $+$ P(PDA) $- P$(cell phone and PDA) $= 0.95 + 0.68 - 0.65 = 0.98$. The probability that an executive owns either device is 98%.

The addition rule can be extended to events with more than two outcomes as illustrated in the following example.

EXAMPLE

If one of the 2072 subjects represented in Table 3-1 is randomly selected, find the probability of getting someone who was given a placebo or was in the control group.

SOLUTION

The probability we seek can be denoted by P(placebo or control). Table 3-1 shows that the subjects who used the placebo and those in the control group are mutually exclusive. That is, there is no overlap between those two groups. Consequently, the total number of people who used a placebo or were in the control group is $665 + 626 = 1291$, and the probability we seek is

$$P(\text{placebo or control}) = \frac{665}{2072} + \frac{626}{2072} = \frac{1291}{2072} = 0.623$$

Table 3-1 Test of Seldane (based on data from Merrell Dow Pharmaceuticals)

	Seldane	Placebo	Control Group	Total
Headache	49	49	24	122
No headache	732	616	602	1950
Total	781	665	626	2072

EXAMPLE

If one of the 2072 subjects represented in Table 3-1 is randomly selected, find the probability of getting someone who used Seldane or did not experience a headache.

SOLUTION

The probability we seek can be denoted by P(Seldane or no headache). Table 3-1 shows that there is overlap between the group of Seldane users and the group of subjects who had no headache. That is, the events are not mutually exclusive, and we must be careful to avoid double counting when we calculate our sums.

The result can be obtained by applying the addition rule as follows.

$$P(\text{Seldane or no headache}) = P(\text{Seldane}) + P(\text{no headache})$$
$$- P(\text{Seldane and no headache})$$
$$= \frac{781}{2072} + \frac{1950}{2072} - \frac{732}{2072} = \frac{1999}{2072} = 0.965$$

We can summarize the key points of this section as follows:

1. To find $P(A \text{ or } B)$, begin by associating *or* with addition.
2. Consider whether events A and B are mutually exclusive; in other words, can they happen at the same time? If they are, $P(A \text{ or } B) = P(A) + P(B)$. If they are not mutually exclusive (that is, if they can happen at the same time), be sure to compensate for double counting by subtracting $P(A \text{ and } B)$. So $P(A \text{ or } B) = P(A) + P(B) - P(A \text{ and } B)$.

Errors made when applying the addition rule often involve double counting. That is, events that are not mutually exclusive are treated as if they were. One indication of such an error is a total probability that exceeds 1; however, errors involving the addition rule do not always cause the total probability to exceed 1.

Complementary Events

In Section 3–2 we defined the complement of event A and denoted it by $\overline{A}$. By definition complementary events must be mutually exclusive because it is impossible

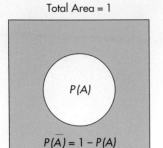

Total Area = 1

$P(A)$

$P(\overline{A}) = 1 - P(A)$

Figure 3-7
Venn Diagram for the
Complement of Event A

for an event both to occur and *not* to occur at the same time. Also, A either does or does not occur. That is, either A or $\overline{A}$ must occur. These observations enable us to apply the addition rule for mutually exclusive events as follows:

$$P(A \text{ or } \overline{A}) = P(A) + P(\overline{A}) = 1$$

We justify $P(A \text{ or } \overline{A}) = P(A) + P(\overline{A})$ by noting that A and $\overline{A}$ are mutually exclusive; we justify the total of 1 by our absolute certainty that A either does or does not occur. This result of the addition rule leads to the following three equivalent forms.

RULE OF COMPLEMENTARY EVENTS

$$P(A) + P(\overline{A}) = 1$$
$$P(\overline{A}) = 1 - P(A)$$
$$P(A) = 1 - P(\overline{A})$$

The first form comes directly from our original result. The second (see Figure 3-7) and third variations involve very simple equation manipulations.

EXAMPLE

If $P(\text{rain}) = 0.4$, find $P(\text{no rain})$.

SOLUTION

Using the rule of complementary events, we get

$$P(\text{no rain}) = 1 - P(\text{rain}) = 1 - 0.4 = 0.6$$

A major advantage of the *rule of complementary events* is that its use can greatly simplify certain problems.

3-3 Exercises A: Basic Skills and Concepts

For each part of Exercises 1 and 2, are the two events mutually exclusive (that is, not overlapping) for a single procedure?

1. a. Selecting a geometric shape
      Selecting an octahedron
   b. Selecting a survey subject who is a member of the Liberal party
      Selecting a survey subject opposed to all welfare plans
   c. Spinning a roulette wheel and getting an outcome of 7
      Spinning a roulette wheel and getting an even number

2. **a.** Buying a new Corvette that is free of defects
    Buying a car with inoperative headlights
  b. Selecting a math course
    Selecting a course that is interesting
  c. Selecting a person with blond hair (natural or otherwise)
    Selecting a person who is bald

3. **a.** If $P(A) = 2/5$, find $P(\overline{A})$.
  b. According to a recent Interac survey, 28% of consumers use their debit cards in supermarkets. Find the probability that they use debit cards only in other stores or not at all.

4. **a.** Find $P(\overline{A})$, given that $P(A) = 0.228$.
  b. Statistics Canada reports that 30% of Canadians own stocks. If you randomly select someone from Canada, what is the probability that he or she does not own stocks?

5. When playing blackjack with a single deck of cards at Casino Regina, you are dealt the first card from the top of a shuffled deck. What is the probability that you get (a) a club or an ace? (b) an ace or a 2?

6. If someone is randomly selected, find the probability that his or her birthday is not October 18, which is National Statistics Day in Japan. Ignore leap years.

7. Refer to Table 3-1 in this section. If one of the 2072 subjects is randomly selected, find the probability of getting someone who used Seldane or a placebo.

8. Refer to Table 3-1 in this section. If one of the 2072 subjects is randomly selected, find the probability of getting someone who used a placebo or experienced a headache.

9. Pollsters are concerned about declining levels of cooperation among persons contacted in surveys. A pollster contacts 84 people in the 18–21 age bracket and finds that 73 of them respond and 11 refuse to respond. When 275 people in the 22–29 age bracket are contacted, 255 respond and 20 refuse to respond (based on data from "I Hear You Knocking but You Can't Come In," by Fitzgerald and Fuller, *Sociological Methods and Research*, Vol. 11, No. 1). Assume that one of the 359 people is randomly selected. Find the probability of getting someone in the 18–21 age bracket or someone who refused to respond.

10. Refer to the data set in Exercise 9, and find the probability of getting someone who is in the 18–21 age bracket or someone who responded.

11. Problems of sexual harassment have received much attention in recent years. In one survey, 420 workers (240 of whom are men) considered a friendly pat on the shoulder to be a form of harassment, whereas 580 workers (380 of whom are men) did not consider that to be a form of harassment (based on data from Bruskin/Goldring Research). If one of the surveyed workers is randomly selected, find the probability of getting someone who does not consider a pat on the shoulder to be a form of harassment.

12. Refer to the data set in Exercise 11, and find the probability of randomly selecting a man or someone who does not consider a pat on the shoulder to be a form of harassment.

13. A survey of 100 businesspeople showed that 25 owned copiers, 50 owned printers, and 13 owned both types of equipment. If a businessperson is selected at random from this sample, find the probability of getting someone who owns a copier or a printer.

14. Refer to the data set in Exercise 13, and find the probability of getting a businessperson who does not own a copier or a printer.

In Exercises 15 and 16, use the data in the accompanying table, which summarizes a sample of 200 times (in minutes) between eruptions of the Old Faithful Geyser in Yellowstone National Park.

Time	Frequency
40–49	8
50–59	44
60–69	23
70–79	6
80–89	107
90–99	11
100–109	1

15. Visitors to Yellowstone naturally want to see Old Faithful erupt, so the interval between eruptions becomes a concern for those with time constraints. If we randomly select one of the times represented in the table, what is the probability that it is at least one hour?

16. If we randomly select one of the times represented in the table, what is the probability that it is at least 70 min or between 60 and 79 min?

In Exercises 17–24, refer to the accompanying figure, which describes the typical blood groups and Rh types of 100 Canadians (based on data from Canadian Blood Services). In each case, assume that 1 of the 100 subjects is randomly selected, and find the indicated probability.

17. P(not group O)

18. P(not type Rh^+)

19. P(group B or type Rh^-)

20. P(group O or group A)

21. P(type Rh^-)

22. P(group A or type Rh^+)

23. P(group AB or type Rh^-)

24. P(group A or B or type Rh^+)

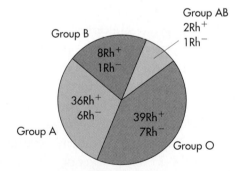

3-3 Exercises B: Beyond the Basics

25. **a.** If $P(A \text{ or } B) = 1/3$, $P(B) = 1/4$, and $P(A \text{ and } B) = 1/5$, find $P(A)$.

  b. If $P(A) = 0.4$ and $P(B) = 0.5$, what is known about $P(A \text{ or } B)$ if A and B are mutually exclusive events?

  c. If $P(A) = 0.4$ and $P(B) = 0.5$, what is known about $P(A \text{ or } B)$ if A and B are not mutually exclusive events?

26. If events A and B are mutually exclusive and events B and C are mutually exclusive, must events A and C be mutually exclusive? Give an example supporting your answer.

27. How is the addition rule changed if the *exclusive or* is used instead of the *inclusive or*? Recall that the *exclusive or* means either one or the other, but not both.

28. Given that $P(A \text{ or } B) = P(A) + P(B) - P(A \text{ and } B)$, develop a formal rule for $P(A \text{ or } B \text{ or } C)$. (*Hint:* Draw a Venn diagram.)

29. One version of DeMorgan's Law states:

$$P(\text{neither } A \text{ nor } B) = P(\overline{A} \text{ and } \overline{B}) = 1 - P(A \text{ or } B)$$

A survey was conducted to gauge support for a dedicated road tax. The respondents for a particular city were divided into those who live in the inner city and those who live in the suburbs. A third category was created for those living outside the city. The percentage of respondents living in the inner city is 12.6% and the percentage living in the suburbs is 79.3%, with the remainder living outside the city forming the remaining percentage of respondents. Responses were classified as support the tax, oppose the tax, and undecided. The percentage overall who support the tax is 45.2%, while the percentage who oppose the tax is 32.3%. The percentage of respondents who live in the inner city and support the tax is 8.5%. The percentage of respondents who live in the inner city and oppose the tax is 1.8%. The percentage of respondents who live in the suburbs and oppose the tax is 27.6%. The percentage of respondents who live outside the city and are undecided is 4.3%.

a. Create a contingency table (similar to Table 3-1) based on these probabilities.
b. Using the addition rule, compute the probability that a respondent lives in the city or does not support the tax.
c. Using DeMorgan's Law, compute the probability from part (b).

 3-4 **Conditional Probabilities**

The Basics

In Section 3–3 we presented the addition rule for finding $P(A \text{ or } B)$, the probability that a trial in a procedure or experiment has an outcome of A or B or both. The main objective in this section is to introduce conditional probability, the probability that one event occurs given that another event occurs or has occurred.

We are accustomed to conditional probabilities in real life. For example, given two male drivers, one aged 22 and the other 54, the male aged 22 will pay more for auto insurance because there is a higher probability that he will be involved in an accident. Another example is in life insurance. Suppose we have two females both aged 40; if one smokes and the other does not smoke, the smoker will pay higher life insurance premiums since there is a higher probability that the smoker will die relatively young. Our assessment of probability for an event can be affected by our knowledge of circumstances. As shown by the previous examples, a conditional probability of an event applies when the probability is affected by other, known circumstances.

The **conditional probability** of B given A, written as $P(B|A)$, is the probability of event B occurring, given that event A occurs or has occurred. It can be found by dividing the probability of events A and B both occurring by the probability of event A.

$$P(B|A) = \frac{P(A \text{ and } B)}{P(A)}$$

Contingency Tables

It is possible to solve conditional probability problems strictly from knowledge of probability theory. However, it is much easier to solve the problems by construction of a **contingency table** (also known as a crosstab), a table of frequencies that correspond to two variables. The most basic contingency table is the 2×2 table, which has 2 rows and 2 columns. To illustrate how to construct a 2×2 table, we return to an earlier example.

EXAMPLE

The probability that an executive owns a cell phone is 95%, the probability that an executive owns a PDA is 68%, and the probability that an executive owns both a cell phone and a PDA is 65%. Construct a contingency table based on this information.

SOLUTION

We represent the event that an executive owns a cell phone by CP. Here is the finished product:

Table 3-2 **Contingency Table of Cell Phone and PDA Owners**

	CP	$\overline{CP}$	Total
PDA	0.65	0.03	0.68
$\overline{PDA}$	0.30	0.02	0.32
Total	**0.95**	**0.05**	**1.00**

Across the top we have one event and its complement (cell phone/no cell phone). Down the side we have the other event and its complement (PDA/no PDA). Since 95% of executives own a cell phone, the 0.95 is in the cell representing the total percentage of executives owning one. This means that 5% of executives do not own a cell phone. Note that the 0.05 is next to the 0.95. Going horizontally, since 68% of executives own a PDA, the 0.68 is in the cell representing the total percentage of executives owning one. This means that 32% of executives do not own a PDA. Note that the 0.32 is next to the 0.68. We position the probabilities this way since they must sum to 100% or 1. Whether we add 0.95 + 0.05 or 0.68 + 0.32, both sum to 1. The other piece of information is that 65% of executives own both a cell phone and PDA. Note the location of 0.65: it is in the crosstab of those who own both a cell phone and a PDA. After this, we fill in the rest of the table by basic arithmetic; for example 0.03 is derived by subtracting 0.65 from 0.68.

The remaining probabilities in the table can be interpreted as follows: 0.03 is the probability that an executive owns a PDA but not a cell phone, 0.3 is the probability that an executive owns a cell phone but not a PDA and 0.02 is the probability that an executive does not own a cell phone and does not own a PDA, i.e., does not own either device.

Now that the contingency table is constructed, we can ask a series of questions.

EXAMPLE

a. What is the probability that an executive owns a PDA, given that the executive owns a cell phone?

b. What is the probability that an executive does not own a PDA, given that the executive owns a cell phone?

c. What is the probability that an executive owns a cell phone, given that the executive does not own a PDA?

SOLUTION

a. We want $P(\text{PDA}|\text{CP}) = P(\text{CP and PDA})/P(\text{CP}) = 0.65/0.95 = 0.684$

b. We want $P(\text{not PDA}|\text{CP}) = P(\text{CP and not PDA})/P(\text{CP}) = 0.3/0.95 = 0.316$

You may notice that the probabilities from parts (a) and (b) sum to 1. The reason is that if an executive owns a cell phone, either the executive owns a PDA or does not own a PDA.

c. We want $P(\text{CP}|\text{not PDA}) = P(\text{CP and not PDA})/P(\text{not PDA}) = 0.30/0.32 = 0.938$.

As with the addition rule, conditional probabilities can be extended to events with more than two outcomes.

Refer to Table 3-3, and base probability calculations on the frequencies shown in the table.

Table 3-3 Contingency Table of Crimes and Victims

Relationship of Criminal to Victim	Homicide	Robbery	Assault	Totals
Stranger	12	379	727	**1118**
Acquaintance or relative	39	106	642	**787**
Relationship unknown	18	20	57	**95**
Totals	**69**	**505**	**1426**	**2000**

a. If one person is randomly selected, what is the likelihood that he or she was victimized by a stranger, given that a robbery victim is selected?

b. Given that an assault victim is selected, what is the likelihood that the criminal is a stranger?

SOLUTION

a. We want P(stranger | robbery). If we assume that the person selected was a robbery victim, we are dealing with the 505 people in the second column of values. Among those 505 people, 379 were victimized by strangers, so

$$P(\text{stranger} \mid \text{robbery}) = \frac{379}{505} = 0.750$$

The same result can be obtained using this formal approach:

$$P(\text{stranger} \mid \text{robbery}) = \frac{P(\text{robbery and stranger})}{P(\text{robbery})}$$

$$= \frac{379/2000}{505/2000} = 0.750$$

b. Here we want P(stranger | assault). If we assume that the person selected was an assault victim, we are dealing with the 1426 people in the third column. Among those 1426 people, 727 were victimized by strangers, so

$$P(\text{stranger} \mid \text{assault}) = \frac{727}{1426} = 0.510$$

Again, the same result can be obtained using a formal approach:

$$P(\text{stranger} \mid \text{assault}) = \frac{P(\text{assault and stranger})}{P(\text{assault})}$$

$$= \frac{727/2000}{1426/2000} = 0.510$$

Interpretation

By comparing the results from parts (a) and (b), we see that the probability of being victimized by a stranger is very different for robberies than for assaults, so there appears to be a dependency between the type of crime and the relationship of the criminal to the victim.

A study on the maximum lifespans of animals shows that, of 15 species selected, 3 live no more than 5 years. Nine of the selected species weigh over 10 kg when full grown. Moreover, 3 species in the study live over 5 years and have weights no more than 10 kg. (These data are based on Data Set 19 in Appendix B.) If a species is selected from those included in this study, what is the probability that it has a maximum lifespan of over 5 years, given that it weighs over 10 kg?

SOLUTION

There are lots of numbers here, but no frequencies are given for animals living over 5 years and having a weight over 10 kg. Yet there are enough data to begin construction of a contingency table; the rest can be filled in later. Table 3-4, in the *non-shaded* cells, summarizes what we have been told. Data for the size of the sample (15) can be placed in the "grand total" on the lower right. Frequency data that apply to just one of the variables, with no reference to other variable, are placed in the Totals at the right or bottom. Table 3-4 shows the lifespan variable at the top, with its category frequencies at the bottom, such as a frequency of 3 for species of lifespan up to 5 years. The weight variable is on the left of the table, with its category frequencies on the right, such as the frequency of 9 for species of weight over 10 kg. There is also one case where we know the frequency for a combination of variable values: 3 species with lifespan over 5 years and weight up to 10 kg. 3 is placed in the appropriate cell.

Next, we complete the table in the manner of a puzzle (the correct answers are shaded in the table). All row frequencies have to add up to their row totals on the right, and all column frequencies have to add up to their column totals on the bottom. The sum of row totals should equal the sum of column totals, which in turn should equal the grand total. At last, we are ready to answer the probability question for this exercise: Of the 9 species whose weight is over 10 kg, 9 species have maximum lifespans over 5 years, so

$$P(\text{over 5 yr.} \mid \text{over 10 kg}) = \frac{9}{9} = 1.00$$

Interpretation

It appears from this sample that all species whose adult weight is over 10 kg. have a maximum lifespan that is over 5 years.

Table 3-4 Contingency Table of Animals' Weights and Lifespans

	Lifespans of Animals (Years Maximum)		
	Up to 5 yr.	Over 5 yr.	Total
Weight up to 10 kg	3	3	6
Weight over 10 kg	0	9	9
Total	3	12	15

Testing for Independence

Two events A and B are **independent** if the occurrence of one does not affect the probability of the occurrence of the other. (Several events are similarly independent if the occurrence of any does not affect the probabilities of the occurrence of the others.) If A and B are not independent, they are said to be **dependent**.

For example, flipping a coin and then tossing a die are *independent* events because the outcome of the coin has no effect on the probabilities of the outcomes of the die. On the other hand, the event of having your car start and the event of driving to class on time are *dependent*, because the outcome of trying to start your car does affect the probability of your getting to class on time (unless you're a dorm resident).

If A is independent of B, this means that $P(B \mid A) = P(B)$, provided $P(A) > 0$. If we expand the left side according to the definition of conditional probability, we get:

$$\frac{P(A \text{ and } B)}{P(A)} = P(B)$$

Cross-multiplying, we get:

$$P(A \text{ and } B) = P(A) \cdot P(B)$$

This equation is known as the test for independence. (It is also known as the multiplication rule for independent events. The multiplication rule is covered in the next section.) If the left side equals the right side, we conclude that A and B are independent events. Otherwise, A and B are not independent. These results are summarized as follows.

Two events A and B are independent if

$$P(B \mid A) = P(B)$$

or

$$P(A \text{ and } B) = P(A) \cdot P(B)$$

Two events A and B are dependent if

$$P(B \mid A) \neq P(B)$$

or

$$P(A \text{ and } B) \neq P(A) \cdot P(B)$$

To illustrate, we return to an earlier example.

EXAMPLE

The probability that an executive owns a cell phone is 95%, the probability that an executive owns a PDA is 68%, and the probability that an executive owns both a cell phone and a PDA is 65%. Does owning one device depend on owning the other?

SOLUTION

The left side of the multiplication rule equation is $P(CP \text{ and } PDA) = 0.65$; the right side is $P(CP)P(PDA) = (0.95)(0.68) = 0.646$. The left side does not equal the right side. Owning a cell phone is not independent of owning a PDA. Therefore, owning one device depends on owning the other.

The other approach is to go back to first principles. Previously, we calculated $P(PDA|CP) = 0.684$. This would be the left side; the right side would be $P(PDA) = 0.68$. Once again, since the left side does not equal the right side, the events are not independent.

3-4 Exercises A: Basic Skills and Concepts

In Exercises 1 and 2, for each given pair of events, classify the two events as independent or dependent. Some of the other exercises are based on concepts from earlier sections of this chapter.

1. **a.** Attending classes in a statistics course
   Passing a statistics course
   b. Getting a flat tire on the way to class
   Sleeping too late for class
   c. Events A and B, where $P(A) = 0.40$, $P(B) = 0.60$, and $P(A \text{ and } B) = 0.20$

2. **a.** Finding your microwave oven inoperable
   Finding your battery-operated smoke detector inoperable
   b. Finding your kitchen light inoperable
   Finding your refrigerator inoperable
   c. Events A and B, where $P(A) = 0.90$, $P(B) = 0.80$, and $P(A \text{ and } B) = 0.72$

3. The probability that a person owns an imported vehicle is 0.43. The probability that a vehicle is more than 5 years old is 0.28. The probability that a vehicle is an import over 5 years old is 0.09.
   a. Create a contingency table based on these probabilities.
   b. What is the probability that a vehicle is more than 5 years old, given that it is an import?
   c. What is the probability that a vehicle is an import, given that it is no more than 5 years old?
   d. What is the probability that a vehicle is no more than 5 years old, given that it is not an import?
   e. Does owning a vehicle more than 5 years old depend on it being an import?

4. Remote sensors are used to control each of two separate and independent valves, denoted by p and q, which must both open to provide water for emergency cooling of a nuclear reactor. Each valve has a 0.95 probability of

opening when triggered. For the given configuration, find the probability that when both sensors are triggered, water will get through the system so that cooling can occur.

5. The probability that a household has an annual income of under $100,000 is 0.72. The probability that a household donates less than $200 per year to charity is 0.63. The probability that a household has an annual income of at least $100,000 and donates at least $200 per year to charity is 0.15.
   a. Create a contingency table based on these probabilities.
   b. What is the probability that a household donates less than $200 per year to charity, given that its annual income is less than $100,000?
   c. What is the probability that a household donates less than $200 per year to charity, given that its annual income is at least $100,000?
   d. Based on the results of parts (b) and (c), does donating less than $200 per year to charity depend on the household's annual income?

6. The probability that a person does his or her own return is 0.85. The probability that a person receives a refund is 0.76. The probability that a person does his or her own return and receives a refund is 0.68.
   a. Create a contingency table based on these probabilities.
   b. Who is more likely to receive a refund: a person who does his or her own return or one who doesn't?

7. When driving to class, a student must pass 2 traffic lights that operate independently. For each light, there is a 0.4 probability that it is green. If she must reach both lights when they are green in order to make class on time, what is the probability that she will be on time?

8. The probability that a student receives an A in a course is 0.1. The probability that a student studies less than 30 hours per week is 0.503. The probability that a student studies less than 30 hours per week and receives less than an A in a course is 0.495.
   a. Create a contingency table based on these probabilities.
   b. What is the probability that a student receives an A in a course, given that the student studies at least 30 hours per week?
   c. What is the probability that a student receives less than an A in a course, given that the student studies less than 30 hours per week?

9. The probability that a person exercises regularly is 0.212. The probability that a person lives to the age of 80 or more is 0.30694. The probability that a person does not exercise regularly and lives to the age of 80 or more is 0.159176.
   a. Create a contingency table based on these probabilities.
   b. What is the probability that a person exercises regularly if the person lives to the age of 80 or more?

c. What is the probability that a person does not exercise regularly if the person does not live to the age of 80?

d. Does living to the age of 80 or more depend on exercising regularly?

10. The probability that a household has an annual income of $75,000 or more is 0.362. The probability that a household recycles regularly is 0.627. The probability that a household has an annual income under $75,000 and does not recycle regularly is 0.289.

   a. Create a contingency table based on these probabilities.
   b. What is the probability that a household has an annual income under $75,000 if it recycles regularly?
   c. Which income group is more likely to recycle regularly?
   d. Based on the results of part (c), what can we conclude about the dependency of recycling regularly on annual household income?

11. A survey of 1000 people was conducted. The following facts were ascertained:
   - 40% of the respondents are male
   - 21% have an annual income under $25,000
   - 57.2% have an annual income of at least $25,000 but less than $50,000
   - 6% are males with an annual income under $25,000
   - 6.3% are females with an annual income of $50,000 or more

   a. Create a contingency table based on these probabilities.
   b. What is the probability that a person has an annual income of $50,000 or more if the person is male?
   c. What is the probability that a person is female if the person has an annual income under $50,000?
   d. Does having an annual income of $50,000 or more depend on the person's gender?

In Exercises 12 and 13, use the data in Table 3-3.

12. a. Find the probability that when 1 of the 2000 subjects is randomly selected, the person chosen was victimized by an acquaintance or relative, given that he or she was robbed.
   b. Find the probability that when 1 of the 2000 subjects is randomly selected, the person chosen was robbed by an acquaintance or relative.
   c. Find the probability that when 1 of the 2000 subjects is randomly selected, the person chosen was robbed or was victimized by an acquaintance or relative.
   d. If two different subjects are randomly selected, find the probability that they were both robbed.

13. a. If one of the crime victims represented in the table is randomly selected, find the probability of getting someone who was victimized by a stranger or who was a homicide victim.

b. If one of the crime victims represented in the table is randomly selected, find the probability of getting someone who was a homicide victim, given that the criminal was a stranger.

c. If one of the crime victims represented in the table is randomly selected, find the probability of getting someone who was victimized by a stranger, given that he or she was a homicide victim.

d. If two different subjects are randomly selected, find the probability that they were both victimized by criminals who were strangers.

In Exercises 14–19, use the following information: Suppose that researchers developed a test to detect a particular disease in the general population. However, the test was not perfect. To determine the accuracy of the test, a study was done on 2000 subjects, 1000 of whom were known to have the disease, and 1000 of whom were known to be free of the disease. Some results are shown in the accompanying table.

Group	Positive	Negative	Total
Subjects with the disease	900		1000
Disease-free subjects			1000
Total		1050	

The researchers indicated that the incidence of the disease in the general population is 0.5%.

14. Based on the study results, what is the probability that if a disease-free person is randomly selected, they will test positive?

15. Based on the study results, what is the probability that if a person with the disease is randomly selected, they will test negative?

16. Based on the study results, what is the probability that if a person without the disease is randomly selected, they will test negative?

17. If a test subject is selected at random, what is the probability that he or she tests negative or is a subject known to be disease-free?

18. If a test subject is selected at random, what is the probability that he or she tests positive or is a subject known to be disease-free?

19. If a test subject is selected at random, what is the probability that he or she tests negative given the subject is known to have the disease?

3-4 Exercises B: Beyond the Basics

20. The probability that a realtor has less than 10 years of service is 0.875. The probability that a realtor has at least 10 but less than 20 years of service is 0.104. The probability that a realtor earns less than $100,000 annually given

less than 10 years of service is 0.984. The probability that a realtor earns less than $100,000 annually given at least 10 but less than 20 years of service is 0.476. The probability that a realtor earns at least $100,000 annually given at least 20 years of service is 0.988.

a. Create a contingency table based on these probabilities.

b. Does a realtor earning at least $100,000 annually depend on having at least 10 years of service?

3-5 Multiplication Rule and Bayes' Theorem

From the previous section, we have $P(B|A) = P(A \text{ and } B)/P(A)$. If we cross-multiply, we derive $P(A \text{ and } B) = P(A)P(B|A)$. This is known as the **multiplication rule** for dependent events.

The multiplication rule is extremely important because it has so many meaningful applications. One area of application involves product testing, as in the following example.

EXAMPLE

A batch of 50 fuel filters is produced by the Windsor Auto Supply Company, and 6 of them are defective. (Optimists would say that 44 are good.) Two of the filters are selected and tested. Find the probability that the first is good and the second is good if the filters are selected (a) with replacement, and (b) without replacement.

SOLUTION

a. If the filters are selected with replacement, the two selections are independent because the second event is not affected by the first outcome. We therefore get

$$P(\text{first good and second good}) = \frac{44}{50} \cdot \frac{44}{50} = 0.774$$

b. If the filters are selected without replacement, the two selections are dependent because the second event is affected by the first outcome. We therefore get

$$P(\text{first good and second good}) = \frac{44}{50} \cdot \frac{43}{49} = 0.772$$

Note that in this case, we adjust the second probability to take into account the selection of a good filter on the first selection. After selecting a good filter the first time, there would be 43 good filters among the 49 that remain. Also note that without replacement, there is a slightly lower chance of getting 2 good filters. If we want to develop a procedure for using samples to test batches of products, we should sample without replacement for two reasons: First, there is a lower chance of getting only good items when some defects are present; second, it doesn't make sense to sample with replacement because it becomes possible to test the same item more than once, and that is a waste.

So far we have discussed two events, but the multiplication rule can be easily extended to several events. In general, the probability of any sequence of independent events is simply the product of their corresponding probabilities. For example, the probability of tossing a coin three times and getting all heads is $0.5 \cdot 0.5 \cdot 0.5 = 0.125$. We can also extend the multiplication rule so that it applies to several dependent events; simply adjust the probabilities as you go along. For example, the probability of getting three aces when three cards are selected without replacement is given by

$$\frac{4}{52} \cdot \frac{3}{51} \cdot \frac{2}{50} = 0.000181$$

In this last example involving three aces, we assumed that the events were dependent because the selections were made without replacement. However, it is a common practice to treat events as independent when *small samples* are drawn from *large populations*. (In such cases, it is rare to select the same item twice.) **A common guideline is to assume independence whenever the sample size is no more than 5% of the size of the population.** When pollsters survey 1200 adults from a population of millions, they typically assume independence, even though they sample without replacement.

BAYES' THEOREM

An important application of the multiplication rule is in Bayes' Theorem. Thomas Bayes (1702–1761) said that probabilities should be revised when we learn more about an event. Here's one form of **Bayes' Theorem:**

$$P(A|B) = \frac{P(A \text{ and } B)}{P(B)} = \frac{P(A \text{ and } B)}{P(A \text{ and } B) + P(\overline{A} \text{ and } B)}$$

$$= \frac{P(A) \cdot P(B|A)}{P(A) \cdot P(B|A) + P(\overline{A}) \cdot P(B|\overline{A})}$$

EXAMPLE

Suppose 60% of a company's computer chips are made in one factory (denoted by A) and 40% are made in its other factory (denoted by B). For a randomly selected chip, the probability it came from factory A is 0.60. Suppose we learn that the chip is defective and the defect rates for the two factories are 35% (for A) and 25% (for B).

1. What is the overall probability of selecting a defective chip?

2. If a chip is defective, what is the probability it is from factory A?

a. We denote a defective chip by D. For each factory, we apply the multiplication rule. The first probability is that a chip is produced by the particular factory. The second probability is the probability that a chip is defective given the factory it is from.

$$P(A \text{ and } D) = P(A)P(D|A) = (0.6)(0.35) = 0.21$$
$$P(B \text{ and } D) = P(B)P(D|B) = (0.4)(0.25) = 0.1$$

If a chip is defective, it is either from factory A or factory B. To find the overall probability that a chip is defective, we add the probability of a defective chip from each factory:

$$P(D) = P(A \text{ and } D) + P(B \text{ and } D) = 0.21 + 0.1 = 0.31$$

This is known as a weighted average. If each factory produced the same percentage of chips, we could simply find the overall probability of a defective chip by adding 0.35 and 0.25 and dividing by 2 to get 0.3. However, since factory A produces more chips than factory B, the overall probability of a defective chip is weighted in favour of factory A.

b. We want $P(A|D) = P(A \text{ and } D)/P(D) = 0.21/0.31 = 0.677$.

We can apply Bayes' Theorem to situations with two weighted averages.

EXAMPLE

The percentage of people living in three regions A, B, and C is 48%, 35%, and 17%. In a recent election, the percentage of people in each region that voted was 42%, 54%, and 63%, respectively. Of those who voted in each region, the percentage of people who voted Liberal was 25%, 33%, and 39%, respectively.

a. What is the overall percentage of people who voted?

b. Of those who voted, what percentage voted Liberal?

We use V to denote those who voted and L to denote those who voted Liberal.

SOLUTION

a. We compute the first weighted average:

$$P(A \text{ and } V) = P(A)P(V|A) = (0.48)(0.42) = 0.2016$$
$$P(B \text{ and } V) = P(B)P(V|B) = (0.35)(0.54) = 0.189$$
$$P(C \text{ and } V) = P(C)P(V|C) = (0.17)(0.63) = 0.1071$$

Then

$$P(V) = P(A \text{ and } V) + P(B \text{ and } V) + P(C \text{ and } V)$$
$$= 0.2016 + 0.189 + 0.1071 = 0.4977 = 49.77\%$$

Notice that the overall percentage who voted is between the lowest percentage of 42% (from region A) and the highest percentage of 63% (from region C).

b. We now compute the second weighted average. In computing this value, we use the percentage in each region that votes from part (a).

$$P(A \text{ and } V \text{ and } L) = P(A \text{ and } V)P(L|A \text{ and } V) = (0.2016)(0.25) = 0.0504$$

$$P(B \text{ and } V \text{ and } L) = P(B \text{ and } V)P(L|B \text{ and } V) = (0.189)(0.33) = 0.06237$$

$$P(C \text{ and } V \text{ and } L) = P(C \text{ and } V)P(L|C \text{ and } V) = (0.1071)(0.39) = 0.041769$$

$$P(V \text{ and } L) = P(A \text{ and } V \text{ and } L) + P(B \text{ and } V \text{ and } L) + P(C \text{ and } V \text{ and } L)$$

$$= 0.0504 + 0.06237 + 0.041769$$

$$= 0.154539 = 15.4539\%$$

This is the percentage of the entire population that voted and voted Liberal. However, we want the percentage who voted Liberal of those who voted. To compute this, we use the two weighted averages:

$$P(L|V) = P(V \text{ and } L)/P(V) = 15.4539/49.77 = 0.3105 = 31.05\%$$

Notice that the overall percentage who voted Liberal is between the lowest percentage of 25% (region A) and the highest percentage of 39% (from region C).

3-5 Exercises A: Basic Skills and Concepts

1. Using funding from Ontario's "Superbuild" Program, a contractor bought a box of decorative screws. Five of the screws are slotted, four are designed for a Phillips-head screwdriver, and three are designed for a square-tipped screwdriver. If two of the screws are randomly selected from the box, find the probability that they are both slotted.
   a. Assume that the first screw is replaced before the second is selected.
   b. Assume that the first screw is not replaced before the second is selected.

2. Find the probability of answering the first 2 questions on a test correctly if random guesses are made and
   a. The first 2 questions are both true/false types.
   b. The first 2 questions are both multiple choice, each with 5 possible answers.

3. Find the probability of getting 4 consecutive aces when 4 cards are drawn without replacement from a shuffled deck.

4. We have noted that when rolling a pair of dice, there are 36 different possible outcomes: 1-1, 1-2, . . . , 6-6.
   a. What is the probability of rolling a 7?
   b. If you have just entered a friendly neighbourhood game of craps and the person who brought the dice rolls eight consecutive 7s, what do you conclude? Why?

5. A classic excuse for a missed test is offered by 4 students who claim that their car had a flat tire. On the makeup test, the instructor asks the students to identify the particular tire that went flat. If they didn't really have a flat tire and randomly select one that supposedly went flat, what is the probability that all 4 select the same tire?

6. Three firms using the same auditor independently and randomly select a month in which to conduct their annual audits. What is the probability that all 3 months are different?

7. In a market survey of Zellers, a researcher is told that 76% of customers are repeat customers. If 10 customers are selected at random, what is the probability that all 10 are repeat customers? Would this be a very common occurrence?

8. Ignoring leap years, find the probability that 3 randomly selected people were all born on September 24.

9. A Statistics Canada study of environmental impacts on human health in 5 provinces indicates that a male has a 0.133 chance of dying in any type of accident. Find the probability that 5 randomly selected males will all eventually die in an accident.

10. Seven financial experts participated in a stock market selection game. Their first choice had to be selected from 8 different resource stocks. All 7 experts picked the same resource stock. If the picks were purely at random, what is the probability of this event occurring?

11. As an insurance claims investigator, you are suspicious of 4 brothers who each reported a stolen car in a different region of Montreal. If Montreal has an annual car-theft rate of 4.5%, find the probability that among 4 randomly selected cars, all are stolen in a given year. What does the result suggest?

12. A blood-testing procedure is made more efficient by combining samples of blood specimens. If samples from 5 people are combined, the mixture will test positive except if all 5 individual samples are negative. Find the probability of a positive result for 5 samples combined into 1 mixture, assuming the probability of an individual blood sample testing positive is 0.015.

13. An employee claims that a new process for manufacturing DVD players is better because the rate of defects is below 5%, the rate of defects in the past. When 20 DVD players are manufactured with the new process, there are no defects. Assuming that the new method has the same 5% defect rate as in the past, find the probability of getting no defects among the 20 DVD players. Based on the result, is there strong evidence to conclude that the new process is better?

14. The probability that a household has an annual income of $80,000 or more is 0.26. The probability that a household owns an SUV, given that it has an

annual income of $80,000 or more is 0.42. The probability that a household owns an SUV, given that it has an annual income under $80,000 is 0.36.

   a. Create a contingency table based on the above probabilities.

   b. What is the overall probability that a household owns an SUV?

   c. What is the probability that a household has an annual income under $80,000, given that it owns an SUV?

15. The market share of three furniture manufacturers is 21.7% for A, 34.6% for B, and 43.7% for C. The percentage of each manufacturer's production that is sold to high-end furniture stores is 89.4%, 60.3%, and 40.7%, respectively for A, B, and C.

   a. What is the overall percentage of the companies' production that is sold to high-end furniture stores?

   b. Of furniture sold to high-end furniture stores, which manufacturer has the highest percentage and what is the percentage for that manufacturer?

   c. Of furniture not sold to high-end furniture stores, which manufacturer has the highest percentage and what is the percentage for that manufacturer?

16. In a survey of those aged 18 or older, 18% of the respondents are aged 18–24, 29% are aged 25–44, and 38% are aged 45–64. The percentage of the respondents in each age group who own their home is 0.2%, 25.4%, 63.2%, and 83.7% for those aged 18–24, 25–44, 45–64, and 65 or older, respectively.

   a. What percentage of the respondents overall do not own their home?

   b. Of those who do not own their home, which age group forms the largest percentage and what is the percentage for that group?

17. Of those with a university degree, 70.2% earn at least $60,000 per year within 10 years of finishing school. Of those without a university degree, the corresponding percentage is 35.4%. Of the entire population, 24.6% have a university degree.

   a. What percentage of the entire population earns at least $60,000 per year within 10 years of finishing school?

   b. Of those who earn less than $60,000 per year within 10 years of finishing school, what percentage don't have a university degree?

3–5 Exercises B: Beyond the Basics

18. Find the probability that of 25 randomly selected people,

   a. No 2 share the same birthday.

   b. At least 2 share the same birthday.

19. Two cards are to be randomly selected without replacement from a shuffled deck. Find the probability of getting a 10 on the first card and a club on the second card.

20. A recycling firm collected recycling statistics for three cities, *A*, *B*, and *C*. The percentage of the total number of residents in each city is 45% in *A*, 30% in *B*, and 25% in *C*. The percentage in each city that recycles is 12% in *A*, 18% in *B*, and 30% in *C*. Of those that recycle, the percentage that recycles milk jugs is 4% in *A*, 8% in *B*, and 40% in *C*.

a. What percentage overall recycles?

b. What percentage overall recycles milk jugs?

c. Of those who recycle, what percentage recycle milk jugs?

3-6 Counting

What are the chances of winning the grand prize in a major lottery like Lotto 6/49? In this lottery, you choose 6 numbers from 49 possible numbers (1 to 49). If you get the same 6-number combination that is randomly drawn, you could potentially win millions of dollars. We could use Rule 2 from Section 3–2 (the classical approach) to find the probability of winning this lottery. That rule, which requires equally likely outcomes, states that the probability of an event *A* can be found using $P(A) = s/n$, where *s* is the number of ways *A* can occur and *n* is the total number of outcomes. With Lotto 6/49, there is only one way to win the grand prize: Choose the same 6-number combination that is drawn in the lottery. Knowing that there is only one way to win, we now need to determine the total number of outcomes; that is, how many 6-number combinations are possible? Writing out all the possible combinations would be an almost impossible task (and boring as well). We could try constructing a **tree diagram**—a branching diagram of the various possibilities and their associated probabilities—but that would require a sheet of paper that would cover Prince Edward Island and probably Newfoundland too! We need a more practical way of finding the total number of possibilities. This section introduces efficient methods of finding such numbers. We will return to this lottery problem after we first present some basic principles. We begin with the **fundamental counting rule**.

FUNDAMENTAL COUNTING RULE

> For a sequence of two events in which the first event can occur *m* ways and the second event can occur *n* ways, the events together can occur a total of *m · n* ways.

For example, if a medical researcher must randomly select 1 of the 2 Rh types (positive, negative) and 1 of the 4 blood groups (A, O, B, AB), the total number of possibilities is $2 \cdot 4 = 8$. We can see the reason for multiplication in Figure 3-8 where we use a tree diagram to depict the different possibilities. The fundamental counting rule easily extends to situations involving more than two events, as illustrated in the following example.

In designing a computer, if a *byte* is defined to be a sequence of 8 bits and each bit must be a 0 or 1, how many different bytes are possible? (A byte is often used to represent an individual character, such as a letter, digit, or punctuation symbol. For example, one coding system represents the letter *A* as 01000001.)

SOLUTION

Because each bit can occur in two ways (0 or 1) and we have a sequence of 8 bits, the total number of different possibilities is given by

$$2 \cdot 2 \cdot 2 \cdot 2 \cdot 2 \cdot 2 \cdot 2 \cdot 2 = 256$$

There are 256 different possible bytes.

EXAMPLE

When designing surveys, pollsters sometimes try to minimize a *lead-in* effect by rearranging the order in which the questions are presented. (A lead-in effect occurs when some questions influence the responses to the questions that follow.) If the Canadian Institute of Public Opinion (CIPO) plans to conduct a consumer survey by asking subjects 5 questions, how many different versions of the survey are required if all possible arrangements are included?

SOLUTION

In arranging any individual survey, there are 5 possible choices for the first question, 4 remaining choices for the second question, 3 choices for the third question, 2 choices for the fourth question, and only 1 choice for the fifth question. The total number of possible arrangements is therefore

$$5 \cdot 4 \cdot 3 \cdot 2 \cdot 1 = 120$$

That is, CIPO would need 120 versions of the survey in order to include every possible arrangement.

Figure 3-8
Tree Diagram of Blood Types/Rh Factors

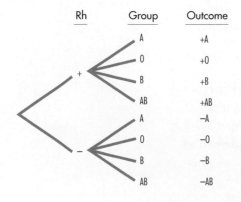

Figure 3-9
Tree Diagram of Routes

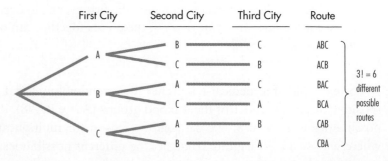

In the preceding example, we found that 5 survey questions can be arranged $5 \cdot 4 \cdot 3 \cdot 2 \cdot 1 = 120$ different ways. This particular solution can be generalized by using the factorial symbol ! and the factorial rule.

NOTATION

The **factorial symbol** ! denotes the product of decreasing positive whole numbers. For example, $4! = 4 \cdot 3 \cdot 2 \cdot 1 = 24$. By special definition, $0! = 1$. (Many calculators have a factorial key.)

FACTORIAL RULE

A collection of n different items can be arranged in order $n!$ different ways. (This **factorial rule** reflects the fact that the first item may be selected n different ways, the second item may be selected $n - 1$ ways, and so on.)

EXAMPLE

Routing problems often involve application of the factorial rule. Bell Canada wants to route telephone calls through the shortest networks. Federal Express wants to find the shortest routes for its deliveries. Suppose a computer salesperson must visit three separate cities denoted by A, B, and C. How many routes are possible?

SOLUTION

Using the factorial rule, we see that the three different cities (A, B, and C) can be arranged in $3! = 6$ different ways. In Figure 3-9 we can see that there are 3 choices for the first city and 2 choices for the second city. This leaves only 1 choice for the third city. The number of possible arrangements for the 3 cities is $3 \cdot 2 \cdot 1 = 6$.

EXAMPLE

In your first assignment as a researcher for Environics, you must conduct a survey in each of the 10 provincial capitals. As you plan your travels, you want to determine the number of different possible routes. How many routes are possible?

SOLUTION

By applying the factorial rule, we know that 10 items can be arranged in order 10! different ways. That is, the 10 provincial capitals can be arranged 10! ways, so the number of different routes is 10! or 3,628,800.

The preceding example is a variation of a classical problem called the *travelling salesman problem*. It is especially interesting because the large number of possibilities (especially if you add a few more stops to the itinerary) means that we can't use a computer to calculate the distance of each route.

Using the factorial counting rule, we determine the number of different possible ways we can arrange a number of items in some type of ordered sequence. The factorial rule tells us how many arrangements are possible when all of n different items are used. Sometimes, however, we want to select only some of the n items.

If we must conduct surveys in provincial capitals, as in the preceding example, but we have time to visit only 4 of the capitals, the number of different possible routes is $10 \cdot 9 \cdot 8 \cdot 7 = 5040$. Another way to obtain this same result is to evaluate

$$\frac{10!}{6!} = 10 \cdot 9 \cdot 8 \cdot 7 = 5040$$

In this calculation, note that the factors in the numerator divide out with the factors in the denominator, except for the factors of 10, 9, 8, and 7 that remain. We can generalize this result by noting that if we have n different items available and we want to select r of them, the number of different arrangements possible is $n!/(n - r)!$, as in $10!/6!$. This generalization is commonly called the **permutations rule**.

Some calculators are designed to automatically calculate values of $_nP_r$. If your calculator lacks such a feature, it is still easy to calculate $n!/(n - r)!$ by using the factorial key identified as !.

It is very important to recognize that the permutations rule requires the following conditions:

- We must have a total of n *different* items available. (This rule does not apply if some of the items are identical to others.)
- We must select r of the n items (without repetition).
- We must consider rearrangements of the same items to be different sequences.

PERMUTATIONS RULE (WHEN ITEMS ARE ALL DIFFERENT)

The number of **permutations** (or sequences) of r items selected from n available items (not allowing repetition) is

$$_nP_r = \frac{n!}{(n - r)!}$$

When we use the terms *permutations*, *arrangements*, or *sequences*, we imply that *order is taken into account* in the sense that different orderings of the same items are counted separately. The letters ABC can be arranged six different ways:

ABC, ACB, BAC, BCA, CAB, CBA. (Later, we will refer to **combinations**, which do not count such arrangements separately.) In the following example, we are asked to find the total number of different sequences that are possible. That suggests use of the permutations rule.

EXAMPLE

A college student planning her fall timetable must select 6 courses from 30 courses offered. The courses must be selected in order of preference. How many different timetables are possible?

SOLUTION

She needs to select $r = 6$ courses from $n = 30$ courses available. Order is relevant because they must be selected in order of preference. Because order counts, we want the number of permutations, which is found as shown:

$$_nP_r = \frac{n!}{(n - r)!} = \frac{30!}{(30 - 6)!} = 427,518,000$$

Because there are 427,518,000 possible timetables, there are too many to consider individually.

The permutations rule can be thought of as an extension of the fundamental counting rule. We can also solve the preceding example by using the fundamental counting rule as follows: With 30 courses available and with a stipulation that 6 are to be selected, we know that there are 30 choices for the first course, 29 choices for the second course, and so on. The total number of possible arrangements is therefore

$$30 \cdot 29 \cdot 28 \cdot 27 \cdot 26 \cdot 25 = 427,518,000$$

but $30 \cdot 29 \cdot 28 \cdot 27 \cdot 26 \cdot 25$ is actually $30! \div 24!$ [or $30! \div (30 - 6)!$].

We sometimes need to find the number of permutations, but some of the items are identical to others. The following variation of the permutations rule applies to such cases.

PERMUTATIONS RULE (WHEN SOME ITEMS ARE IDENTICAL TO OTHERS)

If there are n items with n_1 alike, n_2 alike, . . . , n_k alike, the number of permutations of all n items is

$$\frac{n!}{n_1! \, n_2! \cdots n_k!}$$

The classic examples of the permutations rule are those that show that the letters of the word *Mississippi* can be arranged 34,650 different ways and that the letters of the word *statistics* can be arranged 50,400 ways. We will instead consider the letters *DDDDRRRRR*, which are included in a discussion of the runs test for randomness (Section 12–7). Those letters represent a sequence of diet (*D*) and regular (*R*) colas. How many ways can we arrange the letters *DDDDRRRRR*?

SOLUTION

In the sequence *DDDDRRRRR*, we have $n = 9$ items, with $n_1 = 4$ alike and $n_2 = 5$ others that are alike. The number of permutations is computed as follows:

$$\frac{n!}{n_1!\, n_2!} = \frac{9!}{4!\; 5!} = \frac{362{,}880}{2880} = 126$$

In Section 12–7 we will use the fact that there are 126 different possible sequences of *DDDDRRRRR*, and we can now see how that result is obtained.

The preceding example involved n items, each belonging to one of two categories. When there are only two categories, we can stipulate that x of the items are alike and the other $n - x$ items are alike, so the permutations formula simplifies to

$$\frac{n!}{(n - x)!\; x!}$$

This particular result will be used for binomial experiments, which are introduced in Section 4–3.

When we intend to select r items from n different items but *do not take order into account*, we are really concerned with possible **combinations** rather than permutations. That is, **when different orderings of the same items are to be counted separately, we have a permutation problem, but when different orderings are not to be counted separately, we have a combination problem** and may apply the following rule.

COMBINATIONS RULE

The number of combinations of r items selected from n different items is

$$_nC_r = \frac{n!}{(n - r)!\; r!}$$

Some calculators will automatically evaluate $_nC_r$.

It is very important to recognize that in applying the combinations rule, the following conditions apply:

• We must have a total of n different items available.

- We must select r of the n items (without repetition).
- We must consider rearrangements of the same items to be the same grouping.

Because choosing between the permutations rule and the combinations rule can be confusing, we provide the following example, which is intended to emphasize the difference between them.

EXAMPLE

The Board of Trustees at the author's college has 9 members. Each year, they elect a 3-person committee to oversee buildings and grounds. Each year, they also elect a chairperson, vice chairperson, and secretary.

    a. When the board elects the buildings and grounds committee, how many different 3-person committees are possible?

    b. When the board elects the 3 officers (chairperson, vice chairperson, and secretary), how many different slates of candidates are possible?

SOLUTION

Note that order is irrelevant when electing the buildings and grounds committee. When electing officers, however, different orders are counted separately.

    a. Here we want the number of combinations of $r = 3$ people selected from the $n = 9$ available people. We get

$$_9C_3 = \frac{n!}{(n-r)!r!} = \frac{9!}{(9-3)!\,3!} = \frac{362{,}880}{4320} = 84$$

    b. Here we want the number of sequences (or permutations) of $r = 3$ people selected from the $n = 9$ available people. We get

$$_9P_3 = \frac{n!}{(n-r)!} = \frac{9!}{(9-3)!} = \frac{362{,}880}{720} = 504$$

There are 84 different possible committees of 3 board members, but there are 504 different possible slates of candidates.

The counting techniques presented in this section are sometimes used in probability problems. The following examples illustrate such applications.

EXAMPLE

In Lotto 6/49, a player wins first prize by selecting the correct 6-number combination when 6 different numbers from 1 through 49 are drawn. If a player selects one particular 6-number combination, find the probability of winning. (The player need not select the 6 numbers in the same order that they are drawn, so order is irrelevant).

Because 6 different numbers are selected from 49 different possibilities, the total number of combinations is

$$_{49}C_6 = \frac{49!}{(49 - 6)!6!} = 13{,}938{,}816$$

With only one combination selected, the player's probability of winning is only 1/13,983,816.

In the preceding example, we found that the probability of winning Lotto 6/49 is only 1/13,983,816 or 0.0000000715. In Canada, lightning strikes an average of 60 to 70 people and kills an average of 7 people per year. The probability of being struck and killed by lightning is 7/33,000,000 = 0.000000212. A comparison of these two probabilities shows that in a single year, there is a much better chance of being struck and killed by lightning than of winning Lotto 6/49 in a single try. Of course, you could increase your chances of winning the lottery by buying many tickets, a strategy generally deemed to be unwise.

EXAMPLE

A UPS dispatcher sends a delivery truck to 8 different locations. If the order in which the deliveries are made is randomly determined, find the probability that the resulting route is the shortest possible route.

SOLUTION

With 8 locations there are 8!, or 40,320, different possible routes. Among those 40,320 different possibilities, only two routes will be shortest (actually, the same route in two different directions). Therefore, there is a probability of only 2/40,320, or 1/20,160, or 0.0000496 that the selected route will be the shortest one possible.

This section presented five different counting approaches. When deciding which particular one to apply, you should consider a list of questions that address key issues. The following summary may be helpful.

- Is there a sequence of events in which the first can occur m ways, the second can occur n ways, and so on? If so, use the fundamental counting rule and multiply m, n, and so on.

- Are there n *different* items and are *all* of them to be used in different arrangements? If so, use the factorial rule and find $n!$.

- Are there n *different* items and are only *some* of them to be used in different arrangements? If so, evaluate $_nP_r = \frac{n!}{(n - r)!}$.

- Are there *n* items with some of them *identical* to each other, and is there a need to find the total number of different arrangements of all of those *n* items? If so, use the following expression in which n_1 of the items are alike, n_2 are alike, and so on.

$$\frac{n!}{n_1!\, n_2! \cdots n_k!}$$

- Are there *n* different items, with *some* of them to be selected, and is there a need to find the total number of combinations (that is, is the order irrelevant)? If so, evaluate $_nC_r = \frac{n!}{(n-r)!r!}$.

3-6 Exercises A: Basic Skills and Concepts

In Exercises 1–16, evaluate the given expressions.

1. $6!$
2. $11!$
3. $100!/97!$
4. $85!/82!$
5. $(10-4)!$
6. $(90-87)!$
7. $_6C_4$
8. $_6P_4$
9. $_{12}P_9$
10. $_{10}C_9$
11. $_{40}C_6$
12. $_{40}P_6$
13. $_nC_0$
14. $_nP_0$
15. $_nP_n$
16. $_nC_n$

17. The author uses a home security system that has a code consisting of 4 digits $(0, 1, \ldots, 9)$ that must be entered in the correct sequence. The digits can be repeated in the code.
    a. How many different possibilities are there?
    b. If it takes a burglar 5 seconds to try a code, how long would it take to try every possibility?

18. There are 12 members on the board of directors for Cliffside General Hospital.
    a. If they must elect a chairperson, first vice chairperson, second vice chairperson, and secretary, how many different slates of candidates are possible?
    b. If they must form an ethics subcommittee of 4 members, how many different subcommittees are possible?

19. Each social insurance number (SIN) is a sequence of 9 digits. What is the probability of randomly generating 9 digits and getting your SIN number?

20. A typical "combination" lock is opened with the correct sequence of 3 numbers between 0 and 49 inclusive. How many different sequences are possible? (A number can be used more than once.) Are these sequences combinations or are they actually permutations?

21. In doing a runs test for randomness (Section 12–7), it is found that genders of survey subjects listed in consecutive order are as follows: *MMMMMMM-MMMFFFFFFFFFF*. How many different ways can those letters be arranged?

The Number Crunch
Every so often telephone companies split regions with one area code into regions with two or more area codes because the increased number of area fax and Internet lines has nearly exhausted the possible numbers that can be listed under a single code. A seven-digit telephone number cannot begin with a 0 or 1, but if we allow all other possibilities, we get $8 \cdot 10 \cdot 10 \cdot 10 \cdot 10 \cdot 10 \cdot 10 = 8{,}000{,}000$ different possible numbers! Even so, after surviving for many years with the single area code of 514, the Montreal area was partitioned into the two area codes of 514 and 450. Other regions have also been assigned split area codes.

22. A Federal Express delivery route must include stops at 5 cities.
    a. How many different routes are possible?
    b. If the route is randomly selected, what is the probability that the cities will be arranged in alphabetical order?

23. How many ways can the word *history* be rearranged?

24. a. If a couple plans to have 8 children (it happens), how many different gender sequences are possible?
    b. If a couple has 4 boys and 4 girls, how many different gender sequences are possible?
    c. Based on the results from parts (a) and (b), what is the probability that when a couple has 8 children, the result will consist of 4 boys and 4 girls?

25. Suppose you bought a ticket in a lottery that requires you to select 6 distinct numbers between 1 and 40 inclusive.
    a. How many different selections are possible?
    b. What is the probability of winning with one ticket?
    c. What are the odds against winning such a lottery with one ticket?

26. Refer to the same lottery described in Exercise 25. What is the probability of winning if the rules are changed so that you must select the correct 6 numbers in the same order in which they are drawn?

27. In Denys Parsons' *Directory of Tunes and Musical Themes*, melodies for more than 14,000 songs are listed according to the following scheme: The first note of every song is represented by an asterisk (*), and successive notes are represented by R (for repeat the previous note), U (for a note that goes up), or D (for a note that goes down). Beethoven's Fifth Symphony begins as *RRD*. Classical melodies are represented through the first 16 notes. With this scheme, how many different classical melodies are possible?

28. A Canada Post courier must pick up weather summaries from 11 different locations around the city of Winnipeg. He would like to work out the shortest route. How many routes are possible?

29. M.S. and Associates would like to design a mutual fund portfolio from 50 eligible Canadian equity funds. The portfolio must be made up of 4 funds.
    a. How many different portfolios are possible?
    b. If the funds are selected at random (which we hope they are not), what is the probability of getting the 4 best performing Canadian equity funds?

30. You become suspicious when a genetics researcher randomly selects groups of 20 newborn babies and seems to consistently get 10 girls and 10 boys. The researcher explains that it is common to get 10 boys and 10 girls in such cases.
    a. If 20 newborn babies are randomly selected, how many different gender sequences are possible?

b. How many different ways can 10 boys and 10 girls be arranged in sequence?

c. What is the probability of getting 10 boys and 10 girls when 20 babies are born?

d. Based on the preceding results, do you agree with the researcher's explanation that it is common to get 10 boys and 10 girls when 20 babies are randomly selected?

31. The Halifax Music Company has purchased the rights to 15 different songs and it plans to release a new CD with 8 of them. Recognizing that the order of the songs is important, how many different CDs are possible?

32. A supervisor must visit 8 different distribution locations around the country. She can visit them in any order, but wishes to find the most convenient sequence. How many sequences are possible?

33. a. How many different postal codes are possible if each code is two sequences, the first sequence being letter-number-letter (for example, E1A) and the second sequence being number-letter-number (for example, 3E9)?

b. If a computer randomly generates postal codes, what is the probability that it will produce your postal code?

34. Because of recent retirements, 8 management positions became open at the Red River Finance Company. The jobs were all at different levels of management. A total of 22 employees applied for these management positions. How many different ways could the company fill the 8 vacancies?

35. In an age-discrimination case against Darmin, Inc., evidence showed that among the last 40 applicants for employment, only the 8 youngest were hired. Find the probability of randomly selecting 8 of 40 people and getting the 8 youngest. Based on the result, does it appear that age discrimination is occurring?

36. The following excerpt is from *The Man Who Cast Two Shadows*, by Carol O'Connell: "The child had only the numbers written on her palm in ink . . . all but the last four numbers disappeared in a wet smudge of blood. . . . She would put the coins into the public telephones and dial three untried numbers and then the four she knew. If a woman answered, she would say, 'It's Kathy. I'm lost.'" If it costs Kathy 25¢ for each call and she tries every possibility except those beginning with 0 or 1, what is her total cost?

37. After testing 12 homes for the presence of radon, an inspector is concerned that her test equipment is defective because the measured radon level at each home was higher than the reading at the preceding home. That is, the 12 readings were arranged in order from low to high. If the homes were randomly selected, what is the probability of getting this particular arrangement? Based on the result, is her concern about the test equipment justified?

38. A freight train is to carry 12 coal cars, 5 cars full of lumber, and 4 tankers all carrying kerosene. How many different arrangements are possible?

39. How many 6-digit licence plates are possible? How many plates are possible if no digits can be repeated? How many plates are possible if 1 letter replaces one of the 6 digits (no repeated digits)?

40. The House of Commons would like to set up a committee to investigate foreign investment in Canada. The committee must have 4 members from the government and 4 members from all the opposition parties. If 8 government members and 10 opposition members apply to serve on the committee, how many different ways can it be formed?

3-6 Exercises B: Beyond the Basics

41. Once upon a time, a common computer programming rule was that names of variables must be between 1 and 8 characters long. The first character could be any of the 26 letters, while successive characters could be any of the 26 letters or any of the 10 digits. For example, allowable variable names were A, BBB, and M3477K. How many different variable names were possible?

42. a. Five managers gather for a meeting. If each manager shakes hands with each other manager exactly once, what is the total number of handshakes?
    b. If n managers shake hands with each other exactly once, what is the total number of handshakes?
    c. How many different ways can 5 managers be seated at a round table? (Assume that if everyone moves to the right, the seating arrangement is the same.)
    d. How many different ways can n managers be seated at a round table?

43. Many calculators or computers cannot directly calculate 70! or higher. When n is large, $n!$ can be approximated by $n! = 10^K$ where $K = (n + 0.5) \log n + 0.39908993 - 0.43429448n$.
    a. Evaluate 50! using the factorial key on a calculator and also by using the approximation given here.
    b. Canada Post wanted to find the shortest route for delivering to 300 locations in Winnipeg. There are 300! different possible routes. If 300! is evaluated, how many digits are used in the result?

44. Can computers "think"? According to the *Turing test*, a computer can be considered to think if, when a person communicates with it, the person believes he or she is communicating with another person instead of a computer. In an experiment at Boston's Computer Museum, each of 10 judges communicated with 4 computers and 4 people and was asked to distinguish between them.

a. Assume that the first judge cannot distinguish between the 4 computers and the 4 people. If this judge makes random guesses, what is the probability of correctly identifying the 4 computers and the 4 people?

b. Assume that all 10 judges cannot distinguish between computers and people, so they make random guesses. Based on the result from part (a), what is the probability that all 10 judges make all correct guesses? (That event would lead us to conclude that computers cannot "think" when, according to the Turing criterion, they can.)

VOCABULARY LIST

actual odds against   123
actual odds in favour   123
addition rule   128
Bayes' Theorem   146
classical approach to
    probability   117
combination   155
combinations rule   156
complement of an
    event   121
compound event   129
conditional probability   136
contingency table   136

dependent events   140
event   116
factorial rule   153
factorial symbol   153
fundamental counting
    rule   151
independent events   140
law of large numbers   117
multiplication rule   145
mutually exclusive
    events   129
payoff odds   123
permutation   154

permutations rule   154
probability   117
relative frequency
    approximation
    of probability   117
rule of complementary
    events   132
sample space   116
simple event   116
simulation   118
subjective probability   117
tree diagram   151

REVIEW

This chapter introduced the basic concepts of probability theory. In Section 3–2 we presented the basic definitions and notation, including the representation of events by letters such as A. We defined probabilities of simple events as

$$P(A) = \frac{\text{number of times } A \text{ occurred}}{\text{number of times experiment was repeated}} \quad \text{(relative frequency)}$$

$$P(A) = \frac{\text{number of ways } A \text{ can occur}}{\text{number of different simple events}} = \frac{s}{n} \quad \text{(for equally likely outcomes)}$$

We noted that the probability of any impossible event is 0, the probability of any certain event is 1, and for any event A, $0 \leq P(A) \leq 1$. Also, $\overline{A}$ denotes the complement of event A. That is, $\overline{A}$ indicates that event A does not occur.

After Section 3–2 we proceeded to consider compound events, which involve more than one event. Always keep in mind the following key considerations.

- When conducting one trial, do we want the probability of event *A or B*? If so, use the addition rule, but be careful to avoid counting any outcomes more than once.
- In calculating conditional probabilities, $P(B|A) = \frac{P(A \text{ and } B)}{P(A)}$.
- In calculating Bayes' Theorem probabilities, remember the multiplication rule: $P(A \text{ and } B) = P(A) \cdot P(B|A)$
- In determining if two events are independent or not, $P(A \text{ and } B) = P(A) \cdot P(B)$ if the events are independent. This can be extended to three or more events.

In some probability problems, the biggest obstacle is finding the total number of possible outcomes. The last section of this chapter was devoted to the following counting techniques, which are briefly summarized at the end of Section 3–6:

- Fundamental counting rule
- Factorial rule
- Permutations rule (when items are all different)
- Permutations rule (when some items are identical to others)
- Combinations rule

Most of the material in the following chapters deals with statistical inferences based on probabilities. As an example of the basic approach used, consider a test of someone's claim that a quarter used in a coin toss is fair. If we flip the quarter 10 times and get 10 consecutive heads, we can make one of two inferences from these sample results:

1. The coin is actually fair, and the string of 10 consecutive heads is a fluke.
2. The coin is not fair.

The statistician's decision as to which inference is correct is based on the probability of getting 10 consecutive heads, which, in this case, is so small (1/1024) that the inference of unfairness is the better choice. Here we can see the important role played by probability in the standard methods of statistical inference.

REVIEW EXERCISES

In Exercises 1–8, use the data from Table 3-5, which summarize results of a study to determine if there is a relationship between place of residence and ownership of a foreign car. A random sample of 200 car owners from large cities, 150 from suburbs, and 150 from rural areas was selected.

1. If one of the 500 car owners is randomly selected, find the probability of getting a foreign car owner.

2. If one of the 500 car owners is randomly selected, find the probability of getting a foreign car owner or someone who lives in a large city.

Table 3-5 Car Ownership

	Large City	Suburbs	Rural
Foreign	90	60	25
Domestic	110	90	125

3. If two different car owners are randomly selected, find the probability that both people live in a rural area.

4. If one of the 500 car owners is randomly selected, find the probability of getting a domestic car owner who lives in the suburbs.

5. If one owner is randomly selected, find the probability of getting someone who lives in the suburbs or in a rural area.

6. If three different owners are randomly selected, find the probability that they all own domestic cars.

7. If one owner is randomly selected, find the probability of getting a domestic car owner, given that she or he is from a large city.

8. If one car owner is randomly selected, find the probability of getting someone who lives in a large city, given that her or his car is foreign. Are owning a foreign car and living in large cities independent events? Why or why not?

9. In a census of households, 25% own a high-definition TV, 17% own a camcorder, and 6.2% own both devices. If a household is randomly selected, what is the probability that it owns either device?

10. On the basis of past experience, a commuting student knows that when he speeds on any given day, there is a 2% chance of being ticketed. What is the probability of not getting a speeding ticket if he speeds every one of the 150 days in one year of college? If this student cannot afford the increased insurance costs resulting from a speeding ticket, what does the resulting probability suggest about a course of action?

11. The board of directors for the Southern Alberta Insurance Company has 8 members.
    a. If a committee of 3 is formed by random selection, what is the probability that it consists of the 3 wealthiest members?
    b. If the board must elect a chairperson, vice chairperson, and secretary, how many different slates of candidates are possible?

12. You choose 5 Canadian equity mutual funds for your portfolio. The probability that each fund will rise in value next year is 87.5%. What is the probability that all 5 funds will rise in value next year?

13. When betting on *even* in roulette, there are 38 equally likely outcomes, but only 2, 4, 6, . . . , 36 are winning outcomes.
    a. Find the probability of winning when betting on even.

b. Find the odds against winning with a bet on even.

c. Casinos pay winning bets according to odds described as 1:1. What is your net profit if you bet $5 on even and you win?

14. A question on a history test requires that 5 events be arranged in the proper chronological order. If a random arrangement is selected, what is the probability that it will be correct?

15. A pollster claims that 12 voters were randomly selected from a population of 200,000 voters (30% of whom are Progressive Conservatives), and all 12 were Progressive Conservatives. The pollster claims that this could easily happen by chance. Find the probability of getting 12 Progressive Conservatives when 12 voters are randomly selected from this population. Based on the result, does it seem that the pollster's claim is correct?

16. In a statistics class of 8 women and 8 men, 2 groups of 8 students are formed by random selection. What is the probability that all of the women are in the first group and all of the men are in the second group? (*Hint:* Find the number of ways of arranging *WWWWWWWWMMMMMMMM.*)

■ CUMULATIVE REVIEW EXERCISES

1. Refer to the accompanying frequency table, which describes the age distribution of Canadians who died of cancer (based on data from Statistics Canada).

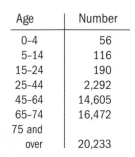

Age	Number
0–4	56
5–14	116
15–24	190
25–44	2,292
45–64	14,605
65–74	16,472
75 and over	20,233

a. Assuming that the class of "75 and over" has a class midpoint of 80, calculate the mean age of the Canadians who died of cancer.

b. Using the same assumption given in part (a), calculate the standard deviation of the ages summarized in the table.

c. If one of the 53,964 ages is randomly selected, find the probability that it is below 15 or above 64.

d. If one of the 53,964 ages is randomly selected, find the probability that it is below 15 or between 5 and 44.

e. If two different ages are randomly selected from those summarized in the table, find the probability that they are both between 0 and 4 years.

2. The accompanying boxplot depicts heights (in inches) of a large collection of randomly selected adult women.

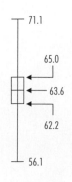

a. If one of these women is randomly selected, find the probability that her height is between 56.1 in. and 62.2 in.

b. If one of these women is randomly selected, find the probability that her height is below 62.2 in. or above 63.6 in.

c. If two women are randomly selected, find the probability that they both have heights between 62.2 in. and 63.6 in.

3. A zookeeper is interested in the lengths of pregnancies of mammals. The gestation periods (in days) of 12 randomly selected mammal species are listed below.

260   42   16   350   280   230   337   113   68   44   13   62

    a. Find the mean gestation period for the sample.

    b. Find the median gestation period for the sample.

    c. Find the standard deviation of the given sample gestation periods.

    d. What is the range of the sample gestation periods?

    e. If one gestation period is randomly selected from this sample, find the probability that it is at least 50 days.

    f. If two gestation periods are randomly selected from this sample, find the probability that they are both at least 50 days.

TECHNOLOGY PROJECT

We can often use computers to find probabilities by simulating an experiment or procedure. A simulation of a procedure is a process that behaves in the same ways as the procedure itself, thus producing similar results. Exercise 18 in Section 3–5 required computation of the probability of getting at least 2 people who share the same birthday when randomly selecting 25 people. Instead of doing theoretical calculations, we will use Excel or STATDISK to simulate the procedure. Instead of generating actual birthdays, we will generate numbers between 1 and 365, which represent the different possible birthdays. (We will ignore leap years.) For example, a generated number of 5 represents January 5, and 364 represents December 30. We can work with the generated numbers; it isn't necessary to identify the actual day and month. After generating 25 such numbers, we can sort them so that it becomes very easy to see if at least 2 of them are the same "birthday."

    Use Excel or STATDISK to generate 25 "birthdays," then sort them and observe whether at least 2 are the same. Record the result. Repeat this experiment until you are reasonably confident that your estimated probability is approximately correct.

EXCEL: Enter the numbers 1 to 365 in Column A and the probability for each of the 365 cells (1/365) in Column B. Select **Data Analysis** under the **Tools** menu. Next, select **Random Number Generation**. In the dialog box, enter 1 for **Number of Variables** and 25 for **Number of Random Numbers**. Choose **Discrete** in the **Distribution** pull-down box. In the box for **Value and Probability Input Range**, enter A1:B365. Select **New Worksheet Ply** for **Output Option** and click **OK**. Twenty-five random numbers between 1 and 365 will appear in Column A of a new worksheet. Sort the numbers ascending and view the sorted results.

First select **Data** from the main menu bar, then choose **Uniform Generator** and use **Num Decimals** to set the number of decimal places to 0 (because we want to generate only whole numbers). Proceed to generate a sample size of 25, with a maximum of 365 and a minimum of 1. Copy the data to the **Sample Editor**, click the **Data Tools** and then **Sort data** to rank and display the 25 simulated birthdays so you can easily see whether at least 2 are the same.

FROM DATA TO DECISION

Drug Testing of Job Applicants

Most U.S. companies now test at least some employees and job applicants for drug use. Some Canadian companies have tried to implement similar policies, particularly in industries where security and workplace safety are major issues. Allyn Clark, a 21-year-old college graduate, applied for a job with a transportation company, took a drug test, and was not offered a job. He suspected that he might have failed the drug test, even though he does not use drugs. In checking with the company's personnel department, he found that the drug test has 99% sensitivity, so only 1% of drug users test negative. Also, the test has 98% specificity, meaning that only 2% of nonusers are incorrectly identified as drug users. Allyn felt relieved by these figures because he believed that they reflected a very reliable test that usually provides good results—but should he be relieved? Can the company feel certain that drug users are not being hired? The accompanying table shows data for Allyn and 1999 other job applicants. Based on those results, find the probability of a "false positive"; that is, find the probability of randomly selecting one of the subjects who tested positive and getting someone who does not use drugs. Also find the probability of a "false negative"; that is, find the probability of randomly selecting someone who tested negative and getting someone who does use drugs. Are the probabilities of these wrong results low enough so that job applicants and the company need not be concerned?

	Drug Users	Nonusers
Positive test result	297	34
Negative test result	3	1666

COOPERATIVE GROUP ACTIVITIES

1. **In-class activity:** Divide into groups of three or four and estimate P(2 girls in 3 births) by using a *simulation* with coins. Describe the exact procedure used and the results obtained.

2. **In-class activity:** Divide into groups of three or four and use actual thumbtacks to estimate the probability that when dropped, a thumbtack will land with the point up. How many trials are necessary to get a result that seems to be reasonably accurate?

3. **In-class activity:** Divide into groups of three or four. In each group, agree on the value of a subjective probability for the event that a woman will be elected as prime minister of Canada in the next federal election. Are the various group values approximately the same, or are they very different? Would agreement among the groups mean that the results are accurate?

4. **In-class activity:** Each student should be given a different page torn from an old telephone directory. Proceed to randomly select 25 simulated "birthdays" by using the last three digits of telephone numbers, ignoring those above 365. After recording the 25 birthdays, check to determine whether any two of them are the same. The class results can be combined to form an estimate of the probability that when 25 people are randomly selected, at least two of them will have the same birthday.

5. **Out-of-class activity:** *The Capture–Recapture Method.* Marine biologists often use the capture–recapture method as a way to estimate the size of a population, such as the number of fish in a lake. This method involves capturing a sample from the population, tagging each member in the sample, then returning them to the population. A second sample is later captured and the tagged members are counted along with the total size of this second sample. As an example, suppose a random sample of 50 fish is captured and tagged. Also suppose that a second random sample (obtained later) consists of 100 fish with 20 of them tagged, suggesting that when a fish is captured, the probability of it being tagged is estimated to be 0.20. That is, 20% of the fish in the population have been tagged. Because the original sample of 50 fish were tagged, we estimate that the population size is 250 fish ($50 \div 20/100 = 250$).

   It's not easy to actually capture and recapture real fish, but we can simulate an experiment using some uniform collection of items such as coloured beads, M&Ms, or Fruit Loop cereal pieces. "Captured" items in the first sample can be "tagged" with a magic marker (or they can be replaced with similar items of a different colour). Illustrate the capture–recapture method by designing and conducting such an experiment. Starting with a large package of M&Ms, for example, collect a sample of 50, then use a magic marker to "tag" each one. Replace the tagged items, mix the whole population, then select a second sample and proceed to estimate the population size. Compare the result to the actual population size obtained by counting all of the items.

6. **In-class activity:** Divide into groups of two. The "Monty Hall" problem has, according to *Chance* magazine, been used to study decision making in business schools. The problem is based on a television game show that was hosted by Monty Hall. Begin by selecting one of the team members to serve as the host. The other team member is the contestant, and there are three doors numbered 1, 2, and 3. The host should *randomly* select one of the doors, and the selection must not be revealed to the contestant. Pretend that the host has

put a prize of a new red Corvette behind the door that was randomly selected, but the other two doors have nothing behind them. The contestant should now select one of the three doors. After the contestant reveals which door has been chosen, the host should select an "empty" door and inform the contestant that this particular door has nothing behind it. The host should now offer the contestant a choice of sticking with the original door or switching to the other door that has not been revealed. After the contestant announces his or her decision to stick or switch, the host should announce that the contestant has (or has not) won the Corvette. Record the result along with the contestant's decision to stick or switch. Repeat the game 20 times with the contestant sticking 10 times and switching 10 times. Then reverse roles and play the game another 20 times. Find the proportion of times the game was won by sticking and find the proportion of times the game was won by switching. Based on the results, which strategy is better: sticking or switching?

7. **In-class activity:** Divide into groups of two for the purpose of doing an experiment designed to show one approach to dealing with sensitive survey questions, such as those related to drug use, stealing, cheating, or sexual activity. For the purposes of this activity, we will use this innocuous question: "Were you born between January 1 and March 31?" We expect that about 1/4 of all responses should be "yes," but let's assume that the question is very sensitive and subjects are usually reluctant to answer honestly. One team member (the "interviewer") should ask the other (the "survey subject") to flip a coin and write "no" on a piece of paper if the survey subject was *not* born between January 1 and March 31 *and* the coin turns up heads; if the subject was born between those dates or if the coin turns up tails, the response of "yes" should be written. Switch roles so that two responses are obtained from each team. Supposedly, respondents tend to be more honest because the coin flip protects their privacy. Combine all results and analyze them to determine the proportion of people born between January 1 and March 31. The accuracy of the results can be checked against the actual birth dates. The experiment can be repeated with a question that is more sensitive, but such a question is not given here because the author already receives enough mail.

INTERNET PROJECT

What's the Probability of Landing on Boardwalk?

Finding probabilities when rolling dice is easy. With one die, there are six possible outcomes, so each outcome, such as rolling a 2, has a probability 1/6. For a card game the calculations are more involved, but they are still manageable. But what about a more complicated game, such as the board game "Monopoly"? What is the probability of landing on a particular space on the board? The probability depends on the space your piece currently occupies, the roll of the dice, the drawing of cards, as well as other factors. Now consider more true-to-life examples, such as the probability of having an auto accident. The number of factors involved there is too large to even consider, yet such probabilities are nonetheless quoted, for example, by insurance companies.

The Internet Project for this chapter considers methods for computing probabilities in complicated situations. Go to either of the following websites:

http://www.mathxl.com   or   http://www.mystatlab.com

Click on Internet Project, then on Chapter 3, and then you will be guided in researching probabilities for a board game. Then you will compute such a probability yourself. Finally, you will compute a health-related probability using empirical data.

4 Discrete Probability Distributions

4-1 Overview

The objectives of this chapter are identified: Random variables and probability distributions are described in general; then, some special probability distributions are examined.

4-2 Random Variables

Discrete and continuous random variables and probability distributions are described in this section. Methods for finding the mean, variance, and standard deviation for a given probability distribution are also described. The expected value of a probability distribution is defined.

4-3 Binomial Probability Distributions

Binomial probability distributions are defined. Probabilities in binomial probability distributions are calculated using the binomial probability formula, a table of binomial probabilities, and statistical software.

4-4 Mean, Variance, and Standard Deviation for the Binomial Distribution

The mean, variance, and standard deviation for a binomial distribution are calculated. Interpretation of those values is also discussed, including the application of the empirical rule for symmetric data.

4-5 The Poisson Distribution

The Poisson distribution is described as another special and important example of a discrete probability distribution. A key characteristic of the Poisson distribution is that it applies to occurrences of some event over a specified interval of time, distance, area, or some similar unit.

4-6 The Hypergeometric Distribution

The hypergeometric distribution is similar to the binomial distribution; its distinction is that the population size is known and is used in calculating probabilities.

Is the number of accidents caused by commercial vehicles just a coincidence?

Much publicity has been given lately to the increasing volumes of traffic on our highways and, in particular, the volume of commercial vehicles, which include trucks, buses, and other vehicles. Along with this increase in volume comes an increase in accidents. Large vehicles pose a large risk to those who travel the roads and highways in smaller, lighter passenger vehicles. In 1997 there were 3,720,000 registered commercial vehicles in Canada, representing about 20% of all vehicles on the road (based on data from Transport Canada, *Canadian Motor Vehicle Traffic Collision Statistics*, 1997).

Commercial vehicles have grown in size as well as numbers. We now see trucks pulling two large trailers behind them. With just-in-time inventory control, there is more emphasis on speed and longer operating hours for these commercial vehicles. Are they safe for our roads? How much of a threat are they to drivers and passengers in smaller vehicles?

Recently, accidents involving trucks have claimed more than a few headlines. The public's anxiety level rises whenever there is a cluster of accidents. Suppose, on a specific stretch of highway during a short period of time, 4 of 7 vehicles involved in accidents were commercial vehicles. Given that commercial vehicles represent 20% of all vehicles on the road, can we conclude from these 7 accident reports that commercial vehicles are not as safe, or are not being operated as safely, as other vehicles? That determination depends on the *probability* that the given events occur by chance. We will consider these two questions:

1. Given that 20% of all vehicles on the road are commercial vehicles, and assuming that the vehicles are as safe as other vehicles and that accidents are independent events that occur at random, what is the probability that 4 of 7 vehicles involved in a cluster of accidents will be commercial vehicles?

2. In deciding whether commercial vehicles are unsafe or the victims of coincidence, is the probability in the preceding question the *relevant* probability? Instead of asking for the probability that exactly 4 of the 7 involved vehicles are commercial, is there another question that better addresses the real issue of whether commercial vehicles are unsafe?

The first question can easily be answered using methods presented in this chapter. The second question is more difficult and requires serious thought, but it is extremely important to correctly identify the event that is key to the issue. We will address both questions later in the chapter.

4-1 Overview

In Chapter 2 we saw that we could explore and describe a set of data by using graphs (such as a histogram or boxplot), measures of central tendency (such as the mean), and measures of variation (such as the standard deviation). In Chapter 3 we discussed the basic principles of probability theory. In this chapter we combine those concepts as we develop probability distributions that describe what will *probably* happen instead of what actually did happen. In Chapter 2 we constructed frequency tables and histograms using *observed* real scores, but in this chapter we will construct probability distributions by presenting possible outcomes along with the relative frequencies we *expect*, given an understanding of the relevant circumstances.

Suppose that a casino manager suspects cheating at a dice table. He or she can compare the relative frequency distribution of the actual sample outcomes to a theoretical model that describes the frequency distribution likely to occur with a fair die. A fair die should have a relative frequency histogram similar to the one shown in Figure 4-1(a), so a relative frequency histogram looking like Figure 4-1(b) may arouse suspicion.

In Figure 4-1(a) we see relative frequencies based not on actual outcomes, but on our knowledge of the probabilities for the outcomes of a fair die. In essence, we can describe the frequency table and histogram for a die rolled an infinite number of times. With this knowledge of the population of outcomes, we are able to determine important characteristics, such as the mean and standard deviation. The remainder of this book and the very core of inferential statistics are based on some knowledge of probability distributions. We begin by examining the concept of a random variable, then we consider important distributions that have many real applications.

Figure 4-1
Histograms of Dice Outcomes for (a) a Fair Die and (b) a Loaded Die

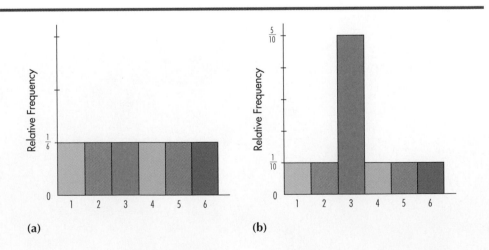

(a)　　　　　　　　　(b)

Random Variables

In this section we discuss random variables, probability distributions, and procedures for finding the mean and standard deviation for a probability distribution. We will see that a random variable has a numeric value for each outcome of a procedure, and a probability distribution associates a probability value with each outcome of a procedure.

DEFINITION

A **random variable** is a variable (typically represented by x) that has a single numerical value (determined by chance) for each outcome of a procedure.

Examples of random variables are the following:

x = The number of women among 10 newly hired employees

x = The number of students absent from statistics class today

x = The height (in cm) of a randomly selected adult male

The word *random* is used to remind us that we don't usually know what the value of x is until we observe or perform a procedure.

EXAMPLE

We randomly select 7 vehicles involved in accidents on public roadways and count how many of these are commercial vehicles. If we let the random variable represent the number of commercial vehicles among 7, this procedure has possible outcomes of 0, 1, 2, 3, 4, 5, 6, 7. (Remember, 0 represents no vehicles are commercial vehicles, 1 represents 1 vehicle is a commercial vehicle, and so on.) The variable is random in the sense that we do not know the value until after the 7 vehicles have been selected.

In Section 1–2 we made a distinction between discrete and continuous data. Random variables may also be discrete or continuous, and the following two definitions are consistent with those given in Section 1–2. This chapter deals with discrete random variables, but the following chapters will deal with continuous random variables.

DEFINITIONS

A **discrete random variable** has either a finite number of values or a countable number of values; that is, they result from a counting process.

A **continuous random variable** has infinitely many values, and those values can be associated with measurements on a continuous scale in such a way that there are no gaps or interruptions.

Figure 4-2
Discrete and Continuous Random Variables

Counter

(a) Discrete Random Variable: Count of the number of movie patrons.

Voltmeter

0   9

(b) Continuous Random Variable: The measured voltage of a smoke detector battery.

EXAMPLE

1. The count of the number of patrons viewing a movie is a whole number and is therefore a discrete random variable. The counting device shown in Figure 4-2(a) is capable of indicating only whole numbers. It can therefore be used to obtain values for a discrete random variable.

2. The measure of voltage for a smoke detector battery can be any value between 0 volts and 9 volts and is therefore a continuous random variable. The voltmeter depicted in Figure 4-2(b) is capable of indicating values on a continuous scale, so it can be used to obtain values for a continuous random variable.

After we identify possible values of a random variable, we can often identify a probability for each of those values. When we know all values of a random variable along with their corresponding probabilities, we have a probability distribution, defined as follows.

DEFINITION

A **probability distribution** gives the probability for each value or range of values of the random variable.

EXAMPLE

Suppose that 20% of all registered motor vehicles are commercial vehicles, and that all vehicles have the same chance of being involved in an accident. If we let the random variable x represent the number of commercial vehicles among 7 randomly selected vehicles involved in accidents, then the probability distribution can be described by Table 4-1. (In Section 4–3 we will see how the probabilities in Table 4-1 are found.) In the table, we see that the probability that 0 of 7 vehicles in accidents is a commercial vehicle is 0.210, the probability that 1 of 7 is commercial is 0.367, and so on. The values denoted by 0+ represent positive probabilities that are so small that they become 0.000 when rounded to three decimal places. We avoid using 0.000 because it incorrectly suggests an impossible event with a probability of 0.

Table 4-1 Probability Distribution for Number of Commercial Vehicles Among Seven Vehicles Involved in Accidents

x	$P(x)$
0	0.210
1	0.367
2	0.275
3	0.115
4	0.029
5	0.004
6	0+
7	0+

Graphs

There are various ways to graph a probability distribution, but we present only the **probability histogram**. Figure 4-3 is a probability histogram that resembles the relative frequency histogram from Chapter 2, but the vertical scale delineates *probabilities* instead of relative frequencies based on actual sample results.

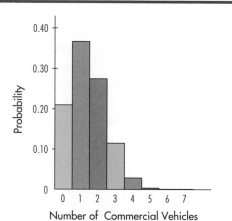

Figure 4-3
Probability Histogram for Number of Commercial Vehicles Among Seven Vehicles Involved in Accidents

In Figure 4-3, note that along the horizontal axis, the values of 0, 1, 2, . . . , 7 are located at the centres of the rectangles. This implies that the rectangles are each 1 unit wide, so the areas of the rectangles are 0.210, 0.367, and so on. When the total area of such a probability histogram is 1, the *probabilities* are equal to the corresponding rectangular *areas*. We will see in Chapter 5 and later chapters that this correspondence between area and probability is very useful in statistics.

For discrete data, every probability distribution must satisfy the following two requirements.

REQUIREMENTS FOR A PROBABILITY DISTRIBUTION (FOR DISCRETE DATA)

1. $\Sigma P(x) = 1$ where x assumes all possible, distinct values
2. $0 \leq P(x) \leq 1$ for every value of x

The first requirement states that the sum of the individual probabilities must equal 1 and is based on the addition rule for mutually exclusive events. The values of the random variable x represent all possible events in the entire sample space, so we are certain (with probability 1) that one of the events will occur. We use simple addition of the values of $P(x)$ because the different values of x correspond to events that are mutually exclusive. In Table 4-1 we can see that the individual probabilities do result in a sum of 1. Also, the probability rule (see Section 3–2) stating that $0 \leq P(A) \leq 1$ for any event A implies that $P(x)$ must be between 0 and 1 for any value of x. Again, refer to Table 4-1 and note that each individual value of $P(x)$ does fall between 0 and 1. Because Table 4-1 does satisfy both of these requirements, it is an example of a probability distribution. A probability distribution may be described by a table, such as Table 4-1, or a graph, such as Figure 4-3, or a formula, as in the following two examples.

EXAMPLE

Can the formula $P(x) = x/5$ (where x can take on the values of 1, 2, 3) determine a probability distribution?

SOLUTION

If a probability distribution is determined, it must conform to the preceding two requirements. But

$$\Sigma P(x) = P(1) + P(2) + P(3)$$
$$= \frac{1}{5} + \frac{2}{5} + \frac{3}{5}$$
$$= \frac{6}{5} \quad \text{(showing that } \Sigma P(x) \neq 1\text{)}$$

Because the first requirement is not satisfied, we conclude that $P(x)$ given in this example cannot be determining a probability distribution.

EXAMPLE

Can the formula $P(x) = x/3$ (where x can be 1 or 2) determine a probability distribution?

SOLUTION

For the given function, we find that $P(1) = 1/3$ and $P(2) = 2/3$ so that

1. $\Sigma P(x) = \dfrac{1}{3} + \dfrac{2}{3} = \dfrac{3}{3} = 1$

2. Each of the $P(x)$ values is between 0 and 1 inclusive.

Because the two requirements are both satisfied, the $P(x)$ function given in this example is a possible probability distribution.

Mean, Variance, and Standard Deviation

In Chapter 2 we saw that three of the extremely important characteristics of data are:

1. *Centre:* Representative value, such as a mean
2. *Variation:* Measure of scattering or variation, such as a standard deviation
3. *Distribution:* Nature or shape of the distribution, such as bell-shaped

The probability histogram can give us insight into the nature or shape of the distribution. Also, we can often find the mean, variance, and standard deviation of data, which provide insight into the other characteristics. The mean, variance, and standard deviation for a probability distribution can be found by applying Formulas 4-1, 4-2, 4-3, and 4-4.

Formula 4-1 $\quad \mu = \Sigma[x \cdot P(x)]$ $\qquad$ Mean for a probability distribution

Formula 4-2 $\quad \sigma^2 = \Sigma[(x - \mu)^2 \cdot P(x)]$ $\qquad$ Variance for a probability distribution

Formula 4-3 $\quad \sigma^2 = [\Sigma x^2 \cdot P(x)] - \mu^2$ $\qquad$ Variance for a probability distribution

Formula 4-4 $\quad \sigma = \sqrt{[\Sigma x^2 \cdot P(x)] - \mu^2}$ $\qquad$ Standard deviation for a probability distribution

Caution: The expression $\Sigma x^2 \cdot P(x)$ is the same as $\Sigma[x^2 \cdot P(x)]$. Evaluate $\Sigma x^2 \cdot P(x)$ by first squaring each value of x, then multiplying each square by the corresponding $P(x)$, then adding.

When using Formulas 4-1 through 4-4, use this rule for rounding results:

ROUND-OFF RULE FOR μ, σ^2, AND σ

Round results by carrying one more decimal place than the number of decimal places used for the random variable x. If the values of x are integers, round μ, σ^2, and σ to one decimal place.

It is sometimes necessary to use a different rounding rule because of special circumstances, such as results that require more decimal places to be meaningful.

When we calculate the mean of a probability distribution, we get the average (mean) value that we would expect to get if the trials could be repeated indefinitely. We *don't* get the value we expect to occur most often. In fact, we often get a mean value that cannot occur in any one actual trial (such as 1.5 girls in 3 births). The standard deviation gives us a measure of how much the probability distribution is spread out around the mean. A large standard deviation reflects considerable spread, whereas a smaller standard deviation reflects lower variability with values relatively closer to the mean. The range rule of thumb (see Section 2–5) may also be helpful in interpreting the value of a standard deviation. According to the range rule of thumb, most values should lie within two standard deviations of the mean; it is unusual for a score to differ from the mean by more than two standard deviations.

EXAMPLE

Table 4-1 represents the probability distribution for the number of commercial vehicles among 7 randomly selected vehicles involved in accidents (assuming that 20% of all registered vehicles are commercial vehicles and that accidents are independent and random events). Using the probability distribution described in Table 4-1, assume that we repeat the procedure of randomly selecting 7 accident vehicles and each time we find the number that are commercial vehicles. Find the mean number of commercial vehicles (among 7), the variance, and the standard deviation. Would it be unusual if 4 of 7 accident vehicles were commercial vehicles?

SOLUTION

In Table 4-2, the two columns at the left describe the probability distribution given earlier in Table 4-1. We create the three columns at the right for the purposes of the calculations required.

Using Formulas 4-1 and 4-3 and the table results, we get

$$\mu = \Sigma[x \cdot P(x)] = 1.398 = 1.4 \text{ commercial vehicles (rounded)}$$

$$\sigma^2 = [\Sigma x^2 \cdot P(x)] - \mu^2$$

$$= 3.066 - 1.398^2 = 1.111596 = 1.1 \text{ commercial vehicles}^2 \text{ (rounded)}$$

Table 4-2 Calculating μ, σ^2, and σ for a Probability Distribution

x	P(x)	x · P(x)	x²	x² · P(x)
0	0.210	0.000	0	0.000
1	0.367	0.367	1	0.367
2	0.275	0.550	4	1.100
3	0.115	0.345	9	1.035
4	0.029	0.116	16	0.464
5	0.004	0.020	25	0.100
6	0+	0.000	36	0.000
7	0+	0.000	49	0.000
Total	1.000	1.398		3.066
	↑	↑		↑
	$\Sigma P(x)$	$\Sigma x \cdot P(x)$		$\Sigma x^2 \cdot P(x)$

The standard deviation is the square root of the variance, so

$$\sigma = \sqrt{1.111596} = 1.054323 = 1.1 \text{ commercial vehicles (rounded)}$$

Interpretation

We now know that among 7 accident vehicles, the mean number of commercial vehicles is 1.4, the variance is 1.1 "commercial vehicles squared" and the standard deviation is 1.1 commercial vehicles. Using the range rule of thumb from Section 2-5, we can conclude that most of the time, there should be between 0 and 3.6 commercial vehicles when 7 accident vehicles are randomly selected. (Recall that with the range rule of thumb, we can find rough estimates of minimum and maximum scores by starting with the mean of 1.4 and adding and subtracting 2.2, which is twice the standard deviation.) If, in fact, there were 4 commercial vehicles among 7 accident vehicles, this would appear to be an unusual result.

Rationale for Formulas 4-1 to 4-4

Why do Formulas 4-1 through 4-4 work? A probability distribution is actually a model of a theoretically perfect population frequency distribution. The probability distribution is like a relative frequency distribution based on data that behave perfectly, without the usual imperfections of samples. Because the probability distribution allows us to predict the population of outcomes, we are able to determine the values of the mean, variance, and standard deviation. Formula 4-1 accomplishes the same task as the formula for the mean of a frequency table. (Recall that f represents class frequency and N represents population size.) Rewriting the formula for the mean of a frequency table so that it applies to a population and then changing its form, we get

$$\mu = \frac{\Sigma(f \cdot x)}{N} = \Sigma\frac{f \cdot x}{N} = \Sigma x \cdot \frac{f}{N} = \Sigma x \cdot P(x)$$

In the fraction f/N, the value of f is the frequency with which the value x occurs and N is the population size, so f/N is the probability for the value of x.

Similar reasoning enables us to take the variance formula from Chapter 2 and apply it to a random variable for a probability distribution; the result is Formula 4-2. Formula 4-3 is a shortcut version that will always produce the same result as Formula 4-2. Although Formula 4-3 is usually easier to work with, Formula 4-2 is easier to understand directly. On the basis of Formula 4-2, we can express the standard deviation as

$$\sigma = \sqrt{\Sigma(x - \mu)^2 \cdot P(x)}$$

or as the equivalent form given in Formula 4-4.

Expected Value

The mean of a discrete random variable is the theoretical mean outcome for infinitely many trials. We can think of that mean as the *expected value* in the sense that it is the average value that we would expect to get if the trials could continue indefinitely. The uses of expected value (also called *expectation* or *mathematical expectation*) are extensive and varied, and they play a very important role in an area of application called *decision theory*.

DEFINITION

The **expected value** of a discrete random variable is denoted by E, and it represents the average value of the outcomes. It is found by finding the value of $\Sigma[x \cdot P(x)]$:

$$E = \Sigma[x \cdot P(x)]$$

From Formula 4-1 we see that $E = \mu$. That is, the mean of a discrete random variable is the same as its expected value. Repeat the experiment of flipping a coin five times and the *mean* number of heads is 2.5; when flipping a coin five times, the *expected value* of the number of heads is also 2.5.

EXAMPLE

Lotteries have become extremely popular with the general public. They might not be so popular with statistics teachers. A ticket in a typical lottery like Lotto 6/49 is usually $2. The probability of winning Lotto 6/49 is 1 in 13,983,816. The payoffs (jackpots) are usually in excess of $1 million. Suppose that you buy one ticket for $2 and the jackpot is $1 million. What is your expected value of gain or loss?

SOLUTION

For this situation there are two simple outcomes: You win or you lose. Your ticket contains one of the 13,983,816 number combinations that could possibly be drawn; your probability of winning is 1/13,983,816 (0.0000000715) and your probability of losing is 13,983,815/13,983,816. Table 4-3 summarizes the situation.

Table 4-3 **The Numbers Game**

Event	x	P(x)	x · P(x)
Win	$999,998	0.0000000715	$0.0715
Lose	−$2	0.9999999285	−$1.9999
Total			−$1.93

From Table 4-3 we can see that when we buy one ticket for $2, our expected value is

$$E = \Sigma x \cdot P(x) = -\$1.93$$

Interpretation
This means that in the long run, for each $2 ticket, we can expect to lose an average of $1.93. This is not a very attractive investment scheme. Even if the jackpot were $10 million, the expected value would be −$1.28.

In this section we learned that a random variable has a numerical value associated with each outcome of some chance experiment, and a probability distribution has a probability associated with each value of a random variable. We examined methods for finding the mean, variance, and standard deviation for a probability distribution. We saw that the expected value of a random variable is really the same as the mean. We also learned that lotteries are lousy investments.

4-2 Exercises A: Basic Skills and Concepts

In Exercises 1–4, identify the given random variable as being discrete or continuous.

1. The weight of a randomly selected textbook

2. The cost of a randomly selected textbook

3. The number of eggs a hen lays

4. The amount of milk obtained from a cow

In Exercises 5–12, determine whether a probability distribution is given. In those cases where a probability distribution is not described, identify the requirement that is not satisfied. In those cases where a probability distribution is described, find its mean, variance, and standard deviation.

5. When a household is randomly selected, the probability distribution for the number x of automobiles owned is as described in the accompanying table.

x	$P(x)$
0	0.011
1	0.394
2	0.380
3	0.215

6. If your college hires the next 4 employees without regard to gender, and the pool of applicants is large with an equal number of men and women, then the probability distribution for the number x of women hired is described in the accompanying table.

x	$P(x)$
0	0.0625
1	0.2500
2	0.3750
3	0.2500
4	0.0625

7. Statistics Canada found that English is the mother tongue of 62.6% of the residents in the Greater Toronto Area (GTA). A small survey of 8 residents in one area of the GTA will be taken, so the accompanying table describes the probability distribution for the number of residents (among the 8 randomly selected residents) whose mother tongue is English.

x	$P(x)$
0	0.000
1	0.002
2	0.012
3	0.053
4	0.147
5	0.261

8. In assessing credit risks, a bank investigates the number of credit cards people have. With x representing the number of credit cards adults have, the accompanying table describes the probability distribution for a population of applicants (based on data from Maritz Marketing Research, Inc.).

x	$P(x)$
0	0.26
1	0.16
2	0.12
3	0.09
4	0.07
5	0.09
6	0.07
7	0.14

9. The Willford Printing Company conducted a survey to monitor daily absenteeism at the plant. The accompanying table describes the probability distribution for one department at the plant, where x represents the number of employees absent in the department.

x	$P(x)$
0	0.15
1	0.50
2	0.25
3	0.10

10. Big Grack's used car dealership reports that the probabilities of selling 0, 1, 2, 3, 4, and 5 cars in one week are 0.256, 0.239, 0.259, 0.71, and 0.061, respectively.

11. A manufacturing process is continually monitored for defective parts. Every hour 5 samples are taken from the process. Listed here are the number of defective parts in a sample of 5 and the probability associated with each.

Defective	Probability
0	0.5905
1	0.3280
2	0.0729
3	0.0081
4	0.0005
5	0.0000

12. A study of gender bias in television news involves the selection of guests and reporters on CBC's *The National*. The people appearing on air during a certain period are randomly selected in groups of 4 and the numbers of women are recorded. The probabilities of getting 0, 1, 2, 3, and 4 women are 0.4979, 0.3793, 0.1084, 0.0138, and 0.0006, respectively (based on data from NewsWatch Canada).

13. When you give a casino $5 for a bet on the number 7 in roulette, you have a 1/38 probability of winning $175 and a 37/38 probability of losing $5. What is your expected value? In the long run, how much do you lose for each dollar bet?

14. Based on past results found in the *Information Please Almanac*, there is a 0.120 probability that a baseball World Series contest will last four games, a 0.253 probability that it will last five games, a 0.217 probability that it will last six games, and a 0.410 probability that it will last seven games. Find the mean and standard deviation for the numbers of games that World Series contests last. Is it unusual for a team to "sweep" by winning in four games?

15. If the probability of being killed by lightning is 0.000002 for a one-year period, what is the expected value of a one-year insurance policy that would pay $1 million if you are killed by lightning?

16. *Reader's Digest* recently ran a sweepstakes in which prizes were listed along with the chances of winning: $5,000,000 (1 chance in 201,000,000), $150,000 (1 chance in 201,000,000), $100,000 (1 chance in 201,000,000), $25,000 (1 chance in 100,500,000), $10,000 (1 chance in 50,250,000), $5000 (1 chance in 25,125,000), $200 (1 chance in 8,040,000), $125 (1 chance in 1,005,000), and a watch valued at $89 (1 chance in 3774).
    a. Find the expected value of the amount won for one entry.
    b. Find the expected value if the cost of entering this sweepstakes is the cost of a postage stamp.

17. The random variable x represents the number of girls in a family of 3 children. (*Hint:* Assuming that boys and girls are equally likely, we get $P(2) = 3/8$ by

examining this sample space: bbb, bbg, bgb, bgg, gbb, gbg, ggb, ggg.) Find the mean, variance, and standard deviation for the random variable x. Also, use the range rule of thumb (from Section 2–5) to approximate the minimum and maximum usual values of x.

18. The random variable x represents the number of boys in a family of 4 children. (See Exercise 17.) Find the mean, variance, and standard deviation for the random variable x. Also, use the range rule of thumb (from Section 2–5) to approximate the usual minimum and maximum values of x.

19. The Don Mills Electronics Company makes switching devices for traffic signals. One batch of 10 switches includes 2 that are defective. If 2 switches are randomly selected from that batch without replacement, let the random variable x represent the number that are defective. Find the mean, variance, and standard deviation for the random variable x.

20. A statistics class includes 3 left-handed students and 24 right-handed students. Two different students are randomly selected for a data collection project and the random variable x represents the number of left-handed students selected. Find the mean, variance, and standard deviation for the random variable x. (*Hint:* Use the multiplication rule of probability to first find $P(0)$ and $P(2)$.)

4-2 Exercises B: Beyond the Basics

21. In each case, determine whether the given function is a probability distribution.
    a. $P(x) = 1/2^x$ where $x = 1, 2, 3, \ldots$
    b. $P(x) = 1/2x$ where $x = 1, 2, 3, \ldots$
    c. $P(x) = 3/[4(3 - x)!x!]$ where $x = 0, 1, 2, 3$
    d. $P(x) = 0.4(0.6)^{x-1}$ where $x = 1, 2, 3, \ldots$

22. The mean and standard deviation for a random variable x are 5.0 and 2.0, respectively. Find the mean and standard deviation if
    a. each value of x is increased by 3
    b. each value of x is multiplied times 3
    c. each value of x is multiplied times 3 and then 4 is added

23. Digits (0, 1, 2, . . . , 9) are randomly selected for telephone numbers in surveys. The random variable x is the selected digit.
    a. Find the mean and standard deviation of x.
    b. Find the z score for each of the possible values of x, then find the mean and standard deviation of the population of z scores.
    c. Will the same mean and standard deviation as you found in part (b) result from every probability distribution?

24. Assume that the discrete random variable x can assume the values 1, 2, . . . , n, and that those values are equally likely.

    a. Show that $\mu = (n + 1)/2$.

    b. Show that $\sigma^2 = (n^2 - 1)/12$.

       [*Hint:* $1 + 2 + 3 + \cdots + n = n(n + 1)/2$ and
$$1^2 + 2^2 + 3^2 + \cdots + n^2 = n(n + 1)(2n + 1)/6.]$$

    c. An experiment consists of randomly selecting a whole number between 1 and 50, and the random variable x is the value of the number selected. Find the mean and standard deviation for x.

25. Let the random variable x represent the number of girls in a family of four children. Construct a table describing the probability distribution, then find the mean and standard deviation. (*Hint:* Begin by listing the different possible outcomes.)

Binomial Probability Distributions

In Section 4–2 we learned that a random variable has a numerical value associated with each outcome of some chance procedure, and a probability distribution has a probability associated with each value or range of values of a random variable. In this section we will focus on one specific type of probability distribution: the *binomial distribution*. Binomial probability distributions are important because they allow us to deal with circumstances in which the outcomes belong to either of *two* relevant categories. In manufacturing, parts either fail or they do not. In medicine, a patient either survives a year or does not. In advertising, a consumer either recognizes a product or does not.

DEFINITION

A **binomial probability distribution** results from a procedure, or **binomial experiment**, that meets all the following requirements:

1. The procedure has a *fixed number of trials* (such as 7 coin tosses, or 20 responses to a survey question).
2. The trials must be *independent.* (The outcome of any individual trial doesn't affect the probabilities in the other trials.)
3. The outcome of each trial must be classifiable into one of *two possible categories.*
4. The probabilities must remain *constant* for each trial.

If a procedure satisfies these four requirements, the distribution of the random variable x is called a *binomial probability distribution* (or *binomial distribution*). The following notation is commonly used.

S and F (success and failure) denote the two possible categories of all outcomes; p and q will denote the probabilities of S and F, respectively, so

$$P(S) = p \qquad (p = \text{probability of a success})$$

$$P(F) = 1 - p = q \qquad (q = \text{probability of a failure})$$

n	denotes the fixed number of trials.
x	denotes a specific number of successes in n trials, so x can be any whole number between 0 and n, inclusive.
p	denotes the probability of success in *one* of the n trials.
q	denotes the probability of failure in *one* of the n trials.
$P(x)$	denotes the probability of getting exactly x successes among the n trials.

The word *success* as used here is arbitrary and does not necessarily describe a desired result. Either of the two possible categories may be called the success S as long as the corresponding probability is identified as p. (The value of q can always be found by subtracting p from 1; if $p = 0.95$, then $q = 1 - 0.95 = 0.05$.) Once a category has been designated as the success S, be sure that p is the probability of a success and x is the number of successes. That is, **be sure that the values of p and x refer to the same category designated as a success.**

One very common application of statistics involves sampling without replacement, as in testing manufactured items or conducting surveys. Strictly speaking, sampling without replacement involves dependent events, which violates the second requirement in the preceding definition. However, the following rule of thumb is based on the fact that if the sample is very small relative to the population size, the difference in results will be negligible if we treat the trials as independent, even though they are actually dependent.

When sampling without replacement, the events can be considered to be independent if the sample size is no more than 5% of the population size. (That is, $n \le 0.05N$.)

EXAMPLE

Five percent of the population of Toronto speaks Italian. Find the probability of getting, in a random sample of 15 Toronto residents, exactly 3 residents who speak Italian.

a. Is this a binomial experiment?

b. If this is a binomial experiment, identify the values of n, x, p, and q.

SOLUTION

a. This experiment does satisfy the requirements for a binomial experiment, as shown below:
   1. The number of trials (15) is fixed.
   2. The trials are independent because one resident speaking Italian doesn't affect the probability of any other randomly selected resident speaking Italian.
   3. Each trial has two categories of outcomes: the resident either speaks Italian or does not.
   4. The probability of selecting an Italian-speaking resident (0.05) remains constant from trial to trial.

b. Having concluded that the experiment is binomial, we now proceed to identify the values of $n, x, p,$ and q:
   1. With 15 residents, we have $n = 15$.
   2. We want 3 residents who speak Italian (successes), so $x = 3$.
   3. The probability of a resident who speaks Italian (success) is 0.05, so $p = 0.05$.
   4. The probability of failure (a resident who does not speak Italian) is 0.95, so $q = 0.95$.

   Again, it is very important to be sure that both x and p refer to the same concept of "success." In this example, we use x to count the desired number of residents who speak Italian, and p is the probability of getting a resident who speaks Italian, so x and p do use the same concept of success.

In this section we present three methods for finding probabilities in a binomial experiment. The first method involves calculations using the *binomial probability formula* and is the basis for the other two methods. The second method involves the use of Table A-1, and the third method involves the use of statistical software. [If your instructor is requiring technology-based solutions (Method 3), we recommend that you also work through some examples using Method 1 and Method 2, to ensure you understand the basis for the calculations.]

METHOD 1: USE THE BINOMIAL PROBABILITY FORMULA   In a binomial distribution, probabilities can be calculated by using the **binomial probability formula**:

Formula 4-5 $$P(x) = \frac{n!}{(n-x)!x!} \cdot p^x \cdot q^{n-x} \quad \text{for } x = 0, 1, 2, \ldots, n$$

where n = number of trials
   x = number of successes among n trials
   p = probability of success in any one trial
   q = probability of failure in any one trial ($q = 1 - p$)

The factorial symbol !, introduced in Section 3–6, denotes the product of decreasing factors. Two examples of factorials are $3! = 3 \cdot 2 \cdot 1 = 6$ and $0! = 1$ (by definition). Many calculators have a factorial key, as well as a key labelled $_nC_r$

that can simplify the computations. For calculators with the $_nC_r$ key, use this version of the binomial probability formula (where n, x, p, and q are the same as in Formula 4-5):

$$P(x) = {_nC_x} \cdot p^x \cdot q^{n-x}$$

n	x	$p = 0.05$
15	0	463
	1	366
	2	135
	3	031
	4	005
	5	001
	6	0+
	7	0+
	8	0+
	9	0+
	10	0+
	11	0+
	12	0+
	13	0+
	14	0+
	15	0+

EXAMPLE

Five percent of the population of Toronto speaks Italian. Use the binomial probability formula to find the probability of getting, in a random sample of 15 Toronto residents, 3 residents who speak Italian. That is, find $P(3)$, given that $n = 15, x = 3, p = 0.05$, and $q = 0.95$.

SOLUTION

Using the given values of n, x, p, and q in the binomial probability formula (Formula 4-5), we get

$$P(3) = \frac{15!}{(15-3)!3!} \cdot 0.05^3 \cdot 0.95^{15-3}$$

$$= \frac{15!}{12!3!} \cdot 0.000125 \cdot 0.5403601$$

$$= (455)(0.000125)(0.5403601) = 0.031$$

The probability that exactly 3 of 15 residents will speak Italian is 0.031.

Binomial probability distribution for $n = 15$ and $p = 0.05$

x	$P(x)$
0	0.463
1	0.366
2	0.135
3	0.031
4	0.005
5	0.001
6	0+
7	0+
8	0+
9	0+
10	0+
11	0+
12	0+
13	0+
14	0+
15	0+

Calculation hint: When computing a probability with the binomial probability formula, it's helpful to get a single number for $_nC_x$, a single number for p^x, and a single number for q^{n-x}, then simply multiply the three factors together. Don't round too much when you find those three factors; round only at the end. Alternatively, if you have a calculator with a display window, it may be advantageous to multiply the three values simultaneously to avoid rounding errors.

METHOD 2: USE TABLE A-1 IN APPENDIX A In some cases, we can easily find binomial probabilities by simply referring to Table A-1 in Appendix A. First locate n and the corresponding value of x that is desired. At this stage, one row of numbers should be isolated. Now align that row with the proper probability of p by using the column across the top. The isolated number represents the desired probability. A very small probability, such as 0.000000345, is indicated by 01.

Shown in the margin is part of Table A-1. When $n = 15$ and $p = 0.05$ in a binomial experiment, the probabilities of 0, 1, 2, . . . , 15 successes are 0.463, 0.366, 0.135, . . . , 0+, respectively.

EXAMPLE

In the preceding example, we used the binomial probability formula to find the probability of 3 successes, given that $n = 15$, $x = 3$, $p = 0.05$, and $q = 0.95$. Use the portion of Table A-1 shown in the margin on the previous page to find

a. The probability of exactly 3 successes

b. The probability of *at least* 3 successes

SOLUTION

a. The display from Table A-1 shows that when $n = 15$ and $p = 0.05$, $P(3) = 0.031$, which is the same value computed with the binomial probability formula in the preceding example.

b. $P(\text{at least } 3) = P(3 \text{ or } 4 \text{ or } 5 \text{ or} \ldots \text{ or } 15)$

$= P(3) + P(4) + P(5) + \cdots + P(15)$

$= 0.031 + 0.005 + 0.001 + \cdots + 0$

$= 0.037$

In part (b) of the preceding solution, if we wanted to find $P(\text{at least } 3)$ by using the binomial probability formula, we would need to apply that formula 13 times to compute 13 different probabilities, which would then be added. (A shortcut would be to calculate $P(0)$, $P(1)$, and $P(2)$ and subtract their sum from 1, but even this shortcut would take much longer than using Table A-1.) Given this choice between the formula and the table, it makes sense to use the table. Note, however, that Table A-1 includes only limited values of n as well as limited values of p, so the table doesn't always work. We must then find the probabilities by using the binomial probability formula or software.

METHOD 3: USE COMPUTER SOFTWARE   Many computer statistics packages and some calculators include an option for generating binomial probabilities. For the details of using Excel or STATDISK, see the Using Technology segment at the end of this section. Your instructor may prefer that you look instead at specific instructions for using Minitab, SPSS, or other such tools.

In general, the following is a good strategy for choosing the best method for finding binomial probabilities:

1. Use computer software or a calculator, if available.

2. If a computer cannot be used, use Table A-1 if possible.

3. If a computer cannot be used and the probabilities can't be found using Table A-1, use the binomial probability formula.

In Section 4–2 we presented Table 4-1 as an example of a probability distribution. The following example illustrates that the probabilities are binomial.

At the beginning of this chapter we noted that commercial vehicles accounted for 20% of all vehicles registered for the road in Canada and considered a situation in which such vehicles constituted 4 of 7 accident vehicles on a certain stretch of road. We assume that vehicle accidents are independent and random events, and also that commercial vehicles are as safe as any other type of vehicle. Find the probability that when there are 7 motor vehicles in accidents, 4 of them are commercial vehicles.

SOLUTION

This is a binomial experiment because of the following:

1. We have a fixed number of trials (7).

2. The trials are assumed to be independent.

3. There are two categories: Each vehicle either is a commercial vehicle or is not.

4. The probability of a selected vehicle being a commercial vehicle (considered a "success" in this example) is 0.20 (because 20% of all vehicles registered are commercial), and it remains a constant for each trial. (We are assuming that accidents are independent and random.)

We begin by identifying the values of n, p, q, and x. We have

$n = 7$	Number of trials (accident vehicles)
$p = 0.20$	Probability of success (an accident vehicle is a commercial vehicle)
$q = 0.80$	Probability of failure (an accident vehicle is not a commercial vehicle)
$x = 4$	Number of commercial vehicles among 7

Referring to Table A-1 with $n = 7$, $p = 0.20$, and $x = 4$, we find that $P(4) = 0.029$. That is, there is a 0.029 probability that among 7 accident vehicles, exactly 4 are commercial vehicles.

Asking the Right Question

The low probability (0.029) that we found in the preceding example might seem to suggest that it is unlikely that commercial vehicles are operating as safely as other vehicles (because they are overrepresented among the accident vehicles). But are we really addressing the correct issue? Is there another question that better addresses the real issue of whether commercial vehicles are unsafe? Here are some possibilities:

1. What is the probability that *at least* 4 out of 7 accident vehicles are commercial vehicles?

2. What is the probability that any single vehicle type (with 20% of total vehicle registrations) will constitute *exactly* 4 out of 7 accident vehicles?

3. What is the probability that *any* single vehicle type (with 20% of total vehicle registrations) will constitute at least 4 out of 7 accident vehicles?

STATISTICAL PLANET

How Likely Is an Asteroid Strike? NASA astronomer David Morrison says that there are about 2000 asteroids with orbits that cross Earth's orbit, yet we have found only 100 of them. It's therefore possible for an undetected asteroid to crash into our planet and cause a global catastrophe that could destroy most life. Morrison says that there is a 1/10,000 probability that within a human lifetime, there will be an asteroid impact large enough to wipe out all agricultural crops for at least a year, with mass starvation following. Some astronomers recommend a 20-year program aimed at observing asteroids with the goal of early detection of those that are dangerous. Early detection could possibly allow us to alter an asteroid's course so that it isn't a fatal threat.

We eliminate the first question because the real issue concerns the likelihood of *any* vehicle type constituting 4 of 7 accident vehicles. To clarify this point, consider your local lottery in which *your* individual chance of winning is extremely small, but the probability of *somebody* winning is fairly high. When someone does win, we don't conclude that this individual had a better chance than anyone else. Similarly, an accident involving a commercial vehicle doesn't necessarily mean that this type of vehicle isn't as safe as others; like winning the lottery, the accident might be the result of random events that affect all vehicles the same way. We therefore need to find the probability of *any* type of vehicle showing up as 4 out of 7 randomly selected accident vehicles, assuming that each type of vehicle represents 20% of total vehicle registrations.

Next, we should eliminate the second question in our list because it refers to *exactly* 4 accident vehicles out of 7. Our real concern is not the likelihood of getting any specific number of accidents. Instead, the issue rests on the likelihood of getting an outcome *at least as extreme* as the one observed. This is a difficult concept, so let's try to clarify it with another example.

Suppose you were flipping a coin to determine whether it favours heads, and suppose 1000 tosses resulted in 501 heads. Intuition should suggest that this is not evidence that the coin favours heads, because it is quite easy to get 501 heads in 1000 tosses. Yet, the probability of getting exactly 501 heads in 1000 tosses is actually quite small: 0.0252. This low probability reflects the fact that with 1000 tosses, *any specific* number of heads will have a very low probability. However, the result of 501 heads among 1000 tosses is not *unusual* because the probability of getting *at least* 501 heads is high: 0.488.

Similarly, we should be asking for the probability of *at least* 4 accidents, not the probability of exactly 4. That is, we need to find the probability that any single vehicle type (with 20% of total vehicle registrations) will constitute at least 4 of 7 vehicles involved in accidents. If that probability is very low (such as 0.05 or less), it is reasonable to conclude that commercial vehicles are not being operated safely, but if it is high (such as greater than 0.05), it is reasonable to conclude that accidents involving commercial vehicles are a coincidence. First we find the probability of commercial vehicles constituting at least 4 out of 7 accident vehicles. We refer to Table A-1 (with $n = 7$, $p = 0.20$, $x = 4, 5, 6, 7$):

$$
\begin{aligned}
P(\text{commercial vehicles constituting at least 4 of 7 accident vehicles}) &= P(4) + P(5) + P(6) + P(7) \\
&= 0.029 + 0.004 + 0^+ + 0^+ \\
&= 0.033
\end{aligned}
$$

But suppose there are five different types of vehicles (denoted by A, B, C, D, E), each with 20% of the total motor vehicle registrations. The probability of *any one* of them constituting 4 of 7 vehicles involved in an accident is found by using the addition rule for mutually exclusive events. (The events are mutually exclusive, because only one type of vehicle could constitute 4 of 7 vehicles selected.)

$$P(A \text{ or } B \text{ or } C \text{ or } D \text{ or } E) = P(A) + P(B) + P(C) + P(D) + P(E)$$
$$= 0.033 + 0.033 + 0.033 + 0.033 + 0.033$$
$$= 0.165$$

This probability suggests that it is hardly rare (it happens over 16% of the time) to have some vehicle type or other constitute 4 of 7 randomly selected accident vehicles, assuming each of the vehicle-type alternatives represents 20% of vehicle registrations. We would not have expected *in advance* that those 4 of 7 would be commercial vehicles (just as I do not really expect to win the next lottery, even though *someone* might win), but it could easily be a coincidence that, yes, some vehicle type constituted 4 of the selected 7 accident vehicles—and it just happened, this time, to be commercial vehicles. (Part of our problem is working with such a small sample.) This analysis is not simple, but it illustrates an important principle: **We should take great care to ask the questions that correctly identify the real issue.**

Rationale for the Binomial Probability Formula

The binomial probability formula is the basis for all three methods presented in this section. Instead of blindly accepting and using that formula, let us see why it works.

In the preceding example we wanted the probability of getting 4 successes among the 7 trials, given a 0.20 probability of success in any one trial. It's correct to reason that for 4 successes among 7 trials, there must be 3 failures. A common error is to find the probability of 4 successes and 3 failures as follows:

$$\overbrace{(0.20 \cdot 0.20 \cdot 0.20 \cdot 0.20)}^{\text{4 successes}} \cdot \overbrace{(0.80 \cdot 0.80 \cdot 0.80)}^{\text{3 failures}} = 0.000819$$

This calculation is *wrong* because it assumes that the *first* 4 outcomes are successes and the *last* 3 are failures. However, the 4 successes and 3 failures can occur in many different sequences, not only the one given above. In fact, there are 35 different sequences of 4 successes and 3 failures, each with a probability of 0.000819, so the correct probability is $35 \cdot 0.000819 = 0.029$, as we have found. In general, the number of ways in which it is possible to arrange x successes and $n - x$ failures is shown in Formula 4-6:

Formula 4-6
$$\frac{n!}{(n - x)!x!}$$
Number of outcomes with exactly x successes among n trials

The expression given in Formula 4-6 is from Section 3–6. (Coverage of Section 3–6 is not required for this chapter.) Combining this counting device (Formula 4-6) with the direct application of the multiplication rule for independent events results in the binomial probability formula:

The number of outcomes with exactly x successes among n trials

The probability of x successes among n trials for any one particular order

$$P(x) = \overbrace{\frac{n!}{(n-x)!x!}} \cdot \overbrace{p^x \cdot q^{n-x}}$$

To keep this section in perspective, remember that the binomial probability formula is only one of many probability formulas that can be used for different situations. It is often used in applications such as quality control, voter analysis, medical research, military intelligence, and advertising. A variation on the binomial distribution is *the negative binomial distribution*. In the binomial distribution, we know the number of trials and compute the probability of a number of successes. In the negative binomial distribution, we know the number of successes and compute the probability of the number of trials needed to attain this number of successes. A particular case of the negative binomial is the *geometric distribution* in which the number of successes is 1 and we compute the probability of the number of trials needed until the first success is attained. See Exercises 34 and 35 for examples of these two distributions.

The Probability of "At Least One"

The rationale for the binomial probability formula can greatly simplify certain types of problems, such as those in which we want to find the probability that among several trials, *at least 1* will result in some specified outcome. In such cases, these two issues of language should be clearly understood:

- "At least 1" is equivalent to "1 or more."
- The complement of getting at least 1 item of a particular type is that you get no items of that type.

EXAMPLE

A survey of households in Canada found that 22.4% of all households with annual incomes over $50,000 subscribed to *The Globe and Mail* (based on data from a NADbank study). Assume that 5 households with incomes over $50,000 are randomly selected. Find the probability that at least 1 of 5 Canadian households with incomes over $50,000 subscribe to *The Globe and Mail*. Assume that newspaper subscriptions are independent and that 77.6% of these households do not subscribe to the newspaper.

SOLUTION

$$P(x \geq 1) = 1 - P(0)$$
$$P(0) = (_5C_0)(0.224^0)(0.776^5) = 0.776^5 = 0.281$$
$$P(x \geq 1) = 1 - 0.281 = 0.719$$

Interpretation

For any five, randomly selected households among Canadian households with an income over $50,000, there is a 71.9% probability that at least one household subscribes to *The Globe and Mail*.

The principle used in the above example can be summarized as follows:

To find the probability of *at least one* of something, calculate the probability of *none*, then subtract that result from 1.

EXCEL: In Excel it is possible—but usually not necessary—to reproduce a full binomial probability distribution table for all values of x, for a given n and p.

EXCEL (prior to 2007): Simply click on *fx* on the toolbar, and select the function category **Statistical** and then the function name **BINOMDIST**. In the dialog box, enter the number of successes x, as well as the values of n (number of trials) and p (the probability of success each trial), plus 0 for the binomial distribution (instead of 1 for the cumulative binomial distribution). The probability value for *exactly* x successes will be displayed. If, instead, you enter a 1 for the cumulative binomial distribution, the value returned is the probability of *up to or including* x successes. (*Hint:* For the probability of "at least x successes," interpret as: $1 -$ [the probability of "*up to or including* $(x - 1)$ successes"].)

EXCEL 2007: Click on **Menus** on the main menu, then the *fx* button, then more functions and select the function category **Statistical** and then the function name **BINOMDIST**. Proceed as above.

MINITAB 15: From the **Calc** menu, choose **Probability Distributions** and then **Binomial**. Suppose $n = 10$ and $p = 0.24$. If you want to calculate $P(x = 2)$, click the **Probability** radio button and enter 10 for the number of trials and 0.24 for the probability of success.

Next, click the **Input constant** button and put 2 in the box. If you want to calculate $P(x \leq 2)$, click the **Cumulative probability** radio button instead, with the rest of the procedure being the same. If you want to calculate $P(x \geq 2)$, keep in mind that $P(x \geq 2) = 1 - P(x \leq 1)$. Click the **Cumulative probability** radio button, and now make the **Input constant** 1.

STATDISK: Select **Analysis** from the main menu, **Probability Distributions**, and then select the **Binomial Probabilities** option. Enter the requested values for n and p, and the entire probability distribution will be displayed. Other columns represent cumulative probabilities obtained by adding the values $P(x)$ as you go down, or up, the column.

EXCEL DISPLAY
Binomial Probabilities

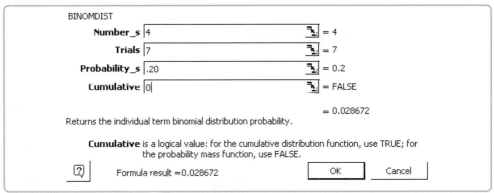

MINITAB DISPLAY
Binomial Probabilities

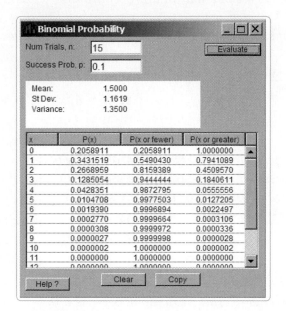

4-3 Exercises A: Basic Skills and Concepts

In Exercises 1–8, determine whether the given experiments are binomial. For those that are not binomial, identify at least one requirement that is not satisfied.

1. Rolling a die 50 times

2. Tossing an unbiased coin 200 times

3. Tossing a biased coin 200 times

4. Surveying 1000 Canadian consumers by asking each one if the brand name Tilley Endurables is recognized

5. Spinning a roulette wheel 500 times

6. Surveying 1067 people by asking each one if he or she voted in the last election

7. Sampling (without replacement) a randomly selected group of 12 different tires from a population of 30 tires that includes 5 that are defective

8. Surveying 2000 television viewers to determine whether they can recall a particular product name after watching a commercial

In Exercises 9–12, assume that in a procedure that yields a binomial distribution, a trial is repeated n times. Find the probability of x successes given the probability p of success on a given trial. (Use the given values of n, x, and p and Table A-1.)

9. $n = 3, x = 2, p = 0.9$ **10.** $n = 2, x = 0, p = 0.6$

11. $n = 8, x = 7, p = 0.99$ **12.** $n = 6, x = 1, p = 0.05$

In Exercises 13–16, assume that in a procedure that yields a binomial distribution, a trial is repeated n times. Find the probability of x successes given the probability p of success on a single trial. Use the given values of n, x, and p and the binomial probability formula or statistical software.

13. $n = 3, x = 2, p = 1/4$ **14.** $n = 6, x = 2, p = 1/3$

15. $n = 10, x = 4, p = 0.35$ **16.** $n = 8, x = 6, p = 0.85$

x	$P(x)$
0	0.021554
1	0.106133
2	0.228640
3	0.281460
4	0.216551
5	0.106631
6	0.032816
7	0.005771
8	0.000444

In Exercises 17–20, refer to the table in the margin for $n = 8$ and $p = 0.381$. When a car buyer is selected at random, there is a 0.381 probability that he or she bought a used car (based on data from a CAA members' survey). In each case, assume that 8 buyers are randomly selected and find the indicated probability.

17. Find the probability that 6 car buyers bought a used car.

18. Find the probability that at least 5 car buyers bought used cars.

19. Find the probability that fewer than 7 car buyers bought used cars.

20. Find the probability that more than 4 car buyers bought used cars.

In Exercises 21–32, find the probability requested.

21. Assume that male and female births are equally likely and that the birth of any child does not affect the probability of the gender of any other children. Find the probability of
   a. Exactly 4 girls in 10 births.
   b. At least 4 girls in 10 births.
   c. Exactly 8 girls in 20 births.

22. According to a market-share study, 13% of televisions in use are tuned to *Hockey Night in Canada* on Saturday night. Assume that an advertiser wants to verify that 13% market share value by conducting its own survey, and a pilot survey begins with 20 households having TV sets in use at the time of a *Hockey Night in Canada* broadcast.
   a. Find the probability that none of the households are tuned to *Hockey Night in Canada*.
   b. Find the probability that at least one of the households is tuned to *Hockey Night in Canada*.

c. Find the probability that at most one household is tuned to *Hockey Night in Canada*.

d. Suppose at most one household is tuned to *Hockey Night in Canada*. Does it appear that the claimed 13% share value is mistaken? Why or why not?

23. Mars, Inc. claims that 20% of its plain M&M candies are red. Find the probability that when 15 plain M&M candies are randomly selected, exactly 20% (or 3 candies) are red.

24. In a market study for Zellers, a researcher found that 76% of customers are repeat customers. If 12 customers are selected at random, find the probability that at least 11 of them are repeat customers. Suppose at least 11 of the 12 are repeat customers; does it appear that the 76% value found in the study is too low? Why or why not?

25. A statistics quiz consists of 10 multiple-choice questions, each with 5 possible answers. For someone who makes random guesses for all of the answers, find the probability of passing if the minimum passing grade is 60%. Is the probability high enough to make it worth the risk of passing by random guesses instead of by studying?

26. A regional airline has a policy of booking as many as 15 persons on an airplane that can seat only 14. (Past studies have revealed that only 85% of the booked passengers actually arrive for the flight.) Find the probability that if the airline books 15 persons, not enough seats will be available.

27. In a survey, the Canadian Automobile Association (CAA) found that 6.1% of its members bought their cars at a used-car lot. If 15 CAA members are selected at random, what is the probability that 4 of them bought their cars at a used-car lot?

28. Deloitte Touche Tohamatsu surveyed British Columbia's top CEOs. According to the survey, 82% of the respondents like to have a computer on their desk. Suppose that a conference is being arranged in Montreal for 9 of these CEOs. Unfortunately there are only 7 computers available for their workstations. Find the probability that more computers will be needed for the CEOs. Is that probability high enough so that plans should be initiated to provide more computers?

29. A quality control manager at the Don Mills Electronics Company knows that his company has been making surge protectors with a 10% rate of defective units. He has instituted several measures designed to lower that defect rate. In a test of 20 randomly selected surge protectors, only one is found to be defective. If the 10% defect rate hasn't changed, find the probability that among 20 units, 1 or none is defective. Based on the result, does it appear that the newly instituted measures are effective?

30. The Telektronic Company purchases large shipments of fluorescent bulbs and uses this *acceptance sampling* plan: Randomly select and test 24 bulbs, then accept the whole batch if there is only 1 or none that doesn't work. If a particular shipment of thousands of bulbs actually has a 4% rate of defects, what is the probability that this whole shipment will be accepted?

31. A survey of college statistics students shows that 30% of the students who show up for their 8:00 a.m. classes are late. One statistics professor reports that among 15 of her students selected at random, only 3 were late for her 8:00 a.m. class. Find the probability that this number will be that low. That is, find the probability of getting 3 or fewer students late for class when 15 are selected at random. Based on the result, is the professor's result likely to occur by chance?

32. Nine percent of men and 0.25% of women cannot distinguish between the colours red and green. This is the type of colour blindness that causes problems with traffic signals. If 6 men are randomly selected for a study of traffic signal perceptions, find the probability that exactly 2 of them cannot distinguish between red and green.

33. A student experiences difficulties with malfunctioning alarm clocks. Instead of using one alarm clock, he decides to use three. What is the probability that at least one alarm clock works correctly if each individual alarm clock has a 98% chance of working correctly?

4-3 Exercises B: Beyond the Basics

34. If a procedure meets all the conditions of a binomial distribution except that the number of trials is not fixed, then the **geometric distribution** can be used. The probability of getting the first success on the xth trial is given by $P(x) = p(1 - p)^{x-1}$ where p is the probability of success on any one trial. Assume that the probability of a defective computer component is 0.2. Find the probability that the first defect is found in the seventh component tested.

35. A more general application of the problem in Question 34 is to use the **negative binomial distribution**, in which we find the probability of getting the kth success on the xth trial to be given by $P(x) = {}_{(x-1)}C_{(k-1)} \cdot p^k \cdot q^{x-k}$, $x \geq k$. Suppose the probability that a salesperson makes a sale is 0.3 and that the person's goal is to make two sales in a week. Find the probability of getting the second sale on the tenth sales call of that week.

36. The binomial distribution applies only to cases involving two types of outcomes, whereas the **multinomial distribution** involves more than two categories.

Suppose we have three types of mutually exclusive outcomes denoted by A, B, and C. Let $P(A) = p_1$, $P(B) = p_2$, and $P(C) = p_3$. In a **multinomial experiment** of n independent trials, the probability of x_1 outcomes of type A, x_2 outcomes of type B, and x_3 outcomes of type C is given by

$$\frac{n!}{(x_1!)(x_2!)(x_3!)} \cdot p_1^{x_1} \cdot p_2^{x_2} \cdot p_3^{x_3}$$

A genetics experiment involves 6 mutually exclusive genotypes identified as A, B, C, D, E, and F, and they are all equally likely. If 20 offspring are tested, find the probability of getting exactly 5 As, 4 Bs, 3 Cs, 2 Ds, 3 Es, and 3 Fs by extending the above expression so that it applies to 6 types of outcomes instead of only 3.

 # 4-4 Mean, Variance, and Standard Deviation for the Binomial Distribution

In Chapter 2 we explored actual collections of real data and identified these three important characteristics: (1) the measure of centre, (2) the measure of variation, and (3) the nature of the distribution. A key point of this chapter is that probability distributions describe what will *probably* happen instead of what actually did happen. In Section 4–2, we learned methods for analyzing probability distributions by finding the mean, the standard deviation, and a probability histogram. Because a binomial distribution is a special type of probability distribution, we could use Formulas 4-1, 4-3, and 4-4 (from Section 4–2) for finding the mean, variance, and standard deviation. Fortunately, those formulas can be greatly simplified for binomial distributions, as shown below.

For Any Probability Distribution:	**For Binomial Distributions:**
Formula 4-1 $\mu = \Sigma[x \cdot P(x)]$	Formula 4-7 $\mu = n \cdot p$
Formula 4-3 $\sigma^2 = [\Sigma x^2 \cdot P(x)] - \mu^2$	Formula 4-8 $\sigma^2 = n \cdot p \cdot q$
Formula 4-4 $\sigma = \sqrt{[\Sigma x^2 \cdot P(x)] - \mu^2}$	Formula 4-9 $\sigma = \sqrt{n \cdot p \cdot q}$

Formula 4-7 for the mean makes sense intuitively. If we were to analyze 100 births, we would expect to get about 50 girls, and $n \cdot p$ becomes $100 \cdot 1/2$, or 50. In general, if we consider p to be the proportion of successes, then the product $n \cdot p$ will give us the actual number of expected successes in n trials. The variance and standard deviation are not so easily justified, and we prefer to omit the complicated algebraic manipulations that lead to Formulas 4-8 and 4-9. Instead, the following example illustrates that for a binomial experiment, Formulas 4-7, 4-8, and 4-9 will produce the same results as Formulas 4-1, 4-3, and 4-4.

In Section 4–3 we confirmed that the chapter problem is really based on the binomial distribution, and Table 4-1 represents a binomial probability distribution for the numbers of commercial vehicles among 7 accident vehicles, randomly selected (assuming that commercial vehicles represent 20% of all registered vehicles on public roads, and that vehicles are "selected" randomly and independently for involvement in accidents). In Section 4-2, we used Formulas 4-1, 4-3, and 4-4 (see Table 4-2) to find these values for the mean, variance, and standard deviation: $\mu = 1.4$ commercial vehicles, $\sigma^2 = 1.1$ "commercial vehicles squared," and $\sigma = 1.1$ commercial vehicles. But given the values we found in Section 4–3 for this distribution ($n = 7$, $p = 0.20$, and $q = 0.80$), we can now use Formulas 4-7, 4-8, and 4-9 to find the mean, variance, and standard deviation. Verify that the results are the same as those obtained by using Formulas 4-1, 4-3, and 4-4.

SOLUTION

Using the values of $n = 7$, $p = 0.20$, and $q = 0.80$, Formulas 4-7, 4-8, and 4-9 provide these results:

$$\mu = n \cdot p = (7)(0.20) = 1.4$$

$$\sigma^2 = n \cdot p \cdot q = (7)(0.20)(0.80) = 1.1 \quad \text{(rounded)}$$

$$\sigma = \sqrt{n \cdot p \cdot q}$$

$$= \sqrt{(7)(0.20)(0.80)} = \sqrt{1.12} = 1.1 \quad \text{(rounded)}$$

These results illustrate that when we have a binomial experiment, Formulas 4-7, 4-8, and 4-9 will produce the same results as Formulas 4-1, 4-3, and 4-4. (There may be small discrepancies in the unrounded values, due to rounding errors in Table 4-2.)

Application of the Empirical Rule for Symmetric Data

In Section 2–5, we introduced the empirical rule for symmetric data in bell-shaped distributions. One of the characteristics of the binomial distribution is that as the number of trials increases, the distribution becomes bell-shaped. Since approximately 99.7% of the distribution lies within 3 standard deviations of the mean, this provides a way to project the minimum and maximum values for the majority of the distribution.

EXAMPLE

The probability that a person orders a salad at a particular fast-food restaurant is 0.18. For a sample of 1000 people, what are the fewest and most number of people who can be expected to order a salad 99.7% of the time, rounding to the nearest whole number?

SOLUTION

Since $n = 1000$ and $p = 0.18$, $\mu = (1000)(0.18) = 180$ and $\sigma = \sqrt{(180)(0.82)} = 12.149$. Then the fewest number would be $\mu - 3\sigma = 180 - 3(12.149) = 144$ and the most would be $\mu + 3\sigma = 180 + 3(12.149) = 216$ approximately 99.7% of the time.

EXAMPLE

Some couples prefer to have baby girls because the mothers are carriers of an X-chromosome-linked recessive disorder that will be inherited by 50% of their sons but none of their daughters. The Ericsson method of gender selection supposedly has a 75% success rate. Suppose 100 couples use the Ericsson method, with the result that among 100 babies, there are 75 girls.

a. Assuming that the Ericsson method has no effect and assuming that boys and girls are equally likely, find the mean and standard deviation for the number of girls in groups of 100 babies. (The assumption that girls and boys are equally likely is not precisely correct, but it will give us very good results.)

b. Interpret the values from part (a) to determine whether the result of 75 girls among 100 babies supports a claim that the Ericsson method is effective.

SOLUTION

a. Let x represent the random variable for the number of girls in 100 births. Assuming that the Ericsson method has no effect and that girls and boys are equally likely, we have $n = 100$, $p = 0.5$, and $q = 0.5$. We can find the mean and standard deviation by using Formulas 4-7 and 4-9 as follows:

$$\mu = n \cdot p = (100)(0.5) = 50$$
$$\sigma = \sqrt{n \cdot p \cdot q} = \sqrt{(100)(0.5)(0.5)} = 5$$

For groups of 100 couples who each have a baby, the mean number of girls is 50 and the standard deviation is 5.

b. We must now interpret the results to determine whether 75 girls among 100 babies is a result that could easily occur by chance, or whether that result is so unlikely that the Ericsson method of gender selection seems to be effective. We will use the range rule of thumb and the empirical rule, both from Section 2-5.

According to the range rule of thumb, rough estimates of the minimum and maximum scores are as follows:

$$\text{minimum "usual" value} \approx (\text{mean}) - 2 \times (\text{standard deviation})$$
$$= 50 - 2(5) = 40$$
$$\text{maximum "usual" value} \approx (\text{mean}) + 2 \times (\text{standard deviation})$$
$$= 50 + 2(5) = 60$$

Interpretation

The range rule of thumb indicates that typical scores are probably between 40 and 60, so 75 girls seems to be a result that is not very likely to occur by chance and may possibly indicate that the Ericsson method is effective.

It is helpful to develop the technical skill to calculate means and standard deviations, but it is especially important to develop an ability to interpret the significance of values of means and standard deviations, as shown in part (b) of the preceding example.

4-4 Exercises A: Basic Skills and Concepts

In Exercises 1–4, find the mean μ, variance σ^2, and standard deviation σ for the given values of n and p. Assume the binomial conditions are satisfied in each case.

1. $n = 64$, $p = 0.5$
2. $n = 150$, $p = 0.4$
3. $n = 1068$, $p = 1/4$
4. $n = 2001$, $p = 0.221$

In Exercises 5–8, find the indicated values.

5. Several students are unprepared for a true/false test with 25 questions, and all of their answers are guesses. Find the mean, variance, and standard deviation for the number of correct answers that would be expected for such students. Would it be unusual for one of these students to get a good mark by guessing at least 20 correct answers? Why or why not?

6. On a multiple-choice test with 50 questions, each question has possible answers of a, b, c, and d, one of which is correct. For students who guess at all answers, find the mean, variance, and standard deviation for the number of correct answers that would be expected. Would it be unusual for one of these students to get a good mark by guessing at least 40 correct answers? Why or why not?

7. The probability of a 7 in roulette is 1/38. In an experiment, the wheel is spun 500 times. If this experiment is repeated many times, find the mean and standard deviation for the number of 7s. Would it be unusual not to win once in 500 trials? Why or why not?

8. The probability of winning Lotto 6/49 is 1/13,983,816. If someone plays 5200 times over 50 years, find the mean and standard deviation for the number of wins. (Express your answer with three significant digits.)

In Exercises 9–16, consider as unusual any result that differs from the mean by more than two standard deviations. That is, unusual values are either less than $\mu - 2\sigma$ or greater than $\mu + 2\sigma$.

9. When surveyed for brand recognition, 95% of consumers recognize Coke (based on data from Total Research Corporation). A new survey of 1200 randomly selected consumers is to be conducted. For such a group of 1200,
   a. Find the mean and standard deviation for the number who recognize the Coke brand name.
   b. Is it unusual to get 1170 consumers who recognize the Coke brand name?

10. A report to Health Canada indicated that the rate of lung cancer development among males is 9.1%. In one region, an intensive education program is used in an attempt to lower that rate. After running the program, a long-term follow-up study of 200 males is conducted.

   a. Assuming the program has no effect, find the mean and standard deviation for the number of lung cancer cases in groups of 200 males.

   b. Among the 200 males in the long-term follow-up study, 7% (or 14 people) test positive for lung cancer. If the program has had no effect, is that rate unusually low? Does this result suggest that the program is effective?

11. *Maclean's* conducted a marketing solutions poll of mutual funds and fund owners. One question asked fund owners what action they took after the October 1997 market drop. Seventeen percent of respondents said they bought more funds.

   a. Find the mean and standard deviation of the number of respondents who bought more funds, if groups of 900 fund owners were polled.

   b. In one of these polls of 900 fund owners, 140 respondents bought more mutual funds. Is this unusual?

12. According to the same *Maclean's* poll as in Exercise 11, fund owners were asked what they believe will happen to the stock market over the next year. In answer to the question, 46% stated that they believe the market will rise. Assume that this poll was conducted among 900 fund owners.

   a. For such groups of 900 fund owners, find the mean and standard deviation of the number of optimistic fund holders.

   b. In a group of 900 fund owners, 450 show optimism about the future of the stock market. Is this result unusual?

13. According to the CBC, 2.539 million Canadians watched the 1997 Grey Cup game. This figure represents a market share of 29%: that is, 29% of televisions were tuned to the game. Assume that this game is being broadcast and 4000 televisions are randomly selected.

   a. For such groups of 4000, find the mean and standard deviation for the number of televisions tuned to that game.

   b. Is it unusual to find that 1372 of the 4000 televisions are tuned to the game? What is the likely cause of a rate that is considerably higher than 29%?

14. A Calgary observatory records daily the mean counting rates for cosmic rays for that day. If a daily mean rate below 3200 is considered low, then "low" values occur about 3.6% of the time (in no particular sequence). Suppose a sample of 90 mean daily counting rates is randomly selected.

   a. Find the mean and standard deviation for the number of "low" daily mean rates in such groups of 90 randomly selected rates.

   b. Suppose that in a 90-day period, 9 "low" values for mean daily counting rates were observed. Is this result unusual?

15. Statistics Canada reports that 27.3% of all deaths are attributable to heart disease.
    a. Find the mean and standard deviation for the number of such deaths that will occur in a typical region with 5000 deaths.
    b. In one region, 5000 death certificates are examined, and it is found that 1500 deaths were attributable to heart disease. Is there cause for concern? Why or why not?

16. One test of extrasensory perception involves the determination of a shape. Fifty blindfolded subjects are asked to identify the one shape selected from the possibilities of a square, circle, triangle, star, heart, and profile of former prime minister Paul Martin.
    a. Assuming that all 50 subjects make random guesses, find the mean and standard deviation for the number of correct responses in such groups of 50.
    b. If 12 of 50 responses are correct, is this result within the scope of results likely to occur by chance? What would you conclude if 12 of 50 responses are correct?

 ## Exercises B: Beyond the Basics

17. The Port Arthur Computer Supply Company knows that 16% of its computers will require warranty repairs within one month of shipment. In a typical month, 279 computers are shipped.
    a. If x is the random variable representing the number of computers requiring warranty repairs among the 279 sold in one month, find its mean and standard deviation.
    b. For a typical month in which 279 computers are sold, what would be an unusually low figure for the number of computers requiring warranty repair within one month? What would be an unusually high figure? (These values are helpful in determining the number of service technicians that are required.)

18. a. If a company makes a product with an 80% yield (meaning that 80% are good), what is the minimum number of items that must be produced to be at least 99% sure that the company produces at least 5 good items?
    b. If the company produces batches of items, each with the minimum number determined in part (a), find the mean and standard deviation for the number of good items in such batches.

4-5 The Poisson Distribution

If you've ever waited in a line at an amusement park, it's likely that your situation could have been analyzed using the Poisson distribution, which is a probability distribution often used as a mathematical model describing arrivals of people in a line. Other applications include the study of vehicle crashes, shoppers arriving at a checkout counter, cars arriving at a gas station, and computer users going onto the Internet. The Poisson distribution is defined as follows.

DEFINITION

> The **Poisson distribution** is a discrete probability distribution that applies to occurrences of some event *over a specified interval*. The random variable x is the number of occurrences of the event in an interval. The interval can be time, distance, area, volume, or some similar unit. The probability of the event occurring x times over an interval is given by Formula 4-10.

Formula 4-10 $P(x) = \dfrac{\mu^x \cdot e^{-\mu}}{x!}$ where $e \approx 2.71828$. (It should be noted that e^x is found on most calculators.)

The Poisson distribution has the following requirements:

- The random variable x is the number of occurrences of an event *over some interval*.
- The occurrences must be *random*.
- The occurrences must be *independent* of each other.
- The occurrences must be *uniformly distributed* over the interval being used.

The Poisson distribution has these parameters:

- The mean is μ.
- The standard deviation is $\sigma = \sqrt{\mu}$.

The Poisson distribution is derived from the binomial distribution. The assumption in the derivation is that n is very large and p is very small. (For those familiar with calculus, starting with the formula for the binomial distribution, we take the limit as n tends to infinity and p tends to zero.) The Poisson distribution differs from the binomial distribution in these important ways:

1. The binomial distribution is affected by the sample size n and the probability p, whereas the Poisson distribution is affected only by the mean μ.

2. In a binomial distribution, the possible values of the random variable x are $0, 1, \ldots, n$, but a Poisson distribution has possible x values of $0, 1, 2, \ldots$ with no upper limit.

There are two scenarios in which we use the Poisson distribution:

SCENARIO 1: n is large and p is small. The first step would be to calculate $\mu = np$, and then we would calculate the probabilities.

SCENARIO 2: n is unknown (and assumed to be large). The first step would be to adjust μ for the question, and then we would calculate the probabilities.

EXAMPLE

The infection rate for a certain rare disease is 1 per 500,000.

a. For a city of 1,200,000 that is exposed to the disease, what is the probability that no more than 1 person contracts the disease?

b. For a town of 4,000, what is the probability that at least 1 person contracts the disease?

c. For a city of 4,000,000, what is the probability that at least 6 but no more than 8 people contract the disease?

SOLUTION

In each of these problems, n and p are given.

a. $\mu = 1{,}200{,}000 \cdot \dfrac{1}{500{,}000} = \dfrac{1{,}200{,}000}{500{,}000} = 2.4$

$P(x \leq 1) = P(0) + P(1)$

$P(0) = \dfrac{e^{-2.4}(2.4^0)}{0!} = e^{-2.4} = 0.0907$

$P(1) = \dfrac{e^{-2.4}(2.4^1)}{1!} = e^{-2.4}(2.4) = 0.2177$

$P(x \leq 1) = 0.0907 + 0.2177 = 0.3084$

There is a 30.84% probability that no more than 1 person in this city will be infected. This means that there is a $100\% - 30.84\% = 69.16\%$ probability that at least 2 people in this city will be infected.

b. $\mu = 4000 \cdot \dfrac{1}{500{,}000} = \dfrac{4{,}000}{500{,}000} = 0.008$

$P(x \geq 1) = 1 - P(0)$

$P(0) = \dfrac{e^{-0.008}(0.008^0)}{0!} = e^{-0.008} = 0.992$

$P(x \geq 1) = 1 - 0.992 = 0.008$

There is only a 0.8% probability that at least 1 person in this town will be infected.

c. $\mu = 4{,}000{,}000 \cdot \dfrac{1}{500{,}000} = \dfrac{4{,}000{,}000}{500{,}000} = 8$

$P(6 \le x \le 8) = P(6) + P(7) + P(8)$

$P(6) = \dfrac{e^{-8}(8^6)}{6!} = 0.1221$

$P(7) = \dfrac{e^{-8}(8^7)}{7!} = 0.1396$

$P(8) = \dfrac{e^{-8}(8^8)}{8!} = 0.1396$

$P(6 \le x \le 8) = 0.1221 + 0.1396 + 0.1396 = 0.4013$

There is a 40.13% probability that at least 6 but no more than 8 people in this city will be infected.

EXAMPLE

An intersection averages 1 accident per 2 weeks.

a. In a 1-week period, what is the probability that the intersection has no more than 1 accident?

b. In a 3-week period, what is the probability that the intersection has at least 1 but no more than 3 accidents?

SOLUTION

Note that n is not provided in this example. The reason is that the intersection could have hundreds or even thousands of vehicles go through it in any given period of time. So, we calculate μ for the time frame in each question.

a. If the intersection averages 1 accident per 2 weeks, this means that it averages 0.5 accidents per week. Thus, $\mu = 0.5$.

$$P(x \le 1) = P(0) + P(1)$$

$$P(0) = \dfrac{e^{-0.5}(0.5^0)}{0!} = e^{-0.5} = 0.6065$$

$$P(1) = \dfrac{e^{-0.5}(0.5^1)}{1!} = 0.3033$$

$$P(x \le 1) = 0.6065 + 0.3033 = 0.9098$$

There is a 90.98% probability that this intersection has no more than 1 accident in a 1-week period.

b. If the intersection averages 0.5 accidents per week, this means it averages 1.5 accidents per 3 weeks.

$$P(1 \leq x \leq 3) = P(1) + P(2) + P(3)$$

$$P(1) = \frac{e^{-1.5}(1.5^1)}{1!} = 0.3347$$

$$P(2) = \frac{e^{-1.5}(1.5^2)}{2!} = 0.2510$$

$$P(3) = \frac{e^{-1.5}(1.5^3)}{3!} = 0.1255$$

$$P(1 \leq x \leq 3) = 0.3347 + 0.2510 + 0.1255 = 0.4405$$

There is a 44.05% probability that the intersection has at least 1 but no more than 3 accidents in a 3-week period.

EXAMPLE

When a bank has 6 customer service reps (CSRs) working, they can collectively serve 1 customer per minute on average.

a. If there are 6 CSRs working, what is the probability that they serve no more than 4 customers in a 5-minute period?

b. If there are 4 CSRs working, what is the probability that they serve at least 3 customers in a 4-minute period?

SOLUTION

This is a situation where we need to adjust μ for both the time period and the number of CSRs who are working.

a. Since there are 6 CSRs in the question, we need to only adjust μ for the time frame. If the CSRs serve 1 customer per minute on average, that means that they serve 5 customers per 5 minutes on average. Thus, $\mu = 5$.

$$P(x \leq 4) = P(0) + P(1) + P(2) + P(3) + P(4)$$

$$P(0) = \frac{e^{-5}(5^0)}{0!} = e^{-5} = 0.0067$$

$$P(1) = \frac{e^{-5}(5^1)}{1!} = 0.0337$$

$$P(2) = \frac{e^{-5}(5^2)}{2!} = 0.0842$$

$$P(3) = \frac{e^{-5}(5^3)}{3!} = 0.1404$$

$$P(4) = \frac{e^{-5}(5^4)}{4!} = 0.1755$$

$$P(x \le 4) = 0.0067 + 0.0337 + 0.0842 + 0.1404 + 0.1755 = 0.4405$$

There is a 44.05% probability that the 6 CSRs can serve no more than 4 customers in a 5-minute period. This means there is a $100\% - 44.05\% = 55.95\%$ probability that they can serve at least 5 customers in the same time frame.

b. If there were 6 CSRs, we would expect that they could serve 4 customers in a 4-minute period. However, since there are 4 CSRs, the number of customers they can expect to serve would be $4 \cdot \frac{4}{6} = \frac{16}{6} = \frac{8}{3}$.

$$P(x \ge 3) = 1 - P(x \le 2) = 1 - [P(0) + P(1) + P(2)]$$

$$P(0) = \frac{e^{(-8/3)}(8/3)^0}{0!} = e^{-8/3} = 0.0695$$

$$P(1) = \frac{e^{(-8/3)}(8/3)^1}{1!} = 0.1853$$

$$P(2) = \frac{e^{(-8/3)}(8/3)^2}{2!} = 0.2471$$

$$P(x \ge 3) = 1 - [0.0695 + 0.1853 + 0.2471] = 1 - 0.5019 = 0.4981$$

There is a 49.81% probability that 4 CSRs can serve at least 3 customers in a 4-minute period.

EXCEL (prior to 2007): Click on *fx* on the toolbar, then select the function category **Statistical** and then **Poisson,** then **OK.** In the dialog box, enter the value of interest for *x* (number of occurrences), as well as the value of μ, and enter 0 for the "Cumulative" option. The probability value for *exactly x* occurrences will be displayed. (Entering 1 for "Cumulative" will return the probability for values of *x up to and including x occurrences.*)

EXCEL 2007: Click on **Menus** on the main menu, then the *fx* button, then more functions and select the function category **Statistical** and then the function name **Poisson.** Proceed as above.

MINITAB 15: From the **Calc** menu, choose **Probability Distributions** and then **Poisson.** Suppose $\mu = 5$. If you want to calculate $P(x = 3)$, click the **Probability** radio button and enter 5 for the mean.

Next, click the **Input constant** button and put 3 in the box. When you click **OK**, the output window gives 0.1404. If you want to calculate $P(x \leq 3)$, click the **Cumulative probability** radio button instead, with the rest of the procedure being the same. The output window gives 0.2650. If you want to calculate $P(x \geq 3)$, keep in mind that $P(x \geq 3) = 1 - P(x \leq 2)$. Click the **Cumulative probability** radio button with an **Input constant** of 2. The output window gives 0.1247. Then, $P(x \geq 3) = 1 - 0.1247 = 0.8753$.

STATDISK: Select **Analysis** from the main menu, **Probability Distributions** and then select the **Poisson Probabilities** option. Enter the requested value for μ, and the entire probability distribution will be displayed. Other columns represent cumulative probabilities obtained by adding the values $P(x)$ as you go down, or up, the column.

 Exercises A: Basic Skills and Concepts

In Exercises 1–4, assume that the Poisson distribution has the indicated mean and use Formula 4-10 to find the probability of the value given for the random variable x.

1. $\mu = 2$, $x = 3$ 2. $\mu = 4$, $x = 1$

3. $\mu = 0.845$, $x = 2$ 4. $\mu = 0.250$, $x = 2$

In Exercises 5–12, use the Poisson distribution to find the indicated probabilities.

5. A new tornado-resistant communications tower is being planned for the area around Regina. The area averages 3.25 tornadoes per year. Find the probability that in a one-year period, the number of tornadoes is
   a. 0   b. 1   c. 4

6. According to data from a Calgary observatory, the mean daily counting rate for cosmic rays falls below 3000 about one day per month. Find the probability that in a randomly selected month, the number of days with a mean counting rate below 3000 is
   a. 0   b. 1   c. 2

7. The Townsend Manufacturing Company experiences a weekly average of 0.2 accidents requiring medical attention. Find the probability that in a randomly selected week, the number of accidents requiring medical attention is
   a. 0   b. 1   c. 2

8. A statistics professor finds that when she schedules an office hour for student help, an average of 2 students arrive. Find the probability that in a randomly selected office hour, the number of student arrivals is

    a. 0   b. 2   c. 5

9. Careful analysis of magnetic computer data tape shows that for each 150 m of tape, the average number of defects is 2.0. Find the probability of more than one defect in a randomly selected length of 150 m of tape.

10. For a recent year, there were 46 aircraft hijackings worldwide (based on data from the FAA). Using one day as the specified interval required for a Poisson distribution, we find the mean number of hijackings per day to be estimated as $\mu = 46/365 = 0.126$. If the United Nations is organizing a single international hijacking response team, there is a need to know about the chances of multiple hijackings in one day. Use $\mu = 0.126$ and find the probability that the number of hijackings (x) in one day is 0 or 1. Is a single response team sufficient?

11. A classic example of the Poisson distribution involves the number of deaths caused by horse kicks of men in the Prussian Army between 1875 and 1894. Data for 14 corps were combined for the 20-year period, and the 280 corps-years included a total of 196 deaths. After finding the mean number of deaths per corps-year, find the probability that a randomly selected corps-year has the following numbers of deaths:

    a. 0   b. 1   c. 2   d. 3   e. 4

The actual results consisted of these frequencies: 0 deaths (in 144 corps-years); 1 death (in 91 corps-years); 2 deaths (in 32 corps-years); 3 deaths (in 11 corps-years); 4 deaths (in 2 corps-years). Compare the actual results to those expected from the Poisson probabilities. Does the Poisson distribution serve as a good device for predicting the actual results?

12. In 1996, there were 572 homicide deaths in Canada (based on data from Statistics Canada). For a randomly selected day, find the probability that the number of homicide deaths is

    a. 0   b. 1   c. 2   d. 3   e. 4

4-5 Exercises B: Beyond the Basics

13. Assume that a binomial experiment has 15 trials, each with a 0.01 probability of success. Find the probability of getting exactly one success among the 15 trials by using (a) Table A-1 and (b) the Poisson distribution as an approximation to the binomial distribution. Note that the rule of thumb requiring that $n \geq 100$ and $np \leq 10$ suggests that the Poisson distribution might not be a good approximation to the binomial distribution.

14. The following is a binomial experiment, but the large number of trials involved creates major problems with many calculators. Overcome that obstacle by approximating the binomial distribution by the Poisson distribution.

If you bet on the number 7 for one spin of a roulette wheel, there is a 1/38 probability of winning. Assume that bets are placed on the number 7 in each of 500 different spins.

a. Find the mean number of wins in such experiments.

b. Find the probability that 7 occurs exactly 13 times.

c. Compare the result to the probability of 0.111, which is found from the binomial probability formula.

The Hypergeometric Distribution

The **hypergeometric distribution** is similar to the binomial distribution in that the outcomes have one of two results. The main distinction between the two distributions is that the population size is known and used in the calculation of hypergeometric probabilities.

Given a population size N, if k items belong to group (or outcome) 1, this implies that $N - k$ items belong to group 2. If n items are selected without replacement, the probability of selecting x items from group 1 is:

Formula 4-11 $$P(x) = \frac{(_kC_x)(_{N-k}C_{n-x})}{_NC_n} \quad x = 0, 1, 2, \ldots, n$$

EXAMPLE

An office has 10 people consisting of 6 women and 4 men. If 3 people are randomly selected for a business trip, what is the distribution of the number of women attending the trip?

SOLUTION

Since $n = 3$, $x = 0, 1, 2,$ or 3.

$$P(0) = \frac{(_6C_0)(_4C_3)}{_{10}C_3} = \frac{(1)(4)}{120} = \frac{4}{120} = 0.0333$$

$$P(1) = \frac{(_6C_1)(_4C_2)}{_{10}C_3} = \frac{(6)(6)}{120} = \frac{36}{120} = 0.3$$

$$P(2) = \frac{(_6C_2)(_4C_1)}{_{10}C_3} = \frac{(15)(4)}{120} = \frac{60}{120} = 0.5$$

$$P(3) = \frac{(_6C_3)(_4C_0)}{_{10}C_3} = \frac{(20)(1)}{120} = \frac{20}{120} = 0.1667$$

Observe how $P(0)$ is calculated. From the 10 office workers, 3 are chosen overall. From the 6 women, none are chosen. This means that of the 4 men, 3 of them must be chosen. Similarly for $P(1)$: if 1 of the 6 women is chosen, this means 2 of the 4 men are chosen, and so on.

Binomial as an Approximation to Hypergeometric

Once N becomes large, we can use the binomial distribution as an approximation to the hypergeometric distribution in which n remains the number of trials and $p = k/N$.

In the previous example, $N = 10$ and $p = 6/10 = 0.6$. We repeat this example with $N = 1000$.

EXAMPLE

A company has 1000 employees of which 600 are women. Based on a random selection of three employees without replacement, compute the distribution of the number of women.

SOLUTION

We first use the hypergeometric distribution:

$$P(0) = \frac{(_{600}C_0)(_{400}C_3)}{_{1000}C_3} = \frac{(1)(10,586,800)}{166,167,000} = \frac{10,586,800}{166,167,000} = 0.0637$$

$$P(1) = \frac{(_{600}C_1)(_{400}C_2)}{_{1000}C_3} = \frac{(600)(79,800)}{166,167,000} = \frac{47,880,000}{166,167,000} = 0.2881$$

$$P(2) = \frac{(_{600}C_2)(_{400}C_1)}{_{1000}C_3} = \frac{(179,700)(400)}{166,167,000} = \frac{71,880,000}{166,167,000} = 0.4326$$

$$P(3) = \frac{(_{600}C_3)(_{400}C_0)}{_{1000}C_3} = \frac{(35,820,200)(1)}{166,167,000} = \frac{35,820,200}{166,167,000} = 0.2156$$

We now recalculate the probabilities using the binomial distribution with $n = 3$ and $p = 0.6$:

$$P(0) = (_3C_0)(0.6^0)(0.4^3) = 0.064$$

$$P(1) = (_3C_1)(0.6^1)(0.4^2) = 0.288$$

$$P(2) = (_3C_2)(0.6^2)(0.4^1) = 0.432$$

$$P(3) = (_3C_3)(0.6^3)(0.4^0) = 0.216$$

The binomial distribution serves as a reasonable approximation to the hypergeometric distribution. It has the advantage in that the calculations are easier to work with.

EXCEL (prior to 2007): Click on *fx* on the toolbar, and select the function category **Statistical** and then the function name **HYPGEOMDIST**. Suppose we revisit the example of 10 people in the office consisting of 6 women and 4 men and we want the probability that, if 3 are randomly chosen, 2 of the 3 are women. In the dialog box, enter Sample $s = 2$, Number sample = 3, Population $s = 6$, and Number pop = 10. Excel gives the probability 0.5.

EXCEL 2007: Click on **Menus** on the main menu, then the *fx* button, then more functions and select the function category **Statistical** and then the function name **HYPGEOMDIST**. Proceed as above.

This chapter presented a variety of different *discrete* probability distributions, including binomial (Sections 4–3 and 4–4), Poisson (Section 4–5), geometric (Exercise 33 in Section 4–3), negative binomial (Exercise 34 in Section 4–3), hypergeometric (this section), and multinomial (Exercise 35 in Section 4–3). In the following chapter we shift our attention to the extremely important *normal* probability distribution, which is *continuous* instead of discrete.

4-6 Exercises A: Basic Skills and Concepts

1. In Lotto 6/49, a bettor selects 6 numbers from 1 to 49 (without repetition), and a winning 6-number combination is later randomly selected. Find the probability of getting
   a. all 6 winning numbers
   b. exactly 5 of the winning numbers
   c. exactly 3 of the winning numbers
   d. no winning numbers

2. A quality control manager uses test equipment to detect defective computer modems. A sample of 3 different modems is to be randomly selected from a group consisting of 12 that are defective and 18 that have no defects. What is the probability that (a) all 3 selected modems are defective, and (b) at least 1 selected modem is defective?

3. A manager can identify employee theft by checking samples of employee shipments. Among 36 employees, 2 are stealing. If the manager checks on 4 different randomly selected employees, find the probability that neither of the thieves will be identified.

4. An approved jury list contains 20 women and 20 men. Find the probability of randomly selecting 12 of these people and getting an all-male jury. Under these circumstances, if the defendant is convicted by an all-male jury, is there strong evidence to suggest that the jury was not randomly selected?

4-6 Exercises B: Beyond the Basics

5. With one method of *acceptance sampling*, a sample of items is randomly selected without replacement and the entire batch is rejected if there is at least one defect. The Niko Electronics Company has just manufactured 5000 CDs, and 3% are defective. If 10 of the CDs are selected and tested, find the probability that the entire batch will be rejected, using:
   a. the hypergeometric distribution
   b. the binomial distribution

VOCABULARY LIST

binomial experiment   186
binomial probability
   distribution   186
binomial probability
   formula   188
continuous random
   variable   175

discrete random
   variable   175
expected value   181
geometric distribution   200
hypergeometric
   distribution   214
multinomial distribution   200

multinomial experiment   201
negative binomial
   distribution   200
Poisson distribution   207
probability distribution   176
probability histogram   177
random variable   175

REVIEW

The central concerns of this chapter were the random variable and the probability distribution. This chapter dealt exclusively with discrete probability distributions. (Chapter 5 will deal with continuous probability distributions.) The following key points were discussed:

- In an experiment yielding numerical results, the *random variable* has numerical values corresponding to different chance outcomes of an experiment.

- A *probability distribution* consists of all values of a random variable, along with their corresponding probabilities. Any probability distribution must satisfy two requirements: $\Sigma P(x) = 1$ and, for each value of x, $0 \le P(x) \le 1$.

- The important characteristics of a *probability distribution* can be investigated by computing its mean (Formula 4-1) and standard deviation (Formula 4-4) and by constructing a probability histogram.

- In a *binomial distribution*, probabilities can be found from Table A-1, they can be calculated with Formula 4-5 or a calculator, or they can be found with software, such as Excel or STATDISK.

- In a *binomial distribution*, the mean and standard deviation can be easily found by calculating the values of $\mu = n \cdot p$ and $\sigma = \sqrt{n \cdot p \cdot q}$.

- A *Poisson probability distribution* applies to occurrences of some event over a specified interval, and its probabilities can be computed with Formula 4-10.

- A *hypergeometric probability distribution* is similar to the binomial distribution in that the outcome can be placed into one of two categories; the distinguishing difference is that the population size N is known. Its probabilities can be computed with Formula 4-11.

- In the solutions to several problems and exercises, this chapter stressed the importance of interpreting results as being unusual outcomes or simply being a typical outcome that one might expect. One approach is to apply the range rule of thumb, where

$$\text{Maximum usual value} = \mu + 2\sigma$$

$$\text{Minimum usual value} = \mu - 2\sigma$$

By this standard, an "unusual" outcome falls outside these bounds. We can also determine whether outcomes in a binomial distribution are unusual by using probability values.

x successes among *n* trials is unusually high if $P(\text{at least } x)$ is very small.

x successes among *n* trials is unusually low if $P(\text{at most } x)$ is very small.

In a similar vein, for the binomial distribution with *n* being large, the empirical rule for symmetric data states that approximately 99.7% of the distribution lies between $\mu - 3\sigma$ and $\mu + 3\sigma$. This allows us to find the lower and upper limits for the vast majority of the distribution.

REVIEW EXERCISE

1. a. What is a random variable?
   b. What is a probability distribution?
   c. An insurance association's study of home smoke detector use involves homes randomly selected in groups of 4. The accompanying table lists values and probabilities for *x*, the number of homes (in groups of 4) that have smoke detectors installed (based on data from the National Fire Protection

Association). Does this table describe a probability distribution? Why or why not?

d. Assuming that the accompanying table does describe a probability distribution, find its mean.

e. Assuming that the accompanying table does describe a probability distribution, find its standard deviation.

x	$P(x)$
0	0.0004
1	0.0094
2	0.0870
3	0.3562
4	0.5470

2. Fifteen percent of sport/compact cars are dark green (based on data from DuPont Automotive). Assume that 50 sport/compact cars are randomly selected.

a. What is the expected number of dark green cars in such a group of 50?

b. In such groups of 50, what is the mean number of dark green cars?

c. In such groups of 50, what is the standard deviation for the number of dark green cars?

d. Is it unusual to get 15 dark green cars in such a group? Why or why not?

e. Find the probability that there are exactly 9 dark green cars in such a group of 50.

3. In a survey, the Canadian Automobile Association (CAA) found that 58% of its members paid cash for their vehicles. Ten CAA members are selected at random.

a. Find the probability that exactly half of the 10 members paid cash for their vehicles.

b. Find the probability that at least half of the 10 members paid cash for their vehicles.

c. If many different groups of 10 CAA members are selected at random, find the mean and standard deviation for the number (among 10) who paid cash for their vehicles.

4. Inability to get along with others is the reason cited in 17% of worker firings (based on data from Robert Half International, Inc.). Concerned about her company's working conditions, the personnel manager at the Drummondville Fabric Company plans to investigate the 5 employee firings that occurred over the past year. Assuming that the 17% rate applies, find the probability that among those 5 employees, the number fired because of an inability to get along with others is

a. 0   b. 4   c. 5   d. at least 3

(Once the actual reasons for the firings have been identified, such probabilities will be helpful in comparing Drummondville Fabric Company to other companies.)

5. Refer to the data given in Exercise 4. Let the random variable x represent the number of fired employees (among 5) who were let go because of an inability to get along with others.

a. Find the mean value of x.

b. Find the standard deviation of the random variable x.

c. Is it unusual to have 4 employees (among 5) fired because of an inability to get along with others? Why or why not?

6. The Western Canada Trucking Company operates a large fleet of trucks. Last year, there were 84 breakdowns.

a. Find the mean number of breakdowns per day.

b. Find the probability that for a randomly selected day, 2 trucks break down.

7. In setting up a manufacturing process for a new computer memory storage device, the initial configuration has a 16% yield: 16% of the devices are acceptable and 84% are defective. If 12 of the devices are made, what is the probability of getting at least 1 that is good? If it is very important to have at least 1 good unit for testing purposes, is the resulting probability adequate?

CUMULATIVE REVIEW EXERCISES

x	f
4	13
5	22
6	20
7	34

1. The Sports Associates Vending Company supplies refreshments at a baseball stadium and must plan for the possibility of a World Series contest. In the accompanying frequency table (based on past results), x represents the number of baseball games required to complete a World Series contest.

a. Construct the corresponding relative frequency table.

b. Does the result from part (a) describe a probability distribution? Why or why not?

c. Based on the past results, what is the probability that the next World Series contest will last at least 5 games?

d. If two different series included in the table are randomly selected, find the probability that they both lasted 7 games.

e. Find the mean number of games for the World Series contests included in the table.

f. Find the standard deviation for the number of games for the World Series contests included in the table.

g. What is the expected number of games for a World Series contest? If a vendor supplies hot dogs to both stadiums involved, and if each stadium averages 30,000 hot dogs sold per game, what is the expected number of hot dogs that will be required?

2. A casino cheat is caught trying to use a pair of loaded dice. At his court trial, physical evidence reveals that some of the black dots were drilled, filled with lead, then repainted to appear normal. In addition to the physical evidence, the dice are rolled in court with these results:

12 8 9 12 12 9 8 7 12 10
12 3 2 12 10 9 12 11 11 12

A probability expert testifies that when fair dice are rolled, the mean should be 7.0 and the standard deviation should be 2.4.

a. Find the mean and standard deviation of the sample values obtained in court.

b. Based on the outcomes obtained in court, what is the probability of rolling a 12? How does this result compare to the probability of 1/36 (or 0.0278) for fair dice?

c. If the probability of rolling a 12 with fair dice is 1/36, find the probability of getting at least one 12 when fair dice are rolled 20 times.

d. If you are the defence attorney, how would you refute the results obtained in court?

TECHNOLOGY PROJECT

An Air Borealis flight from Toronto to Vancouver has seats for 340 passengers. An average of 5% of people with reservations don't show up, so Air Borealis overbooks by accepting 350 reservations for the 340 seats. We can analyze this system by treating it as a binomial experiment with $n = 350$ and $p = 0.95$ (the probability that someone with a reservation does show up).

Find the probability that for a particular flight, there are more passengers than seats. That is, find the probability of at least 341 people showing up with reservations. Because of the value of n, Table A-1 cannot be used and the binomial probability formula would be extremely time consuming and painfully tedious to use. Statistical software would be the best tool.

(For instructions on using Excel or STATDISK for this task, see the Using Technology segment at the end of Section 4–3.)

FROM DATA TO DECISION

Is a Transatlantic Flight Safe with Two Engines?

Statistics is at its best when it is used to benefit humanity in some way. Companies use statistics to become more efficient, increase shareholders' profits, and lower prices. Regulatory agencies use statistics to ensure the safety of workers and clients. This exercise involves a situation in which cost effectiveness and passenger safety are both critical factors. With new aircraft designs and improved engine reliability, airline companies wanted to fly transatlantic routes with twin-engine jets, but the regulations required at least three engines for transatlantic flights. Lowering this requirement was, of course, of great interest to manufacturers of twin-engine jets (such as the Boeing 767). Also, the two-engine jets use about half the fuel of jets with three or four engines. Obviously, the key issue in approving the lowered requirement is the probability of a twin-engine jet making a safe transatlantic crossing. This probability should be compared to that of three- and four-engine jets. Such a study should involve a thorough understanding of the related probabilities.

A realistic estimate for the probability of an engine failing on a transatlantic flight is 1/14,000. Use this probability and the binomial probability formula to find the probabilities of 0, 1, 2, and 3 engine failures for a three-engine jet and the probabilities of 0, 1, and 2 engine failures for a two-engine jet. Because of the numbers involved, carry all results to as many decimal places as your calculator will allow. Summarize your results by entering the probabilities in the two given tables.

Two Engines		Three Engines	
x	$P(x)$	x	$P(x)$
0	?	0	?
1	?	1	?
2	?	2	?
		3	?

Use the results from the tables and assume that a flight will be completed if at least one engine works. Find the probability of a safe flight with a three-engine jet, and find the probability of a safe flight with a two-engine jet. Write a report that outlines the key issues, and include a recommendation. Support your recommendation with specific results.

COOPERATIVE GROUP ACTIVITIES

1. **In-class activity:** Divide into groups of three or four. Let the random variable x be the value of a coin randomly selected from those in possession of a statistics student. On the basis of the coins belonging to the group members, construct a table (similar to Table 4-1) listing the possible values of x along with their probabilities, then find the mean and standard deviation. What is a practical use of the results?

2. **Out-of-class activity:** Divide into groups of three or four. Conduct a survey that includes the following three questions. (Because the first question may be sensitive to some people, use a procedure that provides for anonymity.)
   - How much do you weigh?
   - Enter three random digits (0, 1, 2, 3, 4, 5, 6, 7, 8, 9). A digit may be chosen more than once.
   - Enter the last three digits of your social insurance number.

   First, compile a list consisting of the *last digit* of each weight. If the weights are accurate and precise, we would expect those *last* digits to have a uniform distribution, so each digit has a probability of 1/10. Assuming that each digit has a probability of 1/10, construct a table describing the probability distribution for the last digits. Next, construct a relative frequency table from the list of recorded *last digits*. Compare the probability distribution to the relative frequency table. (This procedure is often used as a test to determine whether the subjects were actually weighed, or whether they simply reported their weights.) Use a similar procedure to analyze the random digits that were selected. Do those digits seem to be random? Finally, use a similar procedure to analyze the digits in the social insurance numbers. Do they seem to be random?

INTERNET PROJECT

Probability Distributions and Simulation

Probability distributions are used to predict the outcomes of the events they model. For example, if we toss a fair coin, the distribution for the outcome is a probability of 1/2 for heads and 1/2 for tails. If we toss the coin ten consecutive times, we expect 5 heads and 5 tails. We might not get this exact result, but in the long run, over hundreds or thousands of tosses, we expect the split between heads and tails to be very close to "50–50". Go to either of the following websites

http://www.mathxl.com   or   http://www.mystatlab.com

and click on Internet Project, and then on Chapter 4, where you will find two explorations. In the first exploration you are asked to develop a probability distribution for a simple experiment, and use that distribution to predict the outcome of repeated trial runs of the experiment. In the second exploration, we will analyze a more complicated situation: the paths of rolling marbles as they move in pinball-like fashion through a set of obstacles. In each case, a dynamic visual simulation will allow you to compare the predicted results with a set of experimental outcomes.

5 Continuous Probability Distributions

5-1 Overview

Whereas Chapter 4 focused on discrete probability distributions, this chapter focuses on continuous distributions. The uniform distribution is introduced as a first example of describing probability as the area under a curve.

5-2 The Standard Normal Distribution

The standard normal distribution is defined as a normal probability distribution having a mean given by $\mu = 0$ and a standard deviation given by $\sigma = 1$. This section presents the basic methods for determining probabilities by using that distribution, as well as for determining standard scores corresponding to given probabilities.

5-3 Normal Distributions: Finding Probabilities

The z score (or standard score) is used to work with normal distributions in which the mean is not 0, the standard deviation is not 1, or both.

5-4 Normal Distributions: Finding Values

Work with nonstandard normal distributions is continued through methods for finding values that correspond to given probabilities.

5-5 The Central Limit Theorem

As the sample size increases, the sampling distribution of sample means is shown to approach a normal distribution with mean μ and standard deviation $\sigma/\sqrt{n}$, where n is the sample size and μ and σ represent the mean and standard deviation of the population, respectively. The ideas of this section form a foundation for the important concepts introduced in Chapters 6 and 7.

5-6 Normal Distribution as Approximation to Binomial Distribution

Use of the normal distribution to estimate probabilities in a binomial experiment is described and illustrated.

5-7 Exponential Distribution

In Chapter 4, the Poisson distribution was used to calculate the probability of a certain number of events over a specified interval such as time. The exponential distribution is the flip side of the Poisson distribution in that we examine the probability of a certain interval, such as time, between events.

CHAPTER PROBLEM

How big are they?

There has been much interest in the last few years, among sports writers and hockey enthusiasts, concerning the ever-increasing size of National Hockey League players. A lot of this interest has been generated by the growing number of injuries, particularly head injuries, that players suffer. In an effort to reduce the number of injuries and amount of lost playing time, the NHL and critics of the game have suggested many changes. One suggestion is to enlarge the rink to give players more room to skate and freewheel. However, enlarging the rink would present an economic setback for team owners, since seats near the ice would have to be removed, causing lost ticket revenue. Of course, there are other less costly ways to protect players, such as rule changes and equipment improvement.

How large are the players in today's NHL? Official rosters of the teams of the 1997-98 season reveal that the average height for the league is 72.9 in. and the average weight is 196.0 lb. The Philadelphia Flyers sit at the top of the height and weight list, with an average height of 74.6 in. and an average weight of 208.3 lb. The Flyers are, on average, 8 lb per player heavier than the second team on the list, the Los Angeles Kings. On the basis of average height per player, the Toronto Maple Leafs are the third smallest team in the NHL. The Leafs' average height is 72.4 in. and their average weight is 198.8 lb.

It is interesting to look back and see what an NHL team looked like in another era. If we check the roster of the 1944-45 Maple Leafs (a Stanley Cup–winning team), the average player's height was 70.2 in. and the average weight was 173.2 lb. The difference over half a century is obvious. Today's Leaf team is 2 in. taller and 25 lb heavier on average.

Anthropometric data, such as height and weight, are important in many aspects of sports, business, and research. The areas in which such data are used extensively include health care, safety research for crash-testing, furniture design, and architecture. This chapter will introduce methods of analyzing data such as heights and weights, and methods for calculating percentages such as the percentage of people within one standard deviation of the average height or weight of a population.

5-1 Overview

In Chapter 4 we introduced the concept of the *random variable* as a variable having a single numerical value (determined by chance) for each outcome of an experiment. We noted that a *probability distribution* gives the probability for each value of the random variable. Chapter 4 was concerned only with *discrete* random variables, such as those in binomial distributions, that have a finite number of possible values. The number of quarters produced by the Royal Canadian Mint each day is an example of a discrete random variables. There are also many different *continuous* probability distributions, such as the weights of the quarters produced at the mint. Distributions can be either discrete or continuous, and they can be described by their *shape*, such as a bell shape.

In Section 4–2 we identified two requirements for a discrete probability distribution: (1) $\sum P(x) = 1$, and (2) $0 \le P(x) \le 1$ for all values of x. Also in Section 4–2, we stated that the graph of a discrete probability distribution is called a *probability histogram*. The graph of a continuous probability distribution, such as Figure 5-1, is called a *density curve*, and it must satisfy two properties similar, but not identical, to the requirements for discrete probability distributions, as listed in the folllowing definition.

DEFINITION

> A **density curve** or **probability density function** is a graph of a continuous probability distribution. It must satisfy the following properties:
>
> 1. The total area under the curve must be 1.
> 2. Every point on the curve must have a vertical height that is 0 or greater.

Figure 5-1 is an example of a density curve, specifically a uniform distribution curve. By setting the height of the rectangle in Figure 5-1 to be 0.2, we force the enclosed area to be $5 \times 0.2 = 1$, as required. This property (area = 1) makes it very easy to solve probability problems, so the following statement is important:

Figure 5-1
Uniform Distribution
of Temperatures

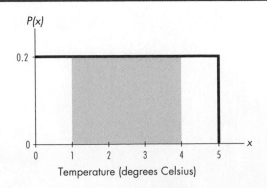

Temperature (degrees Celsius)

Because the total area under the density curve is equal to 1, there is a correspondence between area and probability.

The Uniform Distribution

> A continuous random variable has a **uniform distribution** if its values spread evenly over the full range of possibilities.

In order to compute uniform probabilities, we need a lower limit (called a) and an upper limit (called b). As shown in Figure 5-1, the area under the curve for the uniform distribution is a rectangle with a width of $(b - a)$ and a height of $\frac{1}{(b - a)}$ for a total area of $(b - a) \cdot \frac{1}{(b - a)} = \frac{(b - a)}{(b - a)} = 1$, since the area of a rectangle is width multiplied by height. This basic principle allows us to compute probabilities for the uniform distribution:

For a value k between a and b, $\quad P(X < k) = (k - a) \cdot \dfrac{1}{(b - a)} = \dfrac{(k - a)}{(b - a)}$

For a value k between a and b, $\quad P(X > k) = (b - k) \cdot \dfrac{1}{(b - a)} = \dfrac{(b - k)}{(b - a)}$

For values $k_1 < k_2$ between a and b, $\quad P(k_1 < X < k_2) = (k_2 - k_1) \cdot \dfrac{1}{(b - a)}$

$$= \dfrac{(k_2 - k_1)}{(b - a)}$$

EXAMPLE

One way to visualize the uniform distribution is to picture a straight stretch of highway. Suppose a race is starting at the 100-km mark of the highway and the finish line is the 350-km mark. Suppose that a racer chosen at random has cycled to the 160-km mark at the end of the first day of the race.

a. What percentage of the race has the racer covered so far?

b. What percentage of the race does the racer have left to cover?

c. Suppose the next day the racer departs from the 160-km mark and cycles to the 240-km mark. What percentage of the race did the racer cover that day?

SOLUTION

In this situation, $a = 100$ and $b = 350$. When we draw the rectangle, the width of the rectangle is $350 - 100 = 250$. In order for the rectangle to have an area of 1, the height of the rectangle must be $\frac{1}{250} = 0.004$.

a. Since we want the percentage of the race covered so far, we want:

$$P(X < 160) = \frac{(160 - 100)}{(350 - 100)} = \frac{60}{250} = 0.24$$

Thus, 24% of the trip has been covered so far. This is visualized in Figure 5-2.

Figure 5-2

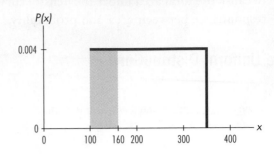

b. For the percentage of the race left to cover, we want:

$$P(X > 160) = \frac{(350 - 160)}{(350 - 100)} = \frac{190}{250} = 0.76$$

The racer has 76% of the race left to cover.

c. We want the percentage of the race covered between the 160 km and 240 km marks. This is:

$$P(160 < X < 240) = \frac{(240 - 160)}{(350 - 100)} = \frac{80}{250} = 0.32$$

So, 32% of the race is covered on that day.

Mean and Standard Deviation of the Uniform Distribution

For the uniform distribution, $\mu = \frac{(a + b)}{2}$ and $\sigma = \frac{(b - a)}{\sqrt{12}}$. The mean is intuitive; it is simply the halfway point between the two limits. The formula for the standard deviation is derived using calculus.

EXAMPLE

The amount of paint contained in a 4-L can is uniformly distributed between 3.98 L and 4.03 L. Suppose that for quality control purposes, the amount of paint should be within 1.5 standard deviations of the mean. What is the probability of achieving this goal?

SOLUTION

We have $\mu = \frac{(3.98 + 4.03)}{2} = 4.05$ and $\sigma = \frac{(4.03 - 3.98)}{\sqrt{12}} = 0.014434$. Then $\mu - 1.5\sigma = 3.983349$ and $\mu + 1.5\sigma = 4.026651$. So

$$P(3.983349 < X < 4.026651) = \frac{(4.026651 - 3.983349)}{(4.03 - 3.98)} = 0.866$$

It should be noted that this problem could also have been solved algebraically:

$$P(\mu - 1.5\sigma < X < \mu + 1.5\sigma) = \frac{[(\mu + 1.5\sigma) - (\mu - 1.5\sigma)]}{(4.03 - 3.98)}$$

$$= \frac{3\sigma}{0.05} = 60\sigma$$

Since $\sigma = 0.014434$, we would multiply it by 60 to obtain 0.866 as above.

5-1 Exercises A: Basic Skills and Concepts

In Exercises 1–5, suppose that the temperature readings for a certain gauge are uniformly distributed between 0°C and 5°C. Find the probability of a randomly selected temperature reading falling in the following range:

1. Greater than 2°C
2. Less than 3°C
3. Between 2°C and 4°C
4. Between 0.8°C and 4.7°C
5. Find the mean and standard deviation for the gauge readings.

In Exercises 6–8, suppose that the amount of paint that goes into a 4-L can is uniformly distributed between 3.85 L and 4.15 L.

6. What is the probability a can has less than 3.9 L?
7. What is the probability a can has more than 4.05 L?
8. Suppose the manufacturer wants the amount of paint in a can to be within 0.5 standard deviations of the mean. Based on the probability of this happening, are these realistic expectations?

5-1 Exercises B: Beyond the Basics

9. For a uniform distribution, show why 100% of the distribution lies within 2 standard deviations of the mean, regardless of the values for *a* and *b* with $a < b$.

5-2 The Standard Normal Distribution

This chapter focuses on normal distributions, which are extremely important because they occur so often in real applications. Heights of adult women, weights of adult men, and third-grade reading test scores are some examples of normally distributed populations.

DEFINITION

A continuous random variable has a **normal distribution** if that distribution has a graph that is symmetric and bell-shaped, as in Figure 5-3, and the distribution fits the equation given as Formula 5-1.

Formula 5-1

$$f(x) = \frac{e^{-\frac{1}{2}\left(\frac{x-\mu}{\sigma}\right)^2}}{\sigma\sqrt{2\pi}}$$

Do not be discouraged by the complexity of Formula 5-1, because it is not really necessary for us actually to use it. What it shows is that any particular normal distribution is determined by two parameters: the mean μ and standard deviation σ. Once specific values are selected for μ and σ, we can graph Formula 5-1 as we would graph any equation relating x and y; the result is a probability distribution with a bell shape. We will see that this normal distribution has many real applications, and we will use it often throughout the remaining chapters.

The Standard Normal Distribution

The density curve of a normal distribution has the more complicated bell shape shown in Figure 5-3, so it's more difficult to find areas; but the basic principle is the same: There is a correspondence between area and probability.

There are many different normal distributions, with each one depending on two parameters: the population mean μ and the population standard deviation σ. Figure 5-4 shows density curves for heights of female and male college students. Because males have a larger mean height, the density curve for males is farther to the right. Because males have a slightly larger standard deviation, the density curve for males is slightly wider. Figure 5-4 shows two different possible normal distributions. There are infinite possibilities, but one is of special interest.

Figure 5-3
The Normal Distribution

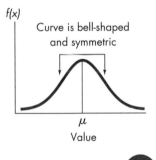

Figure 5-4
Heights of Male and Female College Students

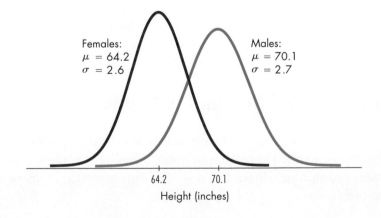

The **standard normal distribution** is a normal probability distribution that has a mean of 0 and a standard deviation of 1. (See Figure 5-5.)

Suppose that somehow we were forced to perform calculations using Formula 5-1. We would quickly see that the most workable values for μ and σ are $\mu = 0$ and $\sigma = 1$. By letting $\mu = 0$ and $\sigma = 1$, mathematicians have calculated areas under the curve. As shown in Figure 5-5, the area under the curve bounded by the mean of 0 and the score of 1 is 0.3413. Remember, the total area under the curve is always 1; this allows us to make the correspondence between area and probability.

Finding Probabilities When Given *z* Scores

Figure 5-5 shows that the area bounded by the curve, the horizontal axis, and the *z scores* of 0 and 1 is an area of 0.3413. Although the figure shows only one area, we can find areas (or probabilities) for many different regions. Such areas can be found by using Table A-2 in Appendix A or by using statistical software. If you are using Table A-2, it is essential to understand the following points.

1. Table A-2 is designed only for the *standard* normal distribution, which has a mean of 0 and a standard deviation of 1.

2. Each value in the body of the table is an area under the curve bounded on the left by a vertical line above the mean of 0 and bounded on the right by a vertical line above a specific positive score denoted by *z*, as illustrated in Figure 5-6.

Figure 5-5   Standard Normal Distribution, with Mean $\mu = 0$ and Standard Deviation $\sigma = 1$

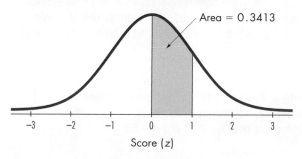

Figure 5-6   The Standard Normal Distribution

The area of the shaded region bounded by the mean of 0 and the positive number *z* can be found in Table A-2.

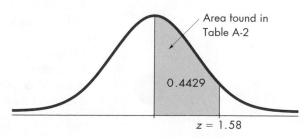

3. When working with a graph, avoid confusing *z* scores and areas.

   z score: *distance* **along the horizontal scale of the standard normal distribution; refer to the leftmost column and top row of Table A-2**

   Area: *region* **under the curve; refer to the values in the body of Table A-2**

4. The part of the *z* score denoting hundredths is found across the top row of Table A-2.

The following example requires that we find the probability associated with a score between 0 and 1.58. Begin with the *z* score of 1.58 by locating 1.5 in the left column; now find the value in the adjoining row of probabilities that is directly below 0.08, as shown in this excerpt from Table A-2.

z	. . .	.08
.		.
.		.
.		.
1.5	. . .	.4429

The area (or probability) value of 0.4429 indicates that there is a probability of 0.4429 of randomly selecting a score between 0 and 1.58. (The following sections will consider cases in which the mean is not 0 or the standard deviation is not 1.)

EXAMPLE

The Precision Scientific Instrument Company manufactures thermometers that are supposed to give readings of 0°C at the freezing point of water. Tests on a large sample of these instruments reveal that at the freezing point of water, some thermometers give readings below 0°C (denoted by negative numbers) and some give readings above 0°C (denoted by positive numbers). Assume that the mean reading is 0°C and the standard deviation of the readings is 1.00°C. Also assume that the frequency distribution of errors closely resembles the normal distribution. If one thermometer is randomly selected, find the probability that, at the freezing point of water, the reading is between 0°C and 1.58°C.

SOLUTION

The probability distribution of the readings is a standard normal distribution because the readings are normally distributed with $\mu = 0$ and $\sigma = 1$. We need to find the area between 0 and *z* (the shaded region) in Figure 5-6 with $z = 1.58$. From Table A-2 we find that this area is 0.4429.

Interpretation

The probability of randomly selecting a thermometer with an error between 0°C and +1.58°C is therefore 0.4429. Another way to interpret this result is to conclude that 44.29% of the thermometers will have errors between 0°C and +1.58°C.

AS

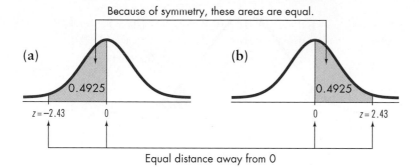

Figure 5-7
Using Symmetry to Find the Area to the Left of the Mean

EXAMPLE

Using the thermometers from the preceding example, find the probability of randomly selecting one thermometer that reads (at the freezing point of water) between $-2.43°C$ and $0°C$.

SOLUTION

We are looking for the region shaded in Figure 5-7(a), but Table A-2 is designed to apply only to regions to the right of the mean (0) as in Figure 5-7(b). By comparing the shaded area in Figure 5-7(a) to the shaded area in Figure 5-7(b), we can see that those two areas are identical because the density curve is symmetric. Referring to Table A-2, we can easily determine that the shaded area of Figure 5-7(b) is 0.4925, so the shaded area of Figure 5-7(a) must also be 0.4925.

Interpretation

The probability of randomly selecting a thermometer with an error between $-2.43°C$ and $0°$ is 0.4925. In other words, 49.25% of the thermometers have errors between $-2.43°C$ and $0°C$.

The above solution illustrates an important principle:

Although a z score can be negative, the area under the curve (or the corresponding probability) can never be negative.

Now recall the empirical rule (presented in Section 2–5) that states that for bell-shaped distributions,

- About 68% of all scores fall within 1 standard deviation of the mean.
- About 95% of all scores fall within 2 standard deviations of the mean.
- About 99.7% of all scores fall within 3 standard deviations of the mean.

If we refer to Figure 5-5 with $z = 1$, Table A-2 shows us that the shaded area is 0.3413. It follows that the proportion of scores between $z = -1$ and $z = 1$ will be $0.3413 + 0.3413 = 0.6826$. That is, about 68% of all scores fall within 1 standard deviation of the mean. A similar calculation with $z = 2$ yields the values of

Figure 5-8
Finding the Area to the
Right of $z = 1.27$

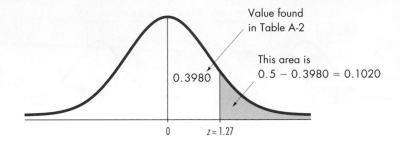

0.4772 + 0.4772 = 0.9544 (or about 95%) as the proportion of scores between $z = -2$ and $z = 2$. Similarly, the proportion of scores between $z = -3$ and $z = 3$ is given by 0.4987 + 0.4987 = 0.9974 (or about 99.7%). These exact values correspond very closely to those given in the empirical rule. In fact, the values of the empirical rule were found directly from the probabilities in Table A-2 and have been slightly rounded for convenience. The empirical rule is sometimes called the *68–95–99 rule*; using exact values from Table A-2, it would be called the *68.26–95.44–99.74 rule*, but then it wouldn't sound as snappy.

Because we are dealing with a density curve for a probability distribution, the total area under the curve must be 1. Now refer to Figure 5-8 and see that a vertical line directly above the mean of 0 divides the area under the curve into two equal parts, each containing an area of 0.5. The following example uses this observation.

EXAMPLE

Once again, make a random selection from the same sample of thermometers. Find the probability that the chosen thermometer reads (at the freezing point of water) greater than $+1.27°C$.

SOLUTION

We are again dealing with normally distributed values having a mean of 0°C and a standard deviation of 1°C. The probability of selecting a thermometer that reads greater than $+1.27°C$ corresponds to the shaded area of Figure 5-8. Table A-2 cannot be used to find that area directly, but we can use the table to find that $z = 1.27$ corresponds to the area of 0.3980, as shown in the figure. We now reason that because the area to the right of zero is one-half of the total area, it has an area of 0.5 and the shaded area is 0.5 − 0.3980, or 0.1020.

Interpretation

We conclude that there is a probability of 0.1020 of randomly selecting one of the thermometers with a reading greater than $+1.27°C$. Another way to interpret this result is to state that if many thermometers are selected and tested, then 0.1020 (or 10.20%) of them will read greater than $+1.27°C$.

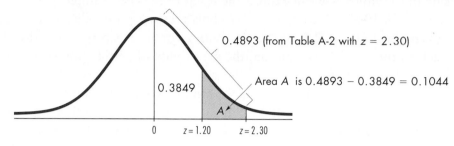

Figure 5-9
Finding the Area Between
$z = 1.20$ and $z = 2.30$

We are able to determine the area of the shaded region in Figure 5-8 by an indirect application of Table A-2. The following example illustrates yet another indirect use.

EXAMPLE

Assuming that one thermometer in our sample is randomly selected, find the probability that it reads (at the freezing point of water) between 1.20°C and 2.30°C.

SOLUTION

The probability of selecting a thermometer that reads between 1.20°C and 2.30°C corresponds to the shaded area of Figure 5-9. However, Table A-2 is designed to provide only for regions bounded on the left by the vertical line above 0. We can use the table to find that $z = 1.20$ corresponds to an area of 0.3849 and that $z = 2.30$ corresponds to an area of 0.4893, as shown in the figure. If we denote the area of the shaded region by A, we can see from Figure 5-9 that

$$0.3849 + A = 0.4893$$

so

$$A = 0.4893 - 0.3849 = 0.1044$$

Interpretation
If one thermometer is randomly selected, the probability that it reads (at the freezing point of water) between 1.20°C and 2.30°C is therefore 0.1044.

The above example concluded with the statement that the probability of a reading between 1.20°C and 2.30°C is 0.1044. Such probabilities can also be expressed with the following notation.

NOTATION

$P(a < z < b)$ denotes the probability that the z score is between a and b.

$P(z > a)$ denotes the probability that the z score is greater than a.

$P(z < a)$ denotes the probability that the z score is less than a.

$P(z = a)$ This probability is always equal to 0.

Using this notation, we can express the result of the last example as $P(1.20 < z < 2.30) = 0.1044$, which states in symbols that the probability of a z score falling between 1.20 and 2.30 is 0.1044. With a continuous probability distribution such as the normal distribution, the probability of getting any single *exact* value is 0. That is, $P(z = a) = 0$.

For example, there is a 0 probability of randomly selecting someone and getting a height of exactly 68.16243357 in. In the normal distribution, any single point on the horizontal scale is represented not by a region under the curve, but by a vertical line above the point. For $P(z = 1.33)$, we have a vertical line above $z = 1.33$, but that vertical line by itself contains no area, so $P(z = 1.33) = 0$. With any continuous random variable, the probability of any one exact value is 0, and it follows that $P(a \leq z \leq b) = P(a < z < b)$. It also follows that the probability of getting a z score of *at most b* is equal to the probability of getting a z score of *less than b*. It is important to interpret correctly key phrases such as *at most, at least, more than, no more than*, and so on. The illustrations in Figure 5-10

Figure 5-10
Interpreting Areas Correctly

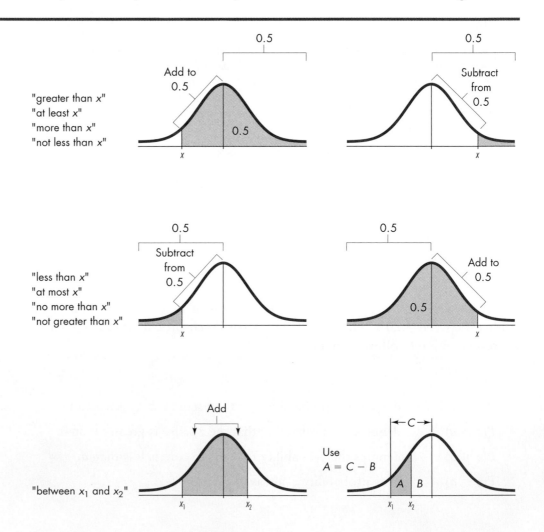

provide an aid to interpreting several of the most common phrases, assuming you will be working with Table A-2.

Finding *z* Scores When Given Probabilities

So far, the examples of this section involving the standard normal distribution have all followed the same format: Given some value(s), we found areas under the curve that represent probabilities. In many other cases, we already know the probability, but we need to find the corresponding *z* score. In such cases, it is very important to avoid confusion between *z* scores and areas. Remember, the numbers Table A-2 shows in the extreme left column and across the top are *z* scores, which are *distances* along the horizontal scale, whereas the numbers in the body of Table A-2 are *areas* (or probabilities). Also, *z* scores to the left of the centre line are always negative (as in Figure 5-7a). If we already know a probability and want to determine the corresponding *z* score using Table A-2, we find it as follows:

1. Draw a bell-shaped curve, draw the centre line, and identify the region under the curve that corresponds to the given probability. If that region is not bounded by the centre line, work with a portion of the curve that *is* bounded, on one side, by the centre line, and, on the other, by a boundary of the area corresponding to the probability.

2. Using the probability representing the area bounded by the centre line, locate the closest probability in the *body* of Table A-2 and identify the corresponding *z* score.

3. If the *z* score is positioned to the left of the centre line, make it negative.

EXAMPLE

Use the same thermometers with temperature readings that are normally distributed with a mean of 0°C and a standard deviation of 1°C. Find the temperature corresponding to P_{95}, the 95th percentile. That is, find the temperature separating the bottom 95% from the top 5%. (See Figure 5-11.)

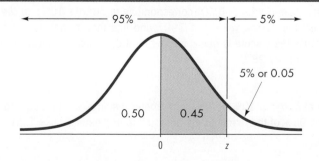

Figure 5-11
Finding the 95th
Percentile

SOLUTION

Figure 5-11 shows the z score that is the 95th percentile, separating the top 5% from the bottom 95%. We must refer to Table A-2 to find that z score, and we must use a region bounded by the centre line (where $\mu = 0$) on one side, such as the shaded region of 0.45 in Figure 5-11. (Remember, Table A-2 is designed to directly provide only those areas that are bounded on the left by the centre line and on the right by the z score.) We first search for the area of 0.45 *in the body of the table* and then find the corresponding z score. In Table A-2 the area of 0.45 is between the table values of 0.4495 and 0.4505, but there's an asterisk with a special note indicating that 0.4500 corresponds to a z score of 1.645. We can now conclude that the z score in Figure 5-11 is 1.645, so the 95th percentile is the temperature reading of 1.645°C.

Interpretation
When tested at freezing, 95% of the readings will be less than or equal to 1.645°C, and 5% of them will be greater than or equal to 1.645°C.

Note that in the preceding solution, Table A-2 led to a z score of 1.645, which is midway between 1.64 and 1.65. When using Table A-2, we can usually avoid interpolation by simply selecting the closest value. There are two special cases involving values that are important because they are used so often in a wide variety of applications (see the accompanying table). Except in these two special cases, we can select the closest value in the table. (If a desired value is midway between two table values, select the larger value.) Also, for z scores above 3.09, we can use 0.4999 as an approximation of the corresponding area.

z score	Area
1.645	0.4500
2.575	0.4950

EXAMPLE

Using the same thermometers, find P_{10}, the 10th percentile. That is, find the temperature reading separating the bottom 10% of all temperatures from the top 90%.

SOLUTION

Refer to Figure 5-12, where the 10th percentile is shown as the z score separating the bottom 10% from the top 90%. Table A-2 is designed for areas bounded by the centre line, so we refer to the shaded area of 0.40 (corresponding to 50% − 10%). *In the body of the table*, we select the closest value of 0.3997 and find that it corresponds to $z = 1.28$. However, because the z score is below the mean of 0, it must be negative. The 10th percentile is therefore −1.28°C.

Interpretation
When tested at freezing, 10% of the thermometer readings will be equal to or less than −1.28°C, and 90% of the readings will be equal to or greater than −1.28°C.

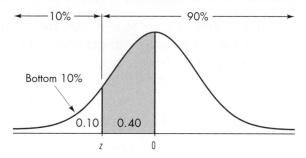

Figure 5-12
Finding the 10th
Percentile

The examples in this section were contrived so that the mean of 0 and the standard deviation of 1 coincided exactly with the parameters of the standard normal distribution described in Table A-2. In reality, it is unusual to find such convenient parameters because typical normal distributions involve means different from 0 and standard deviations different from 1. The next section introduces methods for working with such nonstandard normal distributions.

EXCEL (Prior to 2007): *To find the probability when given a z score:*

Click on the *fx* icon, and select **Statistical, NORMSDIST.** In the dialog box, enter the *z* score for a boundary of the region whose probability you are seeking. Excel returns the cumulative area from the left of the curve up to a vertical line above the specified *z* score. Assuming you input some number as the *z* score, three cases may apply:

1. For $P(z < a)$, the probability is the value returned by Excel.
2. For $P(z > a)$, the probability is 1 − (the value returned by Excel).
3. For $P(a < z < b)$ (assuming you have used Excel twice, using two separate *z* scores), the desired probability is $P(z < b) − P(z < a)$.

To find the z score when given a probability:

Click on the *fx* icon, and select **Statistical, NORMSINV.** In the dialog box, enter the probability corresponding to the area to the left of the *z* score.

EXCEL 2007: Click on **Menus** on the main menu, then the *fx* button, then more functions and select the function category **Statistical** and then either **NORMSDIST** or **NORMSINV.** Proceed as above.

MINITAB 15: *To find the probability when given a z score:*

From the **Calc** menu, choose **Probability Distributions** and then **Normal.** In the dialog box, choose the **Cumulative probability** radio button. The mean and standard deviation are set at 0 and 1 respectively by default. For the probability of a single value, click the **Input constant** radio button, put the value in the box, and click **OK**. Minitab returns the probability to the left of that value. For the probability to the right, subtract the probability from 1.

To find the z score when given a probability:

In the dialog box, choose the **Inverse cumulative probability** button. For a single *z* score, click the **Input constant** radio button and put a value between 0 and 1 in the box. Keep in mind that this value is the left-tail probability of your desired *z* score. For example, if you put 0.025 in the box, Minitab returns -1.96 since $P(z < -1.96) = 0.025$.

STATDISK:

To find the probability when given a z score:

Select **Analysis, Probability Distributions, Normal Distributions.** Enter a *Z* value in the appropriate box and click **Evaluate.** STATDISK provides several outputs, including the probability that Table A-2 would return for that *z* score, as well as the cumulative probabilities to the left, and to the right, of the given *z* score.

To find the z score when given a probability:

Proceed as described above, but enter a left-tail probability in the appropriate box. Read the corresponding *z* score from the right of the display.

EXCEL DISPLAY

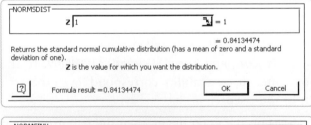

EXCEL DISPLAY

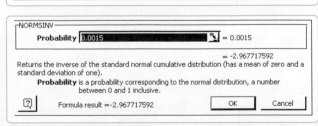

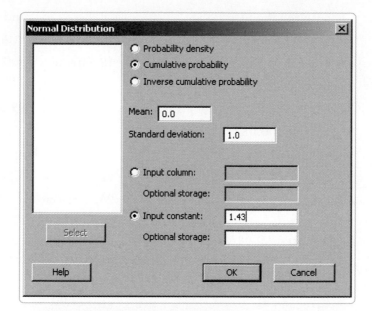

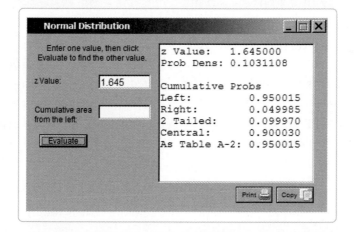

5-2 Exercises A: Basic Skills and Concepts

In Exercises 1–24, assume that the readings on the thermometers are normally distributed with a mean of 0°C and a standard deviation of 1.00°C. A thermometer is randomly selected and tested. In each case, draw a sketch, and find the probability of each reading in degrees.

1. Between 0 and 0.25

2. Between 0 and −0.36

3. Between 0 and 0.89

4. Between 0 and −0.07

5. Between 0 and 3.007

6. Between 0 and 1.96

7. Between 0 and -2.331

8. Between 0 and -1.28

9. Greater than 2.58

10. Less than -1.47

11. Less than -2.09

12. Greater than 0.25

13. Between 1.34 and 2.67

14. Between -1.72 and -0.31

15. Between -2.22 and -1.11

16. Between 0.89 and 1.78

17. Less than 0.08

18. Less than 3.01

19. Greater than -2.29

20. Greater than -1.05

21. Between -1.99 and 2.01

22. Between -0.07 and 2.19

23. Between -1.00 and 4.00

24. Between -5.00 and 2.00

In Exercises 25–28, assume that the readings on the thermometers are normally distributed with a mean of 0°C and a standard deviation of 1.00°C. Find the indicated probability, where z is the reading in degrees.

25. $P(z > 2.33)$

26. $P(2.00 < z < 2.50)$

27. $P(-3.00 < z < 2.00)$

28. $P(z < -1.44)$

In Exercises 29–36, assume that the readings on the thermometers are normally distributed with a mean of 0°C and a standard deviation of 1.00°C. A thermometer is randomly selected and tested. In each case, draw a sketch, and find the temperature reading corresponding to the given information.

29. Find P_{90}, the 90th percentile. This is the temperature reading separating the bottom 90% from the top 10%.

30. Find P_{30}, the 30th percentile.

31. Find Q_1, the temperature reading that is the first quartile.

32. Find D_1, the temperature reading that is the first decile.

33. If 4% of the thermometers are rejected because they have readings that are too high, but all other thermometers are acceptable, find the reading that separates the rejected thermometers from the others.

34. If 8% of the thermometers are rejected because they have readings that are too low, but all other thermometers are acceptable, find the reading that separates the rejected thermometers from the others.

35. A quality control analyst wants to examine thermometers that give readings in the bottom 2%. What reading separates the bottom 2% from the others?

36. If 2.5% of the thermometers are rejected because they have readings that are too high and another 2.5% are rejected because they have readings that are too low, find the two readings that are cutoff values separating the rejected thermometers from the others.

Exercises B: Beyond the Basics

37. Assume that z scores are normally distributed with a mean of 0 and a standard deviation of 1.
    a. If $P(0 < z < a) = 0.3212$, find a.
    b. If $P(2b < z < b) = 0.3182$, find b.
    c. If $P(z > c) = 0.2358$, find c.
    d. If $P(z > d) = 0.7517$, find d.
    e. If $P(z < e) = 0.4090$, find e.

38. For a standard normal distribution, find the percentage of data that are
    a. within 1 standard deviation of the mean
    b. within 1.96 standard deviations of the mean
    c. between $\mu - 3\sigma$ and $\mu + 3\sigma$
    d. between 1 standard deviation below the mean and 2 standard deviations above the mean
    e. Suppose it turns out that the distribution is not exactly normal, but is positively skewed. How does this affect your answers to parts (a), (b), and (d) of this exercise?

39. In a manufacturing plant that makes boxes, the width of a certain type of box is normally distributed. The probability that the width is less than 23.9708 cm is 0.0721 and the probability that the width is more than 24.0404 cm is 0.0217. Find the mean and standard deviation of the box width.

40. In a certain region, annual household incomes are normally distributed. The middle 95% of the incomes are between $72,684 and $78,564. Find the mean and standard deviation of the annual household incomes for this region.

5-3 Normal Distributions: Finding Probabilities

Although Section 5–2 introduced important methods for dealing with normal distributions, the examples and exercises included in that section are generally unrealistic because most normally distributed populations have a nonzero mean, a standard deviation different from 1, or both. In this section we include many real and important nonstandard normal distributions. The basic principle we will be explaining in this section is the following:

> If we convert values to standard scores using Formula 5-2, then procedures for working with all normal distributions are the same as for the standard normal distribution.

Formula 5-2
$$z = \frac{x - \mu}{\sigma}$$

If you use certain calculators or software programs to find probabilities under the normal curve, the conversion to z scores may not be necessary, because the probabilities can be found directly. Regardless of the method used, however, you need to clearly understand the basic principle of this section, because it is an important foundation for concepts introduced in the following chapters.

See Figure 5-13, where we illustrate the important principle that the area bounded by a score and the population mean is the same as the area bounded by the corresponding z score and the mean of 0. Once we convert a nonstandard score to a z score, we can use Table A-2 in the same way it was used in Section 5–2. You can use the following procedure for finding probabilities for values of a random variable with a normal probability distribution:

1. Draw a normal curve, label the mean and the specific x values, then *shade* the region representing the desired probability.

2. For each relevant score x that is a boundary for the shaded region, use Formula 5-2 to find the equivalent z score.

3. Refer to Table A-2 to find the area of the shaded region. This area is the desired probability.

The following example uses these three steps, and it illustrates the relationship between a typical nonstandard normal distribution and the standard normal distribution.

EXAMPLE

The mean value of Canadian imports each year from the Middle East and Africa is $2502 million, according to Statistics Canada data for 1980 to 1999. The standard deviation is $905 million. Assuming there is no trend in the data over time, and that the annual figures are normally distributed, what is the probability that in a randomly selected year from that period, the value of imports from the Middle East and Africa is between $2502 million and $4312 million?

SOLUTION

Step 1: See Figure 5-14, where we enter the mean of 2502 and the x value of 4312, and we shade the area representing the probability we want.
Step 2: To use Table A-2, we must use Formula 5-2 to convert the nonstandard distribution of import values to the standard normal distribution. The import value of 4312 is converted to a z score as follows:

$$z = \frac{x - \mu}{\sigma} = \frac{4312 - 2502}{905} = \frac{1810}{905} = 2.00$$

This result shows that the import value of $4312 million differs from the mean of $2502 million by 2.00 standard deviations.
Step 3: Referring to Table A-2, we find that $z = 2.00$ corresponds to an area of 0.4772.

Figure 5-13
Converting from a Nonstandard Normal Distribution to the Standard Normal Distribution

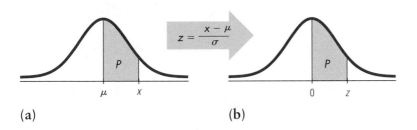

(a)　　　　　　　　　　　　　　　　(b)

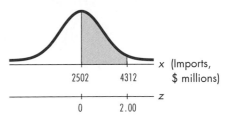

Interpretation

There is a probability of 0.4772 of randomly selecting a year with an import value between $2502 million and $4312 million. This can be expressed in symbols as

$$P(2502 < x < 4312) = P(0 < z < 2.00) = 0.4772$$

Another way to interpret this result is to conclude that 47.72% of years have import values from the Middle East and Africa between $2502 million and $4312 million.

EXAMPLE

Assume that the heights of male college students are normally distributed with a mean of 70.1 in. and a standard deviation of 2.7 in.

a. Find the percentage of male students who fall between the Toronto Maple Leafs' average height of 72.4 in. and the Philadelphia Flyers' average height of 74.6 in.

b. Among 500 randomly selected male college students, how many would you expect to fall between the Toronto Maple Leafs' average height and the Philadelphia Flyers' average height?

SOLUTION

a. The shaded region *B* in Figure 5-15 represents the proportion of male students who fall between the Maple Leafs' average height and the Flyers' average height. We can't find that shaded region directly because Table A-2 isn't designed for such cases, but we can find it indirectly by using the same basic procedures presented in Section 5-2. Find the shaded area *B* by subtracting region *A* from the total area of regions *A* and *B* combined. That is,

$$B = (A \text{ and } B \text{ combined}) - A$$

For the area of regions A and B combined:

$$z = \frac{x - \mu}{\sigma} = \frac{74.6 - 70.1}{2.7} = 1.67$$

Using Table A-2, we find that $z = 1.67$ corresponds to an area of 0.4525.

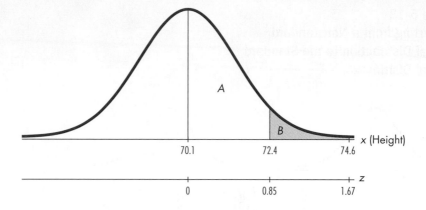

For the area of region A:

$$z = \frac{x - \mu}{\sigma} = \frac{72.4 - 70.1}{2.7} = 0.85$$

Again using Table A-2, we find that $z = 0.85$ corresponds to an area of 0.3023. Region *A* has an area of 0.3023.

For the area of region B, the shaded area is the difference between 0.4525 and 0.3023:

$$\text{Area } B = (\text{areas of } A \text{ and } B, \text{ combined}) - (\text{area } A)$$
$$= 0.4525 - 0.3023 = 0.1502$$

Interpretation
This number indicates that only 15.02% of male college students have heights between the Maple Leafs' average height and the Flyers' average height.

b. Among 500 randomly selected male college students, we expect that 15.02% of them would have heights between the Maple Leafs' average height and the Flyers' average height:

$$500 \cdot 0.1502 = 75.1 \text{ male students}$$

EXAMPLE

Find the percentage of years in which Canadian imports from the Middle East and Africa are between \$339 million and \$8004 million. Again, assume that the annual import values exhibit no trend, and are normally distributed with a mean of \$2502 million and a standard deviation of \$905 million.

SOLUTION

Figure 5-16 shows the normal distribution of import values, with the shaded region representing values between \$339 million and \$8004 million. The method for finding the area of the shaded region involves breaking it up into parts *A* and *B* as shown. We can use

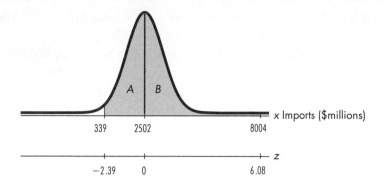

Figure 5-16
Import Values Between
$339 Million and
$8004 Million

Formula 5-2 and Table A-2 to find the areas of those regions separately; then we can add the results.

For area A only:

$$z = \frac{x - \mu}{\sigma} = \frac{339 - 2502}{905} = -2.39$$

We can use Table A-2 to find that $z = -2.39$ corresponds to 0.4916, so the area A is 0.4916.

For area B only:

$$z = \frac{8004 - 2502}{905} = 6.08$$

Table A-2 does not include z scores above 3.09, but it does include a note that for values of z above 3.09, we should use 0.4999 for the area. (If necessary, more accurate results can be obtained by using special tables or software.) Area B is 0.4999.

For areas of regions A and B combined:

$$0.4916 + 0.4999 = 0.9915$$

Interpretation
The proportion of years in which the value of Canadian imports from the Middle East and Africa is between $339 million and $8004 million is 0.9915. That is to say, imports fall within this range 99.15% of all years.

In this section we have extended the concepts of Section 5–2 to include more realistic nonstandard normal probability distributions. However, all of the examples we have considered so far are of the same general type: We are given specific limit values and we must find an area (or probability, or percentage). In many practical and real cases, the probability (or percentage) is known and we must find the relevant value(s). Problems of this type are discussed in the next section.

EXCEL (Prior to 2007): To find the probability in a nonstandard distribution when given an *x* value, use the following procedure in Excel: Click on the *fx* icon, and select **Statistical, NORMDIST**. In the dialog box, enter the value, the mean, the standard deviation, and "true". Excel returns the cumulative area from the left of the curve up to a vertical line above the specified *x* value.

EXCEL 2007: Click on **Menus** on the main menu, then the *fx* button, then more functions and select the function category **Statistical** and then the function name **NORMDIST**. Proceed as above.

MINITAB 15: From the **Calc** menu, choose **Probability Distributions** and then **Normal**. In the dialog box, choose the **Cumulative probability** radio button. The mean and standard deviation are set at 0 and 1 respectively by default. Change these to the mean and standard deviation you want. For the probability of a single value, click the **Input constant** radio button, put the value in the box, and click **OK**. Minitab returns the probability to the left of that value. For the probability to the right, subtract the probability from 1.

STATDISK: Like Table A-2, STATDISK requires that you use Formula 5-2 to convert nonstandard values to standard values; then use the STATDISK procedures described in Using Technology for Section 5–2.

EXCEL DISPLAY

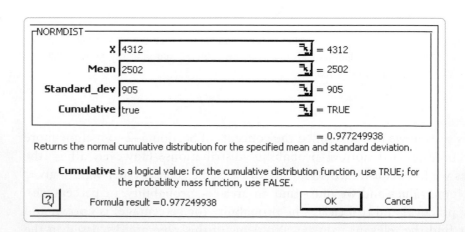

NORMDIST

X	4312		= 4312
Mean	2502		= 2502
Standard_dev	905		= 905
Cumulative	true		= TRUE

= 0.977249938

Returns the normal cumulative distribution for the specified mean and standard deviation.

Cumulative is a logical value: for the cumulative distribution function, use TRUE; for the probability mass function, use FALSE.

Formula result = 0.977249938   OK   Cancel

5-3 Exercises A: Basic Skills and Concepts

In Exercises 1–6, assume that the heights of female students are normally distributed with a mean given by $\mu = 64.2$ in. and a standard deviation given by $\sigma = 2.6$ in. (based on data from a survey of college students). Also assume that a female student is randomly selected. Draw a graph, and find the indicated probability.

1. $P(64.2 \text{ in.} < x < 65.0 \text{ in.})$
2. $P(x < 70.0 \text{ in.})$
3. $P(x > 58.1 \text{ in.})$
4. $P(59.1 \text{ in.} < x < 66.6 \text{ in.})$

5. A fashion agency is looking for females between 65.5 in. and 68.0 in. tall to work as models. Find the probability that a randomly selected female student meets the height requirements to be a model.

6. The Beanstalk Club, a social organization for tall people, has a requirement that women must be at least 70 in. (or 5 ft 10 in.) tall. Suppose you are trying to decide whether to open a branch of the Beanstalk Club at your college with 500 female students.
   a. Find the percentage of female students who are eligible for membership because they meet the minimum height requirement of 70 in.
   b. Among the 500 female students in your college, how many would be eligible for Beanstalk Club membership?
   c. Will you open a branch of the Beanstalk Club?

7. Replacement times for TV sets are normally distributed with a mean of 8.2 years and a standard deviation of 1.1 years (based on data from "Getting Things Fixed," *Consumer Reports*). Find the probability that a randomly selected TV set will have a replacement time of less than 7.0 years.

8. Replacement times for CD players are normally distributed with a mean of 7.1 years and a standard deviation of 1.4 years (based on data from "Getting Things Fixed," *Consumer Reports*). Find the probability that a randomly selected CD player will have a replacement time of less than 8.0 years.

9. Assume that the heights of soldiers in the Canadian Armed Forces are normally distributed with a mean height of 70.3 in. and a standard deviation of 3.4 in. Find the probability of one soldier who is selected at random having a height of 77.0 in. or greater.

10. Based on the sample results in Data Set 18 of Appendix B, assume that human body temperatures are normally distributed with a mean of 36.4°C and a standard deviation of 0.62°C. If we define a fever to be a body temperature above 37.8°C, what percentage of normal and healthy persons would be considered to have a fever? Does this percentage suggest that a cutoff of 37.8°C is appropriate?

11. One classic use of the normal distribution is inspired by a letter to *Dear Abby* in which a wife claimed to have given birth 308 days after a brief visit from her husband, who was serving in the Navy. The lengths of pregnancies are normally distributed with a mean of 268 days and a standard deviation of 15 days. Given this information, find the probability of a pregnancy lasting 308 days or longer. What does the result suggest?

12. Lengths of pregnancies are normally distributed with a mean of 268 days and a standard deviation of 15 days. If we stipulate that a baby is *premature* if born at least three weeks early, what percentage of babies are born prematurely? Why would this information be useful to hospital administrators?

13. Based on daily summaries from a Calgary observatory (for January to November 2000), the mean daily counting rates for cosmic rays are approximately normally distributed, with a mean equal to 3465.5 and a standard deviation of 127.7. If one day is randomly selected, what is the probability that the day's observed mean counting rate is at least 3248?

14. According to the International Mass Retail Association, girls aged 13 to 17 spend an average of $31.20 on shopping trips in a month. Assume that the amounts are normally distributed with a standard deviation of $8.27.

    If a girl in that age category is randomly selected, what is the probability that she spends between $35.00 and $40.00 in one month? Does the assumption of a normal distribution seem plausible for this population?

15. IQ scores are normally distributed with a mean of 100 and a standard deviation of 15. Mensa is an organization for people with high IQs, and eligibility requires an IQ above 131.5.

    a. If someone is randomly selected, find the probability that he or she meets the Mensa requirement.

    b. In a typical region of 75,000 people, how many are eligible for Mensa?

16. An IBM subcontractor was hired to make ceramic substrates that are used to distribute power and signals to and from computer silicon chips. Specifications require resistance between 1.500 ohms and 2.500 ohms, but the population has normally distributed resistances with a mean of 1.978 ohms and a standard deviation of 0.172 ohms. What percentage of the ceramic substrates will not meet the manufacturer's specifications? Does this manufacturing process appear to be working well?

17. The average household expenditure in Canada on postsecondary books is $53.00, with a standard deviation of $18.61 (based on a study by Statistics Canada on family expenditures). If a household is selected at random, find the probability that its expenditure on postsecondary books is between $60.00 and $70.00. Do you expect that the household expenditure on books is normally distributed?

18. Measurements of human skulls from different epochs are analyzed to determine whether they change over time. The maximum breadth is measured for skulls from Egyptian males who lived around 3300 BCE. Results show that those breadths are normally distributed with a mean of 132.6 mm and a standard deviation of 5.4 mm (based on data from *Ancient Races of the Thebaid* by Thomson and Randall-Maciver). An archeologist discovers a male Egyptian skull and a field measurement reveals a maximum breadth of 119 mm. Find the probability of getting a value of 119 or less if a skull is randomly selected from the period around 3300 BCE. Is the newly found skull likely to come from that era?

19. According to a national health survey, the serum cholesterol levels of men aged 18 to 24 are normally distributed with a mean and standard deviation (in mg/100mL) of 178.1 and 40.7, respectively. One criterion for identifying risk of coronary disease is a cholesterol level above 300. If a man aged 18–24 is randomly selected, find the probability that his serum cholesterol level is above 300. Does this probability warrant serious concern?

20. Some vending machines are designed so that their owners can adjust the weights of the quarters that are accepted. If many counterfeit coins are found, adjustments are made to reject more coins, with the effect that most of the counterfeit coins are rejected along with many legal coins. Assume that quarters have weights that are normally distributed with a mean of 5.67 g and a standard deviation of 0.070 g. If a vending machine is adjusted to reject quarters weighing less than 5.50 g or more than 5.80 g, what is the percentage of legal quarters that are rejected?

5-3 Exercises B: Beyond the Basics

In Exercises 21–23, refer to the indicated data set in Appendix B.

  a. *Construct a histogram to determine whether the data set has a normal distribution.*
  b. *Find the sample mean and sample standard deviation s.*
  c. *Use the sample mean as an estimate of the population mean μ, use the sample standard deviation as an estimate of the population standard deviation σ, and use the methods of this section to find the indicated probability.*

21. Use the combined list of 100 weights of M&M plain candies listed in Data Set 11, and estimate the probability of randomly selecting one M&M candy and getting one with a weight greater than 1.000 g.

22. Use the total weights of discarded garbage in Data Set 1, and estimate the probability of randomly selecting a household that discards more than 20.0 lb of garbage in a week.

23. Use the chest measurements of bears in Data Set 3, and estimate the probability of randomly selecting a bear with a chest measuring less than 30 in.

24. When you constructed a histogram for total weights of discarded garbage, for Question 22, was the distribution *exactly* normal? If not, do you believe your probability estimate in Question 22 was likely too low or too high? Explain.

 # Normal Distributions: Finding Values

In this section we consider problems such as this: If companies' revenues per employee are normally distributed with $\mu = \$310{,}000$ and $\sigma = \$147{,}000$, find the revenue per employee separating the bottom 10% from the others. This problem starts from a given probability (0.10). We need to find the appropriate value x. This section reverses the procedure of Section 5–3, where we used a given value to find a probability.

In considering problems of finding values when given probabilities, there are three important cautions to keep in mind.

1. *Don't confuse z scores and areas.* Remember, z scores are *distances* along the horizontal scale, but areas represent *regions* under the normal curve. Table A-2 lists z scores in the left column and across the top row, but areas are found in the body of the table.

2. *Choose the correct (right/left) side of the graph.* A score separating the top 10% from the others will be located on the right side of the graph, but a score separating the bottom 10% will be located on the left side of the graph.

3. *A z score must be negative whenever it is located to the left of the centre line of 0.*

As in Section 5–3, graphs are extremely helpful and they are strongly recommended. Even if you will be using statistical software to find values when given probabilities, you should understand the graphs and procedures that are presented below.

Procedure for Finding Values Using Table A-2 and Formula 5-2

1. Sketch a normal distribution curve; enter the given probability or percentage in the appropriate region of the graph, and identify the x value(s) being sought.

2. Use Table A-2 to find the z score corresponding to the region bounded by x and the centre line of 0. Observe the following cautions:
   • Refer to the *body* of Table A-2 to find the closest area, then identify the corresponding z score.
   • Make the z score *negative* if it is located to the left of the centre line.

3. Using Formula 5-2, enter the values for μ, σ, and the z score found in Step 2, then solve for x. On the basis of the format of Formula 5-2, we can solve for x as follows:

$$x = \mu + (z \cdot \sigma) \qquad \text{(Another form of Formula 5-2)}$$

4. Refer to the sketch of the curve to verify that the solution makes sense in the context of the graph and in the context of the problem.

The following example, which was introduced at the beginning of this section, uses the procedure just outlined. Pay extra attention to Step 2, especially where we make the z score negative because it is to the left of the mean.

EXAMPLE

In its annual lists of the "Top 100 Companies" of Canada, the *Financial Post* uses the companies' revenues as the basis for ranking. There are other company variables which seem, in fact, quite randomly distributed. For example, in 1997, fully two-thirds of the Top 100 companies appeared to have values for the variable "Revenues per employee" that fell into a normal distribution, with a mean and standard deviation (in $1000s) of 310 per employee and 147 per employee, respectively. (The other third of companies truly did distinguish themselves, with values ranging from 3 up to 115 times the mean value for the lower group.) For those lower performing companies, find the value of P_{10}—the revenue per employee separating the bottom 10% from the top 90%.

SOLUTION

Step 1: We begin with the graph shown in Figure 5-17. We have entered the mean of 310, shaded the area representing the bottom 10%, and identified the desired value as x. The area between 310 and x must be 40% of the total area (because the left half of the area must combine to be 50% of the total). Because the total area is 1, the area constituting 40% of the total must be 0.4.

Step 2: We refer to Table A-2, but we look for an area of 0.4000 in the body of the table. (Remember, Table A-2 is designed to list areas only for those regions bounded on the left by the mean and on the right by some value.) The area closest to 0.4000 is 0.3997, and it corresponds to a z score of 1.28. Because the score is to the left of the mean, we make it negative and use $z = -1.28$.

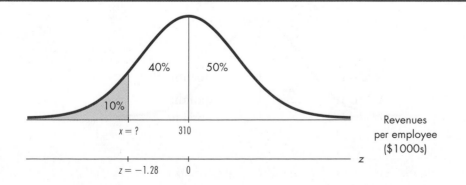

Figure 5-17
Finding P_{10} for Revenues per Employee

Step 3: With $z = -1.28$, $\mu = 310$, and $\sigma = 147$, we solve for x either by using Formula 5-2 directly or by using the following version of Formula 5-2:

$$x = \mu + (z \cdot \sigma) = 310 + (-1.28 \cdot 147) = 121.8$$

Step 4: If we let $x = 121.8$ in Figure 5-17, we see that this solution is reasonable because the 10th percentile should be less than the mean of 310.

Interpretation
A revenue per employee of $121,800 separates the lowest 10% from the highest 90%.

EXAMPLE

Assume that body temperatures of healthy adults are normally distributed with a mean of 36.39°C and a standard deviation of 0.62°C (based on Data Set 18 in Appendix B). If a medical researcher wants to study people in the bottom 2.5% and people in the top 2.5%, find the temperatures separating those limits.

SOLUTION

Step 1: We begin with the graph shown in Figure 5-18. We have shaded the areas representing the bottom 2.5% and top 2.5% (or 0.025). The areas of 0.475 are found by using the fact that the centre line above the mean divides the total area of 1 into two parts, each with area 0.5. We get $0.5 - 0.025 = 0.475$.

Step 2: We refer to Table A-2, but we look for an area of 0.475 in the body of the table. (Remember, Table A-2 is designed to list areas only for those regions bounded on the left by the mean and on the right by some value.) The area of 0.4750 corresponds to $z = 1.96$. For the x value located on the right in Figure 5-18, we use $z = 1.96$; for the x value located on the left we use $z = -1.96$.

Step 3: With $z = 1.96$, $\mu = 36.39$, and $\sigma = 0.62$, we solve for x using a variation of Formula 5-2:

$$x = \mu + (z \cdot \sigma) = 36.39 + (1.96 \cdot 0.62) = 37.61$$

Figure 5-18
Body Temperatures

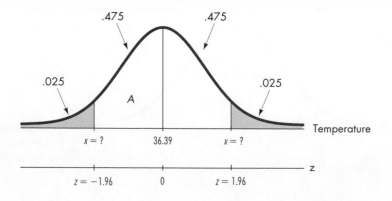

With $z = -1.96$, $\mu = 36.39$, and $\sigma = 0.62$, we solve for x using a variation of Formula 5-2:

$$x = \mu + (z \cdot \sigma) = 36.39 + (-1.96 \cdot 0.62) = 35.17$$

Step 4: If we let $x = 35.17$ and 37.61 in Figure 5-18, we see that our solutions are reasonable.

Interpretation
If the researcher's assumptions about the distribution of body temperatures are correct, then he or she should select people with body temperatures below 35.17°C or above 37.61°C.

EXCEL (prior to 2007): To find a value given a probability: Click on the *fx* icon, and select **Statistical, NORMINV.** In the dialog box, enter the cumulative probability to the left of the given value, the mean, and the standard deviation. Excel returns the corresponding x value.

EXCEL 2007: Click on **Menus** on the main menu, then the *fx* button, then more functions and select the function category **Statistical** and then the function name **NORMINV.** Proceed as above.

MINITAB 15: From the **Calc** menu, choose **Probability distributions** and then **Normal.** In the dialog box, choose the **Inverse cumulative probability** radio button. For a single z score, click the **Input constant** radio button and put a value between 0 and 1 in the box. Keep in mind that this value is the left-tail probability of your desired value. For example, if $\mu = 10$ and $\sigma = 2$ and you put 0.025 in the box, Minitab returns 6.08 since $P(x < 6.08) = 0.025$ when $\mu = 10$ and $\sigma = 2$.

STATDISK: Follow the procedures described in Using Technology for Section 5–2. Like Table A-2, STATDISK provides a z score, which you must convert by use of a formula into the corresponding x value.

EXCEL DISPLAY

NORMINV
Probability 0.025 = 0.025
Mean 36.39 = 36.39
Standard_dev 0.62 = 0.62

= 35.17482413
Returns the inverse of the normal cumulative distribution for the specified mean and standard deviation.

Probability is a probability corresponding to the normal distribution, a number between 0 and 1 inclusive.

Formula result = 35.17482413

OK   Cancel

5-4 Exercises A: Basic Skills and Concepts

In Exercises 1–4, assume that female college students have heights that are normally distributed with a mean of 64.2 in. and a standard deviation of 2.6 in. Find the height for the given percentile.

1. P_{85} **2.** P_{66}

3. P_{15} **4.** P_{35}

5. Replacement times for TV sets are normally distributed with a mean of 8.2 years and a standard deviation of 1.1 years (based on data from "Getting Things Fixed," *Consumer Reports*).

    a. Find the replacement time that separates the top 20% from the bottom 80%.

    b. Find the probability that a randomly selected TV will have a replacement time of less than 5.0 years.

    c. If you want to provide a warranty so that only 1% of the TV sets will be replaced before the warranty expires, what length of time would you recommend for the warranty?

6. Replacement times for CD players are normally distributed with a mean of 7.1 years and a standard deviation of 1.4 years (based on data from "Getting Things Fixed," *Consumer Reports*). Find P_{55}, which is the replacement time separating the top 45% from the bottom 55%.

7. Weights of paper discarded by households each week are normally distributed with a mean of 4.3 kg and a standard deviation of 1.9 kg. Find P_{33}, which is the weight that separates the bottom 33% from the top 67%.

8. Based on the sample results in Data Set 18 of Appendix B, assume that human body temperatures are normally distributed with a mean of 36.4°C and a standard deviation of 0.62°C. What two temperature levels separate the bottom 2% and the top 2%? Could these values serve as reasonable limits that could be used to identify people who are likely to be ill?

9. The durations of pregnancies are normally distributed with a mean of 268 days and a standard deviation of 15 days. If we stipulate that a baby is *premature* if the length of pregnancy is in the lowest 4%, find the duration that separates premature babies from those who are not premature.

10. According to the Opinion Research Corporation, men spend an average of 11.4 min in the shower. Assume that the times are normally distributed with a standard deviation of 1.8 min. Find the values of the quartiles Q_1 and Q_3.

11. IQ scores are normally distributed with a mean of 100 and a standard deviation of 15. If we define a genius to be someone in the top 1% of IQ scores, find the score separating geniuses from the rest of us. Are there any jobs where this score could reasonably be used as one criterion for employment?

12. A subcontractor manufactures ceramic substrates for IBM. These devices have resistances that are normally distributed with a mean of 1.978 ohms and a standard deviation of 0.172 ohms. If the required specifications are to be modified so that 3% of the devices are rejected because their resistances are too low and another 3% are rejected because their resistances are too high, find the cutoff values for the acceptable devices.

13. Scores obtained from the Law School Admission Test (LSAT) are normally distributed with a mean score of 550 and a standard deviation of 110.
    a. If a test score is selected at random, find the probability that it is less than 750.
    b. If the top 16% of the test scores are usually good enough for admission to law school, find the cutoff score for gaining admission.

14. Measurements of human skulls from different epochs are analyzed to determine whether they change over time. The maximum breadth is measured for skulls from Egyptian males who lived around 3300 BCE. Results show that those breadths are normally distributed with a mean of 132.6 mm and a standard deviation of 5.4 mm (based on data from *Ancient Races of the Thebaid* by Thomson and Randall-Maciver).
    a. Find the probability of getting a value greater than 140 mm if a skull is randomly selected from the period of around 3300 BCE.
    b. Find the value that is D_2, the second decile.

15. Based on daily summaries from a Calgary observatory over a period of ten months, the mean daily counting rates for cosmic rays are approximately normally distributed, with a mean equal to 3465.5 and a standard deviation of 127.7. Find the sixth decile D_6. What is the mean daily counting rate that separates the lowest 60% from the highest 40%?

16. Quarters have weights that are normally distributed with a mean of 5.67 g and a standard deviation of 0.070 g.
    a. If a vending machine is adjusted to reject quarters weighing less than 5.53 g or more than 5.81 g, what is the percentage of legal quarters that are rejected?
    b. Find the weights of accepted legal quarters if the machine is readjusted so that the lightest 1.5% are rejected and the heaviest 1.5% are rejected.
    c. If your quarter is rejected from a machine that is set to reject the upper 1.5% and the lower 1.5% of coins, by weight, is it a waste of time to reinsert your coin?

5-4 Exercises B: Beyond the Basics

17. The construction of a histogram for a data set reveals that the distribution is approximately normal and the boxplot is constructed with these quartiles: $Q_1 = 62$, $Q_2 = 70$, $Q_3 = 78$. Estimate the standard deviation.

18. An instructor informs her physics class that a test is very difficult, but the grades will be curved. Scores for the test are normally distributed with a mean of 25 and a standard deviation of 5.

    a. If she curves by adding 50 to each grade, what is the new mean? What is the new standard deviation?

    b. Is it fair to curve by adding 50 to each grade? Why or why not?

    c. If the grades are curved according to the following scheme (instead of adding 50), find the numerical limits for each letter grade.

        A: Top 10%

        B: Scores above the bottom 70% and below the top 10%

        C: Scores above the bottom 30% and below the top 30%

        D: Scores above the bottom 10% and below the top 70%

        F: Bottom 10%

    d. Which method of curving the grades is fairer: Adding 50 to each grade or using the scheme given in part (c)? Explain.

19. According to data from the College Entrance Examination Board, the mean math SAT score is 475, and 17.0% of the scores are above 600. Find the standard deviation, and then use that result to find the 99th percentile. (Assume that the scores are normally distributed.)

20. The College Entrance Examination Board writes that "for the SAT Achievement Tests, your score would fall in a range [between] about 30 points above [and] below your actual ability about two-thirds of the time. This range is called the standard error of measurement (SEM)." Use that statement to estimate the standard deviation for scores of an individual on an SAT Achievement Test. (Assume that the scores are normally distributed.)

5-5 The Central Limit Theorem

This section presents the central limit theorem, which is one of the most important and useful concepts in statistics. It forms a foundation for estimating population parameters and hypothesis testing—topics discussed at length in the following chapters.

We will not present rigorous proofs in this section, but will instead focus on the concepts and how to apply them. You will need to keep in mind the types of data sets we are considering. Instead of sets of *individual* values—the values of a *random variable* (see Section 4–2)—we will work with data sets in which each value is the *mean* of some other sample. Just as a *probability distribution* describes the probability for each value of a random variable *x*, the *sampling distribution of sample means* describes the probability for each value of the sample mean, when drawing samples of a given size from a population.

The **sampling distribution of sample means** is the probability distribution of sample means, with all samples having the sample size n.

One property of the sampling distribution of sample means is key to this section:

As the sample size increases, the sampling distribution of sample means approaches a normal distribution.

In other words, if we collect many samples of the same size from the same population, compute their means, and then draw a histogram of those means, that histogram will tend to have the bell shape of a normal distribution. This is true regardless of the shape of the distribution of the original population. Compare Figure 5-19, which shows a probability distribution for a specific variable, and Figure 5-20, which shows the sampling distribution for sample means for the same variable. Figure 5-21 is a general illustration of the same principle, applied to three different shapes of underlying population distribution. Observations exactly like these led to the formulation of the central limit theorem, which we will now discuss.

The **central limit theorem** involves two different distributions: the distribution of the original population and the distribution of the sample means. As in previous chapters, we use the symbols μ and σ to denote the mean and standard deviation of the original population. We now introduce new notation for the mean and standard deviation of the distribution of sample means.

Figure 5-19
Distribution of 200 Digits from Social Insurance Numbers (Last 4 Digits) of 50 Students

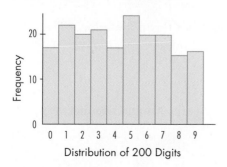

Distribution of 200 Digits

Figure 5-20
Distribution of 50 Sample Means for 50 Students

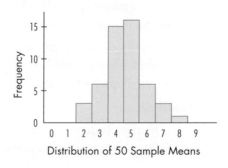

Distribution of 50 Sample Means

Figure 5-21   Normal, Uniform, and Skewed Distributions

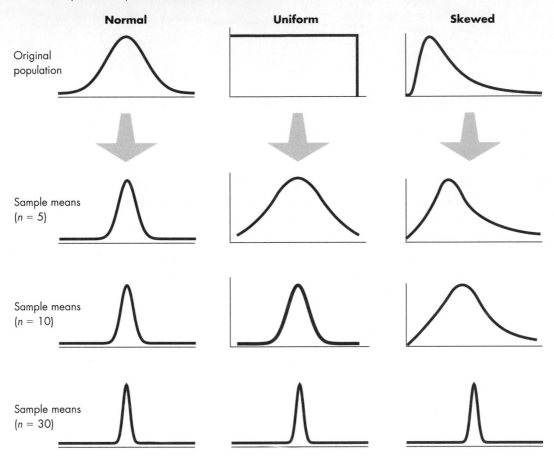

Given:

1. The random variable x has a distribution (which may *or may not* be normal) with mean μ and standard deviation σ.

2. Samples of size n are randomly selected from this population. (The samples are selected so that all possible samples of size n have the same chance of being selected.)

Conclusions:

1. The distribution of sample means $\overline{x}$ will, as the sample size increases, approach a *normal* distribution.

2. The mean of this distribution of sample means will be the population mean μ.

3. The standard deviation of this distribution of sample means will be $\sigma/\sqrt{n}$.

Practical Rules Commonly Used:

1. For samples of size n larger than 30, the distribution of the sample means can be approximated reasonably well by a normal distribution. The approximation gets better as the sample size n becomes larger.

2. If the original population is itself normally distributed, then the sample means will be normally distributed for *any* sample size n (not just the sample sizes n larger than 30).

NOTATION FOR THE CENTRAL LIMIT THEOREM

If all possible random samples of size n are selected from a population with mean μ and standard deviation σ, the mean of the sample means is denoted by $\mu_{\bar{x}}$, so

$$\mu_{\bar{x}} = \mu$$

Also, the standard deviation of the sample means is denoted by $\sigma_{\bar{x}}$, so

$$\sigma_{\bar{x}} = \frac{\sigma}{\sqrt{n}}$$

$\sigma_{\bar{x}}$ is often called the **standard error of the mean**.

EXAMPLE

Table 5-1 illustrates the last four digits of the Social Insurance Numbers of each of 50 students.

The last four digits of Social Insurance Numbers are random. If we combine the four digits from each student into one big collection of 200 numbers, we get a mean of $\bar{x} = 4.5$, a standard deviation of $s = 2.8$, and an approximately uniform distribution with the graph shown in Figure 5-19. Now see what happens when we find the 50 sample means, as shown in Table 5-1. Even though the original collection of data has an approximately *uniform* (that is, not normal) distribution, the sample means have a distribution that is approximately *normal*. This can be a confusing concept, so you should stop right here and study this paragraph until its major point becomes clear: The original set of 200 individual numbers has a uniform distribution (because the digits 0–9 occur with approximately equal frequencies), but the 50 sample means have a normal distribution. It's a truly fascinating and intriguing phenomenon in statistics that by sampling from any distribution, we can create a distribution that is normal or at least approximately normal.

Table 5-1				
SIN Digits				$\bar{x}$
1	8	6	4	4.75
5	3	3	6	4.25
9	8	8	8	8.25
5	1	2	5	3.25
9	3	3	5	5.00
4	2	6	2	3.50
7	7	1	6	5.25
9	1	5	4	4.75
5	3	3	9	5.00
7	8	4	1	5.00
0	5	6	1	3.00
9	8	2	2	5.25
6	1	5	7	4.75
8	1	3	0	3.00
5	9	6	9	7.25
6	2	3	4	3.75
7	4	0	7	4.50
5	7	5	6	5.75
4	1	5	7	4.25
1	2	0	6	2.25
4	0	2	8	3.50
3	1	2	5	2.75
0	3	4	0	1.75
1	5	1	0	1.75
9	7	4	0	5.00
7	3	1	1	3.00
9	1	1	3	3.50
8	6	5	9	7.00
5	6	4	1	4.00
9	3	9	5	6.50
6	0	7	3	4.00
8	2	9	6	6.25
0	2	8	6	4.00
2	0	9	7	4.50
5	8	9	0	5.50
6	5	4	9	6.00
4	8	7	6	6.25
7	1	2	0	2.50
2	9	5	0	4.00
8	3	2	2	3.75
2	7	1	6	4.00
6	7	7	1	5.25
2	3	3	9	4.25
2	4	7	5	4.50
5	4	3	7	4.75
0	4	3	8	3.75
2	5	8	6	5.25
7	1	3	4	3.75
8	3	7	0	4.50
5	6	6	7	6.00

Applying the Central Limit Theorem

Many important and practical problems can be solved with the central limit theorem. When working on such problems, remember the following rules.

- If you are working with a random sample of size $n > 30$, or if the original population is normally distributed, treat the distribution of sample means as a normal distribution.

- Treat the mean μ of the original population as the mean of the distribution of sample means.

- Treat the calculated value $\sigma/\sqrt{n}$ (based on the original population) as the standard deviation of the distribution of sample means.

In the following example, part (a) involves an individual value, so we use the methods presented in Section 5–3; those methods apply to the normal distribution of the random variable x. Part (b), however, involves the mean for a *group* of 36 Canadian Football League (CFL) players, so we must use the central limit theorem in working with the random variable $\bar{x}$. Observe the significant difference between the procedures used in part (a) and part (b).

EXAMPLE

In human engineering and product design, it is often important to consider the weights of people so that airplanes or elevators aren't overloaded, chairs don't break, and other such dangerous or embarrassing mishaps do not occur. Given that the population of players in offensive back positions (including quarterback) in the CFL have weights that are approximately normally distributed, with a mean of 197.5 lb and a standard deviation of 14.2 lb (based on sampling from data on the CFL website), find the probability that

a. if one player is randomly selected, his weight is greater than 200 lb

b. if 36 different players are randomly selected, their mean weight is greater than 200 lb

SOLUTION

a. *Approach: Use the methods presented in Section 5–3* (because we are dealing with an *individual* value from a normally distributed population). We seek the area of the shaded region in Figure 5-22(a).

$$z = \frac{x - \mu}{\sigma} = \frac{200 - 197.5}{14.2} = 0.18$$

We now refer to Table A-2 to find that region A is 0.0714. The shaded region is therefore $0.5 - 0.0714 = 0.4286$. The probability of the player weighing more than 200 lb is 0.4286.

b. *Approach: Use the central limit theorem* (because we are dealing with the *mean for a group* of 36 values, not an individual value). Because we are now dealing with a

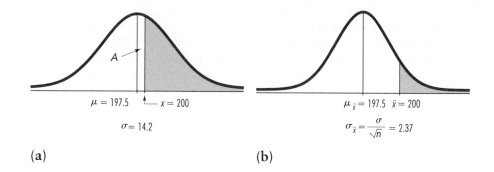

Figure 5-22
Distribution of
(a) Individual Offensive
Backs' Weights and
(b) Means of Samples
of 36 Offensive Backs'
Weights

(a)

$\mu = 197.5$ ⌐— $x = 200$

$\sigma = 14.2$

(b)

$\mu_{\bar{x}} = 197.5$ $\bar{x} = 200$

$\sigma_{\bar{x}} = \dfrac{\sigma}{\sqrt{n}} = 2.37$

distribution of sample means, we must use the parameters $\mu_{\bar{x}}$ and $\sigma_{\bar{x}}$, which are evaluated as follows:

$$\mu_{\bar{x}} = \mu = 197.5$$

$$\sigma_{\bar{x}} = \frac{\sigma}{\sqrt{n}} = \frac{14.2}{\sqrt{36}} = 2.37$$

We want to determine the shaded area shown in Figure 5-22(b), and the relevant z score is calculated as follows:

$$z = \frac{\bar{x} - \mu_{\bar{x}}}{\sigma_{\bar{x}}} = \frac{200 - 197.5}{\dfrac{14.2}{\sqrt{36}}} = \frac{2.5}{2.37} = 1.05$$

Referring to Table A-2, we find that $z = 1.05$ corresponds to an area of 0.3531, so the shaded region is $0.5 - 0.3531 = 0.1469$. The probability that the 36 players have a mean weight greater than 200 lb is 0.1469.

Interpretation

There is a 0.4286 probability that an offensive back in the CFL will weigh more than 200 lb, but there is only a 0.1469 probability that 36 offensive backs (collectively) will have a mean weight of more than 200 lb. It is much easier for an individual to deviate from the mean than it is for a group of 36. A single extreme weight among the 36 weights will lose its impact when it is averaged in with the other 35 weights.

Also, remember that the calculated probabilities are usually approximate, although they are displayed here to four decimal places. For example, the solutions based on using Table A-2 will differ slightly, due to rounding, from solutions based on statistical software. Moreover, these calculations assume (especially when working with single x values) that the population is normal—which may not be *exactly* true for your data.

The next example illustrates another application of the central limit theorem, but carefully examine the conclusion that is reached. This example shows the type of thinking that is the basis for the important procedure of hypothesis testing (discussed in Chapter 7), and illustrates the **rare event rule** for inferential statistics: If, under a given assumption, the probability of a particular observed event is exceptionally small, we conclude that the assumption is probably not correct.

STATISTICAL PLANET

A Professional Speaks About Sampling Error
Daniel Yankelovich, in an essay for *Time*, commented on the sampling error often reported along with poll results. He stated that sampling error refers only to the inaccuracy created by using random sample data to make an inference about a population; the sampling error does not address issues of poorly stated, biased, or emotional questions. He said, "Most important of all, warning labels about sampling error say nothing about whether or not the public is conflict-ridden or has given a subject much thought. This is the most serious source of opinion poll misinterpretation."

The Fuzzy Central Limit Theorem
In *The Cartoon Guide to Statistics* by Gonick and Smith, the authors describe the *fuzzy central limit theorem* as follows: "Data that are influenced by many small and unrelated random effects are approximately normally distributed. This explains why the normal is everywhere: stock market fluctuations, student weights, yearly temperature averages, [test] scores: All are the result of many different effects." People's heights, for example, are the results of hereditary factors, environmental factors, nutrition, health care, geographic region, and other influences that, when combined, produce normally distributed values.

EXAMPLE

Assume that the population of human body temperatures has a mean of 37.0°C, as is commonly believed. Also assume that the population standard deviation is 0.62°C. If a sample of size $n = 108$ is randomly selected, find the probability of getting a mean of 36.4°C or lower. (The value of 36.4°C was actually obtained; see the 108 temperatures in Data Set 18 of Appendix B.)

SOLUTION

We weren't given the distribution of the population, but because the sample size $n = 108$ exceeds 30, we use the central limit theorem and conclude that the distribution of sample means is a normal distribution with these parameters:

$$\mu_{\bar{x}} = \mu = 37.0$$

$$\sigma_{\bar{x}} = \frac{\sigma}{\sqrt{n}} = \frac{0.62}{\sqrt{108}} = 0.059659$$

Figure 5-23 shows the shaded area (see the left tail of the graph) corresponding to the probability we seek. Having already found the parameters that apply to the distribution shown in Figure 5-23, we can now find the shaded area by using the same procedures developed in the preceding section. We first find the z score:

$$z = \frac{\bar{x} - \mu_{\bar{x}}}{\sigma_{\bar{x}}} = \frac{36.4 - 37.0}{0.059659} = -10.06$$

Referring to Table A-2, we find that $z = -10.06$ (or its positive equivalent) is off the chart, but for absolute values of z above 3.09, we use an area of 0.4999. We therefore conclude that region A in Figure 5-23 is 0.4999 and the shaded region is $0.5 - 0.4999 = 0.0001$.

Interpretation

The result shows that if the mean of our body temperatures is really 37°C, then there is an extremely small probability of getting a sample mean of 36.4°C or lower when 108 subjects are randomly selected. Either the population mean really is 37.0°C and the sample represents a chance event that is extremely rare, or the population mean is actually lower than 37.0°C so the sample is typical. Because the probability is so low, it seems more reasonable to conclude that the population mean is lower than 37.0°C. This is the type of reasoning used in hypothesis testing, to be introduced in Chapter 7. For now, we should focus on the use of the central limit theorem for finding the probability of 0.0001, but we should also observe that this theorem will be used later in developing some very important concepts in statistics.

Figure 5-23
Distribution of Sample Mean Body Temperatures ($n = 108$)

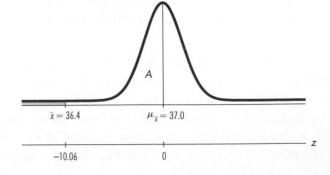

$\bar{x} = 36.4$ $\mu_{\bar{x}} = 37.0$

-10.06 0 z

Correcting for a Finite Population

In applying the central limit theorem, our use of $\sigma_{\bar{x}} = \sigma/\sqrt{n}$ assumes that the population has infinitely many members. When we sample with replacement (that is, put back each selected item before making the next selection), the population is effectively infinite. Yet many realistic applications involve sampling without replacement, so successive samples depend on previous outcomes. In manufacturing, quality control inspectors typically sample items from a finite production run without replacing them. For such a finite population, we may need to adjust $\sigma_{\bar{x}}$. Here is a common rule of thumb:

> **When sampling without replacement and the sample size n is greater than 5% of the finite population size N (that is, $n > 0.05N$), adjust the standard deviation of sample means $\sigma_{\bar{x}}$ by multiplying it by the *finite population correction factor*:**

$$\sqrt{\frac{N - n}{N - 1}}$$

Except for Exercises 15, 20, 21, and 23, the examples and exercises in this section assume that the finite population correction factor does not apply, because the population is infinite or the sample size does not exceed 5% of the population size.

It's very important to note that the central limit theorem applies when we are dealing with the distribution of sample means and either the sample size is greater than 30 or the original population has a normal distribution. Once we have established that the central limit theorem applies, we can determine values for $\mu_{\bar{x}} = \mu$ and $\sigma_{\bar{x}} = \sigma/\sqrt{n}$ and then proceed to use the methods presented in the preceding section.

The central limit theorem is so important because it allows us to use the basic normal distribution methods in a wide variety of different circumstances. In Chapter 6, for example, we will apply the theorem when we use sample data to estimate means of populations. In Chapter 7 we will apply it when we use sample data to test claims made about population means. Such applications of estimating population parameters and testing claims are extremely important uses of statistics, and the central limit theorem makes them possible.

EXAMPLE

An investment club has 500 members. Their annual income is normally distributed with a mean of $125,400 and standard deviation of $27,642.

a. If a random sample of 100 members is taken, what is the probability that their average income is less than $125,000?

b. Suppose the sample is 200 instead of 100. What does the probability in part (a) change to?

SOLUTION

Since $n/N = 100/500 = 20\%$ is greater than 5%, we need to use the finite population correction factor.

a. $P(\bar{X} < 125{,}000)$

$$= P\left(Z < \frac{125{,}000 - 125{,}400}{\dfrac{27{,}642}{\sqrt{100}}\sqrt{\dfrac{500-100}{500-1}}}\right) = P\left(Z < \frac{125{,}000 - 125{,}400}{\dfrac{27{,}642}{\sqrt{100}}\sqrt{\dfrac{400}{499}}}\right)$$

$$= P(Z < -0.16) = 0.5 - 0.0636 = 0.4364$$

b. $P(\bar{X} < 125{,}000)$

$$= P\left(Z < \frac{125{,}000 - 125{,}400}{\dfrac{27{,}642}{\sqrt{200}}\sqrt{\dfrac{500-200}{500-1}}}\right) = P\left(Z < \frac{125{,}000 - 125{,}400}{\dfrac{27{,}642}{\sqrt{200}}\sqrt{\dfrac{300}{499}}}\right)$$

$$= P(Z < \ 0.26) = 0.5 - 0.1026 = 0.3974$$

Notice that the probability in part (b) is less than that in part (a). There are two reasons for this: First, as the sample size increases, $\bar{x}$ becomes closer to μ because of the law of large numbers; second, as the sample size gets closer to the population size, the correction factor becomes smaller and, as a result, the magnitude of the z score becomes greater.

 5-5 **Exercises A: Basic Skills and Concepts**

In Exercises 1–4, assume that female students' heights are normally distributed with a mean given by $\mu = 64.2$ in and a standard deviation given by $\sigma = 2.6$ in.

1. a. If one female student is randomly selected, find the probability that her height is between 64.2 in and 65.2 in.
   b. If 36 female students are randomly selected, find the probability that they have a mean height between 64.2 in and 65.2 in.

2. a. If one female student is randomly selected, find the probability that her height is above 63.0 in.
   b. If 100 female students are randomly selected, find the probability that they have a mean height greater than 63.0 in.

3. a. If one female student is randomly selected, find the probability that her height is above 65.0 in.
   b. If 50 female students are randomly selected, find the probability that they have a mean height greater than 65.0 in.

4. a. If 4 female students are randomly selected, find the probability that their mean height is between 63.0 in and 65.0 in.
   b. Why can the central limit theorem be used in part (a), even though the sample size does not exceed 30?

5. Replacement times for TV sets are normally distributed with a mean of 8.2 years and a standard deviation of 1.1 years (based on data from "Getting Things Fixed," *Consumer Reports*). Find the probability that 40 randomly selected TV sets will have a mean replacement time less than 8.0 years.

6. Replacement times for CD players are normally distributed with a mean of 7.1 years and a standard deviation of 1.4 years (based on data from "Getting Things Fixed," *Consumer Reports*). Find the probability that 45 randomly selected CD players will have a mean replacement time of 6.8 years or less. If 45 CD players were, in fact, selected randomly from the past sales of E-Electronics Edmonton, and had a mean replacement time of 6.8 years, would it appear that this store has been given CD players with lower than average quality?

7. The total weights of garbage that households discard weekly is approximately normally distributed, with a mean of 12.5 kg and a standard deviation of 5.7 kg (based on Data Set 1 in Appendix B).
   a. If 120 households are randomly selected, find the probability that the mean weight of their discarded garbage is over 13.5 kg.
   b. If the town's waste transfer station allocates capacity for 1690 kg of temporary garbage storage per 120 households per week, and the typical garbage route is based on 120 households, what percentage of garbage routes will exceed the allocated capacity for their garbage? Is this an acceptable level or should the town council take corrective action?

8. According to the International Mass Retail Association, girls aged 13 to 17 spend an average of $31.20 on shopping trips in a month. Assume that the amounts have a standard deviation of $8.27. If 85 girls in that age category are randomly selected, what is the probability that their mean monthly shopping expense is between $30.00 and $33.00?

9. For females aged 18–24, systolic blood pressures (in mm Hg) are normally distributed with a mean of 114.8 and a standard deviation of 13.1.
   a. If a woman between the ages of 18 and 24 is randomly selected, find the probability that her systolic blood pressure is above 120.
   b. If 12 women in that age bracket are randomly selected, find the probability that their mean systolic blood pressure is greater than 120.
   c. Given that part (b) involves a sample size that is not larger than 30, why can the central limit theorem be used?

10. The *depth* of a diamond signifies 100 times the ratio of its height to its diameter. For many high-priced diamonds, depth appears to be normally distributed,

with a mean of 60.83 and a standard deviation of 1.97 (based on Data Set 21 in Appendix B). If 40 high-priced diamonds are randomly selected, find the probability that their mean depth is less than 61.2.

11. The ages of U.S. commercial aircraft have a mean of 13.0 years and a standard deviation of 7.9 years (based on data from Aviation Data Services). If the Federal Aviation Administration randomly selects 35 commercial aircraft for special stress tests, find the probability that the mean age of this sample group is greater than 15.0 years.

12. Information provided by Bell Canada in Winnipeg showed that the average monthly bill for basic service was normally distributed with a mean of $15.30 and a standard deviation of $1.25. If 55 phone bills are randomly selected, find the probability that their mean exceeds $15.

13. The typical computer random-number generator yields numbers in a uniform distribution between 0 and 1 with a mean of 0.500 and a standard deviation of 0.289. If 45 random numbers are generated, find the probability that their mean is below 0.565.

14. A study was made of seat-belt use among children who were involved in car crashes that caused them to be hospitalized. It was found that children not wearing any restraints had hospital stays with a mean of 7.37 days and a standard deviation of 0.79 day (based on data from "Morbidity Among Pediatric Motor Vehicle Crash Victims: The Effectiveness of Seat Belts," by Osberg and Di Scala, *American Journal of Public Health,* Vol. 82, No. 3). If 40 such children are randomly selected, find the probability that their mean hospital stay is greater than 7.00 days.

15. As part of his review for a chemistry test, a student randomly selects (without replacement) 40 out of the 102 elements listed on the standard tables, and tries to guess their atomic numbers. Given that the mean of all 102 atomic numbers is 51.794, and the population standard deviation is 29.804, what is the probability that the mean of the 40 atomic numbers selected by the student will be at least 58.3? (*Hint:* See the discussion of the finite population correction factor.)

16. SAT verbal scores are normally distributed with a mean of 430 and a standard deviation of 120 (based on data from the College Board ATP). Randomly selected SAT verbal scores are obtained from the population of students who took a test preparatory course from a training school. Assume that this training course has no effect on test scores.
    a. If one of the students is randomly selected, find the probability that he or she obtained a score greater than 440.
    b. If 100 students of an SAT preparation course achieve a sample mean of 440, does it seem reasonable to conclude that the course is effective because the students perform better on the SAT?

17. The lengths of pregnancies are normally distributed with a mean of 268 days and a standard deviation of 15 days.
    a. If one pregnant female is randomly selected, find the probability that her length of pregnancy is less than 260 days.
    b. If 25 randomly selected females are put on a special diet just before they become pregnant, find the probability that their lengths of pregnancy have a mean that is less than 260 days (assuming that the diet has no effect).
    c. If the 25 females do have a mean of less than 260 days, should the medical supervisors be concerned?

18. A survey by Statistics Canada shows the average household expenditure on post-secondary textbooks is $53.00, with a standard deviation of $18.61. A similar poll randomly selects 150 households to find average expenditure on post-secondary textbooks.
    a. Find the probability that the mean is greater than $57.00.
    b. If a sample of 150 households does yield a mean of $57.00 or greater, is there reason to believe that this sample came from a population with a mean that is higher than $53.00?

19. M&M plain candies have a mean weight of 0.9147 g and a standard deviation of 0.0369 g (based on Data Set 11 in Appendix B). The M&M candies used in Data Set 11 came from a package containing 1498 candies, and the package label stated that the net weight is 48.0 oz, or 1361 g. (If every package has 1498 candies, the mean weight must exceed 1361/1498 = 0.9085 g for the net contents to weigh at least 1361 g.)
    a. If an M&M plain candy is randomly selected, find the probability that it weighs more than 0.9085 g.
    b. If 1498 M&M plain candies are randomly selected, find the probability that their mean weight is at least 0.9085 g.
    c. Given these results, does it seem that Mars is providing M&M consumers with the amount claimed on the label?

20. Weights of players in offensive back positions (including quarterback) in the Canadian Football League are approximately normally distributed, with a mean of 197.5 lb and a standard deviation of 14.2 lb (based on data from the CFL website). An elevator at the hotel that is hosting an awards dinner for all CFL offensive backs (there are 90 such players) posts a limit of 32 occupants, but it will be overloaded if the 32 occupants have a combined weight of 6450 lb or over. If a random selection of 32 of the offensive backs happen to get on the elevator at one time, find the probability that their combined weight will exceed 6450 lb, causing the elevator to be overloaded. Based on the value obtained, is there reason for concern?

21. A *population* consists of these scores: 2, 3, 6, 8, 11, 18.
    a. Find μ and σ.
    b. List all samples of size $n = 2$ that are obtained without replacement.
    c. Find the population of all values of $\bar{x}$ by finding the mean of each sample from part (b).
    d. Find the mean $\mu_{\bar{x}}$ and standard deviation $\sigma_{\bar{x}}$ for the population of sample means found in part (c).
    e. Verify that

$$\mu_{\bar{x}} = \mu \quad \text{and} \quad \sigma_{\bar{x}} = \frac{\sigma}{\sqrt{n}}\sqrt{\frac{N - n}{N - 1}}$$

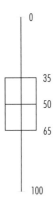

22. An education researcher develops an index of academic interest and obtains scores for a randomly selected sample of 350 college students. The results are summarized in the accompanying boxplot. If 15 of the students are randomly selected, find the probability that their mean score is greater than 55.

23. The finite population correction factor can be ignored when sampling with replacement or when $n \leq 0.05N$. When collecting a sample (without replacement) that is 5% of the population N, what do the values of the finite population correction factor have in common for values of $N \geq 600$? (That is, try calculating the factor for a few values of $N > 600$ and with $n = 0.05N$. What do the results have in common?)

5-6 Normal Distribution as Approximation to Binomial Distribution

In Section 4–3 we introduced the *binomial probability distribution*, which applies to a discrete random variable (rather than to a continuous random variable, as does the normal distribution). We noted that a binomial distribution has these four requirements:

1. The experiment must have a *fixed number of trials*.
2. The trials must be *independent*.
3. Each trial must have all outcomes classified into *two categories*.
4. The probabilities must remain *constant* for each trial.

In Section 4–5, we presented the Poisson distribution, which is derived from the binomial distribution. One scenario in which we use the Poisson distribution is when the number of trials n is large and the probability of success p is small. What if n is large and p is not small? What do we do in a situation like this? The answer is that we can approximate the binomial distribution with the normal distribution.

Suppose we want to determine $P(X = 3)$. If we approximate the binomial distribution with the Poisson distribution, we can still calculate this exact probability since both distributions are discrete. However, if we approximate the binomial distribution with the normal distribution, $P(X = 3) = 0$ under the normal distribution since the normal distribution is continuous. We need to find a way around this problem. The solution is called the continuity correction.

Figure 5-24 shows the histogram of binomial probabilities of $X = 2, 3,$ 4 with $n = 10$ and $p = 0.4$. The way the continuity correction works is that we picture the histogram bar for $X = 3$ stretching from 2.5 to 3.5. Since we now have an area under a curve (specifically a histogram bar), we can now calculate probabilities using the normal distribution.

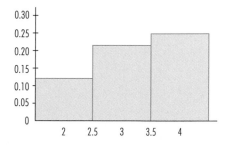

Figure 5-24
Partial Binomial Probabilities for $n = 10$ and $p = 0.4$

Reliability and Validity

The reliability of data refers to the consistency with which results occur, whereas the validity of data refers to how well the data measure what they are supposed to measure. The reliability of an IQ test can be judged by comparing scores for the test given on one date to scores for the same test given at another time. To test the validity of an IQ test, we might compare the test scores to another indicator of intelligence, such as academic performance. Many critics charge that IQ tests are reliable, but not valid; they provide consistent results, but don't really measure intelligence.

When to Use the Normal Approximation

There are principally two situations in which we use the normal approximation to the binomial distribution.

Situation 1: Table A-1 cannot be used. This would arise if either the number of trials n exceeds the table limit of 15 or the probability of success p is not one of the values listed in the table.

Situation 2: Computer software is either unavailable or is inadequate. For example, if $n = 5000$ and $p = 0.2$ and you attempt to calculate the probability of no more than 1000 successes on Excel, depending on the version you have, you may not get a numeric answer.

NORMAL DISTRIBUTION AS APPROXIMATION TO BINOMIAL DISTRIBUTION

If $np \geq 5$ and $nq \geq 5$, and provided p is sufficiently large, then the binomial random variable is approximately normally distributed with the mean and standard deviation given as

$$\mu = np$$
$$\sigma = \sqrt{npq}$$

Figure 5-25
Binomial Histogram for
$n = 20$ and $p = 0.5$

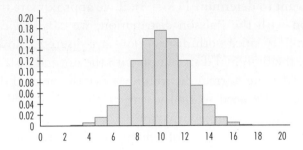

Since the binomial random variable X is approximately normally distributed, we can covert it to Z as we would any other normal random variable:

$$Z = \frac{X - \mu}{\sigma} = \frac{X_B - np}{\sqrt{npq}}$$

where X_B denotes the binomial random variable after the continuity correction.

Observe Figure 5-25 (which applies to a binomial distribution with $n = 20$, $p = 0.5$, and $q = 0.5$) and note that this particular binomial distribution does have a probability histogram with roughly the same shape as that of a normal distribution. The formal justification that allows us to use the normal distribution as an approximation to the binomial distribution results from the central limit theorem, but Figure 5-25 is a visual argument supporting that approximation.

We will now illustrate this normal approximation procedure with an example.

EXAMPLE

Assume that your college has an equal number of qualified male and female applicants, and assume that 64 of the last 100 newly hired employees are men. Estimate the probability of getting *at least* 64 men if each hiring is done independently and with no gender discrimination. (The probability of getting *exactly* 64 men doesn't really tell us anything, because with 100 trials, the probability of any exact number of men is fairly small. Instead, we need the probability of getting a result *at least* as extreme as the one obtained.) Based on the result, does it seem that the college is discriminating on the basis of gender?

SOLUTION

The given problem involves a binomial distribution with a fixed number of trials ($n = 100$), which are presumably independent, two categories (man, woman) of outcome for each trial, and probabilities that presumably remain constant. Since there is an equal number of qualified males and females, $p = 0.5$. We establish that it is reasonable to approximate the binomial distribution by the normal distribution because p is sufficiently large, $np \geq 5$ and $nq \geq 5$, as verified below:

$$np = 100 \cdot 0.5 = 50 \qquad \text{(Therefore } np \geq 5.)$$
$$nq = 100 \cdot 0.5 = 50 \qquad \text{(Therefore } nq \geq 5.)$$

We now proceed to find the values for μ and σ that are needed. We get the following:

$$\mu = np = 100 \cdot 0.5 = 50$$
$$\sigma = \sqrt{npq} = \sqrt{100 \cdot 0.5 \cdot 0.5} = 5$$

In Figure 5-26 we show the discrete value of 64 as being represented by the vertical strip bounded by 63.5 and 64.5. Figure 5-27 shows the area we want: It is the shaded area in the extreme right tail of the graph. This area corresponds to the following z score:

$$z = \frac{x - \mu}{\sigma} = \frac{63.5 - 50}{5} = 2.70$$

Using Table A-2, we find that $z = 2.70$ corresponds to an area of 0.4965, so the shaded area we seek is $0.5 - 0.4965 = 0.0035$.

Assuming that men and women have equal chances of being hired, the probability of getting at least 64 men in 100 new employees is approximately 0.0035. (Using the software packages of Excel or STATDISK, we get the answer of 0.0033, so the approximation is quite good here.)

Interpretation

Because the probability of getting at least 64 men is so small (0.0035), we conclude that either a very rare event has occurred or the assumption that men and women have the same chance is incorrect. It appears that the college discriminates on the basis of gender. The reasoning here will be considered in more detail in Chapter 7 when we discuss formal methods of testing hypotheses. For now, we should focus on the method of finding the probability by using the normal approximation technique.

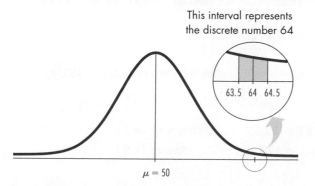

Figure 5-26 **Illustration of Continuity Correction**

This interval represents the discrete number 64

63.5 64 64.5

$\mu = 50$

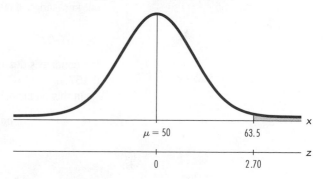

Figure 5-27 **Finding the Probability of "At Least" 64 Men Among 100 New Employees**

$\mu = 50$ 63.5   x

0   2.70   z

Figure 5-28
Identifying the Correct Area

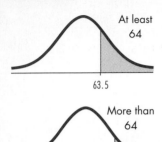

At least 64

63.5

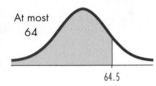

More than 64

64.5

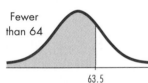

At most 64

64.5

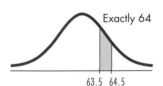

Fewer than 64

63.5

Exactly 64

63.5 64.5

Continuity Corrections

In the previous example, we converted 64 to 63.5, which is the **continuity correction**. Depending on the probability we want, we either add 0.5 or subtract 0.5 from the original value. Using the value of $X = 64$, here is how we would calculate different probabilities using the continuity correction, as follows:

$$P(X \geq 64) = P(X_B > 63.5)$$
$$P(X > 64) = P(X_B > 64.5)$$
$$P(X \leq 64) = P(X_B < 64.5)$$
$$P(X < 64) = P(X_B < 63.5)$$
$$P(X = 64) = P(63.5 < X_B < 64.5)$$

To understand how these corrections work, it may be useful to draw a histogram. For example, $P(X \geq 64)$ is the cumulative probability of 64, 65, 66, and so on up to and including 100. Since the histogram bar containing 64 actually starts at 63.5, we convert 64 to 63.5. Similarly, $P(X > 64)$ is the cumulative probability from 65 through 100. Since we are not including the histogram bar containing 64, we begin at 64.5. Going the other way, $P(X \leq 64)$ is the cumulative probability from 0 through 64 inclusive. Since the histogram bar containing 64 ends at 64.5, we convert 64 to 64.5. Similarly, $P(X < 64)$ is the cumulative probability from 0 through 63 inclusive. Since we are not including the histogram bar containing 64, we end at 63.5. Finally, for $P(X = 64)$, we begin at 63.5 and end at 64.5 since we are interested in just the one histogram bar.

These continuity corrections are illustrated in Figure 5-28.

EXAMPLE

It is common for chemistry data tables to include the atomic masses of the elements. For 15.7% of the elements, however, the atomic masses cannot be known exactly, unless something is known of a specimen's origin. Suppose a chemistry professor is giving to each of his 150 students an individualized test, such that each student receives the name of one of the elements, which they must discuss. A computer randomly selects an element name for each student independently. What is the probability that exactly 30 students will receive the name of an element for which the atomic mass cannot be known exactly without knowledge of origin?

SOLUTION

The conditions described satisfy the criteria for the binomial distribution with $n = 150$, $p = 0.157$, $q = 0.843$, and $x = 30$.
In this problem,

$$np = 150 \cdot 0.157 = 23.55 \qquad \text{(Therefore } np \geq 5.\text{)}$$
$$nq = 150 \cdot 0.843 = 126.45 \qquad \text{(Therefore } nq \geq 5.\text{)}$$

Because p is sufficiently large and np and nq are both at least 5, we conclude that the normal approximation to the binomial distribution is satisfactory.

We obtain the values of μ and σ as follows:

$$\mu = np = 150 \cdot 0.157 = 23.55$$

$$\sigma = \sqrt{npq} = \sqrt{(150)(0.157)(0.843)} = 4.4556313$$

We now convert to Z:

$$P(X = 30) = P(29.5 < X_B < 30.5) = P\left(\frac{29.5 - 23.55}{4.4556313} < Z < \frac{30.5 - 23.55}{4.4556313}\right)$$
$$= P(1.34 < Z < 1.56) = 0.4406 - 0.4099 = 0.0307$$

The probability of exactly 30 students (out of 150) getting names of chemical elements whose atomic mass cannot be exactly known without knowledge of origin is 0.0307.

If we solve the above example using Excel or STATDISK, we get a result of 0.0305, but the normal approximation method resulted in a value of 0.0307. The discrepancy of 0.0002 occurs because the use of the normal distribution results in an *approximate* value that is the area of the shaded region in Figure 5-29, whereas the correct area is a rectangle centred above 30. (Figure 5-29 illustrates this discrepancy.)

Sufficient Size for p

One of the conditions specified to use the normal approximation is for p to be sufficiently large. How large should it be? There are no hard and fast rules but a general guide is that it should be greater than 1%. The following examples illustrate when the Poisson distribution is superior to the normal approximation and vice versa. The probabilities were computed using Excel.

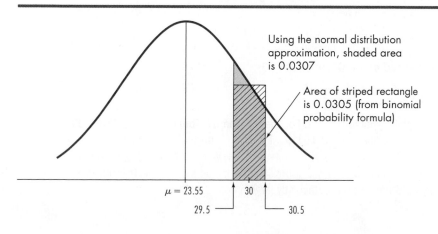

Figure 5-29
Applying a Continuity Correction

Using the normal distribution approximation, shaded area is 0.0307

Area of striped rectangle is 0.0305 (from binomial probability formula)

$\mu = 23.55$ 30

29.5   30.5

EXAMPLE

Given $n = 1,200,000$ and $p = 1/500,000$, find $P(X \leq 1)$.

SOLUTION

Using the binomial distribution, the probability is 0.3084. To use the Poisson distribution, $\mu = (1,200,000)(1/500,000) = 2.4$. The probability is also 0.3084. To use the normal approximation, $\sigma = \sqrt{(1,200,000(1/500,000)(499,999/500,000)} = \sqrt{2.4}$ for all intents and purposes. Then $P(X \leq 1) = P(X_B < 1.5) = P(Z < \frac{1.5 - 2.4}{\sqrt{2.4}}) = P(Z < -0.5809) = 0.2806$. In this case, using the Poisson distribution is better since $np < 5$ and p is small.

EXAMPLE

Given $n = 4,000,000$ and $p = 1/500,000$, find $P(6 \leq X \leq 8) = P(X \leq 8) - P(X \leq 5)$.

SOLUTION

Using the binomial distribution, the probability is 0.4013. For the Poisson distribution, $\mu = (4,000,000)(1/500,000) = 8$ with the resultant probability also being 0.4013. For the normal approximation, $\sigma = \sqrt{8}$. Then $P(X \leq 8) - P(X \leq 5) = P(X_B < 8.5) - P(X_B < 5.5) = P(Z < 0.1768) - P(Z < -0.8839) = 0.3818$. Even though $np \geq 5$, the Poisson distribution is again superior due to p being small.

EXAMPLE

Given $n = 200$ and $p = 0.01$, find $P(X \leq 3)$.

SOLUTION

Using the binomial distribution, the probability is 0.8580. For the Poisson distribution, $\mu = (200)(0.01) = 2$ and the probability is 0.8571. For the normal approximation, $\sigma = \sqrt{(200)(0.01)(0.99)} = \sqrt{1.98}$. Then $P(X \leq 3) = P(X_B < 3.5) = P(Z < 1.066) = 0.8568$. In this situation, even though $np < 5$, the Poisson distribution is only marginally better than the normal approximation since $p = 0.01$.

EXAMPLE

Given $n = 200$ and $p = 0.03$, find $P(X \leq 3)$.

SOLUTION

The binomial probability is 0.1472. For the Poisson distribution, $\mu = (200)(0.03) = 6$ and the probability is 0.1512. For the normal approximation, $\sigma = \sqrt{(200)(0.03)(0.97)} = \sqrt{5.82}$. Then $P(X \leq 3) = P(X_B < 3.5) = P(Z < -1.0363) = 0.15$. In this situation, the normal approximation is superior to the Poisson distribution since (1) $np \geq 5$ and (2) p is sufficiently large.

5-6 Exercises A: Basic Skills and Concepts

In Exercises 1–8, use the continuity correction and describe the region of the normal curve that corresponds to the indicated probability. For example, the probability of "more than 47 successes" corresponds to this area of the normal curve: the area to the right of 47.5.

1. Probability of more than 35 defective parts

2. Probability of at least 175 girls

3. Probability of fewer than 42 correct answers to multiple-choice questions

4. Probability of exactly 65 correct answers to true/false questions

5. Probability of no more than 72 cars with defective brakes

6. Probability that the number of baby girls is between 35 and 45 inclusive

7. Probability that the number of Conservative voters is between 125 and 150 inclusive

8. Probability that the number of patients with group A blood is exactly 34

In Exercises 9–12, do the following:
a. Find the indicated binomial probability using Table A-1 in Appendix A.
b. If $np \geq 5$ and $nq \geq 5$, also estimate the indicated probability by using the normal distribution as an approximation to the binomial distribution; if $np < 5$ or $nq < 5$, then state that the normal approximation is not suitable.

9. With $n = 14$ and $p = 0.50$, find $P(8)$.

10. With $n = 10$ and $p = 0.40$, find $P(7)$.

11. With $n = 15$ and $p = 0.80$, find P(at least 8).

12. With $n = 14$ and $p = 0.60$, find P(fewer than 9).

13. Estimate the probability of getting at least 55 girls in 100 births.

14. Estimate the probability of getting exactly 32 boys in 64 births.

15. Estimate the probability of passing a true/false test of 50 questions if 60% (or 30 correct responses) represents a passing grade and all responses are random guesses.

16. A multiple-choice test consists of 50 questions with possible answers of a, b, c, d, and e. Estimate the probability of getting at most 30% correct if all answers are random guesses.

17. Based on past records, there is a 3.62% chance that any particular CFL player is a punter or kicker. For 140 randomly selected players, estimate the probability that exactly 3 are punters or kickers.

18. The Angus Reid Group surveyed Ontario residents on what they think will be the most valuable type of education to have in the work force ten years

from now. Thirty-five percent of the respondents chose a college diploma in technical occupations. Estimate the probability that if 1000 Ontario residents are selected at random, 350 of them would choose or recommend a college diploma in technical occupations.

19. According to a consumer affairs representative from Mars (the candy company, not the planet), 10% of all M&M plain candies are blue. Data Set 11 in Appendix B shows that among 100 M&Ms chosen, 5 are blue. Estimate the probability of randomly selecting 100 M&Ms and getting 5 or fewer that are blue. Assume that the company's 10% blue rate is correct. Based on the result, is it very unusual to get 5 or fewer blue M&Ms when 100 are randomly selected?

20. John Dukhia plans to place 200 bets of $1 each on the number 7 at roulette. On any one spin there is a probability of 1/38 that 7 will be the winning number. For John to end up with a profit, the number 7 must occur at least 6 times among the 200 trials. Estimate the probability that John finishes with a profit.

21. The Angus Reid Group surveyed Ontario residents on what they think will be the most valuable type of education to have in the work force ten years from now. Eighteen percent of the respondents chose a university degree in science. Estimate the probability that if 400 Ontario residents are selected at random, 70 or fewer would choose or recommend a university degree in science.

22. Air Borealis has been experiencing a 7% rate of no-shows on advance reservations. In a test project that requires passengers to confirm reservations, it is found that among 250 randomly selected advance reservations, there are 4 no-shows. Assuming that the confirmation requirement has no effect so the 7% rate applies, estimate the probability of 4 or fewer no-shows among 250 randomly selected reservations. Based on that result, does it appear that the confirmation requirement is effective?

23. In a survey, the Canadian Automobile Association (CAA) found that 38% of respondents had bought used vehicles. A car dealership surveys 150 people at random and finds that 45 of them have bought used vehicles. Estimate the probability of 45 or more respondents being used-vehicle purchasers. Based on that value, does the CAA survey result seem too high?

24. A successful venture capital firm reports that it finances only 10% of the proposals that it receives. A check of 100 randomly selected proposals shows that 8 of them were financed by the firm. Estimate the probability that among 100 proposals, 8 or fewer were financed by the firm. Based on the result, does it seem that the venture capital firm's claim is correct?

25. A manufacturer of automobile headlights claims that 10% of a batch of these headlights are defective. A quality control inspector tests a random selection of 300 headlights in the batch. The number of defective headlights in this random sample is 35, which is more than expected. Estimate the probability of

getting at least 35 defective lights in a random sample of 300. Based on the result, does it appear that the manufacturer's claim of 10% defectiveness is accurate? Was there a problem with the sample?

26. Some couples have genetic characteristics configured so that one-quarter of all offspring have blue eyes. A study is conducted of 40 couples believed to have those characteristics, with the result that 8 of their 40 offspring have blue eyes. Estimate the probability that among 40 offspring, 8 or fewer have blue eyes. Based on that probability, does it seem that the one-quarter rate is correct?

27. Sunnybrook Hospital is conducting a blood drive because its supply of group O blood is low, and it needs 177 donors of group O blood. If 400 volunteers donate blood, estimate the probability that the number with group O blood is at least 177. Forty-six percent of us have group O blood, according to data provided by Canadian Blood Services.

28. Some companies monitor quality by using a method of acceptance sampling whereby an entire batch of items is rejected if, in a random sample, the number of defects is at least some predetermined number. The Macker Machine Company buys machine bolts in batches of 5000 and rejects a batch if, when 50 of them are sampled, at least 2 defects are found. Estimate the probability of rejecting a batch if the supplier is manufacturing the bolts with a defect rate of 10%. Is this monitoring plan likely to identify the unacceptable rate of defects? Explain.

5-6 Exercises B: Beyond the Basics

29. Replacement times for TV sets are normally distributed with a mean of 8.2 years and a standard deviation of 1.1 years (based on data from "Getting Things Fixed," *Consumer Reports*). Estimate the probability that for 250 randomly selected TV sets, at least 15 of them have replacement times greater than 10.0 years.

30. Assume that a baseball player hits .350, so his probability of a hit is 0.350. (Ignore the complications caused by walks.) Also assume that his hitting attempts are independent of each other.
    a. Find the probability of at least 1 hit in 4 tries in 1 game.
    b. Assuming that this batter gets up to bat 4 times each game, estimate the probability of getting a total of at least 56 hits in 56 games.
    c. Assuming that this batter gets up to bat 4 times each game, find the probability of at least 1 hit in each of 56 consecutive games (Joe DiMaggio's 1941 record).
    d. What minimum batting average would be required for the probability in part (c) to be greater than 0.1?

31. Find the difference between the answers obtained with and without use of the continuity correction in each of the following. What do you conclude from the results?

a. Estimate the probability of getting at least 11 girls in 20 births.

b. Estimate the probability of getting at least 22 girls in 40 births.

c. Estimate the probability of getting at least 220 girls in 400 births.

32. Air Borealis works only with advance reservations and experiences a 7% rate of no-shows. How many reservations could be accepted for an airliner with a capacity of 250 if there is at least a 0.95 probability that all reservation holders who show will be accommodated?

Exponential Distribution

Aside from the uniform and normal distributions, there is one other continuous distribution that is commonly used: the exponential distribution.

The Exponential Distribution

DEFINITION

A continuous random variable X has an **exponential distribution** if the mean, μ, represents the average time, distance, or some other continuous measure between events.

If the Poisson distribution is one side of a coin, the exponential distribution is the other side. In the Poisson distribution, we computed the probability of a certain number of events over a certain interval, such as time. In the exponential distribution, we compute the probability of a certain interval, such as time, between events. An exponential distribution problem will initially appear as a Poisson distribution problem, but, instead of computing the mean of the number of events in the interval, we compute the mean of the interval between events.

Because μ represents the average measure between events, μ is necessarily positive and so are the values of X. Figure 5-30 shows an exponential distribution with $\mu = 2$.

The formula to calculate exponential probabilities is quite simple:

$$\text{For a value } k > 0, \quad P(X > k) = e^{-k/\mu}$$
$$\text{By extension,} \quad P(X < k) = 1 - e^{-k/\mu}$$

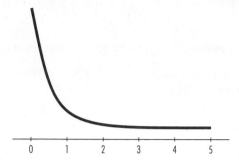

Figure 5-30
Exponential Distribution
with $\mu = 2$

EXAMPLE

A store averages 12 customers per hour.

a. What is the probability the store goes more than 2 minutes between customers?

b. What is the probability the store goes less than half a minute between customers?

You will notice that the opening statement appears to be that of a Poisson problem. However, the two questions refer to the amount of time between customers. This makes the problem exponential.

SOLUTION

In order to solve the problems, we need to calculate μ, the average time between customers. Since the store averages 12 customers per hour (or 60 minutes), this means that the average time between customers is $\frac{60}{12} = 5$ minutes. Thus $\mu = 5$.

a. $P(X > 2) = e^{-2/5} = e^{-0.4} = 0.6703$

b. $P(X < 0.5) = 1 - e^{-0.5/5} = 1 - e^{-0.1} = 1 - 0.9048 = 0.0952$

In the above example, μ was adjusted strictly for the time. There are situations, such as at a store or bank, when the amount of time a person waits also depends on how many staff are working and the number of customers waiting to be served.

EXAMPLE

When a bank has 6 customer service reps (CSRs) working, they collectively serve 1 customer per minute on average.

a. Suppose there are 6 CSRs working and there are 10 people in front of me. What is the probability that I will wait more than 15 minutes to be served?

b. Suppose there are 4 CSRs working and there are 10 people in front of me. What is the probability that I will wait more than 15 minutes to be served?

SOLUTION

In order to solve for μ, we need to adjust it for both the number of CSRs and customers.

a. Since there are 6 CSRs, we need to adjust μ only for the number of customers. If there were only 1 customer in front of me, I would expect to wait 1 minute; since there are 10, I should expect to wait 10 minutes. Thus $\mu = 10$.

$$P(X > 15) = e^{-15/10} = 0.2231$$

b. If there were 6 CSRs, $\mu = 10$ as in the previous example. However, since there are 4 CSRs instead of 6, $\mu = 10 \times 6/4 = 15$.

$$P(X > 15) = e^{-15/15} = 0.3679$$

Memoryless Feature of the Exponential Distribution

We have all had the experience of being on hold on the telephone and wondering how much longer it will be before we are served. The memoryless feature of the exponential distribution examines the probability of the total amount time we will wait given the amount of time we have been waiting already. Suppose the amount of time we have been waiting is t and the total amount of time is $t + s$ where s is the additional amount of time until we are served.

$$P(X > s + t \,|\, X > t) = \frac{P(X > s + t \text{ and } X > t)}{P(X > t)} = \frac{P(X > s + t)}{P(X > t)}$$

$$= \frac{e^{-(s+t)/\mu}}{e^{-t/\mu}} = \frac{e^{-s/\mu} \cdot e^{-t/\mu}}{e^{-t/\mu}} = e^{-s/\mu}$$

In other words, the exponential distribution has no memory of the amount of time you have been already waiting; it is simply concerned with the amount of time left until you are served.

EXAMPLE

The average wait time at a call centre is 25 minutes when there are 50 CSRs.

a. If there are 50 CSRs, what is the probability I will wait more than 20 minutes in total if I have been waiting 15 minutes already?

b. If there are 40 CSRs, what is the probability I will wait less than 20 minutes in total if I have been waiting 15 minutes already?

SOLUTION

As previously, we need to adjust μ for the number of CSRs.

a. Since there are 50 CSRs, $\mu = 25$.

$$P(X > 20 \,|\, X > 15) = P(X > 5) = e^{-5/25} = e^{-0.2} = 0.8187$$

b. Since there are 40 CSRs, $\mu = 25 \times 50/40 = 31.25$

$$P(X < 20 \,|\, X > 15) = 1 - P(X > 20 \,|\, X > 15) = 1 - P(X > 5)$$
$$= 1 - e^{-5/31.25} = 1 - e^{-0.16} = 1 - 0.8521 = 0.1479$$

5-7 Exercises A: Basic Skills and Concepts

In Exercises 1–4, suppose that a gas pipeline averages 1 blemish per 40 km. Find the following probabilities for a section of the pipeline.

1. More than 75 km between successive blemishes

2. Less than 90 km between successive blemishes

3. More than 60 km between successive blemishes if there is at least 40 km between successive blemishes

4. Less than 80 km between successive blemishes if there is at least 35 km between successive blemishes (*Hint:* if there is at least 35 km between blemishes, either there is more than 80 km or less than 80 km.)

In Exercises 5–7, suppose that when a grocery store has 3 cashiers working during its afternoon rush period, a customer waits 8 minutes on average to be served.

5. If there are 3 cashiers working, what is the probability a customer waits more than 10 minutes?

6. If there are 2 cashiers working, what is the probability a customer waits more than 15 minutes in total if the customer has already been waiting 10 minutes?

7. If there are 5 cashiers, what is the probability a customer will wait less than 5 minutes?

5-7 Exercises B: Beyond the Basics

In Exercises 8 and 9, suppose that when a call centre has 25 CSRs, a customer waits 1.5 minutes on average for each person ahead in the queue.

8. Which has the greater probability: (a) waiting more than 20 minutes when there are 25 CSRs and 7 people ahead in the queue or (b) waiting more than 20 minutes when there are 30 CSRs and 10 people ahead in the queue?

9. Suppose there are 10 CSRs and 18 people ahead in the queue. If you have been waiting 40 minutes already, does it seem realistic that you will wait no more than one hour in total to speak with a CSR?

■ VOCABULARY LIST

central limit theorem   **259**

continuity correction   **274**

density curve   **226**

exponential distribution   **280**

finite population correction factor   **265**

normal distribution   **230**

probability density function   **226**

rare event rule   **263**

sampling distribution of sample means   **259**

standard error of the mean   **261**

standard normal distribution   **231**

uniform distribution   **227**

■ REVIEW

Chapter 4 introduced the concept of the probability distribution, but included only *discrete* types. This chapter introduced *continuous* probability distributions, introducing the concept of probability as the area under a curve using the uniform distribution. The chapter focused on the most important category: normal distributions. Normal distributions will be used extensively in the following chapters.

The normal distribution, which appears bell-shaped when graphed, can be described algebraically by an equation, but the complexity of that equation usually forces us to use a table of values (Table A-2) or statistical software instead. Table A-2 gives areas corresponding to specific regions under the standard normal distribution curve, which has a mean of 0 and a standard deviation of 1. Those areas correspond to probability values.

In this chapter we worked with standard procedures for applying Table A-2 to a variety of different situations, including those that involve normal distributions that are nonstandard (with a mean other than 0 or a standard deviation other than 1). Those procedures usually involve use of the standard score: $z = (x - \mu)/\sigma$.

In Section 5–5 we presented the following important points associated with the central limit theorem:

1. The distribution of sample means will, as the sample size n increases, approach a normal distribution.

2. The mean of the sample means is the population mean μ.

3. The standard deviation of the sample means will be $\sigma/\sqrt{n}$.

In Section 5–6 we noted that we can sometimes approximate a binomial probability distribution by a normal distribution. If both $np \geq 5$ and $nq \geq 5$, the binomial random variable x is approximately normally distributed with the mean and standard deviation given as $\mu = np$ and $\sigma = \sqrt{npq}$. Because the binomial probability distribution deals with discrete data and the normal distribution deals with continuous data, we apply the continuity correction, which should be used in normal approximations to binomial distributions.

Finally, in Section 5–7 we presented the exponential distribution. Exponential distribution problems initially appear as Poisson distribution problems, but unlike the Poisson distribution, in which we compute the probability of a certain number of events over an interval such as time, the exponential distribution computes the probability of an interval such as time between events. The memoryless feature of the exponential distribution was also introduced.

REVIEW EXERCISES

In Exercises 1–4, assume that college statistics students have normally distributed heights with a mean of 68.0 in. and a standard deviation of 4.36 in. (based on Data Set 8 in Appendix B).

1. You plan to sell sweat suits in the college bookstore. To minimize start-up costs, you will not stock sweat suits for the tallest 5% and the shortest 5% of male students. Find the minimum and maximum heights of the students for whom sweat suits will be stocked.

2. The Beanstalk Club is a social organization for tall people. Students are eligible for membership if they are at least 74 in. tall. What percentage of students are eligible for membership in the Beanstalk Club?

3. The tallest player on the varsity basketball team is 80 in. tall. The shortest player is 69 in. tall. What percentage of statistics students is in this range?

4. If 45 students are randomly selected, what is the probability that their mean height is between 67.0 in. and 71.0 in.?

5. In a survey, the Canadian Automobile Association (CAA) found that 58% of its members paid cash for their vehicles. Estimate the probability that when 500 CAA members are randomly selected, the number who paid cash is between 250 and 290 inclusive.

6. The Gleason Supermarket uses a scale to weigh produce, and errors are normally distributed with a mean of 0 oz and a standard deviation of 1 oz (The errors can be positive or negative.) One item is randomly selected and weighed. Find the probability that the error is
   a. between 0 and 1.25 oz
   b. greater than 0.50 oz
   c. greater than −1.08 oz
   d. between −0.50 oz and 1.50 oz
   e. between −1.00 oz and −0.25 oz

7. Scores on the biology portion of the Medical College Admissions Test are normally distributed with a mean of 8.0 and a standard deviation of 2.6. Among 600 individuals taking this test, how many are expected to score between 6.0 and 7.0?

8. On the Graduate Record Exam in economics, scores are normally distributed with a mean of 615 and a standard deviation of 107. If a college admissions office requires scores above the 70th percentile, find the cutoff point.

9. According to the *National Hockey League Official Guide and Record Book 1998–1999*, the top goalie in the season let only 6.8% of the shots fired at him into the net for a goal. In a season in which 1739 shots are fired at the goalie, estimate the probability that more than 120 shots will result in a goal.

10. The Chemco Company manufactures car tires that last distances that are normally distributed with a mean of 57,300 km and a standard deviation of 6875 km.

    a. If a tire is randomly selected, what is the probability that it lasts more than 48,000 km?

    b. If 40 tires are randomly selected, what is the probability that they last distances that have a mean greater than 56,300 km?

    c. If the manufacturer wants to guarantee the tires so that only 3% will be replaced because of failure before the guaranteed number of kilometres, for how many kilometres should the tires be guaranteed?

In Exercises 11–13, assume that the print time for a photocopier to copy 20 pages is uniformly distributed between 20.2 and 21.6 seconds. Find the following probabilities for 20 pages:

11. Print time is less than 21 seconds.

12. Print time is more than 20.5 seconds.

13. Print time is between 20.4 and 21.0 seconds.

In Exercises 14–17 suppose that when a company has 5 customer service reps (CSRs) in its call centre at any given time, they collectively serve 40 customers per hour on average.

14. If there are 5 CSRs, what is the probability that a customer waits more than 2 minutes to speak to a CSR?

15. If there are 5 CSRs, what is the probability that a customer waits more than 2 minutes in total after already waiting 30 seconds?

16. Suppose there are only 4 CSRs instead of 5. What is the probability that a customer waits more than 2 minutes?

17. If there are 8 CSRs, what is the probability that a customer will speak to a CSR within 2 minutes after already waiting 30 seconds?

CUMULATIVE REVIEW EXERCISES

1. According to data from Statistics Canada, Chinese is the mother tongue of 8.1% of Toronto residents.
   a. If 3 Toronto residents are randomly selected, find the probability that they all have Chinese as their mother tongue.
   b. If 3 Toronto residents are randomly selected, find the probability that at least one of them has Chinese as his or her mother tongue.
   c. Why can't we solve the problem in part (b) by using the normal approximation to the binomial distribution?
   d. If groups of 100 Toronto residents are randomly selected, what is the mean number of residents whose mother tongue is Chinese in such groups?
   e. If groups of 100 Toronto residents are randomly selected, what is the standard deviation for the number of residents whose mother tongue is Chinese?
   f. Would it be unusual to get 10 people whose mother tongue is Chinese in a randomly selected group of 100 Toronto residents? Why or why not?

2. The sample scores given below are times (in milliseconds) it took the author's disk drive to make one revolution. The times were recorded by a diagnostic software program.

   199.7 200.0 200.1 200.1 200.1 200.3 200.3 200.3 200.3 200.3
   200.3 200.3 200.4 200.4 200.4 200.4 200.4 200.4 200.4 200.4
   200.5 200.5 200.5 200.5 200.5 200.5 200.5 200.5 200.5 200.6
   200.6 200.6 200.6 200.6 200.6 200.7 200.8 201.1 201.2 201.2

   a. Find the mean $\bar{x}$ of the times in this sample.
   b. Find the median of the times in this sample.
   c. Find the mode of the times in this sample.
   d. Find the standard deviation σ of this sample.
   e. Convert the time of 200.5 ms to a z score.
   f. Find the actual percentage of these sample scores that are greater than 201.0 ms.
   g. Assuming a normal distribution, find the percentage of *population* scores greater than 201.0 ms. Use the sample values of $\bar{x}$ and s as estimates of μ and σ.
   h. The specifications require times between 197.0 ms and 202.0 ms. Based on these sample results, does the disk drive seem to be rotating at acceptable speeds?

TECHNOLOGY PROJECT

In this chapter we looked at finding the percentage of male college students with heights between the average height of the Toronto Maple Leafs (the third-shortest team in the NHL), at 72.4 in., and the average height of the Philadelphia Flyers (the tallest team), at 74.6 in. We found that only 15.02% of college males were that tall. The solution, given in Section 5–3, involves theoretical calculations based on the assumptions that the heights of male college students are normally distributed with a mean of 70.1 in. and a standard deviation of 2.7 in. (based on data from a survey of Humber College students). This project describes a different method of solution that is based on a simulation technique: We will use a computer to randomly generate 100 heights of male students (from a normally distributed population with $\mu = 70.1$ and $\sigma = 2.7$), then we will find the percentage of those simulated heights that fall between 72.4 in. and 74.6 in. The Excel, Minitab, and STATDISK procedures are described below.

EXCEL:	Choose **Data Analysis** from the **Tools** menu. Select **Random Number Generation**. In the dialog box, specify 1 for Number of Variables and 100 for **Number of Random Numbers**. Select **Normal** from the pull-down **Distribution** menu, and enter 70.1 for **Mean** and 2.7 for **Standard Deviation**. Choose an output option, and click OK. When the 100 numbers have been generated, sort them by choosing **Sort** from the **Data** menu. Once the numbers are sorted, it is not difficult to count the number of heights between 72.4 in. and 74.6 in. Divide that number by 100 to find the percentage of these simulated heights. Compare the value to the theoretical value of 15.02% that was found in Section 5–3.
MINITAB 15:	From the **Calc** menu, choose **Random Data** and then **Normal**. In the box for the number of rows of data to generate, enter 100 and which column (e.g., C1) you want to store the values in. Enter 70.1 for the mean and 2.7 for the standard deviation. Click **OK**. To sort the data, from the **Data** menu, choose **Sort**. If your data are in C1, enter C1 in the **Sort column(s)** box and the first **By column** box and click the **Original column(s)** radio button. Click **OK**.
STATDISK:	Select **Data** from the main menu bar, then choose the option of **Normal Generator**. Proceed to generate 100 values with a mean of 70.1 and a standard deviation of 2.7. (Use the **Num Decimals** option to specify 1 decimal place.) Next, copy the data to the **Sample Editor**, click **Data Tools**, then **Sort Data**. With this ranked list, it becomes quite easy to count the number of heights between 72.4 and 74.6. Divide that number by 100 to find the percentage of these simulated heights in that range.

FROM DATA TO DECISION

How Can We Outsmart Users of Counterfeit Coins in Vending Machines?

Operators of vending machines are continually devising strategies to combat the use of counterfeit coins and bills.

Let's develop a strategy for minimizing losses from slugs that are counterfeit quarters. In Data Set 13 of Appendix B, we have a random sample of weights (in grams) of legitimate quarters. Based on that sample, let's assume that for the population of all such quarters, the distribution is normal, the mean weight is 5.622 g, and the standard deviation is 0.068 g. Let's also assume that a supply of counterfeit quarters has been collected from vending machines in downtown Calgary. Analysis of those counterfeit quarters shows that they have weights that are normally distributed with a mean of 5.450 g and a standard deviation of 0.112 g. A vending machine is designed so that coins are accepted or rejected according to weight. The limits for acceptable coins can be changed, but the machine currently accepts any coin with a weight between 5.500 g and 5.744 g. Find the percentages of legitimate coins and counterfeit coins accepted by the machine.

Here's a dilemma: If we restrict the weight limits on the coins too much, we reduce the proportion of legitimate coins that are accepted and business is lost. If we don't restrict the weight limits on coins enough, too many counterfeit coins are accepted and losses are incurred. What are the percentages of legitimate and counterfeit coins accepted if the lower weight limit is changed to 5.550 g? Experiment with different weight settings and identify limits that seem reasonable in that the machine does not accept too many counterfeit coins but still does accept a reasonable proportion of legitimate coins. It might be helpful to construct a reasonably accurate normal distribution graph for legitimate coins and another one for counterfeit coins; use the same horizontal scale and position one graph above the other so that numbers on the horizontal scale are aligned. With these graphs, you want to find limits that include the most legitimate coins and the fewest counterfeit coins. (There isn't necessarily a unique answer here; to some extent, the choices depend on subjective judgments.) Write a brief report that includes the settings you recommend, along with specific reasons for your choices.

COOPERATIVE GROUP ACTIVITIES

1. **Out-of-class activity:** Use the reaction timer by following the instructions given below. Collect reaction times for a sample of at least 40 different subjects taken from a homogeneous group, such as right-handed college students. For each subject, measure the reaction time for each hand. Construct a histogram for the right-hand times and another histogram for the left-hand results.

-0.204
-0.202
-0.199
-0.196
-0.194
-0.191
-0.188
-0.185
-0.183
-0.180
-0.177
-0.174
-0.171
-0.168
-0.165
-0.161
-0.158
-0.155
-0.151
-0.148
-0.144
-0.140
-0.137
-0.133
-0.129
-0.125
-0.121
-0.116
-0.112
-0.107
-0.102
-0.097
-0.091
-0.085
-0.079
-0.072
-0.065
-0.056
-0.046
-0.032

Based on the histogram shapes, do the two sets of times each appear to be normally distributed? Calculate the values of the mean and standard deviation for each of the two data sets and compare the two sets of results. Using the right-hand sample mean $\bar{x}$ as an estimate of the population mean μ, and using the right-hand sample standard deviation σ as an estimate of the population standard deviation s, find the quartiles Q_1, Q_2, and Q_3. Repeat that procedure to find Q_1, Q_2, and Q_3 for the left hand. If the right-hand reaction times are to be used to screen job applicants, what time is the cutoff separating the fastest 5% from the slowest 95%?

INSTRUCTIONS FOR USING THE REACTION TIMER

a. Cut the reaction timer along the dashed line.

b. Ask the subject to hold his or her thumb and forefinger horizontally; those fingers should be spread apart by a distance that is the same as the width of the reaction timer.

c. Hold the reaction timer so that the bottom edge is just above the subject's thumb and forefinger.

d. Ask the subject to catch the reaction timer as quickly as possible, then release it after a few seconds.

e. Record the reaction time (in seconds) corresponding to the point at which the subject catches it.

2. **In-class activity:** Use the chalkboard to construct two stem-and-leaf plots with stem values of 0, 1, 2, . . . , 9. For the stem-and-leaf plot at the left, each student should enter the last four digits of his or her Social Insurance Number. For the stem-and-leaf plot at the right, each student should calculate the mean $\bar{x}$ of those same four digits, round the result to the nearest whole number, and enter it on the stem-and-leaf plot. Groups of three or four should now be formed to analyze the results and state how they illustrate the central limit theorem.

3. **In-class activity:** Divide into groups of three or four students. For each group member, find the total value of coins that member has in his or her possession. Next, find the group mean. Share the individual values and the mean value with all of the other groups. What is the distribution of the individual values? What is the distribution of the group means? Do the results illustrate the central limit theorem? How?

4. **In-class activity:** Divide into groups of three or four students. Using a coin to simulate births, each group member should simulate 25 births and record the number of simulated girls. Combine all results in the group and record $n =$ total number of births and $x =$ number of girls. Given batches of n births, compute the mean and standard deviation for the number of girls. Is the simulated result usual or unusual? Why?

INTERNET PROJECT

Exploring the Central Limit Theorem

The central limit theorem is one of the most important results in statistics. It also may be one of the most surprising. Basically, the theorem says that the normal distribution is everywhere. No matter what probability distribution underlies an experiment, there is a corresponding distribution of means that will be approximately normal in shape. The best way to both understand and appreciate the central limit theorem, however, is to see it in action. The Internet Project for this chapter can be found at either of the following websites:

http://www.mathxl.com   or   http://www.mystatlab.com

Click on Internet Project, then on Chapter 5. You will be asked to view, interpret, and discuss a demonstration of the central limit theorem as part of a dice-rolling experiment. In addition, you will be guided in a search through the Internet for other such demonstrations.

6 Estimates and Sample Sizes

6-1 Overview

Chapter objectives are identified. This and the following chapters present topics of *inferential statistics*, characterized by methods of using sample data to form conclusions about population parameters. This chapter focuses on methods of estimating values of population means, proportions, or variances. Methods for determining the sample sizes necessary to estimate those parameters are also presented.

6-2 Estimating a Population Mean: Large Samples

The value of a population mean is approximated with a single value (called a point estimate) and a confidence interval. This section deals with large samples ($n > 30$) only.

6-3 Estimating a Population Mean: Small Samples

This section uses statistics from small samples ($n \leq 30$) to estimate the values of population means. Point estimates and confidence intervals are discussed, and the Student t distribution is introduced.

6-4 Estimating a Population Proportion

The value of a population proportion is estimated with a point estimate and confidence interval. Procedures are described for determining how large a sample must be to estimate a population proportion. This section presents the method commonly used by pollsters to determine how many people must be surveyed in various opinion polls.

6-5 Estimating a Population Variance

The value of a population variance is approximated with a point estimate and confidence interval. The chi-square distribution is introduced.

CHAPTER PROBLEM

Is the mean body temperature really 37.0°C?

Table 6-1 lists 108 body temperatures (from Data Set 18 in Appendix B) obtained courtesy of David Chamberlain and Andrew McRae, Research Institute, Lakeridge Health Oshawa. Using the methods described in Chapter 2, we can obtain the following important characteristics of the sample data set:

- As revealed by a histogram, the distribution of the data is approximately bell-shaped.
- The mean is $\bar{x} = 36.39°C$.
- The standard deviation is $s = 0.62°C$.
- The sample size is $n = 108$.

Most people believe that the mean body temperature is 37.0°C, but the data in Table 6-1 seem to suggest that it is actually 36.39°C. We know that samples tend to vary, so perhaps it is true that the mean body temperature is 37.0°C and the sample mean, $\bar{x} = 36.39°C$, is the result of a chance sample fluctuation. On the other hand, perhaps the sample mean of 36.39°C is correct and the commonly believed value of 37.0°C is wrong. On the basis of an analysis of the sample data in Table 6-1, we will see whether a mean body temperature of 37.0°C seems plausible.

Table 6-1 Body Temperatures (in degrees Celsius) of Adults Accepted for Voluntary Surgery, Measured Before the Surgery

36.4	36.3	36.7	36.3	36.8	36.6	36.6	36.1	36.2	35.6	37.4	36.3
36.2	35.4	36.4	36.3	36.7	36.1	36.0	36.1	36.4	36.4	35.8	36.7
37.3	36.4	37.1	37.1	36.4	36.6	36.6	36.5	36.7	36.5	36.6	36.8
35.5	37.0	36.9	35.2	37.1	36.7	37.4	37.1	37.0	36.4	35.9	36.9
36.3	37.1	36.4	36.8	36.5	37.6	36.1	36.4	36.0	36.7	36.3	36.2
36.4	36.7	35.4	36.1	35.0	39.9	36.6	35.9	36.0	36.0	35.4	36.2
36.0	36.8	37.3	36.3	37.1	35.8	36.0	35.8	36.2	36.9	36.3	36.6
36.0	35.5	36.0	36.1	36.4	35.0	36.5	36.6	35.1	36.4	36.0	36.7
36.3	36.8	36.5	36.5	35.8	36.4	35.9	36.1	36.1	36.2	36.3	36.3

Note that while not all of these temperatures may represent "normal, healthy" temperatures, they are the temperatures of people who were deemed sufficiently nonfevered to undergo voluntary surgery. Temperatures of patients who were operated on but may have had fever are omitted.

6-1 Overview

In Chapter 2 we noted that we use *descriptive statistics* to summarize or describe important characteristics of known population data, but with *inferential statistics* we use sample data to make inferences (or generalizations) about a population. The two major applications of inferential statistics involve the use of sample data to (1) estimate the value of a population parameter and (2) test some claim (or hypothesis) about a population. This chapter presents methods for estimating values of the following population parameters: population means, proportions, and variances. We also present methods for determining the sample sizes necessary to estimate those parameters. The basic methods for testing claims about a population will be introduced in Chapter 7.

As we proceed with methods for using sample data to form inferences about populations, we should recall an extremely important point first made in Chapter 1:

> **Data collected carelessly can be absolutely worthless, even if the sample is quite large.**

The methods used here and in the following chapters require sound sampling procedures.

6-2 Estimating a Population Mean: Large Samples

Consider the 108 body temperatures given in Table 6-1 at the beginning of this chapter. On the basis of those sample values, we want to estimate the mean of *all* body temperatures. We could use a statistic such as the sample median, midrange, or mode as an estimate of this population mean μ, but the sample mean $\bar{x}$ usually provides the best estimate of a population mean. The choice of $\bar{x}$ is based on careful study and analysis of the distributions of the different statistics that could be used as estimators.

DEFINITIONS

An **estimator** is a sample statistic (such as the sample mean $\bar{x}$) used to approximate a population parameter. An **estimate** is a specific value or range of values used to approximate some population parameter.

For example, on the basis of the data in Table 6-1, we might use the *estimator* $\bar{x}$ to conclude that the *estimate* of the mean body temperature of all healthy adults is 36.39°C.

There are two important reasons why a sample mean is a better estimator of a population mean μ than other estimators such as the median or the mode:

1. For many populations, the distribution of sample means $\bar{x}$ tends to be more consistent (i.e., with *less variation*) than the distributions of other sample statistics. (That is, if you use sample means to estimate the population mean μ, those sample means will have a smaller standard deviation than would other sample statistics, such as the median or the mode.)

2. For all populations, we say that the sample mean $\bar{x}$ is an **unbiased estimator** of the population mean μ, meaning that the distribution of sample means tends to centre about the value of the population mean μ. (That is, sample means do not systematically tend to overestimate the value of μ, nor do they systematically tend to underestimate μ. Instead, they tend to target the value of μ itself.)

For these reasons, we will use the sample mean $\bar{x}$ as the best estimate of the population mean μ. Because the sample mean $\bar{x}$ is a single value that corresponds to a point on the number scale, we call it a *point estimate*.

DEFINITION

A **point estimate** is a single value (or point) used to approximate a population parameter.

The sample mean $\bar{x}$ is the best point estimate of the population mean μ.

EXAMPLE

Use the sample body temperatures given in Table 6-1 to find the best point estimate of the population mean μ of all body temperatures.

SOLUTION

The sample mean $\bar{x}$ is the best point estimate of the population mean μ, and for the sample data in Table 6-1 we have $\bar{x} = 36.39°C$. Based on those particular sample values, the best point estimate of the population mean μ of all body temperatures is therefore 36.39°C.

Why Do We Need Confidence Intervals?

In the preceding example we saw that 36.39°C was our best point estimate of the population mean μ, but we had no indication of just how good our best estimate was. If we knew only the first four temperatures of 36.4°C, 36.3°C, 36.7°C, and 36.3°C, the best point estimate of μ would be their mean ($\bar{x} = 36.43°C$), but we wouldn't expect this point estimate to be very good because it is based on such a small sample. Statisticians have therefore developed another type of estimate that does reveal how good the point estimate is. This estimate, called a confidence

STATISTICAL PLANET

TV Sample Sizes
A *Newsweek* article described the use of people meters as a way of determining how many people are watching different television programs. The article stated, "Statisticians have long argued that the household samples used by the rating services are simply too small to accurately determine what America is watching. In that light, it may be illuminating to note that 4000 homes reached by the people meters constitute exactly 0.0045% of the wired nation." This implied claim that the sample size of 4000 homes is too small is not valid. Methods of this chapter can be used to show that a random sample size of 4000 can provide results that are quite good, even though the sample might be only 0.0045% of the population.

interval or interval estimate, consists of a range (or an interval) of values instead of just a single value.

> **DEFINITION**
>
> A **confidence interval** (or **interval estimate**) is a range (or an interval) of values that is *likely* to contain the true value of the population parameter.

A *confidence interval* is associated with a degree of confidence, which is a measure of how certain we are that our interval contains the population parameter. The definition of degree of confidence uses α (lowercase Greek alpha) to describe a probability that corresponds to an area. Refer to Figure 6-1, where the probability α is divided equally between two shaded extreme regions (often called tails) in the standard normal distribution. (We'll describe the role of $z_{\alpha/2}$ later; for now, simply note that α is divided equally between the two tails.)

> **DEFINITION**
>
> The **degree of confidence** is the probability $1 - \alpha$ (often expressed as the equivalent percentage value) that the confidence interval contains the true value of the population parameter. (The degree of confidence is also called the **level of confidence** or the **confidence coefficient**.)

Common choices for the *degree of confidence* are 90% (with $\alpha = 0.10$), 95% (with $\alpha = 0.05$), and 99% (with $\alpha = 0.01$). The choice of 95% is most common because it provides a good balance between precision (as reflected in the width of the confidence interval) and reliability (as expressed by the degree of confidence).

Here's an example of a confidence interval based on the sample data of 108 body temperatures given in Table 6-1:

The 0.95 (or 95%) degree of confidence interval estimate of the population mean μ is 36.27°C $< \mu <$ 36.51°C.

Interpreting a Confidence Interval

We must be careful to interpret confidence intervals correctly. What is the correct interpretation of the confidence estimate 36.27°C $< \mu <$ 36.51°C? We interpret this confidence interval as follows: If we were to select many different samples of size $n = 108$ from the population of all healthy people and construct a similar 95% confidence interval estimate for each sample, in the long run 95% of those intervals would actually contain the value of the population mean μ. We should know that μ is a fixed value, not a random variable, and it is therefore *wrong* to say that there is a 95% chance that μ will fall within the interval. Any particular

Figure 6-1
The Standard Normal Distribution: The Critical Value $z_{\alpha/2}$

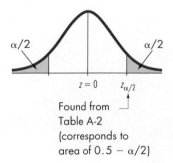

$\alpha/2$ $\alpha/2$

$z = 0$ $z_{\alpha/2}$

Found from Table A-2 (corresponds to area of $0.5 - \alpha/2$)

confidence interval either contains μ or does not, but because μ is fixed, there is no probability that μ falls within an interval. In short, the 95% level refers to the *process* being used to estimate the mean (we have confidence that, 95% of the time, using this process will correctly identify an interval that *contains* the mean); the 95% does *not* refer to the population mean itself.

Critical Values

We know from the central limit theorem that sample means $\bar{x}$ tend to be normally distributed, as in Figure 6-1. Sample means have a relatively small chance of falling in one of the extreme tails of Figure 6-1. Denoting the area of each shaded tail by $\alpha/2$, we see that there is a total probability of α that a sample mean will fall in either of the two tails. By the rule of complements (from Chapter 3), it follows that there is a probability of $1 - \alpha$ that a sample mean will fall within the unshaded region of Figure 6-1. The z score separating the right-tail region is commonly denoted by $z_{\alpha/2}$ and is referred to as a *critical value* because it is on the borderline separating sample means that are likely to occur from those that are unlikely to occur.

NOTATION FOR CRITICAL VALUE

$z_{\alpha/2}$ is the positive z score that is at the vertical boundary separating an area of $\alpha/2$ in the right tail of the standard normal distribution. (The value of $-z_{\alpha/2}$ is at the vertical boundary for the area of $\alpha/2$ in the left tail.)

DEFINITION

A **critical value** is the number on the borderline separating sample statistics that are likely to occur from those that are unlikely to occur. The number $z_{\alpha/2}$ is a critical value that is a z score with the property that it separates an area of $\alpha/2$ in the right tail of the standard normal distribution. (There is an area of $1 - \alpha$ between the vertical borderlines at $-z_{\alpha/2}$ and $z_{\alpha/2}$.)

EXAMPLE

Find the critical value $z_{\alpha/2}$ corresponding to a 95% degree of confidence.

SOLUTION

A 95% degree of confidence corresponds to $\alpha = 0.05$. See Figure 6-2, where we show that the area in each of the shaded tails is $\alpha/2 = 0.025$. We find $z_{\alpha/2} = 1.96$ by noting that the region to its left (and bounded by the mean of $z = 0$) must be $0.5 - 0.025$, or 0.475. Now refer to Table A-2 and find that the area of 0.4750 (found in the *body* of the table) corresponds exactly to a z score of 1.96. For a 95% degree of confidence, the critical value is therefore $z_{\alpha/2} = 1.96$.

STATISTICAL PLANET

Estimating Wildlife Population Sizes
British Columbia's Land Use Coordination Office, working with the Ministry of Forests and the Ministry of Environment, Land, and Parks, developed a management plan to protect the spotted owl, which is considered an endangered species. Biologists and statisticians concluded that survival rates and population sizes were decreasing for the female owls, known to play an important role in species survival. Biologists and statisticians have also studied other wildlife, such as penguins in New Zealand. In the article "Sampling Wildlife Populations" (*Chance*, Vol. 9, No. 2), authors Bryan Manly and Lyman McDonald comment that in such studies, "biologists gain through the use of modelling skills that are the hallmark of good statistics. Statisticians gain by being introduced to the reality of problems by biologists who know what the crucial issues are."

Figure 6-2
Finding $z_{\alpha/2}$ for 95%
Degree of Confidence

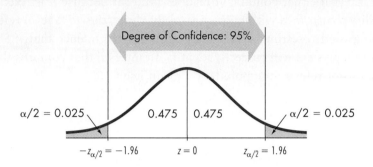

The preceding example showed that a 95% degree of confidence results in a critical value of $z_{\alpha/2} = 1.96$. This is the most common critical value, and it is listed with two other common values in the table that follows:

Degree of Confidence	α	Critical Value $z_{\alpha/2}$
90%	0.10	1.645
95%	0.05	1.96
99%	0.01	2.575

Margin of Error

When we collect a set of sample data, such as the set of 108 body temperatures listed in Table 6-1, we can calculate the sample mean $\bar{x}$ and that sample mean is typically different from the population mean μ. The difference between the sample mean and the population mean can be thought of as an error. This is important because we do not know the true value of μ; and although we assume there is some error, we can't know its exact size. In Section 5–5 we saw that $\sigma/\sqrt{n}$ is the standard deviation of sample means. Using $\sigma/\sqrt{n}$ and the $z_{\alpha/2}$ notation, we now define the margin of error E as follows:

DEFINITION

When data from a random sample are used to estimate a population mean μ, the **margin of error**, denoted by **E,** is the maximum likely (with probability $1 - \alpha$) difference between the observed sample mean $\bar{x}$ and the true value of the population mean μ. The margin of error E is also called the **maximum error of the estimate** and can be found by multiplying the critical value and the standard deviation of sample means, as shown in Formula 6-1.

Formula 6-1
$$E = z_{\alpha/2} \cdot \frac{\sigma}{\sqrt{n}}$$

Given the way that the margin of error E is defined, there is a probability of $1 - \alpha$ that a sample mean will be in error (different from the population mean m) by no more than E, and there is a probability of α that the sample mean will be in error by more than E. The calculation of the margin of error E as given in Formula 6-1 requires that you know the population standard deviation σ, but in reality it's rare to know σ when the population mean μ is not known. The following method of calculation is common practice.

CALCULATING E WHEN σ IS UNKNOWN

> If $n > 30$, we can replace σ in Formula 6-1 by the sample standard deviation s.
>
> If $n \leq 30$, the population must have a normal distribution and we must know σ to use Formula 6-1. [An alternative method for calculating the margin of error E for small ($n \leq 30$) samples will be discussed more fully in the next section.]

On the basis of the definition of the margin of error E, we can now identify the confidence interval for the population mean μ.

CONFIDENCE INTERVAL (OR INTERVAL ESTIMATE) FOR THE
POPULATION MEAN μ (BASED ON LARGE SAMPLES: $n > 30$)

$$\bar{x} - E < \mu < \bar{x} + E \quad \text{where} \quad E = z_{\alpha/2} \cdot \frac{\sigma}{\sqrt{n}}$$

Other equivalent forms for the confidence interval are $\mu = \bar{x} \pm E$ and $(\bar{x} - E, \bar{x} + E)$.

The values $\bar{x} - E$ and $\bar{x} + E$ are called **confidence interval limits.**

**Procedure for Constructing a Confidence Interval for μ
(Based on a Large Sample: $n > 30$)**

1. Find the critical value $z_{\alpha/2}$ that corresponds to the desired degree of confidence. (For example, if the degree of confidence is 95%, the critical value is $z_{\alpha/2} = 1.96$.)

2. Evaluate the margin of error $E = z_{\alpha/2} \cdot \sigma/\sqrt{n}$. If the population standard deviation σ is unknown, use the value of the sample standard deviation s provided that $n > 30$.

3. Using the value of the calculated margin of error E and the value of the sample mean $\bar{x}$, find the values of $\bar{x} - E$ and $\bar{x} + E$. Substitute those values in the general format of the confidence interval:

$$\bar{x} - E < \mu < \bar{x} + E$$
$$\text{or} \qquad \mu = \bar{x} \pm E$$
$$\text{or} \qquad (\bar{x} - E, \quad \bar{x} + E)$$

4. Round the resulting values by using the following round-off rule.

Reporting Poll Results

The following is an excerpt from an Angus Reid press release on a poll taken prior to the Quebec election in 2007. The poll found that the Action Démocratique had 31% support, with the Liberals at 30% and the Parti Québécois at 27%. This example demonstrates the use of the confidence interval in reporting poll results:

"The outcome of this Monday's election in Quebec is too close to call, according to a poll by Angus Reid Strategies. The Canadian province's three main parties attract roughly a third of voters, including those absolutely certain to cast a ballot. In the survey, 31 per cent of decided voters—including leaners—would support the conservative Action démocratique du Québec (ADQ), while 30 per cent would back the governing Liberal Party of Quebec. The sovereignist Parti Québécois (PQ) is third with 27 per cent, followed by the Parti vert (Green Party) with seven per cent, and Québec solidaire with five per cent."

Source: Angus Reid Strategies

Methodology: Online interviews with 838 Quebec adults, conducted on Mar. 22 and Mar. 23, 2007. Margin of error is 3.5 percent, 19 times out of 20. (Note that "19 times out of 20" represents 95% confidence.)

1. When using the *original set of data* to construct a confidence interval, round the confidence interval limits to one more decimal place than is used for the original set of data.

2. When the original set of data is unknown and only the *summary statistics* $(n, \bar{x}, s)$ are used, round the confidence interval limits to the same number of decimal places used for the sample mean.

The following example clearly illustrates the relatively simple procedure for actually constructing a confidence interval. The original data from Table 6-1 use one decimal place and the summary statistics use two decimal places, so the confidence interval limits will be rounded to two decimal places.

EXAMPLE

For the body temperatures in Table 6-1, we have $n = 108$, $\bar{x} = 36.39$, and $s = 0.62$. For a 0.95 degree of confidence, find both of the following:

a. The margin of error E

b. The confidence interval for μ

SOLUTION

a. The 0.95 degree of confidence implies that $\alpha = 0.05$, so $z_{\alpha/2} = 1.96$, as shown in the preceding example. The margin of error E is calculated by using Formula 6-1 as follows. (Note that σ is unknown, but we can use $s = 0.62$ for the value of σ because $n > 30$.)

$$E = z_{\alpha/2} \cdot \frac{\sigma}{\sqrt{n}} = 1.96 \cdot \frac{0.62}{\sqrt{108}} = 0.12$$

b. With $\bar{x} = 36.39$ and $E = 0.12$, we construct the confidence interval as follows:

$$\bar{x} - E < \mu < \bar{x} + E$$
$$36.39 - 0.12 < \mu < 36.39 + 0.12$$
$$36.27 < \mu < 36.51$$

[This result could also be expressed as $\mu = 36.39 \pm 0.12$ or as (36.27, 36.51).]

Interpretation

Based on the sample of 108 body temperatures listed in Table 6-1, the confidence interval for the population mean μ is $36.27°C < \mu < 36.51°C$, and this interval has a 0.95 degree of confidence. This means that if we were to select many different samples of size 108 and construct the confidence intervals as we did here, 95% of them would actually contain the value of the population mean μ.

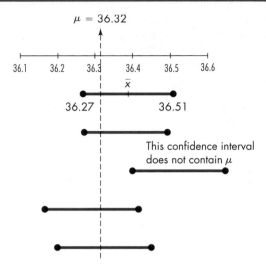

Figure 6-3
Confidence Intervals from Different Samples
The graph shows several confidence intervals, one of which does not contain the population mean μ. For 95% confidence intervals, we expect that among 100 such intervals, 5 will not contain $\mu = 36.32$ while the other 95 will contain it.

Note that the confidence interval limits of 36.27°C and 36.51°C do not contain 37.0°C, the value generally believed to be the mean body temperature. Based on the sample data in Table 6-1, it seems very unlikely that 37.0°C is the correct value of μ.

For interpreting the confidence interval that we derived in the preceding example, Figure 6-3 may be helpful. Suppose that the true mean for body temperatures is really 36.32°C, instead of the 36.39°C that we estimated from our sample. The topmost of the bold line segments in the figure shows the confidence intervals we calculated from our sample. Notice that the interval contains the true mean. The confidence intervals that could have been derived from *other* samples taken from the population are also illustrated. Observe that *most* of the derived confidence intervals contain the population mean—though in a minority of the cases (we expect in about 5% of them) the confidence intervals do not contain the mean.

Basis for the Procedure of Finding the Confidence Interval

The basic idea underlying the construction of confidence intervals relates to the central limit theorem, which indicates that with large ($n > 30$) samples, the distribution of sample means is approximately normal with mean μ and standard deviation $\sigma/\sqrt{n}$. The confidence interval format is really a variation of the equation

$$z = \frac{\overline{x} - \mu}{\dfrac{\sigma}{\sqrt{n}}}$$

Figure 6-4
Distribution of Sample Means

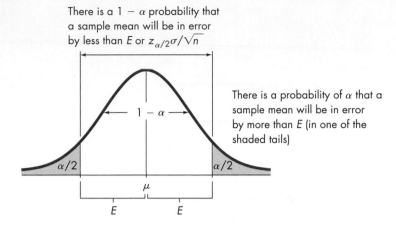

If we solve this equation for μ, we get $\mu = \bar{x} - z\frac{\sigma}{\sqrt{n}}$. Using the positive and negative values for z results in the confidence interval limits we are using.

Let's consider the specific case of a 95% degree of confidence, so $\alpha = 0.05$ and $z_{\alpha/2} = 1.96$. For this case there is a probability of 0.05 that a sample mean will be more than 1.96 standard deviations (or $z_{\alpha/2}\sigma/\sqrt{n}$, which we denote by E) away from the population mean μ. Conversely, there is a 0.95 probability that a sample mean will be within 1.96 standard deviations (or $z_{\alpha/2}\sigma/\sqrt{n}$) of μ. (See Figure 6-4.) If the sample mean $\bar{x}$ is within $z_{\alpha/2}\sigma/\sqrt{n}$ of the population mean μ, then μ must be between $\bar{x} - z_{\alpha/2}\sigma/\sqrt{n}$ and $\bar{x} + z_{\alpha/2}\sigma/\sqrt{n}$; this is expressed in the general format of our confidence interval (with $z_{\alpha/2}\sigma/\sqrt{n}$ denoted as E):
$$\bar{x} - E < \mu < \bar{x} + E.$$

AS

Factors Affecting Confidence Intervals for Means

If we examine the formula for the margin of error, there are three values that make it up. Adjusting any one of these values will change the margin of error and consequently the width of the confidence interval.

Factor 1: The level of confidence   As we increase the level of confidence, the z critical value becomes larger, resulting in a larger margin of error and a wider confidence interval.

Factor 2: The standard deviation   Similarly, a larger standard deviation produces a larger margin of error and a wider confidence interval. It should be noted that a researcher cannot change the standard deviation, since this value is based on the data at hand.

Factor 3: The sample size   As sample size increases, the margin of error becomes smaller, resulting in a narrower confidence interval.

EXAMPLE

We revisit the previous example of the 95% confidence interval with $\sigma = 0.62$ and $n = 108$. In this situation the margin of error is 0.12 and the resulting confidence interval ranges from 36.27 to 36.51.

a. How does the confidence interval change if the level of confidence is increased to 99%?

b. How would the 95% confidence interval change if the standard deviation were 1.2 instead of 0.62?

c. How would the 95% confidence interval change if the sample size were 200 instead of 108?

SOLUTION

a. If the level of confidence is increased to 99%, this means an area of 0.005 in each tail, so the z critical value becomes 2.575. Then:

$$E = 2.575 \cdot \frac{0.62}{\sqrt{108}} = 0.15$$

$$36.39 - 0.15 < \mu < 36.39 + 0.15$$

$$36.24 < \mu < 36.54$$

The margin of error increases by 0.03, resulting in a slightly wider confidence interval.

b. Since the level of confidence is 95%, the z score remains at 1.96. Then:

$$E = 1.96 \cdot \frac{1.2}{\sqrt{108}} = 0.23$$

$$36.39 - 0.23 < \mu < 36.39 + 0.23$$

$$36.16 < \mu < 36.62$$

The margin of error increases by 0.11, again resulting in a wider confidence interval.

c. Since we are increasing the sample size, we should now expect a narrower confidence interval:

$$E = 1.96 \cdot \frac{0.62}{\sqrt{200}} = 0.09$$

$$36.39 - 0.09 < \mu < 36.39 + 0.09$$

$$39.3 < \mu < 39.48$$

The margin of error decreases by 0.03, resulting in a slightly narrower confidence interval.

Determining Sample Size

So far in this section we have discussed ways to find point estimates and interval estimates of a population mean μ. We based our procedures on known sample data. But suppose that we haven't yet collected the sample. How do we know how many members of the population should be selected? For example, suppose we want to estimate the mean first-year income of college graduates. How many such incomes must our sample include? Determining the size of a sample is a very important issue, because samples that are needlessly large waste time and money, and samples that are too small may lead to poor results. In many cases we can find the minimum sample size needed to estimate some parameter, such as the population mean μ.

If we begin with the expression for the margin of error E (Formula 6-1):

$$E = \frac{z_{\alpha/2}\sigma}{\sqrt{n}}$$

Cross-multiplying, we get:

$$\sqrt{n} = \frac{z_{\alpha/2}\sigma}{E}$$

Finally, to solve for the sample size n, we square both sizes and get the following.

SAMPLE SIZE FOR ESTIMATING MEAN μ

Formula 6-2
$$n = \left[\frac{z_{\alpha/2}\sigma}{E}\right]^2$$

where
$z_{\alpha/2}$ = critical z score based on the desired degree of confidence
E = desired margin of error
σ = population standard deviation

Formula 6-2 is quite remarkable because it implies that the sample size does not depend on the size (N) of the population: the sample size depends on the desired degree of confidence, the desired margin of error, and the value of the standard deviation σ.

The sample size must be a whole number, but the calculations for sample size n often result in a value that is not a whole number. When this happens, we observe the following round-off rule that, when rounding is necessary, the required sample size should be rounded *up* so that it is at least adequately large as opposed to slightly too small.

An economist wants to estimate the mean income for the first year of work for a college graduate who has had the profound wisdom to take a statistics course. How many such incomes must be found if we want to be 95% confident that the sample mean is within $500 of the true population mean? Assume that a previous study has revealed that for such incomes, $\sigma = \$6250$.

SOLUTION

We want to find the sample size n given that $\alpha = 0.05$ (from 95% confidence). We want the sample mean to be within $500 of the population mean, so $E = 500$. Assuming that $\sigma = 6250$, we use Formula 6-2 to get

$$n = \left[\frac{z_{\alpha/2}\sigma}{E} \right]^2 = \left[\frac{1.96 \cdot 6250}{500} \right]^2 = 600.25 = 601 \qquad \text{(rounded up)}$$

We should therefore obtain a sample of at least 601 randomly selected first-year incomes of college graduates who have taken a statistics course. With such a sample, we will be 95% confident that the sample mean $\bar{x}$ will be within $500 of the true population mean μ.

Factors Affecting Sample Size

As with the confidence interval for means, there are three values that are used in the computation of the sample size. Changing any one of the values will result in a different sample size.

Factor 1: The level of confidence If we increase the level of confidence, the z critical value also increases resulting in a larger sample size.

Factor 2: The standard deviation Even though a researcher cannot directly change the standard deviation, an increase in this value will also result in a larger sample size.

Factor 3: The margin of error An increase in the margin of error will result in a smaller sample size.

EXAMPLE

We revisit the previous example in which a required sample size of 601 was calculated. Suppose the economist has neither the time nor the money to survey 601 college graduates. Which would result in a smaller sample size: decreasing the level of confidence to 90% while maintaining a margin of error of $500, or maintaining the level of confidence at 95% but increasing the margin of error to $750?

SOLUTION

In the first solution, since the level of confidence is 90%, we have 0.05 in each tail, so the z critical value is 1.645. Then:

$$n = \left[\frac{1.645 \cdot 6250}{500} \right]^2 = 422.8 = 423$$

In the second solution, the z critical value remains at 1.96:

$$n = \left[\frac{1.96 \cdot 6250}{750} \right]^2 = 266.8 = 267$$

The second solution had the greater effect of decreasing the sample size. This will generally be the case; if a researcher has a choice of decreasing the level of confidence or increasing the margin of error, the latter will have a more dramatic effect in decreasing the sample size.

If we are willing to settle for less accurate results by using a larger margin of error such as $1000, the sample size drops to 150.0625, which is rounded up to 151. Doubling the margin of error causes the required sample size to decrease to one-fourth its original value. Conversely, halving the margin of error quadruples the sample size. What this implies is that if you want more accurate results, the sample size must be substantially increased. Because large samples generally require more time and money, there is often a need for a trade-off between the sample size and the margin of error E.

What If σ Is Unknown? Formula 6-2 requires that we substitute some value for the population standard deviation σ, but if it is unknown (as it usually is), we may be able to use a preliminary value obtained from procedures such as these:

1. Using the range rule of thumb (see Section 2–5) to estimate the standard deviation as follows: $\sigma \approx$ typical range/4.

2. Conducting a pilot study by starting the sample process. Based on the first collection of at least 31 randomly selected sample values, calculate the sample standard deviation σ and use it in place of σ. That value can be refined as more sample data are obtained.

3. Estimate the value of σ by using the results of some other study that was done earlier.

   If in doubt when calculating the sample size n, any errors should always be conservative in the sense that they make n too large instead of too small.

In the two examples that follow, the first uses the range rule of thumb to estimate s and the second uses a pilot study consisting of sample data found in Appendix B.

EXAMPLE

You plan to estimate the mean incubation period (in days) for sea birds. How many incubation periods of sea birds must you sample if you want to be 95% confident that the sample mean is within 3 days of the true population mean μ?

SOLUTION

We seek the sample size n given that $\alpha = 0.05$ (from 95% confidence, so $\alpha = 1 - 0.95$), so $z_{\alpha/2} = 1.96$. We want to be within 3 days, so $E = 3$. We do not know the standard deviation σ of incubation periods for all sea birds, but we can estimate σ by using the range rule of thumb: If past experience suggests that these incubation periods typically range from 23 to 78 days (disregarding outliers), the range becomes 55 days so that

$$\sigma \approx \frac{(\text{typical range})}{4} = \frac{(78 - 23)}{4} = 13.75$$

With $z_{\alpha/2} = 1.96$, $E = 3$, and $\sigma \approx 13.75$, we use Formula 6-2 as follows:

$$n = \left[\frac{z_{\alpha/2}\sigma}{E}\right]^2 \approx \left[\frac{1.96 \cdot 13.75}{3}\right]^2 = 80.70 \approx 81 \qquad \text{(rounded up)}$$

We must randomly select 81 incubation periods of seabirds and then find the value of the sample mean $\bar{x}$. We will be 95% confident that the resulting sample mean is within 3 days of the true mean incubation period.

EXAMPLE

If we want to estimate the mean weight of plastic discarded by households in one week, how many households must we randomly select if we want to be 99% confident that the sample mean is within 0.250 lb of the true population mean?

SOLUTION

We seek the sample size n given that $\alpha = 0.01$ (from 99% confidence), so $z_{\alpha/2} = 2.575$. We want to be within 0.250 lb of the true mean, so $E = 0.250$. The value of the population standard deviation σ is unknown, but we can refer to Data Set 1 in Appendix B, which includes the weights of plastic discarded for 62 households. Using that sample as a pilot study, we can calculate the value of the standard deviation to get $s = 1.065$ lb; because this sample is large ($n > 30$), we can use the value of $s = 1.065$ as an estimate of the population standard deviation σ, as shown below:

$$n = \left[\frac{z_{\alpha/2}\sigma}{E}\right]^2 \approx \left[\frac{2.575 \cdot 1.065}{0.250}\right]^2 = 120.3 \approx 121 \qquad \text{(rounded up)}$$

On the basis of the population standard deviation estimated from Data Set 1 in Appendix B, we must sample at least 121 randomly selected households to be 99% confident that the sample mean is within 0.250 lb of the true population mean.

This section dealt with the construction of point estimates and confidence interval estimates of population means and presented a method for determining sample sizes needed to estimate population means to the desired degree of accuracy. All of the examples and exercises in this section involve large ($n > 30$) samples. The following section describes procedures to be used when the samples are small ($n \leq 30$).

EXCEL:

Confidence Intervals: Excel's built-in tools can calculate the margin of error E. You must then subtract this result from $\bar{x}$ and add it to $\bar{x}$ so that you can identify the lower and upper limits of the confidence interval. First, use techniques from Chapter 2 to identify the sample size n and the sample standard deviation s.

EXCEL (Prior to 2007):

Click on the *fx* icon, select the function category **Statistical**, and then select the item **Confidence**. In the dialog box, enter the value of α (called the significance level), the standard deviation, and the sample size. The margin of error E will be displayed.

EXCEL 2007:

Click on **Menus** on the main menu, then the *fx* button, then more functions and select the function category **Statistical** and then the function name **Confidence**. Proceed as above.

Alternatively, you can use the Data Desk XL add-in that is a supplement to this book.

EXCEL (Prior to 2007):

Click on **DDXL** and select **Confidence Intervals**. Under the Function Type options, select **1 Var z Interval**. Click on the pencil icon and enter the Excel range containing the data. Click on **OK**. In the dialog box, set the level of confidence, enter the standard deviation, and click on **Compute Interval**. The confidence interval will be displayed, as shown below.

EXCEL 2007:

Click on the **Add-in** tab on the main menu. You will see **DDXL** to the left. Click on it and proceed as above.

MINITAB 15:

Confidence Intervals: From the Stat menu, choose **Basic Statistics** and then **1-Sample Z**. If you are constructing the confidence interval using summarized data, click that radio button and fill in the necessary boxes such as sample size, mean, and standard deviation. There is an optional box to conduct a hypothesis test. (You would be required to fill in a hypothesis mean, which will be covered in Chapter 7.) The default setting is a 95% level of confidence. If you wish to change it, click the **Options** button and change the level of confidence to what you desire. You can leave the **Alternative** drop-down menu set to not equal.

STATDISK:

Confidence Intervals: First, use techniques from Chapter 2 to identify the sample size n, the sample mean $\bar{x}$ and the sample standard deviation s. Select **Analysis** from the main menu bar, select **Confidence Intervals**, then select **Mean – One Sample**. Proceed to enter the items in the dialog box, then click the **Evaluate** button. Note that if you are using the population standard deviation, enter that in the appropriate box.

Sample Size: Select **Analysis** from the main menu bar at the top, then select **Sample Size Determination**, followed by **Estimate Mean**. You must now enter the confidence level (such as 0.99, for the second example, above), the error E, and the actual or estimated population standard deviation σ. There is also an option that allows you to enter the population size N, if you are sampling without replacement from a finite population (see Exercise 29).

EXCEL DISPLAY

Data Desk® 6.1 Viewer - Untitled

File Edit Data Special Help

z Interval Setup

Press one of the buttons below to set the level of confidence.

| 90% | 95% | 99% |

Confidence Lev

95% Confidence

Type in the population standard deviation:

0.62

Compute Interval

$VAR1 Confidence Interval

Summary Statistics

Count	Mean	Std Dev	Std Dev of the M
108	36.39	0.62	0.0597

Interval Results

Confidence Interval

With 95% Confidence, $36.273 < \mu < 36.507$

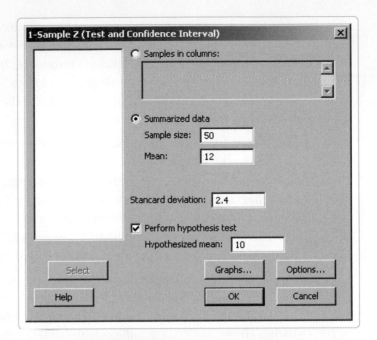

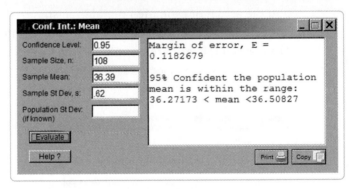

6-2 Exercises A: Basic Skills and Concepts

In Exercises 1–4, find the critical value $z_{\alpha/2}$ that corresponds to the given degree of confidence.

1. 99% **2.** 94% **3.** 98% **4.** 92%

In Exercises 5–8, use the given degree of confidence and sample data to find (a) the margin of error and (b) the confidence interval for the population mean μ.

5. Heights of female students: 95% confidence; $n = 50$, $\bar{x} = 64.2$ in., $s = 2.6$ in.

6. Grade-point averages: 99% confidence; $n = 75$, $\bar{x} = 2.76$, $s = 0.88$

7. Test scores: 90% confidence; $n = 150$, $\bar{x} = 77.6$, $s = 14.2$

8. Cost of basic phone service: 92% confidence; $n = 64$, $\bar{x} = \$15.30$, $s = \$1.25$

9. A sample of 35 skulls is obtained for Egyptian males who lived around 1850 BCE. The maximum breadth of each skull is measured with the result that $\bar{x} = 134.5$ mm and $s = 3.48$ mm (based on data from *Ancient Races of the Thebaid* by Thomson and Randall-Maciver). Using these sample results, construct a 95% confidence interval for the population mean μ. Write a statement that interprets the confidence interval.

10. A sample consists of 75 TV sets purchased several years ago. The replacement times of those TV sets have a mean of 8.2 years and a standard deviation of 1.1 years (based on data from "Getting Things Fixed," *Consumer Reports*). Construct a 90% confidence interval for the mean replacement time of all TV sets from that era. Does the result apply to TV sets currently being sold?

11. A Lyman Electronic Digital Caliper was used to accurately measure the lengths of 40 Triton Quik-Shok 380 ACP 90 grain cartridges (bullets). The results (in millimetres) are listed below. The cartridges are supposed to have a mean length of 0.950 mm and they must be between 0.945 mm long and 0.955 mm, otherwise they could be dangerous when fired. Construct a 90% confidence interval estimate of the mean length, then interpret the results.

0.951	0.954	0.954	0.953	0.953	0.953	0.952	0.955	0.951	0.949
0.950	0.950	0.954	0.950	0.948	0.948	0.945	0.947	0.950	0.954
0.951	0.951	0.952	0.943	0.952	0.952	0.949	0.946	0.949	0.949
0.953	0.953	0.953	0.950	0.949	0.950	0.950	0.949	0.950	0.951

12. An economist is studying the annual revenues of top Canadian companies. The results of a random sample of such revenues (in billions of dollars) are listed below (based on Data Set 6 in Appendix B). Construct a 95% confidence interval estimate of the mean annual revenue for "Top 100" companies. Based on these results, would you expect to find a company on the list with an annual revenue of $350 million?

1.718	6.017	1.681	2.520	1.800	3.678	4.583	2.052	3.795	6.813
3.150	3.149	1.778	2.458	8.509	3.277	2.653	7.312	2.411	4.000
17.161	1.843	2.043	3.076	3.432	3.058	2.683	3.643	3.307	4.148
2.424	1.717	9.512	7.013	33.191					

13. The standard IQ test is designed so that the mean is 100 and the standard deviation is 15 for the population of normal adults. Find the sample size necessary to estimate the mean IQ score of statistics students. We want to be 98% confident that our sample mean is within 1.5 IQ points of the true mean. The mean for this population is clearly greater than 100. The standard

deviation for this population is probably less than 15 because it is a group with less variation than a group randomly selected from the general population; therefore, if we use $\sigma = 15$, we are being conservative by using a value that will make the sample size at least as large as necessary. Assume then that $\sigma = 15$ and determine the required sample size.

14. The McLean Vending Machine Company must adjust its machines to accept only coins with specified weights. We will obtain a sample of quarters and weigh them to determine the mean. How many quarters must we randomly select and weigh if we want to be 99% confident that the sample mean is within 0.025 g of the true population mean for all quarters? If we use the sample of quarters in Data Set 13 of Appendix B, we can estimate that the population standard deviation is 0.068 g.

15. If we refer to the weights (in grams) of quarters listed in Data Set 13 in Appendix B, we will find 50 weights with a mean of 5.622 g and a standard deviation of 0.068 g. Based on this random sample of quarters in circulation, construct a 98% confidence interval estimate of the population mean of all quarters in circulation. Is the claim that quarters are minted to yield a mean weight of 5.670 g consistent with the confidence interval? If not, what is a possible explanation for the discrepancy?

16. Data Set 18 in Appendix B gives the body temperatures in degrees Celsius of adults accepted for voluntary surgery. For that sample, the mean is 36.39°C and the standard deviation is 0.62°C. Construct a 99% confidence interval for the mean body temperature of the general population and interpret the results.

17. A psychologist has developed a new test of spatial perception, and she wants to estimate the mean score achieved by male pilots. How many people must she test if she wants the sample mean to be in error by no more than 2.0 points, with 95% confidence? An earlier study suggests that $\sigma = 21.2$.

18. To plan for the proper handling of household garbage, the city of Fredericton must estimate the mean weight of garbage discarded by households in one week. Find the sample size necessary to estimate that mean if you want to be 96% confident that the sample mean is within 2 lb of the true population mean. For the population standard deviation s, use the value of 12.46 lb, which is the standard deviation of the sample of 62 households included in Data Set 1.

19. A large sample of data ($n = 335$) from a Calgary observatory produced a mean daily counting rate for cosmic rays of 3465.46, and a standard deviation of 127.72. Use these sample data to find the 90% confidence interval for the mean of the daily counting rates. Suppose a scientist claims that the mean of the daily counting rates is actually 3525. Would that claim be consistent with the confidence interval?

20. A random sample of 250 Ontario households using natural gas heating shows that the mean monthly consumption of natural gas is 107.73 cu. ft., with a standard deviation of 13.94 cu. ft. Use these statistics to construct a 94% confidence interval for the mean monthly consumption for the population of all Ontario households.

21. Nielsen Media Research wants to estimate the mean amount of time (in hours) that full-time college students spend watching television each weekday. Find the sample size necessary to estimate that mean with a 0.25 h (or 15 min) margin of error. Assume that a 96% degree of confidence is desired. Also assume that a pilot study showed that the standard deviation is estimated to be 1.87 hours.

22. In deciding whether to attend university, many students are influenced by the increased earnings potential that a university degree is likely to create. A recent census study by Statistics Canada shows that the mean annual income of high-school graduates is $22,846, whereas the mean annual income of university graduates is $42,054. Find the sample size necessary to estimate next year's mean annual income of university graduates. Assume that you want 94% confidence that the sample mean will be within $1000 of the true mean, and assume that the population standard deviation is estimated to be $32,896.

23. You have just been hired by the Stampeder Marketing Company to conduct a survey to estimate the mean amount of money spent by movie patrons (per movie) in Alberta. First use the range rule of thumb to make a rough estimate of the standard deviation of the amounts spent. It is reasonable to assume that typical amounts range from $3 to about $15. Then use that estimated standard deviation to determine the sample size corresponding to 98% confidence and a 25¢ margin of error.

24. Estimate the minimum and maximum ages for typical textbooks currently used in college courses, then use the range rule of thumb to estimate the standard deviation. Next, find the size of the sample required to estimate the mean age (in years) of textbooks currently used in college courses. Assume a 96% degree of confidence that the sample mean will be in error by no more than 0.25 year.

25. Refer to Data Set 1 in Appendix B for the 62 weights (in pounds) of *paper* discarded by households. Using that sample, construct a 92% confidence interval estimate of the mean weight of paper discarded by all households.

26. Refer to Data Set 1 in Appendix B and construct a 92% confidence interval for the mean weight in pounds of discarded plastic goods per household for one week. Compare these results with your solution to Exercise 25, and interpret. Which is more of an ecological problem: discarded paper or discarded plastic?

27. Refer to Data Set 11 in Appendix B and construct a 97% confidence interval for the mean weight of brown M&M plain candies. Can the methods of this section be used to construct a 97% confidence interval for the mean weight of blue M&M plain candies? Why or why not?

28. Refer to Data Set 8 in Appendix B and construct a 98% confidence interval for the mean value of coins in possession of statistics students. Is there reason to believe that this value is different from the mean value of coins in possession of people randomly selected from the general population of adult Canadians?

6-2 Exercises B: Beyond the Basics

29. In Formula 6-1 we assume that the population is infinite, that we are sampling with replacement, or that the population is very large. If we have a relatively small population such that $n > 0.05N$, and sample without replacement, we should modify E to include a *finite population correction* factor as follows:

$$E = z_{\alpha/2} \frac{\sigma}{\sqrt{n}} \sqrt{\frac{N - n}{N - 1}}$$

where N is the population size.

a. Find the 95% confidence interval for the mean of 100 IQ scores if a sample of 31 of those scores produces a mean and standard deviation of 132 and 10, respectively.

b. If you are determining a sample size, and realize that the population size N is relatively small, then the sample-size calculation should be revised to:

$$n = \frac{N\sigma^2 (z_{\alpha/2})^2}{(N - 1)E^2 + \sigma^2(z_{\alpha/2})^2}$$

For a challenge, try solving the expression given at the beginning of this question for n to produce the formula given here. Repeat Exercise 13, assuming that the statistics students are randomly selected without replacement from a population of $N = 200$ statistics students.

30. Test the effect of an outlier as follows: Use the sample data in Table 6-1 to find a 95% confidence interval estimate of the population mean, but change the first entry from 36.4 to 364. Nobody can really have that high a body temperature, but such an error can easily occur when a decimal point is inadvertently omitted when the sample values are entered.

a. Find the 95% confidence interval.

b. By comparing the result from part (a) to the confidence interval found in this section, describe the effect of an outlier on a confidence interval. Are the confidence interval limits sensitive to outliers?

c. Based on part (b), how should you handle outliers when they are found in sample data sets that will be used for construction of confidence intervals?

31. In many of the preceding examples, confidence intervals are used toward the ultimate goal of estimating the value of a population parameter. Confidence intervals can also be used as a data-exploration tool, for describing and comparing data sets. Consider the following descriptive statistics, taken from two samples:

Men:	$n = 100$	$\bar{x} = 1.747$ m	$s = 0.0744$ m
Women:	$n = 100$	$\bar{x} = 1.610$ m	$s = 0.0620$ m

Construct the confidence intervals for the estimates of the population mean of each group. Do the intervals overlap? Based on your answer to the preceding question, would you conclude that the populations of men's and of women's heights are likely to be the same or different? Explain.

32. A 95% confidence interval for the lives (in minutes) of Kodak AA batteries is $430 < \mu < 470$. (See Program 1 of *Against All Odds: Inside Statistics*.) Assume that this result is based on a sample of size 100.
   a. Construct the 99% confidence interval.
   b. What is the value of the sample mean?
   c. What is the value of the sample standard deviation?
   d. If the confidence interval $432 < \mu < 468$ is obtained from the same sample data, what is the degree of confidence?

6-3 Estimating a Population Mean: Small Samples

In Section 6–2 we began our study of inferential statistics by considering point estimates and confidence intervals as methods for estimating the value of a population mean μ. All of the examples and exercises in Section 6–2 involved samples that were large, with sample sizes n greater than 30. Factors such as cost and time often severely limit the size of a sample, so the normal distribution may not be a suitable approximation of the distribution of means from small samples. In this section we investigate methods for estimating a population mean μ when the sample size n is small, where *small* is considered to be 30 or fewer.

Assumptions for This Section

1. $n \leq 30$.

2. The sample has been collected by a valid random-sampling technique.

3. The sample is from a normally distributed population. (This is a loose requirement, which can be met if the population has only one mode and is basically symmetric.)

If the first assumption is not satisfied, we have a large sample, and we can use the methods described in Section 6–2. If the second assumption is not satisfied, we cannot use the methods described in this book, but it might be possible to use more advanced methods. If the third assumption is not satisfied because the population has a distribution that is very nonnormal, we cannot use the methods of this section, but we might be able to use nonparametric methods (see Chapter 12) or bootstrap resampling methods (see the Computer Project at the end of this chapter).

As explained in Section 6–2, *the sample mean $\bar{x}$ is generally the best point estimate of the population mean μ*, because the distribution of sample means $\bar{x}$ tends to be more consistent (with less variation) than distributions of other sample measures of centre, and the sample mean $\bar{x}$ is an *unbiased estimator* of the population mean μ, that is, its distribution tends to centre around the value of the population mean.

As we also saw in Section 6–2, however, the point estimate does not reveal how good that estimate is, and that led us to develop confidence interval estimates for μ. In fact, confidence intervals can be constructed for small samples by using the normal distribution with the same margin of error from the preceding section, **provided that** the original population has a normal distribution and the population standard deviation σ is known (a condition not very common in real applications).

If we have a small ($n \leq 30$) sample but do not know σ, we can sometimes use the Student t distribution, developed by William Gosset (1876–1937), for constructing a confidence interval. Gosset was a Guinness Brewery employee who needed a distribution that could be used with small samples. The Irish brewery where he worked did not allow the publication of research results, so Gosset published under the pseudonym *Student*. As a result of his early experiments and studies of small samples, we can now use the Student t distribution.

STUDENT t DISTRIBUTION

If the distribution of a population is essentially normal (approximately bell-shaped), then the distribution of

$$t = \frac{\bar{x} - \mu}{\frac{s}{\sqrt{n}}}$$

is essentially a **Student t distribution** for all samples of size n. The Student t distribution, often simply referred to as the **t distribution**, is used to find critical values denoted by $t_{\alpha/2}$.

Table A-3 lists values of the t distribution along with areas denoted by α. Values of $t_{\alpha/2}$ are obtained from Table A-3 by locating the proper value for *degrees of freedom* in the left column and then proceeding across the corresponding row until reaching the number directly below the applicable value of α for two tails.

> The number of **degrees of freedom** for a data set corresponds to the number of sample values that can vary after certain restrictions have been imposed on all data values.

For example, if 10 students have quiz scores with a mean of 80, we can freely assign values to the first 9 scores, but the 10th score is then determined. The sum of the 10 scores must be 800, so the 10th score must equal 800 minus the sum of the first 9 scores. Because those first 9 scores can be freely selected to be any values, we say that there are 9 degrees of freedom available. For the applications of this section, the number of degrees of freedom is simply the sample size minus 1:

$$\text{degrees of freedom} = n - 1$$

EXAMPLE

A sample of size $n = 15$ is a simple random sample selected from a normally distributed population. Find the critical value $t_{\alpha/2}$ corresponding to a 95% degree of confidence.

SOLUTION

Because $n = 15$, the number of degrees of freedom is given by $n - 1 = 14$. Using Table A-3, we locate the 14th row by referring to the column at the extreme left. As in Section 6–2, a 95% degree of confidence corresponds to $\alpha = 1 - 0.95 = 0.05$, so we find the column with the heading "0.05 (two tails)." The value in the 14th row and the "0.05 (two tails)" column is 2.145, so $t_{\alpha/2} = 2.145$.

It might seem a bit strange that with a normally distributed population, we sometimes use the t distribution to find critical values, but when σ is unknown, the use of s from a small sample incorporates another source of error. To maintain the desired degree of confidence, we compensate for the additional variability by widening the confidence interval through a process that replaces the critical value $z_{\alpha/2}$ (found from the standard normal distribution values in Table A-2) with the larger critical value of $t_{\alpha/2}$ (found from the t-distribution values in Table A-3).

Important Properties of the Student t Distribution

1. The Student t distribution is different for different sample sizes. (See Figure 6-5 for the cases $n = 3$ and $n = 12$.)
2. The Student t distribution has the same general symmetric bell shape as the standard normal distribution, but it reflects the greater variability (with wider distributions) that is expected with small samples.

Figure 6-5
Student *t* Distributions for *n* = 3 and *n* = 12

The Student *t* distribution has the same general shape and symmetry as the standard normal distribution, but it reflects the greater variability that is expected with small samples.

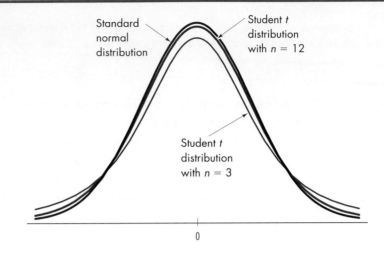

3. The Student *t* distribution has a mean of $t = 0$ (just as the standard normal distribution has a mean of $z = 0$).

4. The standard deviation of the Student *t* distribution varies with the sample size, but it is greater than 1 (unlike the standard normal distribution, which has $\sigma = 1$).

5. As the sample size *n* gets larger, the Student *t* distribution gets closer to the standard normal distribution. For values of $n > 30$, the differences are so small that we can use the critical *z* values instead of developing a much larger table of critical *t* values. (The values in the bottom row of Table A-3 are equal to the corresponding critical *z* values from the standard normal distribution.)

Following is a summary of the conditions indicating use of a *t* distribution instead of the standard normal distribution. (These same conditions will also apply in Chapter 7.)

Conditions for Using the Student *t* Distribution

1. The sample is small ($n \leq 30$); and

2. σ is unknown; and

3. the parent population has a distribution that is essentially normal. (Because the distribution of the parent population is often unknown, we often estimate it by constructing a histogram of sample data.)

We can now determine values for the margin of error *E* in estimating μ when a *t* distribution applies. This margin of error can be used for constructing confidence intervals.

Formula 6-3 $\qquad E = t_{\alpha/2}\dfrac{s}{\sqrt{n}}$ $\quad$ where $t_{\alpha/2}$ has $n - 1$ degrees of freedom

CONFIDENCE INTERVAL FOR THE ESTIMATE OF μ
[BASED ON A SMALL RANDOM SAMPLE ($n \leq 30$) AND UNKNOWN σ
FROM AN APPROXIMATELY NORMALLY DISTRIBUTED POPULATION]

$$\bar{x} - E < \mu < \bar{x} + E$$

where $\qquad\qquad\qquad E = t_{\alpha/2}\dfrac{s}{\sqrt{n}}$

EXAMPLE

With *destructive testing,* sample items are destroyed in the process of testing them. Crash testing of cars is one very expensive example of destructive testing. If you were responsible for such crash tests, there is no way you would want to tell your supervisor that you must crash and destroy more than 30 cars so that you could use the normal distribution. Let's assume that you have crash-tested 12 Dodge Viper sports cars under a variety of conditions that simulate typical collisions. Analysis of the 12 damaged cars results in repair costs having a distribution that appears to be bell-shaped, with a mean of $\bar{x} = \$26{,}227$ and a standard deviation of $s = \$15{,}873$ (based on data from the Highway Loss Data Institute). Find the following:

a. The best point estimate of μ, the mean repair cost for all Dodge Vipers involved in collisions
b. The 95% interval estimate of μ, the mean repair cost for all Dodge Vipers involved in collisions

SOLUTION

a. The best point estimate of the population mean μ is the value of the sample mean $\bar{x}$. In this case, the best point estimate of μ is therefore $26,227.

b. We will proceed to construct a 95% confidence interval by using the t distribution because the following three conditions are met: (1) The sample is small ($n \leq 30$), (2) the population standard deviation σ is unknown, and (3) the population appears to have a normal distribution because the sample data have a bell-shaped distribution.

We begin by finding the value of the margin of error as shown below. Note that the critical value of $t_{\alpha/2} = 2.201$ is found from Table A-3 in the column labelled ".05 two tails" (from 95% confidence) and the row corresponding to 11 degrees of freedom (from $n - 1 = 11$).

$$E = t_{\alpha/2}\frac{s}{\sqrt{n}} = 2.201\frac{15{,}873}{\sqrt{12}} = 10{,}085.29$$

We can now construct the 95% interval estimate of μ by using $E = 10,085.29$ and $\bar{x} = 26,227$. Because the summary statistics are rounded to the nearest dollar, the confidence interval limits will also be rounded to the nearest dollar.

$$\bar{x} - E < \mu < \bar{x} + E$$
$$26,227 - 10,085.29 < \mu < 26,227 + 10,085.29$$
$$\$16,142 < \mu < \$36,312$$

[This result could also be expressed in the format of $\mu = \$26,227 \pm \$10,085$ or as ($\$16,142$, $\$36,312$).]

Interpretation
On the basis of the given sample results, we are 95% confident that the limits of $\$16,142$ and $\$36,312$ actually do contain the value of the population mean μ. These repair costs appear to be quite high. Such information is important to companies that insure Dodge Vipers against collisions.

Choosing the Appropriate Distribution

It is sometimes difficult to decide whether to use Formula 6-1 (and the standard normal distribution) or Formula 6-3 (and the Student t distribution). Figure 6-6 summarizes the traditional approach to making this decision when constructing confidence intervals for estimating μ, the population mean. In Figure 6-6, note that if we have a small ($n \le 30$) sample drawn from a very nonnormal distribution, we can't use the methods described in this chapter. One alternative is to use nonparametric methods (see Chapter 12), and another alternative is to use the computer-oriented bootstrap method, which makes no assumptions about the original population. This method is described in the Computer Project at the end of this chapter.

Another Approach Rather than using *sample size* as the major criterion for choosing between the normal (z) and Student t distributions, some statisticians and software use *knowledge of the population standard deviation σ* as the major criterion—namely, if the sample is from a normally distributed population, then use the *Student t distribution* (regardless of sample size) *unless σ* is known (which is rarely the case). But using the Student t distribution with a large sample size results in a large number of degrees of freedom, so the value of $t_{\alpha/2}$ is, for all practical purposes, the same as the value of $z_{\alpha/2}$. When depending on tables, the normal distribution allows more precision in calculating tail probabilites than does the t table—but if you are using computers, this limitation no longer applies.

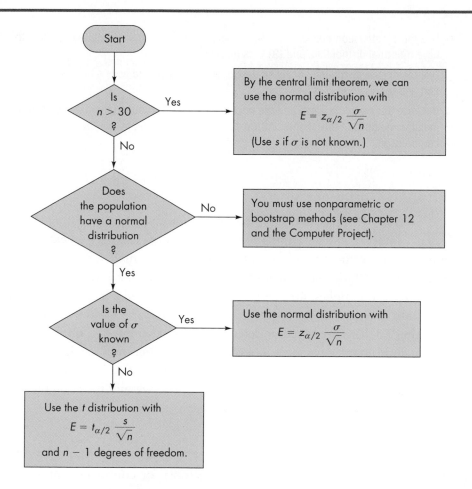

Figure 6-6
Choosing Between Normal (z) and t Distributions to Calculate the Margin of Error E for Estimating μ (Traditional Approach)

Start

Is $n > 30$?

Yes → By the central limit theorem, we can use the normal distribution with

$$E = z_{\alpha/2} \frac{\sigma}{\sqrt{n}}$$

(Use s if σ is not known.)

No ↓

Does the population have a normal distribution ?

No → You must use nonparametric or bootstrap methods (see Chapter 12 and the Computer Project).

Yes ↓

Is the value of σ known ?

Yes → Use the normal distribution with

$$E = z_{\alpha/2} \frac{\sigma}{\sqrt{n}}$$

No ↓

Use the t distribution with

$$E = t_{\alpha/2} \frac{s}{\sqrt{n}}$$

and $n - 1$ degrees of freedom.

EXAMPLE

Assuming that you plan to construct a confidence interval for the population mean μ, use the given data to determine whether the margin of error E should be calculated using the normal distribution, the t distribution, or neither (so that the methods of this chapter cannot be used).

a. $n = 50, \bar{x} = 77.6, s = 14.2$, and the shape of the distribution is skewed.

b. $n = 25, \bar{x} = 77.6, s = 14.2$, and the distribution is bell-shaped.

c. $n = 25, \bar{x} = 77.6, \sigma = 14.2$, and the distribution is bell-shaped.

d. $n = 25, \bar{x} = 77.6, \sigma = 14.2$, and the distribution is extremely skewed.

SOLUTION

Refer to Figure 6-6 and use the flowchart to determine the following.

a. Because the sample is large ($n > 30$), use the normal distribution. When calculating the margin of error E, use Formula 6-1, where the sample standard deviation s is used for σ.

b. Use the t distribution because (1) the sample is small, (2) the population appears to have a normal distribution, and (3) σ is unknown.

c. Use the normal distribution because the population appears to have a normal distribution and the value of σ is known.

d. Because the sample is small and the population has a distribution that is very nonnormal, the methods of this chapter do not apply. Neither the normal nor the t distribution can be used.

AS Section 6–2 presented the three main concepts of point estimate, confidence interval, and determination of sample size. This section extended the concepts of point estimate and confidence interval to small sample cases. This section does not include determination of sample size because with known σ the normal distribution applies, and with unknown σ we need a large sample to justify estimation of σ with s, but the large sample allows us to use the normal distribution again. We therefore base our determinations of sample size on Formula 6-2 only and include no circumstances in which the t distribution is used.

The following section again considers the concepts of point estimate, confidence interval, and determination of sample size, but we will focus on the population proportion instead of the population mean.

USING TECHNOLOGY

EXCEL: *Caution:* Excel's built-in tools for confidence intervals are based on finding critical values of $z_{\alpha/2}$. To obtain confidence intervals that require critical values of $t_{\alpha/2}$, use the Data Desk add-in that is a supplement to this book.

EXCEL (Prior to 2007):   Click on **DDXL**, select **Confidence Intervals**, and select **1 Var t Interval** under the options listed under Function Type. Click on the pencil icon and enter the Excel range containing the data, such as A1:A12. Click on **OK**. In the dialog box, select the level of confidence, then click on **Compute Interval**, and the confidence interval will be displayed.

EXCEL 2007:   Click **Add-Ins** on the main menu and choose **DDXL** from the add-in menu to the left. Proceed as above.

MINITAB 15:   *Confidence Intervals:* From the **Stat** menu, choose **Basic Statistics** and then **1-Sample t**. If you are constructing the confidence interval using raw data, click the **Samples in columns** radio button and select the column with the data.

There is an optional box to conduct a hypothesis test. (You would be required to fill in a hypothesis mean, which will be covered in Chapter 7.) The default setting is a 95% level of confidence. If you wish to change it, click the **Options** button and change the level of confidence to what you desire. You can leave the **Alternative** drop-down menu set to not equal.

STATDISK: Select **Analysis** from the main menu, then select **Confidence Intervals**, followed by **Mean – One Sample**, and enter the items in the dialog box. After you click on **Evaluate**, the confidence interval will be displayed. Note that if you are using the sample standard deviation (which is the case for the t distribution), enter that in the appropriate box.

6-3 Exercises A: Basic Skills and Concepts

In Exercises 1–4, find the critical value $t_{\alpha/2}$ that corresponds to the given degree of confidence and sample size n.

1. 99%; $n = 10$ 2. 95%; $n = 16$
3. 98%; $n = 21$ 4. 90%; $n = 8$

In Exercises 5–8, use the given degree of confidence and sample data to find (a) the margin of error and (b) the confidence interval for the population mean μ. In each case, assume that the population has a normal distribution.

5. Heights of female students: 95% confidence; $n = 10$, $\bar{x} = 64.2$ in., $s = 2.6$ in.

6. Grade-point averages: 99% confidence; $n = 15$, $\bar{x} = 2.76$, $s = 0.88$

7. Test scores: 90% confidence; $n = 16$, $\bar{x} = 77.6$, $s = 14.2$

8. Cost of basic phone service: 98% confidence; $n = 19$, $\bar{x} = \$15.30$, $s = \$1.25$

In Exercises 9–20, be sure to determine correctly whether the confidence intervals are calculated with the normal distribution or the t distribution. (In some cases, you may have a choice.)

9. In crash tests of 15 Honda Odyssey minivans, collision repair costs are found to have a distribution that is roughly bell-shaped, with a mean of $1786 and a standard deviation of $937. Construct the 99% confidence interval for the mean repair cost in all such vehicle collisions.

10. Data Set 18 in Appendix B includes 108 body temperatures. Suppose that we have only the first 10 temperatures given below. For these scores, $\bar{x} = 36.36$

and $s = 0.35$. Construct the 95% confidence interval for the mean of all body temperatures. (A prior study has shown that body temperatures are normally distributed.)

$$36.4 \quad 36.3 \quad 36.7 \quad 36.3 \quad 36.8 \quad 36.6 \quad 36.6 \quad 36.1 \quad 36.2 \quad 35.6$$

11. In a time-use study, 20 randomly selected managers were found to spend a mean of 2.40 h each day on paperwork. The standard deviation of the 20 scores is 1.30 h (based on data from Adia Personnel Services). Also, the sample data appear to have a bell-shaped distribution. Construct the 95% confidence interval for the mean time spent on paperwork by all managers. If the organization requires its managers to have a mean of 5.60 hours available each day (out of 8 hours) for non-desk work, should they be satisfied with the results of the time study? Why or why not?

12. In a study relating the amounts of time required for room-service delivery at a newly opened Radisson Hotel, 20 deliveries had a mean time of 24.2 min and a standard deviation of 8.7 min. The sample data appear to have a bell-shaped distribution. Construct the 90% confidence interval for the mean time of all deliveries.

13. In a study relating physical attractiveness and mental disorders, 231 subjects were rated for attractiveness. The resulting sample mean and standard deviation for that variable were 3.94 and 0.75, respectively. (See "Physical Attractiveness and Self-Perception of Mental Disorder," by Burns and Farina, *Journal of Abnormal Psychology*, Vol. 96, No. 2.) Use these sample data to construct the 95% confidence interval for the population mean.

14. A survey was conducted to estimate the annual income of CEOs of large corporations. Twenty companies were surveyed and the mean income for this group was $220,000 with a standard deviation of $21,000, and the distribution of this group appeared normal. Construct the 99% confidence interval for the mean annual income of all CEOs of large corporations.

15. A manufacturer of AAA batteries wants to estimate the mean life expectancy of the batteries. A sample of 25 such batteries shows that the distribution of life expectancies is roughly normal with a mean of 44.25 h and a standard deviation of 2.25 h. Construct a 98% confidence interval for the mean life expectancy of all the AAA batteries made by this manufacturer.

16. Off-season football players often participate in community events and charity fundraisers. Suppose a firm is bidding to provide T-shirts for players who are representing the CFL in such events. The firm has randomly sampled CFL quarterbacks and running backs, with these results (based on data from the CFL website):

Quarterbacks:	$n = 10$	$\bar{x} = 192.7$ lb.	$s = 11.7$ lb
Running backs:	$n = 9$	$\bar{x} = 199.7$ lb.	$s = 11.8$ lb

Construct a 95% confidence interval for the mean weight of both populations. Assuming that weight is a rough indicator of required T-shirt size, should the firm consider making separate sizes for quarterbacks and running backs? Explain.

17. In a study of the use of hypnosis to relieve pain, sensory ratings were measured for 16 subjects, with the results given below (based on data from "An Analysis of Factors That Contribute to the Efficacy of Hypnotic Analgesia," by Price and Barber, *Journal of Abnormal Psychology*, Vol. 96, No. 1). Use these sample data to construct the 95% confidence interval for the mean sensory rating for the population from which the sample was drawn.

    8.8   6.6   8.4   6.5   8.4   7.0   9.0   10.3
    8.7 11.3   8.1   5.2   6.3   8.7   6.2   7.9

18. Refer to Data Set 11 in Appendix B and use only the sample of *red* M&M plain candies to construct a 95% confidence interval for the mean weight of all M&Ms.

19. Refer to Data Set 11 in Appendix B and use the entire sample of 100 plain M&M candies to construct a 95% confidence interval for the mean weight of all M&Ms. How does the result compare to the confidence interval for Exercise 18?

20. Refer to Data Set 10 in Appendix B.
   a. Construct a 95% confidence interval for the mean annual return over three years of U.S. equity funds.
   b. Construct a 95% confidence interval for the mean annual return over three years of Canadian equity funds and European funds.
   c. Compare and interpret the results of parts (a) and (b).

6-3 Exercises B: Beyond the Basics

21. Assume that a small ($n \leq 30$) sample is randomly selected from a normally distributed population for which σ is unknown. Construction of a confidence interval should use the t distribution, but how are the confidence interval limits affected if the normal distribution is incorrectly used instead?

22. A confidence interval is constructed for a small sample of temperatures (in degrees Celsius) randomly selected from a normally distributed population for which σ is unknown (such as the data set given in Exercise 10).
   a. How is the margin of error E affected if each temperature is converted to the Fahrenheit scale? [$F = \frac{9}{5}C + 32$]
   b. If the confidence interval limits are denoted by a and b, find expressions for the confidence interval limits after the original temperatures have been converted to the Fahrenheit scale.

c. Based on the results from part (b), can confidence interval limits for the Celsius temperatures be found by simply converting the confidence interval limits from the Celsius scale to the Fahrenheit scale?

23. A simple random sample of Boeing 747 aircraft is selected, and the times (in hours) required to test for structural stress fractures are as follows: 8.1, 9.9, 9.5, 6.9, 9.8. Based on these results, a confidence interval is found to be $\mu = 8.84 \pm 1.24$. Find the degree of confidence.

6-4 Estimating a Population Proportion

In this section we apply the same three concepts discussed in Sections 6–2 and 6–3: (1) point estimate, (2) confidence interval, and (3) determining the required sample size. Whereas Sections 6–2 and 6–3 applied these concepts to estimates of a population mean μ, this section applies them to the population proportion p. For example, Nielsen Media Research might need to estimate the proportion of households that tune in to the Grey Cup, a question of great interest to potential television sponsors.

Although this section focuses on the population proportion p, we can also consider a probability or a percentage. Proportions and probabilities are both expressed in decimal or fraction form. When working with percents, convert them to proportions by dropping the percent sign and dividing by 100. For example, the 82.8% rate of Canadians who read newspapers regularly can be expressed in decimal form as 0.828. The symbol p may therefore represent a proportion, a probability, or the decimal equivalent of a percent. We now introduce the new notation of $\hat{p}$ (called "p hat") for the sample proportion.

In previous chapters we stipulated that $q = 1 - p$, so it is natural to note here that $\hat{q} = 1 - \hat{p}$. For example, if you survey 1068 Canadians and find that 673 of them have voicemail, the sample proportion is $\hat{p} = x/n = 673/1068 = 0.630$, and $\hat{q} = 0.370$ (calculated from $1 - 0.630$). In some cases, the value of $\hat{p}$ may be known because the sample proportion or percentage is given directly. If it is reported that 1068 Canadian television viewers are surveyed and 25% of them are college graduates, then $\hat{p} = 0.25$ and $\hat{q} = 0.75$.

NOTATION FOR PROPORTIONS

$p = population$ proportion

$\hat{p} = \dfrac{x}{n} = sample$ proportion of x successes in a sample of size n

$\hat{q} = 1 - \hat{p} = sample$ proportion of failures in a sample of size n

If we want to estimate a population proportion with a single value, the best estimate is the *point estimate* $\hat{p}$.

The sample proportion $\hat{p}$ is the best point estimate of the population proportion p.

We use $\hat{p}$ as the point estimate of p (just as $\bar{x}$ is used as the point estimate of μ) because it is unbiased and is the most consistent of the estimators that could be used. It is unbiased in the sense that the distribution of sample proportions tends to centre about the value of p; that is, sample proportions $\hat{p}$ do not systematically tend to underestimate p, nor do they systematically tend to overestimate p. The sample proportion $\hat{p}$ is the most consistent estimator in the sense that the standard deviation of sample proportions tends to be smaller than the standard deviation of any other unbiased estimators.

As we saw when estimating the mean, the point estimate has a major disadvantage: Although it provides the best single-value estimate, in this case for the population proportion p, it gives no indication of just how good the estimate is. The advantage of a confidence interval estimate is that it does contain an indication of how accurate the point estimate is. We first present the *margin of error* (first defined in Section 6–2), which is used for finding a confidence interval, and then we present the format of the confidence interval itself.

Formula 6-4
$$E = z_{\alpha/2} \sqrt{\frac{\hat{p}\hat{q}}{n}}$$

$$\hat{p} - E < p < \hat{p} + E \qquad \text{where } E = z_{\alpha/2} \sqrt{\frac{\hat{p}\hat{q}}{n}}$$

The confidence interval is sometimes expressed in the following formats:

$$p = \hat{p} \pm E$$

or
$$(\hat{p} - E, \quad \hat{p} + E)$$

In Chapter 3 we rounded probabilities expressed in decimal form to three significant digits. We use that same rounding rule here.

Round the confidence interval limits to three significant digits.

Assumptions for This Section

1. The sample has been collected by a valid random-sampling technique.

2. The four conditions for the binomial distribution are satisfied (see Section 4–3). These are: (1) There is a fixed number of trials; (2) the trials are independent; (3) there are two categories of outcomes; and (4) the probabilities of success remain constant for each trial.

3. The conditions for using the normal distribution to approximate the distribution of sample proportions are satisfied: p is suitably large, $np \geq 5$ and $nq \geq 5$. (Because the population values for p and q are unknown, we use the sample proportion to estimate these values. Also, there are procedures for dealing with situations in which the normal distribution is not a suitable approximation. See Exercise 34.)

EXAMPLE

To estimate the percent of all households in Canada with any children, 19,760 households were selected at random. It was found that 7483 of these households had children (based on data from a NADbank study). Using these sample results, find

a. The point estimate of the population proportion of all households in Canada with any children.

b. The 95% interval estimate of the population proportion of all households in Canada with any children.

SOLUTION

a. The point estimate of p is

$$\hat{p} = \frac{x}{n} = \frac{7483}{19,760} = 0.379$$

b. Construction of the confidence interval requires that we first evaluate the margin of error E. The value of E can be found from Formula 6-4. We use $\hat{p} = 0.379$ (found above), $\hat{q} = 0.621$ (from $\hat{q} = 1 - \hat{p}$), and $z_{\alpha/2} = 1.96$ (from Table A-2, where 95% converts to $\alpha = 0.05$, which is divided equally between the two tails so that $z = 1.96$ corresponds to an area of 0.4750).

$$E = z_{\alpha/2} \sqrt{\frac{\hat{p}\hat{q}}{n}} = 1.96 \sqrt{\frac{(0.379)(0.621)}{19,760}} = 0.0068$$

We can now find the confidence interval by using $\hat{p} = 0.379$ and $E = 0.0068$.

$$\hat{p} - E < p < \hat{p} + E$$
$$0.379 - 0.0068 < p < 0.379 + 0.0068$$
$$0.372 < p < 0.386$$

If we wanted the 95% confidence interval for the true population *percentage*, we could express this result as 37.2% < p < 38.6%. This result is often reported in the following format: "Among Canadian households, the percentage that have children is estimated to be 37.9%, with a margin of error of plus or minus 0.68 percentage points." This is a verbal expression of this format for the confidence interval: $p = 37.9\% \pm 0.68\%$. (The level of confidence should also be reported, but it rarely is in the media. The media typically use a 95% degree of confidence but omit any reference to it.) The confidence interval is also expressed in this format: (0.372, 0.386).

Remember that the population value for p is fixed, and we are not assigning it a probability. In the preceding solution, the "95%" level expresses our confidence that 95% of the time, if we were to randomly select a sample with $n = 19,760$, and construct a confidence interval for p using the procedures shown, the true value of p would in fact be included in the calculated interval.

Basis for the Procedure of Finding the Margin of Error

If the assumptions given for this section are satisfied, then our calculations can be based on a normal approximation for a binomial distribution. Therefore, we can use results from Section 5–6 to conclude that μ and s are given by $\mu = np$ and $\sigma = \sqrt{npq}$. Both of these parameters pertain to n trials, but we convert them to a per-trial basis by dividing by n as follows:

$$\text{Mean of sample proportions} = \frac{np}{n} = p$$

$$\text{Standard deviation of sample proportions} = \frac{\sqrt{npq}}{n} = \sqrt{\frac{npq}{n^2}} = \sqrt{\frac{pq}{n}}$$

The first result may seem trivial because we have already stipulated that the true population proportion is p. The second result is nontrivial and is useful in describing the margin of error E, but we replace the product pq by $\hat{p}\hat{q}$ because we don't yet know the value of p (it is the value we are trying to estimate). Formula 6-4 for the margin of error reflects the fact that $\hat{p}$ has a probability of $1 - \alpha$ of being within $z_{\alpha/2}\sqrt{pq/n}$ of p. The confidence interval for p, as given previously, reflects the fact that there is a probability of $1 - \alpha$ that $\hat{p}$ differs from p by less than the margin of error

$$E = z_{\alpha/2}\sqrt{\frac{\hat{p}\hat{q}}{n}}$$

Factors Affecting Confidence Intervals for Proportions

In Section 6–2, we examined the factors affecting the margin of error and the width of the confidence interval for means. Two of those factors remain the same: the level of confidence and the sample size. The one difference is the magnitude of $\hat{p}$ and $\hat{q}$. As these values approach 0.5, the result is that the margin of error becomes larger and the confidence interval becomes wider. Intuitively, this should make sense: as the values approach 0.5, this indicates more variability in the data. For example, suppose you stand outside the door to a student centre at a college campus and try to predict the gender of the next person to come through the door. Unless you are psychic, you will not perform this task with a great deal of success. Suppose instead that you are standing outside a women's wear store trying to perform the same task. You should have greater success since there should be less variability in the data.

Determining Sample Size

Having discussed point estimates and confidence intervals for p, we will now describe a procedure for determining how large a sample should be when we want to find the approximate value of a population proportion. In the previous section we began with the expression for the margin of error E, then solved for n. Using a similar procedure, we begin with

$$E = z_{\alpha/2} \sqrt{\frac{\hat{p}\hat{q}}{n}}$$

and solve for n to get the sample size, as given in Formula 6-5. Formula 6-5 requires $\hat{p}$ as an estimate of the population proportion p, but if no such estimate is known, we replace both $\hat{p}$ and $\hat{q}$ by 0.5, with the result given in Formula 6-6.

SAMPLE SIZE FOR ESTIMATING PROPORTION p

When an estimate $\hat{p}$ is known: Formula 6-5 $n = \dfrac{[z_{\alpha/2}]^2 \hat{p}\hat{q}}{E^2}$

When no estimate $\hat{p}$ is known: Formula 6-6 $n = \dfrac{[z_{\alpha/2}]^2 \cdot 0.25}{E^2}$

ROUND-OFF RULE FOR DETERMINING SAMPLE SIZE

If the computed sample size is not a whole number, round it up to the next *higher* whole number.

Use Formula 6-5 when reasonable estimates of $\hat{p}$ can be made by using previous samples, a pilot study, or someone's expert knowledge. When no such guess can be made, we can assign the value of 0.5 to each of $\hat{p}$ and $\hat{q}$, so the resulting sample size will be at least as large as it should be. The underlying reason for the assignment of 0.5 is this: The product $\hat{p} \cdot \hat{q}$ has 0.25 as its largest possible value, which occurs when $\hat{p} = 0.5$ and $\hat{q} = 0.5$. (See the accompanying table, which lists some values of $\hat{p}$ and $\hat{q}$.) Note that Formulas 6-5 and 6-6 do not include the population size N, so the size of the population is irrelevant. (*Exception:* When sampling is without replacement from a relatively small finite population. See Exercise 29.)

$\hat{p}$	$\hat{q}$	$\hat{p} \cdot \hat{q}$
0.1	0.9	0.09
0.2	0.8	0.16
0.3	0.7	0.21
0.4	0.6	0.24
0.5	0.5	0.25
0.6	0.4	0.24
0.7	0.3	0.21
0.8	0.2	0.16
0.9	0.1	0.09

EXAMPLE

Insurance companies are becoming concerned that increased use of cellular telephones is resulting in more car crashes, and they are considering implementing higher rates for drivers who use such phones. We want to estimate, with a margin of error of three percentage points, the percentage of drivers who talk on phones while they are driving. Assuming that we want 95% confidence in our results, how many drivers must we survey?

a. Assume that we have an estimate of $\hat{p}$ based on a prior study that showed that 18% of drivers talk on a phone (based on data from *Prevention* magazine).

b. Assume that we have no prior information suggesting a possible value of $\hat{p}$.

SOLUTION

a. The prior study suggests that $\hat{p} = 0.18$, so $\hat{q} = 0.82$ (found from $\hat{q} = 1 - 0.18$). With a 95% level of confidence, we have $\alpha = 0.05$, so $z_{\alpha/2} = 1.96$. Also, the margin of error is $E = 0.03$ (the decimal equivalent of "three percentage points"). Because we have an estimated value of $\hat{p}$, we use Formula 6-5 as follows:

$$n = \frac{[z_{\alpha/2}]^2 \hat{p}\hat{q}}{E^2} = \frac{[1.96]^2(0.18)(0.82)}{0.03^2}$$

$$= 630.0224 = 631 \qquad \text{(rounded up)}$$

We must survey at least 631 randomly selected car drivers.

b. As in part (a), we again use $z_{\alpha/2} = 1.96$ and $E = 0.03$, but with no prior knowledge of $\hat{p}$ (or $\hat{q}$) we use Formula 6-6 as follows:

$$n = \frac{[z_{\alpha/2}]^2 \cdot 0.25}{E^2} = \frac{[1.96]^2 \cdot (0.25)}{0.03^2}$$

$$= 1067.1111 = 1068 \qquad \text{(rounded up)}$$

Interpretation
To be 95% confident that our sample percentage is within three percentage points of the true percentage for all car drivers, we should randomly select and survey 1068 drivers. By comparing this result to the sample size of 631 found in part (a), we can see that if we have no knowledge of a prior study, a larger sample is required to achieve the same results as when the value of $\hat{p}$ can be estimated.

Part (b) of the above example involved application of Formula 6-6, the same formula frequently used by Nielsen, Gallup, and other professional pollsters. Many people incorrectly believe that we should sample some percentage of the population, but Formula 6-6 shows that the population size is irrelevant. (In reality, the population size is sometimes used, but only in cases in which we sample without replacement from a relatively small population. See Exercise 29.) Most of the polls featured in newspapers, magazines, and broadcast media involve polls with sample sizes in the range of 1000 to 2000. Even though such polls may involve a very small percentage of the total population, they can provide results that are quite good. When Nielsen surveys 1068 TV households from a population of 10 million, only 0.01% of the households are surveyed; still, we can be 95% confident that the sample percentage will be within three percentage points of the population percentage.

Polls have become very important and prevalent in Canada. They affect the television shows we watch, the leaders we elect, the legislation that governs us, and the products we consume. An understanding of the concepts of this section removes much of the mystery and misunderstanding surrounding polls.

Factors Affecting Sample Size

As with calculating sample sizes for means, the same factors affect the sample size for proportions.

EXAMPLE

In the previous example, a sample size of 1068 was calculated. Which would reduce the sample size more:

a. reducing the level of confidence to 90% with a 3% margin of error, or

b. maintaining the level of confidence at 95% and increasing the margin of error to 5%?

SOLUTION

a. Since the level of confidence is 90%, the z critical value is 1.645. Then:

$$n = \frac{[1.645]^2(0.18)(0.82)}{0.03^2} = 443.8 = 444$$

b. For the second solution, the z critical value is 1.96:

$$n = \frac{[1.96]^2(0.18)(0.82)}{0.05^2} = 226.8 = 227$$

As with sample size calculations for means, increasing the margin of error has a greater effect on reducing the sample size than reducing the level of confidence.

EXCEL: *Confidence Intervals:* Use the Data Desk XL add-in that is a supplement to this book. First enter the number of successes and the total number of trials in two spreadsheet cells (such as A1 and A2).

EXCEL (Prior to 2007): Click on **DDXL** and select **Confidence Intervals.** Under the Function Type options, select **Summ 1 Var Prop Interval** (which is an abbreviation for "confidence interval for a proportion using summary data for one variable"). Click on the pencil icon for "Num successes" and enter the cell address where you input that data. Similarly, click on the pencil icon for "Num trials" and enter the corresponding Excel address. Click on **OK.** In the dialog box, set the level of confidence, then click on **Compute Interval.** The confidence interval will be displayed, as shown below.

EXCEL 2007: Click **Add-Ins** on the main menu and choose **DDXL** from the add-in menu to the left. Proceed as above.

MINITAB 15: *Confidence Intervals:* From the **Stat** menu, choose **Basic Statistics** and then **1 Proportion.** If you are constructing the confidence interval using summarized data, click that radio button and fill in the necessary boxes such as the number of events (i.e., successes) and trials. There is an optional box to conduct a hypothesis test. (You would be required to fill in a hypothesis proportion, which will be covered in Chapter 7.) The default setting is a 95% level of confidence. If you wish to change it, click the **Options** button and change the level of confidence to what you desire. You can leave the **Alternative** drop-down menu set to not equal.

STATDISK: *Confidence Intervals:* Select **Analysis** from the main menu bar, select **Confidence Intervals,** then select **Proportion One Sample.** Enter the required items in the dialog box, then click the **Evaluate** button.

Sample Size: Select **Analysis** from the main menu bar at the top, then select **Sample Size Determination,** followed by **Estimate Proportion.** Proceed to enter the required items in the dialog box.

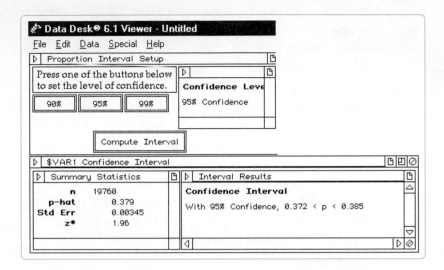

Exercises A: Basic Skills and Concepts

In Exercises 1–4, assume that a sample is used to estimate a population proportion p. Find the margin of error that corresponds to the given values of n and x and the degree of confidence.

1. $n = 800$, $x = 200$, 95%
2. $n = 1400$, $x = 420$, 99%
3. $n = 4275$, $x = 2576$, 98%
4. $n = 887$, $x = 209$, 90%

In Exercises 5–8, use the given sample data and degree of confidence to construct the interval estimate of the population proportion p.

5. $n = 800$, $x = 600$, 95% confidence
6. $n = 2000$, $x = 300$, 99% confidence
7. $n = 2475$, $x = 992$, 90% confidence
8. $n = 5200$, $x = 1024$, 98% confidence

In Exercises 9–12, use the given data to find the minimum sample size required to estimate a population proportion or percentage.

9. Margin of error: 0.02; confidence level: 95%; $\hat{p}$ and $\hat{q}$ unknown
10. Margin of error: 0.01; confidence level: 90%; $\hat{p}$ and $\hat{q}$ unknown
11. Margin of error: four percentage points; confidence level: 99%; $\hat{p}$ is estimated to be 0.20 from a prior study

12. Margin of error: two percentage points; confidence level: 97%; $\hat{p}$ is estimated to be 0.85 from a prior study

13. Wawanesa Mutual Insurance Company wanted to estimate the percentage of drivers who changed tapes or CDs while driving. A random sample of 850 drivers resulted in 544 who changed tapes or CDs while driving (based on data from *Prevention* magazine).
    a. Find the point estimate of the *percentage* of all drivers who change tapes or CDs while driving.
    b. Find a 90% interval estimate of the *percentage* of all drivers who change tapes or CDs while driving.

14. When 160 CEOs in British Columbia are randomly selected and surveyed, it is found that 41 of them possess an MBA.
    a. Find the point estimate of the true proportion of all CEOs in British Columbia who possess an MBA.
    b. Find a 95% confidence interval for the true proportion of all CEOs in British Columbia who possess an MBA.

15. The drug Ziac is used to treat hypertension. In a clinical test, 3.2% of 221 Ziac users experienced dizziness (based on data from Lederle Laboratories).
    a. Construct a 99% confidence interval estimate of the percentage of all Ziac users who experience dizziness.
    b. In the same clinical test, people in the placebo group didn't take Ziac, but 1.8% of them reported dizziness. Based on the results in parts (a) and (b), what can we conclude about dizziness as an adverse reaction to Ziac?

16. In a follow-up article on Y2K problems, a reporter wants to conduct a survey of small businesses, in order to estimate the true proportion of small businesses that took some action regarding the Y2K problem—that is, the then-widespread concern that when the date 2000 arrived, it would confuse software that used two-digit dates (such as 99). The reporter wants 95% confidence that her results have a margin of error of 0.03. How many small businesses must be surveyed?
    a. Assume the reporter will estimate $\hat{p}$ based on a 1998 Industry Canada study, according to which 66% of small businesses were taking action.
    b. Assume, instead, that we do *not* have prior information on which to estimate the value of $\hat{p}$.

17. A random survey of 85 CEOs in British Columbia showed 70 respondents have a computer on their desk. Based on those results, construct a 98% confidence interval for the percentage of all CEOs in British Columbia who have a computer on their desk.

18. Internationally, 53.3% of the candidates who wrote the Certified General Accounting (CGA) exams in December 1995 passed. In Manitoba, out of 781 candidates who wrote those exams, 439 passed (based on data from the

Certified General Accountants Association of Manitoba). Assuming that these 1995 results are representative of years in general, construct a 99% interval estimate of the percentage of all Manitoban candidates of the CGA exams who passed. Is there sufficient evidence to indicate that the pass rate in Manitoba is greater than the international pass rate?

19. The Greybar Tax Company believes that its clients are selected for audits at a rate substantially higher than the rate for the general population. Suppose that the Canada Revenue Agency audits 4.3% of those who earn more than $100,000, but a check of 400 randomly selected Greybar returns with earnings above $100,000 shows that 56 of them were audited. Using a 99% level of confidence, construct a confidence interval for the percentage of Greybar returns with earnings above $100,000 that are audited. Based on the result, does it appear that the high-income Greybar clients are audited at a rate that is substantially higher than the rate for the general population?

20. The Locust Tree Restaurant keeps records of reservations and no-shows. When 150 Saturday reservations are randomly selected, it is found that 70 of them were no-shows (based on data from American Express). Using a 90% degree of confidence, find a confidence interval for the proportion of Saturday no-shows.

21. How many TV households must Nielsen survey to estimate the percentage that are tuned to *The Late Show with David Letterman*? Assume that you want 97% confidence that your sample percentage has a margin of error of two percentage points. Also assume that nothing is known about the percentage of households tuned in to any television shows after 11 p.m.

22. The Western Canada Communications Company is considering a bid to provide long-distance phone service. You are asked to conduct a poll to estimate the percentage of consumers who are satisfied with their current long-distance phone service. You want to be 90% confident that your sample percentage is within 2.5 percentage points of the true population value, and a Roper poll suggests that this percentage should be about 85%. How large must your sample be?

23. A hotel chain gives an aptitude test to job applicants and considers a multiple-choice test question to be easy if at least 80% of the responses are correct. A random sample of 6503 responses to one particular question includes 84% correct responses. Construct the 99% confidence interval for the true percentage of correct responses. Is it likely that this question is really easy?

24. A random sample of 400 people is taken from a large city in Atlantic Canada. They are asked about their views on the amalgamation of the four provinces in Atlantic Canada. Of the respondents, 190 are in favour of this amalgamation. Construct the 98% confidence interval for the true percentage of Atlantic Canada residents who are in favour of amalgamation. Based on the result, does there appear to be much chance that this amalgamation will ever occur?

25. In a study of store checkout scanners, 1234 items were checked and 20 of them were found to be overcharges (based on data from "UPC Scanner Pricing Systems: Are They Accurate?" by Goodstein, *Journal of Marketing,* Vol. 58).
    a. Using the sample data, construct a 95% confidence interval for the proportion of all such scanned items that are overcharges.
    b. Use the sample data as a pilot study and find the sample size necessary to estimate the proportion of scanned items that are overcharges. Assume that you want 99% confidence that the estimate is in error by no more than 0.005.

26. A health study involves 1000 randomly selected deaths, with 273 of them caused by heart disease (based on data from Statistics Canada).
    a. Using the sample data, construct a 99% confidence interval for the proportion of all deaths caused by heart disease.
    b. Use the sample data as a pilot study and find the sample size necessary to estimate the proportion of all deaths caused by heart disease. Assume that you want 98% confidence that the estimate is in error by no more than 0.01.

27. A climatologist claims that half of Canada's largest towns and cities receive more than 200 cm of snowfall each year. Refer to Data Set 4 in Appendix B and construct the 95% confidence interval for the percentage of cities with snowfall over 200 cm. Is the resulting confidence interval consistent with the climatologist's claim?

28. Refer to Data Set 11 in Appendix B and find the sample proportion of M&Ms that are red. Use that result to construct a 95% confidence interval estimate of the population percentage of M&Ms that are red. Is the result consistent with the 20% rate that is reported by the candy maker Mars?

6-4 Exercises B: Beyond the Basics

29. This section presented Formulas 6-5 and 6-6, which are used for determining sample size. In both cases we assumed that the population is infinite or very large, or that we are sampling with replacement. When we have a relatively small population with size N and sample without replacement, we modify E to include the *finite population correction factor* shown here, and we can solve for n to obtain the result shown to the right. Use this result to repeat Exercise 21, assuming that we limit our population to a town of 5000 people.

$$E = z_{\alpha/2} \sqrt{\frac{\hat{p}\hat{q}}{n}} \sqrt{\frac{N - n}{N - 1}} \qquad n = \frac{N\hat{p}\hat{q}[z_{\alpha/2}]^2}{\hat{p}\hat{q}[z_{\alpha/2}]^2 + (N - 1)E^2}$$

30. Consider the following statement: "The results from this poll differ, 19 times out of 20, by no more than 1 percentage point in either direction from the

results that would be obtained by surveying all voters in Canada." Find the sample size suggested by this statement.

31. A newspaper article indicates that an estimate of the unemployment rate involves a survey of 47,000 people. If the reported unemployment rate must have an error no larger than 0.2 percentage points and the rate is known to be about 8%, find the corresponding confidence level.

32. Heights of female students are normally distributed with a mean of 64.2 in and a standard deviation of 2.6 in. How many female students must be surveyed if we want to estimate the percentage who are taller than 5 ft? Assume that we want 98% confidence that the error is no more than 2.5 percentage points.

33. A *one-sided confidence interval* for p can be written as $p < \hat{p} + E$ or $p > \hat{p} - E$, where the margin of error E is modified by replacing $z_{\alpha/2}$ with z_{α}. If Air Borealis wants to report an on-time performance of at least x percent with 95% confidence, construct the appropriate one-sided confidence interval and then find the percent in question. Assume that, from a random sample of 750 flights, 630 are on time.

34. Special tables are available for finding confidence intervals for proportions involving small numbers of cases where the normal distribution approximation cannot be used. For example, given three successes among eight trials, the 95% confidence interval found in *Standard Probability and Statistics Tables and Formulae* (CRC Press) is $0.085 < p < 0.755$. Find the confidence interval that would result if you were to use the normal distribution incorrectly as an approximation to the binomial distribution. Are the results reasonably close?

6-5 Estimating a Population Variance

In this section we consider two concepts discussed earlier in this chapter: (1) point estimates and (2) confidence intervals. Whereas Sections 6–2, 6–3, and 6–4 applied these concepts to estimates of means and proportions, this section applies them to the population variance σ^2 or standard deviation σ. Many real situations, such as quality control in a manufacturing process, require that we estimate values of population variances or standard deviations. In addition to making products with measurements yielding a desired mean, the manufacturer must make products of *consistent* quality that do not run the gamut from extremely good to extremely poor. As this consistency can often be measured by the variance or standard deviation, these become vital statistics in maintaining the quality of products.

Assumptions for This Section

1. The sample has been collected by a valid random-sampling technique.
2. The population must have normally distributed values (even if the sample is large).

The assumption of a normally distributed population was made in earlier sections, but that requirement is more critical here. We describe this sensitivity to a normal distribution by saying that inferences about the population variance σ^2 (or population standard deviation σ) are *not robust*, meaning that the inferences may be very misleading, and may lead to gross errors, if the population does not truly have a normal distribution. Consequently, the normal distribution requirement is much more strict than in previous sections, and we should check the distribution of data first, by constructing histograms and normal probability plots, as described in Section 5–7.

When we considered estimates of means and proportions in Sections 6–2, 6–3, and 6–4, we used the normal and Student t distributions. When developing estimates of variances or standard deviations, we use another distribution, referred to as the *chi-square distribution*. We will examine important features of this distribution before proceeding with the development of confidence intervals.

Chi-Square Distribution

In a normally distributed population with variance σ^2, we randomly select independent samples of size n and compute the sample variance s^2 (see Formula 2-5) for each sample. The sample statistic $\chi^2 = (n - 1)s^2/\sigma^2$ has a distribution called the **chi-square distribution**.

CHI-SQUARE DISTRIBUTION

Formula 6-7
$$\chi^2 = \frac{(n - 1)s^2}{\sigma^2}$$

where
- n = sample size
- s^2 = sample variance
- σ^2 = population variance

We denote chi-square by χ^2, pronounced "kigh square." (The specific mathematical equations used to define this distribution are not given here because they are beyond the scope of this text.) To find critical values of the chi-square distribution, refer to Table A-4. The chi-square distribution is determined by the number of degrees of freedom, and in this chapter we use $n - 1$ degrees of freedom:

degrees of freedom = $n - 1$

In later chapters we will encounter situations in which the degrees of freedom are not $n - 1$, so we should not assume that there are always $n - 1$ degrees of freedom.

Figure 6-7
Chi-Square Distribution

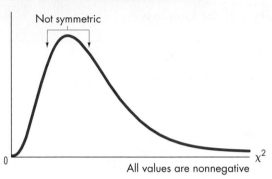

Figure 6-8
Chi-Square Distribution for 10 and 20 Degrees of Freedom (df)

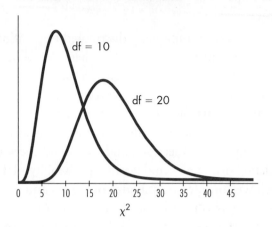

Properties of the Distribution of the Chi-Square Statistic

1. The chi-square distribution is not symmetric, unlike the normal and Student t distributions (see Figure 6-7). (As the number of degrees of freedom increases, the distribution becomes more symmetric, as Figure 6-8 illustrates.)

2. Chi-square values can be zero or positive, but they cannot be negative (see Figure 6-7).

3. The chi-square distribution is different for each number of degrees of freedom (see Figure 6-8), which is determined in this section by df $= n - 1$. As the number of degrees of freedom increases, the chi-square distribution approaches a normal distribution.

Because the chi-square distribution is skewed instead of symmetric, the confidence interval does not fit the familiar format of [**point estimate**] $\pm$ E, and we must do separate calculations for the upper and lower confidence interval limits. There is a different procedure for finding critical values, illustrated in the following example. Note the following essential feature of Table A-4:

In Table A-4, each critical value of χ^2 corresponds to an area given in the top row of the table, and that area represents the ***total region located to the right*** of the critical value.

EXAMPLE

Find the critical values of χ^2 that determine critical regions containing an area of 0.025 in each tail. Assume that the relevant sample size is 10 so that the number of degrees of freedom is $10 - 1$, or 9.

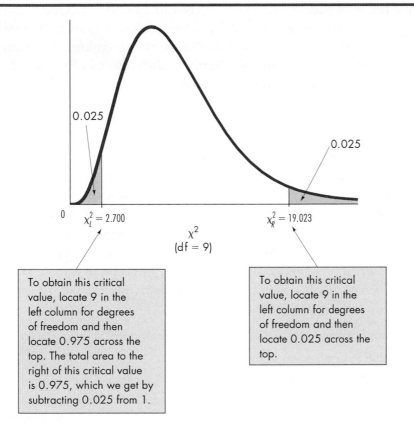

Figure 6-9
Critical Values of the Chi-
Square Distribution

0.025

0.025

0

$\chi_L^2 = 2.700$

$\chi_R^2 = 19.023$

χ^2
(df = 9)

To obtain this critical value, locate 9 in the left column for degrees of freedom and then locate 0.975 across the top. The total area to the right of this critical value is 0.975, which we get by subtracting 0.025 from 1.

To obtain this critical value, locate 9 in the left column for degrees of freedom and then locate 0.025 across the top.

SOLUTION

See Figure 6-9 and refer to Table A-4. The critical value to the right ($\chi^2 = 19.023$) is obtained in a straightforward manner by locating 9 in the degrees-of-freedom column at the left and 0.025 across the top. The critical value of $\chi^2 = 2.700$ to the left once again corresponds to 9 in the degrees-of-freedom column, but we must locate 0.975 (found by subtracting 0.025 from 1) across the top because the values in the top row are always *areas to the right* of the critical value. Refer to Figure 6-9 and see that the total area to the right of $\chi^2 = 2.700$ is 0.975.

Interpretation

Figure 6-9 shows that, for a sample of 10 scores taken from a normally distributed population, the chi-square statistic $(n - 1)s^2/\sigma^2$ has a 0.95 probability of falling between the chi-square critical values of 2.700 and 19.023.

When obtaining critical values of chi-square from Table A-4, note that the numbers of degrees of freedom are consecutive integers from 1 to 30, followed by 40, 50, 60, 70, 80, 90, and 100. When a number of degrees of freedom (such as 52) is not found on the table, you can usually use the closest critical value. For example, if the number of degrees of freedom is 52, refer to Table A-4 and use 50 degrees of freedom. (If the number of degrees of freedom is exactly midway between table values, such as 55, simply find the mean of the two χ^2 values.) For numbers of degrees of freedom greater than 100, use the equation given in Exercise 22, a more detailed table, or a statistical software package.

Estimators of σ^2

Because sample variances s^2 (found by using Formula 2-5) tend to centre on the value of the population variance σ^2, we say that s^2 is an *unbiased estimator* of σ^2. That is, sample variances s^2 do not systematically tend to overestimate the value of σ^2, nor do they systematically tend to underestimate σ^2. Instead, they tend to target the value of σ^2 itself. Also, the values of s^2 tend to produce smaller errors by being closer to σ^2 than do other measures of variation. For these reasons, the value of s^2 is generally the best single value (or point estimate) of the various possible statistics we could use to estimate σ^2.

> **The sample variance s^2 is the best point estimate of the population variance σ^2.**

Because s^2 is the best point estimate of σ^2, it would be natural to expect s to be the best point estimate of σ, but this is not the case, because s is a biased estimator of σ (see Exercise 24 in Section 2–5). If the sample size is large, however, the bias is so small that we can use s as a reasonably good estimate of σ.

> **For large samples, the sample standard deviation s is a reasonably good point estimate of σ (but it is a biased estimate).**

Although s^2 is the best point estimate of σ^2, there is no indication of how good it actually is. To compensate for that deficiency, we develop an interval estimate (or confidence interval) that is more informative.

CONFIDENCE INTERVAL (OR INTERVAL ESTIMATE) FOR THE POPULATION VARIANCE σ^2

$$\frac{(n-1)s^2}{\chi_R^2} < \sigma^2 < \frac{(n-1)s^2}{\chi_L^2}$$

This expression is used to find a confidence interval for variance σ^2, but the confidence interval (or interval estimate) for the standard deviation σ is found by taking the square root of each component, as shown below:

$$\sqrt{\frac{(n-1)s^2}{\chi_R^2}} < \sigma < \sqrt{\frac{(n-1)s^2}{\chi_L^2}}$$

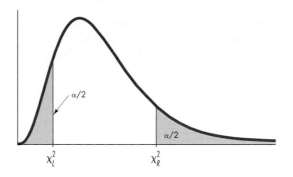

Figure 6-10
Chi-Square Distribution
with Critical Values χ_L^2
and χ_R^2
The critical values χ_L^2 and χ_R^2
separate the extreme areas
corresponding to sample
variances that are unlikely
(with probability α).

The notations χ_R^2 and χ_L^2 in the preceding expressions are described as follows. (Note that some other texts use $\chi_{\alpha/2}^2$ in place of χ_R^2, and they use $\chi_{1-\alpha/2}^2$ in place of χ_L^2.)

NOTATION

With a total area of α divided equally between the two tails of a chi-square distribution, χ_L^2 denotes the left-tailed critical value and χ_R^2 denotes the right-tailed critical value. (See Figure 6-10.)

Confidence interval limits for σ^2 and σ should be rounded by using the following round-off rule, which is really the same basic rule given in Section 6–2.

ROUND-OFF RULE FOR CONFIDENCE INTERVAL ESTIMATES OF σ OR σ^2

1. When using the original set of data to construct a confidence interval, round the confidence interval limits to one more decimal place than is used for the original set of data.

2. When the original set of data is unknown and only the summary statistics (n, s) are used, round the confidence interval limits to the same number of decimal places used for the sample standard deviation or variance.

EXAMPLE

The Qu'appelle Valley Bakery makes doughnuts that are packaged in boxes with labels stating that there are 12 doughnuts weighing a total of 42 oz. If the variation among the doughnuts is too large, some boxes will be underweight (cheating consumers) and others will be overweight (lowering profit). A consumer would not be happy with a doughnut so small that it can be seen only with an electron microscope, nor would a consumer be happy with a doughnut so large that it resembles a tractor tire. The quality control supervisor has found that he can stay out of trouble if the doughnuts have a mean of 3.50 oz and a standard deviation of 0.06 oz or less. Twelve doughnuts are randomly selected from the production line and weighed, with the results given here (in ounces). Construct a 95% confidence interval for σ^2 and a 95% confidence interval for σ, then determine whether the quality control supervisor is in trouble.

$$3.43 \quad 3.37 \quad 3.58 \quad 3.50 \quad 3.68 \quad 3.61 \quad 3.42 \quad 3.52 \quad 3.66 \quad 3.50 \quad 3.36 \quad 3.42$$

SOLUTION

Based on the sample data, the mean of $\bar{x} = 3.504$ seems quite good because it's very close to the desired 3.50 oz. The given scores have a standard deviation of $s = 0.109$, which might seem to be greater than the desired value of 0.06 or less. Let's proceed to find the confidence interval for σ^2.

With a sample of 12 scores, we have 11 degrees of freedom. With a 95% degree of confidence, we divide $\alpha = 0.05$ equally between the two tails of the χ^2 distribution and we refer to the values of 0.975 and 0.025 across the top row. The critical values of χ^2 are $\chi_L^2 = 3.816$ and $\chi_R^2 = 21.920$. Using these critical values, the sample standard deviation of $s = 0.109$, and the sample size of 12, we construct the 95% confidence interval by evaluating the following:

$$\frac{(12-1)(0.109)^2}{21.920} < \sigma^2 < \frac{(12-1)(0.109)^2}{3.816}$$

This becomes $0.006 < \sigma^2 < 0.034$. Taking the square root of each part (before rounding) yields $0.077 < \sigma < 0.185$.

Interpretation

Based on the 95% confidence interval for σ, it appears that the standard deviation is greater than the desired value of 0.06 or less, so the quality control supervisor is in trouble and must take corrective action to make the doughnut weights more consistent.

The confidence interval $0.077 < \sigma < 0.185$ can also be expressed as $(0.077, 0.185)$, but the format of $\sigma = s \pm E$ *cannot* be used because the confidence interval does not have s at its centre.

Basis for the Procedure of Finding the Confidence Interval for σ^2

We now explain why the confidence intervals for σ^2 and s have the forms just given. If we obtain samples of size n from a population with variance σ^2, the distribution of the $(n-1)s^2/\sigma^2$ values will be as shown in Figure 6-10. For a random sample, there is a probability of $1 - \alpha$ that the statistic $(n-1)s^2/\sigma^2$ will fall between the critical values of χ_L^2 and χ_R^2. In other words (and symbols), there is a $1 - \alpha$ probability that both of the following are true:

$$\frac{(n-1)s^2}{\sigma^2} < \chi_R^2 \quad \text{and} \quad \frac{(n-1)s^2}{\sigma^2} > \chi_L^2$$

If we multiply both of the preceding inequalities by σ^2 and divide each inequality by the appropriate critical value of χ^2, we see that the two inequalities can be expressed in the equivalent forms

$$\frac{(n-1)s^2}{\chi_R^2} < \sigma^2 \quad \text{and} \quad \frac{(n-1)s^2}{\chi_L^2} > \sigma^2$$

These last two inequalities can be combined into one inequality:

$$\frac{(n-1)s^2}{\chi_R^2} < \sigma^2 < \frac{(n-1)s^2}{\chi_L^2}$$

There is a probability of $1 - \alpha$ that these confidence interval limits contain the population variance σ^2.

First, ensure that the distribution is sufficiently normal, using the procedures described in Chapter 10.

EXCEL: *Confidence Intervals:* Although Excel has no built-in tool for finding the confidence intervals for σ^2 and σ, a template such as the one illustrated on the next page can be created for this purpose. Place the raw data in column A, and type in the desired σ value in cell C11. All other values will be calculated by the template.

MINITAB 15: *Confidence Intervals:* From the **Stat** menu, choose **Basic Statistics** and then **1 Variance**. If you are constructing the confidence interval using summarized data, click that radio button and fill in the necessary boxes such as sample size and sample standard deviation. There is an optional box to conduct a hypothesis test. (You would be required to fill in a hypothesis standard deviation, which will be covered in Chapter 7.) The default setting is a 95% level of confidence. If you wish to change it, click the **Options** button and change the level of confidence to what you desire. You can leave the **Alternative** drop-down menu set to not equal.

STATDISK: *Confidence Intervals:* Select **Analysis** from the main menu bar, select **Confidence Intervals**, then select **St Dev One Sample**. Enter the required items in the dialog box, then click the **Evaluate** button.

Sample Size: Select **Analysis** from the main menu bar at the top, then select **Sample Size Determination**, followed by **Estimate St Dev**. Proceed to enter the required items in the dialog box.

	E16	▼	=	
	A	B	C	
1	Weight			
2	3.43	Mean	3.504167	
3	3.37	Std.Dev.	0.109084	
4	3.58	Pop.Var.		
5	3.50	Lower Limit	0.005971	
6	3.68	Upper Limit	0.034303	
7	3.61	Pop.Std.Dev.		
8	3.42	Lower Limit	0.077274	
9	3.52	Upper Limit	0.185211	
10	3.66			
11	3.50	alpha	0.050000	
12	3.36	Chi-square Left	3.815742	
13	3.42	Chi-square Right	21.920023	
14		d.f.	11.000000	

Cell	Formula	Explanation
C2	=AVERAGE(A2:A13)	Mean
C3	=STDEV(A2:A13)	Standard deviation
C5	=((C14*C3^2)/C13)	Lower limit of confidence interval for variance
C6	=((C14*C3^2)/C12)	Upper limit of confidence interval for variance
C8	=SQRT(C5)	Lower limit of confidence interval for standard deviation
C9	=SQRT(C6)	Upper limit of confidence interval for standard deviation
C11	{input alpha here}	
C12	=CHIINV(1-C11/2,C14)	Critical chi-square on left tail
C13	=CHIINV(C11/2,C14)	Critical chi-square on right tail
C14	=COUNT(A2:A13)-1	Degrees of Freedom

Use appropriate cell addresses for data

6-5 Exercises A: Basic Skills and Concepts

In Exercises 1–4, find the critical values χ_L^2 and χ_R^2 that correspond to the given degree of confidence and sample size.

1. 95%; $n = 26$ 2. 99%; $n = 17$

3. 90%; $n = 60$ 4. 95%; $n = 50$

In Exercises 5–8, use the given degree of confidence and sample data to find a confidence interval for the population standard deviation σ. In each case, assume that the population has a normal distribution.

5. Heights of female students: 95% confidence; $n = 10$, $\bar{x} = 64.2$ in, $s = 2.6$ in

6. Grade-point averages: 99% confidence; $n = 15$, $\bar{x} = 2.76$, $s = 0.88$

7. Test scores: 90% confidence; $n = 16$, $\bar{x} = 77.6$, $s = 14.2$

8. Cost of basic phone service: 95% confidence; $n = 19$, $\bar{x} = \$15.30$, $s = \$1.25$

Suppose that a sample of customers per hour were randomly taken from a store and that the sample standard deviation of the number of customers per hour was 2.4. Exercises 9–12 are based on this information. Assume that the number of customers per hour is normally distributed.

9. Construct a 95% confidence interval of the population standard deviation based on a sample size of 10.

10. If the sample size were 15 instead of 10, show why the 95% confidence interval would be narrower.

11. If the level of confidence were 99% instead of 95%, with the sample size still 10, show why the confidence interval would be wider than the one in Question 9.

12. Would a 99% confidence interval with a sample size of 15 be wider or narrower than the one in Question 9?

In Exercises 13–20, assume that each sample is obtained by randomly selecting values from a population with a normal distribution.

13. A container of car antifreeze is supposed to hold 3785 mL of the liquid. Realizing that fluctuations are inevitable, the quality control manager wants to be quite sure that the standard deviation is less than 30 mL. Otherwise, some containers would overflow while others would not have enough of the coolant. She randomly selects a sample, with the results given here. Use these sample results to construct the 99% confidence interval for the true value of σ. Does this confidence interval suggest that the fluctuations are at an acceptable level?

3761	3861	3769	3772	3675	3861	$n = 18$
3888	3819	3788	3800	3720	3748	$\bar{x} = 3787.0$
3753	3821	3811	3740	3740	3839	$s = 55.4$

14. Refer to Data Set 21 in Appendix B.
   a. Use the range rule of thumb (see Section 2–5) to estimate σ, the standard deviation of values for the "depth" of high-price diamonds.
   b. Use the listed weights to construct a 98% confidence interval for σ.
   c. Does the confidence interval contain your estimated value for σ?

15. A sample of 35 skulls is obtained for Egyptian males who lived around 1850 BCE. The maximum breadth of each skull is measured with the result that $\bar{x} = 134.5$ mm and $s = 3.5$ mm (based on data from *Ancient Races of the Thebaid* by Thomson and Randall-Maciver). Using these sample results, construct a 95% confidence interval for the population standard deviation s.

16. A sample consists of 75 TV sets purchased several years ago. The replacement times of those TV sets have a mean of 8.2 years and a standard deviation of 1.1 years (based on data from "Getting Things Fixed," *Consumer Reports*). Construct a 90% confidence interval for the standard deviation of replacement times for all TV sets from that era. Does the result apply to TV sets currently being sold?

17. a. The listed values are waiting times (in minutes) of customers at the Humber Valley Credit Union, where customers enter a single waiting line that feeds three windows. Construct a 95% confidence interval for the population standard deviation σ.

   6.5   6.6   6.7   6.8   7.1   7.3   7.4   7.7   7.7   7.7

   b. The listed values are waiting times (in minutes) of customers at the Durham Credit Union, where customers may enter any one of three different lines that have formed at three different teller windows. Construct a 95% confidence interval for σ.

4.2 5.4 5.8 6.2 6.7 7.7 7.7 8.5 9.3 10.0

    c. Interpret the results found in (a) and (b). Do the confidence intervals suggest a difference in the variation among waiting times? Which arrangement seems better: the single-line system or the multiple-line system?

18. Off-season football players often participate in community events and charity fundraisers. Suppose a firm is bidding to provide T-shirts for players who are representing the CFL in such events. The firm has randomly sampled CFL quarterbacks and slotbacks, with these results based on data from the CFL website:

Quarterbacks:	$n = 10$	$\bar{x} = 192.7$ lb	$s = 11.7$ lb
Slotbacks:	$n = 11$	$\bar{x} = 195.0$ lb	$s = 14.3$ lb

    a. Construct a 95% confidence interval estimate for the standard deviation σ for the quarterbacks.

    b. Construct a 95% confidence interval estimate for the standard deviation σ for the slotbacks.

    c. Assume that weight is a rough indicator of required T-shirt size, and that the firm sells a T-shirt with just enough stretchiness to fit the variation in sizes of quarterbacks. Would it be reasonable to assume that this same style of T-shirts could also fit all of the slotbacks? Explain why or why not.

19. Refer to Data Set 4 in Appendix B.

    a. Use the range rule of thumb (see Section 2–5) to estimate s, the standard deviation of the annual snowfall (in centimetres) of the first 31 cities (all except Victoria).

    b. Construct a 99% confidence interval for σ.

    c. Does the confidence interval contain your estimated value of σ?

20. Refer to Data Set 11 in Appendix B.

    a. Use the range rule of thumb (see Section 2–5) to estimate s, the standard deviation of the weights of brown M&M plain candies.

    b. Construct a 98% confidence interval for σ.

    c. Does the confidence interval contain your estimated value of σ?

6-5 Exercises B: Beyond the Basics

21. A journal article includes a graph showing that sample data are normally distributed.

    a. The degree of confidence is inadvertently omitted when this confidence interval is given: $2.8 < \sigma < 6.0$. Find the degree of confidence for these given sample statistics: $n = 20$, $\bar{x} = 45.2$, and $\sigma = 3.8$.

b. This 95% confidence interval is given: $19.1 < \sigma < 45.8$. Find the value of the standard deviation s, which was omitted from the article.

22. In constructing confidence intervals for σ or σ^2, we use Table A-4 to find the critical values χ_L^2 and χ_R^2, but that table applies only to cases in which $n \le 101$, so the number of degrees of freedom is 100 or fewer. For larger numbers of degrees of freedom, we can approximate χ_L^2 and χ_R^2 by using

$$\chi^2 = \frac{1}{2}\left[\pm\ z_{\alpha/2} + \sqrt{2k - 1}\right]^2$$

where k is the number of degrees of freedom. Construct the 95% confidence interval for σ by using the following sample data: The measured heights of 772 men between the ages of 18 and 24 have a standard deviation of 2.8 in.

VOCABULARY LIST

chi-square distribution   339
confidence coefficient   296
confidence interval   296
confidence interval
   limits   299
critical value   297
degree of confidence   296

degrees of freedom   317
estimate   294
estimator   294
interval estimate   296
level of confidence   296
margin of error (E)   298

maximum error of the
   estimate   298
point estimate   295
Student t distribution   316
t distribution   316
unbiased estimator   295

REVIEW

This chapter and the following chapter introduce the fundamental and important concepts of inferential statistics. This chapter focused on *estimates* of parameters as we considered population means, proportions, and variances to develop procedures for each of the following:

- Identifying a point estimate
- Constructing a confidence interval
- Determining the required sample size

We discussed point estimate (or single-valued estimate) and formed these conclusions:

- The best point estimate of μ is $\bar{x}$.
- The best point estimate of p is $\hat{p}$.
- The best point estimate of σ^2 is s^2.

As single values, the point estimates don't convey any real sense of how reliable they are, so we introduced confidence intervals (or interval estimates) as more informative estimates. We also considered ways of determining the sample sizes necessary to estimate parameters to within given tolerance factors. This chapter also introduced the Student t and chi-square distributions. We must be careful to use the correct distribution for each set of circumstances.

It is important to know that all of the confidence interval and sample size procedures in this chapter require that we have a population with a distribution that is approximately normal. If the distribution is very nonnormal, we must use other methods, such as the bootstrap method described in the Computer Project at the end of this chapter.

REVIEW EXERCISES

1. A human resources specialist has compiled annual numbers of person-days of work lost, due to work stoppages, in the water-based transportation industry. These are the results (based on data from Human Resources Canada, Workplace Information Directorate):

   $n = 40$ $\bar{x} = 38,015.7$ median $= 18,240.0$ $s = 49,288.8$

   a. Construct a 96% confidence interval estimate for the population mean of annual numbers of person-days lost.
   b. Why can we not use the methods of Section 6–5 to construct confidence interval estimates of the population standard deviation?

2. You have just been hired by General Motors to tour Canada giving randomly selected drivers test rides in a new Corvette (yeah, right). After giving the test drive, you must ask the rider whether he or she would consider buying a Corvette. How many riders must you survey to be 97% confident that the sample percentage is off by no more than two percentage points?

3. An auto parts supplier wants information about how long car owners plan to keep their cars. A random sample of 25 car owners results in $\bar{x} = 7.01$ years and $s = 3.74$ years, respectively (based on data from a Roper poll). Assuming that the sample is drawn from a normally distributed population, find a 95% confidence interval for the population mean.

4. Using the same sample data from Exercise 3, find a 95% confidence interval for the population standard deviation.

5. The Angus Reid Group randomly selected 1000 Ontario residents ages 18 and older by phone. Fifty-four percent of those surveyed believe the private sector should be responsible for ensuring there are jobs for Ontario residents. Construct the 95% confidence interval for the true proportion

of Ontario residents who have this opinion. Also, what margin of error (in percentage points) should be reported along with the sample percentage of 54%?

6. In a Gallup poll of 1004 adults, 93% indicated that restaurants and bars should refuse service to patrons who have had too much to drink. If you plan to conduct a new poll to confirm that the percentage continues to be correct, how many randomly selected adults must you survey if you want 98% confidence that the margin of error is four percentage points?

7. In designing a new machine to be used on an assembly line at a General Motors plant, an engineer obtains measurements of arm lengths of a random sample of male machine operators. The following values (in centimetres) are obtained. Construct the 95% confidence interval for the mean arm length of all such employees.

| 76.8 | 75.6 | 69.3 | 75.7 | 75.5 | 71.2 | 72.5 | 71.9 |
| 70.9 | 69.4 | 71.7 | 72.5 | 72.2 | 68.5 | 75.9 | 73.0 |

8. Verbal PSAT scores of a random sample of 40 college-bound high-school juniors have a mean of 40.7, a standard deviation of 10.2, and a distribution that is normal (based on data from Educational Testing Service). Find a 99% confidence interval estimate of the population mean.

9. A company that produces electronic ignition systems for an engine developer wants to select a suitable sample size for monitoring the mean defect-free life of the ignition systems. The company wants to be 95% confident that the true mean life of these systems will be estimated within 6300 hours. A test study of these ignition systems provided the data that $s = 1600$ hours. How large must the sample size be to be 95% confident of the true mean?

10. In a Roper survey of 1998 randomly selected adults, 24% included loud commercials among the annoying aspects of television. Construct the 99% confidence interval for the percentage of all adults who are annoyed by loud commercials.

CUMULATIVE REVIEW EXERCISES

1. The college statistics club is reviewing its order for sweat suits for the new fall semester. Its review involves the random sample of heights for 24 students, listed here (in inches):

73.4	66.9	68.2	70.6	67.1	70.4	71.0	65.9
72.5	70.5	71.5	69.5	70.4	68.6	69.7	70.9
72.1	73.3	72.6	67.4	69.4	71.9	69.3	71.6

Find each of the following:

a. mean   b. median   c. mode   d. midrange

e. range   f. variance   g. standard deviation   h. Q_1

i. Q_2 j. Q_3

k. What is the level of measurement of these data? (nominal, ordinal, interval, ratio)

l. Construct a boxplot for the data.

m. Construct a histogram and identify its general shape.

n. Construct a 99% confidence interval for the population mean.

o. Construct a 99% confidence interval for the standard deviation s.

p. Find the sample size necessary to estimate the mean height so that there is 99% confidence that the sample mean is in error by no more than 0.2 in. Use the sample standard deviation s from part (g) as an estimate of the population standard deviation σ.

q. Based on a prior study (see Data Set 8 in Appendix B), the heights of statistics students are approximately normally distributed with a mean of 68.00 in. and a standard deviation of 4.36 in. Do the heights of the new sample of 24 students (as listed above) agree with the population parameters for statistics students? Explain.

2. A genetics expert has determined that for certain couples, there is a 0.25 probability that any child will have an X-linked recessive disorder.

a. Find the probability that among 200 such children, at least 65 have the X-linked recessive disorder.

b. A subsequent study of 200 actual births reveals that 65 of the children have the X-linked recessive disorder. Based on these sample results, construct a 95% confidence interval for the proportion of all such children having the disorder.

c. Based on parts (a) and (b), does it appear that the expert's determination of a 0.25 probability is correct? Explain.

TECHNOLOGY PROJECT

The *bootstrap method* can be used to construct confidence intervals for situations in which traditional methods cannot (or should not) be used. For example, the following sample of 10 scores was randomly selected from a very nonnormal distribution, so the methods previously discussed cannot be used:

2.9   564.2   1.4   4.7   67.6   4.8   51.3   3.6   18.0   3.6

The methods of this chapter require that the population have a distribution that is at least approximately normal. The bootstrap method, which makes no assumptions about the original population, typically requires a computer to build a bootstrap population by replicating (duplicating) a sample many times. We can draw from the sample with replacement, thereby creating an approximation of the original population. In this way, we pull the sample up "by its own bootstraps" to simulate the original population. Using the sample data given above, construct a 95% confidence interval estimate of the population mean μ by using the bootstrap method as described in the following Excel steps.

a. Create 500 new samples, each of size 10, by repeatedly selecting 10 scores with replacement from the 10 sample scores given above. To prepare for this, set up a data block with the raw data in the left column and, in the right column, the assumed probability of occurrence for each value if a random sample were taken. [Since there are 10 scores, we assume a 1/10 (0.10) probability that any one score would be selected.]

b. To begin sampling, select **Data Analysis** from **Tools** on Excel's main menu. From the list of tools, select **Random Number Generation**. For the **Number of Variables**, you want to enter the desired number of bootstrapped samples (500), but Excel will give an error message for such a large number. Instead, enter 250 and run the generator twice. Enter 10 (your intended number of scores per sample) as the **Number of Random Numbers**. For **Distribution**, select **Discrete**. For **Value and Probability Input Range**, enter the Excel range for the input data block that was created in part (a). It is usual to enter 0 for the random seed. (If you enter a different seed number, then you can reproduce the same simulated sample by choosing the same seed number later.) Now select an **Output Option**, as a specified **Output Range** on the same spreadsheet, or on a **New Worksheet Ply**, for which a sheet name can be input. Click on **OK** to obtain the results.

Excel 2007: Choose **Data Analysis** from the **Data** menu. Proceed as above.
A sample of a partial output is shown. Each column represents one of the 250 10-score samples that would be generated. Run the procedure a second time to generate the remaining 250 samples.

4.8	4.8	3.6	51.3	4.7	...
2.9	564.2	2.9	2.9	3.6	...
564.2	18.0	1.4	1.4	67.6	...
18.0	4.7	67.6	2.9	67.6	...
18.0	3.6	51.3	3.6	2.9	...
4.7	4.8	18.0	4.8	67.6	...
1.4	4.8	67.6	4.7	4.7	...
4.7	2.9	1.4	4.8	51.3	...
51.3	3.6	564.2	67.6	4.8	...
3.6	1.4	3.6	4.7	67.6	...

c. Find the means of the 500 bootstrap samples generated in part (b). For example, if the first sample is entered in Cells A1:A10, the formula **=average(A1:A10)** could be put in Cell A11 to obtain the mean of the first sample; this formula could be copied along Row 11 to obtain the means for all the samples of the output.

Because the sample generator had to be run twice in part (b), repeat this process for the second set of samples that were generated.

d. Use the "**Copy\Paste Special\Values**" sequence in Excel to bring all the sample means together in a single block, as shown below. The top row represents the means calculated for the first output of samples, and the bottom row represents the means calculated for the second output of samples.

67.3	61.28	78.16	14.87	34.24
65.9	179.2	125.1	72.31	68.31

e. Find the percentiles $P_{2.5}$ and $P_{97.5}$ for the means that result from the preceding step. To calculate $P_{2.5}$, use the formula **=percentile (____,0.025)**, replacing the blank with the range for the block of means created in part (d). To calculate $P_{97.5}$, use the formula **=percentile(____,0.975)**, replacing the blank with the range for the block of means created in part (d).

Now use the bootstrap method to find a 95% confidence interval for the population standard deviation σ. Use the same steps listed above, but use the formula **=stdev** instead of **=average** in part (c), and use these new values for parts (d) and (e). Compare your result to the interval $318.4 < \sigma < 1079.6$, which was obtained by incorrectly using the methods described in Section 6–5. (Their use is incorrect because the population distribution is very nonnormal.) This incorrect confidence interval for σ does not contain the true value of σ, which is 232.1. Does the bootstrap procedure yield a confidence interval for σ that contains 232.1, verifying that the bootstrap method is effective?

An alternative to using Excel is to use special software designed specifically for bootstrap resampling methods, such as Resampling Stats (available from Resampling Stats, Inc.; see their website, www.resample.com).

FROM DATA TO DECISION

He's Angry, But Is He Right?

The following excerpt is taken from a letter written by a corporation president and sent to the Associated Press:

> When you or anyone else attempts to tell me and my associates that 1223 persons account for our opinions and tastes here in America, I get mad as hell! How dare you! When you or anyone else tells me that 1223 people represent America, it is astounding and unfair and should be outlawed.

The writer then goes on to claim that because the sample size of 1223 people represents 120 million people, his letter represents 98,000 people (120 million divided by 1223) who share the same views.

a. Given that the sample size is 1223 and the degree of confidence is 95%, find the margin of error for the proportion. Assume that there is no prior knowledge about the value of that proportion.

b. The writer of the letter is taking the position that a sample size of 1223 taken from a population of 120 million people is too small to be meaningful. Do you agree or disagree? Write a response that either supports or refutes the writer's position that the sample is too small.

c. The writer also makes the claim that because the poll of 1223 people was projected to reflect the opinions of 120 million, any 1 person actually represents 98,000 other people. As the writer is 1 person, he claims to represent 98,000 other people. Is this claim correct? Explain why or why not.

COOPERATIVE GROUP ACTIVITIES

1. **Out-of-class activity:** Estimate the mean error (in seconds) on wristwatches. First, using a wristwatch that is reasonably accurate, set the time to be exact. Use a radio station, telephone time report, or Internet site (such as www.time.gov) that states that "at the tone, the time is . . ." or otherwise guarantees the exact time. If you cannot set the time to the nearest second, record the error for the watch you are using. Now compare the time on your watch to the time on others. Record the errors with positive signs for watches that are ahead of the actual time and negative signs for those watches that are behind the actual time. Find point estimates and confidence intervals for the mean error and standard deviation of errors. Does the confidence interval for the mean error contain zero? Based on the results, what do you conclude about the accuracy of people's watches? Are the deviations from the correct time random fluctuations, or are there other factors to consider?

2. **In-class activity:** Divide into groups with approximately 10 students in each group. Get the reaction timer from the first Cooperative Group Activity given in Chapter 5 and measure the reaction time of each group member. (Right-handed students should use their right hand, and left-handed students should use their left hand.) Use the methods of this chapter to estimate the mean reaction time for all college students. Construct a 90% confidence interval estimate of that mean. Compare the results to those found in other groups.

3. **In-class activity:** Divide into groups of three or four. Assume that you need to conduct a survey of full-time students at your college with the objective of identifying the percentage of students who are eligible for a student loan. Identify a margin of error that is reasonable, select an appropriate degree of confidence, then find the minimum sample size. Describe a sampling plan that is likely to yield good results.

4. **In-class activity:** Divide into groups with approximately 10 students in each group. First, each group member should write an estimate of the mean amount of cash being carried by students in the group. Next, each group member should report the actual amount of cash being held. (The amounts should be written anonymously on separate sheets of paper that are mixed, so that nobody's privacy is compromised.) Use the reported values to find $\bar{x}$ and s, then construct a 95% confidence interval estimate of the mean μ. Describe the precise population that is being estimated. Which group member came closest to the value of $\bar{x}$? Compare the results with other groups. Are the results consistent, or are there large variations among the group results?

INTERNET PROJECT

Confidence Intervals

The confidence intervals in this chapter illustrate an important point in the science of statistical estimation: Estimations based on sample data are made with certain degrees of confidence. In the Internet Project, you will use confidence intervals to make a statement about the temperature in selected geographical areas. Go to either of the following websites

http://www.mathxl.com   or   http://www.mystatlab.com

and click on Internet Project, and then on Chapter 6. Here you will find instructions on how to use the Internet to find temperature data collected by the weather stations in a wide variety of cities. Once you have collected a set of available weather data, you will construct confidence intervals for temperatures during different time periods and attempt to draw some conclusions about temperature change in the selected area.

7

Hypothesis Testing

7-1 Overview
Chapter objectives are defined. This chapter introduces the basic concepts and procedures used for testing claims made about population parameters. The hypothesis-testing procedures of this chapter and the estimation procedures of Chapter 6 are two of the fundamental and major topics of inferential statistics.

7-2 Fundamentals of Hypothesis Testing
An informal example of a hypothesis test is presented and its important components are described. The types of errors that can be made are also discussed.

7-3 Testing a Claim About a Mean: Large Samples
The traditional approach to hypothesis testing is presented, along with the P-value approach and a third approach based on confidence intervals. The power of the test is also examined. This section is limited to cases involving large ($n > 30$) samples only. It is also noted that, if the population standard deviation σ is known and the population is normally distributed, there is no restriction on the sample size.

7-4 Testing a Claim About a Mean: Small Samples
This section presents the procedure for testing a claim made about the mean of a population, given that the data set is small (with 30 or fewer values), the value of the population standard deviation is not known, and the distribution of the population is assumed to be normal. The Student t distribution will be used in such cases.

7-5 Testing a Claim About a Proportion
The method is described for testing a claim made about a population proportion or percentage. The normal distribution is used for the examples and exercises in this section.

7-6 Testing a Claim About a Standard Deviation or Variance
The method is described for testing a claim made about the standard deviation or variance of a population. The chi-square distribution is used for such tests, and its role is described.

CHAPTER PROBLEM

The mean body temperature is 37°C, right?

When asked, most people will identify the mean body temperature for healthy adults as 37.0°C (or 98.6°F). But "healthy adults" may not always refer to the same population. Table 7-1 lists 108 measured body temperatures found in Data Set 18 of Appendix B (for patients admitted for voluntary surgery at Lakeridge Health Oshawa). These patients had health issues that brought them to surgery; yet their voluntary surgeries would not have proceeded if they were not judged to be essentially healthy overall. These 108 temperatures have a mean of 36.39°C, and a standard deviation of 0.62°C. In Chapter 6 we used the same set of temperatures to estimate μ, the mean body temperature, and to find the 99% confidence interval: $36.24°C < \mu < 36.54°C$ (see Exercise 16 in Section 6-2).

Here's the problem: The confidence interval limits of 36.24°C and 36.54°C do not contain 37.0°C, the value generally believed to be the mean body temperature. Some researchers believe the concept of 37.0°C as a "normal body temperature" is meaningless. Based on the data from Lakeridge Health Oshawa, should we reject the claim that the mean body temperature for those healthy enough for elective surgery is 37.0°C? There is a standard procedure for testing such claims, and this chapter will describe that procedure.

Table 7-1 Body Temperatures (in Degrees Celsius) of Adults Accepted for Voluntary Surgery, Measured Before the Surgery

36.4	36.3	36.7	36.3	36.8	36.6	36.6	36.1	36.2	35.6	37.4	36.3
36.2	35.4	36.4	36.3	36.7	36.1	36.0	36.1	36.4	36.4	35.8	36.7
37.3	36.4	37.1	37.1	36.4	36.6	36.6	36.5	36.7	36.5	36.6	36.8
35.5	37.0	36.9	35.2	37.1	36.7	37.4	37.1	37.0	36.4	35.9	36.9
36.3	37.1	36.4	36.8	36.5	37.6	36.1	36.4	36.0	36.7	36.3	36.2
36.4	36.7	35.4	36.1	35.0	39.9	36.6	35.9	36.0	36.0	35.4	36.2
36.0	36.8	37.3	36.3	37.1	35.8	36.0	35.8	36.2	36.9	36.3	36.6
36.0	35.5	36.0	36.1	36.4	35.0	36.5	36.6	35.1	36.4	36.0	36.7
36.3	36.8	36.5	36.5	35.8	36.4	35.9	36.1	36.1	36.2	36.3	36.3

Note that while not all of these temperatures may represent "normal, healthy" temperatures, they are the temperatures of people who were deemed sufficiently nonfevered to undergo voluntary surgery. Temperatures of patients who were operated on but may have had fever are omitted.

7-1 Overview

In Chapter 6 we introduced a major topic of inferential statistics: estimating values of population parameters based on sample statistics. This chapter introduces another major topic of inferential statistics: testing claims (or *hypotheses*) made about population parameters.

> **DEFINITION**
>
> In statistics, a **hypothesis** is a claim or statement about a property of a population.

The following statements are examples of hypotheses that will be tested by the procedures we develop in this chapter:

- The mean body temperature for patients admitted to elective surgery is not equal to 37.0°C (a claim about a population *mean*).
- Drivers who use cell phones have a car crash rate that is greater than the 13% rate for those who do not use cell phones (a claim about a population *proportion*).
- When new equipment is used to manufacture aircraft altimeters, the variation in the errors is reduced so that the readings are more consistent (a claim about a population *variance*).

Before beginning to study this chapter, you should keep in mind the rare event rule for inferential statistics:

> **If, under a given assumption, the probability of a particular observed event is exceptionally small, we conclude that the assumption is probably not correct.**

Following this rule, we analyze a sample in an attempt to distinguish between results that can easily occur and results that are *highly unlikely*. We can explain the occurrence of highly unlikely results by saying either that a rare event has indeed occurred or that things are not as they are assumed to be. Let us apply that reasoning in the following example (data from Environment Canada):

EXAMPLE

Ground-level ozone is a colourless, unhealthy gas that typically forms in urban areas, from a heated mixture of nitrogen oxide and other air pollutants. In the heavily urbanized corridor from Windsor to Quebec City, it is estimated that 50% of the ozone "smog" is generated locally, and 50% has its origins in the United States. Suppose that Ontario and Quebec take serious steps towards reducing noxious emissions, and that after several years of this program, the proportion of smog generated locally is measured. (Assume that emission rates from the United States remain unchanged.)

Using common sense and no formal statistical methods, what should we conclude about the relative effectiveness of Ontario and Quebec's anti-pollution measures if the new proportion of locally generated smog is:

a. 48%?

b. 2%?

SOLUTION

a. If the two provinces had not lowered their emissions, we would expect about 50% of the smog still to be generated locally. The result of 48% is close to 50%, so we should not conclude that the anti-pollution measures were effective. If the provinces had taken no special measures, it could easily occur by chance that 48% of the observed smog was locally generated.

b. The result that only 2% of the smog is locally generated is extremely unlikely to happen by chance. We could explain the proportion of 2% in one of two ways: Either an *extremely* rare event has occurred by chance, or the provinces' anti-pollution measures were effective in reducing emissions. Given the extremely low probability that 2% of the smog was locally generated, due simply to chance, the more likely explanation is that the program was effective.

Interpretation

We can only conclude that the provinces effectively reduced the percentage of locally generated ozone if they obtain a *significantly* reduced percentage of locally generated ozone compared to what we would expect under normal circumstances. Although the outcomes 48% locally generated and 2% locally generated are both "below average," and less than the expected 50%, the result of 48% is not significant, whereas 2% locally generated ozone does constitute a significant reduction.

This brief example illustrates the basic approach used in testing hypotheses. The formal method involves a variety of standard terms and conditions incorporated into an organized procedure. We recommend that you begin the study of this chapter by first reading Sections 7–2 and 7–3 casually to obtain a general idea of their concepts; then, reread Section 7–2 more carefully to become familiar with the terminology.

7-2 Fundamentals of Hypothesis Testing

In this section we begin with an informal example and then identify the formal components of the standard method of hypothesis testing: null hypothesis, alternative hypothesis, test statistic, critical region, significance level, critical value, type I error, and type II error. After studying this chapter, you should be able to:

- Given a claim, identify the *null hypothesis* and the *alternative hypothesis*, and express them both in symbolic form.
- Given a *significance level*, identify the *critical value(s)*.
- Given a claim and sample data, calculate the value of the *test statistic*.
- State the conclusion of a hypothesis test in simple, nontechnical terms.
- Identify the *type I* and *type II errors* that could be made when testing a given claim.

You should study the following example until you thoroughly understand it. Once you do, you will have captured a major concept of statistics.

EXAMPLE

In the chapter problem, we found that a population of essentially healthy adults does not appear to have a mean body temperature equal to the commonly assumed value of 37.0°C (98.6°F). A random sample was taken from the recorded body temperatures of individuals judged healthy enough to undergo elective surgery at Lakeridge Health Oshawa. The sample has these characteristics: $n = 108$, $\bar{x} = 36.39°$, $s = 0.62$, and the shape of the distribution is approximately normal. Here is the key question: *Do the sample data (with $\bar{x} = 36.39$) constitute sufficient evidence to warrant rejection of the hypothesis that $\mu = 37.0$ for this population?*

We conclude that there is sufficient evidence to warrant rejection of the belief that $\mu = 37.0$ for at least this one population of people who *are* essentially healthy. For if the mean were really 37.0, the probability of getting the sample mean of 36.39 is less than 0.0001, which is too small. (Later, we will show how that probability value of less than 0.0001 is determined.)

If we were to assume that $\mu = 37.0$ and use the central limit theorem (Section 5–5), we know that sample means tend to be normally distributed with these parameters:

$$\mu_{\bar{x}} = \mu = 37.0 \quad \text{(by assumption)}$$

$$\sigma_{\bar{x}} = \frac{\sigma}{\sqrt{n}} \approx \frac{s}{\sqrt{n}} = \frac{0.62}{\sqrt{108}} = 0.06$$

We construct Figure 7-1 by assuming that $\mu = 37.0$ and by using the parameters just shown. Figure 7-1 also shows that if μ is really 37.0, then 99% of all sample means should fall between 36.85 and 37.15. (The values of 36.85 and 37.15 were found by using the methods of Section 5–5. Specifically, 99% of sample means should fall within approximately 2.575 standard deviations of μ. With $\sigma_{\bar{x}} = 0.06$, 99% of sample means should fall within 2.575 × 0.06 ≈ 0.15 of 37.0. Falling within 0.15 of 37.0 is equivalent to falling between 36.85 and 37.15.)

Here are the key points:

- The common belief is that $\mu = 37.0$.
- The sample resulted in $\bar{x} = 36.39$.

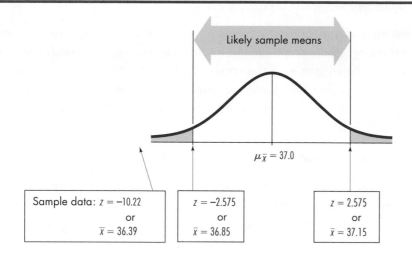

- Considering the distribution of sample means, the sample size, and the magnitude of the discrepancy between 37.0 and 36.39, we find that a sample mean of 36.39 is unlikely (with less than a 1% chance) to occur if μ is really 37.0.

- There are two reasonable explanations for the sample mean of 36.39: Either a very rare event has occurred, or μ is not really 37.0. Because the probability of getting a sample mean of 36.39 (when $\mu = 37.0$) is so low, we go with the more reasonable explanation: The value of μ is not 37.0 for at least one population of healthy people.

The preceding example illustrates well the basic line of reasoning that we will use throughout this chapter. Read it several times until you understand it. Try not to dwell on the details of the calculations. Instead, focus on the key idea that although there is an assumption that $\mu = 37.0$, the sample mean is $\bar{x} = 36.39$. By using the central limit theorem, we determine that if the mean is really 37.0, then the probability of getting a sample with a mean of 36.39 is very small, which suggests that, for the sampled population, the belief that $\mu = 37.0$ should be rejected. Section 7–3 will describe the specific procedure used in hypothesis testing, but we will first describe the components of a **formal hypothesis test** or **test of significance.**

Components of a Formal Hypothesis Test

Testing a claim with a formal hypothesis test is analogous to a trial in a court of law. As you may know, the defendant is assumed to be innocent at the beginning of the trial. It is the responsibility of the Crown to provide sufficient evidence to find the defendant guilty; if it does, the defendant is found guilty, but otherwise the defendant is found not guilty.

There are five basic steps in a formal hypothesis test.

STEP 1: *Null and Alternative Hypotheses*

- The **null hypothesis** (denoted by H_0) is a statement about the value of a population parameter (such as the mean), and it must contain the condition of equality and must be written with the symbol $=$, $\leq$, or $\geq$. (When actually conducting the test, we operate under the assumption that the parameter *equals* some specific value.) For the mean, the null hypothesis will be stated in one of these three possible forms:

$$H_0: \mu = \text{some value} \qquad H_0: \mu \leq \text{some value} \qquad H_0: \mu \geq \text{some value}$$

For example, the null hypothesis corresponding to the common belief that the mean body temperature is 37.0°C is expressed as $H_0: \mu = 37.0$. We test the null hypothesis directly in the sense that we assume it is true and reach a conclusion either to reject H_0 or fail to reject H_0.

- The **alternative hypothesis** (denoted by H_1) is the statement that must be true if the null hypothesis is false. For the mean, the alternative hypothesis will be stated in only one of three possible forms:

$$H_1: \mu \neq \text{some value} \qquad H_1: \mu > \text{some value} \qquad H_1: \mu < \text{some value}$$

Note that H_1 is the opposite of H_0. For example, if H_0 is given as $\mu = 37.0$, then it follows that the alternative hypothesis is given by $H_1: \mu \neq 37.0$.

Note About Using $\leq$ or $\geq$ in H_0: Even though we sometimes express H_0 with the symbol $\leq$ or $\geq$, as in $H_0: \mu \leq 37.0$ or $H_0: \mu \geq 37.0$, we conduct the test by assuming that $\mu = 37.0$ is true. We must have a single fixed value for μ so that we can work with a single distribution having a specific mean. (Some textbooks and some software packages use notation in which H_0 *always* contains only the equals symbol. Where this and many other textbooks might use $\mu \leq 37.0$ and $\mu > 37.0$ for H_0 and H_1, respectively, some others might use $\mu = 37.0$ and $\mu > 37.0$ instead.)

Note About Testing the Validity of a Claim: Sometimes we test the validity of a claim, such as the claim of the Coca Cola Bottling Company that "the mean amount of Coke in cans is at least 355 mL," which becomes the null hypothesis of $H_0: \mu \geq 355$. In this context of testing the validity of a claim, it sometimes becomes the null hypothesis (because it contains equality), and it sometimes becomes the alternative hypothesis (because it does not contain equality).

Analogous to a court of law, the burden of proof is on the alternative hypothesis. If there is sufficient evidence, we reject the null hypothesis; otherwise, we fail to reject the null hypothesis.

In the chapter problem, the null and alternative hypotheses are:

$$H_0: \mu = 37.0$$
$$H_1: \mu \neq 37.0$$

Table 7-2 Type I and Type II Errors

		True State of Nature	
		The null hypothesis is true.	The null hypothesis is false.
Decision	We decide to reject the null hypothesis.	**Type I error** (rejecting a true null hypothesis) α	Correct decision
	We fail to reject the null hypothesis.	Correct decision	**Type II error** (failing to reject a false null hypothesis) β

Before we proceed to the second step, there are some theoretical matters to take care of.

Type I and Type II Errors

In a court of law, there are two errors that the jury may commit in reaching a verdict: finding the defendant guilty when the defendant is innocent or finding the defendant not guilty when the defendant is guilty. These same two errors can also be committed in a formal hypothesis test.

When testing a null hypothesis, we arrive at a conclusion of rejecting it or failing to reject it. Such conclusions are sometimes correct and sometimes not (even if we do everything correctly). Table 7-2 summarizes the different possibilities and shows that we make a correct decision when we either reject a null hypothesis that is false or fail to reject a null hypothesis that is true. However, we make an error when we reject a true null hypothesis or fail to reject a false null hypothesis. These two types of errors are described as follows.

- **Type I error:** *The mistake of rejecting the null hypothesis when it is true.* For the chapter problem, a type I error would be the mistake of rejecting the null hypothesis that the mean body temperature is 37.0 when that mean really is 37.0. The type I error is not a miscalculation or procedural misstep; it is an actual error that can occur when a rare event happens by chance.

- **Type II error:** *The mistake of failing to reject the null hypothesis when it is false.* For the chapter problem, a type II error would be the mistake of failing to reject the null hypothesis ($\mu = 37.0$) when it is actually false (that is, the mean is not really 37.0).

Now that we have defined type I and type II errors, we need to associate probabilities with these types of errors.

- The *maximum probability* of a type I error (rejecting the null hypothesis when it is true) is called the significance level of the test and is denoted by the symbol α (alpha). This is the same α introduced in Section 6–2, where we defined the degree of confidence for a confidence interval to be the probability $1 - \alpha$. Common choices for α are 0.05, 0.01, and 0.10. In Figure 7-1, for example, the shaded regions in the tails combine to be $\alpha = 0.01$. The significance level is described as $\alpha = 0.01$.

- The symbol β (beta) is used to represent the probability of a type II error (not rejecting the null hypothesis when it is false).

- There is one other probability, called the power of the test, which is the probability of rejecting the null hypothesis when it is false. Mathematically, power = $1 - \beta$. Statisticians often use it to gauge the test's effectiveness in recognizing that a null hypothesis is false.

In hypothesis testing, we are generally concerned with the significance level. Analysis of β and the power of the test is done after the initial hypothesis test.

NOTATION

α (alpha) = maximum probability of a type I error (the probability of rejecting the null hypothesis when it is true)

β (beta) = probability of a type II error (failing to reject a false null hypothesis)

STEP 2: *Set Up the Critical Region*

- The critical region is the set of all values of the test statistic (see below) that cause us to reject the null hypothesis. For the chapter problem, the critical region is represented by the shaded part of Figure 7-1 and consists of values of the test statistic less than $z = -2.575$ or greater than $z = 2.575$.

- The acceptance region is the set of all values of the test statistic that cause us to not reject the null hypothesis. For the chapter problem, the acceptance region consists of values of the test statistic between $z = -2.575$ and $z = 2.575$.

- A critical value is any value that separates the critical region (where we reject the null hypothesis) from the acceptance region (where we fail to reject the null hypothesis). The critical values depend on the nature of the null hypothesis, the relevant sampling distribution, and the significance level α. For the chapter problem, the critical values of $z = -2.575$ and $z = 2.575$ separate the shaded critical regions, as shown in Figure 7-1.

STEP 3: *Computing the Test Statistic*

- The test statistic is a value that we compute from the sample data and that we use as a step in deciding whether to reject the null hypothesis. The test statistic converts the sample statistic (such as the sample mean $\bar{x}$) to a score (such as the z score) under the assumption that the null hypothesis is true. The test statistic can therefore be used to gauge whether the discrepancy between the sample and the claim is significant. In this section and the following section, we use only large samples in testing claims made about population means. In such cases, we can use the central limit theorem (Section 5–5) with the test statistic based on the discrepancy between the sample mean $\bar{x}$ and the claimed population mean μ, calculated by using Formula 7-1, which is another form of $z = (\bar{x} - \mu_{\bar{x}})/\sigma_{\bar{x}}$.

Formula 7-1

$$z = \frac{\bar{x} - \mu_{\bar{x}}}{\dfrac{\sigma}{\sqrt{n}}} \qquad \text{Test statistic}$$

Using the sample data from the chapter problem ($n = 108$, $\bar{x} = 36.39$, and $s = 0.62$), and with the assumption that $\mu = 37$, the test statistic is calculated by using Formula 7-1 as shown below (with s used as an estimate of σ).

$$z = \frac{\bar{x} - \mu_{\bar{x}}}{\dfrac{\sigma}{\sqrt{n}}} = \frac{36.39 - 37.0}{\dfrac{0.62}{\sqrt{108}}} = -10.22$$

STEP 4: *Make a Decision to Reject/Not Reject the Null Hypothesis*

- If the test statistic falls in the rejection region, we reject the null hypothesis. Otherwise, we fail to reject the null hypothesis.

In the chapter problem, since the value of the test statistic is -10.22, it falls inside the rejection region (being less than the critical value of -2.575). Therefore, we reject the null hypothesis of $\mu = 37.0$.

It should be noted that since we reject the null hypothesis, there is the possibility that we are committing a type I error of rejecting the null hypothesis when it is, in fact, true.

STEP 5: *State the Final Conclusion*

The conclusion of failing to reject the null hypothesis or rejecting it is fine for those of us with the wisdom to take a statistics course, but we should use simple, nontechnical terms in stating what the conclusion really means. Figure 7-2 shows how to formulate the correct wording of the final conclusion. Note that only one case leads to wording indicating that the sample data actually *support* the conclusion. If you want to support some claim, state it in such a way that it becomes the alternative hypothesis and then hope that the null hypothesis gets rejected. For example, to support the claim that the mean body temperature is different from 37.0°C, make the claim that $\mu \neq 37.0$. This claim will be an alternative hypothesis that will be supported if you reject the null hypothesis of H_0: $\mu = 37.0$.

Figure 7-2 Wording of Conclusions in Hypothesis Tests

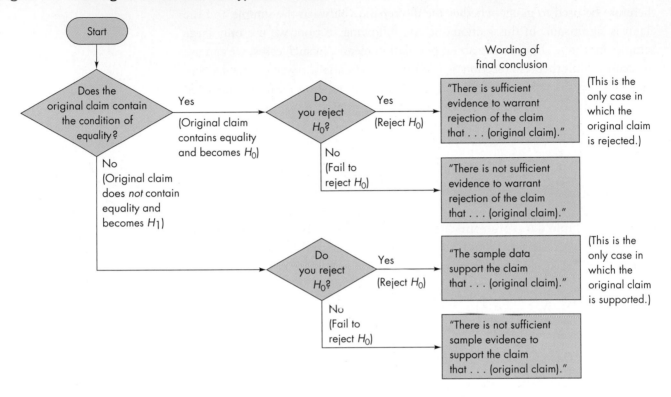

If, on the other hand, you claim that $\mu = 37.0$, you will either reject or fail to reject the claim; in either case, you will not *support* the original claim.

In the original phrasing of the chapter problem, the question was posed whether we should reject the claim that the average body temperature for those healthy enough for elective surgery is 37.0°C. Since we reject the null hypothesis, there is sufficient evidence to reject the claim that the average body temperature is 37.0°C.

Accept/Fail to Reject

Some say "accept the null hypothesis" instead of "fail to reject the null hypothesis." Whether we use the term *accept* or *fail to reject*, we should recognize that *we are not proving the null hypothesis*; we are merely saying that the sample evidence is not strong enough to warrant rejection of the null hypothesis. It's like a jury saying that there is not enough evidence to convict a suspect. The term *accept* is somewhat misleading because it seems to imply incorrectly that the null hypothesis has been proved. The phrase *fail to reject* says more correctly that the available evidence isn't strong enough to warrant rejection of the null hypothesis. In this text we will use the conclusion *fail to reject the null hypothesis*, instead of *accept the null hypothesis*.

Two-Tailed, Left-Tailed, Right-Tailed

The *tails* in a distribution are the extreme regions bounded by critical values. Some hypothesis tests are two-tailed, some are right-tailed, and some are left-tailed.

- **Two-tailed test:** The critical region is in the extreme regions (tails) under both sides of the curve.

- **Right-tailed test:** The critical region is in the extreme region (tail) under the right side of the curve.

- **Left-tailed test:** The critical region is in the extreme region (tail) under the left side of the curve.

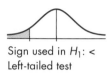

Sign used in H_1: >
Right-tailed test

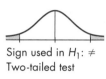

Sign used in H_1: <
Left-tailed test

In two-tailed tests, the level of significance α is divided equally between the two tails that constitute the critical region. For example, in a two-tailed test with a significance level of $\alpha = 0.05$, there is an area of 0.025 in each of the two tails. In right- or left-tailed tests, the area α of the critical region is all in the single tail.

By examining the null hypothesis H_0, we should be able to deduce whether a test is right-tailed, left-tailed, or two-tailed. The tail(s) will correspond to the critical region containing the values that would conflict significantly with the null hypothesis. A useful check is summarized in the margin figures, which show that the inequality sign in H_1 points in the direction of the critical region. The symbol $\neq$ is often expressed in programming languages as $<>$, and this reminds us that an alternative hypothesis such as $\mu \neq 37.0$ corresponds to a two-tailed test.

Sign used in H_1: $\neq$
Two-tailed test

EXAMPLE

Finding critical values: Many passengers on cruise ships wear skin patches that supply dramamine to the body for the purpose of preventing motion sickness. A claim about the mean dosage amount is tested with a significance level of $\alpha = 0.05$. The conditions are such that the standard normal distribution can be used (because the central limit theorem applies). Find the critical value(s) of z if the test is (a) two-tailed, (b) left-tailed, and (c) right-tailed.

SOLUTION

a. In a two-tailed test, the significance level of $\alpha = 0.05$ is divided equally between the two tails, so there is an area of 0.025 in each tail. We can find the critical values in Table A-2 as the values corresponding to areas of 0.4750 (found by subtracting 0.025 from 0.5) to the right and left of the mean. We get critical values of $z = -1.96$ and $z = 1.96$, as shown in Figure 7-3(a).

b. In a left-tailed test, the significance level of $\alpha = 0.05$ is the area of the critical region at the left, so the critical value corresponds to an area of 0.4500 (found from $0.5 - 0.05$) to the left of the mean. Using Table A-2, we get a critical value of $z = -1.645$, as shown in Figure 7-3(b).

c. In a right-tailed test, the significance level of $\alpha = 0.05$ is the area of the critical region to the right, so the critical value corresponds to an area of 0.4500 (found from $0.5 - 0.05$) to the right of the mean. Using Table A-2, we get a critical value of $z = 1.645$, as shown in Figure 7-3(c).

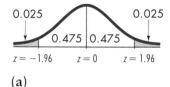

(a)

(b)

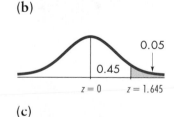

(c)

Figure 7-3
Finding Critical Values

EXAMPLE

Identifying components of a hypothesis test: A concerned citizen is preparing a complaint to the Council of Better Business Bureaus. He has discreetly collected samples from the gasoline pumps of Elite Auto Centre and claims that consumers are being cheated because of this condition: When the meters indicate 1 L, the mean amount of gas actually supplied is less than 1 L.

a. Express in symbolic form the claim that the Elite Auto Centre is cheating consumers.

b. Identify the null hypothesis H_0.

c. Identify the alternative hypothesis H_1.

d. Identify this test as being two-tailed, left-tailed, or right-tailed.

e. Identify the type I error for this test.

f. Identify the type II error for this test.

g. Assume that the conclusion is to reject the null hypothesis. Interpret the conclusion in nontechnical terms; be sure to address the original claim.

h. Assume that the conclusion is failure to reject the null hypothesis. Interpret the conclusion in nontechnical terms; be sure to address the original claim.

SOLUTION

a. The claim that consumers are being cheated is equivalent to claiming that the mean is less than 1 L, which is expressed in symbolic form as $\mu < 1$ L.

b. The original claim of $\mu < 1$ L does not contain equality, as required by the null hypothesis. The original claim is therefore the alternative hypothesis; the null hypothesis is $H_0: \mu \geq 1$ L.

c. See part (b). The alternative hypothesis is $H_1: \mu < 1$ L.

d. This test is left-tailed because the null hypothesis is rejected if the sample mean is significantly less than (or to the left of) 1 L. (As a double check, note that the alternative hypothesis $\mu < 1$ contains the sign $<$, which points to the left.)

e. The type I error (rejection of a true null hypothesis) is to reject the null hypothesis $\mu \geq 1$ L when the population mean is really equal to or greater than 1 L. (This is a serious error, because the business will be charged with cheating consumers when no such cheating is actually happening.)

f. The type II error (failure to reject a false null hypothesis) is to fail to reject the null hypothesis $\mu \geq 1$ L when the population mean is really less than 1 L. (That is, we conclude that there isn't sufficient evidence to charge cheating when cheating is actually taking place.)

g. See Figure 7-2 for interpreting the case where the original claim does not contain equality, but H_0 is rejected. Conclude that there is sufficient evidence to support the claim that the mean is less than 1 L.

h. See Figure 7-2 for interpreting the case where the original claim does not contain equality, and we fail to reject H_0. Conclude that there is not sufficient evidence to support the claim that the mean is less than 1 L.

Controlling Type I and Type II Errors

One step in our procedure for testing hypotheses involves the selection of the significance level α, which is the probability of a type I error. However, we do not select β [P(type II error)]. It would be great if we could always have $\alpha = 0$ and $\beta = 0$, but in reality that is not possible, so we must attempt to manage the α and β error probabilities. Mathematically, it can be shown that α, β, and the sample size n are all related, so when you choose or determine any two of them, the third is automatically determined. The usual practice in research and industry is to select the values of α and n, so the value of β is determined. Depending on the seriousness of a type I error, try to use the largest α that you can tolerate. For type I errors with more serious consequences, select smaller values of α. Then choose a sample size n as large as is reasonable, based on considerations of time, cost, and other such relevant factors. (Sample size determinations were discussed in Section 6–2.) The following practical considerations may be relevant:

1. For any fixed α, an increase in the sample size n will cause a decrease in β. That is, a larger sample will lessen the chance that you make the error of not rejecting the null hypothesis when it's actually false.

2. For any fixed sample size n, a decrease in α will cause an increase in β. Conversely, an increase in α will cause a decrease in β.

3. To decrease both α and β, increase the sample size.

To make sense of these abstract ideas, let's consider M&Ms (produced by Mars, Inc.) and Bufferin brand aspirin tablets (produced by Bristol-Myers Products). The mean weight of the M&M candies should be at least 0.9085 g (in order to conform to the weight printed on the package label). The Bufferin tablets are supposed to contain a mean weight of 325 mg of aspirin. Because M&Ms are candies used for enjoyment whereas Bufferin tablets are drugs used for treatment of health problems, we are dealing with two very different levels of seriousness. If the M&Ms don't have a population mean weight of 0.9085 g, the consequences are not very serious, but if the Bufferin tablets don't have a mean of 325 mg of aspirin, the consequences could be very serious. If the M&Ms have a mean that is too large, Mars will lose some money but consumers will not complain. In contrast, if the Bufferin tablets have too much aspirin, Bristol-Myers could be faced with consumer lawsuits and actions by Health Canada under the Food and Drugs Act. Consequently, in testing the claim that $\mu = 0.9085$ g for M&Ms, we might choose $\alpha = 0.05$ and a sample size of $n = 100$; in testing the claim that $\mu = 325$ mg for Bufferin tablets, we might choose $\alpha = 0.01$ and a sample size of $n = 500$. The smaller significance level α and larger sample size n are chosen because of the more serious consequences associated with testing a commercial drug.

To summarize, the five basic steps of hypothesis testing are:

1. Set up the null and alternative hypotheses.
2. Set up the critical region.
3. Compute the test statistic.
4. Make a decision to reject, or fail to reject, the null hypothesis.
5. State the conclusion in non-technical language.

In Section 7–3 we will examine hypothesis testing for μ using the normal distribution in more detail.

7-2 Exercises A: Basic Skills and Concepts

In Exercises 1–4, what do you conclude? (Do not use formal procedures and exact calculations. Use only the rare event rule described in Section 7–1, and make subjective estimates to determine whether events are likely.)

1. Claim: A coin is fair. Then the coin turns up heads 27 times in 30 tosses.
2. Claim: A die is fair. Then the die turns up a 1 nine times in 60 rolls.
3. Claim: Women who eat blue M&M candies have a better chance of having a baby boy. Then 50 such women give birth to 27 boys and 23 girls.
4. Claim: A roulette wheel is fair. Then the number 7 occurs seven times in seven trials. (A fair roulette wheel has 38 equally likely slots, one of which is a 7.)

In Exercises 5–12, assume that a hypothesis test of the given claim will be conducted. Use μ for a claim about a mean, p for a claim about a proportion, and σ for a claim about variation.

a. *Express the claim in everyday language.*
b. *Identify the null hypothesis H_0.*
c. *Identify the alternative hypothesis H_1.*
d. *Identify the test as being two-tailed, left-tailed, or right-tailed.*
e. *Identify the type I error for the test.*
f. *Identify the type II error for the test.*
g. *Assuming that the conclusion is to reject the null hypothesis, state the conclusion in nontechnical terms; be sure to address the original claim. (See Figure 7-2.)*
h. *Assuming that the conclusion is failure to reject the null hypothesis, state the conclusion in nontechnical terms; be sure to address the original claim. (See Figure 7-2.)*

5. Health Canada claims that a pharmaceutical company makes cold caplets that contain amounts of acetaminophen with a mean different from the 650 mg amount indicated on the label.

6. The college registrar claims that the times students require to earn a college diploma have a mean less than 5 years.

7. The Conservative Party M.P. claims that she is currently favoured by more than half of all voters in her riding.

8. The Home Electronics Supply Company claims that its home circuit breakers trip at levels that have less variation than circuit breakers made by its major competitor, which has variation described by $\sigma = 0.4$ amp.

9. Statistics Canada claims that over 2% of all adults are enrolled in school full time.

10. In explaining the need to attract new business, Antigonish and Guysborough counties claim that, for women, the mean income in that region is $14,000.

11. The Canadian Automobile Association warns that motorists are being seduced by ads to buy higher octane gas than is useful. Test the claim that fewer than 10% of cars on the road have a design that benefits from higher octane gasoline.

12. The buyer for a hospital supply company recommends not buying the new digital thermometers because they vary more than the old thermometers with a standard deviation of 0.33°C.

In Exercises 13–20, find the critical z values for the given conditions. In each case assume that the normal distribution applies, so Table A-2 can be used. Also, draw a graph showing the critical value and critical region.

13. Right-tailed test; $\alpha = 0.05$ 14. Left-tailed test; $\alpha = 0.05$
15. Left-tailed test; $\alpha = 0.01$ 16. Two-tailed test; $\alpha = 0.01$
17. Two-tailed test; $\alpha = 0.10$ 18. Right-tailed test; $\alpha = 0.025$
19. Left-tailed test; $\alpha = 0.025$ 20. Two-tailed test; $\alpha = 0.02$

In Exercises 21–24, use Formula 7-1 to find the value of the test statistic z.

21. From a study of consumer buying: The claim is $\mu = 0.21$, and the sample statistics include $n = 32$, $\bar{x} = 0.83$, and $s = 0.24$.

22. From an article on the ages of humans: The claim is $\mu = 73.4$, and the sample statistics include $n = 35$, $\bar{x} = 69.5$, and $s = 8.7$.

23. From research on seatbelt use: The claim is $\mu < 1.39$, and the sample statistics include $n = 123$, $\bar{x} = 0.83$, and $s = 0.16$.

24. From a study of the amounts of Moosehead Beer in cans: The claim is $\mu = 355$, and the sample statistics include $n = 36$, $\bar{x} = 360.6$, and $s = 3.3$.

7-2 Exercises B: Beyond the Basics

25. When testing a claim that a mean is at least 100, the null hypothesis becomes $\mu \geq 100$. When actually conducting the hypothesis test, why is it necessary to assume $\mu = 100$ instead of $\mu \geq 100$?

26. Someone suggests that in testing hypotheses, you can eliminate a type I error by making $\alpha = 0$. In a two-tailed test, what critical values correspond to $\alpha = 0$? If $\alpha = 0$, will the null hypothesis ever be rejected?

7-3 Testing a Claim About a Mean: Large Samples

In this section we examine hypothesis testing for means using the Z distribution. We will present three methods of testing hypotheses that appear different but are equivalent in the sense that they always lead to similar conclusions. The first procedure is the traditional method; the second is the P-value method; and the third is the confidence interval method, which is based on concepts introduced in Chapter 6.

We begin by identifying the two assumptions that apply to the methods we have already presented in this section.

Assumptions for Testing a Claim About the Mean of a Single Population, from a Large Sample

1. The sample is large ($n > 30$), so the central limit theorem applies and we can use the normal distribution.

2. When applying the central limit theorem, we can use the sample standard deviation s as an estimate of the population standard deviation σ whenever σ is unknown and the sample size is large ($n > 30$).

There is one other scenario in which we could use the Z distribution: namely that the population is known to be normally distributed and the population standard deviation σ is known. This scenario could arise, for example, in a factory that continually samples production for quality control purposes. If this process has proceeded for an extended period of time, then the long-term data could be plotted to determine whether it is normally distributed. (Other methods such as formal normality tests could also be used.) Courtesy of the law of large numbers, the standard deviation from long-term data would be, for all intents and purposes, σ. Under these conditions, the sample size could be under 30.

We also assume that the data have been collected by valid random-sampling techniques. Then, our first step is to *explore* the data set, using methods introduced in Chapter 2 to investigate centre, variation, and distribution, and to identify any outliers.

The Traditional Method of Testing Hypotheses

Figure 7-4 summarizes the steps used in the **traditional** (or **classical**) **method of testing hypotheses**. This procedure uses the components described in Section 7–2 as part of a system to identify a sample result that is *significantly* different from the claimed value. The relevant sample statistic (such as $\bar{x}$) is converted to a test

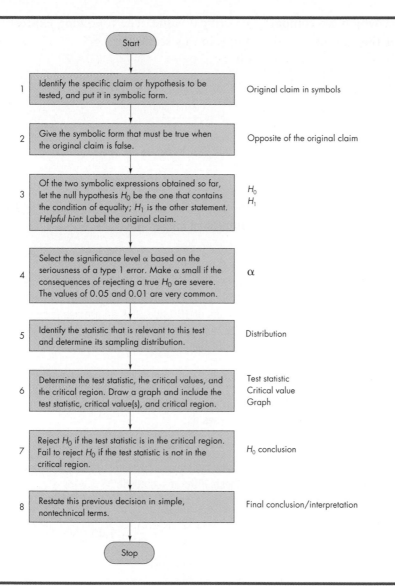

Figure 7-4
Traditional Method of Hypothesis Testing

Start

1. Identify the specific claim or hypothesis to be tested, and put it in symbolic form.

   Original claim in symbols

2. Give the symbolic form that must be true when the original claim is false.

   Opposite of the original claim

3. Of the two symbolic expressions obtained so far, let the null hypothesis H_0 be the one that contains the condition of equality; H_1 is the other statement. *Helpful hint*: Label the original claim.

   H_0
   H_1

4. Select the significance level α based on the seriousness of a type 1 error. Make α small if the consequences of rejecting a true H_0 are severe. The values of 0.05 and 0.01 are very common.

   α

5. Identify the statistic that is relevant to this test and determine its sampling distribution.

   Distribution

6. Determine the test statistic, the critical values, and the critical region. Draw a graph and include the test statistic, critical value(s), and critical region.

   Test statistic
   Critical value
   Graph

7. Reject H_0 if the test statistic is in the critical region. Fail to reject H_0 if the test statistic is not in the critical region.

   H_0 conclusion

8. Restate this previous decision in simple, nontechnical terms.

   Final conclusion/interpretation

Stop

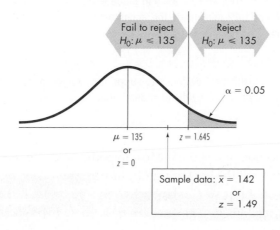

Figure 7-5
Distribution of Means of Textbook Prices Assuming $\mu = 135$

Fail to reject $H_0: \mu \leq 135$ Reject $H_0: \mu \leq 135$

$\alpha = 0.05$

$\mu = 135$
or
$z = 0$

$z = 1.645$

Sample data: $\bar{x} = 142$
or
$z = 1.49$

statistic, which we compare to a critical value. The following expression can be used for the examples and exercises in this section.

TEST STATISTICS FOR CLAIMS ABOUT μ WHEN $n > 30$

$$z = \frac{\bar{x} - \mu_{\bar{x}}}{\dfrac{\sigma}{\sqrt{n}}}$$

Section 7–2 introduced the three types of subtests: right-tailed, left-tailed, and two-tailed. We now present examples of each type.

EXAMPLE

A researcher suspects that the average price of textbooks has significantly increased in the past few years. Based on data from five years previous, textbook prices at the time had an average price of $135. A random sample of 100 titles has an average price of $142 with a standard deviation of $47. Does the evidence provide significant support of the researcher's claim? Test at a 5% level of significance.

SOLUTION

The researcher's claim can be phrased as $\mu > 135$. Since this is a strict inequality, we can set up the null and alternative hypotheses as:

$$H_0: \mu \leq 135 \qquad H_1: \mu > 135$$

Since the sign under the alternative hypothesis is $>$, this is a right-tailed test. Since $\alpha = 0.05$, we reject the null hypothesis if the test statistic > 1.645.
The value of the test statistic is:

$$z = \frac{\bar{x} - \mu}{\dfrac{\sigma}{\sqrt{n}}} = \frac{142 - 135}{\dfrac{47}{\sqrt{100}}} = 1.49$$

Note that since $n = 100$, we can use s in place of σ.
The test statistic of 1.49 falls inside the acceptance region. Therefore, we fail to reject the null hypothesis.

Interpretation
We conclude there is insufficient evidence to support the researcher's claim that the average price of textbooks has significantly increased from $135.

A note on the selection of $\alpha = 0.05$: Most hypothesis testing is conducted between $\alpha = 0.01$ and $\alpha = 0.10$. If the level of significance is set too low, there is an increased risk of a type II error; if it is set too high, there is an increased risk of a type I error. Setting $\alpha = 0.05$ provides a nice balance between these two types of errors.

EXAMPLE

A demographer wants to determine if there has been a significant decrease in the average number of children per household. Based on prior information, the average has been 2.4. A random sample of 400 households has an average of 2.2 with a standard deviation of 0.8. Test at a 1% level of significance.

SOLUTION

Since we want to determine if there has been a significant decrease in the average number of children per household, we want to see if $\mu < 2.4$. The null and alternative hypotheses are:

$$H_0: \mu \geq 2.4 \qquad H_1: \mu < 2.4$$

Since the sign under the alternative hypothesis is $<$, this is a left-tailed test. Since $\alpha = 0.01$, we reject the null hypothesis if the test statistic is less than -2.33. (This critical value is found by noting that $P(0 < Z < 2.33) = 0.4901$, indicating that $P(Z > 2.33) = P(Z < -2.33) \approx 0.01$.)

The value of the test statistic is:

$$z = \frac{\bar{x} - \mu}{\dfrac{\sigma}{\sqrt{n}}} = \frac{2.2 - 2.4}{\dfrac{0.8}{\sqrt{400}}} = -5.00$$

The test statistic of -5.00 falls inside the critical region. Therefore, we reject the null hypothesis.

Interpretation
We conclude there is significant evidence to conclude that the average number of children per household has significantly decreased from 2.4.

You may note that the difference between the mean we would be expecting under the null hypothesis of 2.4 and the mean from the sample of 2.2 is not much. Yet, the magnitude of the test statistic is quite large. There are two reasons for this: The large sample size of 400 and the small standard deviation of 0.8.

The other issue with this example is that of *practical significance* versus that of *statistical significance*. Consider this: Suppose Canada has 11 million households. Note that $2.4 - 2.2 = 0.2$. If we multiply 11 million by 0.2, we get 2.2 million $= 2.2 \times 10^6$. Suppose that when these children become adults, they each earn \$30,000 per year. Multiplying 3×10^4 by 2.2×10^6 yields 6.6×10^{10} or \$66 billion per year. For a program such as Canada Pension Plan, these are the types of issues that the federal government faces.

Figure 7-6
Distribution of Means of Children Per Household Assuming $\mu = 2.4$

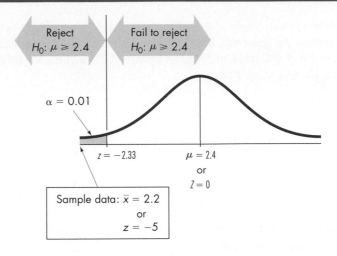

Reject $H_0: \mu \geqslant 2.4$

Fail to reject $H_0: \mu \geqslant 2.4$

$\alpha = 0.01$

$z = -2.33$

$\mu = 2.4$
or
$Z = 0$

Sample data: $\bar{x} = 2.2$
or
$z = -5$

EXAMPLE

A factory tried a new box-cutting machine to determine if it met the specifications of an average box length of 24 cm. In a random sample of 60 boxes, the average was 24.008 cm with a standard deviation of 0.0968 cm. Test at a 10% level of significance.

SOLUTION

If the new machine meets the factory's specifications, then $\mu = 24$. The null hypothesis and alternative hypotheses then become:

$$H_0: \mu = 24 \qquad H_1: \mu \neq 24$$

Since the sign under the alternative hypothesis is $\neq$, this is a two-tailed test. With $\alpha = 0.1$, $\alpha/2 = 0.05$. The critical values are -1.645 for the left tail and 1.645 for the right tail.
The value of the test statistic is:

$$z = \frac{\bar{x} - \mu}{\dfrac{\sigma}{\sqrt{n}}} = \frac{24.008 - 24}{\dfrac{0.0968}{\sqrt{60}}} = 0.64$$

The test statistic of 0.64 falls inside the acceptance region. Therefore, we fail to reject the null hypothesis.

Interpretation
There is insufficient evidence to conclude that the average box length using the new machine is significantly different from 24 cm. We conclude that the new machine meets the factory's specifications.

In presenting the results of a hypothesis test, it is not always necessary to show all of the steps included in Figure 7-4. However, the results should include the null hypothesis, the alternative hypothesis, the calculation of the test statistic, a graph such as Figure 7-6, the initial conclusion (reject H_0 or fail to reject H_0), and

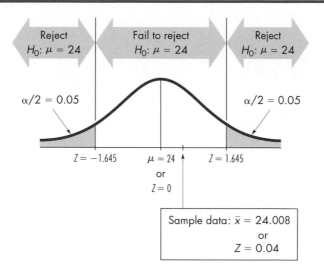

Figure 7-7
**Distribution of Means
of Box Lengths Assuming
$\mu = 24$**

the final conclusion stated in nontechnical terms. The graph should show the test statistic, critical value(s), critical region, and significance level.

The *P*-Value Method of Testing Hypotheses

Many professional articles and software packages use another approach to hypothesis testing that is based on the calculation of a *probability*, or P-value. Given a null hypothesis and sample data, the P-value reflects the likelihood of getting the sample results obtained, assuming that the null hypothesis is actually true.

DEFINITION

> A **P-value** (or **probability value**) is the probability of getting a value of the sample test statistic that is *at least as extreme* as the one found from the sample data, assuming that the null hypothesis is true.

Figure 7-8 outlines the procedure for finding the *P*-value. The figure shows the following:

Right-tailed test: The *P*-value is the area to the right of the test statistic.

Left-tailed test: The *P*-value is the area to the left of the test statistic.

Two-tailed test: The *P*-value is *twice* the area of the extreme region bounded by the test statistic.

STATISTICAL PLANET

Beware of *P*-Value Misuse
John P. Campbell, editor of the *Journal of Applied Psychology*, wrote the following on the subject of *P*-values: "Books have been written to dissuade people from the notion that smaller *P*-values mean more important results or that statistical significance has anything to do with substantive significance. It is almost impossible to drag authors away from their *P*-values, and the more zeros after the decimal point, the harder people cling to them." Although it might be necessary to provide a statistical analysis of the results of a study, we should place strong emphasis on the significance of the results themselves.

Figure 7-8 Finding *P*-Values

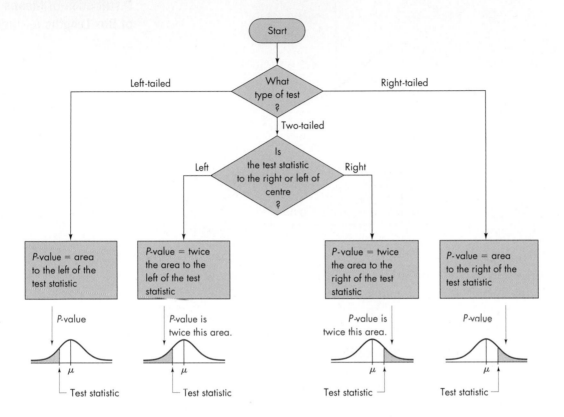

As a way of introducing *P*-values, we revisit the right-tailed example.

EXAMPLE

Recall from the example on page 376 that the null and alternative hypotheses for the mean textbook price are:

$$H_0: \mu \le 135 \qquad H_1: \mu > 135$$

We reject the null hypothesis if the test statistic is greater than 1.645, since $\alpha = 0.05$. Since $z = 1.49$, we fail to reject the null hypothesis.

Suppose that $\alpha = 0.10$ instead of $\alpha = 0.05$. What conclusion would we reach under this condition?

SOLUTION

If $\alpha = 0.10$, the critical value becomes 1.28. (We find this critical value by noting that $P(0 < Z < 1.28) = 0.3997$, indicating that $P(Z > 1.28) \approx 0.10$.) This time, the test statistic of 1.49 falls in the rejection region. Therefore, we reject the null hypothesis and conclude there is significant evidence to conclude that the average textbook price is significantly higher than \$135.

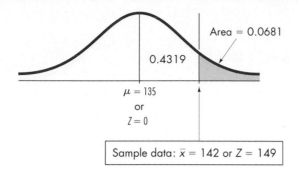

Figure 7-9
P-value of Testing
$H_0: \mu \leq 135$

This presents an interesting situation: When $\alpha = 0.05$, we fail to reject the null hypothesis, but when $\alpha = 0.10$, we do reject the null hypothesis. This raises the question: At which point (i.e., level of significance) does the conclusion change?

This is where the *P*-value comes into play. Looking at Figure 7-8, we see that since we have a right-tailed test, the *P*-value = $P(Z > 1.49) = 0.5 - 0.4319 = 0.0681$. Under the definition of a *P*-value, this is the probability of $\bar{x} > 142$ given that $\mu = 135$, $\sigma = 47$, and $n = 100$. However, there is another definition of a *P*-value, which enables us to answer the above question: A *P*-value is the largest level of significance in which we fail to reject the null hypothesis. Using this definition, the conclusion changes at a 6.81% level of significance.

When $\alpha = 0.05$, we fail to reject the null hypothesis. You may notice that the *P*-value of 0.0681 is greater than the level of significance. On the other hand, when $\alpha = 0.10$, we reject the null hypothesis. This time, the *P*-value is less than the level of significance. This presents an important relationship between the *P*-value and the level of significance:

- *Reject the null hypothesis* if the *P*-value is less than the significance level α.
- *Fail to reject the null hypothesis* if the *P*-value is greater than or equal to the significance level α.

EXAMPLE

Calculate the *P*-value for the left-tailed demographic example on page 377.

SOLUTION

In the left-tailed example, the null and alternative hypotheses are:

$$H_0: \mu \leq 2.4 \qquad H_1: \mu > 2.4$$

With $\alpha = 0.01$, we reject H_0 if the test statistic is less than -2.33. Since $z = -5.00$, we reject the null hypothesis. Using Figure 7-8, since this is a left-tailed test, the *P*-value = $P(Z < -5.00) \approx 0.0001 = 0.01\%$. Since the *P*-value is less than the 1% level of significance, we reject the null hypothesis.

Figure 7-10
***P*-value of Testing**
H_0: $\mu \geq 2.4$

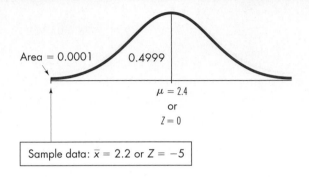

Area = 0.0001

0.4999

$\mu = 2.4$
or
$Z = 0$

Sample data: $\bar{x} = 2.2$ or $Z = -5$

EXAMPLE

Calculate the *P*-value for the two-tailed example of the box factory on page 378.

SOLUTION

In the two-tailed example, the null and alternative hypotheses are:

$$H_0: \mu = 24 \qquad H_1: \mu \neq 24$$

   With $\alpha = 0.10$, we reject the null hypothesis if the test statistic is either less than -1.645 or greater than 1.645. Since $z = 0.64$, we fail to reject the null hypothesis. Looking at Figure 7-8, since this is a two-tailed test and the test statistic is positive, the *P*-value = $2P(Z > 0.64)$. Now, $P(Z > 0.64) = 0.5 - 0.2389 = 0.2611$. Therefore, the *P*-value = $2(0.2611) = 0.5222 = 52.22\%$. Since the *P*-value is greater than the 10% significance level, we fail to reject the null hypothesis.

   Whereas the traditional approach results in a "reject/fail to reject" conclusion, *P*-values measure how confident we are in rejecting a null hypothesis. For example, a *P*-value of 0.0002 would lead us to reject the null hypothesis, but it would also suggest that the sample results are extremely unusual if the claimed

Figure 7-11
***P*-value of Testing**
H_0: $\mu = 24$

Because the test is two-tailed, the *P*-value is *twice* the shaded area.

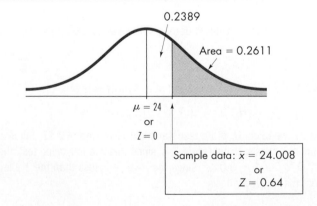

0.2389

Area = 0.2611

$\mu = 24$
or
$Z = 0$

Sample data: $\bar{x} = 24.008$
or
$Z = 0.64$

value of μ is in fact correct. In contrast, given a P-value of 0.40, we fail to reject the null hypothesis because the sample results can *easily* occur if the claimed value of μ is correct.

In the above examples, we compared the P-value to a level of significance. Some statisticians prefer to test a claim without setting a level of significance. If the conclusion is based on the P-value alone, the following guide may be helpful.

P-Value	Interpretation
Less than 0.01	*Unusual* sample results; highly statistically significant Very strong evidence against the null hypothesis
0.01 to 0.05	Statistically significant Adequate evidence against the null hypothesis
Greater than 0.05	Sample results are *not unusual*; insufficient evidence against the null hypothesis

Keep in mind that the above criteria are not set in stone. For example, some statisticians are of the opinion that the P-value should be greater than 0.10 before concluding that there is insufficient evidence against the null hypothesis.

Many statisticians consider it good practice always to select a significance level *before* doing a hypothesis test. This is a particularly good procedure when using P-values because we may be tempted to adjust the significance level based on the results. For example, with a 0.05 level of significance and a P-value of 0.06, we should fail to reject the null hypothesis, but it is sometimes tempting to say that a probability of 0.06 is small enough to warrant rejection of the null hypothesis. Other statisticians believe that prior selection of a significance level reduces the usefulness of P-values. They contend that no significance level should be specified and that the conclusion should be left to the reader. We will use the decision criterion that involves a comparison of a significance level and the P-value.

The next procedure for testing hypotheses is based on confidence intervals and therefore requires the concepts discussed in Section 6–2.

Testing Claims with Confidence Intervals

Confidence intervals can be used to identify results that are highly unlikely, so we can determine whether there is a *significant* difference between sample results and a claimed value of a parameter. For example, let us consider again the hypothesis-testing problem described at the beginning of the chapter. We want to test the claim that the mean body temperature of adults healthy enough to undergo elective surgery is equal to 37.0°C. Sample data consist of $n = 108$

temperatures with mean $\bar{x} = 36.39$ and standard deviation $s = 0.62$. Chapter 6 described methods of constructing confidence intervals. In one exercise, we used the body temperature sample data to construct the following 99% confidence interval:

$$36.24 < \mu < 36.54$$

We are 99% confident that the limits of 36.24 and 36.54 contain the population mean μ. (This means that if we were to repeat the experiment of collecting a sample of 108 body temperatures, 99% of the samples would result in confidence interval limits that actually do contain the value of the population mean μ.) This confidence interval suggests that it is very unlikely that the population mean is equal to 37.0. That is, on the basis of the confidence interval given here, we reject the common belief that the mean body temperature of healthy adults is 37.0°C. We can generalize this procedure as follows: First use the sample data to construct a confidence interval, and then apply the following decision criterion:

A $1 - \alpha$ confidence interval estimate of a population parameter contains the likely values of that parameter. We should therefore reject a claim that the population parameter has a value that is not included in the confidence interval at the level of significance of α.

Using this criterion, we note that the 99% confidence interval given here does not contain the claimed value of 37.0, and we therefore reject the claim that the population mean equals 37.0 at a 1% level of significance. (*Note*: We can make a direct correspondence between a confidence interval and a hypothesis test only when the test is two-tailed. A one-tailed hypothesis test with significance level α corresponds to a confidence interval with degree of confidence $1 - 2\alpha$. For example, a right-tailed hypothesis test with a 0.05 significance level corresponds to a 90% confidence interval.)

EXAMPLE

Construct a 90% confidence interval for the two-tailed example of the box factory. Show why we reach the same conclusion of failing to reject the null hypothesis.

SOLUTION

Note that a 90% level of confidence corresponds to a 10% level of significance, giving a critical value of 1.645. With $\sigma = 0.0968$ and $n = 60$, we first find the margin of error:

$$E = \frac{z_{\alpha/2} \cdot \sigma}{\sqrt{n}} = \frac{1.645 \cdot 0.0968}{\sqrt{60}} = 0.0206$$

With $\bar{x} = 24.008$, we can now construct the 90% confidence interval:

$$\bar{x} - E < \mu < \bar{x} + E$$
$$24.008 - 0.0206 < \mu < 24.008 + 0.0206$$
$$23.9874 < \mu < 24.0286$$

Interpretation

With 90% confidence, the average box length using the new machine is between 23.9874 cm and 24.0286.

Recall the null and alternative hypotheses:

$$H_0: \mu = 24 \qquad H_1: \mu \neq 24$$

Since $\mu = 24$ falls inside the 90% confidence interval, we fail to reject H_0 at a 10% level of significance.

In the remainder of the text, we will apply methods of hypothesis testing to other circumstances, such as those involving claims about proportions or standard deviations or involving more than one population. It is easy to become entangled in a complex web of steps without ever understanding the underlying rationale of hypothesis testing. The key to that understanding lies in the rare event rule for inferential statistics: **If, under a given assumption, the probability of getting the observed sample is exceptionally small, we conclude that the assumption is probably not correct.** When testing a claim, we make an assumption (null hypothesis) that contains equality. We then compare the assumption and the sample results and form one of the following conclusions:

- If the sample results can easily occur when the assumption is true, we attribute any small discrepancy between the assumption and the sample results to chance.
- If the sample results cannot easily occur when the assumption is true, we explain the discrepancy between the assumption and the sample results by concluding that the assumption is not true.

EXCEL: Excel does not include a built-in tool for the one- or two-tailed tests described in this section. Instead, use the Data Desk XL add-in that is a supplement to this book. First enter the sample data into an Excel column. (In the text's data set of body temperatures, the data are found in the range B4:B111.)

EXCEL (Prior to 2007): Select **DDXL** on the main menu bar, then select **Hypothesis Tests**. Under the function type options, select **1 Var z Test**. Click on the pencil icon and enter the range of data values, such as B4:B111. Click on **OK**. Follow the four steps listed in the dialog box. After clicking on **Compute** in Step 4, you will get the *P*-value, test statistic, and conclusion.

EXCEL 2007: Click **Add-Ins** on the main menu and choose **DDXL** from the add-in menu to the left. Proceed as above.

MINITAB 15: From the **Stat** menu, choose **Basic Statistics** and then **1-Sample Z**. If you are testing the hypothesis using summarized data, click that radio button and fill in the necessary boxes such as sample size, mean, and standard deviation. Check the optional box to conduct a hypothesis test and enter the value of the hypothesized mean. The default setting is a two-tailed test at a 5% level of significance. If you wish to change either, click the **Options** button and make the necessary changes.

STATDISK: If working with a list of the original sample values, first find the sample size, sample mean, and sample standard deviation by using the STATDISK procedure described in Section 2–4. After finding the values of n, $\bar{x}$, and s, proceed to select the main menu bar item **Analysis**, then select **Hypothesis Testing**, followed by **Mean-One Sample**. Enter the data in the dialog box. To conduct a Z test, enter the standard deviation in the **Population St. Dev.** box. See the accompanying STATDISK display showing the results from the body temperature example of this section.

EXCEL DISPLAY
Testing a Claim About a
Mean (Large Sample)

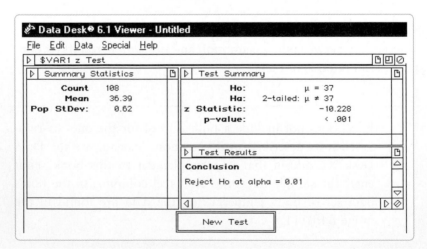

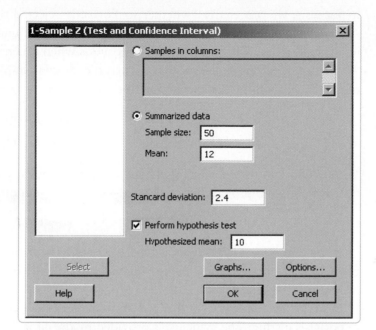

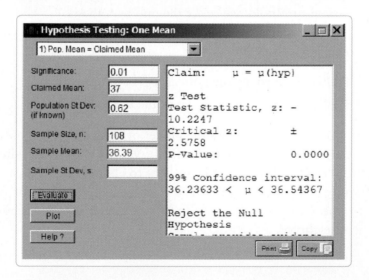

Power of the Test

In a court of law, after the jury has reached a verdict, we can examine the evidence to analyze whether the jury has reached the correct verdict. If the jury found the defendant guilty, we would like to think that the defendant actually is guilty.

In hypothesis testing, we can conduct a similar analysis. If we reject the null hypothesis, we can analyze whether we correctly rejected the null hypothesis.

Recall that there are two types of errors that can be made in hypothesis testing: type I, where we reject the null hypothesis when it is true, and type II, where we do not reject the null hypothesis when the alternative hypothesis is true. Also recall the probabilities associated with these errors:

$$\alpha = P(\text{Type I error}) = P(\text{reject } H_0 \,|\, H_0 \text{ true})$$
$$\beta = P(\text{Type II error}) = P(\text{not reject } H_0 \,|\, H_1 \text{ true})$$

As we saw in Section 7–2, there is a third probability, called the power of the test, which is the probability of correctly rejecting the null hypothesis.

$$\text{Power} = P(\text{reject } H_0 \,|\, H_1 \text{ true})$$

If the alternative hypothesis is true, either we reject the null hypothesis (with probability β) or we don't. Based on this, we can deduce the relationship stated in Section 7–2: Power $= 1 - \beta$.

In a court of law, the verdict is based on the evidence. The same is true for hypothesis testing: The evidence is provided by the data. In the test for one mean, the evidence is provided specifically through the sample mean $\bar{x}$. If $\bar{x}$ is beyond a certain threshold, we reject the null hypothesis.

Because of this criterion, computing the power of the test is essentially a two-step procedure:

STEP 1: Determine the critical region in terms of $\bar{x}$.

STEP 2: Compute the power of the test.

In order to compute the power of the test for one mean, we need an alternative mean (which we will call μ_1). We can choose any alternative mean that falls under H_1. However, it is prudent to choose one that can be considered reasonably realistic. Recall that $\bar{x}$ is an unbiased estimator of μ, meaning that we expect $\bar{x}$ to be equal to μ in theory. For this reason, choosing $\mu_1 = \bar{x}$ may be a wise choice.

EXAMPLE

In the textbook example from page 376, the null and alternative hypotheses are:

$$H_0: \mu \leq 135 \qquad H_1: \mu > 135$$

As well, $n = 100$, $\bar{x} = 142$, and $\sigma = 47$. If we choose $\alpha = 0.10$, the critical region is $Z > 1.28$; since $Z = 1.49$ falls inside the critical region, we reject the null hypothesis.

Note that $\bar{x} = 142$ falls under the alternative hypothesis. For this reason, we choose $\mu_1 = 142$. We can now ask: What is the power of the test with $\alpha = 0.10$, $n = 100$, and $\sigma = 47$ if the actual textbook price is $142?

SOLUTION

The first step is to set up the critical region in terms of $\bar{x}$ with $\alpha = 0.10$.

Reject H_0 if $\quad\quad\quad\quad\quad\quad Z > 1.28$

Since $\quad\quad\quad\quad\quad\quad z = \dfrac{\bar{x} - \mu_0}{\dfrac{\sigma}{\sqrt{n}}}$

Reject H_0 if $\quad\quad\quad\quad\dfrac{\bar{x} - 135}{\dfrac{47}{\sqrt{100}}} > 1.28$

Solving for $\bar{x}$, we get:

Reject H_0 if $\quad\quad\quad\quad \bar{x} > 135 + 1.28 \cdot \dfrac{47}{\sqrt{100}}$

Reject H_0 if $\quad\quad\quad\quad \bar{x} > 141.016$

In other words, when we collect the data from our random sample of 100 titles, if the sample mean is more than $141.016, we reject the null hypothesis and conclude we have sufficient evidence to conclude that the average textbook price is significantly greater than $135.

The power of the test is asking this question: If the actual textbook price is $142, how likely is it that $\bar{x} > 141.016$ based on a random sample of 100 titles and a standard deviation of $47?

$$\text{Power} = P(\text{reject } H_0 | H_1 \text{ true})$$
$$\text{Power} = P(\bar{x} > 141.016 | \mu_1 = 142)$$

Invoking the central limit theorem:

$$\text{Power} = P\left(Z > \dfrac{141.016 - 142}{\dfrac{47}{\sqrt{100}}} \right)$$

$$\text{Power} = P(Z > -0.21) = 0.5 + 0.0832 = 0.5832$$

Interpretation
If the actual textbook price is $142, given a 10% level of significance, sample size of 100, and standard deviation of $47, there is a 58.32% probability of correctly concluding that the average textbook price is significantly greater than $135.

Factors Affecting the Power of the Test

In the previous example, the power of the test at 58.32% is not very high. There are some measures that a researcher can take in order to increase the power.

- Increase the sample size. If your research budget allows you to increase the sample size, this will enable you to have more evidence that, if your alternative hypothesis is true, will support it by both the P-value and the power.

- Choose an alternative mean that is further in the critical region. Care should be taken here. In theory, you could choose any mean that falls under the alternative hypothesis. In the previous example, we could have chosen $1000 as the alternative price of a textbook; however, this is not a realistic expectation for a textbook price.

- Increase the level of significance. This may seem counterintuitive. The reason this works is that type I and type II errors are like pans on a scale; if you increase the probability of one, you decrease the probability of the other. If we increase $\alpha = P(\text{type I error})$, we decrease $\beta = P(\text{type II error})$ and thereby increase power $= 1 - P(\text{type II error})$.

EXAMPLE

In the textbook example, what would be the power of the test if the average textbook price is actually $150?

SOLUTION

We know already that we reject the null hypothesis if $\bar{x} > 141.016$.

$$\text{Power} = P(\bar{x} > 141.016 \,|\, \mu_1 = 150)$$

$$\text{Power} = P\left(Z > \frac{141.016 - 150}{\dfrac{47}{\sqrt{100}}} \right)$$

$$\text{Power} = P(Z > -1.91) = 0.5 + 0.4719 = 0.9719$$

Note that shifting the alternative mean from $142 to $150 increases the power of the test from 58.32% to 97.19%. This is illustrated in Figure 7-12.

EXAMPLE

In the demographic example on page 377 the null and alternative hypotheses are:

$$H_0: \mu \leq 2.4 \qquad H_1: \mu > 2.4$$

We have $n = 400$, $\bar{x} = 2.2$, $\sigma = 0.8$, and $\alpha = 0.01$. Since the test statistic of $z = -5$ falls inside the rejection region, we reject the null hypothesis. What is the power of the test if the average number of children per household is 2.2, given the above information?

SOLUTION

Reject H_0 if $Z < -2.33$. As with the previous example, we set up the critical region in terms of $\bar{x}$.

Reject H_0 if
$$\frac{\bar{x} - 2.4}{\dfrac{0.8}{\sqrt{400}}} < -2.33$$

Reject H_0 if
$$\bar{x} < 2.4 - 2.33 \cdot \frac{0.8}{\sqrt{400}}$$

Reject H_0 if
$$\bar{x} < 2.3068$$

When we collect the data from the 400 random samples, if the average of this sample is less then 2.3068, we reject the null hypothesis. We can now compute the power of the test.

$$\text{Power} = P(reject\ H_0 \,|\, H_1\ true)$$

$$\text{Power} = P(\bar{x} < 2.3068 \,|\, \mu_1 = 2.2)$$

$$\text{Power} = P\left(Z < \frac{2.3068 - 2.2}{\dfrac{0.8}{\sqrt{400}}} \right)$$

$$\text{Power} = P(Z < 2.67) = 0.5 + 0.4962 = 0.9962$$

Interpretation
If the actual average number of children is 2.2, given $\alpha = 0.01$, $\sigma = 0.8$, and $n = 400$, there is a 99.62% probability of correctly rejecting the null hypothesis.

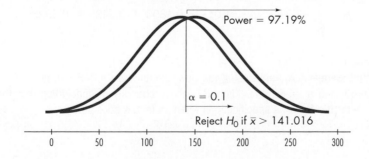

Figure 7-12
Power of the Test with
$\mu_1 = 150$

You may note that even though the level of significance in the above example is low at 1%, the power of the test is extremely high; this is due to the large sample size of 400.

In the box factory example on page 378, the null and alternative hypotheses are:

$$H_0: \mu = 24 \qquad H_1: \mu \neq 24$$

We have $\alpha = 0.1$, $n = 60$, $\bar{x} = 24.008$, and $\sigma = 0.0968$. We reject the null hypothesis if the test statistic is either less than -1.645 or greater than 1.645. Since $Z = 0.64$, we fail to reject the null hypothesis.

Since we do not reject the null hypothesis, there is the possibility that a type II error was made. Rather than computing the power in this case, we will compute β. For the alternative mean, we choose $\mu_1 = 24.008$.

SOLUTION

Since $\beta = P(\text{do not reject } H_0 \mid H_1 \text{ true})$, we set the acceptance region in terms of $\bar{x}$.

Do not reject H_0 if $\qquad\qquad -1.645 < Z < 1.645$

Do not reject H_0 if $\qquad\qquad -1.645 < \dfrac{\bar{x} - 24}{\dfrac{0.0968}{\sqrt{60}}} < 1.645$

Do not reject H_0 if $\quad 24 - 1.645\dfrac{0.0968}{\sqrt{60}} < \bar{x} < 24 + 1.645\dfrac{0.0968}{\sqrt{60}}$

Do not reject H_0 if $\qquad\qquad 23.9794 < \bar{x} < 24.0206$

Now that we know the acceptance region in terms of $\bar{x}$, we can calculate the probability of a type II error.

$$\beta = P(\text{do not reject } H_0 \mid H_1 \text{ true})$$
$$\beta = P(23.9794 < \bar{x} < 24.0206 \mid \mu_1 = 24.008)$$

$$\beta = P\left(\dfrac{23.9794 - 24.008}{\dfrac{0.0968}{\sqrt{60}}} < Z < \dfrac{24.0206 - 24.008}{\dfrac{0.0968}{\sqrt{60}}} \right)$$

$$\beta = P(-2.29 < Z < 1.01) = 0.4890 + 0.3438 = 0.8328$$

Interpretation

If the actual average box length using the new machine is 24.008 cm, given $\alpha = 0.1$, $\sigma = 0.0968$, and $n = 60$, there is a 83.28% probability of concluding that the new machine meets the factory's specifications when, in fact, it does not.

7-3 Exercises A: Basic Skills and Concepts

In Exercises 1–24, test the given claim using the traditional method of hypothesis testing. Clearly identify the following components of the hypothesis test: (a) test statistic, (b) critical value(s), (c) P-value, (d) final conclusion. Check that the P-value method would lead to the same solution as the traditional method. Assume that all samples have been randomly selected.

1. Test the claim that the population mean $\mu = 75$, given a sample of $n = 100$ for which $\bar{x} = 78$ and $s = 15$. Test at the $\alpha = 0.05$ significance level.

2. Test the claim that $\mu > 750$, given a sample of $n = 36$ for which $\bar{x} = 800$ and $s = 100$. Use a significance level of $\alpha = 0.01$.

3. Test the claim that $\mu < 2.50$, given a sample of $n = 64$ for which $\bar{x} = 2.45$ and $s = 0.80$. Use a significance level of $\alpha = 0.02$.

4. Test the claim that a population mean is different from 32.0, given a sample of $n = 75$ for which $\bar{x} = 31.8$ and $s = 0.85$. Use a significance level of $\alpha = 0.01$.

5. In April 1994, during a spring of unusually good hitting in professional baseball, a columnist for the *Toronto Star* hinted at a "conspiracy" by managers in which they "juiced" the baseballs to increase excitement in the game. Suppose that tests of the old balls showed that when dropped 24 ft onto a concrete surface, they bounced an average of 92.14 in. In a test of a sample of 40 new balls, the bounce heights had a mean of 92.67 in. and a standard deviation of 1.79 in. Use a 0.05 significance level to test the claim that the new balls have bounce heights different from 92.14 in. Are these balls "juiced"?

6. The Mackenzie Valley Bottling Company distributes root beer in bottles labelled 32 oz. An inspector randomly selects 50 of these bottles, measures their contents, and obtains a sample mean of 31.8 oz and a sample standard deviation of 0.75 oz. Using a 0.01 significance level, test the inspector's claim that the company is cheating consumers. Should charges be filed?

7. In a study of consumer habits, researchers designed a questionnaire to identify compulsive buyers. For a sample of consumers who identified themselves as compulsive buyers, questionnaire scores have a mean of 0.83 and a standard deviation of 0.24 (based on data from "A Clinical Screener for Compulsive Buying," by Faber and Guinn, *Journal of Consumer Research*, Vol. 19). Assume that the subjects were randomly selected and that the sample size was 32. At the 0.01 level of significance, test the claim that the self-identified compulsive-buyer population has a mean greater than 0.21, the mean for the general population. Does the questionnaire seem to be effective in identifying compulsive buyers?

8. Planners in Ontario's Durham region compiled data for the value of building permits. The mean for all building permits (in thousands of dollars) was 97. But for a random sample of 38 permits in the smaller Durham region communities of Scugog, Brock, Clarington, and Uxbridge, the mean value was 113.6, with a standard deviation of 70.67 (based on data in the *Durham Business News*, June 1998). Using the 0.05 level of significance, test the hypothesis that these smaller communities have a mean value for building permits that is different from the regional mean of 97.

9. Your investment goal is to choose only from mutual fund groups with at least 10% foreign equity. To determine whether the Canadian equity funds group meets your criterion, you randomly sample 33 of the funds in that category. The mean foreign equity percentage in your sample is 8.386, with a standard deviation of 7.126 (based on data from *Mutual Funds Update*, Bank of Montreal, March 31, 1998). Do the Canadian equity funds fail to meet your criterion? What could be the consequence if your conclusion is a type I or type II error?

10. An ambitious young worker claims that older employees have a better chance of being CEOs than younger employees. To confirm this, he sampled the ages of CEOs for 31 companies. The mean age in the sample was found to be 41.77, with a standard deviation of 10.24 (based on data from *Profit*, May 29, 1996). If the mean age for all workers in the population is 37, and if the significance level 0.01 is used, test the hypothesis that the CEOs have a higher mean age than workers in general. Explain how your answer to this question relates to the young worker's original claim—that older employees have a *better chance* of being CEO.

11. The yearly rates of traffic accidents and violations were studied for a group of 137 young adult males who volunteered for treatment for substance abuse. For the five years before receiving treatment, their mean accident rate was 0.123 with a standard deviation of 0.167 (based on data from "Does Treatment for Substance Abuse Improve Driving Safety? A Preliminary Evaluation," by Mann et al., Addiction Research Foundation). At the 0.01 significance level, test the claim that the sample was drawn from the general driving population with a mean accident rate of 0.075. Do the results suggest that the Ministry of Transportation should take an interest in substance-abuse programs?

12. In a study of distances travelled by buses before the first major engine failure, a sampling of 191 buses resulted in a mean of 154,700 km and a standard deviation of 60,000 km (based on data in *Technometrics*, Vol. 22, No. 4). At the 0.05 level of significance, test the manufacturer's claim that mean distance travelled before a major engine failure is more than 144,000 km.

13. A poll of 100 randomly selected car owners revealed that the mean length of time that they plan to keep their cars is 7.01 years and the standard deviation is 3.74 years (based on data from a Roper poll). The president of the Kamloops Car Park is trying to plan a sales campaign targeted at car owners who are ready to buy a different car. Test the claim of the sales manager, who authoritatively states that the mean length of time all car owners plan to keep their cars is less than 7.5 years. Use a 0.05 significance level.

14. When 200 convicted embezzlers were randomly selected, the mean length of prison sentence was found to be 22.1 months and the standard deviation was found to be 8.6 months. Kim Patterson is running for political office on a platform of tougher treatment of convicted criminals. Test her claim that prison terms for convicted embezzlers have a mean of less than 2 years. Use a 0.05 significance level.

15. A nighttime cold medicine bears a label indicating the presence of 600 mg of acetaminophen in each fluid oz of the drug. Suppose that Health Canada randomly selected 65 one-oz samples and found that the mean acetaminophen content was 589 mg, and the standard deviation was 21 mg. Using $\alpha = 0.01$, test the claim of the pharmaceutical company that the population mean is equal to 600 mg. Would you buy this cold medicine?

16. An insurance company is reviewing the driving habits of women aged 16–24 to determine whether they should pay higher premiums than women in a higher age bracket. In a study of 750 randomly selected women drivers aged 16–24, the mean driving distance for one year is 9750 km and the standard deviation is 4750 km. Use a 0.01 significance level to test the claim that the population mean for women in the 16–24 age bracket is less than 11,490 km, which is the known mean for women in the higher age bracket. If women in the 16–24 age bracket drive less, should they be charged lower insurance premiums?

17. A fish-processing company is concerned about the shelf life of its new cat food. A quality manager claims that the mean product shelf life of cans from the Dartmouth facility is less than the advertised mean (of 12 months) for its product's shelf life. Test that claim using the 0.005 level of significance. A sample of 40 items produced in Dartmouth had a mean shelf life of 11 months, and a standard deviation of 1.52 months.

18. The true value of one type of degree or diploma cannot be quantitatively measured, but we can measure its relative impact on starting salary. Graduates from Quebec universities with a B.A. or a B.Sc. degree have a mean annual starting salary of $28,300. Sixty-five Quebec graduates with a civil engineering degree are randomly selected. Their starting salaries have a mean of $36,300 (based

on "Annual Salary Survey," Peat Marwick Stevenson & Kellogg). If the standard deviation is $1670, use a 0.01 level of significance to test the claim that Quebec graduates with a civil engineering degree have a mean starting salary that is greater than the mean for graduates with a B.A. or B.Sc. degree from Quebec. Is the sample size large enough to reach a conclusion?

19. The mean time between failures (in hours) for a Telektronic Company radio used in light aircraft is 420 h. After 35 new radios were modified in an attempt to improve reliability, tests showed that the mean time between failures for this sample is 385 h and the standard deviation is 24 h. Use a 0.05 significance level to test the claim that the modifications improved reliability. (Note that improved reliability should result in a *longer* mean time between failures.)

20. An investment advisor is trying to persuade an investor to switch from her current fund holdings, claiming that the advisor's portfolio recommendations have had a greater one-year return rate than the investor's current funds. Last year, the investor's annual return rate was 35%. Use a 0.01 level of significance to test the claim that the investments that the advisor recommends have a greater one-year return rate than 35%. Base your calculations on the sample of 40 return rates in the one-year return column of Data Set 10 in Appendix B. Is there sufficient evidence to support the advisor's claim? A real investor is not likely to invest in *all* 40 funds in Data Set 10. What additional statistic would help to portray the *risk* of switching to just a portion of the advisor's portfolio?

21. Refer to Data Set 1 in Appendix B for the *total* weights of garbage discarded by households in one week. For that data set, the mean is 27.44 lb and the standard deviation is 12.46 lb. At the 0.01 level of significance, test the claim that the mean weight of all garbage discarded by households each week is less than 35 lb, the amount that can be handled by the town. Based on the result, is there any cause for concern that there might be too much garbage to handle?

22. If we refer to the weights (in grams) of quarters listed in Data Set 13 in Appendix B, we find 50 weights with a mean of 5.622 g and a standard deviation of 0.068 g. Use a 0.01 significance level to test the claim that the mean weight of quarters in circulation is 5.670 g. If the claim is rejected, what is a possible explanation for the discrepancy?

23. Analysis of the last digits in data sometimes reveals whether the data have been accurately measured and reported. If the last digits are uniformly distributed from 0 to 9, then their mean should be 4.5. The last digits in the published numbers of sitting days for federal Parliaments are used to test the claim they come from a population with a mean of 4.5. (The raw data are in Data Set 5 in Appendix B.) If Excel is used to analyze the results, a table similar to the one shown here could be generated. Using a 0.05 significance level, interpret these results.

Mean (Sample)	4.8286
Mean (Null Hyp.)	4.5000
StDev	2.7277
N	35
Std Error	0.4611
Test Statistic z	0.7135
Critical z	1.9600
P-value	0.4755

24. A package of M&M plain candies is labelled as containing 1361 g, and there are 1498 candies, so the mean weight of the individual candies should be 1361/1498, or 0.9085 g. In a test to determine whether consumers are being cheated, a sample of 33 brown M&Ms is randomly selected. (See Data Set 11 in Appendix B.) If Excel is used to analyze the results, a table similar to the one shown here could be generated. Using a 0.05 significance level, for a one-tailed test, interpret these results.

Mean (Sample)	0.91282
Mean (Null Hyp.)	0.90850
StDev	0.03952
N	33
Std Error	0.00688
Test Statistic z	0.62790
Critical z	-1.6450
P-value	0.7350

7-3 Exercises B: Beyond the Basics

25. A journal article reported that a null hypothesis of $\mu = 100$ was rejected because the P-value was less than 0.01. The sample size was given as 62, and the sample mean was given as 103.6. Find the largest possible standard deviation.

26. In Exercise 17, find the *largest* sample mean below 12 months that will support the claim that the mean is less than 12 months. (Use the same sample size and sample standard deviation.)

27. Suppose you are given these hypotheses: H_0: $\mu \leq 25$, H_1: $\mu > 25$; and that the standard deviation from a sample of 100 is 2.4. What is the power of the test at a 5% level of significance if the alternative mean is 25.2?

28. Which would increase the power in Exercise 27 by more: increasing the alternative mean to 25.5 and keeping $\alpha = 0.05$ or keeping the alternative mean at 25.2 and increasing α to 0.1?

29. For the power calculation in Exercise 27, what would the alternative mean need to be at a 5% level of significance if we want the power to be at least 95%?

30. In a certain accounting office, workers have been processing 60 tax returns per day on average with a standard deviation of 10.2. The office introduced a new protocol in order to see if there would be any significant change in the average number of returns processed per day. For a random sample of 200 worker-days, the average was 58. What is the probability of a type II error if the actual average number of returns processed per day is 58 based on a sample of 200?

31. One of the characteristics of the power of the test is that, for the same alternative mean, the power is greater for the appropriate one-tailed test than for the two-tailed test. Suppose we have the same situation as in Exercise 30 but the alternative hypothesis is H_1: $\mu < 60$. Compared to the power that we can calculate from Exercise 30, by how much does the power increase?

32. In some disciplines, both the level of significance and the power are taken into consideration in calculating the sample size. Suppose we have the hypotheses: $H_0: \mu \leq 50$, $H_1: \mu > 50$ and that based on prior information, the standard deviation is 8.5. If we test at a 5% level of significance and we want a minimum power of 99%, what would be the minimum sample size based on an alternative mean of 52?

7-4 Testing a Claim About a Mean: Small Samples

One great advantage of learning the methods of hypothesis testing that we have described in Sections 7–2 and 7–3 is that those same methods can be easily modified for use in many other circumstances. The previous sections' examples and exercises all use large samples in tests of claims about means, but in this section we will deal with small sample cases—that is, cases where the number of sample values is 30 or fewer. The following assumptions apply to the methods described in this section.

Assumptions

1. The data have been collected by valid random sampling.
2. The sample is small ($n < 30$).
3. The value of the population standard deviation σ is unknown. (We seldom know the population standard deviation when we do not know the population mean.)
4. The sample values come from a population with a distribution that is approximately normal.

If these four assumptions are satisfied and we plan to test a hypothesis, then we use the Student t (not the normal) distribution with the test statistic and critical values described as follows.

TEST STATISTIC FOR CLAIMS ABOUT μ WHEN $n \leq 30$ AND σ IS UNKNOWN

If a population is essentially normal, then the distribution of

$$t = \frac{\bar{x} - \mu_{\bar{x}}}{\dfrac{s}{\sqrt{n}}}$$

is essentially a *Student t distribution* for all samples of size *n*. (The Student *t* distribution is often referred to as the *t distribution*.)

Critical Values

1. Critical values are found in Table A-3.
2. Degrees of freedom $= n - 1$.
3. After finding the number of degrees of freedom, refer to Table A-3 and locate that number in the column at the left. With a particular row of t values now identified, select the critical t value that corresponds to the appropriate column heading. If a critical t value is located at the left tail, be sure to make it negative.

Important Properties of the Student t Distribution

1. The Student t distribution is different for different sample sizes (see Figure 6-5 in Section 6–3).
2. The Student t distribution has the same general bell shape as the standard normal distribution; its wider shape reflects the greater variability that is expected with small samples.
3. The Student t distribution has a mean of $t = 0$ (just as the standard normal distribution has a mean of $z = 0$).
4. The standard deviation of the Student t distribution varies with the sample size and is greater than 1 (unlike the standard normal distribution, which has $\sigma = 1$).
5. As the sample size n gets larger, the Student t distribution gets closer to the standard normal distribution. For values of $n > 30$, the differences are so small that we can use the critical z values instead of developing a much larger table of critical t values. (The values in the bottom row of Table A-3 are equal to the corresponding critical z values from the standard normal distribution.)

Choosing the Appropriate Distribution

When testing claims about population means, sometimes the normal distribution applies, sometimes the Student t distribution applies, and sometimes neither applies, so we must use nonparametric methods, which do not require a particular distribution. (Nonparametric methods are discussed in Chapter 13E.)

Figure 7-13 summarizes the decisions to be made in choosing between the normal and Student t distributions. Studying the figure leads to the following observations.

1. According to the central limit theorem, if we obtain large ($n > 30$) samples (from any population with any distribution), the distribution of the sample means can be approximated by the normal distribution.

Figure 7-13

Choosing Between the Normal and Student t Distributions When Testing a Claim About the Population Mean μ

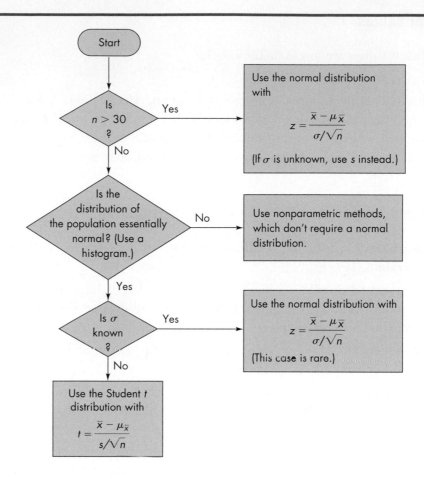

2. When we obtain samples (of any size) from a population with a normal distribution, the distribution of the sample means will be approximately normal with mean μ and standard deviation $\sigma/\sqrt{n}$. In a hypothesis test, the value of μ corresponds to the null hypothesis, and the value of the population standard deviation σ must be known. If σ is unknown and the samples are large, we can use the sample standard deviation s as a substitute for σ because large random samples tend to be representative of the populations from which they come.

3. The conditions for using the Student t distribution are as follows:
   a. The sample is small ($n \leq 30$);
   b. σ is unknown; and
   c. The parent population has a distribution that is essentially normal.

4. If our random samples are small, σ is unknown, and the population distribution is grossly nonnormal, we cannot use the methods of this chapter.

Instead, we can use nonparametric methods, some of which are discussed in Chapter 13E.

The following example involves a small sample drawn from a normally distributed population for which the standard deviation σ is not known. These conditions require that we use the Student t distribution.

EXAMPLE

The seven values listed here are axial loads (in pounds) for the first sample of seven 12-oz aluminum cans (see Data Set 15 in Appendix B). An axial load of a can is the maximum weight supported by its sides, and it must be greater than 165 lb, because that is the maximum pressure applied when the top lid is pressed into place. At the 0.01 level of significance, test the claim of the engineering supervisor that this sample comes from a population with a mean that is greater than 165 lb.

270   273   258   204   254   228   282

SOLUTION

Using the given sample data, we apply procedures of Chapter 2 to find that $n = 7, \bar{x} = 252.7$, and $s = 27.6$. The mean of 252.7 does seem to be well above the required value of 165, but with only seven scores, do we really have enough evidence to support the supervisor's claim? Let's find out by conducting a formal hypothesis test with the same steps outlined in Figure 7-4.

Steps 1, 2, and 3 result in the following null and alternative hypotheses:

$$H_0: \mu \leq 165$$
$$H_1: \mu > 165 \quad \text{(supervisor's claim)}$$

Step 4: The significance level is $\alpha = 0.01$.

Step 5: In this test of a claim about the population mean, the most relevant statistic is the sample mean. Referring to Figure 7-13, we see that the sample is small (because $n = 7$ and does not exceed 30), it's reasonable to conclude that the population distribution is normal (because we're dealing with physical measurements of a product made under standard conditions), and σ is unknown. Figure 7-13 shows that we should use the Student t distribution (not the normal distribution).

Step 6: The test statistic is

$$t = \frac{\bar{x} - \mu_{\bar{x}}}{\frac{s}{\sqrt{n}}} = \frac{252.7 - 165}{\frac{27.6}{\sqrt{7}}} = 8.407$$

The critical value of $t = 3.143$ is found by referring to Table A-3. First locate $n - 1 = 6$ degrees of freedom in the column at the left. Because this test is right-tailed with $\alpha = 0.01$, refer to the column with the heading of 0.01 (one tail). The critical value of $t = 3.143$

Figure 7-14
t-Test of Claim
That $\mu > 165$

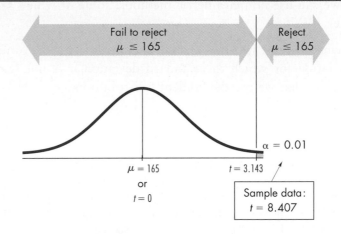

corresponds to 6 degrees of freedom and $\alpha = 0.01$ in one tail. The test statistic and critical value are shown in Figure 7-14.

Step 7: Because the test statistic of $t = 8.407$ falls in the critical region, we reject H_0.

Interpretation
(Refer to Figure 7-2 for help in wording the final conclusion.) There is sufficient evidence to support the supervisor's claim that the sample comes from a population with a mean greater than the required 165 lb.

The critical value in the preceding example was $t = 3.143$, but if the sample had been large ($n > 30$), the critical value would have been $z = 2.33$. The larger Student t critical value shows that with a small sample, the sample evidence must be *more extreme* before we consider the difference to be significant.

P-Values

The preceding example followed the traditional approach to hypothesis testing, but Excel, STATDISK, and much of the literature display *P*-values. Because the t distribution table (Table A-3) includes only selected values of the significance level α, we cannot usually find the specific *P*-value from Table A-3. Instead, we can use that table to identify limits that contain the *P*-value. In the last example we found the test statistic to be $t = 8.407$, and we know that the test is one-tailed with 6 degrees of freedom. By examining the row of Table A-3 corresponding to 6 degrees of freedom, we see that the test statistic of 8.407

exceeds the largest critical value in that row. Although we cannot pinpoint an exact P-value from Table A-3, we can conclude that it must be less than 0.005. That is, we conclude that P-value < 0.005. With a significance level of 0.01 and a P-value less than 0.005, we reject the null hypothesis (because the P-value is less than the significance level), as we did with the traditional method in the preceding example.

EXAMPLE

Use Table A-3 to find the P-value corresponding to these results: The Student t distribution is used in a two-tailed test with a sample of $n = 10$ scores, and the test statistic is found to be $t = 2.567$.

SOLUTION

Refer to the row of Table A-3 with 9 degrees of freedom and note that the test statistic of 2.567 falls between the critical values of 2.821 and 2.262. Because the test is two-tailed, we consider the values of α at the top that are identified with "two tails." The critical values of 2.821 and 2.262 correspond to 0.02 (two tails) and 0.05 (two tails), so we express the P-value as follows.

$$0.02 < P\text{-value} < 0.05$$

With a true null hypothesis, the chance of getting a sample mean (from 10 sample values) that converts to a test statistic of $t = 2.567$ is somewhere between 0.02 and 0.05.

So far, we have discussed tests of hypotheses made about population means only. In the next section we will test hypotheses made about population proportions or percentages, and the last section will consider claims made about standard deviations or variances.

USING TECHNOLOGY

When conducting hypothesis tests using the Student t distribution, the procedures for using Excel and STATDISK are essentially the same as described in Section 7–3, except as noted here.

EXCEL: When choosing the function type option, select **1 Var t Test** instead of **1 Var z Test**.

MINITAB 15: From the **Stat** menu, choose **Basic Statistics** and then **1-Sample t**. If you are testing the hypothesis using raw data, click the **Samples in columns** radio button and select the column that has the data. Check the optional box to conduct a hypothesis test and enter the value of the hypothesized mean. The default setting is a two-tailed test at a 5% level of significance. If you wish to change either, click the **Options** button and make the necessary changes.

STATDISK: STATDISK proceeds exactly as in Section 7–3. To conduct a *t* test, enter the standard deviation in the **Sample St. Dev.** box.

Note that for the example of the seven axial loads, STATDISK estimates the *P*-value = 0.0001; Excel concludes the *P*-value < 0.001.

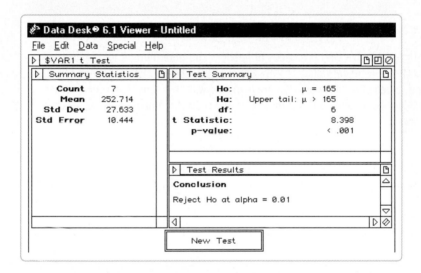

7-4 Exercises A: Basic Skills and Concepts

In Exercises 1 and 2, find the critical t values for the given hypotheses, sample sizes, and significance levels.

1. a. $H_0: \mu = 37.0$ b. $H_0: \mu \geq 100$ c. $H_1: \mu > 32$
      $n = 7$ $n = 12$ $n = 9$
      $\alpha = 0.05$ $\alpha = 0.05$ $\alpha = 0.01$

2. a. $H_0: \mu \leq 1.07$ b. $H_1: \mu > 75.2$ c. $H_1: \mu \neq 64$
      $n = 5$ $n = 14$ $n = 10$
      $\alpha = 0.01$ $\alpha = 0.05$ $\alpha = 0.10$

In Exercises 3 and 4, assume that the sample is randomly selected from a population with a normal distribution. Test the given claim by using the traditional method of testing hypotheses, then re-test using the P-value method.

3. Use a significance level of $\alpha = 0.05$ to test the claim that $\mu \neq 64.8$. The sample data consist of 12 scores for which $\bar{x} = 59.8$ and $s = 8.7$.

4. Use a significance level of $\alpha = 0.01$ to test the claim that $\mu < 927$. The sample data consist of 10 scores for which $\bar{x} = 874$ and $s = 57.3$.

5. A student used just one trade publication to price desktops that have at least a 20 GB hard drive and a 700 MHz processor speed. To justify relying on the one source, he took a sample of prices for such systems, as quoted in dealers' ads in the publication. Assume the industry mean price for such a system is $1750. Use a 0.01 significance level to test the claim that the mean quoted price in the sample is no different from the industry average of $1750. Interpret the results from this table.

Mean (sample)	2062.4
Mean (null hypothesis)	1750
Standard deviation	663.58
n	10
Test statistic t	1.489
Critical t	3.250
P-value	0.171

6. Refer to the display in Exercise 5. If the claim is changed from "is no different from the industry average" to "is greater than the industry average," how are the values in the right column of the table affected?

For hypothesis tests in Exercises 7–24, use the traditional approach summarized in Figure 7-4. Draw a graph showing the test statistic and critical values. In each case, assume that the population has a distribution that is approximately normal and that the sample is randomly selected. Note: Some of the exercises meet the requirements for using the normal distribution (as described in the preceding section) instead of the Student t distribution (as described in this section). Actually, you can also use the t tables for large samples as long as σ is not known (just use the bottom row of the tables); but unless you are using software, the t tables are generally less precise for finding P-values.

7. The top ten punt returners in CIAU football have a mean for total-yards-returned in the season of 377 yards (based on data from the CIAU Football website). To support his arguments for a pay raise, a scout claims that punt returners he has recruited have performed *better* than the mean for top athletes in that position. If five of the punt returners he has recruited have a mean total-yards-returned for a year of 398.8, with a standard deviation of 77.84, test the recruiter's claim, using a 0.05 level of significance.

8. The expense of moving the storage yard for the Consolidated Package Delivery Service (CPDS) is justified only if it can be shown that the daily mean travel distance will be less than 342 km. In trial runs of 12 delivery trucks, the mean and standard deviation are found to be 318 km and 67 km, respectively. At the 0.01 level of significance, test the claim that the mean is less than 342 km. Should the storage yard be moved?

9. Data Set 2 in Appendix B is derived from research into the effects of cold and other factors on human body temperature. Use just the data in the leftmost column, representing the temperatures of 11 subjects after they sat perfectly still for a period of time. Test the claim that the mean body temperature of all healthy adults who have been sitting still is equal to 37.0°C, using a 0.05 level of significance.

10. Refer to Data Set 2 in Appendix B. In two separate analyses, use the data in the two columns labelled "Supine." Both columns represent the temperatures of 11 subjects after they lay face up for a period of time. (In the first experiment, the subjects sat first, before lying down; in the second experiment, they stood before lying down.) For *each* of these two situations, test the hypothesis that the mean body temperature of all healthy adults who have been lying face up (after sitting or standing) is equal to 37.0°C. Use a 0.05 level of significance. (In the lying-after-sitting case, $\bar{x} = 37.02$ and $s = 0.215$; for the lying-after-standing case, $\bar{x} = 36.97$ and $s = 0.249$.) Compare your results with your answer to Exercise 9. What might this suggest about the usual assumptions about a "normal" body temperature, which does not take posture into account?

11. For each of 12 organizations, the cost of operation per client was found. The 12 scores have a mean of $2133 and a standard deviation of $345 (based on data from "Organizational Communication and Performance," by Snyder and Morris, *Journal of Applied Psychology*, Vol. 69, No. 3). At the 0.01 significance level, test the claim of a stockholder who complains that the mean for all such organizations exceeds $1800 per client.

12. The skid properties of a snow tire have been tested, and a mean skid distance of 47 m has been established for standardized conditions. A new, more expensive tire is developed, but tests on a sample of 17 new tires yield a mean skid distance of 45 m with a standard deviation of 3.7 m. Because of the cost involved, the new tires will be purchased only if it can be shown at the $\alpha = 0.005$ significance level that they skid less than the current tires. Based on the sample, will the new tires be purchased?

13. A major bank is concerned about the amount of debt being accrued by customers using its credit cards. The Board of Directors voted to institute an expensive monitoring system if the mean for all of the bank's customers is greater than $2000. The bank randomly selected 50 credit-card holders and determined the amounts they charged. For this sample group, the mean is $2177 and the standard deviation is $1257. Use a 0.025 level of significance to test the claim that the mean amount charged is greater than $2000. Based on the result, will the monitoring system be implemented?

14. When a poultry farmer uses her regular feed, the chickens attain normally distributed weights with a mean of 1.77 kg. In an experiment with an

enriched feed mixture, nine chickens are born that attain the weights (in kg) given below.

1.74 1.77 1.90 1.80 1.88 1.87 1.79 1.81 1.89

a. Use a 0.01 significance level to test the claim that the mean weight is higher with the enriched feed.

b. If the farmer can charge an extra 2 cents for each additional gram that her chickens weigh, and if it will cost an extra $1.17 per chicken to feed them the enriched feed, is it a good decision to switch to the enriched feed?

15. In a study of factors affecting hypnotism, visual analogue scale (VAS) sensory ratings were obtained for 16 subjects. For these sample ratings, the mean is 8.33 while the standard deviation is 1.96 (based on data from "An Analysis of Factors That Contribute to the Efficacy of Hypnotic Analgesia," by Price and Barber, *Journal of Abnormal Psychology*, Vol. 96, No. 1). At the 0.01 level of significance, test the claim that this sample comes from a population with a mean rating of less than 10.00.

16. A critic of Canada Post claims that its financial returns are typically $200 million less than forecast. In five of the years of the 1990s, the mean difference (in millions of dollars) between Canada Post's forecast and actual financial positions was 218.8, with a standard deviation of 118.76 (based on Canada Post data reported by Maclean Hunter, 1996). Use a significance level of 0.05 to test the claim that the mean difference between forecasted returns and actual returns was 200.

17. Because of a merger, a Vancouver executive will commute regularly to San Francisco for important meetings. To plan for travel time, she has compiled, as shown, a sample of flight times, in hours, for flights from Vancouver to San Francisco (based on data from Gary Murphy, chief flight dispatcher, Air Canada). The most convenient flight takes off 2.4 hours before she must reach San Francisco. Test the claim that the mean flight time from Vancouver to San Francisco is less than 2.4 hours. Because of the importance of not being mistaken, use a 0.01 level of significance.

2.30	2.20	1.97	2.22	2.12	$\bar{x} = 2.1564$
1.95	2.03	2.30	2.15	2.33	$s = 0.1312$
2.15					

18. A Calgary observatory records daily the mean counting rates for cosmic rays for that specific day (data from the National Geophysical Data Center). These recorded values are approximately normally distributed. If, for a randomly selected sample of the observatory's records, $\bar{x} = 3455$, and $s = 126.7$, test the claim that the mean for daily counting rates is actually equal to 3465.5.

19. Using the weights of only the *blue* M&Ms listed in Data Set 11 of Appendix B, test the claim that the mean is at least 0.9085 g, the mean value necessary for the 1498 M&Ms to produce a total of 1361 g as the package indicates. Use a 0.05 significance level. For the blue M&Ms, $\bar{x} = 0.9014$ g and $s = 0.0573$ g. Based on the result, can we conclude that the package contents do not agree with the claimed weight printed on the label?

20. Using the weights of only the *brown* M&Ms listed in Data Set 11 of Appendix B, test the claim that the mean is greater than 0.9085 g, the mean value necessary for the 1498 M&Ms to produce a total of 1361 g as the package indicates. Use a 0.05 significance level. For the brown M&Ms, $\bar{x} = 0.9128$ g and $s = 0.0395$ g. Based on the result, can we conclude that the packages contain more than the claimed weight printed on the label?

21. Listed here are the total electric energy consumption amounts (in kWh) for the home of a statistics professor during seven different years:

    11,943   11,463   10,789   9907   9012   9942   11,153

    The utility company claims that the mean annual consumption amount is 11,000 kWh and offers a budget payment plan based on that amount. At the 0.05 significance level, test the utility company's claim that the mean is equal to 11,000 kWh.

22. Given here are the birth weights (in kilograms) of male babies born to mothers on a special vitamin supplement. At the 0.05 level of significance, test the claim that the mean birth weight for all male babies of mothers given vitamins is equal to 3.39 kg, which is the mean for the population of all males. Based on the result, does the vitamin supplement appear to have an effect on birth weight?

    3.73   4.37   3.73   4.33   3.39   3.68   4.68   3.52
    3.02   4.09   2.47   4.13   4.47   3.22   3.43   2.54

23. Rita Gibbons is a stand-up comedian who videotapes her performances and records the total of the times she must wait for audience laughter to subside. Given here are the times (in seconds) for 15 different shows in which she used a new routine. Test the claim that the mean time is greater than 63.2 s, the mean time for her old routine. Based on the results, does her new routine seem to be better than the old one? Use the 0.05 level of significance.

    86   45   44   78   52   79   86   66   61   57   98   44   61   99   87

24. An umbrella salesperson has moved her base from Winnipeg to Atlantic Canada. Believing that the mean annual precipitation for communities in

Atlantic Canada is over twice the annual amount in Winnipeg (526 mm), she expects to do a great business. Refer to the top eight entries for annual precipitation in Data Set 4 in Appendix B. Assuming these data are from a random selection of sites in Atlantic Canada, test the claim that the mean annual precipitation for communities in Atlantic provinces is greater than 1052 mm. Use the 0.05 level of significance.

Suppose the entries from Atlantic Canada in Data Set 4 were *not* selected randomly from all possible sites, but were selected based on population plus convenience of data collection. What would be the implications—theoretical and practical—for using the results of the *t* test?

7-4 Exercises B: Beyond the Basics

25. Refer to the table of summary statistics included with Exercise 5. If the claim were changed from "is not different from the industry average" to "is more than the industry average," how would the values in the right column be affected?

26. Because of certain conditions, a hypothesis test requires the Student *t* distribution, as described in this section. Assume that the standard normal distribution was incorrectly used instead. Does using the standard normal distribution make you more or less likely to reject the null hypothesis, or does it not make a difference? Explain.

27. When finding critical values, we sometimes need significance levels other than those available in Table A-3. Some computer programs approximate critical *t* values by

$$t = \sqrt{df \cdot (e^{A^2/df} - 1)}$$

where df $= n - 1$

$e = 2.718$

$A = z\left(\dfrac{8\,df + 3}{8\,df + 1}\right)$

and z is the critical z score. Use this approximation to find the critical t score corresponding to $n = 10$ and a significance level of 0.05 in a right-tailed case. Compare the results to the critical t value found in Table A-3.

28. Refer to the data in Exercise 23 and assume that you're testing the claim that $\mu > 63.2$ s. Find β (the probability of a type II error), given that the actual value of the population mean is $\mu = 72.1$ s.

7-5 Testing a Claim About a Proportion

In the preceding sections of this chapter we introduced the basic methods of hypothesis testing, but they were used to address claims made about population *means* only. In this section we see how to apply those same basic methods to claims about population *proportions*. Using the methods of this section, we can test a claim about a proportion, percentage, or probability, as illustrated in these examples:

- Based on a sample survey, fewer than 1/4 of all college graduates smoke.
- The percentage of nonmilitary physicians leaving the country who are from Saskatchewan is equal to 5.4%.
- If a driver is fatally injured in a car crash, there is a 0.35 probability that the driver was legally impaired.

Assumptions Used When Testing a Claim About a Population Proportion, Probability, or Percentage

1. As always, we assume that the data have been collected by valid random-sampling techniques.
2. The conditions for a *binomial experiment* are satisfied. That is, we have a fixed number of independent trials having constant probabilities, and each trial has two outcome categories, which we classify as "success" and "failure."
3. The conditions $np \geq 5$ and $nq \geq 5$ are both satisfied, so **the distribution of sample proportions can be derived from the normal approximation of the binomial, with $\mu = np$ and $\sigma = \sqrt{npq}$** (as described in Section 5–6).

If these assumptions are not all satisfied, we may be able to use other methods. In this section, however, we consider only those cases in which the assumptions are satisfied, so the sampling distribution of sample proportions can be approximated by the normal distribution. We will use the following notation and test statistic.

NOTATION

n = number of trials

$\hat{p} = \dfrac{x}{n}$ (*sample* proportion)

p = *population* proportion (used in the null hypothesis)

$q = 1 - p$

$$z = \frac{\hat{p} - p}{\sqrt{\dfrac{pq}{n}}}$$

The Traditional Method

For using the traditional method of testing hypotheses, follow the steps outlined in Figure 7-4, choosing the test statistic z, to be calculated as shown in this section. Find critical z values from Table A-2 (standard normal distribution) by using the procedures described in Section 7–2, or find critical z values with the aid of statistical software. (Note that our method for calculating z is equivalent to finding $z = (x - \mu)/\sigma$ as in Section 2–5, but we have substituted the binomial equivalents for μ and σ, and divided numerator and denominator by n.)

For example, in a two-tailed test with significance level $\alpha = 0.05$, divide α equally between the two tails, then refer to Table A-2 for the z score corresponding to an area of $0.5 - 0.025 = 0.475$; the result is $z = 1.96$, so the critical values are $z = -1.96$ and $z = 1.96$.

When conducting a test of a claim about a population proportion p, be careful to identify correctly the sample proportion $\hat{p}$. The sample proportion $\hat{p}$ is sometimes given directly, as in the statement that "10% of the observed sports cars are red," which is expressed as $\hat{p} = 0.10$. In other cases, we may need to calculate the sample proportion by using $\hat{p} = x/n$. For example, from the statement that "96 surveyed households have cable TV and 54 do not," we can first find the sample size n to be $96 + 54 = 150$, then we can calculate the value of the sample proportion of households with cable TV as follows:

$$\hat{p} = \frac{x}{n} = \frac{96}{150} = 0.64$$

Caution: The value of 0.64 is exact, but when a calculator or computer display of $\hat{p}$ results in many decimal places, use all of those decimal places when evaluating the z test statistic. Large errors can result from rounding $\hat{p}$ too much.

EXAMPLE

In a study of air-bag effectiveness, it was found that in 821 crashes of midsize cars equipped with air bags, 46 of the crashes resulted in hospitalization of the drivers (based on data from the Highway Loss Data Institute). Use a 0.01 significance level to test the claim that the air-bag hospitalization rate is lower than the 7.8% rate for crashes of midsize cars equipped with automatic safety belts but no air bags.

We will use the traditional method of testing hypotheses as outlined in Figure 7-4. Instead of working with percentages, this solution will use the sample proportion of $\hat{p} = 46/821 = 0.0560292$ and the claimed population proportion of $p = 0.078$.

Step 1: The original claim is that the air-bag hospitalization rate is lower than 7.8%. We express this in symbolic form as $p < 0.078$.

Step 2: The opposite of the original claim is $p \geq 0.078$.

Step 3: Because $p \geq 0.078$ contains equality, we have

$$H_0: p \geq 0.078 \quad \text{(null hypothesis)}$$

$$H_1: p < 0.078 \quad \text{(alternative hypothesis and original claim)}$$

Step 4: The significance level is $\alpha = 0.01$.

Step 5: The statistic relevant to this test is $\hat{p} = 46/821 = 0.0560292$. The sampling distribution of sample proportions is approximated by the normal distribution. (The requirements that $np \geq 5$ and $nq \geq 5$ are both satisfied with $n = 821$, $p = 0.078$, and $q = 0.922$.)

Step 6: The test statistic of $z = -2.35$ is found as follows:

$$z = \frac{\hat{p} - p}{\sqrt{\dfrac{pq}{n}}} = \frac{0.0560292 - 0.078}{\sqrt{\dfrac{(0.078)(0.922)}{821}}} = -2.35$$

The critical value of $z = -2.33$ is found from Table A-2. With $\alpha = 0.01$ in the left tail, look for an area of 0.4900 in the body of Table A-2; the closest value is 0.4901 and it corresponds to $z = 2.33$, but it must be negative because it is to the left of the mean. The test statistic and critical value are shown in Figure 7-15.

Step 7: Because the test statistic does fall within the critical region, reject the null hypothesis.

Figure 7-15
Hypothesis Test of Claim
That $p < 0.078$

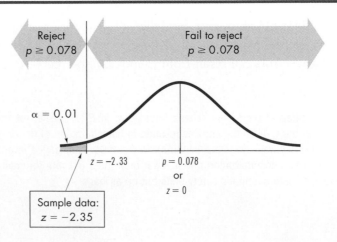

Interpretation

There is sufficient evidence to support the claim that for crashes of midsize cars, the air-bag hospitalization rate is lower than the 7.8% rate for automatic safety belts.

EXAMPLE

In a consumer taste test, 100 regular Pepsi drinkers are given blind samples of Coke and Pepsi; 48 of these subjects preferred Coke. At the 0.05 level of significance, test the claim that Coke is preferred by 50% of Pepsi drinkers who participate in such blind taste tests.

SOLUTION

We summarize the key components of the hypothesis test:

$$H_0: p = 0.5 \quad \text{(from the claim that "Coke is preferred by 50\%")}$$
$$H_1: p \neq 0.5$$

Test statistic:

$$z = \frac{\hat{p} - p}{\sqrt{\dfrac{pq}{n}}} = \frac{0.48 - 0.5}{\sqrt{\dfrac{(0.5)(0.5)}{100}}} = -0.40$$

The test statistic, critical values, and critical region are shown in Figure 7-16. Because the test statistic is not in the critical region, we fail to reject the null hypothesis.

Interpretation

There is not sufficient evidence to reject the claim that 50% of Pepsi drinkers prefer Coke. (Critics of such taste tests claim that the subjects often cannot observe differences and guess when making their choices.)

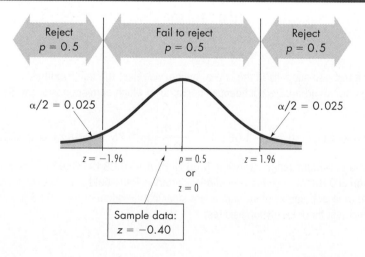

Figure 7-16
Hypothesis Test of $p = 0.5$

The *P*-Value Method

As with the *Z* test for μ with large samples, *P*-values are found in the same way.

Applying this method to the last example, which was two-tailed, the *P*-value is twice the area to the left of the test statistic $z = -0.40$. Table A-2 indicates that the area between $z = 0$ and $z = -0.40$ is 0.1554, so the area to the left of the test statistic ($z = -0.40$) is $0.5 - 0.1554 = 0.3446$. The *P*-value is $2 \times 0.3446 = 0.6892$. Because the *P*-value of 0.6892 is not less than or equal to the significance level of 0.05, we fail to reject the null hypothesis and again conclude that there is not sufficient evidence to reject the claim that 50% of Pepsi drinkers prefer Coke. Again, the *P*-value method is simply another way of arriving at the same conclusion reached using the traditional approach.

Confidence Intervals for Proportions and Two-Tailed Tests

Recall that in a two-tailed test for μ, the $1 - \alpha$ confidence interval can be used to determine whether or not to reject the null hypothesis at the α level of significance; if the hypothesized mean falls outside of the confidence interval, we reject the null hypothesis.

Can we employ the same methodology for a two-tailed test for proportions? The answer is: not always. Consider the following example.

EXAMPLE

Suppose we have these hypotheses:

$$H_0: p = 0.83$$
$$H_1: p \neq 0.83$$

Further suppose that $\hat{p} = 0.75$, $n = 100$, and $\alpha = 0.05$. Under these conditions, we reject the null hypothesis if the test statistic z is less than -1.96 or greater than 1.96. The test statistic is:

$$z = \frac{0.75 - 0.83}{\sqrt{\dfrac{0.83(0.17)}{100}}} = -2.1297$$

Since the test statistic falls in the critical region, we reject the null hypothesis.

Let us now build a 95% confidence interval for p which corresponds to the 5% level of significance:

$$\text{Lower limit} = 0.75 - 1.96\sqrt{\frac{0.75(0.25)}{100}} = 0.75 - 0.0849 = 0.6651$$

So, the 95% confidence interval is $0.6651 < p < 0.8349$. Note that the hypothesis proportion of 0.83 falls inside the confidence interval. This would lead us to conclude incorrectly that the population proportion is not significantly different from 0.83, contradicting the correct results of the hypothesis test.

EXCEL: First enter the number of successes and the total number of trials into separate spreadsheet cells (for example A1 and B1, respectively).

EXCEL (Prior to 2007): Use the Data Disk XL add-in by clicking on **DDXL** on the main menu bar, and then select **Hypothesis Tests**. Under the function type options, select **Summ 1 Var Prop Test** (for testing a claimed proportion for one variable, using summary data). Click on the pencil icon for "Num successes" and enter the appropriate cell address (e.g., A1). Click on the pencil icon for "Num trials" and enter the cell address for this data (e.g., B1). Click **OK**. Follow the four steps listed in the dialog box. After clicking on **Compute** in Step 4, you will get the *P*-value, test statistic, and conclusion.

EXCEL 2007: Click **Add-Ins** on the main menu and choose **DDXL** from the add-in menu to the left. Proceed as above.

MINITAB 15: From the **Stat** menu, choose **Basic Statistics** and then **1 Proportion.** If you are testing the hypothesis using summarized data, click that radio button and fill in the necessary boxes such as number of events (i.e., successes) and number of trials. Check the optional box to conduct a hypothesis test and enter the value of the hypothesized proportion. The default setting is a two-tailed test at a 5% level of significance. If you wish to change either, click the **Options** button and make the necessary changes.

STATDISK: Select **Analysis, Hypothesis Testing, Proportion—One Sample,** then proceed to enter the data in the dialog box.

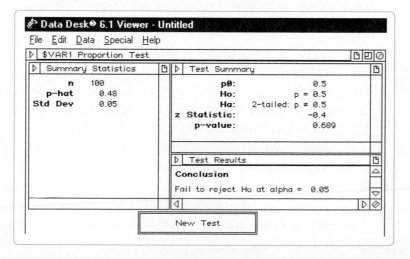

Exercises A: Basic Skills and Concepts

In Exercises 1–20, test the given claim, and draw the appropriate graph. Use the traditional method or the P-value method of testing hypotheses (or your instructor may specify which method to use with assigned exercises). Assume that all samples have been randomly selected.

1. In a study of store checkout scanners, 1234 items were checked and 20 of them were found to be overcharges (based on data from "UPC Scanner Pricing Systems: Are They Accurate?" by Goodstein, *Journal of Marketing*, Vol. 58). Use a 0.05 level of significance to test the claim that with scanners, 1% of sales are overcharges. (Before scanners were used, the overcharge rate was estimated to be about 1%.) Based on these results, do scanners appear to actually increase occurrences of overcharges?

2. Environmental concerns often conflict with modern technology, as is the case with birds that pose a hazard to aircraft during takeoff. An environmental group states that incidents of bird strikes are too rare to justify killing the birds. A pilots' group claims that among aborted takeoffs leading to an aircraft's going off the end of the runway, 10% are due to bird strikes. Use a 0.05 level of significance to test that claim. Sample data consist of 74 aborted takeoffs in which the aircraft overran the runway. Among those 74 cases, five were due to bird strikes (based on data from the Airline Pilots Association and Boeing).

3. The quality control manager at the Telektronic Company considers production of telephone-answering machines to be "out of control" when the overall rate of defects exceeds 4%. Testing of a random sample of 150 machines revealed that nine are defective, so the sample percentage of defects is 6%. The production manager claims that this is only a chance difference, so production is not really out of control and no corrective action is necessary. Use a 0.05 significance level to test the production manager's claim. Does it appear that corrective action is necessary?

4. In a recent year, 64% of all formal complaints to government institutions were resolved. The other complaints were unresolved, discontinued, or not substantiated. A spokesperson for Revenue Canada claims that they had a more positive record. In a sample of 58 complaints received, 42 were resolved (based on Information Commissioner of Canada, Annual Report, 1995–1996). Do the data suggest that more than 64% of complaints to Revenue Canada were resolved? Use the 0.01 level of significance.

5. Ralph Carter is a high-school history teacher who says that if students are not aware of the Holocaust, then the curriculum should be revised to correct that deficiency. A Roper survey of 506 high-school students showed that 268 of them did not know that the term "Holocaust" refers to the Nazi killing of

about 6 million Jews during World War II. Using the sample data and a 0.05 significance level, test the claim that most (more than 50%) students don't know what "Holocaust" refers to. Based on these results, should Ralph Carter seek revisions to the curriculum?

6. Refer to Data Set 8 in Appendix B and consider only those statistics students who are 21 or older. Find the sample percentage of smokers in that age group, then test the claim that statistics students aged 21 or over smoke at a rate that is less than 32%, which is the smoking rate for the general population of persons aged 21 and over. Is there a reason that statistics students 21 and over would smoke at a rate lower than the rate for the general population in that age group?

7. In past years, the percentage of civilian medical doctors leaving Canada who are from Saskatchewan has been equal to 5.4%. In a more recent sample of 120 civilian doctors leaving Canada, 10 were from Saskatchewan (based on data from the Southam Medical Database). At the 0.05 level of significance, test the claim that the percentage of nonmilitary doctors leaving Canada who are from Saskatchewan is now greater than 5.4%.

8. A mutual funds specialist claims that two-year return rates are greater than one-year return rates for 35% of funds on the market. Test that claim using a 0.05 significance level. For sample data, refer to Data Set 10 of Appendix B. *Hint:* You must first identify and count the records that are "successes."

9. Based on seven years of data collected by Transport Canada, 45.8% of all bird strikes on moving aircraft involve gulls. Suppose that the next year, planes are asked to emit gull-repelling sound effects, and that of 364 bird-strike incidents, 150 involve gulls. Test the claim that the sound effects have been effective in reducing the proportion of air-strike incidents that involve gulls. Use the 0.05 level of significance.

10. "Bystanders perform CPR correctly less than half the time," according to a newspaper article that noted that among 662 cases in which bystanders gave CPR, 46% performed the CPR correctly. Use a 0.025 significance level to test the article's claim.

11. Auto insurance companies are beginning to consider raising rates for those who use cell phones while driving. A consumer group claims that the problem really isn't too serious because only 10% of drivers use cell phones. The insurance industry conducts a study and finds that among 500 randomly selected drivers, 90 use cell phones (based on data from *Prevention* magazine). At the 0.02 level of significance, test the consumer group's claim.

12. Many researchers consider that a 75% response rate to a survey is quite good. In spring 1998, Statistics Canada surveyed thousands of representatives of businesses, governments, and institutions regarding their capital spending plans. At the same time, they wished to study the response rates. Of 500 who were

contacted for the spending survey, 395 returned a response (based on data in *The Globe and Mail*, July 23, 1998). At the 0.05 level of significance, test the claim that this survey had a response rate greater than 75%.

13. In clinical studies of the allergy drug Seldane, 70 of the 781 subjects experienced drowsiness (based on data from Merrell Dow Pharmaceuticals, Inc.). A competitor claims that 8% of Seldane users experience drowsiness. Use a 0.10 significance level to test that claim.

14. Late in the campaign for the June 1997 federal election, the Angus Reid Group conducted an opinion poll of public preferences. The poll found the Liberals had the support of 42% of the 3208 voters surveyed. In the election, the Liberals won 38.4% of the popular vote (based on reports by Reuters Limited). Test whether the public support in the poll was greater than on election day. Use the 0.05 level of significance. Did Liberal support drop in between the Angus Reid poll and the election?

15. Dr. Kelly Roberts is dean of a medical school and must plan courses for incoming students. The college president is encouraging her to increase the emphasis on pediatrics, but Dr. Roberts argues that fewer than 10% of medical students prefer pediatrics. She refers to sample data indicating that 64 of 1068 randomly selected medical students chose pediatrics. Is there sufficient sample evidence to support her argument (at the 0.01 significance level)?

16. Several dentist offices were concerned with their 8% rate of no-shows for appointments. The Canadian Dental Association recommended that to reduce these no-shows, the receptionist should call each patient a day before his or her scheduled appointment, as a reminder and confirmation. A study was then made of 125 randomly selected appointments that were made under the new system. If five no-shows were recorded, test the claim that the no-show rate is lower with the new system. Does the new system appear to be effective in reducing no-shows? Use the 0.05 level of significance.

17. In one study of 71 smokers who tried to quit smoking with nicotine patch therapy, 32 were not smoking one year after the treatment. Use a 0.10 level of significance to test the claim that among smokers who try to quit with nicotine patch therapy, the majority are smoking a year after the treatment. Do these results suggest that the nicotine patch therapy is not effective?

18. In a study of the risks to children of automobile accidents, 400 records were randomly selected for the ages of children who died as occupants in motor vehicles. Of these, 285 were teenagers from 15 to 19 years old (based on data from Health Canada). At the 0.03 level of significance, test the claim that more than 2/3 of the children who died in automobile accidents were teenagers from 15 to 19 years old.

19. A research poll of 1012 adults showed that among those who used fruitcakes, 28% ate them and 72% used them for other purposes, such as doorstops,

birdfeed, and landfill. This surprised fruitcake producers, who believe that a fruitcake is an appealing food. The president of the Tombits Food Company claims that the poll results are a fluke and, in reality, half of all adults eat their fruitcakes. Use a 0.01 level of significance to test that claim. Based on the result, does it appear that fruitcake producers should consider changes to make their product more appealing as a food or better suited for its uses as a doorstop, birdfeed, and so on?

20. A *Prevention* magazine article reported on a poll of 1257 adults. The article noted that among those polled, 27% smoke, and 82% of those who smoke have tried to stop at least once. How many of the surveyed people are smokers who have tried to stop at least once? At the 0.05 significance level, test the claim that less than 25% of adults are smokers who have tried to stop at least once.

7-5 Exercises B: Beyond the Basics

21. In a study of the prevalence of cold/flu symptoms in December 2000, 44 residents of Ottawa are tested in December, and five of them have the symptoms (based on data from *The Globe and Mail*, December 20, 2000). We want to use a 0.01 significance level to test the claim that the percentage of Ottawans suffering from cold/flu symptoms in December is greater than the national rate of 11.1%.

   a. Why can't we use the methods of this section?

   b. Assuming that the percentage of Ottawans suffering from cold/flu symptoms in December is equal to the national rate of 11.1%, find the probability that among 44 Ottawans randomly selected in December, five will be suffering from the symptoms.

   c. Based on your result from part (b), what do you conclude?

22. Chemco, a supplier of chemical waste containers, finds that 3% of a sample of 500 units are defective. Being fundamentally dishonest, the Chemco production manager wants to make a claim that the rate of defective units is no more than some specified percentage, and he doesn't want that claim rejected at the 0.05 level of significance if the sample data are used. What is the *lowest* defective rate he can claim under these conditions?

23. A researcher claimed that when 20 small businesses were opened, the success rate was equal to 47%. Describe a basis for rejecting that claim?

24. Refer to Exercise 9. If the true value of p is 0.45, find β, the probability of a type II error. *Hint:* In Step 3 use the values of $p = 0.45$ and $pq/n = (0.45)(0.55)/364$.

7-6 Testing a Claim About a Standard Deviation or Variance

Quality control engineers want to ensure that a product is, on the average, acceptable, but they also want to produce items of *consistent* quality so that there will be few defective products. Consistency is improved by reducing variation. For example, the consistency of dental X-ray tube operation is governed by the Radiation Emitting Devices Regulations (C.R.C., C.1370), administered by Health Canada. Besides requiring that operating X-ray tube voltage must be at least 50 kV (peak value), the regulations specify a maximum allowable deviation: The standard deviation can be no more than 5% of the mean voltage. Even if the mean voltage is perfectly acceptable, an excessively large standard deviation could result in individual voltages that are outside the safe operating range.

While the preceding sections of this chapter introduced methods of testing claims made about population means and proportions, this section uses the same basic procedures to test claims made about a population standard deviation σ or variance σ^2. The following assumptions must be met for such tests.

Assumptions

1. The data have been collected by valid random-sampling techniques.
2. The population has values that are normally distributed.

Other methods of testing hypotheses have also required a normally distributed population, but tests of claims about standard deviations or variances are not as *robust*, meaning that the inferences can be very misleading if the population does not have a normal distribution. Therefore, the condition of a normally distributed population is a much stricter requirement in this section. Given the assumption of a normal distribution, we can use the following notation for the chi-square test statistic.

TEST STATISTIC FOR TESTING HYPOTHESES ABOUT σ OR σ^2

$$\chi^2 = \frac{(n-1)s^2}{\sigma^2}$$

where n = sample size

s^2 = sample variance

σ^2 = population variance (given in the null hypothesis)

Critical Values

1. Critical values are found in Table A-4.

2. Degrees of freedom $= n - 1$.

Do not be confused by the use of both the normal and the chi-square distributions. After verifying that the sample data themselves come from a normally distributed population, we should then shift gears and think "chi-square." The chi-square distribution was introduced in Section 6–5, where we noted the following important properties.

Properties of the Chi-Square Distribution

1. All values of χ^2 are nonnegative, and the distribution is not symmetric (see Figure 7-17).

2. There is a different distribution for each number of degrees of freedom (see Figure 7-18).

3. The critical values are found in Table A-4 using

$$\text{degrees of freedom} = n - 1$$

Critical values are found in Table A-4 by first locating the row corresponding to the appropriate number of degrees of freedom (where df $= n - 1$). Next, the

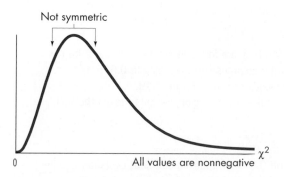

Figure 7-17
Properties of the Chi-Square Distribution

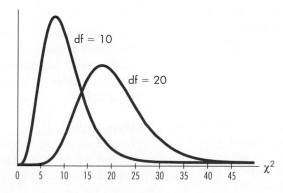

Figure 7-18
Chi-Square Distributions for 10 and 20 Degrees of Freedom
There is a different distribution for each number of degrees of freedom.

significance level α is used to determine the correct column as described in the following list:

Right-tailed test: Locate the area at the top of Table A-4 that is equal to the significance level α.

Left-tailed test: Calculate $1 - \alpha$, then locate the area at the top of Table A-4 that is equal to $1 - \alpha$.

Two-tailed test: Calculate $1 - \alpha/2$ and $\alpha/2$, then locate the area at the top of Table A-4 that equals $1 - \alpha/2$ (for the left critical value) and locate the area at the top equal to $\alpha/2$ (for the right critical value). (See Figure 6-9 and the example on pages 340 and 341.)

EXAMPLE

The Stewart Aviation Products Company has been successfully manufacturing aircraft altimeters with errors normally distributed with a mean of 0 ft (achieved by calibration) and a standard deviation of 43.7 ft. After the installation of new production equipment, 30 altimeters were randomly selected from the new line. This sample group had errors with a standard deviation of $s = 54.7$ ft. Use a 0.05 significance level to test the claim that the new altimeters have a standard deviation different from the old value of 43.7 ft.

SOLUTION

We will use the traditional method of testing hypotheses as outlined in Figure 7-4.

Step 1: The claim is expressed in symbolic form as $\sigma \neq 43.7$ ft.

Step 2: If the original claim is false, then $\sigma = 43.7$ ft.

Step 3: The null hypothesis must contain equality, so we have

$$H_0: \sigma = 43.7 \qquad H_1: \sigma \neq 43.7 \quad \text{(original claim)}$$

Step 4: The significance level is $\alpha = 0.05$.

Step 5: Because the claim is made about σ, we use the chi-square distribution.

Step 6: The test statistic is

$$\chi^2 = \frac{(n-1)s^2}{\sigma^2} = \frac{(30-1)(54.7)^2}{43.7^2} = 45.437$$

The critical values are 16.047 and 45.722. They are found in Table A-4, in the 29th row (degrees of freedom = $n - 1 = 29$) in the columns corresponding to 0.975 and 0.025. See the test statistic and critical values shown in Figure 7-19.

Step 7: Because the test statistic is not in the critical region, we fail to reject the null hypothesis.

Interpretation

There is not sufficient evidence to support the claim that the standard deviation is different from 43.7 ft. (For help in wording the final conclusion, refer to Figure 7-2.) However, it would be wise to continue monitoring and testing the new product line.

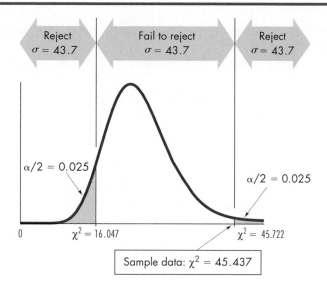

Figure 7-19
Hypothesis Test of Claim
That $\sigma \neq 43.7$

The *P*-Value Method

Instead of the traditional approach to hypothesis testing, we can also use the *P*-value approach summarized in Figure 7-8. If we use Table A-4, we usually cannot find *exact* *P*-values because the chi-square distribution table includes only selected values of α. Instead, we use the table to identify limits that contain the *P*-value. The test statistic from the last example is $\chi^2 = 45.437$, and we know that the test is two-tailed with 29 degrees of freedom. Refer to the 29th row of Table A-4 and see that the test statistic of 45.437 is between the table critical values of 42.557 and 45.722, indicating that the area to the right of the test statistic is between the corresponding areas of 0.05 and 0.025. However, Figure 7-8 shows that in a two-tailed test with a test statistic to the right of the centre, the *P*-value is *twice* the area to the right of the test statistic. The *P*-value is therefore between 0.10 and 0.05, which can be expressed as

$$0.05 < P\text{-value} < 0.10$$

Because the *P*-value is not less than or equal to the significance level of $\alpha = 0.05$, we again fail to reject the null hypothesis. Again, the traditional method and *P*-value method are equivalent in the sense that they always lead to the same conclusion. Note that if statistical software is used, the exact *P*-value (in this example, 0.0533) is usually displayed directly.

Connection Between Confidence Interval and Two-Tailed Test for σ

As with the two-tailed test for μ, we can use the confidence interval as a way of conducting the two-tailed test for σ; if the hypothesized standard deviation falls outside the $1 - \alpha$ confidence interval, we reject the null hypothesis at the α level of significance.

EXAMPLE

Construct the 95% confidence interval of the standard deviation of the aircraft altimeter errors from the previous example.

SOLUTION

$$\text{Lower limit for } \sigma^2 = \frac{29(54.7)^2}{45.722} = \frac{86,770.61}{45.722} = 1897.7868$$

$$\text{Upper limit for } \sigma^2 = \frac{29(54.7)^2}{16.047} = \frac{86,770.61}{16.047} = 5407.2792$$

Thus, $1,897.7868 < \sigma^2 < 5,407.2792$ with 95% confidence. By extension, $43.6 < \sigma < 73.5$ with 95% confidence. Note that the hypothesized standard deviation of 43.7 falls inside the confidence interval. Therefore, we fail to reject the null hypothesis at a 5% level of significance.

USING TECHNOLOGY

STATDISK: Select **Analysis**, then **Hypothesis Testing**, then **StDev—One Sample**. Proceed to provide the required entries in the dialog box, then click on **Evaluate**. STATDISK will display the test statistic, critical value, *P*-value, conclusion, and confidence intervals.

MINITAB 15: From the **Stat** menu, choose **Basic Statistics** and then **1 Variance.** If you are testing the hypothesis using summarized data, click that radio button and fill in the necessary boxes such as sample size and sample standard deviation. Check the optional box to conduct a hypothesis test and enter the value of the hypothesized standard deviation. The default setting is a two-tailed test at a 5% level of significance. If you wish to change either, click the **Options** button and make the necessary changes.

EXCEL: Since there is no Excel Data Analysis or DDXL tool for testing claims about a standard deviation, create a formula-based template, such as the one illustrated below. The template is shown here as being placed in Cells B19 to C27. You must modify the underlying formulas shown in the table based on where the data and the template appear on your own spreadsheet. Enter the right-tailed and left-tailed portions of α. If you are conducting a two-tailed test, then input the value of $\alpha/2$ in both cells. Enter the sample size n, the claimed standard deviation, and the sample standard deviation where indicated. When the underlying formulas are entered into the cells labelled here as C24 to C27, then the test statistic χ^2, lower critical χ^2, upper critical χ^2, and P-value will all be displayed. (For a one-tailed test, the lower or upper critical χ^2 will be omitted, as appropriate.) For a two-tailed test, multiply the displayed P-value by 2.

EXCEL DISPLAY   Testing a Claim About a Standard Deviation

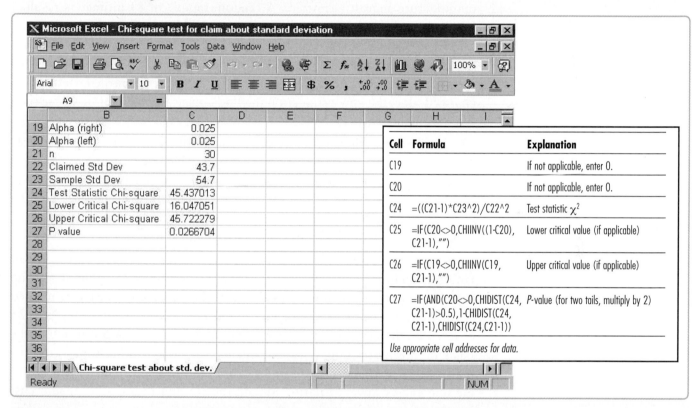

7-6 Exercises A: Basic Skills and Concepts

In Exercises 1 and 2, use Table A-4 to find the critical values of χ^2 based on the given information.

1. a. $H_0: \sigma = 15$ b. $H_1: \sigma > 0.62$ c. $H_1: \sigma < 14.4$

   $n = 10$ $n = 27$ $n = 21$

   $\alpha = 0.01$ $\alpha = 0.01$ $\alpha = 0.05$

2. a. H_1: $\sigma < 1.22$ b. H_1: $\sigma > 92.5$ c. H_0: $\sigma = 0.237$

   $n = 23$ $n = 12$ $n = 16$

   $\alpha = 0.025$ $\alpha = 0.10$ $\alpha = 0.05$

In Exercises 3–14, use the traditional method to test the given hypotheses. Follow the steps outlined in Figure 7-4, and draw the appropriate graph. In all cases, assume that the population is normally distributed and that the sample has been randomly selected.

3. The Stewart Aviation Products Company uses a new production method to manufacture aircraft altimeters. A random sample of 81 altimeters resulted in errors with a standard deviation of $s = 52.3$ ft. At the 0.05 level of significance, test the claim that the new production line has errors with a standard deviation different from 43.7 ft, which was the standard deviation for the old production method. If it appears that the standard deviation has changed, does the new production method appear to be better or worse than the old method?

4. The incubation periods for different, common bird families of the type Hole Nesters have a standard deviation of 6.8 days. The incubation periods for a sample of five different, common bird families of the type Open Nesters have a standard deviation of 2.0. Use a 0.05 significance level to test the claim that Open Nesters have less variation of incubation period than Hole Nesters. Can you suggest a biological explanation for your result?

5. With individual lines at its various windows, the Humber Valley Credit Union found that the standard deviation for normally distributed waiting times on Friday afternoons was 6.2 min. The credit union experimented with a single main waiting line and found that for a random sample of 25 customers, the waiting times have a standard deviation of 3.8 min. On the basis of previous studies, we can assume that the waiting times are normally distributed. Use a 0.05 significance level to test the claim that a single line causes lower variation among the waiting times. Why would customers prefer waiting times with less variation? Does the use of a single line result in a shorter wait?

6. Do men's heights vary more than women's? The nurse at a women's college has collected data from physical exams. The standard deviation of women's heights is found to be 2.5 in. The college has now become coed and the first 20 male students have heights with a standard deviation of 2.8 in. Use a 0.025 significance level to test the claim that men's heights vary more than women's. If that claim is not supported, does this mean that men's heights do not vary more than women's?

7. Xtron Electrics manufactures dental X-ray tubes. Health Canada has warned that they are not fulfilling regulations for Radiation Emitting Devices. For a

sample of 20 X-ray tubes in operation, their mean peak-value voltage ($\bar{x}$ = 52 kV) is acceptable, but it is claimed that the standard deviation is greater than the 2.6 kV allowed (based on Federal Regulation C.R.C., C.1370). The standard deviation for the sample was 2.8 kV. At the 0.05 level of significance, test the claim that Xtron's product is failing to meet the requirement of a maximum standard deviation of 2.6 kV.

8. The St. Lawrence Investment Company finds that if the standard deviation for the weekly downtimes of their computer is 2 h or less, then the computer is predictable and planning is facilitated. Twelve weekly downtimes for the computer were randomly selected, and the sample standard deviation was computed to be 2.85 h. The manager of computer operations claims that computer access times are unpredictable because the standard deviation exceeds 2 h. Test her claim, using a 0.025 significance level. Does the variation appear to be too high?

9. The Medassist Pharmaceutical Company uses a machine to pour cold medicine into bottles in such a way that the standard deviation of the weights is 0.15 oz. A new machine was tested on 71 bottles, and the standard deviation for this sample is 0.12 oz. The Dayton Machine Company, which manufactures the new machine, claims that it fills bottles with a lower variation. At the 0.05 significance level, test the claim made by the Dayton Machine Company. If Dayton's machine is being used on a trial basis, should its purchase be considered?

10. For randomly selected adults, IQ scores are normally distributed with a mean of 100 and a standard deviation of 15. A sample of 24 randomly selected college professors resulted in IQ scores having a standard deviation of 10. A psychologist is quite sure that college professors have IQ scores that have a mean greater than 100. He doesn't understand the concept of standard deviation very well and claims the standard deviation for IQ scores of college professors is equal to 15, the same standard deviation as in the general population. Use a 0.05 level of significance to test that claim. Based on the result, what do you conclude about the standard deviation of IQ scores for college professors?

11. Systolic blood pressure results from contraction of the heart. In comparing systolic blood pressure levels of men and women, Dr. Jane Taylor obtained readings for a random sample of 50 women. The sample mean and standard deviation were found to be 130.7 and 23.4, respectively. If systolic blood pressure levels for men are known to have a mean and standard deviation of 133.4 and 19.7, respectively, test the claim that women have more variation. Use a 0.05 level of significance. (All readings are in millimetres of mercury.)

12. A study examined the effects of a stay in the Arctic on a person's tolerance for cold. After a stay in the Arctic, three subjects were exposed to cold in

experimental conditions and the effects noted. The subjects' mean skin temperatures had a standard deviation of 2.11°C after their exposure to cold (based on data in *International Journal of Biometeorology*, 1996). Supposing the expected standard deviation for mean skin temperature is 1.06°C, test whether the stay in the Arctic has had an effect on the variance of the subjects' mean skin temperature. Test the claim at the 0.05 level of significance.

13. Based on data from a health survey, men aged 25–34 have heights with a standard deviation of 2.9 in. Test the claim that men aged 45–54 have heights with a different standard deviation. The heights of 25 randomly selected men in the 45–54 age bracket are listed below.

66.80	71.22	65.80	66.24	69.62	70.49	70.00	71.46	65.72
68.10	72.14	71.58	66.85	69.88	68.69	72.77	67.34	68.40
68.96	68.70	72.69	68.67	67.79	63.97	67.19		

14. An investor claims that the variability between the two-year return rates of different mutual funds is greater than when she began investing. Refer to the two-year return column in Data Set 10 in Appendix B. If the two-year return rates for different funds had a standard deviation of 10.11 when she began investing, test the claim that the current standard deviation for two-year return rates ($\sigma = 12.845$) is greater. Use the 0.01 level of significance.

7-6 Exercises B: Beyond the Basics

15. Use Table A-4 to find the range of possible *P*-values in the given exercises.

   a. Exercise 3   b. Exercise 5   c. Exercise 9

16. For large numbers of degrees of freedom, we can approximate critical values of χ^2 as follows:

$$\chi^2 = \frac{1}{2}(z + \sqrt{2k - 1})^2$$

Here k is the number of degrees of freedom and z is the critical value, found in Table A-2. For example, if we want to approximate the two critical values of χ^2 in a two-tailed hypothesis test with $\alpha = 0.05$ and a sample size of 150, we let $k = 149$ with $z = -1.96$, followed by $k = 149$ and $z = 1.96$.

   a. Use this approximation to estimate the critical values of χ^2 in a two-tailed hypothesis test with $n = 101$ and $\alpha = 0.05$. Compare the results to those found in Table A-4.

   b. Use this approximation to estimate the critical values of χ^2 in a two-tailed hypothesis test with $n = 150$ and $\alpha = 0.05$.

17. Repeat Exercise 16 using this approximation (with k and z as described in Exercise 16):

$$\chi^2 = k\left(1 - \frac{2}{9k} + z\sqrt{\frac{2}{9k}}\right)^3$$

18. Refer to Exercise 5. Assuming that s is actually 4.0, find β (the probability of a type II error). See Exercise 27 from Section 7–3 and modify the procedure so that it applies to a hypothesis test involving σ instead of μ.

19. When using the hypothesis testing procedure of this section, will the result be dramatically affected by the presence of an outlier? Describe how you arrived at your response.

20. The last digits of sample data are sometimes used in an attempt to determine whether the data have been measured or simply reported by the subject. Reported data often have last digits with disproportionately more 0's and 5's. Measured data tend to have last digits with a mean of 4.5, a standard deviation of about 3, and the digits should occur with roughly the same frequency.

   a. How is the standard deviation of the data affected if there are disproportionately more 0's and 5's?
   b. Why can we not use the methods of this section to test that the last digits of the sample data have a standard deviation equal to 3?

VOCABULARY LIST

α (alpha)   366
β (beta)   366
acceptance region   366
alternative hypothesis   364
classical method of testing
   hypotheses   374
critical region   366
critical value   366

formal hypothesis test   363
hypothesis   360
left-tailed test   369
null hypothesis   364
power   366
probability value   379
P-value   379
right-tailed test   369

significance level   366
test of significance   363
test statistic   366
traditional method of testing
   hypotheses   374
two-tailed test   369
type I error   365
type II error   365

REVIEW

Chapters 6 and 7 introduced two important concepts in using sample data to form inferences about population data: estimating values of population parameters (Chapter 6), and testing claims made about population parameters (Chapter 7). The parameters considered in Chapters 6 and 7 are means, proportions, standard deviations, and variances.

Section 7–2 presented the fundamental concepts of a hypothesis test: null hypothesis, alternative hypothesis, type I error, type II error, test statistic, critical region, critical value, and significance level. We also discussed two-tailed tests, left-tailed tests, right-tailed tests, and the statement of conclusions. Section 7–3 used those components in identifying three different methods for testing hypotheses:

1. The traditional method (summarized in Figure 7-4)

2. The P-value method (summarized in Figure 7-8)

3. Confidence intervals (discussed in Chapter 6)

Sections 7–3 through 7–6 discussed specific methods for dealing with the different parameters. Because it is so important to be correct in selecting the distribution and test statistic, we provide Figure 7-20, which summarizes the key decisions to be made.

REVIEW EXERCISES

In Exercises 1 and 2, find the appropriate critical values. In all cases, assume that the population standard deviation σ is unknown.

1. a. $H_1: \mu < 27.3$ b. $H_1: p \neq 0.5$ c. $H_1: \sigma \neq 15$
    $n = 10$ $n = 150$ $n = 20$
    $\alpha = 0.05$ $\alpha = 0.01$ $\alpha = 0.05$

2. a. $H_0: \mu = 19.9$ b. $H_0: \sigma \leq 0.93$ c. $H_0: p \geq 0.25$
    $n = 40$ $n = 10$ $n = 540$
    $\alpha = 0.10$ $\alpha = 0.025$ $\alpha = 0.01$

In Exercises 3 and 4, respond to each of the following:
    a. *Give the null hypothesis in symbolic form.*
    b. *Is this test left-tailed, right-tailed, or two-tailed?*
    c. *In simple terms devoid of symbolism and technical language, describe the type I error.*
    d. *In simple terms devoid of symbolism and technical language, describe the type II error.*
    e. *Identify the probability of making a type I error.*

3. The claim that statistics consultants have a mean income of $90,000 is to be tested at the 0.05 significance level.

4. The claim that lengths of CBC commercials have a standard deviation less than 15 s is to be tested at the 0.01 significance level.

5. According to the National Population Health Survey for 1996–97, 28% of those with dental insurance did not have their teeth checked during the survey

Figure 7-20 Testing a Claim About a Mean, Proportion, Standard Deviation, or Variance

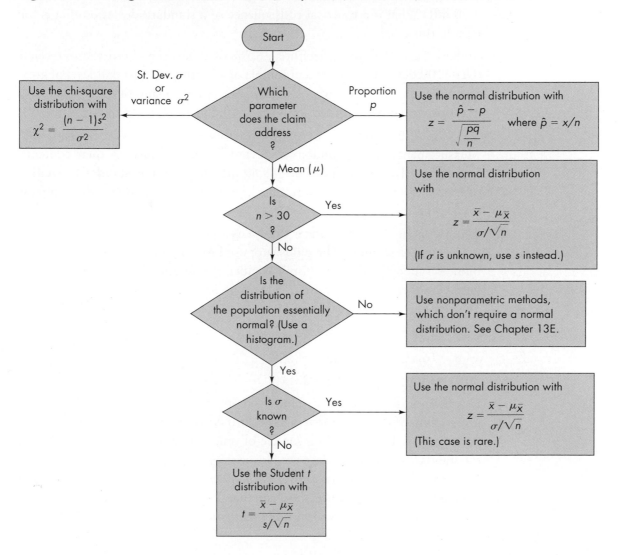

period (based on data from *The Globe and Mail*). Assume that there were 2014 people in the survey group with dental insurance. Test the claim that more than one quarter (25%) of people with dental insurance do not have their teeth checked annually. Use the 0.01 significance level.

6. A Telektronics dental X-ray machine bears a label stating that the machine gives radiation dosages with a mean of less than 5.00 milliroentgens (mR). Sample data consist of 36 randomly selected observations with a mean of 4.13 mR and a standard deviation of 1.91 mR. Using a 0.01 level of significance, test the claim stated on the label.

7. Using the same sample data from Exercise 6, use a 0.01 significance level to test the claim that the radiation dosages have a standard deviation less than 2.50 mR. What is a practical consequence of a standard deviation that is too high in this case?

8. In the 1991 Census of Agriculture, 26.7% of all farms had gross farm receipts over $100,000. In a more recent sample of 75 farms, 23 had that level of gross income. Using the 0.01 level of significance, test the claim that the proportion of farms with gross farm receipts over $100,000 is greater than it was in 1991.

9. The Cape Breton Bottling Company distributes cola in cans labelled 355 mL. Measurement Canada randomly selected 24 cans, measured their contents, and obtained a sample mean of 348.69 mL and a sample standard deviation of 11.21 mL. Use a 0.01 significance level to test the claim that the company is cheating consumers.

10. The Medassist Pharmaceutical Company makes a pill intended for children susceptible to seizures. The pill is supposed to contain 20.0 mg of phenobarbital. A random sample of 20 pills yielded the amounts (in milligrams) listed here. Are these pills acceptable at the $\alpha = 0.01$ significance level?

27.5   26.0   22.9   23.4   23.0

23.9   32.6   20.9   22.9   24.3

24.8   16.1   24.3   17.3   18.9

20.7   33.0   15.6   24.3   23.3

11. Has the number of times the House sits in each session of federal Parliament grown more variable? The overall standard deviation of days the House sits per Parliament is 188.3. But in a sample of nine Parliaments since World War II, the standard deviation of days the House sits per Parliament is 224.8 (based on data from *Duration of Sessions of Parliament*, Library of Parliament). Test, at the 0.05 level of significance, the claim that the standard deviation for the number of sittings per Parliament is no longer equal to 188.3.

CUMULATIVE REVIEW EXERCISES

1. For healthy women aged 18–24, systolic blood pressure readings (in millimetres of mercury) are normally distributed with a mean of 114.8 and a standard deviation of 13.1.

   a. If a healthy woman is randomly selected from the general population of all women aged 18–24, find the probability of getting one with systolic blood pressure greater than 124.23.

   b. If 16 women are randomly selected, find the probability that the mean of their systolic blood pressure readings is greater than 124.23.

2. A medical researcher obtains the systolic blood pressure readings (in millimetres of mercury) in the accompanying list for women aged 18–24 who have a new strain of viral infection. (As in Exercise 1, healthy women in that age group have readings that are normally distributed with a mean of 114.8 and a standard deviation of 13.1.)

| 134.9 | 78.7 | 108.9 | 133.0 | 123.7 | 96.1 | 126.9 | 89.8 |
| 132.0 | 134.7 | 132.1 | 121.7 | 112.3 | 150.2 | 158.3 | 154.4 |

a. Find the sample mean $\bar{x}$ and standard deviation s.

b. Use a 0.05 significance level to test the claim that the sample comes from a population with a mean equal to 114.8.

c. Use the sample data to construct a 95% confidence interval for the population mean m. Do the confidence interval limits contain the value of 114.8, which is the mean for healthy women aged 18–24?

d. Use a 0.05 significance level to test the claim that the sample comes from a population with a standard deviation equal to 13.1, which is the standard deviation for healthy women aged 18–24.

e. Based on the preceding results, does it seem that the new strain of viral infection affects systolic blood pressure?

3. A student majoring in psychology designs an experiment to test for extrasensory perception (ESP). In this experiment, a card is randomly selected from a shuffled deck, and the blindfolded subject must guess the suit (clubs, diamonds, hearts, spades) of the card selected. The experiment is repeated 25 times, with the card replaced and the deck reshuffled each time.

a. For subjects who make random guesses with no ESP, find the expected mean number of correct responses.

b. For subjects who make random guesses with no ESP, find the expected standard deviation for the numbers of correct responses.

c. For subjects who make random guesses with no ESP, find the probability of getting more than 12 correct responses.

d. If a subject gets more than 12 correct responses, test the claim that he or she made random guesses. Use a 0.05 level of significance.

e. You want to conduct a survey to estimate the percentage of adult Canadians who believe that some people have ESP. How many people must you survey if you want 90% confidence that your sample percentage is in error by no more than four percentage points?

4. An important requirement in Sections 7–4 and 7–6 is that the sample data must come from a population that is normally distributed.

a. Refer to the body temperatures of females listed in Data Set 18 of Appendix B and construct a stem-and-leaf plot.

b. Based on the result, does the distribution appear to be normal?

c. What other method could be used to determine whether the distribution is normal?

■ TECHNOLOGY PROJECT

Refer to the body temperatures (in °C) summarized in the given stem-and-leaf plot. (The sample is a subset of the body temperatures listed in Table 7-1.) Use Excel or STATDISK and a 0.05 significance level for the following tests.

35.	02
35.	688
36.	011223
36.	67
37.	1

a. Test the claim that the sample comes from a population with a mean of 37.0°C.

b. Add 5 to each of the original temperatures listed in the stem-and-leaf plot, then test the claim that the sample comes from a population with a mean of 42.0. Compare the results to those found in part (a). In general, what changes when the same constant is added to every score?

c. After multiplying each of the original temperatures by 10, test the claim that the sample comes from a population with a mean of 370. Compare the results to those found in part (a). In general, what changes when every score is multiplied by the same constant?

d. Celsius temperatures can be converted to Fahrenheit temperatures by using $F = (9/5)C + 32$. That is, multiply each Celsius temperature by 9/5, then add 32. Because the conversion from the Celsius scale to the Fahrenheit scale involves adding the same constant (32) and multiplying by the same constant (9/5), can you predict what will happen when the original temperatures are converted and the claim of $\mu = 98.6°F$ is tested? (Note that 37.0°C = 98.6°F.) Verify or disprove your prediction by converting the original temperatures to the Fahrenheit scale and testing the claim that $\mu = 98.6°F$.

e. When testing the claim that the mean weight of men is now greater than the known mean from men who lived 100 years ago, does it make any difference if the weights are measured in pounds or kilograms?

■ FROM DATA TO DECISION

Are Slices of the Venture Capital Pie Getting Smaller?

An increasingly popular way for risk capital companies to raise money is to attract venture capital investment. However, as the number of companies in Canada receiving these moneys has increased, has the mean amount of such funding available to a given recipient decreased? In the first quarter of 1997, the mean amount received was $2150. (All amounts are in thousands of dollars.)

Given the sample data below for the first quarter of 1998 (based on data from "The Venture Capital Survey" in the *Financial Post*, June 27–29, 1998), use a 0.05 significance level to test the claim that the amounts received by venture capital companies in the first quarter of 1998 were less than the mean amount received in the first quarter of 1997. Based on these results, should agents who arrange venture capital investments expect smaller commissions per transaction? Should companies who depend on this type of funding plan on receiving smaller amounts than they might have in previous years?

250	7000	4500	6000	1000	500	250	600	1500	2000	375	5000
3500	750	3000	250	175	500	250	500	2800	150	215	75
1000	150	420	1000	1000	2000	6000	2500	250	50	650	200

COOPERATIVE GROUP ACTIVITIES

1. **Out-of-class activity:** A group activity suggested for Chapter 6 involved estimating the mean error (in seconds) on people's wristwatches. Use the same data collected in that activity to test the claim that the mean error of all wristwatches is equal to 0. Recall that some errors are positive (because the watch reads earlier than the actual time) and some are negative (because the watch reads later than the actual time). Describe the details of the test and include a graph showing the test statistic and critical values. Do we collectively run on time, or are we early or late? Also test the claim that the standard deviation of errors is less than one minute. What are the practical implications of a standard deviation that is excessively large?

2. **In-class activity:** In a group of three or four people, conduct an ESP experiment by selecting one of the group members as the subject.

   Draw a circle on one small piece of paper and draw a square on another sheet of the same size. Repeat this experiment 20 times: Randomly select the circle or the square and place it in the subject's hand behind his or her back so that it cannot be seen, then ask the subject to identify the shape (without looking at it); record whether the response is correct. Test the claim that the subject has ESP because the proportion of correct responses is greater than 0.5.

3. **In-class activity:** Divide into groups of ten to 20 people. Each group member should record the number of heartbeats in one minute. After calculating $\bar{x}$ and s, each group should test the claim that the mean is greater than 59, which is the author's result. (When people exercise, they tend to have lower pulse rates, and the author runs five miles a few times each week. What a guy.)

■ **INTERNET PROJECT**

Hypothesis Testing

This chapter discussed the important method of inferential statistics known as hypothesis testing. This Internet Project will have you conduct tests using a variety of data sets. For each data set, you will be asked to

- Collect data available on the Internet.
- Formulate a hypothesis based on a given question.
- Conduct the test at a specified level of significance.
- Summarize your conclusions.

The Internet Project for this chapter can be found at either of the following websites:

http://www.mathxl.com   or   http://www.mystatlab.com

Click on Internet Project, then on Chapter 7. There you will find instructions outlining tasks that use data from business and economics, education and sports, as well as a classic example from the physical sciences.

8 Inferences from Two Samples

8-1 Overview

Chapters 6 and 7 presented the two major activities of inferential statistics: estimating a population parameter and testing a hypothesis about a population. Whereas those chapters considered cases involving only a single population, this chapter deals with two populations. Methods for hypothesis testing and the construction of confidence intervals are described for cases involving two populations.

8-2 Inferences About Two Means: Dependent Samples (Matched Pairs)

Two dependent samples (consisting of paired data) are used to test hypotheses about the means of two populations. Methods are presented for constructing confidence intervals as estimates of the difference between two population means.

8-3 Inferences About Two Means: Independent and Large Samples

Independent and large ($n > 30$) samples are used to test hypotheses about the means of two populations. Methods are presented for the construction of confidence intervals used to estimate the difference between two population means.

8-4 Comparing Two Variances

A method is presented for testing hypotheses made about two population variances or standard deviations.

8-5 Inferences About Two Means: Independent and Small Samples

This section deals with cases in which two samples are independent and small ($n \leq 30$). Procedures for testing hypotheses about the difference between two population means are discussed. Also discussed are methods for constructing confidence intervals used to estimate the difference between two population means.

8-6 Inferences About Two Proportions

This section deals with methods of inferential statistics applied to two population proportions. Methods are presented for testing claims about the difference between two population proportions. Confidence intervals are constructed for estimating the difference between two population proportions.

CHAPTER PROBLEM

Does El Niño really affect the weather, or is it just a lot of hype with little or no significance?

In Table 8-1 we list temperature and precipitation departures from springtime norms for the strongest El Niño years on record and compare these with data from other randomly selected springs. Temperature departures are in degrees Celsius and precipitation departures are expressed as percents. In Figure 8-1 we use boxplots to compare different stages of El Niño years to other years. The boxplots suggest that there may be some distinct differences, particularly for temperatures in the middle of an El Niño (that is, the second spring). But the number of samples is small (only 7 El Niño samples), so we might question whether these apparent differences are significant. In this chapter, we will consider the data from Table 8-1 as we test the equality of population means.

Table 8-1 Temperature and Precipitation Departures from Seasonal Means for Spring in the Strongest El Niño Years

Spring of El Niño Onset		Second Spring of El Niño		Spring of Other Years	
Departure in Temperature (in degrees Celsius)	Departure in Precipitation (as a percent)	Departure in Temperature (in degrees Celsius)	Departure in Precipitation (as a percent)	Departure in Temperature (in degrees Celsius)	Departure in Precipitation (as a percent)
0.5	−15.9	1.5	−11.1	0.6	−2.3
0.2	−6.8	0.0	−4.5	−0.4	−4.0
−1.2	2.9	0.9	2.0	0.0	28.8
−1.0	4.5	−0.5	10.6	0.0	−5.7
0.6	11.6	1.0	2.2	−0.3	−2.8
1.3	6.4	0.1	3.9	0.8	8.1
−0.4	1.6	3.1	−4.7	1.6	3.0
				1.1	−4.7
				−1.2	−21.3
				−0.3	10.9
				0.0	−7.1
				1.4	9.0
				−0.6	−15.1
				1.1	11.4
				1.0	−0.7
				1.9	26.0
				0.3	−4.5
				−1.8	6.7
				−1.6	−7.0
				2.0	7.1

Based on data from Environment Canada.

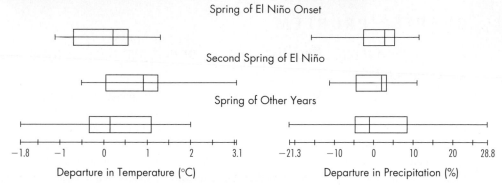

Figure 8-1 Comparisons of El Niño and Non–El Niño Springs

Spring of El Niño Onset

Second Spring of El Niño

Spring of Other Years

Departure in Temperature (°C)

Departure in Precipitation (%)

STATISTICAL PLANET

Research in Twins
Identical twins occur when a single fertilized egg splits in two, so that both twins share the same genetic makeup. There is now an explosion in research focused on those twins. Speaking for the Centre for Study of Multiple Birth, Louis Keith notes that now "we have far more ability to analyze the data on twins using computers with new, built-in statistical packages." A common goal of such studies is to explore the classic issue of "nature versus nurture." For example, Thomas Bouchard, who runs the Minnesota Study of Twins Reared Apart, has found that IQ is 50%–60% inherited, while the remainder is the result of external forces.

Identical twins are matched pairs that provide better results by allowing us to reduce the genetic variation that is inevitable with unrelated pairs of people.

 Overview

Chapter 6 introduced an important activity of inferential statistics: Samples were used to construct confidence intervals, which can be used to estimate values of population parameters. Chapter 7 introduced a second important activity of inferential statistics: Samples were used to test hypotheses about population parameters. In both of those chapters all examples and exercises involved the use of *one* sample to form an inference about *one* population. In reality, however, there are many important and meaningful situations in which it becomes necessary to compare *two* sets of sample data. The following are examples typical of those found in this chapter, which presents methods for using data from two samples so that inferences can be made about the populations from which they came:

• Determine whether there is a difference between the mean amounts of nicotine from filtered cigarettes and nonfiltered cigarettes.

• Determine whether women employees exposed to ethyl glycol ethers have a rate of miscarriage that is different from the rate for women not exposed to those chemicals.

8-2 Inferences About Two Means: Dependent Samples (Matched Pairs)

In this section we consider methods for testing hypotheses and constructing confidence intervals for two samples that are dependent, which means that they are paired or matched. We begin by formally defining *independent* and *dependent*.

DEFINITIONS

> Two samples are **independent** if the sample values selected from one population are not related to or somehow paired with the sample values selected from the other population. If the values in one sample are related to the values in the other sample, the samples are **dependent samples**. Such samples are often referred to as **matched pairs** or **paired samples**.

EXAMPLE

Dependent Samples: Subjects' weights are measured before and after undergoing a regimen of training, with the results in the following table. The sample of pretraining weights and the sample of post-training weights are *dependent samples,* because each pair is matched according to the person involved. "Before/after" data are usually matched and are usually dependent.

Subject	A	B	C	D	E	F
Pretraining weights (kg)	99	62	74	59	70	73
Posttraining weights (kg)	94	62	66	58	70	76

Based on data from the *Journal of Applied Psychology,* Vol. 62, No. 1.

Independent Samples: Weights are taken from two separate groups of subjects. These two samples are *independent* because the sample of females is not related to the sample of males. The data in this table are not matched.

Weights of females (lb)	115	107	110	128	130		
Weights of males (lb)	128	150	160	140	163	155	175

Assumptions

For the hypothesis tests and confidence intervals described in this section, we make the following assumptions:

1. The sample data consist of matched pairs.
2. The samples have been selected by a valid random-sampling technique.
3. If the number of pairs of sample data is small ($n \le 30$), then the population of differences between the paired values must be approximately *normally*

distributed. (If the population of differences departs radically from a normal distribution, we should not use the methods given in this section. We may be able to use nonparametric methods, discussed in Chapter 12.)

Prior to attempting formal inferences with the data, it is advisable to *explore* the data sets, as discussed in Chapter 2. For example, is a data pair an *outlier* that should be removed from further calculations? Have all values been rounded consistently? What are the means and distributions of the two samples?

Hypothesis Tests

x	10	8	5	20
y	7	2	9	20
d	3	6	−4	0

In dealing with two dependent samples, we base our calculations on the differences (d) between the pairs of data, as illustrated in the table in the margin. (Comparing the two sample means $\bar{x}$ and $\bar{y}$ would waste important information about the paired data.) The notation below is based on those differences.

NOTATION FOR TWO DEPENDENT SAMPLES

> μ_d = mean value of the differences d for the *population* of paired data
>
> $\bar{d}$ = mean value of the differences d for the paired *sample* data (equal to the mean of the $x - y$ values)
>
> s_d = standard deviation of the differences d for the paired *sample* data
>
> n = number of *pairs* of data in the sample

We now describe the test statistic to be used in hypothesis tests of claims made about the means of two populations, given that the two samples are dependent. If we have random paired (i.e., dependent) sample data from populations in which the paired differences have a normal distribution, we use the sample differences and proceed with a single test. We use the following test statistic with a Student t distribution.

TEST STATISTIC FOR TWO DEPENDENT SAMPLES (MATCHED PAIRS)

$$t = \frac{\bar{d} - \mu_d}{\frac{s_d}{\sqrt{n}}}$$

where degrees of freedom = $n - 1$

If the number of pairs of data is large ($n > 30$), the number of degrees of freedom will be at least 30, so critical values will in effect be z scores (Table A-2) instead of t scores (Table A-3).

Using a reaction timer similar to the one described in the Cooperative Group Activities of Chapter 5, subjects are tested for reaction times with their left and right hands. (Only right-handed subjects were used.) The results (in thousandths of a second) are given in the accompanying table, and the distribution of the paired differences is approximately normal. Use a 0.05 significance level to test the claim that there is a difference between the mean of the right- and left-hand reaction times. If an engineer is designing a fighter-jet cockpit and must locate the ejection-seat activator to be accessible to either the right or the left hand, does it make a difference which hand she chooses?

Subject	A	B	C	D	E	F	G	H	I	J	K	L	M	N
Right	191	97	116	165	116	129	171	155	112	102	188	158	121	133
Left	224	171	191	207	196	165	177	165	140	188	155	219	177	174
Difference	-33	-74	-75	-42	-80	-36	-6	-10	-28	-86	33	-61	-56	-41

SOLUTION

Using the traditional method of hypothesis testing, we will test the claim that there is a difference between the right- and left-hand reaction times. Because we are dealing with paired data, begin by finding the differences $d = $ right $-$ left. Then follow the steps summarized in Figure 7-4.

Step 1: If there is a difference, we expect the mean of the d values to be different from 0. This is expressed in symbolic form as $\mu_d \neq 0$.

Step 2: If the original claim is not true, we have $\mu_d = 0$.

Step 3: The null hypothesis must contain equality, so we have

$$H_0: \mu_d = 0 \qquad H_1: \mu_d \neq 0 \text{ (original claim)}$$

Step 4: The significance level is $\alpha = 0.05$.

Step 5: Because we are testing a claim about the means of paired dependent data, we use the Student t distribution.

Step 6: Before finding the value of the test statistic, we must first find the values of $\bar{d}$ and s_d. The difference d for each subject is shown in the preceding table. We now use Formulas 2-1 and 2-6 as follows:

$$\bar{d} = \frac{\Sigma d}{n} = \frac{-595}{14} = -42.5$$

$$s_d = \sqrt{\frac{n(\Sigma d^2) - (\Sigma d)^2}{n(n-1)}} = \sqrt{\frac{14(39{,}593) - (-595)^2}{14(14-1)}} = 33.2$$

With these statistics and the assumption that $\mu_d = 0$, we can now find the value of the test statistic:

$$t = \frac{\bar{d} - \mu_d}{\frac{s_d}{\sqrt{n}}} = \frac{-42.5 - 0}{\frac{33.2}{\sqrt{14}}} = -4.790$$

Crest and Dependent Samples
In the late 1950s, Procter & Gamble introduced Crest toothpaste as the first such product with fluoride. To test the effectiveness of Crest in reducing cavities, researchers conducted experiments with several sets of twins. One of the twins in each set was given Crest with fluoride, while the other twin continued to use ordinary toothpaste without fluoride. It was believed that each pair of twins would have similar eating, brushing, and genetic characteristics. Results showed that the twins who used Crest had significantly fewer cavities than those who did not. This use of twins as dependent samples allowed the researchers to control many of the different variables affecting cavities.

The critical values of $t = -2.160$ and $t = 2.160$ are found from Table A-3; use the column for 0.05 (two tails), and use the row with degrees of freedom of $n - 1 = 13$. Figure 8-2 shows the test statistic, critical values, and critical region.

Step 7: Because the test statistic does fall in the critical region, we reject the null hypothesis of $\mu_d = 0$.

Interpretation

There is sufficient evidence to support the claim of a difference between the right- and left-hand reaction times. Because there does appear to be such a difference, an engineer designing a fighter-jet cockpit should locate the ejection-seat activator so that it is readily accessible to the faster hand, which appears to be the right hand with seemingly lower reaction times. (We could require special training for left-handed pilots if a similar test of left-handed pilots shows that their dominant hand is faster.)

The above example is a two-tailed test, but left-tailed tests and right-tailed tests follow the same basic procedure. For example, if we want to test the claim that the right-hand reaction times are lower than the left-hand times, we have a claim that $\mu_d < 0$ (because the values of "right − left" would tend to be negative).

The preceding two-tailed example used the traditional method, but the P-value approach could be used by modifying Steps 6 and 7. In Step 6, use the test statistic of $t = -4.790$ and refer to the 13th row of Table A-3 to find that 4.790 is beyond the largest table value of 3.012. The P-value in the two-tailed test is therefore less than 2(0.005) or 0.01; that is, P-value < 0.01. In Step 7, we again

Figure 8-2
Distribution of Differences Between Right- and Left-Hand Reaction Times

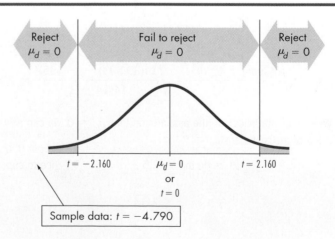

reject the null hypothesis because the *P*-value is less than the significance level of $\alpha = 0.05$.

Confidence Intervals

We can develop a confidence interval estimate of the population mean difference μ_d by using the sample mean $\bar{d}$, the standard deviation of sample means d (which is $s_d/\sqrt{n}$), and the critical value $t_{\alpha/2}$.

The confidence interval estimate of the mean difference μ_d is as follows:

$$\bar{d} - E < \mu_d < \bar{d} + E$$

where $E = t_{\alpha/2}\dfrac{s_d}{\sqrt{n}}$ and degrees of freedom $= n - 1$

EXAMPLE

Use the sample data from the preceding example to construct a 95% confidence interval estimate of μ_d.

SOLUTION

Using the values of $\bar{d} = -42.5$, $s_d = 33.2$, $n = 14$, and $t_{\alpha/2} = 2.160$, we first find the value of the margin of error E:

$$E = t_{\alpha/2}\frac{s_d}{\sqrt{n}} = 2.160\frac{33.2}{\sqrt{14}} = 19.2$$

The confidence interval can now be found:

$$\bar{d} - E < \mu_d < \bar{d} + E$$
$$-42.5 - 19.2 < \mu_d < -42.5 + 19.2$$
$$-61.7 < \mu_d < -23.3$$

The result is sometimes expressed as $\mu_d = -42.5 \pm 19.2$ or as $(-61.7, -23.3)$.

Interpretation

In the long run, 95% of such samples will lead to confidence interval limits that actually do contain the true population mean of the differences. Note that the confidence interval limits do not contain 0, indicating that the true value of μ_d is significantly different from 0. That is, the mean value of the "right $-$ left" differences is different from 0. On the basis of the confidence interval, we conclude that there is sufficient evidence to support the claim that there is a difference between the right- and left-hand reaction times. This conclusion agrees with the conclusion in the preceding example.

EXCEL (Prior to 2007): From the **Tools** menu, select **Data Analysis**. Choose **t-Test: Paired Two Sample for Means**. Enter these details in the dialog box that appears: For **Variable 1 Range**, specify the worksheet range that contains the first variable's data; for **Variable 2 Range**, specify the worksheet range that contains the second variable's data (the data in these two ranges should be arranged in corresponding order). Also enter the appropriate choice for **Alpha**. Then select an **Output Option**, and click on **OK** to obtain the results. Note that both the one-tailed and the two-tailed solutions are provided. (Only positive t values are output, but they may be interpreted as negative where appropriate.)

EXCEL 2007: Click **Data** on the main menu and choose **Data Analysis** from the **Analysis** tab. Proceed as above.

EXCEL DISPLAY
Two Dependent Samples

	A	B	C
1	t-Test: Paired Two Sample for Means		
2			
3		Variable 1	Variable 2
4	Mean	139.57143	182.07143
5	Variance	984.41758	567.45604
6	Observations	14	14
7	Pearson Correlation	0.3020121	
8	Hypothesized Mean Difference	0	
9	df	13	
10	t Stat	-4.793725	
11	P(T<=t) one-tail	0.0001754	
12	t Critical one-tail	1.7709317	
13	P(T<=t) two-tail	0.0003508	
14	t Critical two-tail	2.1603682	

Although Excel cannot calculate confidence intervals directly for paired samples, the Data Desk XL add-in can be used.

EXCEL (Prior to 2007): Click on **DDXL**. Select **Confidence Intervals** and **Paired t Interval**. In the dialog box, click on the pencil icon for the first quantitative column and enter the range for the first sample's data. Then click on the pencil icon for the second quantitative column and enter the range for the second sample's data. Click on **OK**. Follow the steps in the next dialog box that appears.

EXCEL 2007: Click on the **Add-in** tab on the main menu. You will see **DDXL** to the left. Click on it and proceed as above.

MINITAB 15: From the **Stat** menu, choose **Basic Statistics** and then **Paired t**. If the test is being conducted with raw data, select the appropriate columns. The default is a two-tailed test at a 5% level of significance with a hypothesized difference of zero. If you wish to change any of these details, click the **Options** button and make the necessary changes.

STATDISK: Select **Analysis**, then **Hypothesis Testing**, then **Mean-Matched Pairs**. The result will be the display shown here. (STATDISK automatically provides confidence interval limits for two-tailed hypothesis tests.)

We used Table A-3 to conclude that the *P*-value is less than 0.01, but from the Excel and STATDISK displays we can see that the *P*-value is actually 0.0004.

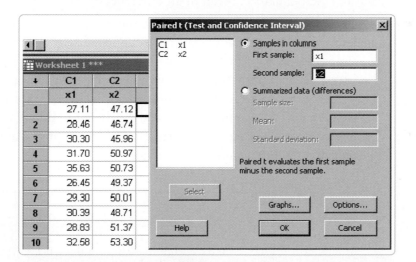

MINITAB DISPLAY
Two Dependent Samples

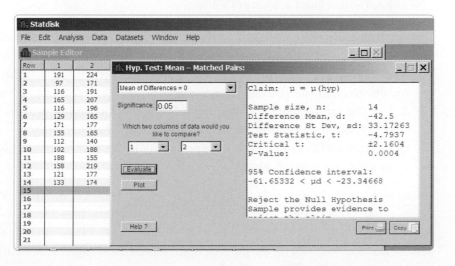

STATDISK DISPLAY
Two Dependent Samples

8-2 Exercises A: Basic Skills and Concepts

In Exercises 1 and 2, assume that you want to test the claim that the paired sample data come from a population for which the mean difference is $\mu_d = 0$. Assuming a 0.05 level of significance, find (a) $\bar{d}$, (b) s_d, (c) the t test statistic, and (d) the critical values.

1. x	8	8	6	9	7
y	3	2	6	4	9

2. x	20	25	27	27	23	29	30	26
y	20	24	25	29	20	29	32	29

3. Using the sample paired data in Exercise 1, construct a 95% confidence interval for the population mean of all differences $x - y$.

4. Using the sample paired data in Exercise 2, construct a 99% confidence interval for the population mean of all differences $x - y$.

5. Do strong El Niño phenomena change the weather as much as people claim?
   a. Using the data in the following table (aggregate temperature departures from spring norms in degrees Celsius), test the claim that the progression of a strong El Niño, from the spring of onset to the next spring, affects Canadian temperature departures from seasonal norms. Use a 0.05 level of significance.
   b. Construct a 95% confidence interval for the mean difference of the "onset" minus "second spring" departures from norms. Write a statement that interprets the resulting confidence interval.

El Niño Cycle	A	B	C	D	E	F	G
Spring of El Niño Onset	0.5	0.2	−1.2	−1	0.6	1.3	−0.4
Second Spring of El Niño	1.5	0.0	0.9	−0.5	1.0	0.1	3.1

Based on data from Environment Canada.

6. Captopril is a drug designed to lower systolic blood pressure. When subjects were tested with this drug, their systolic blood pressure readings (in millimetres of mercury) were measured before and after the drug was taken, with the results given in the accompanying table.
   a. Use the sample data to construct a 99% confidence interval for the mean difference between the before and after readings.
   b. Is there sufficent evidence to support the claim that Captopril is effective in lowering systolic blood pressure?

Subject	A	B	C	D	E	F	G	H	I	J	K	L
Before	200	174	198	170	179	182	193	209	185	155	169	210
After	191	170	177	167	159	151	176	183	159	145	146	177

Based on data from "Essential Hypertension: Effect of an Oral Inhibitor of Angiotensin-Converting Enzyme" by MacGregor et al., *British Medical Journal*, Vol. 2.

7. Does past growth in a company translate to continued growth in the future? In 1988, *Profit* magazine identified the 50 fastest-growing companies in Canada. In 1998, it revisited these companies. Of the companies for which relevant data were available, 10 have been randomly selected. Compare their sales in 1987 and 1997 (in millions of dollars).

   a. At the 0.05 significance level, test the claim that fast-growing companies do grow in sales after 10 years. Does the fact that some companies in the 1987 survey went out of business or would not report their sales for 1997 affect the validity of the sample? (Consider the issues raised in Section 1–3.)

   b. Construct a 95% confidence interval for the mean of the difference between 1987 sales and 1997 sales.

Company	A	B	C	D	E	F	G	H	I	J
1987 Sales	1.0	1.0	3.1	2.7	4.2	7.8	7.6	2.0	7.0	6.0
1997 Sales	5.0	2.5	1.7	4.5	19.0	17.0	17.0	1.2	64.0	8.0

Based on data from *Profit*, Vol. 17, No. 3.

8. According to a report in *The Globe and Mail*, the Canadian oil and gas industry "ramped up" investment spending in the period from 2000 to 2001 ("Energy Sector Ramps Up Spending," *The Globe and Mail*, January 29, 2001, p. B14). The accompanying table shows the spending levels (in millions of dollars) in each of those years for nine top oil and gas companies.

   a. Using a 0.01 level of significance, is there sufficient evidence to support the claim that spending is greater in 2001 than in 2000?

   b. Construct a 98% confidence interval for the mean of the differences. Do the confidence interval limits contain 0, indicating that there is not a significant difference between the two years' spending levels?

Company	A	B	C	D	E	F	G	H	I
2000 spending	1685	1153	1190	1415	1125	1203	954	957	1998
2001 spending	1940	1800	1710	1500	1500	1410	1320	1200	935

9. A study was conducted to examine differences in starting salaries for university graduates in Ontario and Quebec. Because starting salaries are in different ranges based on type of degree, the Ontario and Quebec results are paired with respect to degree earned in the adjacent table.

   a. At the 0.05 significance level, test the claim that starting salaries in Ontario and Quebec are equal. What do you conclude about the effect of province on the starting salary for graduates?

   b. Construct a 95% confidence interval for the mean of the differences between Ontario and Quebec starting salaries for various types of degrees earned.

Degree	Ontario	Quebec
B.A. or B.Sc.	$28,900	$28,300
B. Comm.	$29,500	$28,500
M.A. or M.Sc.	$34,600	$31,900
M.B.A.	$36,600	$34,100
Chem. Eng.	$32,900	$34,400
Civil Eng.	$33,700	$36,300
Elec. Eng.	$34,000	$32,300
Mech. Eng.	$33,700	$34,300
Structural Eng.	$33,500	$35,100

Based on "Annual Salary Survey," Peat Marwick Stevenson & Kellogg.

10. In May 1997, Calgary-based Bre-X Minerals collapsed when its gold assets were revealed to be dramatically less than had been claimed. Did this collapse affect the prices of other mining stocks?

a. Using the price data below, construct a 98% confidence interval for the mean difference of their pre-collapse prices minus their prices three months later.

b. At the 0.01 level of significance, does the evidence support a claim that the prices of mining stocks dropped during the period?

Stock	A	B	C	D	E
Before the collapse	1.4	1.2	1.15	0.280	2.73
After the collapse	0.7	0.5	0.41	0.225	2.06

11. Listed below are the costs (in dollars) of flights from New York (JFK) to San Francisco for US Air, Continental, Delta, United, American, Alaska, and Northwest. Use a 0.01 significance level to test the claim that flights scheduled one day in advance cost more than flights scheduled 30 days in advance. What strategy appears to be effective in saving money when flying?

Flight scheduled 1 day in advance	456	614	628	1088	943	567	536
Flight scheduled 30 days in advance	244	260	264	264	278	318	280

12. If a *t* test were run on Excel at the 0.01 confidence level, using the data in Exercise 10, the following output would be produced.

	A	B	C
1	t-Test: Paired Two Sample for Means		
2			
3		Variable 1	Variable 2
4	Mean	1.352	0.779
5	Variance	0.77857	0.54203
6	Observations	5	5
7	Pearson Correlation	0.951423	
8	Hypothesized Mean Difference	0	
9	df	4	
10	t Stat	4.4084749	
11	P(T<=t) one-tail	0.0058072	
12	t Critical one-tail	3.7469363	
13	P(T<=t) two-tail	0.0116144	
14	t Critical two-tail	4.6040805	

Could a person who ran this test validly reject a null hypothesis that there is no difference in means? Is this the same answer as is reached in Exercise 10(b)? If not, explain how these answers could be different.

13. Refer to Data Set 2 in Appendix B. Use the paired data consisting of the body temperatures of "initially standing" subjects in the standing posture (column STAND1) and after lying down (column SUPINE2).

   a. Construct a 95% confidence interval for the mean difference of the STAND1 temperatures minus the SUPINE2 temperatures.

   b. Using a 0.05 significance level, test the claim that for those temperatures the mean difference is 0. Based on the result, do the body temperatures in the two postures appear to be about the same?

14. A company offers a daily list of stock picks to paid subscribers. Its "Track Record for TSE" web page shows the prices of stocks it has recommended (1) on the day the "buy" recommendation was made, and (2) on a subsequent day within the next week or two. The accompanying table is a sample from these data, for November to December 2000. (Based on data from Active Trading Global.)

   a. At the 0.05 significance level, do the data support the company's claim that, for stocks bought and sold at the prices cited, their prices show an increase during the period?

   b. Suppose the company recommends good stock buys, but leaves it up to the client to decide *when actually to sell* the stocks. Is the displayed data sufficient to support a claim that the actual clients of this company made money? Explain.

Stock Pick No.	1	2	3	4	5	6	7	8	9	10	11	12
Entry price	4.87	22.85	14.05	13.80	22.70	10.00	25.45	18.90	20.90	4.50	31.30	14.00
Exit price	5.35	23.65	14.25	14.80	25.25	9.65	26.70	20.50	25.25	9.15	41.25	14.40

8-2 Exercises B: Beyond the Basics

15. Re-examine the data presented in Exercise 7. Is there any data value that may possibly be an outlier? How would the results be affected if that value were excluded from the analysis? In general, can an outlier have a dramatic effect on the hypothesis test and confidence interval?

16. Exercise 13 used temperatures given in degrees Celsius. Suppose we express all of the temperatures in their equivalent Fahrenheit value. Is the hypothesis test affected by such a change in units? Is the confidence interval affected by such a change in units? How?

17. The 95% confidence interval for a collection of paired sample data is $0.0 < \mu_d < 1.2$. Based on this confidence interval, the traditional method of hypothesis testing leads to the conclusion that the claim of $\mu_d > 0$ is supported. What is the smallest possible value of the significance level of the hypothesis test?

8-3 Inferences About Two Means: Independent and Large Samples

Section 8–2 described methods of using inferential statistics for dependent (paired) data. In this section our two samples are *independent*, that is, the sample selected from one population is not related to the sample selected from the other population. This section is also restricted to large ($n > 30$) samples. For the hypothesis tests and confidence intervals described in this section, we make the following assumptions.

Assumptions

1. The two samples are *independent*.
2. The two sample sizes are large. That is, $n_1 > 30$ and $n_2 > 30$.
3. Both samples have been selected by valid random-sampling techniques.

Exploring the Data Sets

Before we jump into a formal hypothesis test or construction of a confidence interval, it is often enlightening to *explore* the two samples, using the techniques described in Chapter 2. For example, Figure 8-3 shows the value of boxplots for comparing the axial loads of a sample of cans 0.0109 in. thick and another sample of cans 0.0111 in. thick. (Based on Data Set 15 in Appendix B; an axial load of a can is the maximum weight supported by its sides.)

The axial load of 504 lb (for the cans that are 0.0111 in. thick) stands out as a possible outlier, and indeed it is very far away from all of the other values. If that outlier is deleted, the long right tail (or "whisker") on the bottom boxplot is shortened to the value identified as 317. A comparison of the two boxplots shows that the spread of the data is about the same, but the values for the 0.0111 in. cans appear to be farther to the right, suggesting that those cans have larger axial loads and are stronger. We can now use a hypothesis test to determine whether that apparent difference is significant.

Hypothesis Tests

In testing a claim about the difference $\mu_1 - \mu_2$, we use the following test statistic.

TEST STATISTIC FOR TWO MEANS: INDEPENDENT AND LARGE SAMPLES

$$z = \frac{(\overline{x}_1 - \overline{x}_2) - (\mu_1 - \mu_2)}{\sqrt{\dfrac{\sigma_1^2}{n_1} + \dfrac{\sigma_2^2}{n_2}}}$$

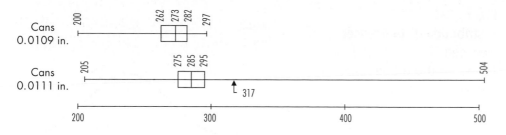

Figure 8-3
**Boxplots of Axial Loads
(in Pounds) of Cans
0.0109 in. Thick and
0.0111 in. Thick**

Except for the form of this test statistic, the method of hypothesis testing is essentially the same as that described in Chapter 7. (Later in this section, we will discuss the rationale for the form of the given test statistic.) As in Chapter 7, if the values of σ_1 and σ_2 are not known, we can use s_1 and s_2 in their places, provided that both samples are large. If σ_1 and σ_2 are known, we use those values in calculating the test statistic, but realistic cases will usually require the use of s_1 and s_2.

EXAMPLE

Use a 0.01 significance level to test the claim that cans 0.0109 in. thick have a lower mean axial load than cans that are 0.0111 in. thick. The original data are listed in Data Set 15 of Appendix B, and the summary statistics are listed in the margin.

SOLUTION

Step 1: The claim can be expressed symbolically as $\mu_1 < \mu_2$.

Step 2: If the original claim is false, then $\mu_1 \geq \mu_2$.

Step 3: The null hypothesis must contain equality, so we have

$$H_0: \mu_1 \geq \mu_2 \qquad H_1: \mu_1 < \mu_2 \text{ (original claim)}$$

We now proceed with the assumption that $\mu_1 = \mu_2$, or $\mu_1 - \mu_2 = 0$.

Step 4: The significance level is $\alpha = 0.01$.

Step 5: Because we have two independent and large samples and we are testing a claim about the two population means, we use a normal distribution with the test statistic given earlier in this section.

Step 6: The values of σ_1 and σ_2 are unknown, but the samples are both large, so we can use the sample standard deviations as estimates of the population standard deviations, and the test statistic is calculated as follows:

$$z = \frac{(\bar{x}_1 - \bar{x}_2) - (\mu_1 - \mu_2)}{\sqrt{\dfrac{\sigma_1^2}{n_1} + \dfrac{\sigma_2^2}{n_2}}} = \frac{(267.1 - 281.8) - 0}{\sqrt{\dfrac{22.1^2}{175} + \dfrac{27.8^2}{175}}} = -5.48$$

Because we are using a normal distribution, the critical value of $z = -2.33$ is found from Table A-2. (With $\alpha = 0.01$ in the left tail, find the z score corresponding to an area of $0.5 - 0.01$, or 0.4900.) The test statistic, critical value, and critical region are shown in Figure 8-4.

Step 7: Because the test statistic does fall within the critical region, reject the null hypothesis $\mu_1 \geq \mu_2$.

Axial Loads (lb)
of 0.0109 in. Cans

$n_1 = 175$

$\bar{x}_1 = 267.1$
$s_1 = 22.1$

Axial Loads (lb)
of 0.0111 in. Cans

$n_2 = 175$

$\bar{x}_2 = 281.8$
$s_2 = 27.8$

Figure 8-4
Distribution of Differences Between Means of Axial Loads for 0.0109 in. Cans and 0.0111 in. Cans

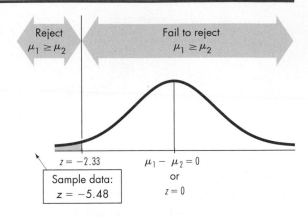

Interpretation

There is sufficient evidence to support the claim that the 0.0109 in. cans have a mean axial load that is lower than the mean for the 0.0111 in. cans. We might have expected the thinner cans to be weaker, but we now know that the difference is statistically significant. The *practical* significance of that difference is another issue not addressed by this hypothesis test, and to assess practical significance would require additional information—for example, about the *actual* loads typically applied to such cans, and whether these are at all close to the cans' *maximum* loads.

P-Values

Determination of *P*-values is easy here because the test statistics are z scores from the standard normal distribution. Simply follow the method outlined in Figures 7-7 and 7-8. For the preceding left-tailed hypothesis test, the *P*-value is the area to the left of the test statistic $z = -5.48$. Referring to Table A-2, we find that the area to the left of $z = -5.48$ is 0.0001. (Because the z score is beyond 3.09, we use an area of 0.4999 as indicated in the footnote to Table A-2.) Because the *P*-value of 0.0001 is less than the significance level of $\alpha = 0.01$, we again reject the null hypothesis and conclude that the 0.0109 in. cans have a mean that is lower than the mean for the 0.0111 in. cans.

Confidence Intervals

We can construct confidence interval estimates of the difference between two population means, which we denote as $\mu_1 - \mu_2$.

The confidence interval estimate of the difference $\mu_1 - \mu_2$ is as follows:

$$(\bar{x}_1 - \bar{x}_2) - E < (\mu_1 - \mu_2) < (\bar{x}_1 - \bar{x}_2) + E$$

where

$$E = z_{\alpha/2} \sqrt{\frac{\sigma_1^2}{n_1} + \frac{\sigma_2^2}{n_2}}$$

As with hypothesis testing, if σ_1 and σ_2 are not known, use s_1 and s_2 in their places, provided that both samples are large.

EXAMPLE

Using the sample data given in the preceding example, construct a 99% confidence interval estimate of the difference between the means of the axial loads of the 0.0109 in. cans and the 0.0111 in. cans.

SOLUTION

We first find the value of the margin of error E:

$$E = z_{\alpha/2} \sqrt{\frac{\sigma_1^2}{n_1} + \frac{\sigma_2^2}{n_2}} = 2.575 \sqrt{\frac{22.1^2}{175} + \frac{27.8^2}{175}} = 6.9$$

We now find the desired confidence interval as follows:

$$(\bar{x}_1 - \bar{x}_2) - E < (\mu_1 - \mu_2) < (\bar{x}_1 - \bar{x}_2) + E$$
$$(267.1 - 281.8) - 6.9 < (\mu_1 - \mu_2) < (267.1 - 281.8) + 6.9$$
$$-21.6 < (\mu_1 - \mu_2) < -7.8$$

Interpretation
We are 99% confident that the limits of -21.6 and -7.8 actually do contain the difference between the two population means. This result could be more clearly presented by stating that μ_2 exceeds μ_1 by an amount that is between 7.8 lb and 21.6 lb. Because those limits do not contain 0, it is very unlikely that the two population means are equal.

Rationale: Why Do the Test Statistic and Confidence Interval Have the Particular Forms Presented?

Both forms are based on a normal distribution of $\bar{x}_1 - \bar{x}_2$ values with mean $\mu_1 - \mu_2$ and standard deviation $\sqrt{\sigma_1^2/n_1 + \sigma_2^2/n_2}$. This follows from the central limit theorem (introduced in Section 5–5). The central limit theorem tells us that sample means $\bar{x}$ are normally distributed with mean μ and standard deviation $\sigma/\sqrt{n}$. Also, when samples have a size of 31 or larger, the normal distribution

STATISTICAL PLANET

The Placebo Effect
It has been a common belief that when patients are given a placebo (a treatment with no medicinal value), about one-third of them show some improvement. However, a more recent study of 6000 patients showed that for those with mild medical problems, the placebos seemed to result in improvement in about two-thirds of the cases. The placebo effect seems to be strongest when patients are very anxious and they like their physicians. Because it could cloud studies of new treatments, the placebo effect is minimized by using a *double-blind* experiment in which neither the patient nor the physician knows whether the treatment is a placebo or a real medicine.

serves as a reasonable approximation to the distribution of sample means. By similar reasoning, the values of $\bar{x}_1 - \bar{x}_2$ also tend to approach a normal distribution with mean $\mu_1 - \mu_2$. When both samples are large, the following property of variances leads us to conclude that the values of $\bar{x}_1 - \bar{x}_2$ will have a standard deviation of

$$\sqrt{\frac{\sigma_1^2}{n_1} + \frac{\sigma_2^2}{n_2}}$$

The variance of the *differences* between two independent random variables equals the variance of the first random variable *plus* the variance of the second random variable.

That is, the variance of sample values $\bar{x}_1 - \bar{x}_2$ will tend to equal

$$\sigma_{\bar{x}_1}^2 + \sigma_{\bar{x}_2}^2$$

provided that $\bar{x}_1$ and $\bar{x}_2$ are independent. (See Exercise 21.) For large random samples, the standard deviation of sample means is $\sigma/\sqrt{n}$, so the variance of sample means is σ^2/n. If we combine the additive property of variances with the central limit theorem's expression for variance of sample means, we get the following:

$$\sigma_{\bar{x}_1 - \bar{x}_2}^2 = \sigma_{\bar{x}_1}^2 + \sigma_{\bar{x}_2}^2 = \frac{\sigma_1^2}{n_1} + \frac{\sigma_2^2}{n_2}$$

Taking the square root, we see that the standard deviation of the differences $\bar{x}_1 - \bar{x}_2$ can be expressed as

$$\sqrt{\frac{\sigma_1^2}{n_1} + \frac{\sigma_2^2}{n_2}}$$

Because z is a standard score that corresponds in general to

$$z = \frac{\text{(sample statistic)} - \text{(population mean)}}{\text{(standard deviation of sample statistics)}}$$

we get

$$z = \frac{(\bar{x}_1 - \bar{x}_2) - (\mu_1 - \mu_2)}{\sqrt{\frac{\sigma_1^2}{n_1} + \frac{\sigma_2^2}{n_2}}}$$

by noting that the sample values of $\bar{x}_1 - \bar{x}_2$ will have a mean of $\mu_1 - \mu_2$ and the standard deviation previously given.

EXCEL (Prior to 2007): From the **Tools** menu, select **Data Analysis**. Choose **z-Test: Two Sample for Means**. Enter these details in the dialog box that appears: For **Variable 1 Range**, specify the worksheet range that contains the first variable's data; for **Variable 2 Range**, specify the worksheet range that contains the second variable's data. Also enter the appropriate choice for **Alpha** (in this example, 0.01). Enter the **Variances** for the two variables (which you must calculate separately, for example, by using the **=VAR()** function) Now select an **Output Option**, and click on **OK** to obtain the results.

EXCEL 2007: Click **Data** on the main menu and choose **Data Analysis** from the **Analysis** tab. Proceed as above.

EXCEL DISPLAY
Hypothesis Test

	Variable 1	Variable 2
z-Test: Two Sample for Means		
Mean	267.11429	281.80571
Known Variance	488.9524	771.4333
Observations	175	175
Hypothesized Mean Difference	0	
z	-5.474334	
P(Z<=z) one-tail	2.201E-08	
z Critical one-tail	2.3263419	
P(Z<=z) two-tail	4.402E-08	
z Critical two-tail	2.5758345	

Note that both the one-tailed and the two-tailed solutions are provided, although you should determine in advance which version will apply. As in a printed z table, only positive critical z values are output, but they may be interpreted as negative where appropriate.

Although Excel cannot calculate a confidence interval directly for the data, the Data Desk XL add-in can be used.

EXCEL (Prior to 2007): Click on **DDXL**. Select **Confidence Intervals** and **2 Var t Interval**. In the dialog box, click on the pencil icon for the first quantitative column and enter the range for the first sample's data. Then click on the pencil icon for the second quantitative column and enter the range for the second sample's data. Click on **OK**. Follow the steps in the next dialog box that appears. (In Step 1, select **2 Sample**.)

EXCEL 2007: Click on the **Add-in** tab on the main menu. You will see **DDXL** to the left. Click on it and proceed as above.

STATDISK: Select **Analysis**, then **Hypothesis Testing**, then **Mean-Two Independent Samples**. To conduct a Z test, enter the standard deviations in the **Population St Dev** boxes. Confidence interval limits are included with hypothesis test results.

STATDISK DISPLAY
Hypothesis Test and
Confidence Interval

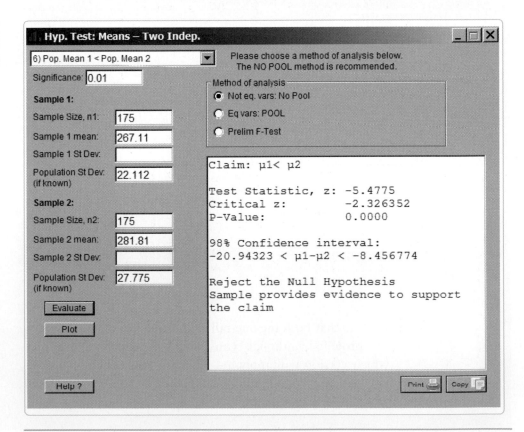

8-3 Exercises A: Basic Skills and Concepts

In Exercises 1 and 2, use a 0.05 significance level to test the claim that the two samples come from populations with the same mean. In each case, the two samples are independent and have been randomly selected.

1.

Control Group	Experimental Group
$n_1 = 50$	$n_2 = 100$
$\bar{x}_1 = 75$	$\bar{x}_2 = 73$
$s_1 = 15$	$s_2 = 14$

2.

Math Majors	English Majors
$n_1 = 60$	$n_2 = 75$
$\bar{x}_1 = 8.75$	$\bar{x}_2 = 9.66$
$s_1 = 2.05$	$s_2 = 2.88$

3. Using the sample data given in Exercise 1, construct a 95% confidence interval estimate of the difference between the two population means. Do the confidence interval limits contain 0? What do you conclude about the difference between the control group and experimental group?

4. Using the sample data given in Exercise 2, construct a 95% confidence interval estimate of the difference between the two population means. Do the confidence interval limits contain 0? What do you conclude about the difference between the math majors and the English majors?

5. For comparison purposes, mutual funds are often grouped into various categories. Do Canadian equity funds and Canadian balanced equity funds differ with respect to percentage of foreign holdings within the funds?

   a. Using the data shown, and a 0.01 level of significance, test the claim that there is no difference in foreign content between these fund groups.

   b. Construct a 99% confidence interval for $\mu_1 - \mu_2$, where μ_1 is the mean foreign content (in percent) in Canadian equity funds. (Based on data from *Mutual Funds Update*, published by the Bank of Montreal.)

Canadian Equity Funds
$n_1 = 33$
$\bar{x}_1 = 8.386$
$s_1 = 7.126$

Canadian Balanced Equity Funds
$n_2 = 33$
$\bar{x}_2 = 9.233$
$s_2 = 7.454$

6. Provincial governments across Canada have been re-examining their delivery of services, such as health care. A researcher compared the number of beds in long-term care centres in New Brunswick and in Nova Scotia. Her sample results are summarized in the margin, based on data in *Guide to Canadian Healthcare Facilities 1998–1999*, Vol. 6 (Canadian Healthcare Association, Ottawa). Testing at the 0.05 significance level, is there sufficient evidence to support a claim that the mean number of beds per centre is greater in Nova Scotia than in New Brunswick? Do the results support a claim that there are more long-term care beds per person in Nova Scotia than in New Brunswick? Why or why not?

New Brunswick	Nova Scotia
$n_1 = 31$	$n_2 = 38$
$\bar{x}_1 = 37.054$	$\bar{x}_2 = 60.973$
$s_1 = 49.046$	$s_2 = 76.602$

7. The Medassist Pharmaceutical Company wants to test Dozenol, a new cold medicine intended for night use. Tests for such products often include a "treatment group" of people who use the drug and a "control group" of people who don't use the drug. Fifty people with colds are given Dozenol, and 100 others are not. The systolic blood pressure is measured for each subject, and the sample statistics are as given in the margin. The head of research at Medassist claims that Dozenol does not affect blood pressure—that is, the treatment population mean μ_1 and the control population mean μ_2 are equal.

Treatment Group	Control Group
$n_1 = 50$	$n_2 = 100$
$\bar{x}_1 = 203.4$	$\bar{x}_2 = 189.4$
$s_1 = 39.4$	$s_2 = 39.0$

a. Test that claim using a significance level of 0.01. Based on the result, would you recommend advertising that Dozenol does not affect blood pressure?

b. Construct a 99% confidence interval for $\mu_1 - \mu_2$, where μ_1 and μ_2 represent the mean for the treatment group and the control group, respectively.

8. A group of college students randomly selected 217 student cars and found that they had ages with a mean of 7.89 years and a standard deviation of 3.67 years. They also randomly selected 152 faculty cars and found that they had ages with a mean of 5.99 years and a standard deviation of 3.65 years.

a. Use a 0.05 significance level to test the claim that student cars are older than faculty cars.

b. Construct a 95% confidence interval for the mean $\mu_1 - \mu_2$, where μ_1 is the mean age of student cars.

9. In April 1996, *The Globe and Mail* listed the salaries of the hundreds of University of Toronto staff who earn at least $100,000 annually. On studying this article, several disgruntled staff (whose last names begin with S, V, and Q) concluded that top staff whose names begin with S to Z earn less than their counterparts.

Names A to R	Names S to Z
$n_1 = 40$	$n_2 = 40$
$\bar{x}_1 = 118.980$	$\bar{x}_2 = 117.580$
$s_1 = 27.050$	$s_2 = 15.796$

a. Based on the sample data shown, construct appropriate null and alternate hypotheses, and test the employees' claim, at the 0.05 significance level. Mean salaries are in $1000s.

b. Construct a 95% confidence interval for $\mu_1 - \mu_2$, where μ_1 is the mean salary of the reported staff with last names starting with the letters A to R. Do the confidence interval limits contain 0? Does this indicate that there is or is not a significant difference between the two means?

10. A fish-processing company is concerned about the shelf life of its new cat food. A sample examined in Halifax, Nova Scotia ($n = 35$) had an acceptable mean shelf life of 13 months, with a standard deviation of 1.44 months. But a second sample, examined in Dartmouth, Nova Scotia ($n = 40$), had a mean shelf life of only 11 months, with a standard deviation of 1.52 months.

The manager in Dartmouth claims that there is no significant difference between these two means.

a. Test her claim at the 0.05 level of significance. Based on the result, should the operation in Dartmouth be subjected to a special review?

b. Construct a 95% confidence interval for the difference between the population means $(\mu_1 - \mu_2)$, where μ_1 is the mean shelf life of cat food in Halifax and μ_2 is the mean shelf life of cat food in Dartmouth. Do the confidence interval limits contain 0? Does this suggest that there is not a significant difference between the two population means?

11. Data Set 9 in Appendix B contains measured levels of serum cotinine (in ng/mL) for two samples of nonsmokers: nonsmokers with no environmental tobacco smoke exposure at home or at work, and nonsmokers who *are* exposed to environmental tobacco smoke at home or at work.

a. Testing at the 0.02 level of significance, do the data support a claim that nonsmokers with exposure to environmental tobacco smoke have higher levels of serum cotinine than nonsmokers without such exposure?

b. Construct a 96% confidence interval for $\mu_1 - \mu_2$ where μ_1 is the mean for exposed nonsmokers and μ_2 is the mean for nonexposed nonsmokers. Do the confidence interval limits contain 0? Assuming that raised levels of serum cotinine are undesirable, do the results suggest that exposure to environmental tobacco smoke poses a possible health hazard?

12. Data Set 11 in Appendix B contains weights of 100 randomly selected M&M plain candies. Those weights have a mean of 0.9147 g and a standard deviation of 0.0369 g. A previous edition of this book used a different sample of 100 M&M plain candies, with a mean and standard deviation of 0.9160 g and 0.0433 g, respectively. Is the discrepancy between the two sample means significant?

13. Two samples are used to compare building permit values for larger communities versus smaller ones in the regional municipality of Durham, in Ontario. For both samples n is 38. Expressed in thousands of dollars, the mean value per building permit for the larger communities is 102.008, with a standard deviation of 25.546. The mean value per permit for smaller communities is 113.637, with a standard deviation of 70.670. At the 0.05 significance level, test the claim that the population of smaller communities has a higher mean value for the building permits issued (based on data in the *Durham Business News,* June 1998).

14. As an aid to farmers, the Alberta government records freezing dates for over a hundred locations around the province. Based on average last-frost dates for spring, a researcher has proposed dividing the province into early last-frost locations and late last-frost locations. Refer to Data Set 17 of Appendix B.

Use the day-of-the-year columns to test, at the 0.01 significance level, the claim that the late last-frost locations have a different mean date for last spring frost than the early last-frost locations.

a. Conduct the hypothesis test using the traditional method.

b. Conduct the hypothesis test using the *P*-value method.

c. Conduct the hypothesis test by constructing a 99% confidence interval for the difference $\mu_1 - \mu_2$.

d. Compare the results from parts (a), (b), and (c).

15. Refer to the pulse rates in Data Set 8 from Appendix B. First identify and delete the two outliers that are unrealistic pulse rates for living and reasonably healthy statistics students.

a. Use a 0.05 level of significance and the traditional method of testing hypotheses to test the claim that male and female statistics students have the same mean pulse rate.

b. Use the *P*-value method to test the claim given in part (a).

c. Construct a 95% confidence interval for the difference $\mu_1 - \mu_2$, where μ_1 and μ_2 are the mean pulse rates for male and female statistics students, respectively.

d. Compare the results from parts (a), (b), and (c).

16. When testing for a difference between the means of a treatment group and a group given a placebo, the following Excel display is obtained. Using a 0.05 significance level, is there sufficient evidence to support the claim that the treatment group (Variable 1) comes from a population with a mean that is less than the mean for the placebo population? Explain.

EXCEL DISPLAY

	A	B	C
1	z-Test: Two Sample for Means		
2			
3		Variable 1	Variable 2
4	Mean	152.0739	154.9669
5	Known variance	438.4388	239.1461
6	Observations	50	50
7	Hypothesized mean difference	0	
8	z	-0.7858	
9	P(Z<=z) one-tail	0.215991	
10	z Critical one-tail	1.644853	
11	P(Z<=z) two-tail	0.431982	
12	z Critical two-tail	1.959961	

17. When testing for a difference between the means of a treatment group and a group given a placebo, the following Excel display is obtained. Using a 0.02 significance level, is there sufficient evidence to support the claim that the treatment group (Variable 1) comes from a population with a mean that is different from the mean for the placebo population? Explain.

	A	B	C
1	z-Test: Two Sample for Means		
2			
3		Variable 1	Variable 2
4	Mean	75.34798	73.81763
5	Known variance	1.55383	11.65292
6	Observations	50	50
7	Hypothesized mean difference	0	
8	z	2.199422	
9	P(Z<=z) one-tail	0.013924	
10	z Critical one-tail	1.644853	
11	P(Z<=z) two-tail	0.027848	
12	z Critical two-tail	1.959961	

8-3 Exercises B: Beyond the Basics

18. Refer to Exercise 5. If the actual difference between the population means is 1.5, find β, the probability of a type II error. (See Section 7–2.) *Hint:* In Step 1, replace $\bar{x}$ by $(\bar{x}_1 - \bar{x}_2)$, replace $\mu_{\bar{x}}$ by 0, and replace $\sigma_{\bar{x}}$ by

$$\sqrt{\frac{\sigma_1^2}{n_1} + \frac{\sigma_2^2}{n_2}}$$

19. The examples in this section used the axial loads of a sample of cans 0.0109 in. thick and a second sample of cans 0.0111 in. thick. (The original data are listed in Data Set 15 of Appendix B.) We noted that the second sample includes an outlier of 504 lb. How are the boxplots of Figure 8-3, the hypothesis test, and the confidence interval affected if we decide that the outlier is actually an error that should be deleted?

20. The examples in this section used the axial loads of cans, where the loads were given in pounds. How are the boxplots, hypothesis test, and confidence interval affected if the loads are all converted to kilograms?

21. a. Find the variance for this *population* of x scores: 5, 10, 15. (See Section 2–5 for the variance σ^2 of a population.)

    b. Find the variance for this *population* of y scores: 1, 2, 3.

    c. List the *population* of all possible differences $x - y$, and find the variance of this population.

    d. Use the results from parts (a), (b), and (c) to verify that $\sigma_{x-y}^2 = \sigma_x^2 + \sigma_y^2$. (This principle is used to derive the test statistic and confidence interval for this section.)

    e. How is the *range* of the differences $x - y$ related to the range of the x values and the range of the y values?

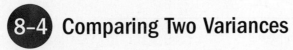

8-4 Comparing Two Variances

Because the characteristic of variation among data is extremely important, this section presents a method for using two samples to compare the variances of the populations from which the samples are drawn. The method we use requires the following assumptions.

Assumptions

1. The two populations are *independent* of each other. (Recall from Section 8–2 that two samples are independent if the sample selected from one population is not related to the sample selected from the other population.)
2. The two populations are each *normally distributed*. (This assumption is important, because the test statistic is extremely sensitive to departures from normality.)

Exploring the Data Sets

As in previous sections of this chapter, it is important to explore the data sets before undertaking formal hypothesis tests or constructing confidence intervals. In this section, because the requirement of a normal distribution is quite strict, procedures such as the construction of *normal probability plots* to test for normality (discussed more fully in Section 10–4) become especially important, as does the identification and removal of outliers.

Hypothesis Tests

The computations of this section will be greatly simplified if we stipulate that s_1^2 represents the *larger* of the two sample variances. It doesn't really make any difference which sample is designated as sample 1, so we use the following notation.

s_1^2 = *larger* of the two sample variances

n_1 = size of the sample with the *larger* variance

σ_1^2 = variance of the population from which the sample with the larger variance was drawn

The symbols s_2^2, n_2, and σ_2^2 are used for the other sample and population.

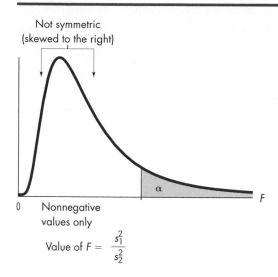

Not symmetric
(skewed to the right)

0   Nonnegative
      values only

Value of $F = \dfrac{s_1^2}{s_2^2}$

Figure 8-5 *F* Distribution
There is a different *F distribution* for each different pair of degrees of freedom for numerator and denominator.

For two normally distributed populations with equal variances (that is, $\sigma_1^2 = \sigma_2^2$), the sampling distribution of the following test statistic is the **F distribution** shown in Figure 8-5 with critical values listed in Table A-5. If you continue to repeat an experiment of randomly selecting samples from two normally distributed populations with equal variances, the distribution of the ratio s_1^2/s_2^2 of the sample variances is the *F* distribution. In Figure 8-5, note these properties of the *F* distribution:

- The *F* distribution is not symmetric.

- Values of the *F* distribution cannot be negative.

- The exact shape of the *F* distribution depends on two different degrees of freedom.

TEST STATISTIC FOR HYPOTHESIS TESTS WITH TWO VARIANCES

$$F = \frac{s_1^2}{s_2^2}$$

If the two populations really do have equal variances, then $F = s_1^2/s_2^2$ tends to be close to 1 because s_1^2 and s_2^2 tend to be close in value. But if the two populations have radically different variances, s_1^2 and s_2^2 tend to be very different numbers. Denoting the larger of the sample variances by s_2^2, we see that the ratio s_1^2/s_2^2 will be a large number whenever s_2^2 and s_2^2 are far apart in value. Consequently, a value of F near 1 will be evidence in favour of the conclusion that $\sigma_1^2 = \sigma_2^2$, but a large value of F will be evidence against the conclusion of equality of the population variances.

This test statistic applies to a claim made about two variances, but we can also use it for claims about two population standard deviations. Any claim about two population standard deviations can be restated in terms of the corresponding variances.

Critical Values

Using Table A-5, we obtain critical F values that are determined by the following three values:

1. **The significance level α** (Table A-5 has six pages of critical values for $\alpha = 0.01$, 0.025, and 0.05.)

2. **Numerator degrees of freedom** $= n_1 - 1$

3. **Denominator degrees of freedom** $= n_2 - 1$

To find a critical value, first refer to the part of Table A-5 corresponding to α (for a one-tailed test) or $\alpha/2$ (for a two-tailed test), then intersect the column representing the degrees of freedom for s_1^2 with the row representing the degrees of freedom for s_2^2. Because we are stipulating that the larger sample variance is s_1^2, all one-tailed tests will be right-tailed and all two-tailed tests will require that we find only the critical value located to the right. Good news: We have no need to find a critical value separating a left-tailed critical region. (If you would like to try anyway, see Exercise 15.)

We often have numbers of degrees of freedom that are not included in Table A-5. We could use linear interpolation to approximate the missing values, but in most cases that's not necessary because the F test statistic is either less than the lowest possible critical value or greater than the largest possible critical value. For example, Table A-5 shows that given $\alpha = 0.025$ in the right tail, 20 degrees of freedom for the numerator, and 34 degrees of freedom for the denominator, the critical F value is between 2.0677 and 2.1952. Any F test statistic below 2.0677 will result in failure to reject the null hypothesis, any F test statistic above 2.1952 will result in rejection of the null hypothesis, and interpolation is necessary only if the F test statistic falls between 2.0677 and 2.1952. (See Exercise 13 in this section.) The use of a statistical software package such as Excel or STATDISK eliminates this problem by providing critical values or P-values.

Axial Loads (lb) of 0.0109 in. Cans	Axial Loads (lb) of 0.0111 in. Cans
$n = 175$	$n = 175$
$\bar{x} = 267.1$	$\bar{x} = 281.8$
$s = 22.1$	$s = 27.8$

EXAMPLE

In the preceding section we used a sample of 175 aluminum cans that are 0.0109 in. thick and a second sample of 175 aluminum cans (this time we do not remove the outlier) that are 0.0111 in. thick. Data Set 15 lists the axial loads of the cans in those samples, and the summary statistics are listed in the margin. Use a 0.05 significance level to test the claim that the samples come from populations with the same variance.

SOLUTION

Because we stipulate in this section that the larger variance is denoted by s_1^2, we reverse the subscript notation used in the preceding section, and we let $s_1^2 = 27.8^2$, $n_1 = 175$, $s_2^2 = 22.1^2$, and $n_2 = 175$. We now proceed to use the traditional method of testing hypotheses as outlined in Figure 7-4:

Step 1: The claim of equal population variances is expressed symbolically as $\sigma_1^2 = \sigma_2^2$.

Step 2: If the original claim is false, then $\sigma_1^2 \neq \sigma_2^2$.

Step 3: Because the null hypothesis must contain equality, we have

$$H_0: \sigma_1^2 = \sigma_2^2 \text{ (original claim)} \qquad H_1: \sigma_1^2 \neq \sigma_2^2$$

Step 4: The significance level is $\alpha = 0.05$.

Step 5: Because this test involves two population variances, we use the F distribution.

Step 6: The test statistic is

$$F = \frac{s_1^2}{s_2^2} = \frac{27.8^2}{22.1^2} = 1.5824$$

For the critical values, first note that this is a two-tailed test with 0.025 in each tail. As long as we are stipulating that the larger variance is placed in the numerator of the F test statistic, we need to find only the right-tailed critical value. From Table A-5 we get a critical value of $F = 1.4327$, which corresponds to 0.025 in the right tail, with 174 degrees of freedom for the numerator and 174 degrees of freedom for the denominator. (Actually, the table does not include 174 degrees of freedom, so we chose the closest value of 120 degrees of freedom. If the difference in sample variances is significant based on sample sizes of 121, that difference would certainly be significant with sample sizes of 175.)

Step 7: Figure 8-6 shows that the test statistic $F = 1.5824$ falls within the critical region, so we reject the null hypothesis.

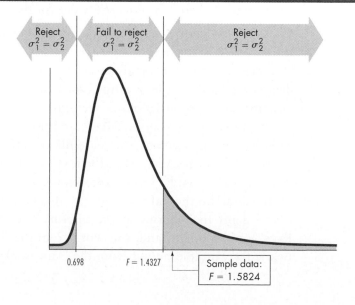

Figure 8-6
Distribution of s_1^2/s_2^2 for Axial Loads of 0.0111 in. Cans and 0.0109 in. Cans

Interpretation

There is sufficient evidence to warrant rejection of the claim that the two variances are equal. However, we should recognize that the F test is extremely sensitive to distributions that are not normally distributed—and the distributions of axial loads are by no means perfectly normal. Therefore, the apparent result of this test (that the variances are not equal) cannot be accepted as definitive.

In the preceding example we used a two-tailed test for the claim of equal variances. A right-tailed test of the claim that the 0.0111 in. cans have a larger variance would yield the same test statistic of $F = 1.5824$, but a different critical value of $F = 1.3519$ (found from Table A-5 with $\alpha = 0.05$). We would again have sufficient evidence to support the claim of a larger variance.

We have described the traditional method of testing hypotheses made about two population variances. Exercise 14 below deals with the P-value approach, and Exercise 16 deals with the construction of confidence intervals.

USING TECHNOLOGY

EXCEL (Prior to 2007): From the **Tools** menu, select **Data Analysis**. Choose **F-Test Two-Sample for Variances**. Enter these details in the dialog box that appears: For **Variable 1 Range**, specify the worksheet range that contains the data for the variable with the larger variance; for **Variable 2 Range**, specify the worksheet range that contains the other variable's data. Also enter the appropriate choice for the right-tail **Alpha** (in this example, 0.025). For a two-tailed hypothesis test, enter $\alpha/2$ at this point. Then select an **Output Option**, and click on **OK** to obtain the results. The Data Analysis tool returns the critical F value for the right tail.

EXCEL 2007: Click **Data** on the main menu and choose **Data Analysis** from the **Analysis** tab. Proceed as above.

Excel formulas can be also used for constructing confidence intervals for the F ratio, based on the calculations described in Exercise 16 of this section.

The following figure is Excel's solution to the solved example for this section. There are minor differences from the text version because (1) in the text-based solution, 174 degrees of freedom had to be approximated by 120 degrees of freedom in the tables, and (2) the text version has rounded off the s terms in the intermediate calculations. Excel, on the other hand, can work from the full data set. Below the example output from the Data Analysis tool are sample formulas for finding the left-tail critical F and the confidence interval.

MINITAB 15: From the **Stat** menu, choose **Basic Statistics** and then **2 Variances**. If the test is being conducted with raw data, select the appropriate columns. The default is a 5% level of significance. If you wish to change this, click the **Options** button and make the necessary changes.

STATDISK: Select **Analysis** from the main menu, then select **Hypothesis Testing**, then **St. Dev.-Two Samples**.

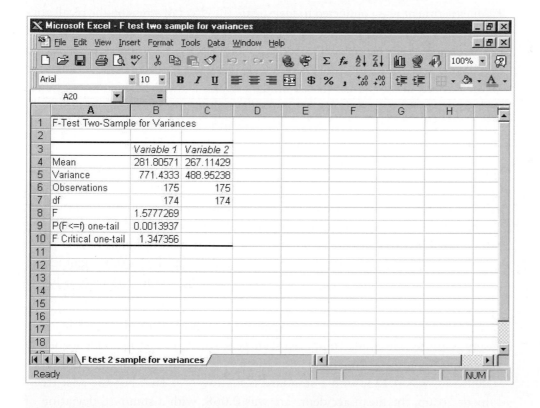

Cell	Formula	Result	Explanation
B12	5FINV(1-0.025,B7,C7)	0.742193862	Critical F, left tail, for two-tail test
B13	5B5/C5*(1/B10)	1.170979939	Lower bound of confidence interval
B14	5B5/C5*(1/B12)	2.125761115	Upper bound of confidence interval

Use appropriate cell addresses for data. Formula in B12 assumes $\alpha = 0.05$.

8-4 Exercises A: Basic Skills and Concepts

In Exercises 1 and 2, test the given claim. Use a significance level of $\alpha = 0.05$, and assume that all populations are normally distributed. Use the traditional method of testing hypotheses outlined in Figure 7-4, and draw the appropriate graphs.

1. Claim: Populations A and B have different variances ($\sigma_1^2 \neq \sigma_2^2$).
   Sample A: $n = 25$, $\overline{x} = 175$, $s^2 = 900$
   Sample B: $n = 31$, $\overline{x} = 200$, $s^2 = 2000$

2. Claim: Population A has a larger variance than population B.
   Sample A: $n = 50$, $\overline{x} = 77.4$, $s = 18.1$
   Sample B: $n = 15$, $\overline{x} = 75.7$, $s = 7.0$

Nonstress	Stress
$n_1 = 40$	$n_2 = 40$
$\overline{x}_1 = 53.3$	$\overline{x}_2 = 45.3$
$s_1 = 11.6$	$s_2 = 13.2$

3. An experiment was conducted to investigate the effect of stress on the recall ability of police eyewitnesses. The experiment involved a nonstressful interrogation of a cooperative suspect and a stressful interrogation of an uncooperative and belligerent suspect. The numbers of details recalled a week after the incident are summarized in the margin (based on data from "Eyewitness Memory of Police Trainees for Realistic Role Plays," by Yuille et al., *Journal of Applied Psychology,* Vol. 79, No. 6). Use a 0.10 level of significance to test the claim that the samples come from populations with different standard deviations.

4. Customer waiting times are studied at the Humber Valley Credit Union. When 25 randomly selected customers enter any one of several waiting lines, their times have a mean of 6.896 min and a standard deviation of 3.619 min. When 20 randomly selected customers enter a single main waiting line that feeds the individual teller stations, their waiting times have a mean of 7.460 min and a standard deviation of 1.841 min. Use a 0.01 significance level to test the claim that waiting times for the single line have a lower standard deviation.

5. A study was done of 137 young adult males who had been treated for substance abuse. The study compared their yearly rates of traffic accidents before treatment and after treatment. For the pre-treatment group, the mean accident rate was 0.123, with a standard deviation of 0.167. For the post-treatment group, the mean accident rate was 0.068, with a standard deviation of 0.223 (based on data from "Does Treatment for Substance Abuse Improve Driving Safety? A Preliminary Evaluation," by Mann et al., Addiction Research Foundation). At the 0.05 significance level, test the claim that the standard deviation for the post-treatment group is greater than that for the pre-treatment group.

6. Canadian equity funds and Canadian balanced equity funds can be compared for their percentages of foreign holdings. At the 0.05 significance level, test the claim that both types of funds have the same variance.

Canadian Equity Funds	Canadian Balanced Equity Funds
$n_1 = 33$	$n_2 = 33$
$\bar{x}_1 = 8.386$	$\bar{x}_2 = 9.233$
$s_1 = 7.126$	$s_2 = 7.454$

Based on data from *Mutual Funds Update*, published by the Bank of Montreal, March 31, 1998.

7. A fish-processing company is concerned about the shelf life of its new cat food. A sample examined in Halifax ($n = 35$) had an acceptable mean shelf life of 13 months, with a standard deviation of 1.44 months. But a second sample, examined in Dartmouth ($n = 40$), had a mean shelf life of only 11 months, with a standard deviation of 1.52 months. At the 0.05 level of significance, test the claim that shelf lives at both plants have the same standard deviation.

8. An investor is comparing the variability of estimates by stock analysts, when they analyze the earnings potentials of U.S. and Canadian firms. For a sample of estimates (by different analysts) of the earning potentials of Canadian firms, the standard deviation is 0.0204, whereas the standard deviation for a sample of earnings estimates for U.S. firms is 0.0076 (based on data from "The Seasonal Impact of Institutional Investors (The January Effect)", Vol. 11, *Canadian Investment Review*, 09-01-1998, pp. 28–31). Suppose that n (the number of analysts making the projections) is 12 for each sample. At the 0.01 significance level, test the claim that the variability in analysts' earnings forecasts is different for firms in the two countries. If you base your investment decisions on earnings forecasts, would it be wiser to listen to a forecaster of Canadian or of U.S. firms? Explain.

9. Because of wind patterns, Air Canada flights between Vancouver and San Francisco can take different times in each direction. Sample times for flights between these cities, from March 8 to March 18, 1997, are shown. Test the claim that both sets of times come from populations with the same variance. Use a 0.05 level of significance.

Vancouver to San Francisco					San Francisco to Vancouver			
2.30	2.20	1.97	2.22	2.12	1.80	1.68	2.03	1.83
1.95	2.03	2.30	2.15	2.33	1.77	1.98	1.85	
2.15								

Based on data from Gary Murphy, chief flight dispatcher, Air Canada.

10. Provincial governments across Canada have had to re-examine their delivery of services, such as health care, to avoid expensive duplication, and yet provide fair access to their citizens. The following table is based on one random sample of hospitals in New Brunswick and another in Nova Scotia, and shows the total number of beds in each hospital (*Guide to Canadian Healthcare Facilities*, 1998–99, Vol. 6, Canadian Healthcare Association, Ottawa). At the 0.05 significance level, test the claim that the two given samples come from populations with the same variance. If the variances are different, could this have implications for the equity of services around each province?

Explain why it would be useful to know the population served by each hospital in order fully to answer the preceding question.

New Brunswick		Nova Scotia	
23	47	15	13
153	47	27	136
397	15	8	26
12	500	85	311
15	56	12	132
398			

11. Sample data were collected to compare the ages of CEOs of top growth companies in western Canada and Quebec. Because of small sample sizes, techniques introduced later in this chapter would be needed to compare group means. One such method assumes equal variances for the two groups. At the 0.05 significance level, test the claim that these two sample groups come from populations with the same variance.

Western Canada	38	54	35	40	41	34	59	37	35	48	35	59
Quebec	50	31	38	49	34	36	35	33	39	38		

Based on data from *Profit*, May 29, 1996.

12. Refer to Data Set 11 in Appendix B and use a 0.10 level of significance to test the claim that red and yellow M&M plain candies have weights with different standard deviations. If you were responsible for controlling production so that red and yellow M&Ms have weights with the same amount of variation, would you take corrective action?

Exercises B: Beyond the Basics

13. A hypothesis test of the claim $\sigma_1^2 = \sigma_2^2$ is being conducted with a 0.05 level of significance. What do you conclude in each of the following cases?
    a. Test statistic is $F = 2.0933$; $n_1 = 50$; $n_2 = 35$
    b. Test statistic is $F = 1.8025$; $n_1 = 50$; $n_2 = 35$
    c. Test statistic is $F = 2.3935$; $n_1 = 40$; $n_2 = 20$

14. To test a claim about two population variances by using the *P*-value approach, first find the *F* test statistic, then refer to Table A-5 to determine how it compares to the critical values listed for $\alpha = 0.01$, $\alpha = 0.025$, and $\alpha = 0.05$. For example, if the *F* test statistic (for a one-tailed test) is greater than the table value for $\alpha = 0.05$ but *not* greater than the value for $\alpha = 0.025$, then $0.025 < p < 0.05$. That is, *p* is between 0.025 and 0.05. Also recall that for two-tailed tests, the significance level in the actual problem is halved for purposes of using the *F* distribution tables. Therefore, in these cases, if the test statistic

exceeds the critical value for F on a table, the p indicated is *2 times* the α shown at the top of that table. For example, an F just above the critical F on the $\alpha = 0.025$ table represents (for a two-tailed test) $p < 0.05$. Referring to Exercises 4 and 13(a), what can be concluded about the P-values?

15. For hypothesis tests in this section that were two-tailed, we found only the upper critical value. (Recall that for two-tailed tests, the significance level that you look up in the F distribution tables is in fact $\alpha/2$.) Let's denote that value by F_R, where the subscript suggests the critical value for the right tail. The lower critical value F_L (for the left tail) can be found as follows: First interchange the degrees of freedom, and then take the reciprocal of the resulting F value found in Table A-5. (F_R is often denoted by $F_{\alpha/2}$, and F_L is often denoted by $F_{1-\alpha/2}$.) Find the critical values F_L and F_R for two-tailed hypothesis tests based on the following values.

a. $n_1 = 10$, $n_2 = 10$, $\alpha = 0.05$ b. $n_1 = 10$, $n_2 = 7$, $\alpha = 0.05$
c. $n_1 = 7$, $n_2 = 10$, $\alpha = 0.05$ d. $n_1 = 25$, $n_2 = 10$, $\alpha = 0.02$
e. $n_1 = 10$, $n_2 = 25$, $\alpha = 0.02$

16. In addition to testing claims involving σ_1^2 and σ_2^2, we can also construct interval estimates of the ratio σ_1^2/σ_2^2 using the following expression:

$$\left(\frac{s_1^2}{s_2^2} \cdot \frac{1}{F_R} \right) < \frac{\sigma_1^2}{\sigma_2^2} < \left(\frac{s_1^2}{s_2^2} \cdot \frac{1}{F_L} \right)$$

Here F_L and F_R are as described in Exercise 15. Construct the 95% confidence interval estimates for the ratios between group variances for the data in (a) Exercise 9 and (b) Exercise 11.

17. Sample data consist of temperatures recorded for two different groups of items that were produced by two different production techniques. A quality control specialist plans to analyze the results. She begins by testing for equality of the two population standard deviations.

a. If she adds the same constant to every temperature from both groups, how is the value of the test statistic F affected?

b. If she multiplies every score from both groups by the same constant, how is the value of the test statistic F affected?

c. If she converts all temperatures from the Celsius scale to the Fahrenheit scale, how is the value of the test statistic F affected?

8-5   Inferences About Two Means: Independent and Small Samples

In this section we present methods of inferential statistics for situations involving the means of two independent populations, when (unlike Section 8–3) at least one of the two samples is small (with $n \leq 30$). We describe methods for testing hypotheses and constructing confidence intervals for situations in which the following assumptions apply.

Assumptions

1. The two samples are *independent*.
2. The two samples are *randomly selected from normally distributed* populations.
3. At least one of the two samples is small ($n \leq 30$).

When these conditions are satisfied, we use one of three different procedures corresponding to the following cases:

Case 1: The values of both population variances are known. (In reality, this case rarely occurs.)

Case 2: The two populations appear to have equal variances. (That is, $\sigma_1^2 = \sigma_2^2$.)

Case 2: The two populations appear to have unequal variances. (That is, $\sigma_1^2 \neq \sigma_2^2$.)

Case 1: Both Population Variances Are Known

In reality, Case 1 doesn't occur very often. Usually, the population variances are computed from the known population data, and if we could find σ_1^2 and σ_2^2, we should be able to find the values of μ_1 and μ_2, so there would be no need to test claims or construct confidence intervals. If some strange set of circumstances allows us to know the values of σ_1^2 and σ_2^2 but not μ_1 and μ_2, then we can use the same methods as described in Section 8–3 for large samples. Because this case of known population variances is so unlikely to occur, we will not pursue it here.

Choosing Between Cases 2 and 3: How Do We Decide That Two Populations Have Equal or Unequal Variances?

If the three assumptions listed at the beginning of this section are satisfied, inferences about the two population means are made with the procedures we will describe in Cases 2 and 3. Case 2 applies when the two populations have equal variances, and Case 3 applies when the two populations have *different* variances. But how do we determine whether the two population variances are equal?

In Section 8–4 we described a procedure for testing the claim that two populations have equal variances, and this approach, which we will now call the *preliminary F test approach*, is preferred by some statisticians. Not all statisticians agree: Some argue that if we apply the F test with a certain significance level and then do a t test at the same level, the overall result will not be at the same level of significance. Also, the F statistic is much more sensitive to departures from normal distributions than is the t statistic—so there are samples for which the t test is appropriate, but the F test is not appropriate for the preliminary analysis. (For one argument against the preliminary F test, see "Homogeneity of Variance in the Two-Sample Means Test," by Moser and Stevens, *The American*

Statistician, Vol. 46, No. 1.) Different textbooks address this difficulty in different ways. In this text, we provide some *tools* that different strategies can utilize, and your instructor may recommend a specific approach. Here are two options:

1. *Preliminary F test approach*:

   Use the *F* test described in Section 8–4 to test the null hypothesis that $\sigma_1^2 = \sigma_2^2$. Use the conclusion of that test as follows:
   - *Fail to reject $\sigma_1^2 = \sigma_2^2$*: Treat the populations as if they have equal variances (Case 2).
   - *Reject $\sigma_1^2 = \sigma_2^2$*: Treat the populations as if they have unequal variances (Case 3).

2. *Assume unequal variances*: Assume that the two populations have *unequal* variances $\sigma_1^2 \neq \sigma_2^2$ and use the methods of Case 3. (Remember, this assumption applies only to cases in which we have two independent samples, at least one of which is small, and the population variances are not known.)

You might also consider making inferences using *both* methods (for equal variances and for unequal variances). If the two sets of results agree, your conclusion is probably OK.

Case 2: The Two Populations Appear to Have Equal Variances

If we are working on the assumption of equal variances, we calculate a **pooled estimate of σ^2** that is common to both populations; that pooled estimate is denoted by s_p^2 and is a weighted average of s_1^2 and s_2^2, as shown in the box. (Excel is one of many systems using the "pooled" option of this case.) Hypotheses about μ_1 and μ_2 can be tested and confidence interval estimates of $\mu_1 - \mu_2$ can be constructed by using the following methods.

EXAMPLE

Refer to the data listed in Table 8-1 in the chapter problem. Use a 0.05 significance level to test the claim that, for the initial and for the subsequent spring seasons of strong El Niño events, the mean departures in temperatures from seasonal norms were equal. (Temperature departures are in degrees Celsius.)

TEST STATISTIC (SMALL INDEPENDENT SAMPLES AND EQUAL VARIANCES)

$$t = \frac{(\bar{x}_1 - \bar{x}_2) - (\mu_1 - \mu_2)}{\sqrt{\dfrac{s_p^2}{n_1} + \dfrac{s_p^2}{n_2}}}$$

where

$$s_p^2 = \frac{(n_1 - 1)s_1^2 + (n_2 - 1)s_2^2}{(n_1 - 1) + (n_2 - 1)}$$

and the degree of freedom is given by df $= n_1 + n_2 - 2$

$$(\bar{x}_1 - \bar{x}_2) - E < (\mu_1 - \mu_2) < (\bar{x}_1 - \bar{x}_2) + E$$

where $\qquad E = t_{\alpha/2}\sqrt{\dfrac{s_p^2}{n_1} + \dfrac{s_p^2}{n_2}}$

and s_p^2 is as given in the test statistic.

SOLUTION

In Figure 8-1 we used boxplots to compare weather deviations at different stages of El Niño events. The boxplots representing temperature deviations seem to suggest an increase in temperature deviations from the norm in the second spring of an El Niño compared to the initial spring, but let's use a formal hypothesis of the claim that the two mean amounts are equal. We will use the sample statistics listed in the margin.

The samples are independent (because they are separate and are not matched), the sample sizes are small, and we don't know the values of s_1 and s_2, so we are dealing with either Case 2 or Case 3. If we use a preliminary F test (described in Section 8–4) to choose between Case 2 and Case 3, we test $H_0 : \sigma_1^2 = \sigma_2^2$ with the test statistic

	Temperature Departures (°C)
Second Spring	Spring of Onset
$n_1 = 7$	$n_2 = 7$
$\bar{x}_1 = 0.88$	$\bar{x}_2 = 0$
$s_1 = 1.20$	$s_2 = 0.91$

$$F = \frac{s_1^2}{s_2^2} = \frac{1.20^2}{0.91^2} = 1.7389$$

With $\alpha = 0.05$ in a two-tailed F test and with 6 degrees of freedom for both the numerator and denominator, we use Table A-5 to find the critical F value of 5.8198. Because the computed test statistic of $F = 1.7389$ does *not* fall within the critical region, we fail to reject the null hypothesis of equal variances and proceed by using the approach outlined in Case 2.

In using the Case 2 approach, we test the claim that $\mu_1 = \mu_2$ with the following null and alternative hypotheses:

$$H_0 : \mu_1 = \mu_2 \ (\text{or } \mu_1 - \mu_2 = 0) \qquad H_1 : \mu_1 \neq \mu_2$$

The test statistic for this case of equal variances requires the value of the pooled variance s_p^2, so we find that value first:

$$s_p^2 = \frac{(n_1 - 1)s_1^2 + (n_2 - 1)s_2^2}{(n_1 - 1) + (n_2 - 1)} = \frac{(7 - 1) \cdot 1.20^2 + (7 - 1) \cdot 0.91^2}{(7 - 1) + (7 - 1)}$$

$$= 1.134$$

We can now find the value of the test statistic:

$$t = \frac{(\bar{x}_1 - \bar{x}_2) - (\mu_1 - \mu_2)}{\sqrt{\dfrac{s_p^2}{n_1} + \dfrac{s_p^2}{n_2}}} = \frac{(0.88 - 0) - 0}{\sqrt{\dfrac{1.134}{7} + \dfrac{1.134}{7}}} = 1.546$$

The critical values of $t = -2.179$ and $t = 2.179$ are found from Table A-3 by referring to the column for $\alpha = 0.05$ (two tails) and to the row for df $= 12$ (the value of $7 + 7 - 2$). Figure 8-7 shows the test statistic, critical values, and critical region. We see that the test statistic does not fall within the critical region, so we fail to reject the null hypothesis that $\mu_1 = \mu_2$.

Interpretation

Despite localized effects and subjective reports about El Niño, there is not sufficient evidence to warrant rejection of the claim that the mean temperature deviations in Canada are the same during the first and second springs of El Niños. The tentative conclusion we formed by intuitively analyzing Figure 8-1 was, in this case, not upheld by a formal hypothesis test. If instead of using the preliminary F test approach, we used the approach of automatically assuming *unequal* variances (Case 3), we would again fail to reject the null hypothesis of equal population means.

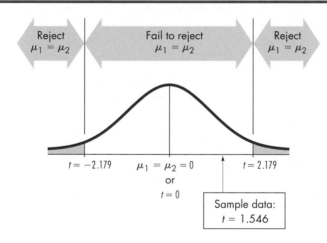

Figure 8-7
Distribution of Differences Between Means of Temperature Deviations

EXAMPLE

Using the data in the preceding example, construct a 95% confidence interval estimate of $\mu_1 - \mu_2$.

SOLUTION

It is easy to find that $\bar{x}_1 - \bar{x}_2 = 0.88 - 0 = 0.88$. Next, we find the value of the margin of error E:

$$E = t_{\alpha/2}\sqrt{\frac{s_p^2}{n_1} + \frac{s_p^2}{n_2}} = 2.179\sqrt{\frac{1.134}{7} + \frac{1.134}{7}} = 1.240$$

With $\bar{x}_1 - \bar{x}_2 = 0.88$ and with $E = 1.240$, we proceed to construct the confidence interval as follows:

$$(\bar{x}_1 - \bar{x}_2) - E < (\mu_1 - \mu_2) < (\bar{x}_1 - \bar{x}_2) + E$$
$$0.88 - 1.240 < (\mu_1 - \mu_2) < 0.88 + 1.240$$
$$-0.36 < (\mu_1 - \mu_2) < 2.12$$

This confidence interval contains zero, which means that the apparent difference in means for temperature deviations, for the first and second springs of El Niños, could be the result of chance. We fail to reject the hypothesis that the difference in means is zero.

Case 3: The Two Populations Appear to Have Unequal Variances

If we are working on the assumption of unequal variances, there is no exact method for testing equality of means and constructing confidence intervals. An *approximate* method is to use the following test statistic and confidence interval.

TEST STATISTIC (SMALL INDEPENDENT SAMPLES AND UNEQUAL VARIANCES)

$$t = \frac{(\bar{x}_1 - \bar{x}_2) - (\mu_1 - \mu_2)}{\sqrt{\dfrac{s_1^2}{n_1} + \dfrac{s_2^2}{n_2}}}$$

where df = smaller of $n_1 - 1$ and $n_2 - 1$

CONFIDENCE INTERVAL (SMALL INDEPENDENT SAMPLES AND UNEQUAL VARIANCES)

$$(\bar{x}_1 - \bar{x}_2) - E < (\mu_1 - \mu_2) < (\bar{x}_1 - \bar{x}_2) + E$$

where
$$E = t_{\alpha/2}\sqrt{\frac{s_1^2}{n_1} + \frac{s_2^2}{n_2}}$$

and df = smaller of $n_1 - 1$ and $n_2 - 1$

This test statistic and confidence interval give the number of degrees of freedom as the smaller of $n_1 - 1$ and $n_2 - 1$, but this is a more conservative and simplified alternative to computing the number of degrees of freedom by using Formula 8-1.

Formula 8-1

$$df = \frac{(A + B)^2}{\dfrac{A^2}{n_1 - 1} + \dfrac{B^2}{n_2 - 1}}$$

where

$$A = \frac{s_1^2}{n_1} \text{ and } B = \frac{s_2^2}{n_2}$$

More exact results are obtained by using Formula 8-1, but they continue to be only approximate. (*Note:* Excel and STATDISK use Formula 8-1 instead of the "smaller of $n_1 - 1$ and $n_2 - 1$," so their results will be a little more precise than those included in this text.)

EXAMPLE

Refer to the companies in Data Set 6 in Appendix B, which are taken from the *Financial Post* list of top 500 Canadian companies for 1997. Conglomerates (marked "f") are clearly major employers, some having as many as 50,000 and 83,000 employees. Wholesale distributors (marked "dd") do not appear on the list until row 60, and appear to have fewer employees. Use a 0.05 significance level to test the claim that for *all* 500 companies on this list, the mean number of employees in conglomerates is greater than the mean number of employees in wholesale distributors.

SOLUTION

A visual scan of Data Set 6 in Appendix B suggests that conglomerates are bigger employers than wholesale distributors, but let's justify that observation with a formal hypothesis test based on the sample statistics listed in the margin. Note that these samples were drawn randomly from the *complete* list of top 500 companies, not just those in Data Set 6.

The two samples are independent (because they are separate and are not matched), the sample sizes are small, and we don't know the values of σ_1 and σ_2, so we are dealing with either Case 2 or Case 3. If we use a preliminary F test to choose between Case 2 and Case 3, and we test $H_0: \sigma_1^2 = \sigma_2^2$ with the test statistic

$$F = \frac{s_1^2}{s_2^2} = \frac{143.643^2}{29.003^2} = 24.5292$$

	Employees (00s)	
	Conglomerates	Wholesale Distributors
	$n_1 = 10$	$n_2 = 11$
	$\bar{x}_1 = 105.77$	$\bar{x}_2 = 17.77$
	$s_1 = 143.643$	$s_2 = 29.003$

With $\alpha = 0.05$ in a two-tailed F test and with 9 and 10 degrees of freedom for the numerator and denominator, respectively, we use Table A-5 to find the critical F value of 3.7790. Because the computed test statistic of $F = 24.5292$ does fall within the critical region, we reject the null hypothesis of equal variances and proceed by using the approach outlined in Case 3.

In using the Case 3 approach, we test the claim that $\mu_1 > \mu_2$ with the following null and alternative hypotheses:

$$H_0: \mu_1 \leq \mu_2 \text{ (or } \mu_1 - \mu_2 \leq 0) \qquad H_1: \mu_1 > \mu_2 \qquad \text{(original claim)}$$

The test statistic for this case of unequal variances is

$$t = \frac{(\bar{x}_1 - \bar{x}_2) - (\mu_1 - \mu_2)}{\sqrt{\dfrac{s_1^2}{n_1} + \dfrac{s_2^2}{n_2}}} = \frac{(105.77 - 17.77) - 0}{\sqrt{\dfrac{143.643^2}{10} + \dfrac{29.003^2}{11}}} = 1.902$$

The critical value of $t = 1.833$ is found from Table A-3 by referring to the column for $\alpha = 0.05$ (one tail) and to the row for df $= 9$ (the smaller of $10 - 1$ and $11 - 1$). Because this right-tailed test has a test statistic of $t = 1.902$ and a critical value of $t = 1.833$, the test statistic does fall within the critical region, so we reject the null hypothesis that $\mu_1 \le \mu_2$.

Interpretation

There is sufficient evidence to support the claim that the mean number of employees for top conglomerates in Canada is greater than the mean number of employees for top wholesale distribution companies. (If we use the same data with the methods of Case 2 for equal variances, we again reject the null hypothesis, and reach the same conclusion, that $\mu_1 > \mu_2$.)

EXAMPLE

Using the data in the preceding example, construct a 95% confidence interval estimate of $\mu_1 - \mu_2$.

SOLUTION

With $\bar{x}_1 - \bar{x}_2 = 105.77 - 17.77 = 88.0$, we proceed to find the value of the margin of error E:

$$E = t_{\alpha/2}\sqrt{\frac{s_1^2}{n_1} + \frac{s_2^2}{n_2}} = 2.262\sqrt{\frac{143.643^2}{10} + \frac{29.003^2}{11}} = 104.636$$

With $\bar{x}_1 - \bar{x}_2 = 88.0$ and with $E = 104.636$, we now construct the confidence interval as follows:

$$(\bar{x}_1 - \bar{x}_2) - E < (\mu_1 - \mu_2) < (\bar{x}_1 - \bar{x}_2) + E$$
$$88.0 - 104.636 < (\mu_1 - \mu_2) < 88.0 + 104.636$$
$$-16.6 < (\mu_1 - \mu_2) < 192.6$$

Interpretation

Based on this result, we are 95% confident that the mean number of employees in conglomerates (in hundreds) exceeds the mean number of employees in wholesale distributors (in hundreds) by an amount that is between 192.6 and *minus* 16.6. That is, there *may* actually be a higher mean for the wholesale distributors. Note the apparent discrepancy of this result with the previous example, which concluded, based on a one-tailed test, in favour of conglomerates having the greater mean number of employees. The one-tailed test achieved, just barely, the required 0.05 level of significance. A confidence-interval approach corresponds more closely with a two-tailed test: The critical region under each tail is only $\alpha/2$ (0.025). It is important to decide *before* calculating results which approach will be considered to represent sufficient evidence for a difference in means, if the test is passed.

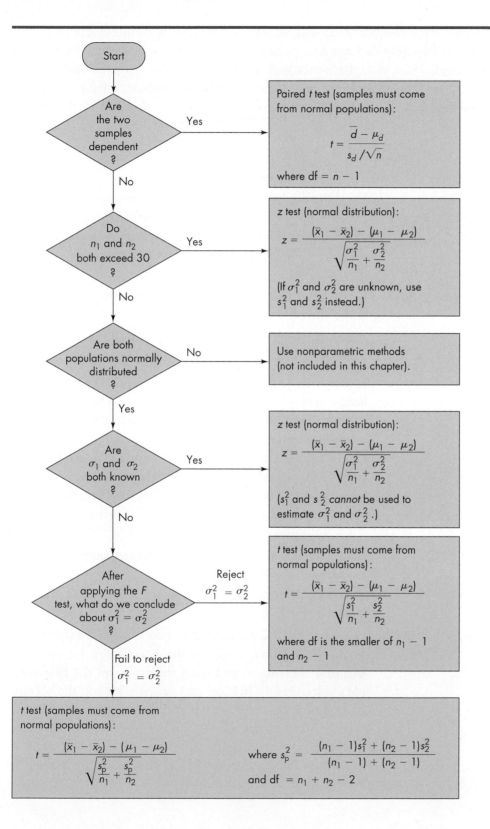

Figure 8-8
Testing Hypotheses Made
About the Means of Two
Populations

Start

Are the two samples dependent ?

Yes

Paired t test (samples must come from normal populations):

$$t = \frac{\overline{d} - \mu_d}{s_d / \sqrt{n}}$$

where df $= n - 1$

No

Do n_1 and n_2 both exceed 30 ?

Yes

z test (normal distribution):

$$z = \frac{(\overline{x}_1 - \overline{x}_2) - (\mu_1 - \mu_2)}{\sqrt{\dfrac{\sigma_1^2}{n_1} + \dfrac{\sigma_2^2}{n_2}}}$$

(If σ_1^2 and σ_2^2 are unknown, use s_1^2 and s_2^2 instead.)

No

Are both populations normally distributed ?

No

Use nonparametric methods (not included in this chapter).

Yes

Are σ_1 and σ_2 both known ?

Yes

z test (normal distribution):

$$z = \frac{(\overline{x}_1 - \overline{x}_2) - (\mu_1 - \mu_2)}{\sqrt{\dfrac{\sigma_1^2}{n_1} + \dfrac{\sigma_2^2}{n_2}}}$$

(s_1^2 and s_2^2 *cannot* be used to estimate σ_1^2 and σ_2^2.)

No

After applying the F test, what do we conclude about $\sigma_1^2 = \sigma_2^2$?

Reject $\sigma_1^2 = \sigma_2^2$

t test (samples must come from normal populations):

$$t = \frac{(\overline{x}_1 - \overline{x}_2) - (\mu_1 - \mu_2)}{\sqrt{\dfrac{s_1^2}{n_1} + \dfrac{s_2^2}{n_2}}}$$

where df is the smaller of $n_1 - 1$ and $n_2 - 1$

Fail to reject $\sigma_1^2 = \sigma_2^2$

t test (samples must come from normal populations):

$$t = \frac{(\overline{x}_1 - \overline{x}_2) - (\mu_1 - \mu_2)}{\sqrt{\dfrac{s_p^2}{n_1} + \dfrac{s_p^2}{n_2}}}$$

where $s_p^2 = \dfrac{(n_1 - 1)s_1^2 + (n_2 - 1)s_2^2}{(n_1 - 1) + (n_2 - 1)}$

and df $= n_1 + n_2 - 2$

This section, Section 8–2, and Section 8–3 all deal with inferences about the means of two populations. Determining the correct procedure can be difficult because we must consider issues such as the independence of the samples, the sizes of the samples, and whether the two populations appear to have equal variances. Figure 8-8 is designed to simplify the decisions that lead to the correct procedure.

EXCEL: Excel works from the raw data for the two samples, and requires that you first determine whether Case 2 or Case 3 applies. If you intend to use a *preliminary F test* for that decision, then begin with the Excel procedures described in Section 8–4.

EXCEL (Prior to 2007): Next, from the **Tools** menu, select **Data Analysis**, and then select one of:

t-test: Two-Sample Assuming Equal Variances

t-test: Two-Sample Assuming Unequal Variances

In either case, the same dialog box appears. For **Variable 1 Range**, specify the worksheet range that contains the data for one of the two variables; for **Variable 2 Range**, specify the worksheet range that contains the other variable's data. Unlike for the Excel *F* test, the appropriate choice for **Alpha** is the *full* intended α, not just the right-tailed portion. Next, select an **Output Option**, and click on **OK** to obtain the results. The illustration shows the results for Case 2 and Case 3 types of output.

EXCEL 2007: Click **Data** on the main menu and choose **Data Analysis** from the **Analysis** tab. Proceed as above.

To generate confidence intervals, the Data Desk XL add-in can be used.

EXCEL (Prior to 2007): Click on **DDXL**. Select **Confidence Intervals** and **2 Var t Interval**. In the dialog box, click on the pencil icon for the first quantitative column and enter the range for the first sample's data. Then click on the pencil icon for the second quantitative column and enter the range for the second sample's data. Click on **OK**. Follow the steps in the next dialog box that appears. (In Step 1, select **2 Sample** if assuming unequal population variances, or **Pooled** if assuming equal variances.)

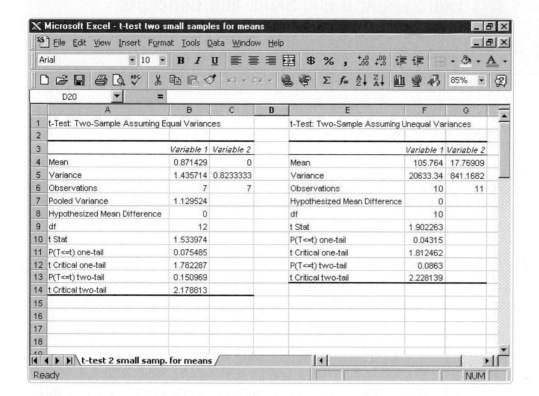

	A	B	C	D	E	F	G
1	t-Test: Two-Sample Assuming Equal Variances				t-Test: Two-Sample Assuming Unequal Variances		
2							
3		Variable 1	Variable 2			Variable 1	Variable 2
4	Mean	0.871429	0		Mean	105.764	17.76909
5	Variance	1.435714	0.8233333		Variance	20633.34	841.1682
6	Observations	7	7		Observations	10	11
7	Pooled Variance	1.129524			Hypothesized Mean Difference	0	
8	Hypothesized Mean Difference	0			df	10	
9	df	12			t Stat	1.902263	
10	t Stat	1.533974			P(T<=t) one-tail	0.04315	
11	P(T<=t) one-tail	0.075485			t Critical one-tail	1.812462	
12	t Critical one-tail	1.782287			P(T<=t) two-tail	0.0863	
13	P(T<=t) two-tail	0.150969			t Critical two-tail	2.228139	
14	t Critical two-tail	2.178813					

EXCEL 2007: Click on the **Add-in** tab on the main menu. You will see **DDXL** to the left. Click on it and proceed as above.

MINITAB 15: From the **Stat** menu, choose **Basic Statistics** and then **2-Sample t.** If the test is being conducted with raw data, select the appropriate columns. There is an optional checkbox to assume equal variances. The default is a two-tailed test at a 5% level of significance with a hypothesized difference of zero. If you wish to change any of these, click the **Options** button and make the necessary changes.

STATDISK: Unlike Excel, STATDISK works from summary data. Begin by selecting the menu items **Analysis, Hypothesis Testing,** and **Mean-Two Independent Samples.** Choose the appropriate test to pool or not pool based on a preliminary F test. There is an optional radio button to conduct the F test which is then used for the appropriate t test. The output from STATDISK includes both the test results and a confidence interval.

Exercises A: Basic Skills and Concepts

In Exercises 1 and 2, test the given claim. Use a significance level of $\alpha = 0.05$, and assume that all populations are normally distributed. Use the traditional method of testing hypotheses outlined in Figure 7-4.

1. Claim: Populations A and B have the same mean ($\mu_1 = \mu_2$).
   Sample A: $n_1 = 10$, $\bar{x}_1 = 75$, $s_1 = 15$
   Sample B: $n_2 = 10$, $\bar{x}_2 = 80$, $s_2 = 12$

2. Claim: Populations A and B have the same mean ($\mu_1 = \mu_2$).
   Sample A: $n_1 = 15$, $\bar{x}_1 = 150$, $s_1 = 40$
   Sample B: $n_2 = 5$, $\bar{x}_2 = 140$, $s_2 = 10$

3. Using the sample data in Exercise 1, construct a 95% confidence interval estimate of the difference between the means of the two populations.

4. Using the sample data in Exercise 2, construct a 95% confidence interval estimate of the difference between the means of the two populations.

5. The equality of Air Canada flight times (in hours) between Vancouver and San Francisco is tested using Excel. A preliminary *F* test shows that there is no significant difference between the two sample variances, so, as shown in the accompaning figure ($\alpha = 0.05$), the *t* test assuming equal variances was used.
   a. Based on the displayed results below, construct a 95% confidence interval for the difference $\mu_1 - \mu_2$, where μ_1 denotes the mean flight time from Vancouver to San Francisco. Do the confidence intervals contain 0? What can you conclude about equality of the mean flight times in both directions?
   b. Use elements from the Excel display on the next page to answer these questions: At the 0.05 significance level, test the claim that flights from Vancouver to San Francisco take a longer time than flights in the other direction. What is the test statistic? What is the critical value? What is the *P*-value?

Vancouver to San Francisco	San Francisco to Vancouver
2.30 2.20 1.97 2.22 2.12	1.80 1.68 2.03 1.83
1.95 2.03 2.30 2.15 2.33	1.77 1.98 1.85
2.15	

Based on data from Gary Murphy, chief flight dispatcher, Air Canada.

```
X  Microsoft Excel - t-test two small samples for means--equal

  File  Edit  View  Insert  Format  Tools  Data  Window  Help

  Arial                    ▾  10  ▾   B  I  U   ≡ ≡ ≡ ⊞

  □ ⊡ ⊟ ⊜ ⊡ ᴬᴮᶜ  ✕ ⊟ ⊟ ◇   ⟲ ▾ ⟳ ▾   ⊜

       A20          ▾        =

              A                    |    B    |    C
 1  t-Test: Two-Sample Assuming Equal Variances
 2
 3                              | Variable 1 | Variable 2
 4  Mean                        | 2.1560606  |   1.85
 5  Variance                    | 0.0173485  | 0.0148148
 6  Observations                |    11      |    7
 7  Pooled Variance             | 0.0163984  |
 8  Hypothesized Mean Difference|    0       |
 9  df                          |    16      |
10  t Stat                      | 4.943293   |
11  P(T<=t) one-tail            | 7.338E-05  |
12  t Critical one-tail         | 1.7458842  |
13  P(T<=t) two-tail            | 0.0001468  |
14  t Critical two-tail         | 2.1199048  |
```

6. Employees who work with electronic components at Canadian Automated Electronics are required to undergo special training. Training includes completing tasks under simulated conditions and recording completion times. A supervisor claims he has found some star performers among a new group of trainees.

   a. Based on the data shown, test the claim at the 0.01 significance level that the star-performers group has a mean completion time less than the other group (based on data from Canadian Automated Electronics). The assumption of equal variances for the groups has been rejected ($p < 0.01$).

   b. Construct a 99% confidence interval estimate of the difference between the mean task completion time for the two groups.

   Star performers: $n = 15$, $\bar{x} = 286$, $s = 71.93$
   Other trainees: $n = 10$, $\bar{x} = 873.7$, $s = 259.26$

7. In a sample of the driving records of 137 young adult males who underwent treatment for substance abuse in a Toronto clinic, the mean accident rate after treatment was 0.068, with a standard deviation of 0.223 (based on data from "Does Treatment of Substance Abuse Improve Driving Safety? A Preliminary Evaluation," by Mann et al., Addiction Research Foundation). Suppose a competing clinic claims that those who complete its substance-abuse program have better driving records. In a random sample of 20 of its

past clients, the mean accident rate after treatment was 0.061, with a standard deviation of 0.210.

a. At the 0.05 level of significance, test the claim that the past clients of both clinics have the same mean accident rate. Based on this result, is the competing clinic entitled to make the assertion that it does?

b. Construct a 95% confidence interval for the difference between the mean accident rates for the client groups from the two clinics.

8. Assume that Data Set 21 in Appendix B represents sample prices and characteristics of high-quality diamonds sold at auction. A vendor claims that near-colourless diamonds sell at a higher price than yellower ones. Based on the data summarized in the margin (from Data Set 21, with colours 5–9 interpreted as "More Yellow"), use a 0.05 level of significance to test the vendor's claim. Assume unequal variances for the two groups. Should a buyer for the auction house avoid buying the yellower diamonds? Is more information needed to answer that question? Explain.

9. Assume that Data Set 21 in Appendix B represents sample prices and characteristics of high quality diamonds sold at auction. A vendor claims that nearly flawless diamonds sell at a higher price than noticeably flawed ones. Based on the data summarized in the margin (from Data Set 21, with clarity 4–9 interpreted as "More Flawed"), use a 0.05 level of significance to test the vendor's claim. Assume equal variances for the two groups. Should a buyer for the auction house avoid buying the noticeably flawed diamonds? Is more information needed to answer that question? Explain.

10. A simple random sample of 21 peanut M&M candies is selected, and each candy is weighed. The mean is 2.4658 g and the standard deviation is 0.3127 g. The 100 g M&M plain candies listed in Data Set 11 in Appendix B have a mean of 0.9147 g and a standard deviation of 0.0369 g. Is there sufficient evidence to support the claim that there is a difference between the mean weight of peanut M&Ms and the mean weight of plain M&Ms?

11. Data Set 11 in Appendix B includes a sample of 21 red M&Ms with weights having a mean of 0.9097 g and a standard deviation of 0.0275 g. The data set also includes a sample of 8 orange M&Ms with weights having a mean of 0.9251 g and a standard deviation of 0.0472 g. Use a 0.05 level of significance to test for equality of the two population means. If you were responsible for controlling production so that red and orange M&Ms have weights with the same mean, would you take corrective action?

12. Refer to Data Set 11 in Appendix B and test the claim that yellow M&Ms and green M&Ms have the same mean weight. If you were part of a quality control team with responsibility for ensuring that yellow M&Ms and green M&Ms have the same mean weight, would you take corrective action?

13. Refer to Data Set 4 in Appendix B. If the Ontario–Manitoba border is taken as the boundary between East and West, test the claim that in comparing

Less Yellow

$n_1 = 13$

$\bar{x}_1 = \$30{,}638.31$

$s_1 = \$28{,}346.06$

More Yellow

$n_2 = 17$

$\bar{x}_2 = \$14{,}541.00$

$s_2 = \$11{,}297.64$

Less Flawed

$n_1 = 13$

$\bar{x}_1 = \$27{,}083.23$

$s_1 = \$25{,}644.35$

More Flawed

$n_2 = 17$

$\bar{x}_2 = \$17{,}259.59$

$s_2 = \$17{,}660.96$

provincial averages, there is no difference in mean annual precipitation between western and eastern provinces. Use the 0.05 level of significance.

14. Provincial governments have been re-examining their delivery of services, such as health care. The accompanying table is based on one random sample of hospitals in New Brunswick and another in Nova Scotia, and shows the total number of beds in each hospital (*Guide to Canadian Healthcare Facilities,* 1998–99, Vol. 6, Canadian Healthcare Association, Ottawa). At the 0.05 significance level, test the claim that the two given samples come from populations with the same mean (assume unequal variances). If the means are different, what are the possible implications for the equity of services between these two provinces? If the means are different, do we have sufficient information to conclude that the numbers of hospital beds per person are different in the two provinces? Explain.

New Brunswick		Nova Scotia	
23	47	15	13
153	47	27	136
397	15	8	26
12	500	85	311
15	56	12	132
398			

15. Sample data were collected to compare the ages of CEOs of top growth companies in western Canada and Quebec, shown in the table below. Exercise 11 in Section 8–4 looked at the variances for these two groups. Assuming the populations have equal variances, test at the 0.05 significance level the claim that these two samples come from populations with the same mean.

Western Canada	38	54	35	40	41	34	59	37	35	48	35	59
Quebec		50	31	38	49	34	36	35	33	39	38	

Based on data from *Profit,* May 29, 1996.

16. An experiment was conducted to test the effects of alcohol. The errors were recorded in a test of visual and motor skills for a treatment group of people who drank ethanol and another group given a placebo. The results are shown in the accompanying table. Use a 0.05 significance level to test the claim that the two groups come from populations with the same mean. Do these results support the common belief that drinking is hazardous for drivers, pilots, ship captains, and so on?

Treatment Group	Placebo Group
$n_1 = 22$	$n_2 = 22$
$\bar{x}_1 = 4.20$	$\bar{x}_2 = 1.71$
$s_1 = 2.20$	$s_2 = 0.72$

Based on data from "Effects of Alcohol Intoxication on Risk Taking, Strategy, and Error Rate in Visuomotor Performance," by Streufert et al., *Journal of Applied Psychology,* Vol. 77, No. 4.

17. An experiment was conducted to test the effects of alcohol. The breath alcohol levels were measured for a treatment group of people who drank ethanol and another group given a placebo. The results are given in the accompanying table. Use a 0.05 significance level to test the claim that the two groups come from populations with the same mean.

Treatment Group	Placebo Group
$n_1 = 22$	$n_2 = 22$
$\bar{x}_1 = 0.049$	$\bar{x}_2 = 0.000$
$s_1 = 0.015$	$s_2 = 0.000$

Based on data from "Effects of Alcohol Intoxication on Risk Taking, Strategy, and Error Rate in Visuomotor Performance," by Streufert et al., *Journal of Applied Psychology*, Vol. 77, No. 4.

18. Assume that two samples have the same standard deviation and that both are independent, small, and randomly selected from normally distributed populations. Also assume that we want to test the claim that the samples come from populations with the same mean.
    a. Is it necessary to conduct a preliminary *F* test?
    b. If both samples have standard deviation *s*, what is the value of s_p^2 expressed in terms of *s*?

19. Refer to the two examples of this section that used the number of employees in key industry groups. (See the Case 3 subsection.) How are the hypothesis test and confidence interval affected if the number of degrees of freedom is calculated with Formula 8-1, instead of using the smaller of $n_1 - 1$ and $n_2 - 1$?

20. Suppose that the sample data for the ages of CEOs displayed for Exercise 15 had omitted one data point because it was viewed as an outlier. This extra data value was an 11th CEO from Quebec, who is 80. How are the hypothesis test and confidence interval affected if this apparently extreme value is re-included?

Inferences About Two Proportions

Although this section appears last in the chapter, it is arguably the most important because this is where we describe methods for using two sample proportions to make inferences (hypothesis testing and confidence interval construction) about two population proportions.

When testing a hypothesis made about two population proportions—such as the proportions of cured patients in a population given some treatment and a second population given a placebo—or when constructing a confidence interval for the difference between two population proportions, we make the following assumptions and use the following notation.

Assumptions

1. We have two *independent* sets of randomly selected sample data.
2. For both samples, the conditions $np \geq 5$ and $nq \geq 5$ are satisfied. That is, there are at least 5 successes and 5 failures in each of the two samples. (In many cases, we will test the claim that two populations have equal proportions so that $p_1 - p_2 = 0$. Because we assume that $p_1 - p_2 = 0$, it is not necessary to specify the particular value that p_1 and p_2 have in common. In such cases, the conditions $np \geq 5$ and $nq \geq 5$ can be checked by replacing p by the estimated pooled proportion $\bar{p}$, which will be described later.)

NOTATION FOR TWO PROPORTIONS

For population 1 we let

$p_1 = population$ proportion

$n_1 = $ size of the sample

$x_1 = $ number of successes in the sample

$\hat{p}_1 = \dfrac{x_1}{n_1} = sample$ proportion

$\hat{q}_1 = 1 - \hat{p}_1$

The corresponding meanings are attached to p_2, n_2, x_2, $\hat{p}_2$, and $\hat{q}_2$, which come from population 2.

How do we find the values of x_1 and x_2?   Sometimes we are told directly the numbers of successes x_1 and x_2. In other cases, we may have to calculate the values of x_1 and x_2, which can be found by noting that $\hat{p}_1 = x_1/n_1$ implies that

$$x_1 = n_1 \cdot \hat{p}_1$$

For example, if we know that sample 1 consists of 500 treated patients and 30% of them were cured, the actual number of cured patients is $x_1 = 500 \cdot 0.30 = 150$. In this case, $n_1 = 500$, $\hat{p}_1 = 0.30$, and $x_1 = 150$.

Hypothesis Tests

In Section 7–5 we discussed tests of hypotheses made about a single population proportion. We will now consider tests of hypotheses made about two population proportions, but *we will be testing only claims that $p_1 = p_2$* and we will use the following pooled (or combined) estimate of the value that p_1 and p_2 have in common. (For claims that the difference between p_1 and p_2 is equal to a nonzero constant, see Exercise 17 of this section.)

POOLED ESTIMATE OF p_1 AND p_2

The **pooled estimate of p_1 and p_2** is denoted by $\overline{p}$ and is given by

$$\overline{p} = \frac{x_1 + x_2}{n_1 + n_2}$$

We denote the complement of $\overline{p}$ by $\overline{q}$, so $\overline{q} = 1 - \overline{p}$.

TEST STATISTIC FOR TWO PROPORTIONS

The following test statistic applies to null and alternative hypotheses that fit one of these three formats:

$$H_0: p_1 = p_2 \quad H_0: p_1 \leq p_2 \quad H_0: p_1 \leq p_2$$
$$H_1: p_1 \neq p_2 \quad H_1: p_1 < p_2 \quad H_1: p_1 > p_2$$

$$z = \frac{(\hat{p}_1 - \hat{p}_2) - (p_1 - p_2)}{\sqrt{\dfrac{\overline{p}\,\overline{q}}{n_1} + \dfrac{\overline{p}\,\overline{q}}{n_2}}}$$

where $p_1 - p_2 = 0$ (assumed in the null hypothesis), and

$$\hat{p}_1 = \frac{x_1}{n_1} \text{ and } \hat{p}_2 = \frac{x_2}{n_2}$$

$$\overline{p} = \frac{x_1 + x_2}{n_1 + n_2}$$

$$\overline{q} = 1 - \overline{p}$$

The following example will help clarify the roles of x_1, n_1, $\hat{p}_1$, $\overline{p}$, and so on. In particular, you should recognize that under the assumption of equal proportions, the best estimate of the common proportion is obtained by pooling both samples into one big sample, so that $\overline{p}$ becomes a more obvious estimate of the common population proportion.

Johns Hopkins researchers conducted a study of pregnant IBM employees. Among 30 employees who worked with glycol ethers, 10 (or 33.3%) had miscarriages, but among 750 who were not exposed to glycol ethers, 120 (or 16.0%) had miscarriages. At the 0.01 significance level, test the claim that the miscarriage rate is greater for women exposed to glycol ethers.

SOLUTION

For notation purposes, we stipulate that sample 1 is the group that worked with glycol ethers and sample 2 is the group not exposed, so the sample statistics can be summarized as shown here:

Exposed to Glycol Ethers	Not Exposed to Glycol Ethers
$n_1 = 30$	$n_2 = 750$
$x_1 = 10$	$x_2 = 120$

$$\hat{p}_1 = \frac{10}{30} = 0.333 \qquad \hat{p}_2 = \frac{120}{750} = 0.160$$

We will now use the traditional method of hypothesis testing, as summarized in Figure 7-4.

Step 1: The claim of a greater miscarriage rate for women exposed to glycol ethers can be represented by $p_1 > p_2$.

Step 2: If $p_1 > p_2$ is false, then $p_1 \leq p_2$.

Step 3: Because our claim of $p_1 > p_2$ does not contain equality, it becomes the alternative hypothesis, and we have

$$H_0: p_1 \leq p_2 \qquad H_1: p_1 > p_2 \quad \text{(original claim)}$$

Step 4: The significance level is $\alpha = 0.01$.

Step 5: We will use the normal distribution (with the test statistic previously given) as an approximation to the binomial distribution. We have two independent samples, and the conditions $np \geq 5$ and $nq \geq 5$ are satisfied for each of the two samples. To check this, we note that in conducting this test, we assume that $p_1 = p_2$, where their common value is the pooled estimate $\bar{p}$, calculated as

$$\bar{p} = \frac{x_1 + x_2}{n_1 + n_2} = \frac{10 + 120}{30 + 750} = 0.1667$$

With $\bar{p} = 0.1667$, it follows that $\bar{q} = 1 - 0.1667 = 0.8333$. We verify that $np \geq 5$ and $nq \geq 5$ for both samples as follows:

Sample 1	Sample 2
$n_1 p = (30)(0.1667) = 5 \geq 5$	$n_2 p = (750)(0.1667) = 125 \geq 5$
$n_1 q = (30)(0.8333) = 25 \geq 5$	$n_2 q = (750)(0.8333) = 625 \geq 5$

Step 6: We can now find the value of the test statistic:

$$z = \frac{(\hat{p}_1 - \hat{p}_2) - (p_1 - p_2)}{\sqrt{\dfrac{\bar{p}\,\bar{q}}{n_1} + \dfrac{\bar{p}\,\bar{q}}{n_2}}}$$

$$= \frac{(0.333 - 0.160) - 0}{\sqrt{\dfrac{(0.1667)(0.8333)}{30} + \dfrac{(0.1667)(0.8333)}{750}}} = 2.49$$

The critical value of $z = 2.33$ is found by observing that we have a right-tailed test with $\alpha = 0.01$. The value of $z = 2.33$ is found from Table A-2 as the z score corresponding to an area of $0.5 - 0.01 = 0.4900$.

Step 7: In Figure 8-9 we see that the test statistic falls within the critical region, so we reject the null hypothesis of $p_1 \leq p_2$.

Interpretation

We conclude that there is sufficient evidence to support the claim that the miscarriage rate is greater for women exposed to ethyl glycol. Johns Hopkins researchers concluded that women employees exposed to glycol ethers "have a significantly increased risk of miscarriage." On the basis of these results, IBM warned its employees of the danger, notified the Environmental Protection Agency, and greatly reduced its use of glycol ethers.

(These results could also be reached with the P-value approach. In Step 6, instead of finding the critical value of z, we would find the P-value by using the procedure summarized in Figures 7-7 and 7-8. With a test statistic of $z = 2.49$ and a right-tailed test, we get

$$P\text{-value} = (\text{area to the right of } z = 2.49) = 0.0064$$

Again, we reject the null hypothesis because the P-value of 0.0064 is less than the significance level of $\alpha = 0.01$.)

Figure 8-9

Distribution of Differences Between Proportions for Group Exposed to Glycol Ethers and Group Not Exposed

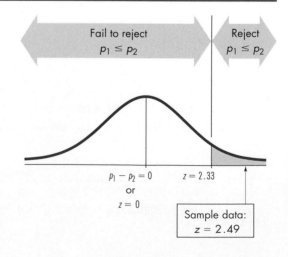

Confidence Intervals

In the preceding example we found that there appears to be a larger miscarriage rate for women exposed to glycol ethers, so we now proceed to estimate the amount of the difference. Is the difference large enough to justify extensive and expensive changes in manufacturing procedures?

With the same assumptions given at the beginning of this section, a confidence interval for the difference between population proportions $p_1 - p_2$ can be constructed by evaluating the following:

CONFIDENCE INTERVAL ESTIMATE OF $p_1 - p_2$

The confidence interval estimate of the difference $p_1 - p_2$ is:

$$(\hat{p}_1 - \hat{p}_2) - E < (p_1 - p_2) < (\hat{p}_1 - \hat{p}_2) + E$$

where

$$E = z_{\alpha/2}\sqrt{\frac{\hat{p}_1\hat{q}_1}{n_1} + \frac{\hat{p}_2\hat{q}_2}{n_2}}$$

EXAMPLE

Use the sample data given in the preceding example to construct a 99% confidence interval for the difference between the two population proportions.

SOLUTION

With a 99% degree of confidence, $z_{\alpha/2} = 2.575$ (from Table A-2). We first calculate the value of the margin of error E as shown below:

$$E = z_{\alpha/2}\sqrt{\frac{\hat{p}_1\hat{q}_1}{n_1} + \frac{\hat{p}_2\hat{q}_2}{n_2}} = 2.575\sqrt{\frac{(0.333)(0.667)}{30} + \frac{(0.160)(0.840)}{750}}$$
$$= 0.2242$$

With $\hat{p}_1 = 0.333$, $\hat{p}_2 = 0.160$ and $E = 0.2242$, the confidence interval is evaluated as follows:

$$(\hat{p}_1 - \hat{p}_2) - E < (p_1 - p_2) < (\hat{p}_1 - \hat{p}_2) + E$$
$$(0.333 - 0.160) - 0.2242 < (p_1 - p_2) < (0.333 - 0.160) + 0.2242$$
$$-0.0512 < (p_1 - p_2) < 0.3972$$

Interpretation

The confidence interval limits contain 0, suggesting that there is not a significant difference between the two proportions. Similarly to what occurred in the example of employment numbers by industry in Section 8-5, the confidence-interval results appear to contradict the conclusions from the related hypothesis-testing procedure. In this instance, the discrepancy is

STATISTICAL PLANET

Polio Experiment
In 1954 an experiment was conducted to test the effectiveness of the Salk vaccine as protection against the devastating effects of polio. Approximately 200,000 children were injected with an ineffective salt solution, and 200,000 other children were injected with the vaccine. The experiment was "double blind" because the children being injected didn't know whether they were given the real vaccine or the placebo, and the doctors giving the injections and evaluating the results didn't know either. Only 33 of the 200,000 vaccinated children later developed paralytic polio, whereas 115 of the 200,000 injected with the salt solution later developed paralytic polio. Statistical analysis of these and other results led to the conclusion that the Salk vaccine was indeed effective against paralytic polio.

attributable to two factors: (1) The hypothesis test was right-tailed with a 0.01 significance level, so it corresponds to a 98% confidence interval (but the above confidence interval has a 99% degree of confidence); (2) the variance used for the hypothesis test (based on an assumption of equal proportions) is different from the variance used for the confidence interval (where the proportions are not assumed to be equal).

As always, we should be careful when interpreting confidence intervals. Because p_1 and p_2 have fixed values and are not variables, it is wrong to state that there is a 99% chance that the value of $p_1 - p_2$ falls between -0.0512 and 0.3972. It is correct to state that if we repeat the same sampling process and construct 99% confidence intervals, in the long run 99% of the intervals will actually contain the value of $p_1 - p_2$.

Rationale for the Procedures Used in This Section

The test statistic given for hypothesis tests is justified by the following:

1. With $n_1 p_1 \geq 5$ and $n_1 q_1 \geq 5$, the distribution of $\hat{p}_1$ can be approximated by a normal distribution with mean p_1, standard deviation $\sqrt{p_1 q_1 / n_1}$, and variance $p_1 q_1 / n_1$. These conclusions are based on Sections 5–6 and 6–4 and they also apply to the second sample.

2. Because $\hat{p}_1$ and $\hat{p}_2$ are each approximated by a normal distribution, $\hat{p}_1 - \hat{p}_2$ will also be approximated by a normal distribution with mean $p_1 - p_2$ and variance

$$\sigma^2_{(\hat{p}_1 - \hat{p}_2)} = \sigma^2_{\hat{p}_1} + \sigma^2_{\hat{p}_2} = \frac{p_1 q_1}{n_1} + \frac{p_2 q_2}{n_2}$$

(In Section 8–2 we established that the variance of the differences between two independent random variables is the sum of their individual variances.)

3. Because the values of p_1, p_2, q_1, and q_2 are typically unknown and from the null hypothesis we assume that $p_1 = p_2$, we can pool (or combine) the sample data. The pooled estimate of the common value of p_1 and p_2 is $\overline{p} = (x_1 + x_2)/(n_1 + n_2)$. If we replace p_1 and p_2 by $\overline{p}$ and replace q_1 and q_2 by $\overline{q} = 1 - \overline{p}$, the variance from Step 2 leads to the following standard deviation:

$$\sigma_{(\hat{p}_1 - \hat{p}_2)} = \sqrt{\frac{\overline{p}\,\overline{q}}{n_1} + \frac{\overline{p}\,\overline{q}}{n_2}}$$

4. We now know that the distribution of $p_1 - p_2$ is approximately normal, with mean $p_1 - p_2$ and standard deviation as given, so that the z test statistic has the form given earlier.

The form of the confidence interval requires an expression for the variance different from the one given in Step 3. In Step 3 we are assuming that $p_1 = p_2$, but if we don't make that assumption (as in the construction of a confidence interval), we estimate the variance of $\hat{p}_1 - \hat{p}_2$ as

$$\sigma^2_{(\hat{p}_1 - \hat{p}_2)} = \sigma^2_{\hat{p}_1} + \sigma^2_{\hat{p}_2} = \frac{\hat{p}_1 \hat{q}_1}{n_1} + \frac{\hat{p}_2 \hat{q}_2}{n_2}$$

and the standard deviation becomes

$$\sqrt{\frac{\hat{p}_1 \hat{q}_1}{n_1} + \frac{\hat{p}_2 \hat{q}_2}{n_2}}$$

In the test statistic

$$z = \frac{(\hat{p}_1 - \hat{p}_2) - (p_1 - p_2)}{\sqrt{\frac{\hat{p}_1 \hat{q}_1}{n_1} + \frac{\hat{p}_2 \hat{q}_2}{n_2}}}$$

let z be positive and negative (for two tails) and solve for $p_1 - p_2$. The results are the limits of the confidence interval given earlier.

EXCEL: In Excel, you must use the **Data Disk XL** add-in, which is a supplement to this book. In four spreadsheet cells, input these numbers, respectively: the number of successes in Sample 1, the number of trials for Sample 1, the number of successes in Sample 2, and the number of trials for Sample 2.

EXCEL (Prior to 2007): Click on **DDXL**. For the hypothesis test, select **Hypothesis Tests** and **Summ 2 Var Prop Test**; or for the confidence interval, select **Confidence Intervals** and **Summ 2 Var Prop Interval**. In the dialog box, click on the each of the four pencil icons in turn, and enter the corresponding cell addresses (for example, in the first input box, enter the cell address containing the number of successes in Sample 1). Click **OK**. Proceed to complete the new dialog box.

EXCEL 2007: Click **Add-Ins** on the main menu and choose **DDXL** from the add-ins menu to the left. Proceed as above.

MINITAB 15: From the Stat menu, choose **Basic Statistics** and then **2 Proportions**. If the test is being conducted with summarized data, click that button and fill in the boxes with the number of events (i.e., successes) and trials. The default is a two-tailed test at a 5% level of significance with a hypothesized difference of zero and unpooled proportions. If you wish to change any of these, click the **Options** button and make the necessary changes.

STATDISK: Select **Analysis**, then **Hypothesis Testing**, then **Proportion-Two Samples**. Enter the required items in the dialog box. Confidence interval limits are included with the hypothesis test results.

8-6 Exercises A: Basic Skills and Concepts

In Exercises 1 and 2, assume that you plan to use a significance level of $\alpha = 0.05$ to test the claim that $p_1 = p_2$. Use the given sample sizes and numbers of successes to find (a) the pooled estimate $\bar{p}$, (b) the z test statistic, (c) the critical z values, and (d) the P-value.

1. Sample 1	Sample 2		2. Sample 1	Sample 2
$n_1 = 50$	$n_2 = 100$		$n_1 = 500$	$n_2 = 200$
$x_1 = 25$	$x_2 = 55$		$x_1 = 300$	$x_2 = 150$

In Exercises 3 and 4, find the number of successes x suggested by the given statement.

3. From Health Canada: Of 2429 industrial radiographers whose exposures to radiation were monitored for a year, 2.0% of them had radiation doses above 20 mSv.

4. Based on a list of major bridges (BlackDog Media): Of a random selection of 10 major continuous truss bridges, 10% are in Canada.

5. The Composting Council of Canada requires that type AA compost must comprise less than 1% foreign matter. A ministry inspector uses random sampling to test two commercial composting facilities against the standards. Of 50 samples collected at Facility A, 15 had less than half of the allowable level of foreign matter. Facility B attained this quality in 7 of 55 samples. Test the claim that Facility A has a greater proportion of product containing less than half of the allowable level of foreign matter than Facility B. Use the 0.05 significance level. If the claim is true, does it follow that Facility A consistently produces better quality compost?

6. According to the *Annual Report of the Correctional Investigator* (1995–1996), this agency was able to resolve or provide assistance for 88 of the 280 complaints received about the prison grievance procedure. On the other hand, they were able to resolve or provide assistance for 55 of 232 complaints received about private family visits. At the 0.05 level of significance, and assuming that data for the years covered in the report are representative of future years, test the claim that the Correctional Investigator has more success in resolving or assisting complaints about the grievance procedure than complaints about private visits. Is it reasonable to treat the reported data as a sample of a larger population of future years or years in general?

7. Based on a random survey of 1000 adult Canadians conducted in January 2001, a headline in *The Globe and Mail* (February 5, 2001) concluded that "fewer Canadians support free trade." In the survey, 64% of the respondents supported the North American free trade agreement, but in a 1999

survey 70% supported the agreement. Testing at the 0.01 level of significance, is there sufficient evidence to support the claim in the headline? (Assume that the number of Canadians surveyed in 1999 was 1000.) We are not told in the article how many Canadians were surveyed in 1999. Could it make a difference to our conclusion if our assumption of $n = 1000$ is incorrect? Explain.

8. In initial tests of the Salk vaccine, 33 of 200,000 vaccinated children later developed polio. Of 200,000 children vaccinated with a placebo, 115 later developed polio. At the 0.01 level of significance, test the claim that the Salk vaccine is effective in lowering the polio rate. Does it appear that the vaccine is effective?

9. A random sample was taken from the *Financial Post* list of top 500 companies for 1998. Of 26 food- and beverage-related industries in the sample, 12 had at least 10% foreign ownership. Of 29 wholesale distribution firms in the sample, 15 had at least 10% foreign ownership.
   a. Use a 0.01 significance level to test the claim that both industry groups sampled had the same proportion of companies with a foreign ownership level of at least 10%.
   b. Construct the 99% confidence interval for the difference between proportions of companies that are significantly (at least 10%) foreign-owned, in top food- and beverage-related companies versus top wholesale distribution firms.

10. During the campaign for the June 1997 federal election, many opinion polls were conducted. Early in the campaign, a poll found that 47.6% of voters supported the Liberals. Then in May 1997 an Angus Reid poll found that of 3208 voters surveyed, 42% supported the Liberals (based on a report conducted for the CTV television network and the Southam newspaper chain). Some analysts suggested that this difference was within the margin of error for the polls.
   a. Assuming that 3100 voters participated in the earlier poll, test the claim that there was no change in support for the Liberals between the two polls. Based on the result, did the Liberals need to change campaign strategies?
   b. Construct a 95% confidence interval for the difference between the proportions of voters who supported the Liberals in the earlier poll and those who supported the Liberals in the May poll.

11. Karl Pearson, who developed many important concepts in statistics, collected crime data in 1909. Of those convicted of arson, 50 were drinkers and 43 abstained. Of those convicted of fraud, 63 were drinkers and 144 abstained. Use a 0.01 significance level to test the claim that the proportion of drinkers among convicted arsonists is greater than the proportion of drinkers convicted of fraud. Does it seem reasonable that drinking might have an effect on the type of crime? Why?

12. An initial public offering (IPO) of a company's stock is considered "under-priced" if there is a large percentage difference between its issuing price and its closing price after one day of trading. A Canadian study based on a sample of 399 IPOs, over 25 years, showed that approximately 8% of those stocks were underpriced when they were issued ("The Mixed Results of Canadian IPOs," *Canadian Investment Review*, Vol. 10, 12-01-1997, pp. 22–26). Suppose that a study of 430 U.S. IPOs had shown that 66 of those issues were underpriced.

    a. Use the 0.05 level of significance to test the claim that fewer Canadian IPOs have been underpriced than U.S. IPOs.

    b. "Day traders" like underpriced IPOs, because if bought early and sold late on their day of issuance, the trades make a profit. Do the results from (a) support a claim that day traders who knew which IPOs were under-priced could have made more money by investing in U.S. IPOs? What additional information, if any, is needed to answer the question?

    c. Construct a 95% confidence interval for the difference in proportions of underpriced IPOs for Canadian and U.S. stocks.

13. Professional pollsters are becoming concerned about the growing rate of refusals among potential survey subjects. In analyzing the problem, there is a need to know if the refusal rate is universal or if there is a difference between the rates for central-city residents and those not living in central cities. Specifically, it was found that when 294 central-city residents were surveyed, 28.9% refused to respond. A survey of 1015 residents not in a central city resulted in a 17.1% refusal rate (based on data from "I Hear You Knocking But You Can't Come In," by Fitzgerald and Fuller, *Sociological Methods and Research*, Vol. 11, No. 1). At the 0.01 significance level, test the claim that the central-city refusal rate is the same as the refusal rate in other areas.

14. A mining company conducted a safety analysis for workers at a Saskatchewan site. Of 450 sample records in January, 16 revealed undesirable safety conditions. Of 660 sample records in February/March, 10 showed undesirable conditions. Is there sufficient sample evidence to support the site manager's claim, at the 0.02 level of significance, that safety conditions improved between the two sample periods?

15. When games were sampled from throughout a season, it was found that the home team won 127 of 198 professional *basketball* games, and the home team won 57 of 99 professional *football* games (based on data from "Predicting Professional Sports Game Outcomes from Intermediate Game Scores," by Cooper et al., *Chance*, Vol. 5, No. 3–4). Construct a 95% confidence interval for the difference between the proportions of home wins. Do the confidence interval limits contain 0? Based on the results, is there a significant difference between the proportions of home wins? What do you conclude about the home field advantage?

16. Ziac is a Lederle Laboratories drug developed to treat hypertension. Lederle Laboratories reported that when 221 people were treated with Ziac, 3.2% of them experienced dizziness. It was also reported that among the 144 people in the placebo group, 1.8% experienced dizziness.
    a. Can you use the methods of this section to test the claim that there is a significant difference between the two rates of dizziness? Why or why not?
    b. Can the given information be correct? Why or why not?

17. To test the null hypothesis that the difference between two population proportions is equal to a nonzero constant c, use the test statistic

$$z = \frac{(\hat{p}_1 - \hat{p}_2) - c}{\sqrt{\dfrac{\hat{p}_1(1 - \hat{p}_1)}{n_1} + \dfrac{\hat{p}_2(1 - \hat{p}_2)}{n_2}}}$$

As long as n_1 and n_2 are both large, the sampling distribution of the test statistic z will be approximately the standard normal distribution. Refer to the sample data included with the example presented in this section, and use a 0.05 significance level to test the claim of a medical expert that when women are exposed to glycol ethers, their percentage of miscarriages is 10 percentage points more than the percentage for women not exposed to glycol ethers.

18. Sample data are randomly drawn from three independent populations, each of size 100. The sample proportions are $\hat{p}_1 = 40/100$, $\hat{p}_2 = 30/100$, and $\hat{p}_3 = 20/100$.
    a. At the 0.05 significance level, test H_0: $p_1 = p_2$.
    b. At the 0.05 significance level, test H_0: $p_2 = p_3$.
    c. At the 0.05 significance level, test H_0: $p_1 = p_3$.
    d. In general, if hypothesis tests lead to the conclusions that $p_1 = p_2$ and $p_2 = p_3$ are reasonable, does it follow that $p_1 = p_3$ is also reasonable? Why or why not?

19. The *sample size* needed to estimate the difference between two population proportions to within a margin of error E with a confidence level of $1 - \alpha$ can be found as follows. In the expression

$$E = z_{\alpha/2}\sqrt{\frac{p_1 q_1}{n_1} + \frac{p_2 q_2}{n_2}}$$

replace n_1 and n_2 by n (assuming that both samples have the same size) and replace each of $p_1, q_1, p_2,$ and q_2 by 0.5 (because their values are not known). Then solve for n.

Use this approach to find the size of each sample if you want to estimate the difference between the proportions of men and women who own cars. Assume that you want 95% confidence that your error is no more than 0.03.

VOCABULARY LIST

denominator degrees of
freedom **466**
dependent samples **441**
F distribution **465**
independent samples **441**

matched pairs **441**
numerator degrees of
freedom **466**
paired samples **441**

pooled estimate of p_1
and p_2 **490**
pooled estimate of σ^2 **475**
significance level **466**

REVIEW

Chapters 6 and 7 introduced two major concepts of inferential statistics: the estimation of population parameters and the methods of testing hypotheses made about population parameters. Whereas those chapters considered only cases involving a single population, this chapter considered two samples drawn from two populations.

- Section 8–2 considered inferences made about the means of (dependent) population data consisting of matched pairs.

- Section 8–3 considered inferences made about two independent populations, but the involved samples were both large ($n > 30$).

- Section 8–4 presented methods for testing claims about two population standard deviations or variances.

- Section 8–5 described procedures for making inferences about two independent populations, with at least one of the samples being small ($n \leq 30$). Sections 8–2, 8–3, and 8–5 dealt with a variety of different cases for making inferences about two population means; those cases were summarized in Figure 8-8 (for hypothesis tests).

- Section 8–6 discussed the procedures for making inferences about two population proportions.

REVIEW EXERCISES

1. In a study of people who stop to help drivers with disabled cars, researchers hypothesized that more people would stop to help someone if they first saw another driver with a disabled car getting help. In one experiment, 2000 drivers first saw a woman being helped with a flat tire and then saw a second woman who was alone, farther down the road, with a flat tire; 2.90% of those 2000 drivers stopped to help the second woman. Among 2000 other drivers who did not see the first woman being helped, only 1.75% stopped to help (based on data from "Help on the Highway," by McCarthy, *Psychology Today*). At the 0.05 significance level, test the claim that the

percentage of people who stop after first seeing a driver with a disabled car being helped is greater than the percentage of people who stop without first seeing someone else being helped.

2. Twelve Dozenol tablets are tested for solubility before and after being stored for one year. The indexes of solubility are given in the table below.

Before	472	487	506	512	489	503	511	501	495	504	494	462
After	562	512	523	528	554	513	516	510	524	510	524	508

   a. At the 0.05 significance level, test the claim that the Dozenol tablets are more soluble after the storage period.

   b. Construct a 95% confidence interval estimate of the mean difference of after − before.

3. In clinical tests of adverse reactions to the drug Viagra, 7% of the 734 subjects in the treatment group experienced dyspepsia (indigestion), but 2% of the 725 subjects in the placebo group experienced dyspepsia (based on data from Pfizer Pharmaceuticals).

   a. Is there sufficient evidence to support the claim that dyspepsia occurs at a higher rate among Viagra users than those who do not use Viagra?

   b. Construct a 95% confidence interval estimate of the difference between the dyspepsia rate for Viagra users and the dyspepsia rate for those who use a placebo. Do the confidence interval limits contain 0, and what does this suggest about the two dyspepsia rates?

4. Having just been charged more to fly from Toronto to Saint John than from Toronto to Britain, an annoyed passenger claims there is no difference in price between domestic and international flights. Using ads in the *Toronto Star*, his travelling companion compiles a random sample of prices for domestic flights from Toronto and flights from Toronto to Europe. Based on his results, shown below, test the disgruntled passenger's claim that there is no difference in price between these two types of flight. Use the 0.02 level of significance. Assume unequal standard deviations for the two samples.

Domestic	International
$n_1 = 12$	$n_2 = 16$
$\bar{x}_1 = 354.8$	$\bar{x}_2 = 522.6$
$s_1 = 76.8$	$s_2 = 179.3$

5. To conduct the means test in Exercise 4, it was assumed that the two samples had unequal variances. Test whether the evidence supports that assumption, at the 0.02 level of significance.

6. A test question is considered good if it discriminates between prepared and unprepared students. The first question on a test was answered correctly by 62 of 80 prepared students and by 23 of 50 unprepared students. At the 0.05 level of significance, test the claim that this question was answered correctly by a greater proportion of prepared students.

7. In a study of techniques used to measure lung volumes, physiological data were collected for 10 subjects. The values given in the accompanying table are in litres, representing the measured forced vital capacities of the 10 subjects in a sitting position and in a supine (lying) position. At the 0.05 significance level, test the claim that the position has no effect, so the mean difference is zero.

Subject	A	B	C	D	E	F	G	H	I	J
Sitting	4.66	5.70	5.37	3.34	3.77	7.43	4.15	6.21	5.90	5.77
Supine	4.63	6.34	5.72	3.23	3.60	6.96	3.66	5.81	5.61	5.33

Based on data from "Validation of Esophageal Balloon Technique at Different Lung Volumes and Postures," by Baydur et al., *Journal of Applied Physiology,* Vol. 62, No. 1.

8. A major airline is experimenting with its training program for flight attendants. With the traditional six-week program, a random sample of 60 flight attendants achieves competency test scores with a mean of 83.5 and a standard deviation of 16.3. With a new ten-day program, a random sample of 35 flight attendants achieve competency test scores with a mean of 79.8 and a standard deviation of 19.2. At the 0.01 significance level, test the claim that the ten-day program results in scores with a lower mean.

9. Researchers are testing commercial air-filtering systems made by the Winston Industrial Supply Company and the Barrington Filter Company. Random samples are tested for each company, and the filtering efficiency is scored on a standard scale, with the results shown in the margin. (Higher scores correspond to better filtering.) At the 0.05 level of significance, test the claim that both systems have the same mean.

Winston	Barrington
$n = 18$	$n = 24$
$\bar{x} = 85.7$	$\bar{x} = 80.6$
$s = 2.8$	$s = 9.7$

10. Data Set 18 in Appendix B shows body temperatures (in degrees Celsius) of adults accepted for voluntary surgical procedures in an urban hospital. Sample data for men are summarized by these statistics: $n = 12$, $\bar{x} = 36.275$, $s = 0.561$. Sample data for women are summarized by these statistics: $n = 12$, $\bar{x} = 36.533$, $s = 0.365$.

   a. At the 0.05 significance level, test the claim that both men and women who have been accepted for voluntary surgical procedures have the same mean temperature. Use a preliminary F test to determine whether the two populations have equal variances.

   b. Construct a 95% confidence interval for the difference between the two population means.

CUMULATIVE REVIEW EXERCISES

1. The data in the table to the right were obtained through a survey of randomly selected subjects.

   a. If one of the survey subjects is randomly selected, find the probability of getting someone ticketed for speeding.

   b. If one of the survey subjects is randomly selected, find the probability of getting a male or someone ticketed for speeding.

   c. Find the probability of getting someone ticketed for speeding, given that the selected person is a man.

   d. Find the probability of getting someone ticketed for speeding, given that the selected person is a woman.

   e. Use a 0.05 level of significance to test the claim that the percentage of women ticketed for speeding is less than the percentage for men. Can we conclude that men generally speed more than women?

	Ticketed for Speeding Within the Last Year?	
	Yes	No
Men	26	224
Women	27	473

Based on data from R.H. Bruskin Associates.

2. The Newton Scientific Instrument Company manufactures scales with its day shift and night shift. Random samples are routinely tested with errors recorded as positive amounts for readings that are too high or negative for readings that are too low.

   a. A random sample of 40 scales made by the day shift is tested and the errors are found to have a mean of 1.2 g and a standard deviation of 3.9 g. Use a 0.05 significance level to test the claim that the sample comes from a population with a mean error equal to 0.

   b. A random sample of 33 scales made by the night shift is tested and the errors are found to have a mean of -1.4 g and a standard deviation of 4.3 g. Use a 0.05 significance level to test the claim that the sample comes from a population with a mean error equal to 0.

   c. Use a 0.05 significance level to test the claim that both shifts manufacture scales with the same mean error.

   d. Construct a 95% confidence interval for the mean error of the day shift.

   e. Construct a 95% confidence interval for the mean error of the night shift.

   f. Construct a 95% confidence interval for the difference between the means of errors for the day shift and night shift.

   g. If you want to estimate the mean error for the day shift, how many scales must you check if you want to be 95% confident that your mean is off by no more than 0.5 g? (*Hint:* Use the sample standard deviation in part (a) as an estimate of the population standard deviation σ.)

Refer to Data Set 1 in Appendix B and use Excel or STATDISK to retrieve the weights of discarded metal and plastic, which are already stored on a disk. Use Excel or STATDISK for each of the following.

a. Test the claim that the weights of discarded metal and the weights of discarded plastic have the same mean. Record the test statistic and P-value, and state your conclusion. Also construct a 95% confidence interval for the difference $\mu_m - \mu_p$, where μ_m is the mean weight of discarded metal and μ_p is the mean weight of discarded plastic.

b. The weights used in part (a) are given in pounds. They can be converted to kilograms by multiplying each value by 0.4536. Convert all of those weights from pounds to kilograms and then repeat part (a). Compare the results to those originally found in part (a). In general, are the results affected if a different scale is used?

c. The first weight of discarded metal is 1.09 lb. Change that value to 109 lb by deleting the decimal point. The value of 109 lb is clearly an outlier, but it is only one wrong entry among the 62 metal weights. After making that change, repeat part (a). How are the results affected by this outlier?

■ FROM DATA TO DECISION

Does Aspirin Help Prevent Heart Attacks?

In a recent study of 22,000 male physicians, half were given regular doses of aspirin while the other half were given placebos. The study ran for six years at a cost of $4.4 million. Among those who took the aspirin, 104 suffered heart attacks. Among those who took the placebos, 189 suffered heart attacks. (The figures are based on data from *Time* and the *New England Journal of Medicine*, Vol. 318, No. 4.) Do these results show a statistically significant decrease in heart attacks among the sample group who took aspirin? The issue is clearly important because it can affect many lives.

Use the methods of this chapter to determine whether the use of aspirin seems to help prevent heart attacks. Write a report that summarizes your findings. Include any relevant factors you can think of that might affect the validity of the study. For example, is it noteworthy that the study involved only male physicians? Is it noteworthy that aspirin sometimes causes stomach problems?

■ COOPERATIVE GROUP ACTIVITIES

1. **Out-of-class activity:** Are estimates influenced by anchoring numbers? Refer to the related Chapter 2 Cooperative Group Activity. In Chapter 2 we noted that, according to author John Rubin, when people must estimate a value,

their estimate is often "anchored" to (or influenced by) a preceding number. In that Chapter 2 activity, subjects were asked to quickly estimate the value of $8 \times 7 \times 6 \times 5 \times 4 \times 3 \times 2 \times 1$, while others are asked to quickly estimate the value of $1 \times 2 \times 3 \times 4 \times 5 \times 6 \times 7 \times 8$.

In Chapter 2, we compared the two sets of results by using statistics (such as the mean) and graphs (such as boxplots). The methods of Chapter 8 now allow us to compare the results with a formal hypothesis test. Collect sample data and test the claim that when we begin with larger numbers ($8 \times 7 \times 6 \times \ldots$), our estimates tend to be larger.

2. **In-class activity:** In Exercise 15 from Section 8–3, we used pulse rates in Data Set 8 to test the claim that male and female statistics students have the same mean pulse rate. Divide into groups according to gender, with about 10 or 12 students in each group. Each group member should record his or her pulse rate by counting the number of heartbeats in one minute, and the group statistics ($n, \bar{x}, s$) should be calculated. The groups should exchange their results and test the same claim given in Exercise 15 from Section 8–3. Is there a difference? Why?

3. **In-class activity:** Divide into groups of about 10 or 12 students and use the same reaction timer included with the Chapter 5 Cooperative Group Activities. Each group member should be tested for right-hand reaction time and left-hand reaction time. Using the group results, test the claim that there is no difference between the right-hand and left-hand reaction times. Compare the conclusion to the conclusion reached by other groups. Is there a difference?

INTERNET PROJECT

Hypothesis Testing

Whereas Chapter 7 presented methods for testing hypotheses about a single population, the methods of this chapter extend those methods to two populations. Similarly, the Internet Project for this chapter differs from that of Chapter 7 through the use of two data sets in each of the hypothesis tests. The project can be found at either of the following websites:

http://www.mathxl.com   or   http://www.mystatlab.com

After choosing "Internet Projects" and then "Chapter 8" from the menus, you will find several hypothesis-testing problems that examine population demographics, salary fairness, and a traditional superstition. In each case you will formulate the problem as a hypothesis test using the notation of the chapter, seek out appropriate data, conduct the test, and then interpret the results with a summarizing conclusion.

9

Analysis of Variance

9-1 Overview

Objectives are identified for this chapter, which introduces methods for testing the hypothesis that three or more populations have the same mean. In Chapter 8 we tested for equality between the means of two populations and used test statistics having normal or Student t distributions. The methods of this chapter, however, are very different.

9-2 One-Way ANOVA

This section introduces the basic method of analysis of variance (ANOVA), which is used to test the claim that three or more populations have the same mean. One-way ANOVA involves different samples that are categorized according to a single characteristic. The setup of the data is called a completely randomized design since the data are randomly sampled from each population independently.

9-3 Two-Way ANOVA

While the samples in Section 9-2 are independent, such is not the case for the data in this section. First we examine the randomized block design (also known as two-way ANOVA without replication), in which the random sampling is done with the blocks. This approach is common in consumer surveys. Second, we examine the factorial design (also called two-way ANOVA with replication), in which there are two factors and multiple observations for each factor combination. This approach is common in scientific experiments.

CHAPTER PROBLEM

Is there a significant difference in the average price per litre between grades of gasoline?

Each week, MJ Ervin & Associates Inc. does a survey of pump prices across Canada. Table 9-1 contains a random sample of prices for each of regular, mid-grade, premium and diesel before taxes are added, which were published on July 15, 2008. The four samples have means of 107.0625, 109.3, 113.325, and 120.9875, respectively, so it appears that there may not be a significant difference in the average price of regular and mid-grade, but that may not hold true for the rest. Figure 9-1 shows boxplots representing the four samples, and these also suggest that the samples come from populations with non-equal means. Considering, however, that each sample consists of only 8 values, are those differences statistically significant?

Chapter 8 included methods for testing equality between *two* population means. We now want to test for equality among *three or more* population means. Specifically, we want to test the claim that $\mu_{regular} = \mu_{mid\text{-}grade} = \mu_{premium} = \mu_{diesel}$. This chapter introduces methods for testing such claims, and we will consider the data included in Table 9-1.

Table 9-1 Gas Pump Prices by Grade

Regular	Mid-grade	Premium	Diesel
108.9	105.7	113.6	124.5
103.8	106.8	119.0	124.2
106.8	105.9	112.8	119.4
106.3	113.7	110.7	121.6
103.9	110.0	109.9	118.8
105.6	112.2	117.5	118.5
113.5	113.5	111.0	119.4
107.7	106.6	112.1	121.5
$n_1 = 8$	$n_2 = 8$	$n_3 = 8$	$n_4 = 8$
$\bar{x}_1 = 107.0625$	$\bar{x}_2 = 109.3$	$\bar{x}_3 = 113.325$	$\bar{x}_4 = 120.9875$
$s_1 = 3.132$	$s_2 = 3.463$	$s_3 = 3.284$	$s_4 = 2.370$

MJ Ervin & Associates, www.mjervin.com.

Figure 9-1 Comparison of Gas Prices for Various Grades

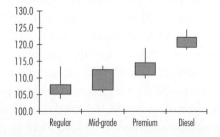

MJ Ervin & Associates, www.mjervin.com.

9-1 Overview

In Sections 8–2, 8–3, and 8–5 we developed procedures for testing the hypothesis that *two* population means are equal ($H_0: \mu_1 = \mu_2$). Section 9–2 introduces a procedure for testing the hypothesis that *three or more* population means are equal. (The methods used in this chapter can also be used to test for equality between two population means, but the methods used in Chapter 8 are more efficient.) A typical null hypothesis in Section 9–2 will be $H_0: \mu_1 = \mu_2 = \mu_3 = \mu_4$. The alternative hypothesis is H_1: At least two means are different. The method we use is based on an analysis of sample variances.

> **DEFINITION**
>
> **Analysis of variance (ANOVA)** is a method of testing the equality of three or more population means by analyzing sample variances.

ANOVA is used in applications such as the following:

- When three different groups of people (such as smokers, nonsmokers exposed to environmental tobacco smoke, and nonsmokers not so exposed) are measured for cotinine (a marker of nicotine), we can test to determine whether they have the same level.

- When the Canadian Nuclear Safety Commission is concerned with exposures of uranium miners to ionizing radiation, it can test to determine whether there are differences in mean annual exposures for various categories of mine workers.

Why bother with a new procedure when we can test for equality of two means by using the methods presented in Chapter 8, where we developed tests of $H_0: \mu_1 = \mu_2$? For example, if we want to use the sample data from Table 9-1 to test the claim (at the $\alpha = 0.05$ level) that the four populations have the same mean, why not simply pair them off and do two at a time by testing $H_0: \mu_1 = \mu_2$, then $H_0: \mu_2 = \mu_3$, and so on? This approach (doing two at a time) requires six different hypothesis tests, so the degree of confidence could be as low as 0.95^6 (or 0.7351). In general, as we increase the number of individual tests of significance, we increase the likelihood of finding a difference by chance alone. The risk of a type I error—finding a difference when no such difference actually exists—is higher than is implied by the selected significance level for each test. The method of analysis of variance lets us avoid that particular pitfall (rejecting a true null hypothesis) by using one test for equality of several means.

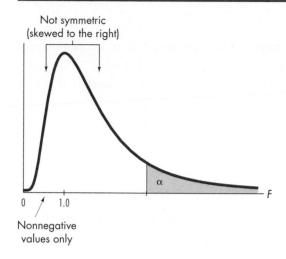

Not symmetric
(skewed to the right)

0   1.0 F

Nonnegative
values only

α

Figure 9-2
The *F* Distribution

There is a different *F* distribution
for each different pair of degrees
of freedom for numerator and
denominator.

F Distribution

The ANOVA methods use the *F* distribution, which was first introduced in
Section 8–4. In Section 8–4 we noted that the *F* distribution has the following
important properties (see Figure 9-2):

1. The *F* distribution is not symmetric; it is skewed to the right.

2. The values of *F* can be 0 or positive, but they cannot be negative.

3. There is a different *F* distribution for each pair of degrees of freedom for the
   numerator and denominator.

Critical values of *F* are given in Table A-5.

Analysis of variance (ANOVA) is based on a comparison of two different estimates of the variance common to the different populations. Those estimates (the
variance between samples and the *variance within samples*) will be described in
Section 9–2. The term *one-way* is used because the sample data are separated
into groups according to one characteristic or factor. For example, the prices
listed in Table 9-1 are separated into four different groups according to the one
characteristic (or factor) of the grade from which they were observed. Section 9–3
will introduce two-way analysis of variance, which allows us to compare populations separated into categories using two characteristics (or factors). For
example, we might separate lengths of movies that are broadcast on television
by using the following two criteria: (1) their star ratings (1 star, 2 stars, 3 stars,
and 4 stars) and (2) the Canadian classification system for violence in programming (FAM, PA, 14+, and 18+).

There is a different *F* distribution for each different pair of degrees of freedom
for numerator and denominator.

9-2 One-Way ANOVA

In this section we consider tests of hypotheses that three or more population means are all equal, for example, $H_0: \mu_1 = \mu_2 = \mu_3$.

The following assumptions apply when testing the hypothesis that three or more samples come from populations with the same mean.

Assumptions

1. The populations have distributions that are approximately normal.

2. The populations have the same variance σ^2 (or standard deviation σ).

3. The samples are random and independent of each other.

4. The different samples are from populations that are categorized in only one way.

The setup of the data is known as a completely randomized design, because the data for each population are sampled independently of the other populations. The requirements of normality and equal variances are somewhat relaxed, as the methods in this section work reasonably well unless a population has a distribution that is very nonnormal or the population variances differ by large amounts. Statistician George E.P. Box showed that as long as the sample sizes are equal (or nearly equal), the variances can differ by amounts where the largest is up to nine times the smallest and the results of ANOVA will continue to be essentially reliable. (If the samples are independent but the distributions are very nonnormal, we can use the Kruskal-Wallis test presented in Section 12–5.)

The method we use is called **one-way analysis of variance** (or **single-factor analysis of variance**) because we use a single property, or characteristic, for categorizing the populations. This characteristic is sometimes referred to as a *treatment* or *factor*.

DEFINITION

A **treatment** (or **factor**) is a property, or characteristic, that allows us to distinguish the different populations from one another.

For example, the prices listed in Table 9-1 are sample data drawn from four different populations that are distinguished according to the treatment (or factor) of the grade from which they were observed. The term *treatment* is used because early applications of analysis of variance involved agricultural experiments in which different plots of farmland were treated with different fertilizers, seed types, insecticides, and so on.

Rationale

The method of analysis of variance is based on this fundamental concept:

Total variation of all data = Variation between groups
+ Variation within groups

To illustrate, examine this set of data:

Group 1	Group 2	Group 3
1	2	3
1	2	3
1	2	3

With the data arranged this way, there is more variation between the groups than within the groups. Thus we conclude the means of the three groups are different.

Let's now rearrange the data:

Group 1	Group 2	Group 3
1	1	1
2	2	2
3	3	3

This time, there is less variation between the groups than within the groups. Thus we conclude the means of the three groups are equal.

It should be noted that the total variation of all the data is the same for both data sets since we just rearranged the data. So, the left side of the above equation is the same for both data sets; what changes as we rearrange the data is the right side. If the variation between the groups increases, that means the variation within the groups decreases and vice versa. Granted, this is an oversimplified example but it serves to illustrate the key principle behind analysis of variance:

> **The more variation there is between the groups, the more likely we are to conclude that there is a significant difference between the means of at least two of these groups.**

With the assumption that the populations all have the same variance σ^2, we estimate the common value of σ^2 using two different approaches:

1. The **variance between samples** (also called **variation due to treatment**) is an estimate of the common population variance σ^2 that is based on the variability among the sample *means*.

2. The **variance within samples** (also called **variation due to error**) is an estimate of the common population variance σ^2 based on the sample *variances*.

What we want to do is compare the variation between the groups to the variation within the groups. Recall from Section 8–4 that the F distribution is used for the hypothesis test in which we compare two variances. That is why we use the F distribution for analysis of variance.

Construction of the ANOVA Table

In order to test the hypothesis using the F distribution, we need the numerator and denominator degrees of freedom (in order to find the critical value and subsequent critical region) and the value of the test statistic. We find these values by constructing an ANOVA table.

To begin, we introduce some basic terminology:

x = observed data value

$\bar{x}$ = mean of all sample scores combined

k = number of population means being compared

n_i = number of values in the ith sample

n = total number of values in all samples combined

$\bar{x}_i$ = mean of values in the ith sample

s_i^2 = variance of values in the ith sample

For the chapter problem, this is the ANOVA table:

Source	df	SS	MS	F
between	3	899.28625	299.7621	31.3837
within	28	267.4425	9.5515	
Total	31	1166.72875		

The table has five columns: source of variation (usually just labelled source), degrees of freedom (df), sum of squares (SS), mean square (MS), and the F statistic.

First the sources of variation: Note that the between and within are above the addition line and that total is below it. This is to drive home the point that the total variation is the sum of the between and within sources of variation.

Next the degrees of freedom:

$$\text{total df} = n - 1 = 32 - 1 = 31$$
$$\text{between df} = k - 1 = 4 - 1 = 3$$
$$\text{within df} = \text{total df} - \text{between df} = 31 - 3 = 28$$

The calculations for the SS column are the most complicated part of the ANOVA table. To illustrate how the SS column is constructed, we provide a condensed derivation of the equations needed. We begin with this equation:

$$x = \bar{x} + \bar{x}_i - \bar{x} + x - \bar{x}_i$$

Note that the two sides balance. Next we subtract $\bar{x}$ from both sides and group terms:

$$(x - \bar{x}) = (\bar{x}_i - \bar{x}) + (x - \bar{x}_i)$$

Then we square both sides and sum over all values of x:

Formula 9-1 $\qquad \Sigma(x - \bar{x})^2 = \Sigma n_i(\bar{x}_i - \bar{x})^2 + \Sigma(x - \bar{x}_i)^2$

Recall from Chapter 2 the formula for s^2:

$$s^2 = \frac{\Sigma(x - \overline{x})^2}{n - 1}$$

If we cross-multiply:

$$(n - 1)s^2 = \Sigma(x - \overline{x})^2$$

Note that the right side of the above equation is equivalent to the left side of Formula 9-1 and similar to the second term of the right side. Substituting, we have:

$$(n - 1)s^2 = \Sigma n_i(\overline{x}_i - \overline{x})^2 + \Sigma(n_i - 1)s_i^2$$

These terms are known as sums of squares. The term on the left side represents the total variation of all the data. The terms on the right side are, respectively, the variation between the groups and the variation within the groups. These terms are named as follows:

SS(TOTAL), OR TOTAL SUM OF SQUARES, IS A MEASURE OF THE TOTAL VARIATION (AROUND $\overline{x}$) IN ALL OF THE SAMPLE DATA COMBINED.

$$\text{SS(total)} = (n - 1)s^2$$

SS(BETWEEN) IS A MEASURE OF THE VARIATION BETWEEN THE SAMPLE MEANS. [IN ONE-WAY ANOVA, SS(BETWEEN) IS SOMETIMES REFERRED TO AS SS(FACTOR) OR SS(TREATMENT).]

$$\begin{aligned}\text{SS(between)} &= n_1(\overline{x}_1 - \overline{x})^2 + n_2(\overline{x}_2 - \overline{x})^2 + \cdots + n_k(\overline{x}_k - \overline{x})^2 \\ &= \Sigma n_i(\overline{x}_i - \overline{x})^2\end{aligned}$$

Fortunately, there is a shortcut formula for SS(between):

$$\text{SS(between)} = \Sigma\left(\frac{T_i^2}{n_i}\right) - \frac{T^2}{n}$$

In this formula, T_i is the total of the values in group i and T is the total of all the data.

SS(WITHIN) IS A SUM OF SQUARES REPRESENTING THE VARIABILITY THAT IS ASSUMED TO BE COMMON TO ALL THE POPULATIONS BEING CONSIDERED. [IN ONE-WAY ANOVA, SS(WITHIN) IS SOMETIMES REFERRED TO AS SS(ERROR).]

$$\text{SS(within)} = (n_1 - 1)s_1^2 + (n_2 - 1)s_2^2 + \cdots + (n_k - 1)s_k^2 = \Sigma(n_i - 1)s_i^2$$

We can rewrite Formula 9-1 as:

$$SS(total) = SS(between) + SS(within)$$

First we compute SS(total). If we put all the data into a calculator, we find that $n = 32$, $s = 6.134852182$ and $s^2 = 37.63641129$. Then SS(total) = (31)(37.63641129) = 1166.72875.

To compute SS(between), $n_1 = n_2 = n_3 = n_4 = 8$, $T_1 = 856.5$, $T_2 = 874.4$, $T_3 = 906.6$, $T_4 = 967.9$, and $T = 3605.4$. Then SS(between) = $856.5^2/8$ + $874.4^2/8$ + $906.6^2/8$ + $967.9^2/8 - 3605.4^2/32$ = 899.28625.

To compute SS(within), $s_1 = 3.132$, $s_2 = 3.463$, $s_3 = 3.284$, and $s_4 = 2.37$. Then SS(within) = $7(3.132^2) + 7(3.463^2) + 7(3.284^2) + 7(2.37^2)$ = 68.65875 + 83.96 + 75.515 + 39.30875 = 267.4425. (The calculations were done with a calculator and not with the above rounded values.) Note that SS(between) + SS(within) = 899.28625 + 267.4425 = 1,166.72875 = SS(total).

We now turn our attention to the mean square (MS) column. The mean square (MS) column is calculated by dividing the value in the SS column by its respective degrees of freedom. For one-way ANOVA, there are two MS values to calculate, **MS(between)** and **MS(within)**:

$$MS(between) = \frac{SS(between)}{k - 1} = \frac{899.28625}{3} = 299.7621$$

$$MS(within) = \frac{SS(within)}{N - k} = \frac{267.4425}{28} = 9.5515$$

It should be noted that MS(within) is commonly called **MS(error)** and abbreviated **MSE**. (MS(between) is also known as **MS(treatment)**.) As mentioned earlier, one of the assumptions of ANOVA is that the populations share a common variance σ^2. MS(error) serves as an estimate of that variance. This fact will be used in later computations.

Finally, the F test statistic is calculated by dividing the MS(between) by the MS(within):

$$F = \frac{MS(between)}{MS(within)} = \frac{299.7621}{9.5515} = 31.3837$$

To find the critical value from the F distribution, we need the numerator and denominator degrees of freedom:

$$numerator\ df = between\ df = 3$$
$$denominator\ df = within\ df = 28$$

Unlike tests for two populations in which the critical region can be in the right tail, left tail, or divided between both tails, the critical region for ANOVA is always in the right tail. The reason is based on the rationale given earlier: If the population means (μ_1, μ_2, . . . , μ_k) are not equal, we would expect more variation between the groups than within, resulting in a larger SS(between) and a larger

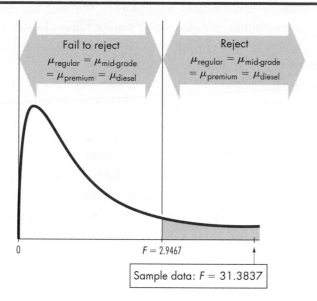

Figure 9-3
F Distribution for Gasoline
Grade Data

Fail to reject
$\mu_{regular} = \mu_{mid\text{-}grade}$
$= \mu_{premium} = \mu_{diesel}$

Reject
$\mu_{regular} = \mu_{mid\text{-}grade}$
$= \mu_{premium} = \mu_{diesel}$

0

$F = 2.9467$

Sample data: $F = 31.3837$

test statistic. On the other hand, if there is no significant difference between the population means, we would expect less variation between the groups than within, resulting in a smaller SS(between) and a smaller test statistic.

If we test at a 0.05 level of significance, the critical value is 2.9467. The null and alternative hypotheses are:

$$H_0: \mu_1 = \mu_2 = \mu_3 = \mu_4$$

H_1: at least two means are different

We reject the null hypothesis if the test statistic is greater than 2.9467. The value of the test statistic from the ANOVA table is 31.3837. Therefore, we reject the null hypothesis and conclude that at least two of the means are significantly different.

Interpretation   We conclude that the average prices of at least two of the grades of gasoline are significantly different at a 0.05 level of significance.

A suggested strategy for constructing a one-way ANOVA table is to compute SS(total) first, then SS(between), and afterward use basic arithmetic to construct the rest of the table.

Multiple Comparisons of Means

In conducting ANOVA, if we fail to reject the null hypothesis, no further analysis is required. If we do reject the null hypothesis, we know that at least two means are significantly different, but which ones? Further analysis is required to determine this.

There are several other tests that can be used to make such identifications, and those procedures are called **multiple comparison procedures**. Comparison of confidence intervals, the Scheffé test, the extended Tukey test, and the Bonferroni test are four common multiple comparison procedures. We will use **Tukey's method of simultaneous confidence intervals,** as described below. The choice of Tukey over Scheffé and Bonferroni is mainly due to Tukey's method producing narrower confidence intervals.

If we are comparing k population means, we can construct $_kC_2$ confidence intervals to compare pairs of means. For example, if $k = 4$, we can construct $_4C_2 = 6$ individual confidence intervals: $\mu_1 - \mu_2, \mu_1 - \mu_3, \mu_1 - \mu_4, \mu_2 - \mu_3, \mu_2 - \mu_3$, and $\mu_3 - \mu_4$. If the level of confidence for each one is 95%, the combined level of confidence for all the intervals is $0.95^6 = 0.7351 = 73.51\%$. The advantage of constructing simultaneous intervals is that the level of confidence is maintained (say at 95%) for all the intervals simultaneously. The main shortcoming is that the intervals must be wider in order for them to share the same level of confidence simultaneously.

For groups i and j, the margin of error is:

Formula 9-2
$$E = Q\sqrt{\frac{\text{MSE}}{2n_i} + \frac{\text{MSE}}{2n_j}}$$

Q is from Table A-11, in which $r = k$ and df = error degrees of freedom. MSE is the MS(error).

If $n_i = n_j = m$, then
$$E = Q\sqrt{\frac{\text{MSE}}{m}}$$

The confidence interval for $\mu_i - \mu_j$ is:
$$(\bar{x}_i - \bar{x}_j) - E < \mu_i - \mu_j < (\bar{x}_i - \bar{x}_j) + E$$

Suppose we are testing the hypothesis
$$H_0: \mu_i = \mu_j$$
$$H_1: \mu_i \neq \mu_j$$

at a level of significance of α. Recall from Chapter 8 that if $\mu_i - \mu_j = 0$ falls outside the $1 - \alpha$ confidence interval, we reject the null hypothesis and conclude that $\mu_i \neq \mu_j$. So, if we reject the null hypothesis in ANOVA, the simultaneous confidence intervals can help us determine which pairs of means are significantly different.

EXAMPLE

Construct 95% simultaneous confidence intervals of the mean difference in price between the grades of gasoline. Use these intervals to determine which pairs of means are significantly different.

SOLUTION

To find Q, we have $k = 4$ and the error degrees of freedom at 28. Using Table A-11 with $\alpha = 0.05$, we use $Q = 3.845$ since 30 is the closest value of degrees of freedom to 28.

Both groups have the same sample size of 8 and the MSE is 9.5515 from the ANOVA table. The margin of error for the intervals is:

$$E = 3.845\sqrt{\frac{9.5515}{8}} = 4.2013$$

The sample means are $\bar{x}_1 = 107.0625$ (regular), $\bar{x}_2 = 109.3$ (mid-grade), $\bar{x}_3 = 113.325$ (premium), and $\bar{x}_4 = 120.9875$ (diesel). Since most of us are more comfortable subtracting smaller values from larger ones, this is how we will proceed. We show how the confidence interval is computed for the first pair of mid-grade and regular. The other pairs would follow the same procedure. The results are summarized in this table:

Groups	Difference of Means	Confidence Interval
mid-grade, regular	$\bar{x}_2 - \bar{x}_1 = 2.2375$	$2.2375 - 4.2013 < \mu_2 - \mu_1 <$ $2.2375 + 4.2013$ $-1.9638 < \mu_2 - \mu_1 < 6.4388$
premium, regular	$\bar{x}_3 - \bar{x}_1 = 6.2625$	$2.0612 < \mu_3 - \mu_1 < 10.4638$
diesel, regular	$\bar{x}_4 - \bar{x}_1 = 13.925$	$9.7237 < \mu_4 - \mu_1 < 18.1263$
premium, mid-grade	$\bar{x}_3 - \bar{x}_2 = 4.025$	$-0.1763 < \mu_3 - \mu_2 < 8.2263$
diesel, mid-grade	$\bar{x}_4 - \bar{x}_2 = 11.6875$	$7.4862 < \mu_4 - \mu_2 < 15.8888$
diesel, premium	$\bar{x}_4 - \bar{x}_3 = 7.6625$	$3.4612 < \mu_4 - \mu_3 < 11.8638$

From the six pairs, we see that the only intervals that contain zero are for mid-grade and regular as well as for premium and mid-grade. Therefore, at a 0.05 level of significance, we fail to reject the null hypothesis for these pairings, but reject the null hypothesis for all the other pairings.

Interpretation
We conclude that there is no significant difference in the mean prices of mid-grade and regular gasoline and of premium and mid-grade gasoline, and that there is a significant difference in mean price between the other pairings.

	A	B	C	D	E	F	G
1	Anova: Single Factor						
2							
3	SUMMARY						
4	Groups	Count	Sum	Average	Variance		
5	Summer 1997	19	86.7	4.563158	29.09023		
6	Summer 1998	19	105.7	5.563158	27.86023		
7	Fall 1999	19	108.1	5.689474	31.22322		
8							
9							
10	ANOVA						
11	Source of Variation	SS	df	MS	F	P-value	F crit
12	Between Groups	14.46877	2	7.234386	0.246141	0.782684	3.168246
13	Within Groups	1587.126	54	29.39123			
14							
15	Total	1601.595	56				
16							

EXCEL DISPLAY
One-Way ANOVA

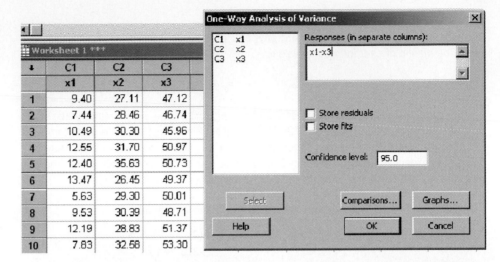

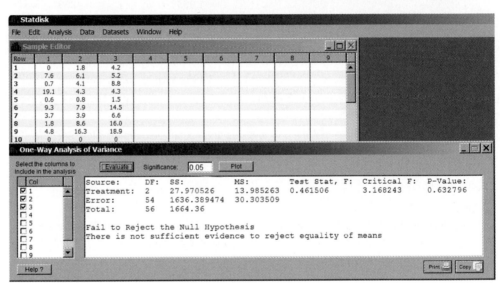

USING TECHNOLOGY

On the previous page and above, we illustrated sample computer displays. Here we describe how to generate those displays.

EXCEL: Organize the raw data into an Excel range, with the data for each population in neighbouring columns (or rows).

EXCEL (Prior to 2007): Select **Data Analysis** from the **Tools** menu. Choose **Anova: Single Factor**. Enter these details: For **Input Range**, specify the range that contains the data; input the desired **alpha**; and specify whether the input range is organized into columns or rows. Then select an **Output Option**, and click on **OK** to obtain the results.

EXCEL 2007:	Click **Data** on the main menu, then **Data Analysis** from the **Analysis** tab. Proceed as above.
MINITAB 15:	Organize the raw data into separate columns. You might want to put labels for each group in the label row. From the **Stat** menu, choose **ANOVA** and then **One-Way (Unstacked)**. Select the columns with the data and click **OK**.
STATDISK:	Select **Analysis** from the main menu bar, then select **One-Way Analysis of Variance**. Select the columns you want. Click **Evaluate**.

As efficient and reliable as computer programs may be, they are worthless if you don't understand the relevant concepts. The methods in this section are used to test the claim that several samples come from populations with the same mean. These methods require normally distributed populations with the same variance, and the samples must be independent. We reject or fail to reject the null hypothesis of equal means by analyzing these two estimates of variance: the variance between samples and the variance within samples. MS(between) or MS(treatment) is an estimate of the variation between samples, and MS(within) or MS(error) is an estimate of the variation within samples. If MS(between) is significantly greater than MS(within), we reject the claim of equal means; otherwise, we fail to reject that claim.

Even if we reject the hypothesis of equal means, we cannot be absolutely sure that the given factor is responsible for the differences. We must identify other relevant factors that could be causing the effect. In general, good results require that the experiments that generate the data be carefully designed and executed.

9-2 Exercises A: Basic Skills and Concepts

1. Analysis of variance results generated by Excel are given on the next page. The sample results are measured consumer reactions to television commercials for different products. Assume that we want to use a 0.05 significance level in testing the null hypothesis that the different products have commercials with the same mean reaction score.
   a. How many different samples are included in this study?
   b. Identify the value of the test statistic.
   c. Find the critical value from Table A-5.
   d. Identify the *P*-value.

e. Based on the preceding results, what do you conclude about the equality of the population means?

Source of Variation	SS	df	MS	F	P-value
Between groups	50.00	5	10.00	3.33	0.032
Within groups	72.00	24	3.00		
Total	122.00	29			

2. Repeat Exercise 1 assuming that the Excel results are as shown below:

Source of Variation	SS	df	MS	F	P-value
Between groups	20.68	2	10.34	6.68	0.011
Within groups	18.59	12	1.55		
Total	39.27	14			

3. The following table is based on data in Data Sets 2 and 18 in Appendix B. Samples of human body temperatures in three contexts are compared. (Assume for this exercise that all three samples are independent.)
   a. Construct the ANOVA table for these data.
   b. Find the F critical value based on a 0.05 level of significance. What do you conclude about the claim that the three populations have the same mean body temperature?
   c. Using Tukey's method of simultaneous confidence intervals, determine which pairs of means, if any, are significantly different.

Three Samples of Body Temperatures (in degrees Celsius)

Awaiting Surgery	Sitting in Lab	Standing in Lab
36.0	37.61	37.63
35.0	37.07	37.17
36.0	37.20	36.93
36.0	37.23	37.35
36.7	37.38	37.73
36.5	37.28	37.39
36.3	37.12	37.57
35.5	37.55	37.89
36.1	37.04	37.50
36.7	37.78	37.98
37.1	37.59	37.36
$n_1 = 11$	$n_2 = 11$	$n_3 = 11$
$\bar{x}_1 = 36.17$	$\bar{x}_2 = 37.35$	$\bar{x}_3 = 37.50$
$s_1 = 0.588$	$s_2 = 0.249$	$s_3 = 0.307$

4. A large company is deciding between transportation methods for shipping its finished goods. Since the possibility of labour stoppages is a concern, a manager studied Data Set 20 in Appendix B for annual person-days lost through work stoppages in different modes of transportation. For each of the transportation modes shown in the table below, she independently selected a random sample of 14 years' data. (The table shows annual

person-days lost, in thousands.) At the 0.05 level of significance, test the claim that the four populations of annual person-days lost have the same mean. Based on the results, does it appear that the four modes are equally reliable, in terms of possible loss of service due to labour stoppage? Does any single category seem to be particularly reliable in this regard? Do we have enough information, from these results, to decide on the best mode of transportation to use? Explain.

Air	Rail	Water	Truck
$n_1 = 14$	$n_2 = 14$	$n_3 = 14$	$n_4 = 14$
$\bar{x}_1 = 60.805$	$\bar{x}_2 = 58.036$	$\bar{x}_3 = 12.631$	$\bar{x}_4 = 19.896$
$s_1 = 63.303$	$s_2 = 71.558$	$s_3 = 13.627$	$s_4 = 33.131$

For Exercises 5–7, use the data shown in the margin (based on the Financial Post list of top 500 companies for 1998). Revenues, assets, and numbers of employees are compared for samples of companies from three industry groups: wholesale distributors (1), food production and sales industries (2), and conglomerates (3). Revenues and assets are in millions of dollars.

5. At the 0.05 significance level, test the claim that the means of the revenues are the same in all three industry groups. Does any industry group seem to have higher revenues? If so, is this apparent difference significant?

6. At the 0.05 significance level, test the claim that the means of the assets are the same in all three industry groups. Does any industry group seem to have greater assets?

7. At the 0.05 significance level, test the claim that the means of the numbers of employees are the same in all three industry groups. Does any industry group seem to have greater numbers of employees?

8. A disgruntled University of Toronto employee reads the list (published in *The Globe and Mail*) of U of T staff who earn over $100,000 annually. The employee claims that for staff in this category, mean salaries are different based on the starting letters of their last names. A random sample from the full list follows. (Salaries are in thousands of dollars.) At the 0.05 level of significance, test the claim that the means are not equal for the three groups shown.

	Revenues	Assets	Employees
1	2,639	326	462
1	2,411	934	2,352
1	1,667	762	2,098
1	1,574	726	10,000
1	478	301	2,800
1	405	159	721
1	327	141	300
1	322	30	29
1	188	86	400
2	6,813	1,506	16,050
2	4,720	1,157	31,000
2	3,432	726	6,500
2	2,458	616	19,700
2	1,442	824	4,428
2	813	278	4,200
2	676	327	3,200
2	461	232	3,400
2	278	144	1,492
3	11,212	5,846	43,000
3	10,008	50,258	27,000
3	4,000	2,200	18,000
3	2,043	872	1,335
3	1,036	724	3,784
3	913	600	5,820
3	504	243	600
3	410	587	4,700
3	349	1,331	670

Names		
A to I	J to R	S to Z
113.7	108.0	114.7
106.4	109.2	111.0
120.3	100.9	121.5
104.7	105.4	156.1
143.8	103.0	107.7
118.3	102.5	112.1
166.7	102.7	108.1
116.7		102.0
		111.1
		158.1
		111.3

Laboratory				
1	2	3	4	5
2.9	2.7	3.3	3.3	4.1
3.1	3.4	3.3	3.2	4.1
3.1	3.6	3.5	3.4	3.7
3.7	3.2	3.5	2.7	4.2
3.1	4.0	2.8	2.7	3.1
4.2	4.1	2.8	3.3	3.5
3.7	3.8	3.2	2.9	2.8
3.9	3.8	2.8	3.2	
3.1	4.3	3.8	2.9	
3.0	3.4	3.5		
2.9	3.3			

9. Flammability tests were conducted on children's sleepwear. The Vertical Semirestrained Test was used, in which pieces of fabric were burned under controlled conditions. After the burning stopped, the length of the charred portion was measured and recorded. Results are given in the margin for the same fabric tested at different laboratories. Because the same fabric was used, the different laboratories should have obtained the same results. Did they? Use a 0.05 significance level to test the claim that the different laboratories have the same population mean.

10. The values in the list below are measured maximum breadths of male Egyptian skulls from different epochs (in mm, based on data from *Ancient Races of the Thebaid*, by Thomson and Randall-Maciver). Changes in head shape over time suggest that interbreeding occurred with immigrant populations. Use a 0.05 significance level to test the claim that the mean is the same for the different epochs.

4000 BCE	1850 BCE	150 CE
131	129	128
138	134	138
125	136	136
129	137	139
132	137	141
135	129	142
132	136	137
134	138	145
138	134	137

11. A political scientist claims that since Confederation, the number of times the House of Commons sits in each session of Parliament has been increasing. The table shows, in three samples grouped by Parliament number, how many times the House of Commons has sat in each Parliament. Parliaments lasting fewer than 1.5 years were not sampled. At the 0.05 level of significance, test the claim that all three groups have the same mean number of sittings.

Parliament		
1 to 12	13 to 24	25 to 35
460	507	447
283	355	418
568	605	687
303	292	405
254	366	707
262	565	767
458	390	590

Based on data from *Duration of Sessions of Parliament*, Library of Parliament.

12. The political scientist from Exercise 11 also claims that for Parliaments of 4, 4.5, and 5 years in length, the mean number of times the Senate has sat during a session is equal. The table shows, in three samples grouped by length in years, the number of times that the Senate has sat in each Parliament.
    a. At the 0.05 level of significance, test the claim that all three groups have the same mean number of Senate sittings.
    b. Based on Tukey's method of simultaneous confidence intervals, between which groups, if any, is there a significant difference?

Length of Parliament in Years		
4	4.5	5
201	478	241
158	339	283
383	352	366
217		357
257		255
261		363

Based on data from *Duration of Sessions of Parliament*, Library of Parliament.

13. Refer to Data Set 21 in Appendix B. At the 0.05 significance level, test the claim that the mean weight (in carats) of diamonds is the same for each of the four clarity levels (2, 3, 4, or 5) shown. If you need to save money by compromising on clarity, does this necessarily limit the carat of the diamond that you choose?

14. Refer to Data Set 9 in Appendix B. Use a 0.01 significance level to test the claim that the mean cotinine level is different for these three groups: nonsmokers not exposed to environmental tobacco smoke, nonsmokers who are exposed to tobacco smoke, and people who smoke. What do the results suggest about second-hand smoke?

9-2 Exercises B: Beyond the Basics

15. Repeat Exercise 3 after adding 1°C to each temperature listed in the "Awaiting Surgery" group. Compare these results to those found in Exercise 3.

16. If Exercise 3 is repeated after changing the temperature readings from the Celsius scale to the Fahrenheit scale, are the results affected? In general, how is the analysis of variance F test statistic affected by the scale used?

17. How are analysis of variance results affected in each of the following cases?
    a. The same constant is added to every sample score.
    b. Every sample score is multiplied by the same constant.
    c. The order of the samples is changed.

18. Five independent samples of 50 scores each are randomly drawn from populations that are normally distributed with equal variances. We wish to test the claim that $\mu_1 = \mu_2 = \mu_3 = \mu_4 = \mu_5$.

   a. If we used only the methods given in Chapter 8, we would test the individual claims $\mu_1 = \mu_2$, $\mu_1 = \mu_3$, ..., $\mu_4 = \mu_5$. How many ways could we pair off 5 means?

   b. Assume that for each test of equality between two means, there is a 0.95 probability of not making a type I error. If all possible pairs of means are tested for equality, what is the probability of making no type I errors? (Although the tests are not actually independent, assume that they are.)

   c. If we use analysis of variance to test the claim that $\mu_1 = \mu_2 = \mu_3 = \mu_4 = \mu_5$ at the 0.05 level of significance, what is the probability of not making a type I error?

   d. Compare the results of parts (b) and (c). Which approach is better in the sense of giving us a greater chance of not making a type I error?

19. In this exercise you will verify that when you have two sets of sample data, the t test for independent samples (assuming equal variances) and the ANOVA method of this section are equivalent. Refer to the data used in Exercises 5–7, but exclude the data on conglomerates.

   a. Use a 0.05 level and the method of Section 8–5 to test the claim that wholesale distributors and companies in the food industry have the same revenues.

   b. Use a 0.05 level and the ANOVA method of this section to test the claim made in part (a).

   c. Verify that the squares of the t test statistic and critical value from part (a) are equal to the F test statistic and critical value from part (b).

20. Complete the following ANOVA table if it is known that there are three samples with sizes of 5, 7, and 7, respectively.

Source of Variation	Sum of Squares (SS)	Degrees of Freedom	Mean Square (MS)	Test Statistic
Treatments	?	?	?	$F = ?$
Error	100.00	?	?	
Total	123.45	?		

9-3 Two-Way ANOVA

Section 9–2 illustrated the use of analysis of variance to decide whether three or more populations have the same mean. We called those procedures *one*-way analysis of variance (or single-factor analysis of variance) because the data are categorized into groups according to a *single* factor (or treatment). As such, each sample is independent of the others.

In **two-way analysis of variance,** the samples are dependent. There are two types of designs we will examine: the randomized block design (also known as two-way ANOVA without replication) and the factorial design (also called two-way ANOVA with replication).

When we use the one-way (or single factor) analysis of variance technique and conclude that the differences among the means are significant, we cannot necessarily conclude that the given factor is responsible for the differences. It is possible that the variation of some other unknown factor is responsible for the differences. One way to reduce the effect of the extraneous factors is to design the experiment so that it has a **completely randomized design,** in which each element is given the same chance of belonging to the different categories or treatments. Another way to reduce the effect of extraneous factors is to use a **rigorously controlled design,** in which elements are carefully chosen so that all other factors have no variability. That is, select elements that are the same in every characteristic except for the single factor being considered.

Yet another way to control for extraneous variation is to use a **randomized block design,** in which a measurement is obtained for each treatment on each of several individuals that are matched according to similar characteristics.

Randomized Block Design

Consider the following data set in which daily sales were drawn from three stores of the same type for the same series of dates:

	Store 1	Store 2	Store 3
March 2	5000	4800	4700
March 6	5200	5000	4900
March 10	4800	4900	4500
March 14	5200	5200	4600
March 18	5600	5500	5000
March 22	5100	4700	5100
March 26	5400	4800	4700
March 30	5300	5200	4800

In the randomized block design, the sampling is done through the blocks. In this situation, the dates are known as a systematic random sample in which a random starting date is chosen and then each fourth date after that is chosen. If the stores constitute the treatment (or factor) samples, these samples are dependent since the daily sales figure for each store depends on the date from which it is pulled.

The date is known as a blocking variable. This goes back to when agricultural experiments were first conducted; the land was divided into blocks and a variety of fertilizers were applied to each block. (It should be noted that there is other terminology for blocks, such as second factor, categories, or row variable.)

Since each observed value depends on both the treatment (in this case, the store) and the block, the equation for the breakdown of the variation is slightly different than that of one-way ANOVA:

Total variation of all data = Variation due to treatments
+ Variation due to blocks + Error

There is a slight change in the terminology; the variation between the groups is now the variation due to treatments and the variation within the groups is now the error source of variation. However, the computation of the sum of squares for these sources of variation are the same, i.e., SS(treatment) = SS(between) and SS(error) = SS(within). As before, the mean square (MS) column is calculated by dividing entries in the SS column by entries in the degrees of freedom (df) column.

Because there are three terms on the right side (compared to two for one-way ANOVA), the ANOVA table for the randomized block design has an extra row for the blocks. Here is the ANOVA table for this data set:

Source	df	SS	MS	F
Treatment	2	682,500	341,250	10.0052
Block	7	740,000	105,714.2857	3.0995
Error	14	477,500	34,107.1429	
Total	23	1,900,000		

In constructing the row for the block, we use b to represent the number of blocks. The block degrees of freedom is $b - 1 = 8 - 1 = 7$. We use B_i to represent the total of each block. Then,

$$SS(block) = \sum \left(\frac{B_i^2}{b} \right) - \frac{T^2}{n}$$

where T is the total of all the observations, as with one-way ANOVA, and n is the total sample size. In our example, $B_1 = 14,500$, $B_2 = 15,100$, . . . , $B_8 = 15,300$, and $T = 120,000$.

$$SS(block) = 14,500^2/3 + 15,100^2/3 \ + \ ... \ + 15,300^2/3 - 120,000^2/24$$
$$= 740,000$$

As with the treatment and error sources of variation, MS(block) is calculated by dividing SS(block) by its degrees of freedom:

$$MS(block) = \frac{SS(block)}{b - 1} = \frac{740,000}{7} = 105,714.2857$$

The F test statistic for the blocks is calculated by dividing the MS(block) by the MS(error):

$$F\ (block) = \frac{MS(block)}{MS(error)} = \frac{105,714.2857}{34,107.1429} = 3.0995$$

As mentioned earlier, the math for the total variation, treatment variation, and error sources of variation is the same as for one-way ANOVA. The suggested method to construct a two-way ANOVA table is to calculate SS(total), SS(treatment), SS(block) from the data and then use basic arithmetic to compute the rest.

Hypothesis Testing for Two-Way ANOVA

The following assumptions apply when testing with two-way analysis of variance.

Assumptions

1. For each cell, or factor/block combination, the sample values come from a population with a distribution that is approximately normal.
2. The populations have the same variance σ^2 (or standard deviation σ).
3. The sample within each treatment level and block combination is random.
4. The sample values are categorized in two ways.
5. All of the cells have the same number of sample values. (This is called a *balanced* design.)

One of the reasons that two-way ANOVA is called such is that there are two hypotheses we can test: one for the treatments and one for the blocks. You will notice in the two-way ANOVA table that there are two F test statistics, one for each hypothesis. We will first examine the hypothesis test involving the treatments and then the one involving the blocks.

EXAMPLE

Is there any significant difference between the stores in average daily sales? Test at a 0.05 level of significance.

SOLUTION

The null and alternative hypotheses are:

$$H_0: \mu_1 = \mu_2 = \mu_3$$
H_1: at least two treatment means are different

For this hypothesis test, the critical value is found the same way as with one-way ANOVA.

$$\text{numerator df} = \text{treatment df} = 2$$
$$\text{denominator df} = \text{error df} = 14$$

With $\alpha = 0.05$, the critical value is 3.7389. From the ANOVA table, we find the F test statistic of 10.0052 in the treatment row. Since the test statistic is greater than the critical value, we reject the null hypothesis.

Interpretation
We conclude there is a significant difference in average daily sales between at least two of the stores.

Is there any significant difference between the dates in average daily sales? Test at a 0.05 level of significance.

SOLUTION

Since there are 8 dates, the null and alternative hypotheses are:

$$H_0: \mu_1 = \mu_2 = \mu_3 = \mu_4 = \mu_5 = \mu_6 = \mu_7 = \mu_8$$

H_1: at least two block means are different

In order to find the critical value, we need the block degrees of freedom:

$$\text{numerator df} = \text{block df} = 7$$

$$\text{denominator df} = \text{error df} = 14$$

With $\alpha = 0.05$, the critical value is 2.7642. The F test statistic of 3.0995 is found in the block row of the ANOVA table. Since the test statistic is greater than the critical value, we reject the null hypothesis at a 0.05 level of significance.

Interpretation

We conclude there is a significant difference between at least 2 dates in average daily sales.

As with one-way ANOVA, we can use Tukey's method of simultaneous confidence intervals to determine which pairs of means are significantly different. Generally speaking, researchers are more concerned with differences between the treatments than with differences between the blocks. For example, if a firm conducted a survey of 400 people to compare how much they spend at three competing grocery stores in their area, they would be more concerned with finding a significant difference between the stores than between any pairs of respondents.

EXAMPLE

Between which stores is there a significant difference?

SOLUTION

We have $r = 3$ and df $= 14$. From Table A-11, $Q = 3.701$. Since each sample size is 8 and the MS(error) $= 34,107.1429$,

$$E = 3.701\sqrt{\frac{34,107.1429}{8}} = 241.6$$

From the data, $\bar{x}_1 = 5200, \bar{x}_2 = 5012.5$, and $\bar{x}_3 = 4787.5$. The following table summarizes the confidence intervals:

Groups	Difference of Means	Confidence Interval
Store 1/store 2	187.5	$-54.1 < \mu_1 - \mu_2 < 429.1$
Store 1/store 3	412.5	$170.9 < \mu_1 - \mu_3 < 654.1$
Store 2/store 3	225	$-16.6 < \mu_1 - \mu_2 < 466.6$

The only stores that are significantly different in average daily sales are store 1 and store 3.

In the following example, we use four different cars to test the mileage produced by three different grades (regular, extra, premium) of gasoline. Using a randomized block design as a way to control differences among cars, we burn each grade of gas in each of the four cars, randomly selecting the order in which this is done. We have three treatments corresponding to the three grades of gas; we have four blocks corresponding to the four cars used.

EXAMPLE

When we burn each of three different grades of gas in each of four different cars, we obtain the results shown in Table 9-2. Use two-way analysis of variance to test the claim that the grade of gasoline does not affect mileage, and test the claim that the different cars do not affect mileage.

Table 9-2 **Mileage (L/100 km) for Different Grades of Gasoline**

		Block			
		Car 1	Car 2	Car 3	Car 4
	Regular gas	12.39	7.13	10.23	8.72
Treatment	Extra gas	12.39	6.92	9.05	8.12
	Premium gas	10.7	6.04	9.05	6.92

SOLUTION

A portion of the Excel display for the data in Table 9-2 is shown below.

EXCEL DISPLAY
ANOVA for Randomized Block Design. (The labels in parentheses were added after Excel generated the table.)

	A	B	C	D	E	F	G
1	ANOVA						
2	Source of Variation	SS	df	MS	F	P-value	F crit
3	Rows (Gas)	4.2792	2	2.1396	14.2327	0.0053	5.143253
4	Columns (Cars)	43.9654	3	14.6551	97.4860	1.76E-05	4.757063
5	Error	0.9020	6	0.1503			
6							
7	Total	49.1466	11				

Considering the treatment (row) effects, we calculate the test statistic:

$$F = \frac{MS(Gas)}{MS(error)} = \frac{2.1396}{0.1503} = 15.2327$$

Because the displayed P-value for the gas factor is less than the significance level, we reject the null hypothesis. The grade of gas does seem to have an effect on the mileage.

For the block (column) effects,

$$F = \frac{MS(Car)}{MS(error)} = \frac{14.6551}{0.1503} = 97.486$$

This value is significant because the P-value for the cars factor (0.0000176, displayed by Excel in scientific notation) is much less than the specified alpha. We therefore reject the null hypothesis of equal block means at a 0.05 level of significance.

Interpretation

We conclude that the grade of gas affects mileage, and that the different cars have different mileage values. (If we interchange the roles played by the blocks and treatments by transposing Table 9-2, we obtain the same results.)

Factorial Design

In the randomized block design, there is only one observation for each treatment/block combination. This is known as two-way ANOVA without replication. However, there are situations (particularly with scientific experiments) in which there are two factors with multiple observations for each factor combination. This is known as **factorial design** (or two-way ANOVA with replication).

In addition to examining the equality of the means for each of the factors, we can also examine if there is significant interaction between the factors.

DEFINITION

> There is an **interaction** between two factors if the effect of one of the factors changes for different categories of the other factor.

As a result, the equation breaking down the total variation of the data becomes:

Total variation of data = Variation due to factor 1 + Variation due to factor 2 + Interaction + Error

Consider the following table:

Table 9-3 Lengths (in min) of Movies on Television Categorized by Two Factors: Star Ratings and the Canadian Classification System for Violence in Programming

	2.0-2.5 Stars	3.0-3.5 Stars	4.0 Stars
CTR Rating: FAM/PA/14+	98	93	103
	100	94	193
	123	94	168
	92	105	88
	99	111	121
CTR Rating: 18+	110	115	72
	114	133	120
	96	106	106
	101	93	104
	155	129	159

In using ANOVA for the data of Table 9-3, we will consider the effect of an *interaction* between star ratings and CTR ratings, as well as the effects of star ratings and the effects of CTR ratings on movie length. The calculations are quite involved, so *we will assume that a software package is being used*. The Excel display shows results from the data in Table 9-3. (Excel can do two-way ANOVA, but STATDISK cannot. The Excel procedures are described at the end of this section.)

	A	B	C	D	E	F	G	H
1	ANOVA							
2	Source of Variation	SS	df	MS	F	P-value	F crit	
3	Sample (CTR)	32.0333333	1	32.0333333	0.0481825	0.828113	4.259675	
4	Columns (Stars)	1582.06667	2	791.033333	1.18982201	0.321594	3.402832	
5	Interaction	2256.06667	2	1128.03333	1.69671597	0.20454	3.402832	
6	Within (error)	15956	24	664.833333				
7								
8	Total	19826.1667	29					

You will notice the extra row in the ANOVA table for the interaction. The interaction degrees of freedom are calculated as $(k - 1)(b - 1) = (3 - 1)(2 - 1) = 2$. As with the other sources of variation, MS(interaction) is calculated by dividing SS(interaction) by its degrees of freedom:

$$MS(\text{interaction}) = \frac{SS(\text{interaction})}{(k - 1)(b - 1)} = \frac{2256.0667}{2} = 1128.0333$$

You will notice the third F test statistic in the interaction row. This is used to test if there is significant interaction between the star rating and the CTR rating. It is calculated by dividing the MS(interaction) by the MS(error):

$$F(\text{interaction}) = \frac{MS(\text{interaction})}{MS(\text{error})} = \frac{1128.0333}{664.8333} = 1.6967$$

EXAMPLE

Is there significant interaction between star rating and the CTR rating? Test at a 0.05 level of significance.

SOLUTION

The null and alternative hypotheses are:

H_0: There is not significant interaction between star rating and CTR rating.

H_1: There is significant interaction between star rating and CTR rating.

To find the critical value, we use the interaction degrees of freedom:

$$\text{numerator df} = \text{interaction df} = 2$$

$$\text{denominator df} = \text{error df} = 24$$

The critical value is 3.4028. Since the F test statistic of 1.6967 is not greater than the critical value, we fail to reject the null hypothesis at a 0.05 level of significance.

There is not sufficient evidence to conclude that movie length is affected by an interaction between star rating and CTR rating.

It should be noted that the *P*-value from the ANOVA table of 0.2045 is greater than the 0.05 level of significance, which leads to the same conclusion.

An aid in determining if there is interaction between a pair of factors is an interaction plot. In order to create one, we compute the mean for each set of replicates. The following table summarizes the means for each star/CTR rating combination:

Table 9-4 Means for Star/CTR Combinations

	2.0–2.5 Stars	3.0–3.5 Stars	4.0 Stars
CTR Rating: FAM/PA/14+	102.4	99.4	134.6
CTR Rating: 18+	115.2	115.2	112.2

For example, the average of 102.4 for the factor combination of 2.0–2.5 stars and Fam/PA/14+ is computed by taking the average of the values 98, 100, 123, 92, and 99. Here is the interaction plot derived from these data:

Figure 9-4
Interaction Plot of
Star/CTR Ratings

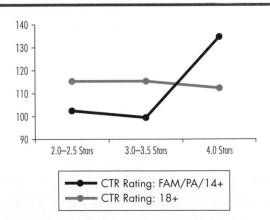

By and large, there is no interaction between the rating systems; they only cross when the average time for the CTR rating of FAM/PA/14+ increases from 99.4 minutes for 3.0–3.5 stars to 134.6 minutes for 4.0 stars. We now turn our attention to the other two hypothesis tests. We can use SS(Stars) as a measure of variation among the star-rating means; this is SS for "Columns" in the Excel display. We use SS(CTR) as a measure of the variation among the CTR means; Excel labels this SS for the "Sample."

If we fail to reject the null hypothesis of no interaction between factors, then we should proceed to test the following two hypotheses:

H_0: There are no effects from the row factor (that is, the row means arc equal).
H_0: There are no effects from the column factor (that is, the column means are equal).

In other words, if there is no apparent interaction between factors, there may still be an effect based on one or both of the factors, individually.

As shown previously, we did fail to reject the null hypothesis of no interaction between factors, so we proceed with the next two hypothesis tests.

For the row factor (CTR),

$$F = \frac{MS(CTR)}{MS(error)} = \frac{32.0333}{664.8333} = 0.0482$$

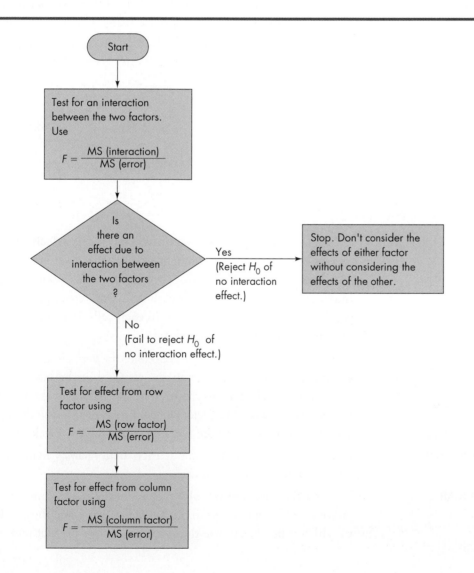

**Figure 9-5
Procedure for Two-Way
ANOVA**

Interpretation Because the corresponding *P*-value for the CTR factor (0.8281) exceeds the significance level, we fail to reject the null hypothesis of no effects from CTR. The CTR rating does not appear to have an effect on movie length.

For the column factor (Stars),

$$F = \frac{\text{MS(Stars)}}{\text{MS(error)}} = \frac{791.0333}{664.8333} = 1.1898$$

Interpretation This value is not significant because the *P*-value for the stars factor (0.3216) exceeds the specified significance level. Thus, we fail to reject the null hypothesis of no effects from the star rating. The star rating of a movie does not appear to have an effect on the movie's length.

USING TECHNOLOGY

EXCEL: *For two-way tables with more than one entry per cell:* Entries from the same cell must be listed down a column, and not across a row. Enter the labels corresponding to the data set in column A and row 1, as in this example, which corresponds to Table 9-3.

A	B	C	D
	2.0–2.5 Stars	3.0–3.5 Stars	4.0–4.5 Stars
FAM/PA/14+	98	93	103
FAM/PA/14+	100	94	193
FAM/PA/14+	123	94	168
:	:	:	:
18+	101	93	104
18+	155	129	159

EXCEL (Prior to 2007): After entering the sample data and labels, select **Tools** from the main menu bar, then **Data Analysis**, then **Anova: Two-Factor With Replication**. In the dialog box, enter the input range (including row and column headings). For "rows per sample," enter the number of values in each cell (5, in the example shown in Table 9-3). Click **OK**.

For two-way tables with exactly one entry per cell: Enter the sample data as they appear in the table—but the labels are not required. Select **Tools** from the main menu bar, then **Data Analysis**, then **Anova: Two-Factor Without Replication**. In the dialog box, enter the input range (*excluding* headings). Click **OK**.

EXCEL 2007: Click **Data** on the main menu, then **Data Analysis** from the **Analysis** tab. Proceed as above.

MINITAB 15: To enter the data, put the observed values in one column, the factor to which it belongs in the second column, and the block to which it belongs in the third column. From the **Stat** menu, choose **ANOVA** and then **Two-Way**. Select the appropriate columns as shown in the figure.

STATDISK: From **Analysis**, choose **Two-Way Analysis of Variance**. Enter the number of categories for both the row and column variables and the number of values in each cell. Click **Continue**. In the spreadsheet to the right, enter the appropriate values. Click **Evaluate**.

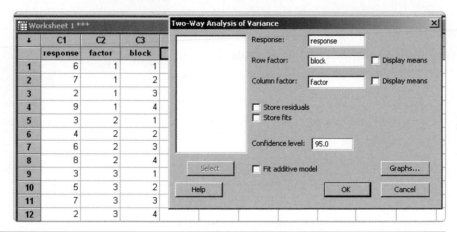

MINITAB DISPLAY
Two-Way ANOVA

9-3 Exercises A: Basic Skills and Concepts

In Exercises 1–4, use the Excel display below, which corresponds to the data in the accompanying table.

	A	B	C	D	E	F	G	H
1	ANOVA							
2	*Source of Variation*	SS	df	MS	F	P-value	F crit	
3	Sample (*CTR*)	14.0625	1	14.0625	0.02023563	0.890398	5.317645	
4	Columns (*Stars*)	1049.1875	3	349.729167	0.50325269	0.690607	4.06618	
5	Interaction	3790.1875	3	1263.39583	1.81799922	0.221873	4.06618	
6	Within (*error*)	5559.5	8	694.9375				
7								
8	Total	10412.9375	15					

Lengths (in min) of Movies Categorized
by Star Ratings and the Canadian
Classification System for Violence in Programming

	Poor 0.0–1.5 Stars	Fair 2.0–2.5 Stars	Good 3.0–3.5 Stars	Excellent 4.0 Stars
CTR Rating: FAM/ PA/ 14+	108	98	93	103
	91	100	94	193
CTR Rating: 18+	105	110	115	72
	96	114	133	120

1. Identify the indicated values.
   a. MS(interaction)   b. MS(error)
   c. MS(Stars)   d. MS(CTR)

2. Find the test statistic and critical value for the null hypothesis of no interaction between star rating and CTR rating. What do you conclude?

3. Assume that the length of a movie is not affected by an interaction between its star rating and CTR rating. Find the test statistic and critical value for the null hypothesis that star rating has no effect on movie length. What do you conclude?

4. Assume that the length of a movie is not affected by an interaction between its star rating and CTR rating. Find the test statistic and critical value for the null hypothesis that CTR rating has no effect on movie length. What do you conclude?

In Exercises 5 and 6, use only the first value from each of the eight cells in the table used for Exercises 1–4. When we use only these first values, the Excel display includes the following.

	A	B	C	D	E	F	G	H
12	ANOVA							
13	*Source of Variation*	*SS*	*df*	*MS*	*F*	*P-value*	*F crit*	
14	Rows	0	1	0	0	1	10.12796	
15	Columns	459	3	153	0.574468	0.669941	9.276619	
16	Error	799	3	266.3333				
17								
18	Total	1258	7					

5. Assuming that there is no effect on movie length from the interaction between star rating and CTR rating, test the null hypothesis that CTR rating has no effect on movie length. Identify the test statistic and critical value, and state the conclusion. Use a 0.05 significance level.

6. Assuming that there is no effect on movie length from the interaction between star rating and CTR rating, test the null hypothesis that star rating has no effect on movie length. Identify the test statistic and critical value, and state the conclusion. Use a 0.05 significance level.

Exercises 7 and 8 refer to the sample data in the following table and the corresponding Excel display. The table entries are the numbers of support beams manufactured by four different operators using each of three different machines. In the Excel results, the operators are represented by rows and the machines are represented by columns.

		Machine		
		1	2	3
	1	66	74	67
Operator	2	58	67	68
	3	65	71	65
	4	60	64	66

	A	B	C	D	E	F	G	H	
12	ANOVA								
13	Source of Variation	SS	df	MS	F	P-value	F crit		
14	Rows	59.58333	3	19.86111	2.474048	0.158993	4.757055		
15	Columns	93.16667	2	46.58333	5.802768	0.039583	5.143249		
16	Error	48.16667	6	8.027778					
17									
18	Total	200.9167	11						

7. Using a 0.05 significance level, test the claim of the hypothesis that the four operators have the same mean production output. Identify the test statistic and critical value, and state the conclusion.

8. Using a 0.05 significance level, test the claim that the choice of machine has no effect on the production output. Identify the test statistic and critical value, and state the conclusion.

9. Refer to Data Set 8 in Appendix B and construct a table with pulse rates categorized according to the two factors of gender and whether the individual smokes. Select 9 scores for each cell and test the null hypothesis of no interaction between gender and smoking.

10. Use the same data collected for Exercise 9, assume that pulse rates are not affected by an interaction between gender and smoking, and test the null hypothesis that gender has no effect on pulse rates.

11. Use the same data collected for Exercise 9, assume that pulse rates are not affected by an interaction between gender and smoking, and test the null hypothesis that smoking has no effect on pulse rates.

9-3 Exercises B: Beyond the Basics

12. Use a statistics software package, such as Excel or SPSS/PC, that can produce results for two-way analysis of variance. First enter the data in the table used for Exercises 1–4 and verify that the results are as given in this section. Then transpose the table by making the CTR rating the column factor and making the star rating the row factor. Obtain the computer display for the transposed table, and compare the results to those previously given.

13. Refer to the data in the table used for Exercises 1–4 and subtract 10 from each table entry. Using a statistics software package with a two-way analysis of variance capability, determine the effects of subtracting 10 from each entry.

14. Refer to the data in the table used for Exercises 1–4 and multiply each table entry by 10. Using a statistics software package with a two-way analysis of variance capability, determine the effects of multiplying each entry by 10.

15. In analyzing Table 9-3, we concluded that movie length is not affected by an interaction between star rating and CTR rating; it is not affected by star rating; and it is not affected by CTR rating.

    a. Change the table entries so that there is an effect from the interaction between star rating and CTR rating.

    b. Change the table entries so that there is no effect from the interaction between star rating and CTR rating, and there is no effect from star rating, but there is an effect from CTR rating.

    c. Change the table entries so that there is no effect from the interaction between star rating and CTR rating, and there is no effect from CTR rating, but there is an effect from star rating.

■ VOCABULARY LIST

analysis of variance
(ANOVA) **508**
completely randomized
design **525**
factor **510**
factorial design **530**
interaction **530**
MS(between) **514**
MS(error) **514**
MS(treatment) **514**
MS(within) **514**
MSE **514**
multiple comparison
procedures **515**

one-way analysis of
variance **510**
randomized block
design **525**
rigorously controlled
design **525**
single-factor analysis
of variance **510**
SS(between) **513**
SS(error) **513**
SS(factor) **513**
SS(total) **513**
SS(treatment) **513**
SS(within) **513**

treatment **510**
Tukey's method of
simultaneous
confidence
intervals **516**
two-way analysis of
variance **525**
variance between
samples **511**
variance within
samples **511**
variation due to error **511**
variation due to
treatment **511**

■ REVIEW

In this chapter we used analysis of variance (or ANOVA) to test for equality of population means. This method requires (1) normally distributed populations, (2) populations with the same standard deviation (or variance), and (3) random samples that are independent of each other.

In Section 9–2 we considered one-way analysis of variance, characterized by sample data categorized according to a single factor. The following are key features of one-way analysis of variance:

- The F test statistic is based on the ratio of two different estimates of the common population variance σ^2, as shown below.

$$F = \frac{\text{variance between samples}}{\text{variance within samples}} = \frac{\text{MS(between)}}{\text{MS(within)}}$$

- Critical values of F are found in Table A-5, with the degrees of freedom (df) found as follows:

$$\text{df for numerator} = k - 1 \quad \text{(where } k = \text{number of samples)}$$
$$\text{df for denominator} = N - k \quad \text{(where } N = \text{the total number of values in all samples combined)}$$

In Section 9–3 we considered two-way analysis of variance, characterized by data categorized according to two different factors. The method for two-way analysis of variance is summarized in Figure 9-5. We also considered randomized block design (two-way analysis of variance with one observation per cell) and factorial design (two-way analysis of variance with multiple observations per cell).

REVIEW EXERCISES

1. In a study of the effects of drinking and driving, 3 groups of adult men were randomly selected for an experiment designed to measure their blood alcohol levels after consuming 5 drinks. Members of group A were tested after 1 hour, members of group B were tested after 2 hours, and members of group C were tested after 4 hours. The results are given in the accompanying table; the ANOVA table for these data is also shown. At the 0.05 level of significance, test the claim that the 3 groups have the same mean level.

A	B	C
0.11	0.08	0.04
0.10	0.09	0.04
0.09	0.07	0.05
0.09	0.07	0.05
0.10	0.06	0.06
		0.04
		0.05

ANOVA

Source of Variation	SS	df	MS	F
Between groups	0.007657143	2	0.003829	46.9
Within groups	0.001142857	14	8.16E–05	
Total	0.0088	16		

2. Is the mean cost of food items in a "typical" basket of products the same across Canada? The accompanying list shows the costs of the same items when purchased in four different cities. Do these sample data support the claim that the mean food costs are the same? Use a 0.05 significance level.

	St. John's	Moncton	Regina	Victoria
Food item 1	5.16	4.37	3.26	3.67
Food item 2	9.83	8.04	8.62	9.68
Food item 3	0.88	1.07	0.94	1.38
Food item 4	3.64	2.98	2.91	3.19
Food item 5	2.87	2.65	2.79	2.76
Food item 6	1.65	1.29	0.96	1.18
Food item 7	3.34	2.94	3.39	3.47
Food item 8	0.56	0.62	0.53	0.60

Based on data from Statistics Canada.

3. Twelve different 4-cylinder cars were tested for fuel consumption (in litres per 100 km) after being driven under identical highway conditions; the results are listed in the table and accompanying ANOVA table. At the 0.05 significance level, test the claim that fuel consumption is not affected by an *interaction* between engine size and transmission type.

	Highway Fuel Consumption (L/100 km) of Different 4-Cylinder Compact Cars		
	Engine Size (litres)		
	1.5	2.2	2.5
Automatic transmission	9.1, 8.8	10.1, 10.9	9.1, 12.3
Manual transmission	8.6, 7.8	8.6, 9.4	10.5, 8.1

ANOVA

Source of Variation	SS	df	MS	F	P-value	F crit
Sample	4.440833333	1	4.440833	2.958912	0.136196	5.987374
Columns	4.631666667	2	2.315833	1.543032	0.287956	5.143249
Interaction	0.331666667	2	0.165833	0.110494	0.897172	5.143249
Within	9.005	6	1.500833			
Total	18.40916667	11				

4. Refer to the same data used in Exercise 3 and assume that fuel consumption is not affected by an interaction between engine size and type of transmission. Use a 0.05 level of significance to test the claim that fuel consumption is not affected by engine size.

5. Refer to the same data used in Exercise 3 and assume that fuel consumption is not affected by an interaction between engine size and type of transmission. Use a 0.05 level of significance to test the claim that fuel consumption is not affected by type of transmission.

6. Transport Canada is concerned about the braking performance of snow-mobiles. The following table shows the distances in metres that snowmobiles took to stop when the brakes were applied. The results are grouped by two factors: the speed of the snowmobile and the brand of the snowmobile. If you have access to a computer that can perform two-way analysis of variance, test the claim that the interaction of these two variables does affect mean brake performance, at the 0.01 significance level. What is the *P*-value for this test?

Snowmobile Braking Distances (in metres)
with Respect to Brand of Snowmobile and Speed

	Brand A	Brand B	Brand C	Brand D
10 km/h	3.0	4.4	9.5	1.1
	3.3	7.7	7.6	4.3
	4.8	7.0	6.8	3.5
20 km/h	15.8	24.7	18.0	13.7
	15.2	25.1	21.5	10.1
	19.5	35.7	20.3	12.6
30 km/h	37.9	66.4	48.9	32.2
	37.5	71.1	52.5	28.9
	31.6	66.4	55.5	28.7

CUMULATIVE REVIEW EXERCISES

1. The North of Superior Brewing Company plans to launch a major media campaign. Three advertising companies prepared trial commercials in an attempt to win a $2 million contract. The commercials were tested on randomly selected consumers, whose reactions were measured; the results are summarized in the accompanying table. (Higher scores indicate more positive reactions to the commercial.)

Aurora Advertising Co.				Solomon & Campbell Advertising				Gagné and Turner Advertising			
52	68	75	40	69	73	82	59	42	73	69	53
77	63	55	72	66	84	75	70	57	61	73	74

a. Construct a boxplot for each of the three samples. Use the same scale so that the boxplots can be compared. Do the boxplots reveal any notable differences?

b. Find the mean and standard deviation for each of the three sets of sample data.

c. Use the methods of Section 8–5 to test the claim that the population of Aurora scores has a mean that is equal to the population mean for Solomon & Campbell scores. Use a 0.05 significance level.

d. For each of the three samples, construct a 95% confidence interval estimate of the population mean μ. Do the results suggest any notable differences?

e. At the 0.05 significance level, test the claim that the three populations have the same mean reaction score. If you were responsible for advertising at this brewery, which company would you select on the basis of these results? Why?

2. The tread life in kilometres of a certain model of radial tire is normally distributed, with a mean of 60,000 km and a standard deviation of 5000 km (based on data from Goodyear).

a. If one radial tire is randomly selected, what is the probability that it will have a tread life of more than 62,210 km?

b. If 16 radial tires are randomly selected, what is the probability that their mean tread life is more than 62,210 km?

   c. What is the probability that each of the next 3 tires selected will have a tread life greater than 60,000 km?

3. Refer to the numbers of Newfoundland males in the labour force, in different SOC job categories, as listed in Data Set 14 in Appendix B.

   a. Find the mean.

   b. Find the standard deviation.

   c. Construct a boxplot.

   d. Identify any outliers.

   e. Construct a histogram.

   f. Assume that you want to test the null hypothesis that the mean number of workers per job category is the same for males in all provinces. Can you use one-way ANOVA? Why, or why not?

   g. Based on the sample data, estimate the probability that a randomly selected male member of the labour force in Newfoundland works in the SOC job category "B5."

TECHNOLOGY PROJECT

For Canadian prime ministers, popes, and British monarchs after 1690, the accompanying table lists the numbers of years that they lived after they first came to office or were crowned. Use boxplots and analysis of variance to determine whether the survival times for the different groups differ. Conduct the analysis of variance by running Excel, STATDISK, or some other statistical software package. Obtain printed copies of the computer displays and write your observations and conclusions.

Prime Ministers		Popes		Kings and Queens	
Macdonald	24	Alexander VIII	2	James II	17
Mackenzie	19	Innocent XII	9	Mary II	6
Abbott	1	Clement XI	21	William III	13
Thompson	2	Innocent XIII	3	Anne	12
Bowell	23	Benedict XIII	6	George I	13
Tupper	19	Clement XII	10	George II	33
Laurier	23	Benedict XIV	18	George III	59
Borden	26	Clement XIII	11	George IV	10
Meighen	40	Clement XIV	6	William IV	7
Mackenzie King	29	Pius VI	25	Victoria	63
Bennett	17	Pius VII	23	Edward VII	9
St. Laurent	25	Leo XII	6	George V	25
Diefenbaker	22	Pius VIII	2	Edward VIII	36

(Continued)

Prime Ministers		Popes		Kings and Queens	
Pearson	9	Gregory XVI	15	George VI	15
Trudeau	32	Pius IX	32		
		Leo XIII	25		
		Pius X	11		
		Benedict XV	8		
		Pius XII	17		
		Pius XIII	19		
		John XXIII	5		
		Paul VI	15		
		John Paul I	0		
		John Paul II	26		

Based on data from *Computer-Interactive Data Analysis*, by Lunn and McNeil, John Wiley & Sons.

FROM DATA TO DECISION

Was the Experiment Designed Correctly?

An experiment was designed to compare the effects of weight loss to the effects of aerobic exercise training on coronary artery disease risk factors in men who are healthy but sedentary, obese, and middle-aged or older. (See "Effects of Weight Loss vs. Aerobic Exercise Training on Risk Factors for Coronary Disease in Healthy, Obese, Middle-Aged and Older Men," by Katzel, Bleecker, et al., *Journal of the American Medical Association*, Vol. 274, No. 24.) The study involved 170 men divided into three groups: (1) 73 men were assigned to lose weight, (2) 71 men were assigned to aerobic exercise, and (3) 26 men were in a control group instructed not to lose weight or change the level of physical activity. In such a study, it is important to begin with groups that are similar in "baseline" characteristics of age, weight, body mass, body fat, and so on. In a report of the study, it was claimed that "there were no significant differences among the three groups in any of the baseline characteristics." The mean weights for the three sample groups are 94.3 kg, 93.9 kg, and 88.3 kg. Are these differences significant? Is there sufficient evidence to warrant rejection of the claim of no significant differences among the three groups? Analyze the data in the following list, form a conclusion, and write a report stating your results.

Weights (in kg) of men assigned to the weight loss group

95	82	94	87	84	111	98	101	85	117	111	86	114	85	98	102	97
92	76	95	88	107	89	81	93	77	91	106	97	107	78	101	88	92
101	110	96	83	92	106	107	106	110	102	105	84	93	69	96	70	80
108	87	86	93	78	98	86	93	101	94	85	107	74	101	105	93	89
90	103	103	80	116												

Weights (in kg) of men assigned to the aerobic exercise group

90	101	89	76	93	110	87	108	109	105	83	98	100	82	71	67	126
114	102	78	101	79	97	100	106	84	80	100	103	80	71	99	108	84
78	92	92	101	107	95	106	84	111	108	109	79	116	98	77	110	93
95	85	86	104	93	92	86	93	74	74	89	85	91	104	97	97	80
104	95	109														

Weights (in kg) of men assigned to the control group

91	71	74	86	97	78	81	82	78	106	79	110	91	93	87	72	95
88	83	96	98	77	84	95	103	102								

 COOPERATIVE GROUP ACTIVITIES

1. **In-class activity:** Divide into groups of five or six students. This activity is a contest of reaction times to determine which group is fastest, if there is a "fastest" group. Use the same reaction timer included with the Cooperative Group Activities in Chapter 5. Test and record the reaction time using the dominant hand of each group member. (Only one try per person.) Each group should calculate the values of n, $\bar{x}$, and s and record those summary statistics on the chalkboard along with the original list of reaction times. After all groups have reported their results, identify the group with the fastest mean reaction time. But is that group actually fastest? Use ANOVA to determine whether the means are significantly different. If they are not, there is no real "winner."

2. **Out-of-class activity:** Divide into groups of three or four students. Each group should survey other students at the same college by asking them to identify their major and gender. You might include other factors, such as employment (none, part-time, full-time) and age (under 21, 21–30, over 30). For each surveyed subject, determine the accuracy of the time on his or her wristwatch. First set your own watch to the correct time using an accurate and reliable source ("At the tone, the time is . . .") For watches that are ahead of the correct time, record positive times. For watches that are behind the correct time, record negative times. Use the sample data to address questions such as these:

   • Does gender appear to have an effect on the accuracy of the wristwatch?

   • Does major have an effect on wristwatch accuracy?

   • Does an interaction between gender and major have an effect on wristwatch accuracy?

INTERNET PROJECT

Analysis of Variance

Go to either of the following websites:

http://www.mathxl.com   or   http://www.mystatlab.com

Click on Internet Project, then on Chapter 9. There you will find instructions for finding data sets. Each data set can be divided naturally into samples from different populations so that ANOVA methods may be applied. In applying ANOVA methods, you will be conducting analyses in areas as varied as the functioning of the human body, consumer product labelling, and performance in sports.

10 Goodness-of-Fit and Contingency Tables

10-1 Overview

Chapter objectives are identified. This chapter deals with sample data consisting of frequency counts arranged in one row with at least three categories (one-way frequency table), or a table with at least two rows and at least two columns (two-way frequency table).

10-2 Goodness-of-Fit

The goodness-of-fit procedure of hypothesis testing is used to test claims that observed sample frequencies fit, or conform to, a particular distribution. We present the procedures for discrete uniform, multinomial, binomial, and Poisson distributions.

10-3 Contingency Tables: Independence and Homogeneity

A standard method is presented for testing claims that in a contingency table, the row variable and the column variable are independent. Also presented is a method for a test of homogeneity, in which we test a claim that

different populations have the same proportions of some characteristics.

10-4 Tests of Normality

Some hypothesis tests are based on the assumption that the population is normally distributed. We present a number of procedures, including formal goodness-of-fit hypothesis tests, to determine the validity of these assumptions.

CHAPTER PROBLEM

Is the percentage of married couples with children the same across Canada?

In the span of a generation, the percentage of married couples with children has changed. Once upon a time, it was not uncommon for a family to have 4, 5, or even more children. Table 10-1 is a sample based on the 2006 census. It shows by region the number of married couples (rounded to the nearest thousand) with children and without children. From the table we can calculate that the Arctic has the highest percentage of couples with children at 69.23% while Quebec has the lowest percentage at 51.94%. We want to determine if the circumstance of a married couple having children or not significantly depends on the region of residence. Section 10-3 will consider this particular set of data.

Table 10-1 Number of Married Couples With and Without Children (in Thousands) by Region

	B.C.	Prairies	Ontario	Quebec	Maritimes	Arctic
With children	457	604	1522	601	251	9
Without children	388	474	1009	556	230	4

Adapted from the Statistics Canada website, http://www12.statcan.ca/english/census06/data/highlights/households.

10-1 Overview

In this chapter we examine two different types of hypothesis tests. In Section 10–2, we explore goodness-of-fit tests, which have the purpose of determining if a set of data follows a certain distribution. In Section 10–3 we will consider tests of independence or homogeneity based on contingency tables, which consist of frequency counts arranged in a table such as Table 10-1. In Section 10–4, we examine a specific goodness-of-fit test: namely, whether a data set follows a normal distribution. We will examine a variety of methods including two different formal hypothesis tests.

Most of the tests require the original data to be divided into categories. In order to conduct hypothesis tests on data arranged this way, we will use the same χ^2 (chi-square) test statistic that was introduced in Section 6–5. Recall the following important properties of the chi-square distribution:

1. Unlike the normal and Student t distributions, the chi-square distribution is not symmetric. (See Figure 10-1.)

2. The values of the chi-square distribution can be 0 or positive, but they cannot be negative. (See Figure 10-1.)

3. The chi-square distribution is different for each number of degrees of freedom. (See Figure 10-2.)

Critical values of the chi-square distribution are found in Table A-4.

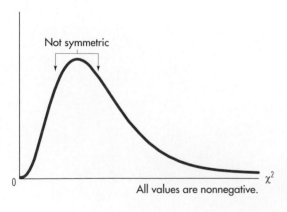

Figure 10-1
The Chi-Square Distribution

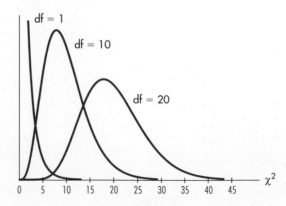

Figure 10-2
Chi-Square Distributions for 1, 10, and 20 Degrees of Freedom

10-2 Goodness-of-Fit

Each data set in this section consists of data that have been separated into different categories. We aim to determine whether the distribution agrees with or "fits" some claimed distribution.

The following assumptions apply when we test a hypothesis that the population proportion for each of the categories is as claimed.

Assumptions

1. The data have been randomly selected.

2. The sample data consist of frequency counts for each of the different categories.

3. For each category, the *expected* frequency is at least 5. (The expected frequency for a category is the expected *mean* of the frequencies for that category that would occur if random samples of the same size were repeatedly selected from the same population. There is *no* minimum-size requirement for the *observed frequencies* in each category of the actual sample being analyzed.)

In this section we present a method for testing a claim that the frequencies observed in the different categories fit a particular distribution. Because we test for how well an observed frequency distribution conforms to some theoretical distribution, this method is often called a *goodness-of-fit test*.

DEFINITION

A **goodness-of-fit test** is used to test the hypothesis that an observed frequency distribution fits (or conforms to) some claimed distribution.

EXAMPLE

Examine the *last digits* of each of the company revenues listed in Data Set 6 in Appendix B. If dollar amounts are counted precisely, we usually expect that the last digits will occur with relative frequencies (or probabilities) that are roughly the same. In contrast, estimated or rounded-off values tend to have 0 or 5 occurring much more often as last digits. Using the methods of this section, we could test the hypothesis that the data (that is, the last digits of the company revenues) fit a uniform distribution, with all of the digits being equally likely.

Our goodness-of-fit tests will incorporate the following notation.

NOTATION

O represents the *observed frequency* of an outcome.

E represents the *expected frequency* of an outcome.

k represents the *number of different categories* or outcomes.

n represents the total *number of trials*.

Finding Expected Frequencies

In the typical situation requiring a goodness-of-fit test, we have *observed* frequencies (denoted by O) and must use the claimed distribution to determine the *expected* frequencies (denoted by E). In many cases, an expected frequency can be found by multiplying the probability p for a category by the number of different trials n, so

$$E = np$$

For example, if we test the claim that a die is fair by rolling it 60 times, we have $n = 60$ (because there are 60 trials) and $p = 1/6$ (because a die is fair if the six possible outcomes are equally likely with the same probability of 1/6). The expected frequency for each category or cell is therefore

$$E = np$$
$$= (60)(1/6) = 10$$

We know that sample frequencies typically deviate somewhat from the values we theoretically expect, so we now present the key question: Are the differences between the actual *observed* values O and the theoretically *expected* values E statistically significant? To answer this question we use the following test statistic, which measures the discrepancy between observed and expected frequencies.

TEST STATISTIC FOR GOODNESS-OF-FIT TESTS

$$\chi^2 = \sum \frac{(O - E)^2}{E}$$

Critical Values

1. Critical values are found in Table A-4 by using $k - 1$ degrees of freedom, where k = number of categories.

2. Goodness-of-fit hypothesis tests are always right-tailed.

The form of the χ^2 test statistic is such that *close agreement* between observed and expected values will lead to a *small* value of χ^2. A large value of χ^2 will indicate strong disagreement between observed and expected values. A significantly large value of χ^2 will thus cause rejection of the null hypothesis of no difference between observed and expected frequencies. Our test is therefore always right-tailed because the critical value and critical region are located at the extreme right of the distribution.

Once we know how to find the values of the test statistic and critical value, we can test hypotheses by using the same procedure introduced in Chapter 7 and summarized in Figure 7-4.

Suppose a die is rolled 60 times in order to determine if it is a fair die. The results are summarized in Table 10-2. Test at a 0.05 level of significance.

SOLUTION

As mentioned previously, if this is a fair die, we would expect each value to appear 10 times. Table 10-2 shows the actual number of observations for each value.

Table 10-2	Results of Tossing a Die 60 Times					
Die roll	1	2	3	4	5	6
Frequency	13	8	9	9	6	15

The null hypothesis must contain the condition of equality, so we have

$$H_0: p_1 = p_2 = p_3 = p_4 = p_5 = p_6 \text{ (a uniform distribution)}$$

H_1: At least one of the probabilities is different from the others.

Using the observed frequencies O listed in Table 10-2 and the expected frequencies E (equal to 10), we compute the value of the χ^2 test statistic, as shown in Table 10-3. The test statistic is $\chi^2 = 5.6$. The critical value is $\chi^2 = 11.071$ (found in Table A-4 with $\alpha = 0.05$ in the right tail and degrees of freedom equal to $k - 1 = 5$). The test statistic and critical value are shown in Figure 10-3.

Because the test statistic does not fall within the critical region, there is insufficient evidence to reject the null hypothesis.

Interpretation
There is insufficient evidence to conclude that the probability of one of the die values is significantly different from the others. We conclude that the die is fair.

Table 10-3 Calculating the χ^2 Test Statistic for Die Roll Test

Die Roll	Observed Frequency O	Expected Frequency E	$O - E$	$(O - E)^2$	$\dfrac{(O - E)^2}{E}$
1	13	10	3	9	0.9
2	8	10	-2	4	0.4
3	9	10	-1	1	0.1
4	9	10	-1	1	0.1
5	6	10	-4	16	1.6
6	15	10	5	25	2.5
	60	60		$\chi^2 = \sum \dfrac{(O - E)^2}{E} = 5.6$	

Figure 10-3
Goodness-of-Fit Test of
$p_1 = p_2 = \cdots = p_6$

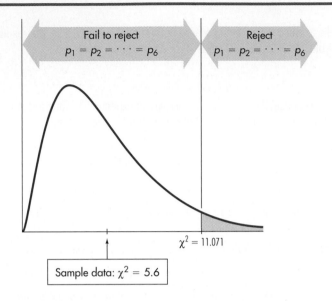

The techniques in this section can be used to test whether an observed frequency distribution conforms to some theoretical frequency distribution. The preceding example used a goodness-of-fit test to decide whether the observed frequencies conformed to a uniform distribution. Because many statistical analyses require a normally distributed population, we can use the chi-square test in this section to determine whether given samples are drawn from normally distributed populations (see Exercise 22).

The preceding example dealt with the null hypothesis that the probabilities for the different categories are all equal. The methods of this section can also be used when the hypothesized probabilities (or frequencies) are different, as in the next example.

EXAMPLE

Mars, Inc. claims that its M&M plain candies are distributed with the colour percentages of 30% brown, 20% yellow, 20% red, 10% orange, 10% green, and 10% blue. The colours of the M&Ms listed in Data Set 11 of Appendix B are summarized in Table 10-4. Using the sample data and a 0.05 significance level, test the claim that the colour distribution is as claimed by Mars, Inc.

Table 10-4 **Frequencies of M&M Plain Candies**

	Brown	Yellow	Red	Orange	Green	Blue
Observed frequency	33	26	21	8	7	5
Expected frequency	30	20	20	10	10	10

SOLUTION

We extended Table 10-4 to include the expected frequencies, which are calculated as follows. For n, we use the total number of trials (100), which is the total number of M&Ms observed in the sample. For the probabilities, we use the decimal equivalents of the claimed percentages (30%, 20%, . . . , 10%).

$$\text{Brown: } E = np = (100)(0.30) = 30$$
$$\text{Yellow: } E = np = (100)(0.20) = 20$$
$$\vdots$$
$$\text{Blue: } E \;\;= np = (100)(0.10) = 10$$

In testing the given claim, we have the following hypotheses:

H_0: $p_{br} = 0.3$ and $p_y = 0.2$ and $p_r = 0.2$ and $p_o = 0.1$ and $p_g = 0.1$ and $p_{bl} = 0.1$
H_1: At least one of the above proportions is different from the claimed value.

The test statistic is calculated from Table 10-5.

The test statistic is $\chi^2 = 5.950$. The critical value of χ^2 is 11.071, and it is found in Table A-4 (using $\alpha = 0.05$ in the right tail with $k - 1 = 5$ degrees of freedom). The test statistic and critical value are shown in Figure 10-4. Because the test statistic does not fall within the critical region, there is not sufficient evidence to warrant rejection of the null hypothesis.

Interpretation

There is not sufficient evidence to warrant rejection of the claim that the colours are distributed with the percentages given by Mars, Inc.

In Figure 10-5, we graph the claimed proportions of 0.30, 0.20, . . . , 0.10, along with the observed proportions of 33/100, 26/100, 21/100, 8/100, 7/100, 5/100, so that we can visualize the discrepancy between the distribution that was claimed and the frequencies that were observed. The points along the blue line represent the claimed proportions,

Table 10-5 Calculating the χ^2 Test Statistic for M&M Data

Colour Category	Observed Frequency O	Expected Frequency $E = np$	$O - E$	$(O - E)^2$	$\dfrac{(O - E)^2}{E}$
Brown	33	30	3	9	0.3000
Yellow	26	20	6	36	1.8000
Red	21	20	1	1	0.0500
Orange	8	10	−2	4	0.4000
Green	7	10	−3	9	0.9000
Blue	5	10	−5	25	2.5000
	100	100		$\chi^2 = \sum \dfrac{(O - E)^2}{E} = 5.9500$	

Figure 10-4
Goodness-of-Fit
Test of $p_{br} = 0.3$ and $p_y = 0.2$ and $p_r = 0.2$ and $p_o = 0.1$ and $p_g = 0.1$ and $p_{bl} = 0.1$

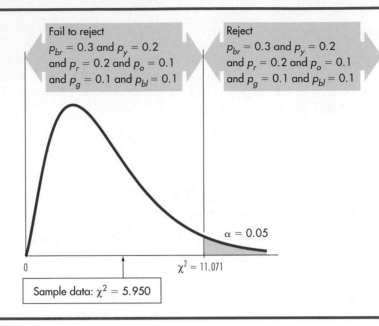

Fail to reject
$p_{br} = 0.3$ and $p_y = 0.2$
and $p_r = 0.2$ and $p_o = 0.1$
and $p_g = 0.1$ and $p_{bl} = 0.1$

Reject
$p_{br} = 0.3$ and $p_y = 0.2$
and $p_r = 0.2$ and $p_o = 0.1$
and $p_g = 0.1$ and $p_{bl} = 0.1$

$\alpha = 0.05$

0    $\chi^2 = 11.071$

Sample data: $\chi^2 = 5.950$

Figure 10-5
Comparison of Claimed and Observed Proportions

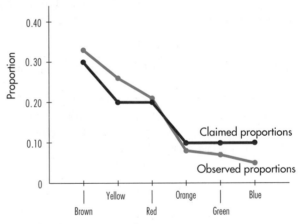

Claimed proportions

Observed proportions

and the points along the green line represent the observed proportions. The corresponding pairs of points are all fairly close, showing that all of the expected frequencies are reasonably close to the corresponding observed frequencies. In general, graphs such as Figure 10-5 are helpful in visually comparing expected frequencies and observed frequencies, as well as suggesting which categories result in the major discrepancies.

In the preceding examples, calculating the expected values was straightforward since the probabilities were the same for each category (as in the die example) or given to us (as in the M&M example). However, for goodness-of-fit tests involving the binomial and Poisson distributions, there are a couple of complicating factors: first, we need to calculate the probability for each category; second, the expected value for a category may be less than 5, requiring us to combine categories.

Binomial Distribution

Recall that we use the binomial distribution when the outcome is either succcss or failure, the number of trials n is known, the long-term probability of success p is known, and the trials are independent of each other.

EXAMPLE

A factory receives panels for wide-screen televisions in crates of 100. A random sample of 5 panels is taken to test that the percentage of defective panels does not exceed 20%. In the course of a year, the factory received 500 shipments. In the 500 samples, the number of defective panels is summarized in Table 10-6.

Table 10-6 Distribution of Defective Panels

Value	0	1	2	3	4	5
Frequency	200	212	70	15	3	0

Do the data follow a binomial distribution with $n = 5$ and $p = 0.2$? Test at a 0.05 level of significance.

SOLUTION

Note that the distribution is binomial since a panel is either defective or not defective, the panels are independent of each other, we know the number of trials, namely 5, and the expected defect rate is 0.2. We first state the null and alternative hypotheses:

H_0: the distribution is binomial with $n = 5$ and $p = 0.2$

H_1: the distribution is not binomial with $n = 5$ and $p = 0.2$

In order to compute the expected values, we need $p = P(x) = (_5C_x)(0.2^x)(0.8^{5-x})$ for each value of $x = 0, 1, 2, 3, 4,$ and 5. We can proceed as before to calculate $E = np = 500p$. The results are summarized in Table 10-7.

Table 10-7 Computation of Expected Values for a Binomial Distribution With $n = 5$ and $p = 0.2$

Value of x	0	1	2	3	4	5
$P(x)$	0.3277	0.4096	0.2048	0.0512	0.0064	0.00032
$E = 500p$	163.84	204.8	102.4	25.6	3.2	0.16

You will notice that the expected values for $x = 4$ and $x = 5$ are each less than 5. If we combine these two categories, their combined expected value of $3.2 + 0.16 = 3.36$ is still less than 5. However, the expected value for $x = 3$ at 25.6 is well above 5. Therefore, we combine $x = 3, x = 4,$ and $x = 5$ into one category, which we label $3+$. The expected value of this category is $25.6 + 3.2 + 0.16 = 28.96$. Because of this, we also add the

Table 10-8 Calculating the χ^2 Test Statistic for Binomial Data

Binomial Category	Observed Frequency O	Expected Frequency $E = np$	$O - E$	$(O - E)^2$	$\dfrac{(O - E)^2}{E}$
0	200	163.84	36.16	1307.5456	7.9806
1	212	204.8	7.2	51.84	0.2531
2	70	102.4	−32.4	1049.76	10.2516
3+	18	28.96	−10.96	120.1216	4.1478
	500	500			$\chi^2 = \sum \dfrac{(O - E)^2}{E} = 22.6331$

observed values of 15, 3, and 0 to get 18. We now have 4 categories instead of the original 6. The calculation of the χ^2 test statistic is summarized in Table 10-8.

Since $k = 4$, there are 3 degrees of freedom. With $\alpha = 0.05$, we reject the null hypothesis if the χ^2 test statistic is greater than 7.815. Since the test statistic of 22.6331 is greater than 7.815, we reject the null hypothesis.

Interpretation

We have sufficient evidence to conclude that the distribution of defective wide-screen television panels does not follow a binomial distribution with $n = 5$ and $p = 0.2$.

This example raises the question as to what is an appropriate value for p in this binomial distribution. If we compare the observed values to the expected values, we will note that for $x = 0$ and $x = 1$, the observed values are greater than the expected values, while for $x = 2$ and $x = 3+$, the opposite is true. This would suggest that our original choice of $p = 0.2$ is too high. Experimenting with $p = 0.15$, $p = 0.16$, and $p = 0.17$, their respective test statistics are 5.1691, 1.9626, and 2.5302. Keeping in mind that a lower test statistic value indicates greater agreement between what we observe and what we expect, the best conclusion based on the data is that the distribution is binomial with $n = 5$ and $p = 0.16$.

Poisson Distribution

Recall that we use the Poisson distribution when we need the probability of some event occurring over a specified interval of time, distance, area, or some similar unit.

EXAMPLE

Sheet metal is examined to determine the number of blemishes per 1000 m. In 400 samples, the number of blemishes are summarized in Table 10-9.

In a Poisson distribution, the mean is also the mode. Since $x = 2$ occurs the most often, it is reasonable to think that $\mu = 2$. Do the number of blemishes follow a Poisson distribution with $\mu = 2$? Test at a 0.05 level of significance.

Table 10-9 Distribution of Blemishes

Value	0	1	2	3	4	5	6	7	8
Frequency	52	102	115	70	35	17	5	3	1

SOLUTION

We first state the null and alternative hypotheses:

H_0: The distribution is Poisson with $\mu = 2$.

H_1: The distribution is not Poisson with $\mu = 2$.

As with the binomial example, we need to calculate $p = P(x) = e^{-2}\frac{2^x}{x!}$ for $x = 0, 1, 2, \ldots$. Since there is no upper limit on the values of x, but $P(x)$ decreases as x increases, our strategy will be to calculate expected values until they fall below 5. The results are summarized in Table 10-10.

Table 10-10   Computation of Expected Values for a Poisson Distribution with $\mu = 2$

Value of x	0	1	2	3	4	5	6
$p = P(x)$	0.1353	0.2707	0.2707	0.1804	0.0902	0.0361	0.0120
$E = 400p$	54.1341	108.2682	108.2682	72.1788	36.0894	14.4358	4.8119

The expected values drop below 5 when $x = 6$. Since the values of x, in theory, continue beyond this to infinity, the last category is 6+. What is the expected frequency for this class? We know that all the expected frequencies must sum to 400. If we add the expected frequencies for $x = 0$, $x = 1$, $x = 2$, $x = 3$, $x = 4$, and $x = 5$, they sum to 392.3745. Therefore, the expected frequency for 6+ is $400 - 392.3745 = 7.6255$. As with the binomial example, we also add the observed frequencies for 6+: $5 + 3 + 1 = 9$. We have $k = 7$ classes to compute our test statistic, as summarized in Table 10-11.

Since $k = 7$, there are 6 degrees of freedom. With $\alpha = 0.05$, we reject the null hypothesis if the χ^2 test statistic is greater than 12.592. Since the test statistic of 1.6212 is less than the critical value, we fail to reject the null hypothesis.

Interpretation

We conclude that the average number of blemishes per 1000 m of sheet metal follows a Poisson distribution with $\mu = 2$.

Table 10-11 Calculating the χ^2 Test Statistic for Poisson Data

Poisson Category	Observed Frequency O	Expected Frequency $E = np$	$O - E$	$(O - E)^2$	$\dfrac{(O - E)^2}{E}$
0	52	54.1341	-2.1341	4.5544	0.0841
1	102	108.2682	-6.2682	39.2903	0.3629
2	115	108.2682	6.7318	45.3171	0.4186
3	70	71.1788	-1.1788	1.3896	0.0195
4	35	36.0894	-1.0894	1.1868	0.0329
5	17	14.4358	2.5642	6.5751	0.4555
6+	9	7.6255	1.3745	1.8893	0.2478
	400	400			$\uparrow$

$$\chi^2 = \sum \frac{(O - E)^2}{E} = 1.6212$$

Rationale for the Test Statistic

The preceding examples should be helpful in developing a sense for the role of the χ^2 test statistic. It should be clear that we want to measure the amount of disagreement between observed and expected frequencies. Simply summing the differences between observed and expected values does not result in an effective measure because that sum is always 0, as shown below:

$$\Sigma(O - E) = \Sigma O - \Sigma E = n - n = 0$$

Squaring the $O - E$ values provides a more useful statistic, which reflects the differences between observed and expected frequencies. (The reasons for squaring the $O - E$ values are essentially the same reasons for squaring the $x - \bar{x}$ values in the formula for standard deviation.) The value of $\Sigma(O - E)^2$ measures only the magnitude of the differences, but we need to find the magnitude of the differences relative to what was expected. This relative magnitude is found through division by the expected frequencies, as in the test statistic.

The theoretical distribution of $\Sigma(O - E)^2/E$ is a discrete distribution because the number of possible values is limited. The distribution can be approximated by a chi-square distribution, which is continuous. This approximation is generally considered acceptable, provided that all values of E are at least 5. We included this requirement with the assumptions that apply to this section. In Section 5–6 we saw that the continuous normal probability distribution can reasonably approximate the discrete binomial probability distribution, provided that np and nq are both at least 5. We now see that the continuous chi-square distribution can reasonably approximate the discrete distribution of $\Sigma(O - E)^2/E$, provided that all values of E are at least 5. (There are ways of circumventing the problem of an expected frequency that is less than 5, such as combining categories so that all expected frequencies are at least 5.)

The number of degrees of freedom reflects the fact that we can freely assign frequencies to $k - 1$ categories before the frequency for every category is determined. Although we say that we can "freely" assign frequencies to $k - 1$ categories, we cannot have negative frequencies, nor can we have frequencies so large that their sum exceeds the total of the observed frequencies for all categories combined.

P-Values

The examples in this section used the traditional approach to hypothesis testing, but the P-value approach can also be used. P-values can be obtained by using the same methods described in Sections 7–3 and 7–6. For instance, the preceding example resulted in a test statistic of $\chi^2 = 1.6212$. That example had $k = 7$ categories, so there were $k - 1 = 6$ degrees of freedom. Referring to Table A-4, we see that for the row with 6 degrees of freedom, the test statistic of 1.6212 is less than the lowest right-tailed critical value of 10.645, so the P-value is greater than 0.10. (If the preceding example is run on a computer, you can find a more precise P-value of 0.9510.) The high P-value suggests that the null hypothesis should not be rejected. Remember, we reject the null hypothesis only when the P-value is less than the significance level.

EXCEL: In Excel, you can conduct goodness-of-fit tests using the template shown here. Type in the column and row headings shown in bold. Manually enter the data for the categories, observed frequencies, and expected frequencies. (If more rows are required, adjust the cell addresses in the formulas.) Enter the indicated formulas in Cells D4, E4, and F4. Select Cells D4:F4 then, under the **Edit** menu, select **Copy**; select Cells D5:D9 then, under the **Edit** menu, select **Paste**. This completes the upper portion of the template. In the bottom portion of the template, enter the desired alpha. The n value can be input manually or else calculated by the **=COUNT** function. The formulas in Cells B13 to B15 complete the analysis.

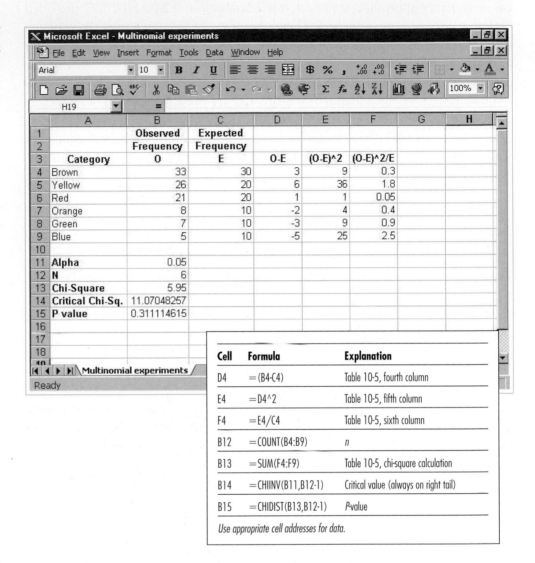

STATDISK: Select **Analysis** from the main menu bar, then select the option **Goodness-of-Fit**. Choose between "equal expected frequencies" and "unequal expected frequencies." For equal expected frequencies, choose the column with the observed frequencies. For unequal expected frequencies, choose the column with the observed frequencies and the column with the expected frequencies. Then click **Evaluate**. It should be noted that if any expected values are less than 5, a warning box comes up and the data will not be evaluated.

10-2 Exercises A: Basic Skills and Concepts

1. One common way to test for authenticity of data is to analyze the frequencies of digits. When people are weighed and their weights are rounded to the nearest pound, we expect the last digits 0, 1, 2, . . . , 9 to occur with about the same frequency. In contrast, if people are asked how much they weigh, the digits 0 and 5 tend to occur at higher rates. The author randomly selected 80 students, obtained their weights, and recorded only the last digits, with the results shown in the following table. At the 0.01 significance level, test the claim that the last digits occur with the same frequency. Based on the results, does it appear that the students were actually weighed or were they asked to report their weights?

Last digit	0	1	2	3	4	5	6	7	8	9
Frequency	35	0	2	1	4	24	1	4	7	2

2. Excel can be used to generate random integers between 0 and 9 inclusive. Suppose the following results are obtained. Use a 0.05 significance level to test the claim that the computer generates digits that are uniformly distributed.

Digit	0	1	2	3	4	5	6	7	8	9
Frequency	32	29	37	21	41	24	33	32	25	26

3. One of the authors observed 500 spins of the roulette wheel at the Mirage Resort and Casino. For each spin, the ball can land in any one of the 38 different slots that are supposed to be equally likely. When computer software (STATDISK) was used to test the claim that the slots are in fact equally likely, the test statistic $\chi^2 = 38.232$ was obtained.
   a. Find the critical value, assuming that the significance level is 0.10.
   b. STATDISK displayed a P-value of 0.41331, but suppose you are working only with Table A-4. Given the known value for χ^2 and the number of categories (38), what would be your estimate for the P-value, based on the table?
   c. Write a conclusion about the claim that the 38 results are equally likely.

4. Such measures of inflation as the consumer price index can also be used to compare prices in different regions. A standardized "basket" of grocery or consumer items can be purchased in different cities and stores, and their prices compared. Suppose that 51 stores are randomly selected, from all grocery and convenience stores in the four cities shown that charged over $90 for a standard basket of consumer goods. At the 0.05 level of significance,

test the claim that stores charging over $90 for the same grocery and convenience items are found in equal numbers in the four cities.

City	St. John's	Moncton	Regina	Victoria
Number	20	3	11	17

Based on data from Statistics Canada.

5. A labour specialist is analyzing the numbers of work stoppages in the transportation industry. She claims that, in general, 10% of stoppages involve air transport, 10% involve rail transport, 20% water transport, and 60% road-based transport (including trucks, buses, and taxis). Based on the table below, drawn from a random selection of years from 1961 to 1999, test the specialist's claim. Use the 0.05 level of significance.

Mode of transport	Air	Rail	Water	Road
Number of stoppages	84	66	123	426

Based on data from Human Resources Canada.

6. A study was made of 147 industrial accidents that required medical attention. Among those accidents, 31 occurred on Monday, 42 on Tuesday, 18 on Wednesday, 25 on Thursday, and 31 on Friday (based on results from "Counted Data CUSUM's," by Lucas, *Technometrics*, Vol. 27, No. 2). Test the claim that accidents occur with equal proportions on the five workdays. If they are not the same, what factors might explain the differences?

7. Use a 0.05 significance level and the industrial accident data from Exercise 6 to test the claim of a safety expert that accidents are distributed on workdays as follows: 30% on Monday, 15% on Tuesday, 15% on Wednesday, 20% on Thursday, and 20% on Friday. Does rejection of that claim provide any help in correcting the industrial accident problem?

8. The Gleason Supermarket's manager must decide how much of each ice-cream flavour he should stock so that customer demands are satisfied but unwanted flavours don't result in waste. The ice-cream supplier claims that among the four most popular flavours, customers have these preference rates: 62% prefer vanilla, 18% prefer chocolate, 12% prefer neapolitan, and 8% prefer vanilla fudge. A random sample of 200 customers produces the results below. At the $\alpha = 0.05$ significance level, test the claim that the supplier has correctly identified customer preferences.

Flavour	Vanilla	Chocolate	Neapolitan	Vanilla Fudge
Customers	120	40	18	22

Data are based on results from the International Association of Ice Cream Manufacturers.

9. The number π is an irrational number with the property that when we try to express it in decimal form, it requires an infinite number of decimal places and there is no pattern of repetition. In the decimal representation of π, the first 100 digits occur with the frequencies described in the table below. At the 0.05 significance level, test the claim that the digits are uniformly distributed.

Digit	0	1	2	3	4	5	6	7	8	9
Frequency	8	8	12	11	10	8	9	8	12	14

10. The number 22/7 is similar to π in the sense that they both require an infinite number of decimal places. However, 22/7 is a rational number because it can be expressed as the ratio of two integers, whereas π cannot. When rational numbers such as 22/7 are expressed in decimal form, there is a pattern of repetition. In the decimal representation of 22/7, the first 100 digits occur with the frequencies described in the table below. At the 0.05 significance level, test the claim that the digits are uniformly distributed. How does the result differ from that found in Exercise 9?

Digit	0	1	2	3	4	5	6	7	8	9
Frequency	0	17	17	1	17	16	0	16	16	0

11. In a "Scratch & Save" promotion, Zellers promised all customers a discount ranging from 5% to 50%, depending on the value that appeared on their "Scratch & Save" card. In the accompanying table, the bottom row shows the expected distribution of discounts for every 1000 cards submitted. Suppose that the manager of one branch says the actual distribution of discounts for 1000 cards that were submitted in her store was as shown in the middle row. At the 0.05 level of significance, test the claim that this actual distribution of discounts is consistent with the distribution that Zellers promised.

Percent discount	5%	10%	15%	20%	25%	30%	50%
Actual quantity	203	715	35	22	8	14	3
Expected quantity	185	725	40	20	15	10	5

Based on promotional materials from Zellers.

12. In analyzing hits by V-1 buzz bombs in World War II, South London was subdivided into regions, each with an area of 0.25 km². Use the values listed here to test the claim that the actual frequencies fit a Poisson distribution. Use a 0.05 level of significance.

Number of bomb hits	0	1	2	3	4 or more
Actual number of regions	229	211	93	35	8
Expected number of regions (from Poisson distribution)	227.5	211.4	97.9	30.5	8.7

13. Refer to the Old Faithful geyser data in Data Set 16 of Appendix B. Test the claim that the time intervals are uniformly distributed among the five categories of 55–64, 65–74, 75–84, 85–94, 95–104.

14. Refer to Data Set 8 in Appendix B and record only the last digits of the pulse counts. Test for authenticity of the pulse counts by testing the claim that the last digits occur with equal frequency. Do the pulse counts appear to be authentic? (See Exercise 1.)

15. Refer to Data Set 10 in Appendix B and categorize the listed one-year returns for mutual funds as poor (up to 24%), fair (above 24% to 36%), good (above 36% to 48%), or excellent (above 48%). Test the claim that in the year sampled, the returns were evenly distributed among the four categories. Use the 0.05 level of significance.

16. Many lottery enthusiasts search for patterns that they hope will indicate winning combinations. Yuri thinks that sums of numbers in Lotto 6/49 draws are not evenly distributed. Refer to Data Set 12 in Appendix B and find the sum for each row of numbers (*not* including the bonus number). Place these numbers into categories by this procedure: Divide each sum by 20, then ignore the decimal portion of the result. (For example, the sum 57 would be placed in the category 2, since $57/20 = 2.85$, or 2 with the decimal portion dropped.) At the 0.05 level of significance, test the claim that the categories based on sums of the numbers occur with the same frequency. Do the results imply that the lottery is not fair?

10-2 Exercises B: Beyond the Basics

17. In doing a test for the goodness-of-fit as described in this section, does an outlier have much of an effect on the value of the χ^2 test statistic? Test for the effect of an outlier by repeating Exercise 4 after changing the frequency for Victoria from 17 to 170. Describe the general effect of an outlier.

18. In doing a test for goodness-of-fit as described in this section, suppose that we multiply each observed frequency by the same positive integer greater than 1. How is the critical value affected? How is the test statistic affected?

19. In this exercise we will show that a hypothesis test involving a binomial experiment is equivalent to a hypothesis test for a proportion (Section 7–5). Assume that a particular experiment has only two possible outcomes, A and B, with observed frequencies of f_1 and f_2, respectively.
    a. Find an expression for the χ^2 test statistic, and find the critical value for a 0.05 significance level. Assume that we are testing the claim that both categories have the same frequency, $(f_1 + f_2)/2$.

b. The test statistic

$$z = \frac{\hat{p} - p}{\sqrt{\dfrac{pq}{n}}}$$

is used to test the claim that a population proportion is equal to some value p. With the claim that $p = 0.5$, $\alpha = 0.05$, and

$$\hat{p} = \frac{f_1}{f_1 + f_2}$$

show that z^2 is equivalent to χ^2 [from part (a)]. Also show that the square of the critical z score is equal to the critical χ^2 value from part (a).

20. An observed frequency distribution is as follows:

Number of successes	0	1	2	3
Frequency	89	133	52	26

a. Assuming a binomial distribution with $n = 3$ and $p = 1/3$, use the binomial probability formula to find the probability corresponding to each category of the table.

b. Using the probabilities found in part (a), find the expected frequency for each category.

c. Use a 0.05 level of significance to test the claim that the observed frequencies fit a binomial distribution for which $n = 3$ and $p = 1/3$.

21. In a recent year, customers of Victoria Fridge and Stove (which is open 7 days a week) returned 146 appliances. If the frequencies of returns on different days conform to a Poisson distribution, they will be as shown on the bottom row of the following table. (For example, on about 245 days there will be no returns at all.) Use a 0.05 significance level to test the claim that the actual frequencies fit a Poisson distribution. (*Caution:* Not all of the expected frequencies are at least 5, as assumed by the hypothesis test.)

Number of returns on that day	0	1	2	3	4
Actual number of days	247	90	24	3	1
Expected number of days	244.7	97.9	19.6	2.6	0.2

22. An observed frequency distribution of sample IQ scores is as follows:

IQ score	Less than 80	80–95	96–110	111–120	More than 120
Frequency	20	20	80	40	40

a. Assuming a normal distribution with $\mu = 100$ and $\sigma = 15$, use the methods given in Chapter 5 to find the probability of a randomly selected subject belonging to each class. (Use class boundaries of 79.5, 95.5, 110.5, 120.5.)

b. Using the probabilities found in part (a), find the expected frequency for each category.

c. Use a 0.01 level of significance to test the claim that the IQ scores were randomly selected from a normally distributed population with $\mu = 100$ and $\sigma = 15$.

10-3 Contingency Tables: Independence and Homogeneity

The examples and exercises in Section 10–2 involved frequencies according to a certain distribution. This section examines frequencies in a contingency table such as Table 10-1. Contingency tables were introduced in Chapter 3 as an aid to solving conditional probability problems. Contingency tables are especially important because they are frequently used to analyze survey results. Consequently, the methods presented in this section are among those used most often.

This section presents two types of hypothesis testing based on contingency tables. We first consider tests of independence, used to determine whether a contingency table's row variable is independent of its column variable. We then consider tests of homogeneity, used to determine whether different populations have the same proportions of some characteristic. Good news: Both types of hypothesis testing use the *same* basic methods. We begin with tests of independence.

Test of Independence

DEFINITION

A **test of independence** tests the null hypothesis that the row variable and the column variable in a contingency table are not related. (The null hypothesis is the statement that the row and column variables are independent.)

It is very important to recognize that in this context, the word *contingency* refers to dependence, but this is only a statistical dependence and cannot be used to establish a direct cause-and-effect link between the two variables in question. For example, after analyzing the data in Table 10-1, we might conclude that there is a relationship between region of residence and the likelihood of a married couple having children, but that doesn't mean that the region a married couple lives in directly affects the causes for having children.

When testing the null hypothesis of independence between the row and column variables in a contingency table, the following assumptions apply. (Note that these assumptions do not require that the parent population have a normal distribution or any other particular distribution.)

You may recall a test for independence from Chapter 3. Given two variables A and B, we would conclude that A and B are independent if $P(A \text{ and } B) = P(A) \cdot P(B)$.

EXAMPLE

Given this contingency table, show that A and B are independent.

	A	$\overline{A}$	Total
B	10	30	40
$\overline{B}$	40	120	160
Total	50	150	200

SOLUTION

The left side is $P(A \text{ and } B) = 10/200 = 0.05$ and the right side is $P(A) \cdot P(B) = (50/200)(40/200) = (0.25)(0.2) = 0.05$. Since the left side equals the right side, A and B are independent.

The problem with this approach to determining if two events are independent is that, in order to achieve this conclusion, the percentages in each column must be exactly equal as is the situation with this particular contingency table. Note that in this example, $P(B|A) = 10/50 = 0.2$ and $P(B|\overline{A}) = 30/150 = 0.2$. In a real-world situation, this is not a realistic expectation. In the hypothesis test approach to independence, we will conclude two variables are independent if the percentages in each column are *approximately* equal.

Assumptions

1. The sample data are randomly selected.

2. The null hypothesis H_0 is the statement that the row and column variables are *independent*; the alternative hypothesis H_1 is the statement that the row and column variables are dependent.

3. For every cell in the contingency table, the *expected* frequency E is at least 5. (There is no requirement that every *observed* frequency must be at least 5.)

Our test of independence between the row and column variables uses the following test statistic and critical values.

$$\chi^2 = \Sum \frac{(O - E)^2}{E}$$

Critical Values

1. The critical values are found in Table A-4 by using

$$\text{degrees of freedom} = (r - 1)(c - 1)$$

where r is the number of rows and c is the number of columns.

2. In a test of independence with a contingency table, the critical region is located in the *right tail only*.

The test statistic allows us to measure the degree of disagreement between the frequencies actually observed and those that we would theoretically expect when the two variables are independent. Small values of the χ^2 test statistic result from close agreement between observed frequencies and frequencies expected with independent row and column variables. Large values of the χ^2 test statistic are to the right of the chi-square distribution, and they reflect significant differences between observed and expected frequencies. In repeated large samplings, the distribution of the test statistic χ^2 can be approximated by the chi-square distribution, provided that all expected frequencies are at least 5.

The number of degrees of freedom $(r - 1)(c - 1)$ reflects the fact that because we know the total of all frequencies in a contingency table, we can freely assign frequencies to only $r - 1$ rows and $c - 1$ columns before the frequency for every cell is determined. [However, we cannot have negative frequencies or frequencies so large that any row (or column) sum exceeds the total of the observed frequencies for that row (or column).]

In the preceding section we knew the corresponding probabilities and could easily determine the expected values, but the typical contingency table does not come with the relevant probabilities. Consequently, we need to devise a method for obtaining the corresponding expected values. We will first describe the procedure for finding the values of the expected frequencies, and then we will justify that procedure. For each cell in the frequency table, the expected frequency E can be calculated by using the following equation.

EXPECTED FREQUENCY FOR A CONTINGENCY TABLE

$$\text{Expected frequency} = \frac{(\text{row total})(\text{column total})}{(\text{grand total})}$$

Table 10-12 Observed Frequencies (and Expected Frequencies) (in 1000s)

	BC	Prairies	Ontario	Quebec	Maritimes	Arctic	Row totals
			Region				
With children	457 (476.6880)	604 (608.1297)	1522 (1427.8074)	601 (652.6958)	251 (271.3455)	9 (7.3337)	3444
Without children	388 (368.3120)	474 (469.8703)	1009 (1103.1926)	556 (504.3042)	230 (209.6545)	4 (5.6663)	2661
Column totals:	845	1078	2531	1157	481	13	Grand total: 6105

Here *grand total* refers to the total of all observed frequencies in the table. For example, the expected frequency for the upper left cell of Table 10-12 (a duplicate of Table 10-1 with expected frequencies inserted in parentheses) is 476.688. It is found by noting that the total of all frequencies for that row is 3444, the total of the column frequencies is 845, and the total of all frequencies in the table is 6105, so we get an expected frequency of

$$E = \frac{\text{(row total)(column total)}}{\text{(grand total)}} = \frac{(3444)(845)}{6105} = 476.688$$

EXAMPLE

The expected frequency for the upper left cell of Table 10-12 is 476.688. Find the expected frequency for the lower left cell, assuming independence between whether a married couple has children or not and the region of residence.

SOLUTION

The lower left cell lies in the second row (with total 2661) and the first column (with total 845). The expected frequency is

$$E = \frac{\text{(row total)(column total)}}{\text{(grand total)}} = \frac{(2661)(845)}{6105} = 368.312$$

Interpretation
To interpret this result in the lower left cell, we can say that although 388,000 married couples in B.C. do not have children, we would have expected 368,312 if the circumstance of whether a married couple has children or not is independent of that couple's region of residence. There is a discrepancy between $O = 388$ and $E = 368.312$ (in thousands), and such discrepancies are key components of the test statistic.

It would be very helpful to stop here and verify the other expected frequencies.

Rationale for the Procedure for Finding Expected Frequencies

As mentioned earlier, if two events A and B are independent, then $P(A \text{ and } B) = P(A) \cdot P(B)$. In theory, if the circumstance of a married couple having children is independent of the region of residence,

$$P(\text{B.C. and children}) = P(\text{B.C.}) \cdot P(\text{children}) = \left(\frac{845}{6105}\right)\left(\frac{3444}{6105}\right)$$

Having found an expression for the probability of being in the upper left cell, we now proceed by finding the *expected value* for that cell, which we get by multiplying the probability for that cell by the total number of married couples, as shown in the following equation:

$$E = p \cdot n = \left[\frac{845}{6105} \cdot \frac{3444}{6105}\right] \cdot 6105 = 476.688$$

The form of this product suggests a general way to obtain the expected frequency of a cell:

$$\text{expected frequency } E = \frac{(\text{row total})}{(\text{grand total})} \cdot \frac{(\text{column total})}{(\text{grand total})} \cdot (\text{grand total})$$

This expression can be simplified to

$$E = \frac{(\text{row total}) \cdot (\text{column total})}{(\text{grand total})}$$

We can now proceed to use contingency table data for testing hypotheses, as in the following example, which uses the data given in the chapter problem.

AS

EXAMPLE

At the 0.05 significance level, use the data in Table 10-1 to test whether a married couple has children or not significantly depends on the region of residence.

SOLUTION

The null hypothesis and alternative hypothesis are as follows:

H_0: A married couple having children is independent of the region of residence.

H_1: A married couple having children depends on the region of residence.

The significance level is $\alpha = 0.05$.

Because the data are in the form of a contingency table, we use the χ^2 distribution with this test statistic, together with the expected values from Table 10-12:

$$\chi^2 = \sum \frac{(O - E)^2}{E}$$

$$= \frac{(457 - 476.688)^2}{476.688} + \frac{(604 - 608.1297)^2}{608.1297} + \frac{(1522 - 1427.8074)^2}{1427.8074}$$

$$+ \frac{(601 - 652.6958)^2}{652.6958} + \frac{(251 - 271.3455)^2}{271.3455} + \frac{(9 - 7.3337)^2}{7.3337}$$

$$+ \frac{(388 - 368.312)^2}{368.312} + \frac{(474 - 469.8703)^2}{469.8703} + \frac{(1009 - 1103.1926)^2}{1103.1926}$$

$$+ \frac{(556 - 504.3042)^2}{504.3042} + \frac{(230 - 209.6545)^2}{209.6545} + \frac{(4 - 5.6663)^2}{5.6663}$$

$$= 0.8131 + 0.028 + 6.2139 + 4.0945 + 1.5255 + 0.3786 + 1.0524$$

$$+ 0.0363 + 8.0423 + 5.2993 + 1.9744 + 0.4900$$

$$= 29.9485$$

The critical value is $\chi^2 = 11.071$, and it is found from Table A-4 by noting that $\alpha = 0.05$ in the right tail and the number of degrees of freedom is given by $(r - 1)(c - 1) = (2 - 1)(6 - 1) = 5$. The test statistic and critical value are shown in Figure 10-6. Because the test statistic falls within the critical region, we reject the null hypothesis that whether a married couple has children or not is independent of the region of residence.

Interpretation
There is enough evidence to conclude that whether a married couple has children or not significantly depends on the region of residence.

P-Values

The preceding example used the traditional approach to hypothesis testing, but we can easily use the *P*-value approach. Excel and STATDISK both provide the same *P*-value of 0.000015. Using the criterion given in Section 7–3, we reject the null hypothesis because the *P*-value is less than the significance level of $\alpha = 0.05$.

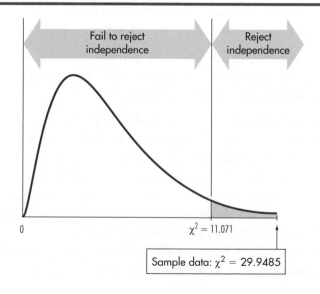

Fail to reject independence

Reject independence

$\chi^2 = 11.071$

Sample data: $\chi^2 = 29.9485$

0

Figure 10-6
Test of Independence Between a Couple Having Children and the Region of Residence

If we don't have a suitable calculator or statistical software package, we can estimate P-values by using the same methods introduced earlier. The preceding example resulted in a test statistic of $\chi^2 = 29.9485$, and the critical value is based on 5 degrees of freedom. Refer to Table A-4 and note that for the row with 5 degrees of freedom, the test statistic of $\chi^2 = 29.9485$ is greater than the 0.005 critical value of 16.750, indicating that the P-value < 0.005. On the basis of this relatively small P-value, we again reject the null hypothesis and conclude that there is sufficient sample evidence to warrant rejection of the null hypothesis that whether a married couple has children or not is independent of the region of residence. If the P-value had been greater than the significance level of 0.05, we would have failed to reject the null hypothesis of independence.

Test of Homogeneity

In the preceding example, we illustrated a test of independence by using a sample of married couples drawn from a single population of married Canadian couples. In some cases samples are drawn from different populations, and we want to determine whether those populations have the same proportions of the characteristics being considered.

DEFINITION

> In a **test of homogeneity**, we test the claim that different populations have the same proportion of some characteristics.

Because a test of homogeneity uses data sampled from different populations, we have predetermined totals for either the rows or the columns in the contingency table. Consequently, a test of homogeneity involves random selections made in such a way that either the row totals are predetermined or the column totals are predetermined. In trying to distinguish between a test for homogeneity and a test for independence, we can therefore pose the following question:

> **Were predetermined sample sizes used for different populations (test of homogeneity), or was one big sample drawn so both row and column totals were determined randomly (test of independence)?**

As an example of a test of homogeneity, suppose we want to test the claim that the proportion of voters who are Conservative supporters is the same in British Columbia, Alberta, Saskatchewan, and Manitoba. If we choose to find the political party registrations for 250 British Columbians, 200 Albertans, 100 Saskatchewanians, and 80 Manitobans, then the contingency table summarizing the results will have either the row totals or the column totals (whichever represent the different provinces) predetermined as 250, 200, 100, 80.

In conducting a test of homogeneity, we can use the same procedures already presented in this section, as illustrated in the following example.

Does a pollster's gender have an effect on poll responses by responses of men? In an article about polls, it was stated that people often give "acceptable" rather than honest responses on sensitive issues. The gender or race of the interviewer may influence the answers. To support that claim, data were provided for a poll in which surveyed men were asked if they agreed with this statement: "Abortion is a private matter that should be left to the woman to decide without government intervention." We will analyze the effect of gender on male survey subjects only. Table 10-13 shows the surveyed men (based on data from *U.S. News and World Report*). Assume that the survey was designed so that male interviewers were instructed to obtain 800 responses from male subjects, and female interviewers were instructed to obtain 400 responses from male subjects. Using a 0.05 significance level, test the claim that the proportions of agree/disagree responses are the same for the subjects interviewed by men and the subjects interviewed by women.

Table 10-13 **Gender and Survey Responses**		
	Gender of Interviewer	
	Man	Woman
Men who agree	560	308
Men who disagree	240	92

SOLUTION

Because we have predetermined column totals of 800 subjects interviewed by men and 400 subjects interviewed by women, we test for homogeneity with these hypotheses:

H_0: The proportions of agree/disagree responses are the same for the subjects interviewed by men as the subjects interviewed by women.

H_1: The proportions are different.

EXCEL DISPLAY
Contingency Table for Test of Homogeneity

The significance level is $\alpha = 0.05$. We use the same χ^2 test statistic described earlier, and it is calculated by using the same procedure. Instead of listing the details of that calculation, we provide the accompanying Excel display that results from the data in Table 10-13.

The Excel display shows the expected frequencies of 578.67, 289.33, 221.33, and 110.67. The display also includes the test statistic of $\chi^2 = 6.529$ and the P-value of 0.0106. Using the P-value approach to hypothesis testing, we reject the null hypothesis of equal (homogeneous) proportions. There is sufficient evidence to warrant rejection of the claim that the proportions are the same. It appears that response and the gender of the interviewer are dependent. It seems that men are influenced by the gender of the interviewer, although this conclusion is a statement of causality that is not justified by the statistical analysis.

It should be noted that this example could have been solved using the Z test for two proportions. In this case, the pooled proportion would be:

$$\bar{p} = \frac{560 + 308}{800 + 400} = \frac{868}{1200} = 0.723\dot{3}$$

The subsequent test statistic would be:

$$Z = \frac{0.77 - 0.7}{\sqrt{\dfrac{(0.7233)(0.2767)}{800} + \dfrac{(0.7233)(0.2767)}{400}}} = 2.5553$$

One of the relationships between Z and χ^2 is that a χ^2 random variable with 1 degree of freedom is equal to a squared Z (standard normal) random variable. Note that $2.5553^2 = 6.529$, the value of the test statistic from the above test. Similarly, the P-value of the Z test would be $2P(Z > 2.5553) = 2(0.0053) = 0.0106$, the P-value from the above test.

EXAMPLE

A city conducted a survey to determine support for a dedicated road tax. According to census figures, 25% of those working in the city live in the inner city, 60% live in the suburbs and 15% live in neighbouring bedroom communities. A stratified random sample of 1000 respondents was based on these figures. The survey had the following results:

	Inner City	Suburbs	Bedroom Communities	Total
For	98	202	52	352
Against	140	360	89	589
Undecided	12	38	9	59
Total	250	600	150	1000

Are the percentages of how people stand on the tax equivalent for the three regions? Test at a 0.05 level of significance.

SOLUTION

If we designate inner city as region 1, suburbs as region 2, and bedroom communities as region 3, we can write the null and alternative hypotheses as:

$$H_0: p_1 = p_2 = p_3$$
$$H_1: \text{not all the percentages are equal}$$

The expected values are as follows:

	Inner City	Suburbs	Bedroom Communities
For	88	211.2	52.8
Against	147.25	353.4	88.35
Undecided	14.75	35.4	8.85

The value of the χ^2 test statistic is:

$$\chi^2 = \sum \frac{(O - E)^2}{E} = \frac{(98 - 88)^2}{88} + \frac{(202 - 211.2)^2}{211.2} + \cdots + \frac{(9 - 8.85)^2}{8.85}$$

$$= 2.7405$$

The number of degrees of freedom for the critical value is $(3 - 1)(3 - 1) = 4$. Using Table A-4, with $\alpha = 0.05$, the critical value is 9.488. Since the test statistic value of 2.7405 is less than this, we fail to reject the null hypothesis.

Interpretation

We conclude the percentages of how people stand on the tax is equivalent for the three regions.

The P-value for this test is $P(\chi^2 > 2.7405) = 0.6021$. Since this is well above the level of significance of 0.05, we fail to reject the null hypothesis.

USING TECHNOLOGY

EXCEL: In Excel, you can use the following template to find the test statistic, critical value, and P-value. Adjust the displayed formulas based on where you set up the template and whether you need to add extra rows or columns of data. First, set up an area for entering the row and column data. In Cell H2, input the formula shown to compute the row total; copy to Cell H3. In Cell D4, input the formula shown to compute the column total; copy to Cells E4:G4. In Cell H5, enter the formula for the grand total. Create a second area to find the expected frequencies for each cell. Input the formulas as shown in Cells D8 and D9, then copy to Cells E8:G8 and E9:G9, respectively. In the template area for subtotals, copy the formula shown for Cell D11 into the range D11:G12. In the bottom portion of the template, enter the desired alpha and the formulas for calculating the degrees of freedom test statistic, critical value, and P-value.

MINITAB 15: Enter the data in the spreadsheet part of Minitab as you would for Excel. From the **Stat** menu, choose **Tables** and then **Chi-Square Test (Two-Way Table in Worksheet)**. In the dialog box, select the appropriate columns and click **OK**.

STATDISK: The STATDISK display shows the results obtained for the data in Table 10-1. If you are using STATDISK, select **Analysis** from the main menu bar, then select **Contingency Tables** and check the boxes of the required columns. The STATDISK results include the test statistic, critical value, *P*-value, and conclusion.

EXCEL DISPLAY

Contingency Table for Test of Independence (or Test of Homogeneity)

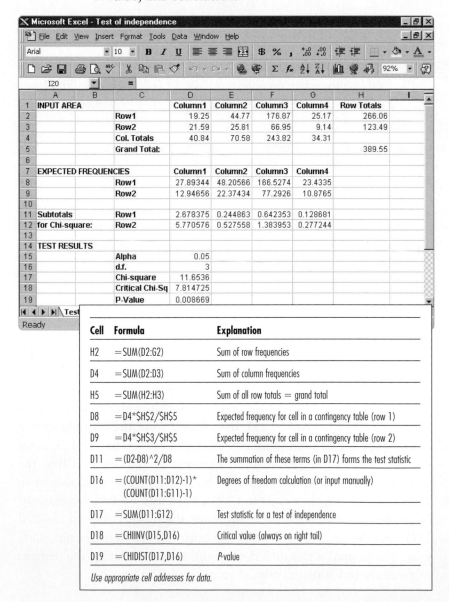

Cell	Formula	Explanation
H2	=SUM(D2:G2)	Sum of row frequencies
D4	=SUM(D2:D3)	Sum of column frequencies
H5	=SUM(H2:H3)	Sum of all row totals = grand total
D8	=D4*$H$2/$H$5	Expected frequency for cell in a contingency table (row 1)
D9	=D4*$H$3/$H$5	Expected frequency for cell in a contingency table (row 2)
D11	=(D2-D8)^2/D8	The summation of these terms (in D17) forms the test statistic
D16	=(COUNT(D11:D12)-1)*(COUNT(D11:G11)-1)	Degrees of freedom calculation (or input manually)
D17	=SUM(D11:G12)	Test statistic for a test of independence
D18	=CHIINV(D15,D16)	Critical value (always on right tail)
D19	=CHIDIST(D17,D16)	*P*-value

Use appropriate cell addresses for data.

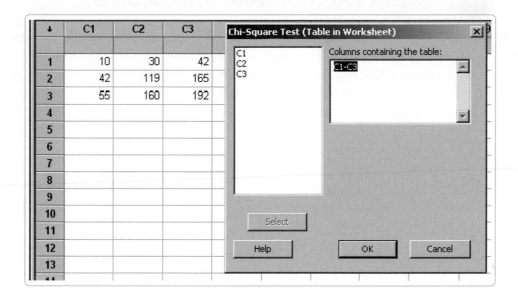

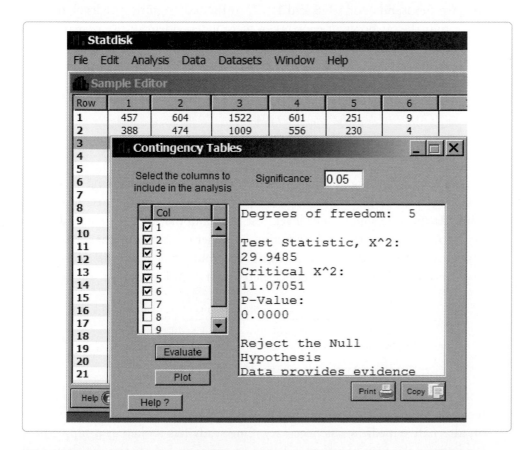

Exercises A: Basic Skills and Concepts

1. Table 10-13 summarizes data for male survey subjects, but the accompanying table summarizes data for a sample of women. Using a 0.01 significance level, and assuming that the sample sizes of 800 men and 400 women are predetermined, test the claim that the proportions of agree/disagree responses are the same for the subjects interviewed by men and the subjects interviewed by women.

	Gender of Interviewer	
	Man	Woman
Women who agree	512	336
Women who disagree	288	64

2. The number of overnight trips to Canada declined from 1996 to 1997, perhaps partly in response to the world financial conditions of 1997. Were these changes from year to year independent of the purposes of the trips made to Canada? Refer to the accompanying table showing the numbers of overnight trips (in thousands) for 1996 and 1997. At the 0.05 significance level, test the claim that the purposes of trips taken were independent of the year the trips were taken. Were some kinds of trips more affected than others?

	Business	Visit	Pleasure	Other
1996	671	1171	2360	174
1997	769	1204	2080	181

Based on data in *Industry Canada: Characteristics of International Travellers, 1997.*

3. Nicorette is a chewing gum designed to help people stop smoking cigarettes. Tests for adverse reactions yielded the results given in the accompanying table. At the 0.05 significance level, test the claim that the treatment (drug or placebo) is independent of the reaction (whether or not mouth or throat soreness was experienced). If you are thinking about using Nicorette as an aid to stop smoking, should you be concerned about mouth or throat soreness?

	Drug	Placebo
Mouth or throat soreness	43	35
No mouth or throat soreness	109	118

Based on data from Merrell Dow Pharmaceuticals, Inc.

4. Both proponents of and skeptics about health-store pills and nutritional supplements agree that to make valid claims about them, more clinical research on alternative medicines is necessary. One study has looked at a supplement for fighting the pain of arthritis. Based on the table below, test the claim that

for those with arthritis, glucosamine hydrochloride has no effect on whether the subject feels improvement. Use the 0.05 level of significance. Does this study show that glucosamine hydrochloride is effective in reducing the pain of arthritis? If the study had a larger sample size and similar ratios were found within the contingency table, could this affect the outcome of the statistical test?

	Glucosamine Hydrochloride	Placebo
Felt improvement	22	24
Did not feel improvement	23	29

Based on data from the *Toronto Star*, September 19, 1998.

5. Firms that sponsor or participate in sporting events to gain marketing exposure like to represent winning teams. The accompanying table shows the numbers of cars of different makes in the top 14 and second 14 final positions in the 1996 Molson Indy Toronto. Does the evidence support a claim that the car model and the final position are independent? *Warning:* Not all expected frequencies are at least 5. If all the frequencies in the table had been five times greater than shown, the above warning would not be needed. In that scenario, would the *P*-value be greater than or less than the *P*-value that you get by using the original numbers in the table?

	Ford	Honda	Mercedes	Other
Top 14	5	3	6	0
Second 14	6	3	2	3

Based on data from Molson Indy Toronto.

6. In a study of store checkout scanning systems, samples of purchases were used to compare the scanned prices to the posted prices. The accompanying table summarizes results for a sample of 819 items. When stores use scanners to check out items, are the error rates the same for regular-priced items as they are for advertised-special items? How might the behaviour of consumers change if they believe that disproportionately more overcharges occur with advertised-special items?

	Regular-Priced Items	Advertised-Special Items
Undercharge	20	7
Overcharge	15	29
Correct price	384	364

Based on data from "UPC Scanner Pricing Systems: Are They Accurate?" by Ronald Goodstein, *Journal of Marketing*, Vol. 58.

7. The Boeing Commercial Airplane Group conducted a survey of 133 planes being flown or on order from three airline companies based in different countries. The sample data are summarized in the accompanying table. Is there

sufficient evidence to support the claim that the distribution of planes flying and planes ordered is independent of the airline company? Use a 0.01 significance level.

	United	British	Singapore
Boeing 777s flying	36	22	14
Boeing 777s ordered	16	23	22

8. A market researcher investigated cars and trucks sold in Canada to determine whether the proportions of sales for these two vehicle types were the same for all manufacturers. He examined randomly selected records for 50 sales of GM vehicles, 50 of Fords, 50 of Chryslers, 30 of Toyotas, and 20 of Volkswagens. (All sales were in December 1996.) The results are displayed below. At the 0.05 level of significance, test the claim that all these manufacturers sell cars and trucks in the same proportions.

	GM	Ford	Chrysler	Toyota	Volkswagen
Cars	24	14	16	24	19
Trucks	26	36	34	6	1

Based on data from *The Globe and Mail,* January 7, 1997.

9. A sports-medicine clinic is investigating the relation between types of sports injuries and the ages of patients involved. The accompanying table summarizes randomly selected sample data for patients with sports-based injuries. At the 0.05 significance level, test the claim that the sport in which the accident occurred (basketball or other sport) is independent of the patient's age (under 10 or 10 and over). Does it follow from the results that playing basketball is safer at lower ages? Explain your answer.

	Sport Where Injury Occurred	
	Basketball	Other Sports
Under 10	23	124
10 and over	427	490

Based on data from *Chronic Diseases in Canada,* Vol. 16, No. 3.

10. A study of randomly selected car accidents and drivers who use cellular phones provided the following sample data. At the 0.05 level of significance, test the claim that the occurrence of accidents is independent of the use of cellular phones. Based on these results, does it appear that the use of cellular phones affects driving safety?

	Had Accident in Last Year	Had No Accident in Last Year
Cellular phone user	23	282
Not cellular phone user	46	407

Based on data from AT&T.

11. A provider of uniforms for CFL players is examining players' weights. The accompanying table compares the weights of 32 randomly selected offensive backs (including quarterbacks) in the league. At the 0.05 level of significance, do the data indicate that the weight distribution for offensive backs is independent of position played?

	Quarterback	Other Offensive Back
Less than 200 lb	7	9
At least 200 lb	3	13

Based on data from the CFL website.

12. Winning team data were collected for teams in different sports, with the results given in the accompanying table. Use a 0.10 level of significance to test the claim that home/visitor wins are independent of the sport.

	Basketball	Baseball	Hockey	Football
Home team wins	127	53	50	57
Visiting team wins	71	47	43	42

Based on data from "Predicting Professional Sports Game Outcomes from Intermediate Game Scores," by Copper, DeNeve, and Mosteller, *Chance*, Vol. 5, No. 3–4.

13. The accompanying table lists sample data that statistician Karl Pearson used in 1909. Does the type of crime appear to be related to whether the criminal drinks or abstains? Are there any crimes that appear to be associated with drinking?

	Arson	Rape	Violence	Stealing	Coining	Fraud
Drinker	50	88	155	379	18	63
Abstainer	43	62	110	300	14	144

14. A study of people who refused to answer survey questions provided the randomly selected sample data in the accompanying table. At the 0.01 significance level, test the claim that the cooperation of the subject (response, refusal) is independent of the age category. Does any particular age group appear to be particularly uncooperative?

	Age					
	18–21	22–29	30–39	40–49	50–59	60 and over
Responded	73	255	245	136	138	202
Refused	11	20	33	16	27	49

Based on data from "I Hear You Knocking But You Can't Come In," by Fitzgerald and Fuller, *Sociological Methods and Research*, Vol. 11, No. 1.

15. Refer to Data Set 8 in Appendix B and test the claim that the gender of statistics students is independent of whether they smoke.

16. Refer to Data Set 8 in Appendix B and test the claim that whether statistics students exercise is independent of gender.

10-3 Exercises B: Beyond the Basics

17. The chi-square distribution is continuous, whereas the test statistic used in this section is discrete. Some statisticians use *Yates' correction for continuity* in cells with an expected frequency of less than 10 or in all cells of a contingency table with two rows and two columns. With Yates' correction, we replace

$$\sum \frac{(O - E)^2}{E} \quad \text{with} \quad \sum \frac{(|O - E| - 0.5)^2}{E}$$

Given the contingency table in Exercise 10, find the value of the χ^2 test statistic with and without Yates' correction. In general, what effect does Yates' correction have on the value of the test statistic?

18. **a.** For a contingency table with two rows and two columns and frequencies of a and b in the first row and frequencies of c and d in the second row, verify that the test statistic becomes

$$\chi^2 = \frac{(a + b + c + d)(ad - bc)^2}{(a + b)(c + d)(b + d)(a + c)}$$

b. Let $\hat{p}_1 = a/(a + c)$ and let $\hat{p}_2 = b/(b + d)$. Show that the test statistic

$$z = \frac{(\hat{p}_1 - \hat{p}_2) - 0}{\sqrt{\dfrac{\overline{p}\,\overline{q}}{n_1} + \dfrac{\overline{p}\,\overline{q}}{n_2}}}$$

where $\qquad \overline{p} = \dfrac{a + b}{a + b + c + d}$

and $\qquad \overline{q} = 1 - \overline{p}$

is such that $z^2 = \chi^2$ [the same result as in part (a)]. (This result shows that the chi-square test involving a 2×2 table is equivalent to the test for the difference between two proportions, as described in Section 8–6.)

10-4 Tests of Normality

Many important statistical methods require that the sample data were collected from a population that has a *normal* distribution. It is therefore very important to be able to determine whether we are working with data that are normally distributed. In this section, we present a number of **normality test** procedures, including formal goodness-of-fit hypothesis tests, to determine the validity of this assumption.

Informal Procedures

There are a handful of procedures that can be used without the rigour of a formal hypothesis test to determine the normality of the data:

1. *Histogram:* Construct a histogram. Reject normality if the histogram departs dramatically from a bell shape.

2. *Outliers:* Identify outliers. Reject normality if there is more than one outlier present. (Just one outlier could be an error or the result of chance variation; but be careful, because even a single outlier can have a dramatic effect on results.)

3. *Normal probability plot:* If the histogram is basically symmetric and there is at most one outlier, construct a normal probability plot, using the following steps.
   a. First sort the data by arranging the values in order from lowest to highest, and use the sorted data to form the left column of a table (see Figure 10-7).
   b. With a sample of size n, each value in the left column represents a proportion of $1/n$ of the sample. The procedure to fill the right column is based on finding the z-score which corresponds to the centre of each successive area under the assumed normal curve. There are two steps:
      i. First calculate, for each successive portion of the curve, the area under the normal curve that it represents—that is, the area extending from the

(sorted) Imports to Canada from the Middle East and Africa	Middle z-score for the expected, corresponding portion of the normal curve
1387	−1.960
1490	−1.440
1590	−1.150
1598	−0.935
1691	−0.755
1774	−0.598
1815	−0.454
1937	−0.319
1992	−0.189
2039	−0.063
2283	0.063
2554	0.189
2571	0.319
2840	0.454
3372	0.598
3446	0.755
3503	0.935
3989	1.150
4057	1.440
4103	1.960

Figure 10-7
Table for Constructing a Probability Plot

Figure 10-8
Normal Probability Plot

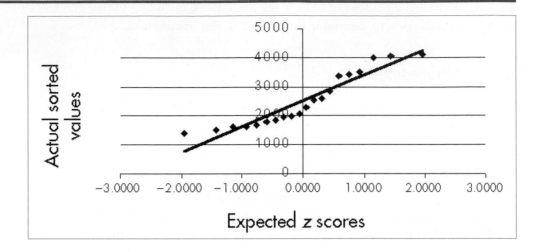

extreme left to a line over the z score at the centre of that portion of the curve. For example, the first z score represents an area of $0.5/n$, the next z score represents an area of $1.5/n$, then $2.5/n$, $3.5/n$, and so on.

ii. Use Table A-2 (or statistical software) to find the z scores that correspond to each of these areas under the curve. That is, for each portion, find z such that $P(Z < z)$ is the value you calculated in step (a).

c. Plot the points (x, y) where the x's are the expected middle z scores for the first, second, third, . . . portions of the normal curve, and the y's are the corresponding values in the original sample. (See Figure 10-8.)

(*Note:* Some details for constructing normal probability plots can vary among different textbooks or software packages—for example, whether to assign the original or the expected values to the x-axis, or whether to work with z scores (as we do here) or with t scores (which is recommended if you are working with small samples.)

To interpret the results of this procedure, compare the plot you have created with the ideal of a straight line. If the distribution of the sample is normal, we expect the points to exhibit only a small amount of random variation from a straight-line relationship.

EXAMPLE

The *F* test for comparing variances (Chapter 8) has a strong requirement that each population being compared have a normal distribution. Suppose the data in the rightmost column of Data Set 22 in Appendix B, for imports to Canada from the Middle East and Africa over 20 years, represent one of the samples to be compared. Although the data were collected as a time series (one total is given per year), would it be reasonable to assume that the results are normally distributed?

Step 1: We begin by constructing the histogram of the original values, with the result shown in the accompanying Excel (DDXL) display. An examination of the histogram shows that the data appear to be somewhat skewed to the right.

Step 2: There do not appear to be any outliers.

Step 3: We construct the normal probability plot, as shown in Figure 10-8.

Interpretation

Both the histogram and the normal probability plot suggest that these data are, at best, loosely approximated by a normal curve. For example, in the normal probability plot, every actual value in the middle part of the data set is below the expected value for that position. Therefore, it would not be valid to apply a technique such as the *F* test for comparing sample variances, since the test requires a good fit to normality.

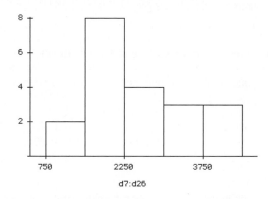

DDXL Output

DDXL HISTOGRAM

Formal Procedures

In the preceding informal procedures, judging whether a data set follows a normal distribution is at the discretion of the researcher. In the following formal procedures, a critical value and test statistic provide a degree of rigour in determining normality.

χ^2 Goodness-of-Fit Test for Normality

This test is similar to the χ^2 goodness-of-fit tests presented earlier in the chapter. There are two features that distinguish the test for normality:

1. The previous distributions were discrete, making it easier to distinguish the categories. Since the normal distribution is continuous, we must arbitrarily decide

the categories, keeping in mind that the expected number of observations for each category must still be at least 5.

2. In the previous tests, we specifically laid out the parameters of the distribution, such as n and p for the binomial distribution and μ for the Poisson distribution. This will not necessarily be the case for the normality test.

Because of the restriction of expecting at least 5 observations per category, the χ^2 test usually requires a fair number of observations. A recommended minimum number is 20.

EXAMPLE

In a milk bottling plant, the amount of milk going into a 1 L carton is expected to be normally distributed with a mean of 1 L and standard deviation of 0.05 mL. For a random sample of 25 cartons, the measurements (in mL) are given in Table 10-14.

Table 10-14	Measurements of 25 cartons of 1 L milk (in mL)			
1000.005	1000.1067	999.9855	999.9141	1000.0037
999.9561	999.9108	999.9355	999.9916	1000.0015
1000.0322	1000.0083	999.9277	999.9998	1000.0871
1000.0838	1000.0748	999.9711	1000.0267	1000.0157
1000.0799	999.9657	999.9813	1000.0017	1000.0107

Does the amount of milk in the carton follow a normal distribution with $\mu = 1000$ and $\sigma = 0.05$? Test at a 0.05 level of significance.

SOLUTION

We first state the null and alternative hypotheses:

H_0: The data follow a normal distribution with $\mu = 1000$ and $\sigma = 0.05$.

H_1: The data do not follow a normal distribution with $\mu = 1000$ and $\sigma = 0.05$.

The first step is to set up the categories so that we can expect at least 5 observations per category. For example, if we let X be the amount of milk, one common approach to setting up the categories is $X < \mu - \sigma, \mu - \sigma < X < \mu, \mu < X < \mu + \sigma$, and $X > \mu + \sigma$. In our situation, the lowest category would be $X < 1000 - 0.05$ or $X < 999.95$. Since 999.95 is one standard deviation away from the mean, $P(X < 999.95) = P(Z < -1) = 0.5 - 0.3413 = 0.1587$. The expected number of observations for this category would be $25(0.1587) = 3.9675$, which is below the minimum threshold of 5. Given the few number of observations, perhaps 3 categories is optimal with approximately $p = 0.33$ in each category. From Table A-2, we can calculate that $P(-0.44 < Z < 0) = P(0 < Z < 0.44) = 0.17$. Then $P(Z < -0.44) = P(Z > 0.44) = 0.5 - 0.17 = 0.33$ and $P(-0.44 < Z < 0.44) = 0.17 + 0.17 = 0.34$. The expected number of observations for the lowest and highest categories is $25(0.33) = 8.25$ and for the middle category is $25(0.34) = 8.5$.

What we now need is the actual number of observations per category. Starting with the lowest category,

$$-0.44 = \frac{X - 1000}{0.05}$$

from which we can solve $X = 1000 - 0.44(0.05) = 999.978$. Similarly, for the upper category, $X = 1000 + 0.44(0.05) = 1000.022$. The categories are $X < 999.978$, $999.978 < X < 1000.022$, and $X > 1000.022$. If we sort the data from lowest to highest, the numbers of observations in each category are 7, 11, and 7, respectively. We can now construct the χ^2 test statistic:

Normal Category	Observed Frequency O	Expected Frequency $E = np$	$O - E$	$(O - E)^2$	$\dfrac{(O - E)^2}{E}$
$X < 999.978$	7	8.25	-1.25	1.5625	0.1894
$999.978 < X <$ 1000.022	11	8.5	2.5	6.25	0.7353
$X > 1000.022$	7	8.25	-1.25	1.5625	0.1894
	25	25		$\chi^2 = \sum \dfrac{(O - E)^2}{E} = 1.1141$	

Since $k = 3$, the number of degrees of freedom is 2. With $\alpha = 0.05$, we reject the null hypothesis if the χ^2 test statistic is greater than 5.991, from Table A-4. Since the test statistic of 1.1141 is not greater than this, we fail to reject the null hypothesis.

Interpretation
The amount of milk in the 1 L carton follows a normal distribution with $\mu = 1000$ mL and $\sigma = 0.05$ mL.

In the previous example, μ and σ were specified. However, there are times when the researcher has no preconceived notion as to what these parameters should be. In this case, they are estimated by their sample counterparts, $\overline{x}$ and s. This is illustrated in the following example.

EXAMPLE

A dollar store wanted to determine if its sales were normally distributed. From a sample of 200 sales, it found $\overline{x} = 12.3053$ and $s = 5.0112$. The categories used were $X < 7.2941$, $7.2941 < X < 12.3053$, $12.3053 < X < 17.3165$, and $X > 17.3165$, which delineate the category boundaries by one standard deviation. The observed frequencies for these categories are 27, 86, 53, and 34. Test at a 0.05 level of significance.

SOLUTION

The null and alternative hypotheses are:

H_0: Sales are normally distributed.

H_1: Sales are not normally distributed.

To find the expected frequencies, we will work on the assumption that $\mu = 12.3053$ and $\sigma = 5.0112$. Then $P(X < 7.2941) = P(Z < -1) = 0.5 - 0.3413 = 0.1587$. Similarly $P(X > 17.3165) = P(Z > 1) = 0.1587, P(7.2941 < X < 12.3053) = P(-1 < Z < 0) = 0.3413$ and $P(12.3053 < X < 17.3165) = P(0 < Z < 1) = 0.3413$. The expected values for the lowest and highest categories are $200(0.1587) = 31.74$ while those for the two middle categories are $200(0.3413) = 68.26$. We now construct the χ^2 test statistic:

Normal Category	Observed Frequency O	Expected Frequency $E = np$	$O - E$	$(O - E)^2$	$\dfrac{(O - E)^2}{E}$
$X < 7.2941$	27	31.74	−4.74	22.4676	0.7079
$7.2941 < X < 12.3053$	86	68.26	17.74	314.7076	4.6104
$12.3053 < X < 17.3165$	53	68.26	−15.26	232.8676	3.4115
$X > 17.3165$	34	31.74	2.26	5.1076	0.1609
	200	200		$\chi^2 = \sum \dfrac{(O - E)^2}{E} = 8.8907$	↑

Normally, since $k = 4$, there would be 3 degrees of freedom. However, since both μ and σ were not specified in the null hypothesis, we must subtract 1 degree of freedom for each of these parameters. Therefore, the actual number of degrees of freedom is $4 - 1 - 1 - 1 = 1$. Using Table A-4 and $\alpha = 0.05$, the critical value is 3.841. Since the test statistic value of 8.8907 is greater than this, we reject the null hypothesis.

Interpretation
We conclude that the dollar store's sales are not normally distributed.

Lilliefors Test for Normality

Whereas the χ^2 test for normality requires a fair number of observations, the Lilliefors test for normality does not; it can be conducted with a sample size as small as four. It also does not require μ and σ to be specified. This makes it ideal for situations such as determining if the data are normally distributed as a precursor to hypothesis tests such as the t test for one mean. The null and alternative hypotheses are:

H_0: Data are normally distributed.

H_1: Data are not normally distributed.

The approach to this test is not unlike that for normal probability plots in that we examine the difference between the distribution we have (called the empirical distribution) and the normal distribution we expect. The test statistic for the Lilliefors test is the largest difference between what we have and what we expect. Lilliefors developed a table of critical values (Table A-12). If the test statistic is greater than the critical value, we reject the null hypothesis and conclude that the data are not normally distributed.

If each value occurs only once, the procedure to finding the test statistic is as follows:

1. Sort the data from lowest to highest.

2. Calculate $\bar{x}$ and s. Convert each data value to a Z value using the formula:

$$Z = \frac{X - \bar{x}}{s}$$

3. Calculate $F(z)$, the left-tail probability of each Z value.

4. Calculate $S(z)$, the empirical cumulative probability for each Z value, which is equal to (number of observations $\leq x$ value) $\div n$. Also calculate $S'(z) = S(z) - 1/n$, which is also equal to (number of observations $< x$ value) $\div n$.

5. Compute $|F(z) - S(z)|$ and $|F(z) - S'(z)|$ for each Z value. The test statistic is the largest of these differences.

If there are values that occur more than once, refer to the Using Technology section.

EXAMPLE

In Chapter 7, a t test was conducted on a random sample of seven 12-oz aluminum cans to determine if the mean axial load was greater than 165 pounds. We reproduce the data:

270   273   258   204   254   228   282

One of the assumptions of the t test is that the data follow a normal distribution. Based on these data, can we conclude that axial loads of 12-oz aluminum cans are normally distributed? Test at a 0.05 level of significance.

SOLUTION

The null and alternative hypotheses are:

H_0: The axial loads are normally distributed.

H_1: The axial loads are not normally distributed.

To find the critical value from Table A-12, we look in the 0.95 column (since $1 - 0.05 = 0.95$) and the $n = 7$ row. The critical value is 0.300.

To find $F(z)$ for each Z value, we will work from the assumption that we are using Table A-2. As such, we will round the Z values to 2 decimals. From the data, we find $\bar{x} = 252.7143$ and $s = 27.6328$. The following table summarizes the results:

X	Z	F(z)	S(z)	$S'(z) = S(z) - 1/7$	$\|F(z) - S(z)\|$	$\|F(z) - S'(z)\|$
204	−1.76	0.0392	$1/7 = 0.1429$	0.0000	0.1037	0.0392
228	−0.89	0.1867	$2/7 = 0.2857$	0.1429	0.0990	0.0438
254	0.05	0.5199	$3/7 = 0.4286$	0.2857	0.0913	0.2342
258	0.19	0.5753	$4/7 = 0.5714$	0.4286	0.0039	0.1467
270	0.63	0.7357	$5/7 = 0.7143$	0.5714	0.0214	0.1643
273	0.73	0.7673	$6/7 = 0.8571$	0.7143	0.0898	0.0530
282	1.06	0.8554	$7/7 = 1.0000$	0.8571	0.1446	0.0017

To illustrate how the values are found, we will use the first row of $X = 204$. To convert this to Z, we have:

$$Z = \frac{204 - 252.7143}{27.6328} = -1.76$$

Then $F(-1.76) = P(Z < -1.76) = 0.5 - 0.4608 = 0.0392$. Since this is the first Z value, $S(z) = 1/7 = 0.1429$. Therefore, $S'(z) = 1/7 - 1/7 = 0$. Then $|F(z) - S(z)| = |0.0392 - 0.1429| = 0.1037$ and $|F(z) - S'(z)| = |0.0392 - 0| = 0.0392$. We repeat this process for all seven values. Of all the computed differences, the largest is 0.2342 for $X = 254$. Since the test statistic is less than the critical value of 0.300, we fail to reject the null hypothesis.

Interpretation

We conclude that the axial loads of 12-oz aluminum cans are normally distributed.

EXCEL: The **Data Desk XL** add-in (Figure 10-9) can be used to generate a *normal probability plot*. First enter the sample values into an Excel column. (There is no need to sort the data.)

EXCEL (Prior to 2007): Click on **DDXL** on the main menu bar. Select **Charts and Plots**, then select the function type of **Normal Probability Plot**. Click on the pencil icon for **Quantitative Variable**, then enter the range of values, such as A3:A22. Click **OK**.

EXCEL 2007: Click **Add-Ins** on the main menu and then **DDXL** from the add-in menu to the left. Proceed as above.

MINITAB 15: From the **Stat** menu, choose **Basic Statistics** and then **Normality Test**. Select the variable which you wish to test for normality. You have a choice of three normality tests. Click **OK**.

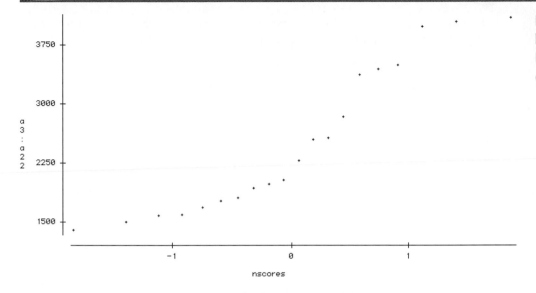

Figure 10-9
Normal Probability Plot
Produced by Excel
(DDXL add-in)

MINITAB DISPLAY
Normality Test

INTERNET: There are a wide variety of normality tests available on the Internet. A search on the phrase "normality test Excel" will produce a variety of tools (some freeware) that you can use with Excel to test a data set for normality.

10-4 Exercises A: Basic Skills and Concepts

1. In a survey at a mall during August, people were asked how much they spent that day. For a sample of 8 people, these were the results:

    50   75   100   120   140   150   240   1350

    Are the sales normally distributed? Test at a 0.05 level of significance.

2. In the previous question, 1350 is an outlier. If we remove this value, show why the sales without outliers are normally distributed at a 0.05 level of significance.

3. A factory wanted to institute a quality control program. In theory, the diameter of a certain plastic part to be used in build-it-yourself furniture should be normally distributed with a mean of 15 mm and standard deviation of 0.002 mm. Based on a sample of 1000 pieces, this was the distribution:

$X < 14.996$	$14.996 < X < 14.998$	$14.998 < X < 15$	$15 < X < 15.002$	$15.002 < X < 15.004$	$X > 15.004$
20	138	352	327	140	23

   Are the diameters normally distributed with a mean of 15 mm and standard deviation of 0.002 mm? Test at a 0.05 level of significance.

4. Refer to the height data in Data Set 8. Use normal probability plots to confirm the assumption of normality. (Be sure to identify and remove the outlier in the data set—you can't miss it.)

10-4 Exercises B: Beyond the Basics

5. In the χ^2 goodness-of-fit test, suppose the total number of observations is n. If the number of observations in each category is multiplied by a common factor N, show why the value of the test statistic is also multiplied by N.

VOCABULARY LIST

goodness-of-fit test   **549**   test of homogeneity   **572**
normality test   **583**   test of independence   **566**

REVIEW

Section 10–2 described the goodness-of-fit test for determining whether a set of data has some claimed distribution. Section 10–3 described contingency tables and the test of independence between the row and column variables, as well as the test of homogeneity, in which different populations have the same proportions of some characteristics. Section 10–4 examined ways of determining the normality of a data. The following are some key elements of the methods discussed in this chapter.

- *Section 10–2 (Test for goodness-of-fit):*

Test statistic is
$$\chi^2 = \Sigma \frac{(O - E)^2}{E}$$

Test is right-tailed with $k - 1$ degrees of freedom. All expected frequencies must be at least 5.

- *Section 10–3 (Contingency table test of independence or homogeneity):*

Test statistic is
$$\chi^2 = \Sigma \frac{(O - E)^2}{E}$$

Test is right-tailed with $(r - 1)(c - 1)$ degrees of freedom. All expected frequencies must be at least 5.

REVIEW EXERCISES

1. The latest product of a cat food manufacturer is apparently well liked and nutritious, but cans are reportedly going bad before the end of their stated shelf life. Tests are now being conducted with an improved product using a revised set of additives. Based on the accompanying results, and using a 0.05 level of significance, test the claim that the shelf life of their product is independent of the additives that are used.

	Below Standard	Acceptable	Above Standard
Current additives	34	25	5
Revised additives	7	44	10

2. In recent years, El Niño phenomena have been popularly blamed for a variety of weather effects. The accompanying table compares the number of springs with precipitation levels at least 2% above seasonal norms. Fourteen springs were sampled for each of three time periods related to El Niño cycles: (a) springs prior to a major El Niño cycle (either one or two years prior), (b) the springs within a major El Niño cycle (there are two springs per cycle), and (c) other springs, unrelated to an El Niño cycle. At the 0.05 level of significance, test the claim that the proportions of springs

with above-normal precipitation are the same regardless of which part of the El Niño cycle is occurring.

	Prior Springs	El Niño Springs	Other Springs
Springs with above-normal precipitation	5	8	8
Springs without above-normal precipitation	9	6	6

Based on data from Environment Canada.

3. Based on seven years of data collected by Transport Canada, 72% of all bird strikes on moving aircraft involve gulls, sparrows, swallows, or snow buntings. The accompanying table is based on a random selection of bird strikes on planes, by other species than those just mentioned. Test the claim that the all the other bird species are involved in bird strikes with equal frequencies. Use the 0.05 level of significance. Would your results support a policy of focusing bird-strike prevention programs on particular species of birds? Why or why not?

Species	Hawks	Starlings	Ducks	Plovers	Owls	Sandpipers	Larks	Kestrels
Frequency	17	13	15	11	10	9	8	7

4. Refer to Data Set 18 in Appendix B, which shows the body temperatures for people who were sufficiently healthy to be accepted for voluntary surgery. Below, the temperature ranges of females and males in the sample are compared. Test the hypothesis that the patients' temperatures and genders are independent. Use the 0.05 level of significance.

	Female	Male
Below 37°C	43	51
At least 37°C	11	3

5. Clinical tests of the allergy drug Seldane yielded results summarized in the accompanying table. At the 0.05 significance level, test the claim that the occurrence of headaches is independent of the group (Seldane, placebo, control). Based on these results, should Seldane users be concerned about getting headaches?

	Seldane Users	Placebo Users	Control
Headache	49	49	24
No headache	732	616	602

Based on data from Merrell Dow Pharmaceuticals, Inc.

CUMULATIVE REVIEW EXERCISES

Table 10-15

	A	B	C	D
x	66	80	82	75
y	77	89	94	84

1. Assume that Table 10-15 lists test scores for four people, where the x score is from a test of memory and the y score is from a test of reasoning. Test the claim that both sets of scores are normally distributed at a 0.05 level of significance.

2. Assume that Table 10-15 lists test scores for four people, where the x score is from a pretest taken before a training session on memory improvement and the y score is from a test taken after the training. Test the claim that the training session is effective in raising scores.

3. Assume that in Table 10-15, the letters A, B, C, and D represent the choices on the first question of a multiple-choice quiz. Also assume that x represents men and y represents women, and the table entries are frequency counts, so that 66 men chose answer A, 77 women chose answer A, 80 men chose answer B, and so on. Test the claim that men and women choose the different answers in the same proportions.

4. Assume that in Table 10-15, the letters A, B, C, D represent different versions of the same test of reasoning. The x scores were obtained by four randomly selected men and the y scores were obtained by four randomly selected women. Test the claim that men and women have the same mean score.

TECHNOLOGY PROJECT

Use Excel to randomly generate 100 digits between 0 and 9 inclusive, then use STATDISK to randomly generate another 100 digits between 0 and 9 inclusive. Record the results in the accompanying table, then test the claim that the digit selected is independent of whether you used Excel or STATDISK. Test the claim by using Excel or STATDISK, and obtain a printed copy of the display. Interpret the computer results and write your conclusions.

Digit	0	1	2	3	4	5	6	7	8	9
Frequency in Excel										
Frequency in STATDISK										

FROM DATA TO DECISION

In response to serious public allegations by a staff psychiatrist at a British Columbia hospital, the government of British Columbia ordered an independent investigation ("B.C. Probes Big Increase in Shock Treatments," *The Globe and Mail*, December 13, 2000). It was claimed that following a change in health policy, allowing doctors to be paid $62 for administering electroshock treatments to geriatric patients, the numbers of such treatments actually administered rose dramatically. The accompanying table shows the numbers of these treatments given during three consecutive time periods—one period *before* the policy of paying for treatments, and two subsequent periods.

ANALYZE THE RESULTS

Analyze the sample data and draw conclusions. Is there a significant difference in the numbers of electroshock treatments administered during the three time periods shown? If there does appear to be an upward change in the numbers of treatments, try to identify one factor *besides* the new payment policy that could be contributing to the higher numbers. Make a recommendation that might reduce any tendency for the payment system itself to be an incentive for ordering additional treatments.

	Time Period		
	1996/1997	1997/1998	1998/1999
Number of electroshock treatments	689	1249	1533

COOPERATIVE GROUP ACTIVITIES

1. **Out-of-class activity:** Divide into groups of four or five students. Each group member should survey at least 15 male students and 15 female students at the same college by asking two questions: (1) Which political party does the subject favour most? (2) If the subject were to make up an absence excuse of a flat tire, which tire would he or she say went flat if the instructor asked? Ask the subject to write the two responses on an index card, and also record the gender of the subject and whether the subject wrote with the right or left hand. Use the methods of this chapter to analyze the data collected.

2. **Out-of-class activity:** In random sampling, each member of the population has the same chance of being selected. When the author asked 60 students to "randomly" select three digits each, the results listed below were obtained. Use this sample of 180 digits to test the claim that students select digits randomly.

213	169	812	125	749	137	202	344	496	348	714	765
831	491	169	312	263	192	584	968	377	403	372	123
493	894	016	682	390	123	325	734	316	357	945	208
115	776	143	628	479	316	229	781	628	356	195	199
223	114	264	308	105	357	333	421	107	311	458	007

Were the students successful in choosing their own random numbers? Select your own sample of students and ask each one to randomly select three digits. Test your results for randomness. Also test your results to determine whether they agree with those listed on next page. Write a report describing your experiment and clearly state your major conclusions.

3. **In-class activity:** Divide into groups of three or four students. Each group should be given a die along with the instruction that it should be tested for "fairness." Is the die fair or is it biased? Describe the analysis and results.

4. **In-class activity:** Divide into groups of six or more students. Each group member should provide the digits of his or her Social Insurance Number. (The digits should be rearranged so that privacy is maintained.) Construct a combined list of those digits and test the claim that the digits 0, 1, 2, . . . , 9 occur with the same frequency.

INTERNET PROJECT

Contingency Tables

In all the examples in Section 10–3, data were sorted into categories formed by the rows and columns of a contingency table. We used contingency tables along with the associated chi-square test to form conclusions regarding qualitative population data. Characteristics such as gender, race, and geographic location become fair game for formal hypothesis testing procedures. Go to the Internet Project for this chapter, located at either of the following websites:

http://www.mathxl.com   or   http://www.mystatlab.com

   Click on Internet Project, then on Chapter 10. After collecting demographic data for a number of populations, you will apply the techniques of this chapter to form conclusions about the independence of interesting pairings of characteristics.

---

11 Correlation and Regression

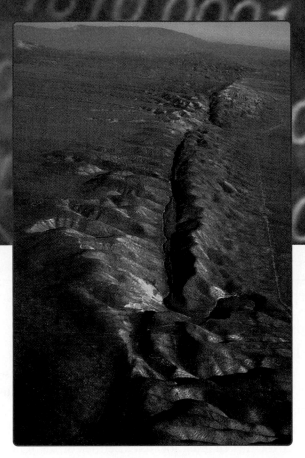

11-1 Overview

This chapter describes methods for dealing with relationships between two variables. The important concepts of correlation and regression are discussed.

11-2 Correlation

The relationship between two variables is investigated with a graph (called a scatter diagram) and a measure (called the linear correlation coefficient).

11-3 Regression

Linear relationships between two variables are described using the equation and graph of a straight line, called the regression line. A method for determining predicted values of a variable is also presented. The general concept of mathematical modelling is introduced. Residual plots are introduced as a tool for determining model sufficiency.

11-4 Variation and Prediction Intervals

A method is presented for analyzing the differences between the predicted values of a variable and the actual observed values. Regression ANOVA is presented to determine model significance as well as tests and confidence intervals for the regression slope. Prediction intervals, which are confidence interval estimates of predicted values, are constructed as well as confidence intervals for the mean value of *y* (the dependent variable).

11-5 Multiple Regression

Methods are presented for finding a linear equation that relates three or more variables. The multiple coefficient of determination is presented as a measure of how well the sample points fit that linear equation. Because of the calculations involved, this section emphasizes the use of computer software and the interpretation of computer displays.

CHAPTER PROBLEM

Do earthquakes at all latitudes have the same magnitude?

To most Canadians, earthquakes are probably viewed as rare occurrences, which usually strike somewhere else, such as in California. If a tremor is noticeable nearby (as has happened a few times in the Toronto area), it is treated as a newsworthy event. In fact, "seismic activity" or earthquakes are quite common. In just one month in 2001, Natural Resources Canada recorded 215 separate earthquakes that affected Canada, from Vancouver Island, B.C., to Wager Bay, Nunavut, to the Labrador Sea.

Not all earthquakes have the same strength either at their centre or in how they spread their energies. The commonly known Richter Scale (officially the Magnitude Local, or ML, scale) is one system for measuring earthquake magnitudes; but other systems may be preferred in certain circumstances.

The 215 earthquakes referred to above were spread across the whole of Canada. But was their magnitude the same in all locations? In particular, were the magnitudes greater at higher latitudes?

The data in Table 11-1 represent 13 randomly selected cases of seismic activity (earthquakes) from among those recorded between January 9 and February 8, 2001. (Only events whose magnitudes are recorded in the ML scale were selected.) For reasons of clarity, we use this abbreviated data set, but better results could be obtained if we used a more complete set of sample values. Based on these data, does there appear to be a relationship between the recorded magnitude and the recorded latitude of seismic activity? If so, what is that relationship? If we learn that an earthquake has occurred at a latitude of 64°N, how can we use that information to predict the magnitude of the event? These questions will be addressed in this chapter.

Table 11-1 Latitudes and Magnitudes of Seismic Activity

x Latitude (°N)	60.0	77.5	50.7	65.6	48.2	63.5	49.2	60.3	52.6	52.8	64.3	49.3	48.3
y Magnitude (ML)	4.1	4.0	2.6	2.8	0.9	2.2	3.0	4.1	1.2	1.1	5.5	2.7	0.9

Based on data from Natural Resources Canada.

11-1 Overview

The main focus of this chapter is to form inferences based on sample data that come in *pairs*. Given such paired data, we want to determine whether there is a relationship between the two variables and, if so, identify what the relationship is. For example, using the data in Table 11-1, we want to determine whether there is a relationship between the latitude of recorded seismic activity and its magnitude. If such a relationship exists, we want to identify it with an equation so that we can predict the magnitude of seismic activity if we are informed of the latitude where it occurred.

We begin in Section 11–2 by considering the concept of correlation, which is used to determine whether there is a statistically significant relationship between two variables. We investigate correlation using the scatter diagram (a graph) and the linear correlation coefficient (a measure of the strength of linear association between two variables). Section 11–3 investigates regression analysis, where we construct an explanation that mathematically describes the relationship between two variables. We show how to use that equation to predict values of a variable.

In Section 11–4, we analyze the differences between predicted values and actual observed values of a variable. Sections 11–2 through 11–4 deal with relationships between two variables, but Section 11–5 uses concepts of multiple regression to describe the relationship among three or more variables. Throughout this text we deal only with linear (or straight-line) relationships between two or more variables.

11-2 Correlation

In this section we address the issue of determining whether there appears to be a relationship between two variables. In statistics, we refer to such a relationship as a *correlation*.

DEFINITION

> A **correlation** exists between two variables when one of them is related to the other in some way.

Table 11-1, for example, consists of paired data (sometimes called **bivariate data**). We will determine whether there is a correlation between the variable x (length) and the variable y (weight).

As we work with sample data and develop methods of forming inferences about populations, we make the following assumptions.

Assumptions

1. The sample of paired (*x*, *y*) data is a *random* sample.
2. The *y* data have a *normal distribution*. (Normal distributions are discussed in Chapter 5, but this assumption basically requires that the values of *y* have a distribution that is bell-shaped. The second assumption can be checked using the normality test presented in Chapter 10.)

Exploring the Data

Before working with the more formal computational methods of this section, we should first explore the data set to see what we can learn. We can often see a relationship between two variables by constructing a graph called a *scatterplot*, or *scatter diagram*.

DEFINITION

> A **scatterplot** (or **scatter diagram**) is a graph in which the paired (*x*, *y*) sample data are plotted with a horizontal *x*-axis and a vertical *y*-axis. Each individual (*x*, *y*) is plotted as a single point.

As an example, see the Excel display of the 13 pairs of data listed in Table 11-1. When we examine such a scatterplot, we should study the overall pattern of the plotted points. If there is a pattern, we should note its direction: That is, as one variable increases, does the other seem to increase or decrease? We should observe whether

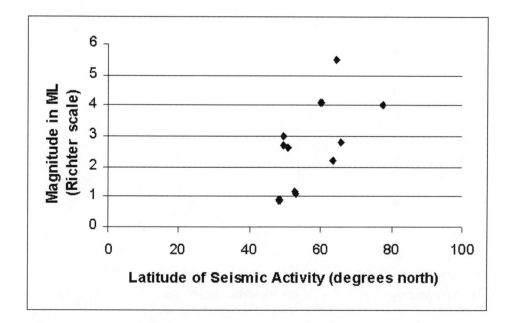

EXCEL DISPLAY
Scatter Diagram Showing Correlation

there are any outliers, which, in this context, means points that lie very far away from all the other points. The Excel-generated display does seem to reveal a pattern showing that larger recorded magnitudes for seismic activity tend to go along with higher recorded latitudes for the activity. There do not appear to be any outliers.

Other examples of scatter diagrams are shown in Figure 11-1. The graphs in Figure 11-1(a), (b), and (c) depict a pattern of increasing values of y that correspond to increasing values of x. As you proceed from (a) to (c), the dot pattern

Figure 11-1
Scatter Diagrams

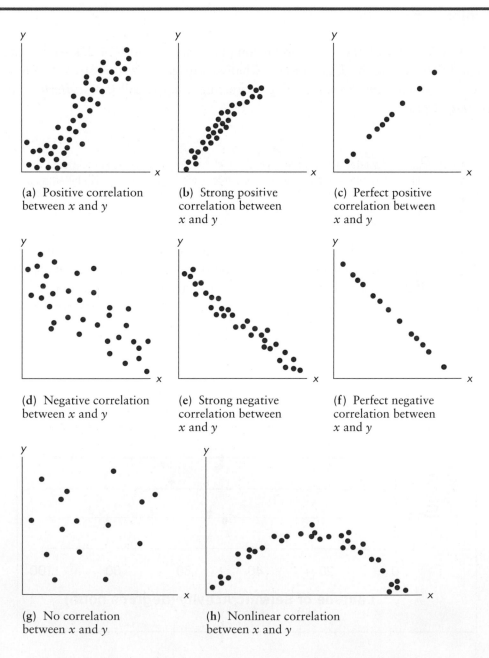

(a) Positive correlation between x and y

(b) Strong positive correlation between x and y

(c) Perfect positive correlation between x and y

(d) Negative correlation between x and y

(e) Strong negative correlation between x and y

(f) Perfect negative correlation between x and y

(g) No correlation between x and y

(h) Nonlinear correlation between x and y

becomes closer to a straight line, suggesting that the relationship between x and y becomes stronger. The scatter diagrams in (d), (e), and (f) depict patterns in which the y values decrease as the x values increase. Again, as you proceed from (d) to (f), the relationship becomes stronger. In contrast to the first six graphs, the scatter diagram of (g) shows no pattern and suggests that there is no correlation (or relationship) between x and y. Finally, the scatter diagram of (h) shows a pattern, but it is not a straight-line pattern.

Linear Correlation Coefficient

Because conclusions drawn from scatter diagrams are largely subjective, we need more precise and objective methods. We will use the linear correlation coefficient, which is useful for detecting straight-line patterns.

DEFINITION

> The **linear correlation coefficient** r (often called simply the **correlation coefficient**) measures the strength of the linear relationship between the paired x and y values in a *sample*. Its value is computed by using Formula 11-1, which follows. [The linear correlation coefficient is also sometimes referred to as **Pearson's product moment correlation coefficient** in honour of Karl Pearson (1857–1936), who originally developed it.]

Formula 11-1
$$r = \frac{n\Sigma xy - (\Sigma x)(\Sigma y)}{\sqrt{n(\Sigma x)^2 - (\Sigma x)^2}\sqrt{n(\Sigma y^2) - (\Sigma y^2)}}$$

Because r is calculated using sample data, it is a sample statistic used to measure the strength of the linear correlation between x and y. If we had every pair of population values for x and y, the result of Formula 11-1 would be a population parameter, represented by ρ.

We will describe how to compute and interpret the linear correlation coefficient r given a list of paired data, but we will first identify the notation relevant to Formula 11-1. Later in this section we will present the underlying theory that led to the development of Formula 11-1.

Rounding the Linear Correlation Coefficient r

Round the linear correlation coefficient r to three decimal places (so that its value can be directly compared to critical values in Table A-6). In the social sciences r is often rounded off to two decimal points. When calculating r and other statistics in this chapter, rounding in the middle of a calculation often creates substantial errors, so try using your calculator's memory to store intermediate results and round off only at the end. Many inexpensive calculators have Formula 11-1 built in so that you can automatically evaluate r after entering the sample data.

STATISTICAL PLANET

Cell Phones and Crashes
Because some countries have banned the use of cell phones in cars while other countries are considering such a ban, researchers studied the issue of whether the use of cell phones while driving increases the chance of a crash. A sample of 699 drivers was obtained. Members of the sample group used cell phones and were involved in crashes. Subjects completed questionnaires and their telephone records were checked. Telephone usage was compared to the time interval immediately preceding a crash to a comparable time period the day before. Conclusion: Use of a cell phone was associated with a crash risk that was about four times as high as the risk when a cell phone was not used. (See "Association between Cellular-Telephone Calls and Motor Vehicle Collisions," by Redelmeier and Tibshirani, *New England Journal of Medicine*, Vol. 336, No. 7.)

n	represents the number of pairs of data present.
Σ	denotes the addition of the items indicated.
Σx	denotes the sum of all x values.
Σx^2	indicates that each x value should be squared and then those squares added.
$(\Sigma x)^2$	indicates that the x values should be added and the total then squared. It is extremely important to avoid confusing Σx^2 and $(\Sigma x)^2$.
Σxy	indicates that each x value should be first multiplied by its corresponding y value. After obtaining all such products, find their sum. (*Note*: each product of corresponding x and y values is called a **cross product**.)
r	represents the linear correlation coefficient for a *sample*.
ρ	represents the linear correlation coefficient for a *population*.

EXAMPLE

Using the data in Table 11-1, find the value of the linear correlation coefficient r. (A later example will use this value to determine whether there is a relationship between the recorded latitudes and magnitudes of seismic activity.)

SOLUTION

For the sample paired data in Table 11-1, $n = 13$ because there are 13 pairs of data. The other components required in Formula 11-1 are found from the calculations in Table 11-2. Note how this vertical format makes the calculations easier.

Using the calculated values and Formula 11-1, we can now evaluate r as follows:

$$r = \frac{n\Sigma xy - (\Sigma x)(\Sigma y)}{\sqrt{n(\Sigma x^2) - (\Sigma x)^2}\sqrt{n(\Sigma y^2) - (\Sigma y)^2}}$$

$$= \frac{13(2100.84) - (742.3)(35.1)}{\sqrt{13(43,344.79) - (742.3)^2}\sqrt{13(119.87) - (35.1)^2}}$$

$$= \frac{1256.19}{\sqrt{12,472.98}\sqrt{326.3}} = 0.623$$

These calculations get quite messy with large data sets, so it's fortunate that the linear correlation coefficient can be found automatically with many different calculators and computer programs. For example, see "Using Technology" at the end of this section.

Table 11-2 Finding Statistics Used to Calculate r

Latitude (°N)	Magnitude (ML)			
x	y	x · y	x^2	y^2
60.0	4.1	246.00	3600.00	16.81
77.5	4.0	310.00	6006.25	16.00
50.7	2.6	131.82	2570.49	6.76
65.6	2.8	183.68	4303.36	7.84
48.2	0.9	43.38	2323.24	0.81
63.5	2.2	139.70	4032.25	4.84
49.2	3.0	147.60	2420.64	9.00
60.3	4.1	247.23	3636.09	16.81
52.6	1.2	63.12	2766.76	1.44
52.8	1.1	58.08	2787.84	1.21
64.3	5.5	353.65	4134.49	30.25
49.3	2.7	133.11	2430.49	7.29
48.3	0.9	43.47	2332.89	0.81
Total 742.3	35.1	2100.84	43,344.79	119.87
↑	↑	↑	↑	↑
Σx	Σy	Σxy	Σx^2	Σy^2

Interpreting the Linear Correlation Coefficient

We need to interpret a calculated value of r, such as the value of 0.623 found in the preceding example. Given the way that Formula 11-1 was developed, the value of r must always fall between -1 and $+1$ inclusive. If r is close to 0, we conclude that there is no significant linear correlation between x and y, but if r is close to -1 or $+1$, we conclude that there is a significant linear correlation between x and y. Interpretations of "close to" 0 or 1 or -1 are vague, so we use the following very specific decision criterion:

> If the absolute value of the computed value of r exceeds the value in Table A-6, conclude that there is a significant linear correlation. Otherwise, there is not sufficient evidence to support the conclusion of a significant linear correlation.

When there really is no actual linear correlation between x and y, Table A-6 lists values that are "critical" in this sense: They separate *usual* values of r from those that are *unusual*. For example, Table A-6 shows us that with $n = 13$ pairs of sample data, the critical values are 0.553 (for $\alpha = 0.05$) and 0.684 (for $\alpha = 0.01$). Critical values and the role of α are carefully described in Chapters 6 and 7. Here's how we interpret those numbers: With 13 pairs of data and no actual linear correlation between x and y, there is a 5% chance that the absolute value of the computed linear correlation coefficient r will exceed 0.553. With $n = 13$ and no actual linear correlation, there is a 1% chance that $|r|$ will exceed 0.684.

EXAMPLE

Given the sample data in Table 11-1 for which $r = 0.623$, refer to Table A-6 to determine whether there is a significant linear correlation between the recorded latitudes and magnitudes of seismic activity. In Table A-6, use the critical value for $\alpha = 0.05$. (With $\alpha = 0.05$, we will conclude that there is a significant linear correlation only if the sample is unlikely in the sense that such a value of r occurs less than 5% of the time.)

SOLUTION

Referring to Table A-6, we locate the row for which $n = 13$ (because there are 13 pairs of data). That row contains the critical values of 0.553 (for $\alpha = 0.05$) and 0.684 (for $\alpha = 0.01$). Using the critical value for $\alpha = 0.05$, we see that there is less than a 5% chance that with no linear correlation, the absolute value of the computed r will exceed 0.553. Because $r = 0.623$, its absolute value does exceed 0.553, so we conclude that there is a significant linear correlation between recorded latitudes and magnitudes of seismic activity.

We have already noted that the format of Formula 11-1 requires that the calculated value of r always falls between -1 and $+1$ inclusive. We list that property along with other important properties.

PROPERTIES OF THE LINEAR CORRELATION COEFFICIENT r

1. *The value of r is always between -1 and 1. That is,*
$$-1 \leq r \leq 1$$
2. *The value of r does not change if all values of either variable are converted to a different scale. For example, if the latitudes in Table 11-1 are given in minutes instead of degrees, the value of r will not change.*
3. *The value of r is not affected by the choice of x or y. Interchange all x and y values and the value of r will not change.*
4. *r measures the strength of a linear relationship. It is not designed to measure the strength of a relationship that is not linear.*

Common Errors Involving Correlation

We now identify three of the most common errors made in interpreting results involving correlation.

1. *A common source of error involves concluding that correlation implies causality.* One study showed a correlation between the salaries of statistics

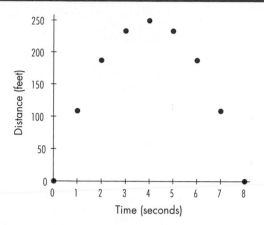

professors and per capita beer consumption, but those two variables are affected by the state of the economy, a third variable lurking in the background. (A **lurking variable** is formally defined as one that affects the variables being studied, but is not included in the study.)

2. *Another source of error arises with data based on rates or averages.* Averages suppress the variation among the individuals or items, and this may lead to an inflated correlation coefficient. One study produced a 0.4 linear correlation coefficient for paired data relating income and education among individuals, but the linear correlation coefficient became 0.7 when regional averages were used.

3. *A third source of error involves the property of linearity.* The conclusion that there is no significant linear correlation does not mean that x and y are not related in any way. The data depicted in Figure 11-2 result in a value of $r = 0$, which is an indication of no *linear* correlation between the two variables. However, we can easily see from Figure 11-2 that there is a pattern reflecting a very strong *nonlinear* relationship. (Figure 11-2 is a scatter diagram that depicts the relationship between distance above ground and time elapsed for an object thrown upward.)

Formal Hypothesis Test (requires coverage of Chapter 7)

We present two methods (summarized in Figure 11-3) for using a formal hypothesis test to determine whether there is a significant linear correlation between two variables. Some instructors prefer Method 1 because it reinforces concepts introduced in earlier chapters. Others prefer Method 2 because it involves easier calculations.

Figure 11-3 shows that the null and alternative hypotheses will be expressed as follows:

$$H_0: \rho = 0 \qquad \text{(no significant linear correlation)}$$
$$H_1: \rho \neq 0 \qquad \text{(significant linear correlation)}$$

Figure 11-3
Testing for a Linear Correlation

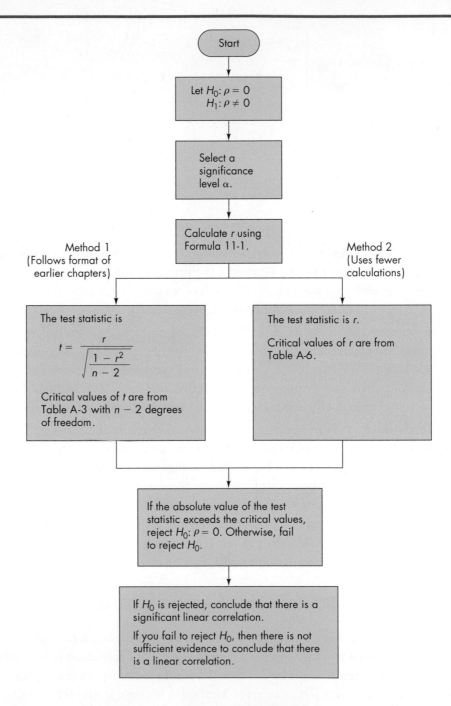

For the test statistic, we use one of the following methods.

METHOD 1: TEST STATISTIC IS t This method follows the format presented in earlier chapters. It uses the Student t distribution with a test statistic having the form $t = (r - \mu_r)/s_r$, where μ_r and s_r denote the claimed value of the mean and the sample standard deviation of r values. Because we assume that $\rho = 0$, it follows that

$\mu_r = 0$. Also, it can be shown that s_r, the standard deviation of linear correlation coefficients, can be expressed as $\sqrt{(1 - r^2)/(n - 2)}$. We can therefore use the following test statistic.

TEST STATISTIC t FOR LINEAR CORRELATION

$$t = \frac{r}{\sqrt{\dfrac{1 - r^2}{n - 2}}}$$

Critical values: Use Table A-3 with degrees of freedom $= n - 2$.

METHOD 2: TEST STATISTIC IS r This method requires fewer calculations. Instead of calculating the test statistic just given, we use the computed value of r as the test statistic. Critical values are found in Table A-6.

TEST STATISTIC r FOR LINEAR CORRELATION

Test statistic: r

Critical values: Refer to Table A-6.

Figure 11-3 shows that the decision criterion is to reject the null hypothesis of $\rho = 0$ if the absolute value of the test statistic exceeds the critical values; rejection of $\rho = 0$ means that there is sufficient evidence to support a claim of a linear correlation between the two variables. If the absolute value of the test statistic does not exceed the critical values, then we fail to reject $\rho = 0$; that is, there is not sufficient evidence to conclude that there is a linear correlation between the two variables.

EXAMPLE

Using the sample data in Table 11-1, test the claim that there is a linear correlation between the recorded latitudes and magnitudes of seismic activity. For the test statistic, use both (a) Method 1 and (b) Method 2.

SOLUTION

Refer to Figure 11-3. To claim that there is a significant linear correlation is to claim that the population linear correlation ρ is different from 0. We therefore have the following hypotheses:

$H_0: \rho = 0$ (no significant linear correlation)
$H_1: \rho \neq 0$ (significant linear correlation)

No significance level α was specified, so use $\alpha = 0.05$.

In a preceding example we already found that $r = 0.623$. With that value, we now find the test statistic and critical value, using each of the two methods just described.

Figure 11-4 Testing H_0: $\rho = 0$ with Method 1

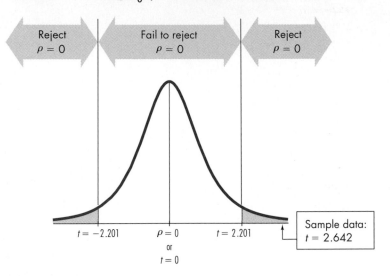

Reject $\rho = 0$ Fail to reject $\rho = 0$ Reject $\rho = 0$

$t = -2.201$ $\rho = 0$ or $t = 0$ $t = 2.201$

Sample data: $t = 2.642$

Figure 11-5 Testing H_0: $\rho = 0$ with Method 2

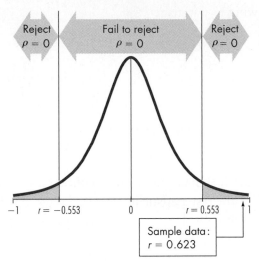

Reject $\rho = 0$ Fail to reject $\rho = 0$ Reject $\rho = 0$

-1 $r = -0.553$ 0 $r = 0.553$ 1

Sample data: $r = 0.623$

a. *Method 1:* The test statistic is

$$t = \frac{r}{\sqrt{\dfrac{1 - r^2}{n - 2}}} = \frac{0.623}{\sqrt{\dfrac{1 - 0.623^2}{13 - 2}}} = 2.642$$

The critical values of $t = -2.201$ and $t = 2.201$ are found in Table A-3, where 2.201 corresponds to 0.05 divided between two tails (with 0.025 in each tail) and the number of degrees of freedom is $n - 2 = 11$. See Figure 11-4 for the graph that includes the test statistic and critical values.

b. *Method 2:* The test statistic is $r = 0.623$. The critical values of $r = -0.553$ and $r = 0.553$ are found in Table A-6 with $n = 13$ and $\alpha = 0.05$. See Figure 11-5 for a graph that includes this test statistic and critical values.

Using either of the two methods, we find that the absolute value of the test statistic does exceed the critical value (Method 1: 2.642 > 2.201; Method 2: 0.623 > 0.553); that is, the test statistic does fall within the critical region. We therefore reject H_0: $\rho = 0$.

Interpretation
There is sufficient evidence to support the claim of a linear correlation between the recorded latitudes and magnitudes of seismic activity. Seismic activity detected and recorded at higher latitudes tends to have a higher magnitude than activity at lower latitudes.

One-tailed tests: The preceding example and Figures 11-4 and 11-5 illustrate a two-tailed hypothesis test. The examples and exercises in this section will generally involve only two-tailed tests, but one-tailed tests can occur with a claim of a positive linear correlation or a claim of a negative linear correlation. In such cases, the hypotheses will be as shown here.

Claim of Negative Correlation (Left-tailed test)	Claim of Positive Correlation (Right-tailed test)
$H_0: \rho \geq 0$	$H_0: \rho \leq 0$
$H_0: \rho < 0$	$H_0: \rho > 0$

For these one-tailed tests, Method 1 can be handled as in earlier chapters. For Method 2, either calculate the critical value as described in Exercise 25 or modify Table A-6 by replacing the column headings of $\alpha = 0.05$ and $\alpha = 0.01$ by the one-sided critical values of $\alpha = 0.025$ and $\alpha = 0.005$, respectively.

Explanation for formula for calculating r:

We have presented Formula 11-1 for calculating r and have illustrated its use; we will now give a justification for it. Formula 11-1 simplifies the calculations used in this equivalent formula:

$$r = \frac{\Sigma(x - \overline{x})(y - \overline{y})}{(n - 1)s_x s_y}$$

We will temporarily use this latter version of Formula 11-1 because its form relates more directly to the underlying theory. We will consider the following paired data, which are depicted in the scatter diagram shown in Figure 11-6:

x	1	1	2	4	7
y	4	5	8	15	23

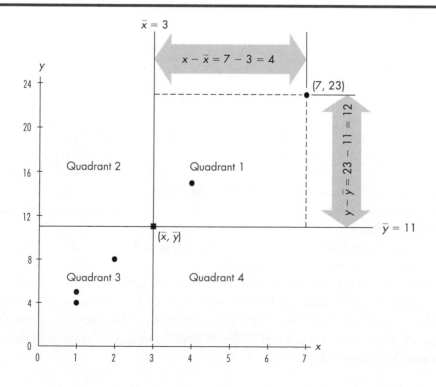

Figure 11-6
Scatter Diagram
Partitioned into Quadrants

Figure 11-6 includes the point $(\bar{x}, \bar{y}) = (3, 11)$, which is called the centroid of the sample points.

DEFINITION

Given a collection of paired (x, y) data, the point $(\bar{x}, \bar{y})$ is called the **centroid**.

The statistic r, sometimes called the Pearson product moment, was first developed by Karl Pearson. It is based on the product of the moments $(x - \bar{x})$ and $(y - \bar{y})$; that is, Pearson based his measure of scattering on the statistic $\Sigma(x - \bar{x})(y - \bar{y})$. In any scatter diagram, vertical and horizontal lines through the centroid $(\bar{x}, \bar{y})$ divide the diagram into four quadrants, as in Figure 11-6. If the points of the scatter diagram tend to approximate an uphill line (as in the figure), individual values of the product $(x - \bar{x})(y - \bar{y})$ tend to be positive because most of the points are found in the first and third quadrants, where the products of $(x - \bar{x})$ and $(y - \bar{y})$ are positive. If the points of the scatter diagram approximate a downhill line, most of the points are in the second and fourth quadrants, where $(x - \bar{x})$ and $(y - \bar{y})$ are opposite in sign, so $\Sigma(x - \bar{x})(y - \bar{y})$ is negative. Points that follow no linear pattern tend to be scattered among the four quadrants, so the value of $\Sigma(x - \bar{x})(y - \bar{y})$ tends to be close to 0.

The sum $\Sigma(x - \bar{x})(y - \bar{y})$ depends on the magnitude of the numbers used. For example, if you change x from inches to feet, that sum will change. To make r independent of the particular scale used, we include the sample standard deviations as follows:

$$r = \frac{\Sigma(x - \bar{x})(y - \bar{y})}{(n - 1)s_x s_y}$$

This expression can be algebraically manipulated into the equivalent form of Formula 11-1.

Preceding chapters discussed methods of inferential statistics by addressing methods of hypothesis testing, as well as methods for constructing confidence interval estimates. A similar procedure may be used to find confidence intervals for ρ. However, because the construction of such confidence intervals involves somewhat complicated transformations, that process is presented in Exercise 26 (Beyond the Basics).

We can use the linear correlation coefficient to determine whether there is a linear relationship between two variables. Using the data in Table 11-1, we have concluded that there is a linear correlation between the recorded latitudes and magnitudes of earthquakes. Having concluded that a relationship exists, we would like to determine what that relationship is so that we can calculate the weight of a bear when we know only its length. This next stage of analysis is addressed in Section 11–3.

EXCEL: Excel has a function that calculates the value of the linear correlation coefficient. First enter the paired sample data in two adjacent columns (such as A1:A13 for x and B1:B13 for y).

EXCEL (Prior to 2007): Click on the *fx* function key located on the main menu bar. Select the function category **Statistical** and the function name **CORREL**, then click **OK**. In the dialog box, enter the cell range of values for x, and the cell range of values for y.

To obtain a scatterplot, click on the Chart Wizard icon, and then select the chart type identified as **XY(Scatter)**. In the dialog box, enter the input range of the data, such as A1:B13. Click **Next** and proceed to use the dialog boxes to modify the graph as desired.

EXCEL 2007: Click on **Menus** on the main menu, then the *fx* button, then more functions and select the function category **Statistical** and then the function name **CORREL**. Proceed as above.

MINITAB 15: From the **Stat** menu, choose **Basic Statistics** and then **Correlation**. Select the columns for which you want to compute the correlation. There is an optional checkbox to display P-values. Click **OK**.

STATDISK: Select **Analysis** from the main menu bar, then use the option **Correlation and Regression**. Select the columns with the x and y data. Click on the **Evaluate** button. The display will include the value of the linear correlation coefficient along with the critical value of r, the conclusion of a hypothesis test, and other results to be discussed in later sections. A scatterplot can also be obtained by clicking on the **Plot** button.

11-2 Exercises A: Basic Skills and Concepts

In Exercises 1 and 2, assume that a sample of n pairs of data results in the given value of r. Is there is a significant linear correlation between x and y? Explain.

1. a. $n = 32, r = 0.992$
   b. $n = 50, r = -0.333$
   c. $n = 17, r = 0.456$

2. a. $n = 22, r = -0.087$
   b. $n = 40, r = 0.299$
   c. $n = 25, r = -0.401$

In Exercises 3 and 4, (a) use a scatterplot to determine whether there is a significant linear correlation between x and y, and (b) find the values of n, Σx, Σx^2, $(\Sigma x)^2$, Σxy, and the linear correlation coefficient r.

3.

x	2	3	5	5	10
y	6	9	14	16	30

4.

x	2	3	5	5	10
y	6	0	15	5	2

In Exercises 5–18,
    a. *Construct the scatterplot.*
    b. *Find the value of the linear correlation coefficient r.*
    c. *Determine whether there is a significant linear correlation between the two variables. (Use $\alpha = 0.05$.)*
    d. *Save your work because some of the same data will be used in the next section.*

5. When bears were anesthetized, researchers measured the distances (in inches) around their chests and they weighed the bears (in pounds). The results are given below for 8 male bears. Based on the results, does a bear's weight seem to be related to its chest size? Do the results change if the chest measurements are converted to feet, with each of those values divided by 12?

x Chest (in.)	26	45	54	49	41	49	44	19
y Weight (lb)	90	344	416	348	262	360	332	34

Based on data from Gary Alt and Minitab.

6. The accompanying table lists the number of Christmas trees (in thousands) exported from New Brunswick in one year and the dollar value (also in thousands) of those exports. Based on the results, can New Brunswick depend on receiving a greater dollar value for tree exports when it exports more trees? Do the results change if the tree numbers are entered as 358,000, 341,000, 364,000, and 367,000?

x Trees	358	341	364	367
y Dollars	3978	3729	4073	4564

Based on data from Statistics Canada.

7. The accompanying table lists weights (in kilograms) of plastic discarded by a sample of households, along with the sizes of the households (adapted from Data Set 1). Is there a significant linear correlation? This issue is important because the presence of a correlation implies that Statistics Canada could predict population size by analyzing discarded garbage.

Plastic	0.12	0.64	0.99	1.28	0.99	0.82	0.39	1.38
Household size	2	3	3	6	4	2	1	5

8. The paired data below consist of weights (in kilograms) of discarded paper and sizes of the households.

Paper	1.09	3.43	4.33	4.00	3.96	3.16	3.10	5.18
Household size	2	3	3	6	4	2	1	5

9. A study was conducted to compare the incubation periods and fledgling periods (in days) of different bird species. Sample data are given below for randomly selected species.
   a. Do the incubation and fledgling periods appear to be related?
   b. Each of the numbers shown in the table is itself a mean—for the incubation or fledgling periods *of an entire species*. Suppose the data gave the incubation and fledgling periods of randomly selected *individual* birds. Would you expect r to be larger or smaller than you obtained in part (a)? Explain.

Incubation	23	13	53	23	30	17	78	49	15	12	18
Fledgling	35	11	360	30	53	28	280	49	17	22	28

10. An investor, considering whether to add a mining stock to his portfolio as a long-term investment, compared, for the years 1995–2000, the year's closing price for a Toronto-based mining firm and the year's closing value for the TSE 300 index, an indicator of general price trends for stocks traded on the Toronto Stock Exchange (based on data from *Management Information Circular*, January 26, 2001, Twin Mining Corporation, Toronto).
    a. Does the stock price appear to be related to the TSE 300 index?
    b. These data were collected during a long period of general increase in TSE 300 values. If the TSE 300 were to begin to drop, would it be reasonable to rely on the results found in part (a)? Why or why not?

Mining stock	0.450	0.600	0.210	0.160	0.335	0.400
TSE 300	9397.97	12061.95	13868.54	13648.84	17977.46	19309.36

11. To investigate the relationship between a person's education level and his or her salary, these variables were compared for people in selected occupations. Salaries are shown in thousands of dollars. Based on these results, does salary appear to depend on years of education?

Years of Education	Salary
16	61.10
10	26.53
11	29.57
11	28.38
13	37.62
13	27.22
12	36.70
10	14.77
12	22.23
12	24.82
14	32.89
13	36.94
15	35.99
11	23.11
14	62.56
10	21.31

Based on data from Statistics Canada.

12. A stock-market investor is seeking clues to pick the most profitable companies. The accompanying table lists, for each of 11 companies, its profit margin (as a percent) versus its ratio of current assets to current liabilities. Are higher profit margins associated with higher ratios of current assets to current liabilities? Would it be advisable for the investor to rely heavily on this indicator?

Profit margin	3.9	2.9	9.4	6.1	−0.6	3.6	2.7	1.0	2.2	8.3	3.9
Assets/liabilities	1.3	1.3	1.1	1.5	1.7	1.5	1.6	1.5	1.0	1.1	0.9

Based on data from Statistics Canada.

13. Refer to Data Set 16 in Appendix B.
    a. Use the paired data for durations and intervals after eruptions of the Old Faithful geyser. Is there a significant linear correlation, suggesting that the interval after an eruption is related to the duration of the eruption?
    b. Use the paired data for intervals after eruptions and heights of eruptions of the Old Faithful geyser. Is there a significant linear correlation, suggesting that the interval after an eruption is related to the height of the eruption?
    c. Assume you want to develop a method for predicting the time interval to the next eruption. Based on the results from parts (a) and (b), which factor would be more relevant: eruption duration or eruption height? Why?

14. Refer to Data Set 4 in Appendix B and use the paired columns for annual temperature and precipitation. Based on the result, does there appear to be a significant linear correlation between annual temperature and precipitation? If so, can researchers reduce their expenses by measuring only one of these two variables?

15. Refer to Data Set 21 in Appendix B.
    a. Use the paired data consisting of price and carat (weight). Is there a significant linear correlation between the price of a diamond and its weight in carats?
    b. Use the paired price/colour data. Assuming the colour data is recorded in an interval scale, is there a significant linear correlation between the price of a diamond and its colour?
    c. Assume that you are planning to buy a diamond engagement ring. In considering the value of a diamond, which characteristic should you consider to be more important: the carat weight or the colour? Why?

16. Refer to Date Set 4 in Appendix B and use the paired columns for annual temperature and snowfall. Based on the results, does there appear to be a significant correlation between annual temperature and snowfall? If so, can researchers reduce their expenses by measuring only one of these two variables?

17. Unlike Data Set 4, Data Set 7 in Appendix B records precipitation levels and temperatures in terms of *deviations from seasonal norms*, rather than in absolute terms. Use the columns in Data Set 7 in Appendix B for temperature departures and precipitation departures to test whether there appears to be a significant linear correlation between annual spring temperature departures and precipitation departures from seasonal norms. Is there a contradiction between the results from this exercise and the results from Exercise 14?

18. The data on the next page are extracted from the top ten entries in the *Financial Post* list of top 500 companies. The number of employees is given in thousands and the revenues for 1997 are given in billions of dollars. Is there a significant linear correlation between these companies' revenues and

their numbers of employees? Would you necessarily expect that your results for these top companies would apply to all companies on the list, or to those that did not make the list?

Employees	29	122	24	30	16	3	83	50	43	33
Revenues	34.2	33.2	27.9	17.2	16.7	14.2	13.9	12.1	11.2	10.8

In Exercises 19–22, describe the error in the stated conclusion. (See the list of common errors included in this section.)

19. *Given:* The paired sample data of the ages of subjects and their scores on a test of reasoning result in a linear correlation coefficient very close to 0.
    Conclusion: Younger people tend to get higher scores.

20. *Given:* There is a significant linear correlation between personal income and years of education.
    Conclusion: More education causes a person's income to rise.

21. *Given:* Subjects take a test of verbal skills and a test of manual dexterity, and those pairs of scores result in a linear correlation coefficient very close to 0.
    Conclusion: Scores on the two tests are not related in any way.

22. *Given:* There is a significant linear correlation between provincial average tax burdens and provincial average incomes.
    Conclusion: There is a significant linear correlation between individual tax burdens and individual incomes.

Exercises B: Beyond the Basics

23. How is the value of the linear correlation coefficient r affected in each of the following cases?
    a. Each x value is switched with the corresponding y value.
    b. Each x value is multiplied by the same nonzero constant.
    c. The same constant is added to each x value.

24. In addition to testing for a linear correlation between x and y, we can often use *transformations* of data to explore for other relationships. For example, we might replace each x value by x^2 and use the methods of this section to determine whether there is a linear correlation between y and x^2. Given the paired data in the accompanying table, construct the scatter diagram and then test for a linear correlation between y and each of the following. Which case results in the largest value of r?
    a. x b. x^2 c. $\log x$ d. $\sqrt{x}$ e. $1/x$

x	1.3	2.4	2.6	2.8	2.4	3.0	4.1
y	0.11	0.38	0.41	0.45	0.39	0.48	0.61

25. The critical values of r in Table A-6 are found by solving

$$t = \frac{r}{\sqrt{\dfrac{1 - r^2}{n - 2}}}$$

for r to get

$$r = \frac{t}{\sqrt{t^2 + n - 2}}$$

where the t value is found from Table A-3 by assuming a two-tailed case with $n - 2$ degrees of freedom. Table A-6 lists the results for selected values of n and α. Use the formula for r given here and Table A-3 (with $n - 2$ degrees of freedom) to find the critical values of r for the given cases.

a. H_0: $\rho = 0$, $n = 50$, $\alpha = 0.05$
b. H_1: $\rho \neq 0$, $n = 75$, $\alpha = 0.10$
c. H_0: $\rho \geq 0$, $n = 20$, $\alpha = 0.05$
d. H_0: $\rho \leq 0$, $n = 10$, $\alpha = 0.05$
e. H_1: $\rho > 0$, $n = 12$, $\alpha = 0.01$

26. Given n pairs of data from which the linear correlation coefficient r can be found, use the following procedure to construct a confidence interval about the population parameter ρ.

Step a. Use Table A-2 to find $z_{\alpha/2}$ that corresponds to the desired degree of confidence.

Step b. Evaluate the interval limits w_L and w_R:

$$w_L = \frac{1}{2} \ln \left(\frac{1 + r}{1 - r} \right) - z_{\alpha/2} \cdot \frac{1}{\sqrt{n - 3}}$$

$$w_R = \frac{1}{2} \ln \left(\frac{1 + r}{1 - r} \right) - z_{\alpha/2} \cdot \frac{1}{\sqrt{n - 3}}$$

Step c. Now evaluate the confidence interval limits in the expression below:

$$\frac{e^{2w_L} - 1}{e^{2w_L} + 1} < \rho < \frac{e^{2w_R} - 1}{e^{2w_R} + 1}$$

Use this procedure to construct a 95% confidence interval for ρ, given 50 pairs of data for which $r = 0.600$.

27. Refer to Table 11-1, used for the chapter problem. This table includes paired data representing the recorded latitudes and magnitudes of seismic activity (earthquakes). These data can only be recorded if an earthquake is picked up by monitors, and the great majority of these are located in Canada's lower latitudes. If it is possible that the widely spaced monitors in northern Canada are failing to detect lower-magnitude earthquakes in those latitudes, what effect could this have on our interpretation of this chapter's correlation results?

11-3 Regression

In Section 11–2 we analyzed paired data with the goal of determining whether there is a significant linear correlation between two variables. We now want to describe the relationship by finding the graph and equation of the straight line that represents the relationship. This straight line is called the *regression line*, and its equation is called the *regression equation*. Sir Francis Galton (1822–1911) studied the phenomenon of heredity and showed that when tall or short couples have children, the heights of those children tend to *regress*, or revert to a more typical mean height. We continue to use his terminology.

DEFINITIONS

Given a collection of paired sample data, the **regression equation**

$$\hat{y} = b_0 + b_1 x$$

algebraically describes the relationship between the two variables. The graph of the regression equation is called the **regression line** (or *line of best fit*, or *least-squares line*).

This definition expresses a relationship between x (called the **independent variable** or **predictor variable**) and $\hat{y}$ (called the **dependent variable** or **response variable**). In the preceding definition, the typical equation of a straight line ($y = mx + b$) is expressed in the format of $\hat{y} = b_0 + b_1 x$, where b_0 is the y-intercept and b_1 is the slope. The following notation box shows that b_0 and b_1 are sample statistics used to estimate the population parameters β_0 and β_1.

NOTATION FOR REGRESSION EQUATION

	Population Parameter	Sample Statistic
y-intercept of regression equation	β_0	b_0
Slope of regression equation	β_1	b_1
Equation of the regression line	$y = \beta_0 + \beta_1 x$	$\hat{y} = b_0 + b_1 x$

For the regression methods given in this section, we make these assumptions:

Assumptions

1. We are investigating only *linear* relationships.

2. For each x value, y is a random variable having a normal (bell-shaped) distribution. All of these y distributions have the same variance. Also, for a given value of x, the distribution of y values has a mean that lies on the regression line. (Results are not seriously affected if departures from normal distributions and equal variances are not too extreme.)

An important goal of this section is to use sample paired data to estimate the regression equation. Using only sample data, we can't find the exact values of the population parameters β_0 and β_1, but we can use the sample data to estimate them with b_0 and b_1, which are found by using Formulas 11-2 and 11-3:

Formula 11-2

$$b_0 = \frac{(\Sigma y)(\Sigma x^2) - (\Sigma x)(\Sigma xy)}{n(\Sigma x^2) - (\Sigma x)^2} \qquad \text{(y-intercept)}$$

Formula 11-3

$$b_1 = \frac{n(\Sigma xy) - (\Sigma x)(\Sigma y)}{n(\Sigma x^2) - (\Sigma x)^2} \qquad \text{(slope)}$$

These formulas might look intimidating, but they are programmed into many calculators and computer programs, so the values of b_0 and b_1 can be easily found (see Using Technology at the end of this section). In those cases when we must use formulas instead of a calculator or computer, the required computations will be much easier if we keep the following facts in mind:

1. If the linear correlation coefficient r has been computed using Formula 11-1, the values of Σx, Σy, Σx^2, and Σxy have already been found and they can be used again in Formula 11-3. (Also, the numerator for r in Formula 11-1 is the same numerator for b_1 in Formula 11-3; the denominator for r includes the denominator for b_1. If the calculation for r has been set up carefully, the calculation for b_1 requires the simple division of one known number by another.)

2. If you find the slope b_1 first, you can use Formula 11-4 to find the y-intercept b_0. [The regression line always passes through the centroid $(\overline{x}, \overline{y})$ so that $\overline{y} = b_0 + b_1\overline{x}$ must be true, and this equation can be expressed as Formula 11-4. It is usually easier to find the y-intercept b_0 by using Formula 11-4 than by using Formula 11-2.]

Formula 11-4 $$b_0 = \overline{y} - b_1\overline{x}$$

Once we have evaluated b_0 and b_1, we can identify the estimated regression equation, which has the following special property: *The regression line fits the sample points best.* (The specific criterion used to determine which line fits "best" is the least-squares property, which will be described later.) We will now briefly discuss rounding and then illustrate the procedure for finding and applying the regression equation.

Rounding the *y*-Intercept b_0 and the Slope b_1

It's difficult to provide a simple universal rule for rounding values of b_0 and b_1, but we usually try to round each of these values to *three significant digits* or use the values provided by a statistical software package or calculator. Because these values are very sensitive to rounding at intermediate steps of calculations, you should try to carry at least six significant digits (or use exact values) in the intermediate steps. Depending on how you round, this book's answers to examples and exercises may be slightly different from your answers.

EXAMPLE

In Section 11–2 we used the Table 11-1 data (x = latitudes of seismic activity; y = magnitudes of seismic activity) to find that the linear correlation coefficient of $r = 0.623$ indicates that there is a significant linear correlation. Now find the regression equation of the straight line that relates x and y.

SOLUTION

We will find the regression equation by using Formulas 11-3 and 11-4 and these values already found in Table 11-2 from Section 11–2:

$$n = 13 \qquad\qquad \Sigma x = 742.3 \qquad\qquad \Sigma y = 35.1$$

$$\Sigma x^2 = 43{,}344.79 \qquad\qquad \Sigma y^2 = 119.87 \qquad\qquad \Sigma xy = 2100.84$$

First find the slope b_1 by using Formula 11-3:

$$b_1 = \frac{n(\Sigma xy) - (\Sigma x)(\Sigma y)}{n(\Sigma x^2) - (\Sigma x)^2}$$

$$= \frac{13(2100.84) - (742.3)(35.1)}{13(43{,}344.79) - (742.3)^2} = \frac{1256.19}{12{,}472.98} = 0.100713$$

$$= 0.101 \quad \text{(rounded)}$$

Next, find the y-intercept b_0 by using Formula 11-4:

$$b_0 = \bar{y} - b_1\bar{x}$$

$$= \frac{35.1}{13} - (0.100713)\frac{742.3}{13} = -3.05 \quad \text{(rounded)}$$

Knowing the slope b_1 and y-intercept b_0, we can now express the estimated equation of the regression line as

$$\hat{y} = -3.05 + 0.101x$$

We should realize that this equation is an *estimate* of the true regression equation $y = \beta_0 + \beta_1 x$. This estimate is based on one particular set of sample data listed in Table 11-1, but another sample drawn from the same population would probably lead to a somewhat different equation.

The Excel display below shows the regression line plotted on the scatter diagram. We can see that this line fits the data reasonably well. (The data points tend to be distributed uniformly

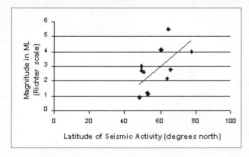

EXCEL DISPLAY

Scatter Diagram with
Regression Line

above and below the line.) Note that even though the regression line "fits the data," it does not actually pass through any of the original data points. This is common, although in a minority of cases, the regression line might pass through one or more of the sample data points.

Using the Regression Equation for Predictions

Regression equations can be helpful when used in *predicting* the value of one variable, given some particular value for the other variable. If the regression line fits the data quite well, then it makes sense to use its equation for predictions, provided that we don't go beyond the scope of the available scores. However, *we should use the equation of the regression line only if r indicates that there is a significant linear correlation. In the absence of a significant linear correlation, we should not use the regression equation for projecting or predicting; instead, our best estimate of the second variable is simply its sample mean.*

In predicting a value of y based on some given value of x:

1. If there is *not* a significant linear correlation, the best predicted y value is $\bar{y}$.

2. If there *is* a significant linear correlation, the best predicted y value is found by substituting the x value into the regression equation.

Figure 11-7 summarizes this process, which is easier to understand if we think of r as a measure of how well the regression line fits the sample data.

Figure 11-7
Predicting the Value of a Variable

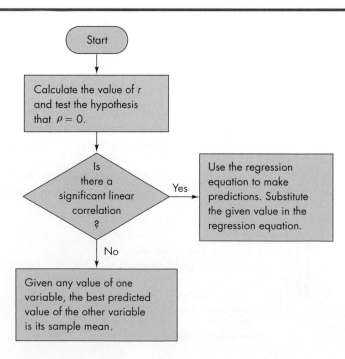

If r is near -1 or $+1$, then the regression line fits the data well, but if r is near 0, then the regression line fits poorly (and should not be used for predictions).

EXAMPLE

Using the sample data in Table 11-1, we found that there is a significant linear correlation between latitudes of earthquakes and their magnitudes, and we also found that the regression equation is $\hat{y} = -3.05 + 0.101x$. If an earthquake is recorded at latitude 64°N, predict its magnitude.

SOLUTION

There's a strong temptation to jump in and substitute 64.0 for x in the regression equation, but we should first consider whether there is a significant linear correlation that justifies the use of that equation. In this example, we do have a significant linear correlation (with $r = 0.623$), so our **predicted value**—the value of a dependent variable found by using values of independent variables in a regression equation of x based on the model—is found as follows:

$$\hat{y} = -3.05 + 0.101x$$
$$= -3.05 + 0.101(64.0) = 3.41$$

The predicted magnitude of an earthquake recorded at 64°N is 3.41 ML. (If there had not been a significant linear correlation, our best predicted magnitude would have been $\bar{y} = 2.7$ ML.)

EXAMPLE

The Excel scatter diagram on the next page compares, for each of the first 35 Parliaments, the number of days it was in session versus its approximate duration in years. Using Data Set 5 in Appendix B, find the best predicted number of days in session for a parliament that lasts 4.5 years.

SOLUTION

The diagram clearly shows that Assumption 2 for regressions (stated near the beginning of Section 11-3) does not apply to the relation between Parliament duration and number of days in session. There is *not* a consistent variance of y values around all parts of the trend line. For Parliaments lasting up to 3 years, it may be reasonable to find a regression equation, but the equation would not apply to Parliaments of at least 4 years' duration. Since there is no significant linear correlation between duration and days in session for Parliaments lasting 4 or more years, the mean number of days in session for these longer Parliaments would be a better predictor. That mean is $\Sigma(\text{days})/n = 21,619/25 = 864.8$.

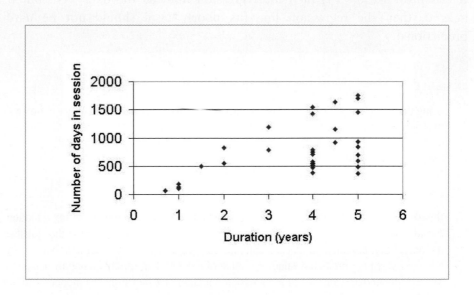

Carefully compare the solutions to the preceding two examples and note that we used the regression equation when there was a significant linear correlation, but in the absence of such a correlation, the best predicted value of *y* is simply the value of the sample mean $\bar{y}$.

Interpreting Predicted Values   We should understand that when we determine a predicted value of *y* based on some given value of *x*, the predicted value will not necessarily be the exact result. There are other factors, such as random variation and characteristics not included in the study, that can also affect *y*. Based on the sample data in Table 11-1, we see that for an earthquake recorded at latitude 64°N, the best predicted magnitude (where the ML scale applies) is 3.41 ML. This does not mean that every recorded earthquake or other seismic activity at latitude 64°N will have a magnitude of exactly 3.41 ML; some will be less and some will be more. But the regression equation tells us that, on the whole, 3.41 ML is the single best estimate of magnitude whenever seismic activity is recorded at latitude 64°N. A common error is to use the regression equation when there is no significant linear correlation. That error violates the first of the following guidelines.

Guidelines for Using the Regression Equation

1. *If there is no significant linear correlation, don't use the regression equation to make predictions.*

2. *When using the regression equation for predictions, stay within the scope of the available sample data.* If you find a regression equation that relates women's heights and shoe sizes, it's absurd to predict the shoe size of a woman who is 10 ft tall.

3. *A regression equation based on old data is not necessarily valid now.* The regression equation relating used car prices and ages of cars is no longer usable if it's based on data from the 1970s.

4. *Don't make predictions about a population that is different from the population from which the sample data were drawn.* If we collect sample data from men and develop a regression equation relating age and TV remote control usage, the results don't necessarily apply to women.

Similarly, the regression for the chapter problem is based on data for which the Magnitude Local (ML) scale is appropriate. Magnitudes for events recorded by the Magnitude Nuttli (MN) scale (which involves different assumptions and calculations) cannot be predicted by the same regression equation.

Interpreting the Regression Equation: Marginal Change

We can use the regression equation to see the effect on one variable when the other variable changes by some specific amount.

DEFINITION

In working with two variables related by a regression equation, the **marginal change** in a variable is the amount that it changes when the other variable changes by exactly one unit.

The slope b_1 *in the regression equation represents the marginal change resulting when x changes by one unit.* For the earthquake data of Table 11-1, we can see that an increase in x of one unit will cause $\hat{y}$ to change by 0.101 units. That is, if an earthquake is recorded at 1° more north, its predicted magnitude increases by 0.101 ML.

Outliers and Influential Points

A correlation/regression analysis of bivariate data should include an investigation of *outliers* and *influential points*, defined as follows.

DEFINITIONS

In a scatterplot, an **outlier** is a point lying far away from the other data points. Paired sample data may include one or more **influential points,** which are points that strongly affect the graph of the regression line.

An outlier is easy to identify: Examine the scatterplot and identify a point that is far away from the others. To determine whether a point is an influential point, graph the regression line resulting from the data with the point included, then graph the regression line resulting from the data with the point excluded. If the graph changes by a considerable amount, the point is influential. Influential points are often found by identifying those outliers that are *horizontally* far away from the other points.

For example, refer to the Excel display on page 621. If we include an additional earthquake at latitude 77°N that has a 9.0 ML magnitude, we have an outlier, because that point would fall in the upper-right corner of the graph and would be far away from the other points. This point would not be an influential point because the regression line would not change its position very much. If Canada annexed an island at 20°N, which then registered a 9.0 ML earthquake, it would become an influential point because the graph of the regression line would change considerably, as shown by the following Excel display.

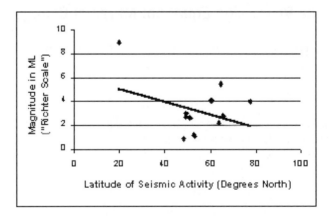

Residuals and the Least-Squares Property

We have stated that the regression equation represents the straight line that fits the data "best," and we will now describe the criterion used in determining the line that is better than all others. This criterion is based on the vertical distances between the original data points and the regression line. Such distances are called *residuals*.

DEFINITION

For a sample of paired (x, y) data, a **residual** is the difference $(y - \hat{y})$ between an observed sample y value and the value of $\hat{y}$, which is the value of y that is predicted from the regression equation.

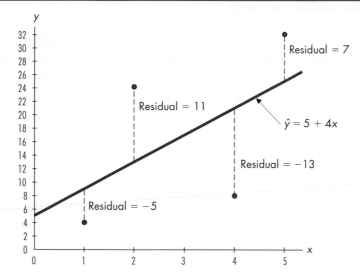

Figure 11-8
**Scatter Diagram with
Regression Line and
Residuals**

This definition might seem as clear as tax form instructions, but you can easily understand residuals by referring to Figure 11-8, which corresponds to the paired sample data in the margin. In Figure 11-8, the residuals are represented by the dashed lines. For a specific example, see the residual indicated as 7, which is directly above $x = 5$. If we substitute $x = 5$ into the regression equation $\hat{y} = 5 + 4x$, we get a predicted value of $\hat{y} = 25$. When $x = 5$, the *predicted* value of y is $\hat{y} = 25$, but the actual *observed* sample value is $y = 32$. The difference of $y - \hat{y} = 32 - 25 = 7$ is a residual.

x	1	2	4	5
y	4	24	8	32

The regression equation represents the line that fits the points "best" according to the *least-squares property*.

DEFINITION

A straight line satisfies the **least-squares property** if the sum of the squares of the residuals is the smallest sum possible.

From Figure 11-8, we see that the residuals are -5, 11, -13, and 7, so the sum of their squares is

$$(-5)^2 + 11^2 + (-13)^2 + 7^2 = 364$$

Any straight line different from $\hat{y} = 5 + 4x$ will result in residuals with a sum of squares that is greater than 364.

Fortunately, we need not deal directly with the least-squares property when we want to find the equation of the regression line. Calculus has been used to build the least-squares property into Formulas 11-2 and 11-3. Because the derivations of these formulas require calculus, we don't include them in this text.

Other Mathematical Models

In many cases, there is a relationship between two variables that is not linear. A **mathematical model** is similar to a regression, in that it is a mathematical function that "fits" or describes real-world data, and which can be used to predict the value of a dependent variable, given values of one or more independent variables.

For example, inspection of the table shown here shows that each y value is the square of the corresponding x value, so that the two variables are related by the equation $y = x^2$ instead of a linear equation, which has the form $y = b_0 + b_1 x$.

x	2	5	4	8	10
y	4	25	16	64	100

With specialized statistical software and calculators, it is possible to find directly the best curve-fitting equation for data, from a variety of complex options. Excel and STATDISK can be coaxed to assist in this process (see Using Technology), but the process is not so automatic.

The following are some common structures of equations that can be the basis of mathematical models:

Linear: $\quad y = a + bx$

Quadratic: $\quad y = ax^2 + bx + c$

Logarithmic: $\quad y = a + b \ln x$

Exponential: $\quad y = ab^x$

Power: $\quad y = ax^b$

Logistic: $\quad y = \dfrac{c}{1 + ae^{-bx}}$

These common models are illustrated on the following page. The particular model that you select depends on the nature of the sample data; a scatterplot can be very helpful in making that determination.

If you have the tools to assist you, there are three basic rules for developing a good mathematical model:

1. *Look for a pattern in the graph.* Examine the graph of the plotted points and compare the basic pattern to the known generic graphs illustrated on the next page. When trying to select a model, consider only those functions that visually appear to fit the observed points reasonably well.

2. *Find and compare values of R^2.* If your software can generate values of R^2, these can be interpreted as indicating degree of fit, similar to regression; select functions that result in larger values of R^2. However, don't place much importance on small differences, such as the difference between $R^2 = 0.984$ and $R^2 = 0.989$. (Another measurement used to assess the quality of a model is the sum of squares of the residuals.)

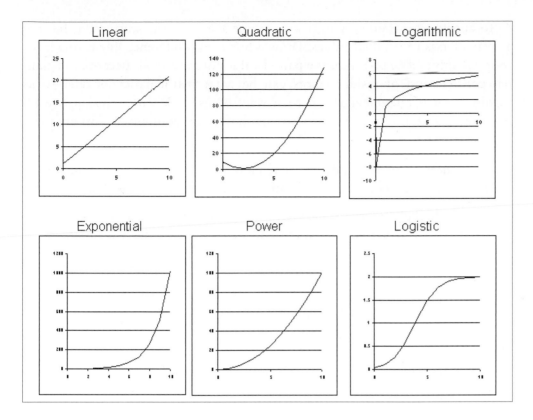

3. *Think*. Use common sense. Don't use a model that leads to predicted values known to be totally unrealistic. Use the model to calculate future values, past values, and values for missing years, then determine whether the results are realistic.

Residual Plots

In this section and the preceding section we listed simplified requirements for the effective analyses of correlation and regression results. We noted that we should always begin with a scatterplot, and we should verify that the pattern of points is approximately a straight-line pattern. We should also consider outliers. A *residual plot* can be another helpful device for analyzing correlation and regression results and for checking the requirements necessary for making inferences about correlation and regression.

DEFINITION

A **residual plot** is a scatterplot of the (x, y) values after each of the y-coordinate values has been replaced by the residual value $y - \hat{y}$ (where $\hat{y}$ denotes the predicted value of y). That is, a residual plot is a graph of the points $(x, y - \hat{y})$.

To construct a residual plot, use the same x-axis as the scatterplot, but use a vertical axis of residual values. Draw a horizontal reference line through the residual value of 0, then plot the paired values of $(x, y - \hat{y})$. Because the manual construction of residual plots can be what mathematicians refer to as "tedious," it is recommended that software be used. When analyzing a residual plot, look for a pattern in the way the points are configured, and use these criteria:

If a residual plot does not reveal any pattern, the regression equation is a good representation of the association between the two variables.

If a residual plot reveals some systematic pattern, the regression equation is not a good representation of the association between the two variables.

EXAMPLE

Construct a residual plot for the chapter problem.

SOLUTION

Here is the plot of the residuals vs latitude constructed in Excel:

EXCEL DISPLAY
Residuals vs Latitude

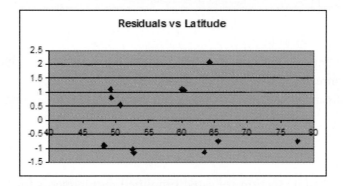

There is no systematic pattern in the residuals as the latitude values increase. This indicates that the linear regression model is adequate.

EXAMPLE

The density of wood is being used to predict its stiffness. A random sample of 11 readings is taken:

Density	21.7	15.2	23.4	15.4	14.5	16.7	15	25.6	15	24.4	7
Stiffness	47,661	28,028	104,170	25,312	22,148	49,499	25,319	96,305	26,222	72,594	5304

Construct a residual plot of this data.

SOLUTION

We begin with a scatterplot of the data:

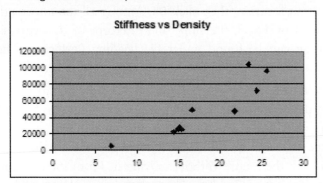

EXCEL DISPLAY
Stiffness vs Density

There is a definite non-linear trend to the data. This should be reflected in the the the plot of the residuals vs density constructed in Excel:

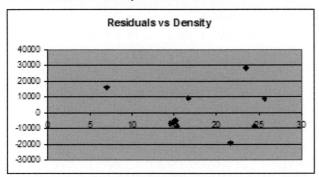

EXCEL DISPLAY
Residuals vs Density

There is a fan-shaped pattern in the residuals as the density values increase. This indicates that the linear regression model is not adequate and that a different mathematical model should be constructed. In the next section, we present a simple procedure to help determine what type of model should be built.

Another common residual plot has the $\hat{y}$ values along the x-axis, with the residuals still along the y-axis. If there is only one independent variable, there is no advantage to this particular plot; it becomes useful when there are two or more independent variables.

It should be mentioned that there are times when analyzing residual plots is not unlike reading tea leaves; discerning whether there is a pattern or not can be at the discretion of the researcher. If in doubt, experiment with various models and use the methods presented in the next section.

USING TECHNOLOGY

Because of the messy calculations involved, the linear correlation coefficient r and the slope and y-intercept of the regression line are usually found by using a calculator or computer software.

EXCEL (Prior to 2007):   *Regression:* Use Excel's Data Analysis add-in by selecting **Tools** from the main menu, then selecting **Data Analysis** and **Regression**, then clicking **OK.** In the spaces provided, enter Excel ranges for the columns containing the y and x values, respectively. There is an optional checkbox for residual plots. Specify an output location, then click **OK**. Among all the information provided by Excel, the slope and intercept of the regression equation can be found under the heading **Coefficient**. The r value is found to the right of the heading **Multiple R**.

Trend Line: Once you have generated a scatter diagram for the x and y data (see Using Technology, Section 11–2), you can easily insert a trend line: Aim the mouse pointer at one of the plotted points, and left-click the mouse to highlight all the points. Then right-click the mouse without moving it. From the menu that appears, select **Add Trendline**. Then select the type **Linear**, and click on **OK**. (There are also some nonlinear options you can explore.)

Mathematical Model: With patience, built-in Excel functions can be used to transform data, so the results can be analyzed by the linear regression techniques of this chapter. For instance, if your model is that $y = x^2$, then the square root of each y should approximately equal its corresponding x value. Thus, use a conventional regression analysis to compare the transformed y values (that is, after taking the square root) with their corresponding x's. (See Exercise 25 in this section and Exercise 17 in Section 11–5 for more detailed examples.)

EXCEL 2007:

Click **Data** on the main menu, then **Data Analysis** from the analysis tab. Proceed as above.

MINITAB 15:

Regression: From the **Stat** menu, choose **Regression** and then **Regression** again. In the dialog box, choose the column with the dependent variable and the column with the independent (**Predictor**) variable. To generate residual plots, click the **Graphs** button, choose the **Regular** radio button under the **Residuals for Plots** section, and then check the box for the type of residual plot you want. Finally, select the variable (either x or y) that you wish to plot the residuals against.

Mathematical Model: Use the Minitab calculator to transform the required variable. For example, suppose the model is $y = x^2$. If the y variable is in C1 and the x variable is in C2, to generate the square root of y choose **Calculator** from the **Calc** menu, choose the square root function, select C1 as the argument, and put the results in C3. You can then proceed to do regression with C3 as the dependent variable and C2 as the independent variable.

STATDISK: Select **Analysis** from the main menu bar, then use the option **Correlation and Regression**. Select the columns with the x and y data. Enter a value for the significance level. Click on the **Evaluate** button. The display will include the value of the linear correlation coefficient along with the critical value of r, the conclusion about correlation, and the intercept and slope of the regression equation, as well as some other results. Click on **Plot** to get a graph of the scatterplot with the regression line included.

STATDISK can also assist in testing mathematical models. Select **Data** from the main menu bar, then select the option of **Sample Transformations**, in order to transform columns of data mathematically. As described for Excel, transform the data in such a way that the results can be tested by regression.

11-3 Exercises A: Basic Skills and Concepts

In Exercises 1–4, use the given data to find the equation of the regression line.

1.

x	1	2	4	5
y	3	5	9	11

2.

x	5	3	2	1	0	2
y	-2	0	1	2	3	1

3.

x	2	3	5	5	10
y	6	9	14	16	30

4.

x	2	3	5	5	10
y	6	0	15	5	2

In Exercises 5–18, find the regression equation; unless the problem suggests otherwise, let the first variable be the independent (x) variable. Caution: When finding predicted values, be sure to follow the prediction procedure described in this section.

5.

x Chest (in.)	26	45	54	49	41	49	44	19
y Weight (lb)	90	344	416	348	262	360	332	34

Find the best predicted weight of a bear with a chest size of 52 in.

6.

Trees exported (1000s)	358	341	364	367
Export value ($1000s)	3978	3729	4073	4564

Find the best predicted export value when 362,000 Christmas trees are exported.

7.

Plastic (kg)	0.12	0.64	0.99	1.28	0.99	0.82	0.39	1.38
Household size	2	3	3	6	4	2	1	5

What is the best predicted size of a household that discards 0.25 kg of plastic?

8.

Paper (kg)	1.09	3.43	4.33	4.00	3.96	3.16	3.10	5.18
Household size	2	3	3	6	4	2	1	5

What is the best predicted size of a household that discards 5.00 kg of paper?

9.

Incubation period (days)	23	13	53	23	30	17	78	49	15	12	18
Fledgling period (days)	35	11	360	30	53	28	280	49	17	22	28

Data from www.swishweb.com/Animal_Kingdom/Birds/animal03b.htm.

What is the best predicted fledgling period for a bird species whose incubation period is 36 days?

10.

TSE 300	9397.97	12,061.95	13,868.54	13,648.84	17,977.46	19,309.36
Mining stock	0.450	0.600	0.210	0.160	0.335	0.400

Data from *Management Information Circular*, January 26, 2001, TWIN Mining Corporation, Toronto.

What is the best predicted value for this mining stock, if the TSE 300 closes at 15,445?

11. The salaries in the accompanying table are shown in thousands of dollars. What is the best predicted salary level for a person with 13 years of education? Does everyone with 13 years of education actually make that figure?

Years of Education	Salary
16	61.10
10	26.53
11	29.57
11	28.38
13	37.62
13	27.22
12	36.70
10	14.77
12	22.23
12	24.82
14	32.89
13	36.94
15	35.99
11	23.11
14	62.56
10	21.31

Based on data from Statistics Canada.

12. What is the best predicted profit margin (as a percent) for a company whose ratio of current assets to current liabilities is 1.4?

Profit margin	3.9	2.9	9.4	6.1	−0.6	3.6	2.7	1.0	2.2	8.3	3.9
Assets/liabilities	1.3	1.3	1.1	1.5	1.7	1.5	1.6	1.5	1.0	1.1	0.9

13. Refer to Data Set 16 in Appendix B.
   a. Use the paired duration and interval data for eruptions of the Old Faithful geyser. What is the best predicted time before the next eruption if the previous eruption lasted for 210 seconds?
   b. Use the paired height and interval data for eruptions of the Old Faithful geyser. What is the best predicted time before the next eruption if the previous eruption had a height of 275 ft?
   c. Which predicted time is more reliable, the result from part (a) or the result from part (b)? Why?

14. Refer to Data Set 4 in Appendix B and use the paired columns for average annual temperature and precipitation. What is the best predicted amount of precipitation when the average annual temperature is 7.1°C?

15. Refer to Data Set 21 in Appendix B.
   a. Use the paired data consisting of price and carat (weight). What is the best predicted price of a diamond with a weight of 1.5 carats?

b. Use the paired price/colour data. Assuming the colour data is recorded in an interval scale, what is the best predicted price of a diamond with a colour rating of 3?

c. Which prediction is better, the result from part (a) or the result from part (b)? Why?

16. Refer to Data Set 4 in Appendix B and use the paired columns for average annual temperature and snowfall. What is the best predicted amount of snowfall when the average annual temperature is 7.1°C?

17. Refer to Data Set 7 of Appendix B and use the paired data consisting of temperature departures and precipitation departures from seasonal norms. What is the best predicted precipitation departure in a spring whose temperature was 0.9°C *below* the norm?

18.

Employees	29	122	24	30	16	3	83	50	43	33
Revenues	34.2	33.2	27.9	17.2	16.7	14.2	13.9	12.1	11.2	10.8

What is the best predicted revenue for a top company that employs 10,000 people? (Note that in the table the number of employees is given in thousands and revenues in billions of dollars.)

19. In each of the following cases, find the best predicted value of y given that $x = 3.00$. The given statistics are summarized from paired sample data.

a. $r = 0.931$, $\bar{y} = 7.00$, $n = 10$, and the equation of the regression line is $\hat{y} = 4.00 + 2.00x$.

b. $r = -0.033$, $\bar{y} = 2.50$, $n = 80$, and the equation of the regression line is $\hat{y} = 5.00 - 2.00x$.

20. In each of the following cases, find the best predicted value of y given that $x = 2.00$. The given statistics are summarized from paired sample data.

a. $r = -0.882$, $\bar{y} = 3.57$, $n = 15$, and the equation of the regression line is $\hat{y} = 23.00 - 8.00x$.

b. $r = 0.187$, $\bar{y} = 9.33$, $n = 60$, and the equation of the regression line is $\hat{y} = 4.00 + 8.00x$.

21. Refer to these data for Parliaments of up to 3 years' duration. If there were an additional Parliament to include, which lasted 1 year and was in session for 364 days, would the new point be an outlier? Would it be an influential point?

Duration (years)	0.7	1	1	1	1	1.5	2	2	3	3
Days in session	67	111	132	177	177	491	550	828	788	1189

22. Refer to the 13 pairs of data listed in Table 11-1. If we include a 14th data point, for an earthquake at 85°N latitude with magnitude 75ML, is the new point an outlier? Is it an influential point?

23. Large numbers, such as those in the accompanying table, often cause computational problems. First use the given data to find the equation of the regression line, then find the equation of the regression line after each x value has been divided by 1000. How are the results affected by the change in x? How would the results be affected if each y entry were also divided by 1000?

x	924,736	832,985	825,664	793,427	857,366
y	142	111	109	95	119

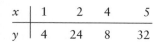

24. According to the least-squares property, the regression line minimizes the sum of the squares of the residuals. We noted that with the paired data in the margin, the regression equation is $\hat{y} = 5 + 4x$ and the sum of the squares of the residuals is 364. Show that the equation $\hat{y} = 8 + 3x$ results in a sum of squares of residuals that is greater than 364.

25. If the scatterplot reveals a nonlinear (not a straight line) pattern that you recognize as another type of curve, you may be able to apply the methods of this section. For the data given in the margin, find the linear equation $(y = b_0 + b_1 x)$ that best fits the sample data, and find the logarithmic equation $(y = a + b \ln x)$ that best fits the sample data. (*Hint:* Begin by replacing each x value with $\ln x$.) Which of these two equations fits the data better? Why?

26. Explain why a test of the null hypothesis H_0: $\rho = 0$ is equivalent to a test of the null hypothesis H_0: $\beta_1 = 0$, where ρ is the linear correlation coefficient for a population of paired data, and β_1 is the slope of the regression line for that same population.

27. See Data Set 19 in Appendix B. If you draw a scatter plot to compare (y) the maximum lifespans of the first six animals listed with (x) their metabolic rates, there appears to be a nonlinear relationship. For those data, find the linear equation $(y = b_0 + b_1 x)$ that best fits the sample data. Then find the exponential equation $(y = a \cdot e^{bx}$, where e represents the constant 2.7182818) that best fits the sample data. [*Hint:* Begin by replacing each y (maxlife) value with $\ln y$. Perform the linear regression versus x (metabolic rate). To transform the results of the regression: a for the exponential equation equals the constant e raised to the power b_0 (from the linear regression), and b for the exponential equation equals b_1.] Which of these two equations fits the data better? Explain.

11-4 Variation and Prediction Intervals

We have said that for paired variables, the actual y values rarely lie exactly on the regression line. There is usually variation between the actual and predicted values for y. We now examine this variation for these two major applications:

1. To determine the proportion of the variation in y that can be explained by the linear relationship between x and y.

2. To construct confidence interval estimates of predicted y-values. Such confidence intervals are called *prediction intervals*.

Explained and Unexplained Variation

In Section 11–2 we introduced the concept of correlation and used the linear correlation coefficient r in determining whether there is a significant linear correlation between two variables denoted by x and y. In addition to serving as a measure of the linear correlation between two variables, the value of r can also provide us with additional information about the variation of sample points about the regression line. We begin with a sample case, which leads to an important definition.

Suppose we have a large collection of paired data, which yields the following results:

- There is a significant linear correlation.
- The equation of the regression line is $\hat{y} = 3 + 2x$.
- The mean of the y values is given by $\bar{y} = 9$.
- One of the pairs of sample data is $x = 5$ and $y = 19$.
- Figure 11-9 shows that the point $(5, 13)$ is one of the points on the regression line, because substitution of $x = 5$ into the regression equation yields the following:

$$\hat{y} = 3 + 2x = 3 + 2(5) = 13$$

Figure 11-9 shows that the point $(5, 13)$ lies on the regression line, but the point $(5, 19)$, which is from the original data set, does not lie on the regression line because it does not satisfy the regression equation. Take time to examine Figure 11-9 carefully and note the differences defined as follows.

DEFINITIONS

Assume that we have a collection of paired data containing the particular point (x, y), that $\hat{y}$ is the predicted value of y (obtained by using the regression equation), and that the mean of the sample y values is $\bar{y}$.

The **total deviation** (from the mean) of the particular point (x, y) is the vertical distance $y - \hat{y}$, which is the distance between the point (x, y) and the horizontal line passing through the sample mean $\bar{y}$.

Figure 11-9
Unexplained, Explained, and Total Deviation

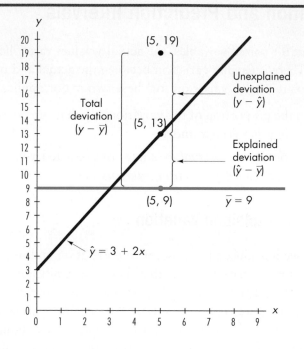

The **explained deviation** is the vertical distance $\hat{y} - \bar{y}$, which is the distance between the predicted y value and the horizontal line passing through the sample mean $\bar{y}$.

The **unexplained deviation** is the vertical distance $y - \hat{y}$, which is the vertical distance between the point (x, y) and the regression line. (The distance $y - \hat{y}$ is also called a *residual*, as defined in Section 11-3.)

For the specific data under consideration, we get these results:

total deviation of $(5, 19) = y - \bar{y} = 19 - 9 = 10$

explained deviation of $(5, 19) = \hat{y} - \bar{y} = 13 - 9 = 4$

unexplained deviation of $(5, 19) = y - \hat{y} = 19 - 13 = 6$

If we were totally ignorant of correlation and regression concepts and wanted to predict a value of y given a value of x and a collection of paired (x, y) data, our best guess would be $\bar{y}$. But we are not totally ignorant of correlation and regression concepts: We know that in this case (with a significant linear correlation) the way to predict the value of y when $x = 5$ is to use the regression equation, which yields $\hat{y} = 13$, as calculated before. We can explain the discrepancy between $\bar{y} = 9$ and $\hat{y} = 13$ by simply noting that there is a significant linear correlation best described by the regression line. Consequently, when $x = 5$, y should be 13 and not the mean y value of 9. But whereas y should be 13, it is 19. The discrepancy between 13 and 19 cannot be explained by the regression line and is called an *unexplained deviation* or a *residual*. The specific case illustrated in Figure 11-9 can be generalized as follows:

(total deviation) = (explained deviation) + (unexplained deviation)

or $\quad (y - \bar{y}) \quad = \quad (\hat{y} - \bar{y}) \quad + \quad (y - \hat{y})$

This last expression applies to a particular point (x, y), but it can be further generalized and modified to include all of the pairs of sample data, as shown in Formula 11-5. In that formula, the **total variation** is expressed as the sum of the squares of the total deviation values, the **explained variation** is the sum of the squares of the explained deviation values, and the **unexplained variation** is the sum of the squares of the unexplained deviation values.

Formula 11-5

(total variation) = (explained variation) + (unexplained variation)

or $\quad \Sigma(y - \bar{y})^2 \quad = \quad \Sigma(\hat{y} - \bar{y})^2 \quad + \quad \Sigma(y - \hat{y})^2$

Coefficient of Determination

The components of Formula 11-5 are used in the following important definition.

The **coefficient of determination** is the amount of the variation in y that is explained by the regression line. It is computed as

$$r^2 = \frac{\text{explained variation}}{\text{total variation}}$$

We can compute r^2 by using the definition just given with Formula 11-5, or we can simply square the linear correlation coefficient r, which is found by using the methods described in Section 11–2. As an example, if $r = 0.8$, then the coefficient of determination is $r^2 = 0.64$, which means that *64% of the total variation in y can be explained by the regression line. It follows that 36% of the total variation in y remains unexplained.*

EXAMPLE

Referring to the earthquake data in Table 11-1, find the percentage of the variation in y (magnitude) that can be explained by the regression line.

SOLUTION

Recall that Table 11-1 contains 13 pairs of sample data, with each pair consisting of the recorded latitude and magnitude of seismic activity. In Section 11–2 we found that the linear correlation coefficient is $r = 0.623$. The coefficient of determination is $r^2 = 0.623^2 = 0.388$, indicating that the ratio of explained variation in y to total variation in y is 0.388.

Interpretation
38.8% of the total variation in y can be explained by the regression line. That is, while 38.8% of the total variation in earthquake magnitudes can be explained by the variation in their recorded latitudes; most of the variation (61.2%) is attributable to other factors.

As mentioned in Section 11–3, a data set should always be examined for outliers. The following example shows how to use r^2 as part of that analysis.

EXAMPLE

The size of a home (in thousands of square feet) was used to predict the selling price (in thousands of dollars). A sample of homes gave the following results:

Size	2.5	1.5	3.1	1.0	1.1	1.3	1.0	1.3
Price	451.0	256.5	626.7	204.6	237.2	257.4	575.9	203.5

Identify the outlying data and the influence exerted on r^2.

SOLUTION

Most of the prices seem in line with the size of the home except for the seventh pair in which a home with 1000 square feet has a selling price of $575,900. For this data set, $r^2 = 39.89\%$. If we remove this pair of data, $r^2 = 96.31\%$, a vast improvement.

The coefficient of determination can also be useful in determining if a mathematical model should be constructed as illustrated in the following example.

EXAMPLE

The density of wood is being used to predict its stiffness. A random sample of 11 readings is taken:

Density	21.7	15.2	23.4	15.4	14.5	16.7	15	25.6	15	24.4	7
Stiffness	47,661	28,028	104,170	25,312	22,148	49,499	25,319	96,305	26,222	72,594	5304

Using r^2, which is the better transformation of stiffness: natural logarithm or square root?

SOLUTION

A plot of the data indicates that the relationship is nonlinear. If we construct a linear model with no transformation, $r^2 = 82.16\%$. A natural log transformation of stiffness produces $r^2 = 91.41\%$ while a square root transformation has $r^2 = 90.47\%$. While both transformations improve r^2, the natural log transformation is the better of the two.

Model Significance

If we have a linear regression model, we would like to know if the model is useful for predicting. In Section 11–2, we examined a formal hypothesis test for correlation to determine if the correlation of the dependent and independent variable is significant. Clearly, if the correlation is significant, we can safely conclude that the model is useful for predicting. This is fine if there is only one independent variable. We now introduce another formal hypothesis test that can be used regardless of the number of independent variables.

If we let k represent the number of independent variables, the theoretical linear regression model is:

$$y = B_0 + B_1 x_1 + B_2 x_2 + \cdots + B_k x_k$$

where B_0 is the theoretical intercept, B_1 is the theoretical coefficient of x_1, and so on. What if $B_1 = B_2 = \cdots = B_k = 0$? How useful would the model be for predicting? If each of the variable coefficients were zero, the model would become $y = B_0$ and regardless of the values of the independent variables, the dependent variable y would always be equal to the theoretical intercept. Such a model would be useless for prediction. At least one of the coefficients should not equal zero in order to have a linear regression model that is useful for prediction.

To that end, we present the following null and alternative hypotheses:

$$H_0: B_1 = B_2 = \cdots = B_k = 0$$

H_1: at least one coefficient is not equal to zero

This hypothesis appears similar to the null and alternative hypotheses for ANOVA, as we saw in Chapter 9. Just as we constructed an ANOVA table to test the hypothesis of the equality of several means, we now use **regression ANOVA**, constructing a special type of table, called a regression ANOVA table, to test the null hypothesis that all of the variable coefficients equal zero. If we reject the null hypothesis, we say that the model is significant; that is, that at least one variable is useful for prediction.

To build the table, we need the sources of variation, the sum of squares (SS), the mean square (MS), and the F test statistic. In Formula 11-5, we presented that the total variation in y is equal to the sum of the explained variation (provided by the linear regression model) and the unexplained variation (provided by the sum of the squared residuals). These are the sources of variation for the regression ANOVA table.

EXAMPLE

Is the model constructed for the chapter problem significant? Test at a 0.05 level of significance.

SOLUTION

Since $k = 1$, the null and alternative hypotheses are:

$$H_0: B_1 = 0$$

$$H_1: B_1 \neq 0$$

Here is the regression ANOVA table for this particular model:

Source	df	SS	MS	F
Regression	1	9.7319	9.7319	6.9658
Error	11	15.3681	1.3971	
Total	12	25.1		

The total number of degrees of freedom is $n - 1 = 13 - 1 = 12$. The regression degrees of freedom $= k = 1$. The residual (error) degrees of freedom is calculated by simple arithmetic: $12 - 1 = 11$.

In Formula 11-5, SS(total) is presented as $\Sigma(y - \bar{y})^2$. If we take advantage of the formula for s^2, SS(total) can be rewritten as $(n - 1)s_y^2$. Using a calculator, $s_y = 1.44626$ leading to $s_y^2 = 2.091\dot{6}$. Therefore, SS(total) $= (12)(2.091\dot{6}) = 25.1$.

In Formula 11-5, SS(regression) is presented as $\Sigma(\hat{y} - \bar{y})^2$, which would require the calculation of each predicted value for each data pair, a tedious process. There is a simpler way. We know that r^2 is the ratio of the explained variation in y to the total variation in y. Using our sum of squares notation, we can write:

$$r^2 = \frac{SS(\text{regression})}{SS(\text{total})}$$

from which we can solve for SS(regression) $=$ SS(total) $\cdot r^2$.

We earlier found that $r^2 = 0.388$ after rounding. If we use the value from a calculator, we find SS(regression) $= (25.1)(0.387724606) = 9.7319$ if we round to 4 decimals.

SS(error) $=$ SS(total) $-$ SS(regression) $= 25.1 - 9.7319 = 15.3681$. It should be noted that SS(error) could also be calculated by computing the residual for each data pair and summing the squared residuals.

The MS values are calculated in the standard manner by dividing the SS value by its respective degrees of freedom.

$$MS(\text{regression}) = \frac{SS(\text{regression})}{\text{regression df}} = \frac{9.7319}{1} = 9.7319$$

$$MS(\text{error}) = \frac{SS(\text{error})}{\text{error df}} = \frac{15.3681}{11} = 1.3971$$

The F test statistic is calculated by dividing the MS(regression) by the MS(error):

$$F = \frac{MS(\text{regression})}{MS(\text{error})} = \frac{9.7319}{1.3971} = 6.9658$$

To find the F critical value, the numerator df $=$ regression df $= 1$ and the denominator df $=$ error df $= 11$. With $\alpha = 0.05$, the critical value from Table A-5 is 4.8443. We reject the null hypothesis if the F test statistic is greater than 4.8443. Since the test statistic has a value of 6.9658, we reject the null hypothesis.

Interpretation

We conclude that the model is significant and that latitude is a useful predictor of magnitudes of seismic activity.

In Section 11–2, there was a formal hypothesis test for the correlation ρ using the t distribution to determine whether it was significant. There is a parallel test for B_1. In the correlation test, the number of degrees of freedom for the t critical value was given as df $= n - 2$. Note that this is equal to the error df from the regression ANOVA table. This is also the df for the test for B_1. The test statistic is:

$$t = \frac{b_1}{s_{b_1}} = \frac{b_1}{\sqrt{\dfrac{MS(error)}{(n-1)s_x^2}}}$$

EXAMPLE

Test the significance of B_1 for the chapter problem using the t test.

SOLUTION

The null and alternative hypotheses remain the same:

$$H_0 : B_1 = 0$$
$$H_1 : B_1 \neq 0$$

The df for the t critical value is the error degrees of freedom of 11. As before, the critical values are $t = -2.201$ and $t = 2.201$. To construct the test statistic, we have $b_1 = 0.100713$, MS(error) $= 1.3971$ from the regression ANOVA table and $(n-1)s_x^2 = (12)(8.9418)^2 = 959.46$. The test statistic is:

$$t = \frac{0.100713}{\sqrt{\dfrac{1.3971}{959.46}}} = \frac{0.100713}{0.038159292} = 2.6393$$

This is equivalent to the test statistic from the correlation significance test (aside from rounding errors). As before, we reject the null hypothesis and conclude that B_1 is significant.

In a computer output, such as Excel, you will find a $1 - \alpha$ confidence interval for B_1. Using the data from the t test for B_1, we can construct such an interval:

$$b_1 - t \cdot s_{b_1} < B_1 < b_1 - t \cdot s_{b_1}$$

where t is the critical value from the two-tailed test for B_1.

Construct a 95% confidence interval for B_1 for the chapter problem.

We have $b_1 = 0.100713$, $t = 2.201$ and $s_{b_1} = 0.038159292$. Then:

$$0.100713 - (2.201)(0.038159292) < B_1 < 0.100713 + (2.201)(0.038159292)$$
$$0.100713 - 0.083989 < B_1 < 0.100713 + 0.083989$$
$$0.0167 < B_1 < 0.1847$$

Interpretation
With 95% confidence, for each additional degree of latitude, the magnitude of an earthquake will increase by an amount between 0.0167 ML and 0.1847 ML.

We can use this confidence interval to test the null hypothesis of $B_1 = 0$ versus the alternative hypothesis of $B_1 \neq 0$. The theory is: If $B_1 = 0$ falls outside the $1 - \alpha$ confidence interval of B_1, we reject the null hypothesis at the α level of significance. Since $B_1 = 0$ falls outside the 95% confidence interval of B_1, we reject the null hypothesis at a 5% level of significance.

Depending on the sign of b_1, we can also test the null hypothesis of $B_1 \leq 0$ versus the alternative hypothesis of $B_1 > 0$ if $b_1 > 0$ or $B_1 \geq 0$ versus $B_1 < 0$ if $b_1 < 0$. In the chapter problem, since $b_1 = 0.100713$ is positive, the null and alternative hypotheses would be:

$$H_0: B_1 \leq 0$$
$$H_1: B_1 > 0$$

The test statistic would still be $t = 2.6393$. With $\alpha = 0.05$ and the degrees of freedom at 11, the t critical value from Table A-3 would be 1.796. We reject the null hypothesis if the t test statistic is greater than 1.796. Since the test statistic is greater than the critical value, we reject the null hypothesis.

Interpretation We conclude that $B_1 > 0$. Therefore, there is a significant positive relationship between latitude and earthquake magnitude.

Prediction and Confidence Intervals for the Dependent Variable

In Section 11–3, we used the Table 11-1 sample data to find the regression equation $\hat{y} = -3.05 + 0.101x$, where $\hat{y}$ represents the predicted magnitude and x the recorded latitude of the seismic activity. We then used that equation to predict the y value, given that $x = 64°N$. We found that the best predicted magnitude of an earthquake recorded at 64°N is 3.41 ML. Because 3.41 is a single value, it is referred to as a *point estimate*. In Chapter 6, we saw that point estimates have the disadvantage of not conveying any sense of how accurate they might be. Here, we know that 3.41 is the best predicted value, but we don't know how accurate that value is. In Chapter 6 we developed confidence interval estimates to overcome that disadvantage, and in this section we follow that precedent. We present two types of intervals: a **prediction interval**, which is a confidence interval estimate of a *predicted value* of y, and a **confidence interval for** μ_y, the mean value of y.

The development of these intervals requires a measure of the spread of sample points about the regression line. Recall that the unexplained deviation (or residual) is the vertical distance between a sample point and the regression line, as illustrated in Figure 11-9. The *standard error of estimate* is a collective measure of the spread of the sample points about the regression line; it is formally defined as follows.

DEFINITION

The **standard error of estimate**, denoted by s_e, is a measure of the differences (or distances) between the observed sample y values and the predicted values $\hat{y}$ that are obtained using the regression equation. It is given as

$$s_e = \sqrt{\frac{\Sigma(y - \hat{y})^2}{n - 2}}$$

where $\hat{y}$ is the predicted y value.

Formula 11-6 can also be used to compute the standard error of estimate s_e. It is algebraically equivalent to the expression in the definition, but this form is generally easier to work with because it doesn't require that we compute each of the predicted values y by substitution in the regression equation. However, Formula 11-6 does require that we find the y-intercept b_0 and the slope b_1 of the estimated regression line.

Formula 11-6 $\qquad s_e = \sqrt{\dfrac{\Sigma y^2 - b_0 \Sigma y - b_1 \Sigma xy}{n - 2}}$ (standard error of estimate)

It should be noted if you have constructed a regression ANOVA table, the standard error of estimate is equal to the square root of the MS(error), that is:

$$s_e = \sqrt{\text{MS(error)}}$$

In Excel, Minitab and STATDISK, the value of the standard error of estimate is displayed as part of the output when their regression procedures are executed.

Rationale for the s_e Definition

The development of the standard error of estimate s_e closely parallels that of the ordinary standard deviation introduced in Chapter 2. Just as the standard deviation is a measure of how scores deviate from their mean, the standard error of estimate s_e is a measure of how sample data points deviate from their regression line. The reasoning behind dividing by $n - 2$ is similar to the reasoning that led to division by $n - 1$

for the ordinary standard deviation, and we will not pursue the complex details. It is important to note that smaller values of s_e reflect points that stay close to the regression line, and larger values are from points farther away from the regression line.

EXAMPLE

Use Formula 11-6 to find the standard error of estimate s_e for the earthquake measurement data listed in Table 11-1.

SOLUTION

In Section 11-2 we used the Table 11-1 data to find that

$$n = 13 \quad \Sigma y^2 = 119.87 \quad \Sigma y = 35.1 \quad \Sigma xy = 2100.84$$

In Section 11-3 we used the Table 11-1 data to find the y-intercept and slope of the regression line. Those values are given here with extra decimal places for greater precision:

$$b_0 = -3.05071 \qquad b_1 = 0.100713$$

We can now use these values to find the standard error of estimate s_e:

$$s_e = \sqrt{\frac{\Sigma y^2 - b_0 \Sigma y - b_1 \Sigma xy}{n - 2}}$$

$$= \sqrt{\frac{119.87 - (-3.05071)(35.1) - (0.100713)(2100.84)}{13 - 2}}$$

$$= 1.18199 = 1.18 \quad \text{(rounded)}$$

Lastly, as a check, MS(error) = 1.3971 from the regression ANOVA table in the previous section. Then $s_e = \sqrt{1.3971} = 1.18$. We can measure the spread of the sample points about the regression line with the standard error of estimate $s_e = 1.18$.

We can use the standard error of estimate s_e to construct interval estimates that will help us see how dependable our point estimates of y really are. Assume that for each fixed value of x, the corresponding sample values of y are normally distributed about the regression line, and those normal distributions have the same variance. The following interval estimate applies to an *individual y*.

PREDICTION INTERVAL FOR AN INDIVIDUAL y

Given the fixed value x_0, the *prediction interval for an individual y* is

$$\hat{y} - E < y < \hat{y} + E$$

where the margin of error E is

$$E = t_{\alpha/2} s_e \sqrt{1 + \frac{1}{n} + \frac{n(x_0 - \bar{x})^2}{n(\Sigma x^2) - (\Sigma x)^2}}$$

and x_0 represents the given value of x, $t_{\alpha/2}$ has $n - 2$ degrees of freedom, and s_e is found from Formula 11-6.

EXAMPLE

Refer to the Table 11-1 sample data listing earthquake latitudes (x) along with their magnitudes (y). In previous sections we have shown that

- There is a significant linear correlation (at the 0.05 significance level).
- The regression equation is $\hat{y} = -3.05 + 0.101x$.
- When $x = 64.0$, the predicted y value is 3.41 ML, but a more accurate value of 3.39 ML is obtained if we calculate the predicted value using unrounded values of slope and intercept.

Construct a 95% prediction interval for the magnitude of an earthquake that is recorded at 64°N. This will provide a sense of how reliable the estimate of 3.39 ML really is.

SOLUTION

We have already used the Table 11-1 sample data to find the following values:

$$n = 13 \quad \bar{x} = 57.1 \quad \Sigma x = 742.3 \quad \Sigma x^2 = 43344.79 \quad s_e = 1.18199$$

From Table A-3 we find $t_{\alpha/2} = 2.201$. (We used $13 - 2 = 11$ degrees of freedom with $\alpha = 0.05$ in two tails.) We can now calculate the margin of error E by letting $x_0 = 64.0$, because we want the prediction interval of y for $x = 64.0$.

$$E = t_{\alpha/2} s_e \sqrt{1 + \frac{1}{n} + \frac{n(x_0 - \bar{x})^2}{n(\Sigma x^2) - (\Sigma x)^2}}$$

$$= (2.201)(1.18199) \sqrt{1 + \frac{1}{13} + \frac{13(64.0 - 57.1)^2}{13(43,344.79) - (742.3)^2}}$$

$$= (2.201)(1.18199)(1.06139) = 2.76$$

With $\hat{y} = 3.39$ and $E = 2.76$, we get the prediction interval as follows:

$$\hat{y} - E < y < \hat{y} + E$$
$$3.39 - 2.76 < y < 3.39 + 2.76$$
$$0.63 < y < 6.15$$

That is, for seismic activity recorded at latitude 64°N, we have 95% confidence that its true magnitude is between 0.63 ML and 6.15 ML (considering only events recorded in the ML scale). For earthquakes, that is a very large range. (One factor contributing to the large range is the small sample size of 13.)

Interpretation

Although we have an equation to predict a magnitude (3.39 ML) for seismic activity at 64°N, we have discovered that the estimate is not very reliable. The 95% prediction interval shows that the magnitude estimate can vary considerably—from a very minor tremor up to a serious earthquake.

To construct the *confidence interval* for the *mean value* of y, the margin of error is slightly different than that of the prediction interval:

$$E = t_{\alpha/2}s_e\sqrt{\frac{1}{n} + \frac{n(x_0 - x)^2}{n(\Sigma x^2) - (\Sigma x)^2}}$$

As you may observe, the only difference is the absence of the "1 +" before $1/n$ under the square root. Since this is a confidence interval for the mean value of y, the interval is:

$$\hat{y} - E < \mu_y < \hat{y} + E$$

using the above margin of error.

EXAMPLE

Construct a 95% confidence interval of the mean magnitude based on a latitude of 64.0.

SOLUTION

As with the prediction interval, $y = 3.41$ ML. The margin of error is similar to that constructed for the prediction interval:

$$E = t_{\alpha/2}s_e\sqrt{\frac{1}{n} + \frac{n(x_0 - x)^2}{n(\Sigma x^2) - (\Sigma x)^2}}$$

$$= (2.201)(1.18199)\sqrt{\frac{1}{13} + \frac{13(64.0 - 57.1)^2}{13(43{,}344.79) - (742.3)^2}}$$

$$= (2.201)(1.18199)(0.355731273)$$

$$= 0.93$$

Note that the margin of error of 0.93 for the confidence interval of the mean value of y is smaller than the margin of error of 2.76 for the prediction interval of y. The reason is that the confidence interval of the mean value of y is predicated on numerous observations of the same value of y whereas the prediction interval is predicted on a single observation of y.

With $\hat{y} = 3.39$ and $E = 0.93$, we can now construct the confidence interval of the mean value of y:

$$\hat{y} - E < \mu_y < \hat{y} + E$$
$$3.39 - 0.93 < \mu_y < 3.39 + 0.93$$
$$2.46 < \mu_y < 4.32$$

Interpretation
With 95% confidence, the mean magnitude based on a latitude of 64°N ranges from 2.46 ML to 4.32 ML.

EXCEL: For procedures, see Using Technology in Section 11–3. Outputs include b_0 and b_1 (for the regression equation), the standard error estimate s_e, and the coefficient of determination (labelled as **R square**).

MINITAB 15: For procedures, see Using Technology in Section 11–3. Outputs include b_0 and b_1 (for the regression equation), the standard error estimate s_e, and the coefficient of determination (labelled as **R-Sq**).

STATDISK: For procedures, see Using Technology in Section 11–3. Outputs include the linear correlation coefficient r, the equation of the regression line, the standard error estimate s_e, the total, explained, and unexplained variation, and the coefficient of determination.

Exercises A: Basic Skills and Concepts

In Exercises 1–4, use the value of the linear correlation coefficient r to find the coefficient of determination and the percentage of the total variation that can be explained by the regression line.

1. $r = 0.2$
2. $r = -0.6$
3. $r = -0.225$
4. $r = 0.837$

In Exercises 5–8, find the (a) explained variation, (b) unexplained variation, (c) total variation, (d) coefficient of determination, and (e) standard error of estimate s_e.

5. The accompanying table lists numbers x of patio tiles and costs y (in dollars) of having them manually cut to fit. (The equation of the regression line is $\hat{y} = 2 + 3x$.)

x	1	2	3	5	6
y	5	8	11	17	20

6. The paired data below consist of the chest sizes (in inches) and weights (in pounds) of a sample of male bears. (The equation of the regression line is $\hat{y} = -187.462 + 11.2713x$.)

x Chest (in.)	26	45	54	49	41	49	44	19
y Weight (lb)	90	344	416	348	262	360	332	34

7. The paired data below consist of the weights (in kilograms) of discarded plastic and sizes of households. (The equation of the regression line, for predicting household size from the weight of discarded plastic, is $\hat{y} = 0.5486 + 3.2694x$.)

Plastic (kg)	0.12	0.64	0.99	1.28	0.99	0.82	0.39	1.38
Household size	2	3	3	6	4	2	1	5

8. The paired data below consist of annual closing values of the TSE 300 stock index and of corresponding closing prices for a Canadian mining stock. (The equation of the regression line, for predicting the stock closing value from the TSE value, is $\hat{y} = 0.5013 - 0.000009886x$.)

TSE 300	9397.97	12,061.95	13,868.54	13,648.84	17,977.46	19,309.36
Mining stock	0.450	0.600	0.210	0.160	0.335	0.400

9. Refer to the data given in Exercise 5 and assume that the necessary conditions of normality and variance are met.
   a. For $x = 4$, find $\hat{y}$, the predicted value of y.
   b. How does the value of s_e affect the construction of the 95% prediction interval of y for $x = 4$?

10. Refer to Exercise 6 and assume that the necessary conditions of normality and variance are met.
   a. For a bear with a chest size of 52 in., find $\hat{y}$, the predicted weight.
   b. Find the 99% prediction interval of y for $x = 52$.

11. Refer to the data given in Exercise 7 and assume that the necessary conditions of normality and variance are met.
   a. Find the predicted size of a household that discards 1.25 kg of plastic.
   b. Find the 95% prediction interval for the size of a household that discards 1.25 kg of plastic.

12. Refer to the data given in Exercise 8, and assume that the necessary conditions of normality and variance are met.
   a. For a year that the TSE 300 closes at 14,447.99, find the predicted closing price for the mining stock.
   b. Find the 99% prediction interval for the closing price of the mining stock, given that the TSE closed at 14,447.99.
   c. For this data set, does the answer from part (b) appear very useful? Why or why not?

In Exercises 13–16, refer to the data in Exercise 11 in Section 11–3. Let x represent the number of years of a person's education, and let y represent his or her salary (in thousands of dollars). Construct a prediction interval estimate of the salary of a person who has the given years of education. Use the given degree of confidence. (See the example in this section.)

13. 12 years; 95% confidence

14. 14 years; 90% confidence

15. 13.5 years; 90% confidence

16. 13 years; 99% confidence

17. Confidence intervals for the y-intercept β_0 of a regression line $(y = \beta_0 + \beta_1 x)$ can be found by evaluating the limits in the intervals below:

$$b_0 - E < \beta_0 < b_0 + E$$

where

$$E = t_{\alpha/2} s_e \sqrt{\frac{1}{n} + \frac{\bar{x}^2}{\Sigma x^2 - \frac{(\Sigma x)^2}{n}}}$$

In these expressions, the y-intercept b_0 is found from the sample data and $t_{\alpha/2}$ is found from Table A-3 by using $n - 2$ degrees of freedom. Using the earthquake data in Table 11-1, find the 95% confidence interval estimates of β_0.

18. a. If a collection of paired data includes at least three pairs of values, what do you know about the linear correlation coefficient if $s_e = 0$?
    b. If a collection of paired data is such that the total explained variation is 0, what do you know about the slope of the regression line?

19. a. Find an expression for the unexplained variation in terms of the sample size n and the standard error of estimate s_e.
    b. Find an expression for the explained variation in terms of the coefficient of determination r^2 and the unexplained variation.
    c. Suppose we have a collection of paired data for which $r^2 = 0.900$ and the regression equation is $\hat{y} = 3 - 2x$. Find the linear correlation coefficient.

20. In constructing a confidence interval for the mean value of y for a mathematical model based on a natural log transformation of y, the initial confidence interval is:

$$\hat{y} - E < \ln(\mu_y) < \hat{y} + E$$

which we can then transform back to the original units:

$$e^{(\hat{y} - E)} < \mu_y < e^{(\hat{y} + E)}$$

In the example of the wood density being used to predict stiffness, we reproduce the data:

Density	21.7	15.2	23.4	15.4	14.5	16.7	15	25.6	15	24.4	7
Stiffness	47,661	28,028	104,170	25,312	22,148	49,499	25,319	96,305	26,222	72,594	5304

Adapted from *Probability and Statistics for Engineers and Scientists*, 7e, by Ronald E. Walpole. Published by Prentice Hall Inc.

In that example, it was shown that a natural log transformation of stiffness had $r^2 = 91.41\%$. If we construct a mathematical model based on this transformation, for a density of 20, construct a 95% confidence interval of the mean stiffness, rounding the transformed limits to the nearest hundred.

21. An alternate approach to calculating standard errors and confidence intervals for regression is to apply them to the calculated b_0 and x coefficient(s), instead of to the y estimate. For example, this is part of Excel's regression output for comparing the weights and lengths of bears.

This shows that the b_0 term, which the regression formula would estimate as -352, could, in fact, range from -663 to -40 in the 95% confidence interval. Similarly, the x coefficient, which is estimated as 9.66, could range from 4.9 to 14.4.

Refer to the data in Exercise 11 in Section 11–3. Let x represent the number of years of a person's education, and let y represent his or her salary (in thousands of dollars). Use Excel to find the confidence limits for the intercept and x coefficient for the regression between education and salary.

 # **11-5 Multiple Regression**

So far, we have examined relationships between exactly *two* variables. In this section we develop a method of **multiple regression** for analyzing relationships involving *more than two* variables. For example, in addition to considering the relationship between the magnitude of an earthquake and its recorded latitude, we might also include variables such as the depth of focus (hypocentre) of the earthquake, or its recorded longitude. We will focus on four key elements: the value of adjusted R^2, the regression ANOVA P-value, individual t tests for variable significance, and the presence of multicollinearity. As in the preceding sections, we will work with linear relationships only.

Multiple Regression Equation

DEFINITION

A **multiple regression equation** expresses a linear relationship between a dependent variable y and two or more independent variables $(x_1, x_2, \ldots, x_k)$. The general form of a multiple regression equation is $\hat{y} = b_0 + b_1x_1 + b_2x_2 + \cdots + b_kx_k$.

We will use the following notation, which was briefly introduced in the discussion of regression ANOVA in Section 11–3.

$$\hat{y} = b_0 + b_1x_1 + b_2x_2 + \cdots + b_kx_k \quad \text{(General form of the estimated multiple regression equation)}$$

n = sample size

k = number of *independent* variables. (The independent variables are also called predictor variables or x variables.)

$\hat{y}$ = predicted value of the dependent variable y (computed by using the multiple regression equation)

$x_1, x_2, \ldots, x_k$ are the independent variables

β_0 = the y-intercept, or the value of y when all of the predictor variables are 0. (This value is a population parameter.)

b_0 = estimate of β_0 based on the sample data (b_0 is a sample statistic)

$\beta_1, \beta_2, \ldots, \beta_k$ are the coefficients of the independent variables $x_1, x_2, \ldots, x_k$

$b_1, b_2, \ldots, b_k$ are the sample estimates of the coefficients $\beta_1, \beta_2, \ldots, \beta_k$

The calculations of the preceding sections of this chapter can all be done with any scientific calculator. The computations required for multiple regression are so complicated that a statistical software package must be used. We will focus here on *interpreting* the outputs of a statistical computer display. (See Using Technology in this section on how to generate such displays in Excel, Minitab, or STATDISK.)

Our goal in constructing a multiple regression equation (or model) is to find the "best" model with the fewest number of independent variables using the four elements outlined above. The reason the word "best" is in quotation marks is because there are more elements, beyond the scope of this textbook, that could be used. Depending on the software being used, these four elements are easy to find and can be compared to their counterparts from other models.

EXAMPLE

Table 11-3 is a hypothetical data set with the following variables:

$$x_1 = \text{gender with male} = 0 \text{ and female} = 1$$

$$x_2 = \text{monthly income recorded in thousands}$$

$$y = \text{amount spent per month eating out}$$

Construct a multiple regression model in which gender and monthly income are used to predict how much a person spends per month eating out on average. Interpret the coefficients.

Table 11-3	Hypothetical Multiple Regression Data		
Sample	y	x_1	x_2
1	120	1	2.5
2	170	0	3.2
3	95	1	1.8
4	220	0	4.5
5	190	1	3.2
6	195	0	3.5

SOLUTION

Before constructing the model, it is important to understand the raw data. We will use sample 1 to illustrate.

Variable x_1 is what is known as a **dummy** (or **qualitative**) **variable**. It is used when we believe a qualitative variable (such as gender) may be useful in predicting the dependent variable. It is generally at 2 levels. For ease of computations, the values 0 and 1 are used. In this example, 0 is used to represent male, 1 for female. For the 6 cases, we have 3 males and 3 females. The person in sample 1 is female.

Variable x_2 is a coded variable, representing income coded in thousands of dollars. This is done for ease of data entry and to reduce the magnitude of the coefficients. In sample 1, 2.5 represents a monthly income of $2,500.

EXCEL DISPLAY
Multiple Regression Using
Gender and Income

1	SUMMARY OUTPUT						
2							
3	*Regression Statistics*						
4	Multiple R	0.9574					
5	R Square	0.9166					
6	Adjusted R Square	0.8609					
7	Standard Error	17.8850					
8	Observations	6					
9							
10	ANOVA						
11		*df*	*SS*	*MS*	*F*	*Significance F*	
12	Regression	2	10540.3846	5270.19231	16.4760	0.0241	
13	Residual	3	959.6154	319.87179			
14	Total	5	11500.0000				
15							
16		*Coefficients*	*Standard Error*	*t Stat*	*P-value*	*Lower 95%*	*Upper 95%*
17	Intercept	1.1538	49.4459	0.0233	0.9828	-156.2051	158.5128
18	gender	4.0385	21.6434	0.1866	0.8639	-64.8406	72.9175
19	income	51.9231	12.9524	4.0088	0.0278	10.7027	93.1435

To construct the multiple regression model, we look at the bottom of the output under the coefficients column. The intercept $b_0 = 1.1538$, the coefficient for gender $b_1 = 4.0385$, and the coefficient for income $b_2 = 51.9231$. The multiple regression model is: $\hat{y} = 1.1538 + 4.0385x_1 + 51.9231x_2$. If we have a male who earns $3,000 per month, $x_1 = 0$ and $x_2 = 3$ from which we get $\hat{y} = 1.1538 + 4.0385(0) + 51.9231(3) = 156.92$ after rounding to the nearest cent. This is the average amount spent per month eating out by a male earning \$3,000 per month. If we have a female who earns \$3,000 per month, $x_1 = 1$ and $\hat{y} = 1.1538 + 4.0385(1) + 51.9231(3) = 160.96$. Notice that the difference in the average amount spent eating out between females and males with the same income is $160.96 - 156.92 = 4.04$, which is the coefficient for gender after rounding to 2 decimals. Therefore for a male and female with the same monthly income, a female spends \$4.04 more on average per month eating out. In a similar manner, for a male who earns \$4,000 per month, $x_2 = 4$ and $\hat{y} = 1.1538 + 4.0385(0) + 51.9231(4) = 208.85$. The difference between this male and the one earning \$3,000 per month in the average amount spent per month eating out is $208.85 - 156.92 = 51.92$ (adjusting for rounding), which is the coefficient for income after rounding to 2 decimals. Therefore, for two people of the same gender, for each \$1,000 difference in income, the average difference in the amount spent per month eating out is \$51.92.

In this model, we have two independent variables. The question is whether this is the best model to predict how much a person spends per month eating out. To answer this question, we proceed through each of the four elements introduced earlier.

Adjusted R^2

R^2 denotes the **multiple coefficient of determination**, which is a measure of how well the multiple regression equation fits the sample data. A perfect fit would result in $R^2 = 1$. A very good fit results in a value near 1. A very poor fit results in a value of R^2 close to 0. The value of $R^2 = 0.9166$ in the Excel display indicates that 91.66% of the variation in the amount spent per month eating out can be explained by gender and monthly income. *The multiple coefficient of determination R^2 is a measure of how well the regression equation fits the sample data, but it has a serious flaw:* As more variables are included, R^2 increases. (Actually, R^2 could remain the same, but it usually increases.) Although the largest R^2 is thus achieved by simply including all of the available variables, the best multiple regression equation does not necessarily use all of the available variables. Consequently, it is better to use the *adjusted coefficient of determination* when comparing different multiple regression equations, because it adjusts the R^2 value based on the number of variables and the sample size.

DEFINITION

> The **adjusted coefficient of determination** is the multiple coefficient of determination R^2 modified to account for the number of variables and the sample size. It is calculated by using Formula 11-7.

STATISTICAL PLANET

Statistics in Court
Owners of a five-building apartment complex filed a lawsuit because of extensive damage to bricks. The damage occurred when water was absorbed by the brick face, followed by freezing and thawing cycles, causing part of the brick face to separate. With about 750,000 bricks, it was not practical to inspect each one, so methods of sampling were used instead. Statisticians used regression methods to predict the total number of damaged bricks. The independent variables included which of the five buildings was used, direction of wall exposure, height, and whether the wall faced an interior courtyard or was an exterior wall. The estimate of total damage appears to have strongly influenced the final settlements. (See "Bricks, Buildings, and the Bronx: Estimating Masonry Deterioration," by Fairley, Izenman, and Whitlock, *Chance*, Vol. 7, No. 3.)

Formula 11-7 $\text{Adjusted } R^2 = 1 - \dfrac{(n-1)}{[n-(k+1)]}(1-R^2)$

where

n = sample size
k = number of independent (x) variables

The Excel display shows the adjusted coefficient of determination as 0.8609. If we use Formula 11-7 with the R^2 value of 0.9166, $n = 6$, and $k = 2$, we find that the adjusted R^2 value is 0.8609, confirming Excel's displayed value.

In using this element to find the best model, we will choose the one with the highest adjusted R^2.

Regression ANOVA *P*-Value

Recall from Section 11–4 that the purpose of regression ANOVA is to determine whether the model as a whole is significant. Since $k = 2$, the null and alternative hypotheses are:

$H_0: B_1 = B_2 = 0$
$H_1:$ At least one coefficient does not equal zero.

Under the ANOVA section, the Excel display shows a *P*-value (labelled "Significance *F*") of 0.0241. This is based on an *F* distribution with the numerator df = 2 (since $k = 2$) and the denominator df = 3. The value of the *F* test statistic is 16.476. Thus, the *P*-value = $P(F > 16.476) = 0.0241$. This value is a measure of the overall significance of the multiple regression equation. In this case, if we test at a 0.05 level of significance, the value of 0.0241 suggests that the multiple regression equation has good overall significance and is usable for predictions. That is, it makes sense to predict the average amount spent per month eating out based on gender and monthly earnings. Like the adjusted R^2, this *P*-value is a good measure of how well the equation fits the sample data. Rejection of the null hypothesis implies that at least one of the coefficients is not zero.

In using this element to find the best model, we cannot use the value of the *F* test statistic directly because another model may not have the same number of independent variables, which would affect the numerator degrees of freedom for the *F* distribution. However, the further the test statistic is into the critical region, the smaller the *P*-value is and the more significant the model becomes. Therefore, we will choose the model with the smallest regression ANOVA *P*-value.

Individual *t* Tests for Variable Significance

Recall from ANOVA in Chapter 9 that if we reject the null hypothesis, further analysis is required to determine which pairs of means are significantly different. We face a similar situation in regression ANOVA. If we reject the null hypothesis, we know that at least one coefficient does not equal zero, meaning that at least one variable is useful in prediction, thus the purpose of the individual t tests.

For variable x_i, $i = 1$ through k, the null and alternative hypotheses are:

$$H_0 : B_i = 0$$
$$H_1 : B_i \neq 0$$

Recall that if $k = 1$, the t test is equivalent to the regression ANOVA test. However, if $k > 1$, this is not the case due to the algebra involved.

Since $k = 2$, we have two null hypotheses to test: H_0: $B_1 = 0$ and H_0: $B_2 = 0$. If we look in the section where we found the coefficients, we see in the next column that each variable has its own standard deviation (or standard error). As with the prior t test, we find the value of the t test statistic by dividing the coefficient by its standard error.

$$\text{gender } t \text{ test statistic} = \frac{\text{gender coefficient}}{\text{gender standard error}} = \frac{4.0385}{21.6434} = 0.1866$$

$$\text{income } t \text{ test statistic} = \frac{\text{income coefficient}}{\text{income standard error}} = \frac{51.9231}{12.9524} = 4.0088$$

Notice that the column for the t test statistic is to the right of the standard error column labelled **t Stat**.

The degrees of freedom for the t distribution is the error degrees of freedom from the regression ANOVA table, which is 3. If we choose $\alpha = 0.05$, the upper critical value from Table A-3 would be 3.182. However, this is not necessary. To the right of the test statistic column is the P-value column. If the P-value of any variable is less than the level of significance, we reject the null hypothesis for that variable and conclude that variable is useful in prediction. The P-value for gender is $2P(t > 0.1866) = 0.8639$ while that of income is $2P(t > 4.0088) = 0.0278$. Based on the decision criterion, we conclude that gender is not useful in predicting the amount spent per month eating out, while income is.

Variable significance is a term used to mean that, at a certain level of significance, a variable is useful in prediction. In other words, if a variable is useful in prediction, we say that variable is significant. In using this element to find the best model, we will choose a model in which all of the variables are significant.

Presence of Multicollinearity

While it is desirable for an independent variable to be highly correlated with the dependent variable (as indicated by the variable being significant), it is not desirable for the independent variables to be highly correlated with each other. If some or all independent variables *are* highly correlated, we have the presence of **multicollinearity**. There are two main symptoms of multicollinearity:

SYMPTOM 1: THE MODEL AS A WHOLE IS SIGNIFICANT BUT NO VARIABLES ARE SIGNIFICANT.

SYMPTOM 2: THE COEFFICIENT FOR A VARIABLE CAN HAVE THE OPPOSITE SIGN OF WHAT IT SHOULD. For example, if a variable is positively correlated with the dependent variable, its coefficient should be positive; in the presence of multicollinearity, it could be negative instead.

There is a statistic that is used to detect if a variable suffers from multi-collinearity called the Variance Inflation Factor (VIF). For independent variable x_j, $j = 1$ through k, the VIF is found by first regressing x_j against the other independent variables; that is, by constructing a multiple regression model with x_j as the dependent variable. As with any multiple regression model, we can measure the coefficient of determination R_j^2, with the subscript j referring to the fact that x_j is the dependent variable. From this, we can calculate the VIF for x_j:

$$VIF_j = \frac{1}{1 - R_j^2}$$

If x_j is independent of the other independent variables, R_j^2 should be close to zero and its VIF should be close to the minimum value of 1. However, as it becomes more correlated with the other independent variables, R_j^2 should increase and, subsequently, so will its VIF. A common threshold used is that if the VIF > 10 for any variable, the model suffers from multicollinearity.

In our example, since there are only two variables, gender and income, it does not matter which one we make the dependent variable to find R^2 and the subsequent VIF: R^2 will be the same for both models, namely, 0.5448. The VIF for both variables is:

$$VIF = \frac{1}{1 - 0.5448} = \frac{1}{0.4552} = 2.2$$

Therefore, the model does not suffer from multicollinearity.

A couple of notes: First, if there is only one independent variable, VIF = 1 by default. Second, if a model with more than one independent variable does not suffer from multicollinearity, any subsequent model built with a subset of the variables from the initial model will also not suffer from multicollinearity.

In using this element to find the best model, we will choose a model which does not suffer from multicollinearity.

Finding the Best Multiple Regression Equation

In the initial model, we found that gender is not significant while income is significant. While there are no hard and fast rules, if a variable is not significant in one model, it is unlikely that it would be significant in other models. So, for the second model, we choose only income as the independent variable. Here is the Excel output:

1	SUMMARY OUTPUT					
2						
3	*Regression Statistics*					
4	Multiple R	0.9569				
5	R Square	0.9156				
6	Adjusted R Square	0.8945				
7	Standard Error	15.5784				
8	Observations	6				
9						
10	ANOVA					
11		*df*	*SS*	*MS*	*F*	*Significance F*
12	Regression	1	10529.2479	10529.2479	43.3859	0.0028
13	Residual	4	970.7521	242.6880		
14	Total	5	11500			
15						
16		*Coefficients*	*Standard Error*	*t Stat*	*P-value*	*Lower 95%*	*Upper 95%*
17	Intercept	8.7326	24.5620	0.3555	0.7401	-59.4626	76.9278
18	income	50.1393	7.6121	6.5868	0.0028	29.0047	71.2739

The adjusted $R^2 = 0.8945$ and the regression ANOVA P-value is 0.0028. Recall from the previous section that if there is only one independent variable, the P-value for both regression ANOVA and the t test would be the same, which is the case here. If we use $\alpha = 0.05$ to determine if income is significant, since $0.0028 < 0.05$, we conclude that it is. Finally, VIF $= 1$ by default since there is only one independent variable.

Finally, we construct a model using only gender. Here is the Excel output:

1	SUMMARY OUTPUT					
2						
3	*Regression Statistics*					
4	Multiple R	0.6852				
5	R Square	0.4696				
6	Adjusted R Square	0.3370				
7	Standard Error	39.0512				
8	Observations	6				
9						
10	ANOVA					
11		*df*	*SS*	*MS*	*F*	*Significance F*
12	Regression	1	5400	5400	3.5410	0.1330
13	Residual	4	6100	1525		
14	Total	5	11500			
15						
16		*Coefficients*	*Standard Error*	*t Stat*	*P-value*	*Lower 95%*	*Upper 95%*
17	Intercept	195	22.5462	8.6489	0.0010	132.4014	257.5986
18	gender	-60	31.8852	-1.8818	0.1330	-148.5277	28.5277

The adjusted $R^2 = 0.337$ and both the regression ANOVA and t test P-value are equal at 0.1330 indicating that the model is not significant and, of course, neither is the variable. As with the last model, its VIF $= 1$.

Now that we have constructed the models, we can summarize the results:

Table 11-4 Searching for the Best Multiple Regression Equation

Model	Adjusted R^2	ANOVA P-value	t Tests (all significant— yes/no)	Multicollinearity (yes/no)
#1: Gender, Income	0.8906	0.0241	no	no
#2: Income	0.8945	0.0028	yes	no
#3: Gender	0.3370	0.1330	no	no
Best model	#2	#2	#2	tie

One possible method to choose the best overall model is to go through each of the elements and see which model performs best. In our example, model #2 using only income appears to be the best model since it has the highest adjusted R^2 = 89.45% and the lowest regression ANOVA P-value of 0.28%, its variable is significant, and it does not suffer from multicollinearity.

The initial model in the previous example has no presence of multicollinearity. We now present a data set with multicollinearity.

EXAMPLE

Table 11-5 is drawn from data in Data Set 3 in Appendix B, which contains measurements from anaesthetized bears. Using all the independent variables x_2 through x_7, find the VIF for each one.

Table 11-5 Data from Anesthetized Male Bears

Sample Number	y Weight	x_2 Age	x_3 Headlen	x_4 Headwth	x_5 Neck	x_6 Length	x_7 Chest
1	80	19	11.0	5.5	16.0	53.0	26
2	344	55	16.5	9.0	28.0	67.5	45
3	416	81	15.5	8.0	31.0	72.0	54
4	348	115	17.0	10.0	31.5	72.0	49
5	262	56	15.0	7.5	26.5	73.5	41
6	360	51	13.5	8.0	27.0	68.5	49
7	332	68	16.0	9.0	29.0	73.0	44
8	34	8	9.0	4.5	13.0	37.0	19

SOLUTION

The VIFs for the variables are as follows:

Age	Headlen	Headwth	Neck	Length	Chest
8.78	38.4877	17.9902	163.5921	11.8115	47.4805

To find the VIF of 8.78 for Age, we regress this variable against the remaining five independent variables. We find that $R^2 = 0.8861$, from which we get VIF $= 1/(1 - 0.8861) = 8.78$. We would follow the same procedure for the remaining variables.

Aside from the Age, all the other variables suffer from multicollinearity. This stands to reason since we would expect any one of a bear's dimensions to be proportional to the others. It may be enlightening to view the Excel output for the model using all the variables:

1	SUMMARY OUTPUT						
2							
3	Regression Statistics						
4	Multiple R	0.9997					
5	R Square	0.9994					
6	Adjusted R Square	0.9957					
7	Standard Error	9.1474					
8	Observations	8					
9							
10	ANOVA						
11		df	SS	MS	F	Significance F	
12	Regression	6	136564.3249	22760.7208	272.0132	0.0464	
13	Residual	1	83.6751	83.6751			
14	Total	7	136648				
15							
16		Coefficients	Standard Error	t Stat	P-value	Lower 95%	Upper 95%
17	Intercept	-216.4152	29.2156	-7.4075	0.0854	-587.6335	154.8031
18	Age	-1.3720	0.3041	-4.5122	0.1388	-5.2356	2.4915
19	Headlen	-4.4894	7.5519	-0.5945	0.6586	-100.4446	91.4659
20	Headwth	6.8528	7.9249	0.8647	0.5461	-93.8421	107.5477
21	Neck	19.4721	6.4019	3.0416	0.2022	-61.8709	100.8150
22	Length	-2.9307	0.9155	-3.2013	0.1927	-14.5631	8.7016
23	Chest	6.7196	1.9631	3.4229	0.1810	-18.2244	31.6637

Examining the results of the regression ANOVA table, if we use $\alpha = 0.05$, we find that the model as a whole is significant with a P-value of 0.0464, though barely so. However, when we examine the P-values of each of the independent variables, Age has the lowest one at 0.1388. Again using $\alpha = 0.05$, we conclude that none of the variables is significant. This is due to the presence of multicollinearity.

Furthermore, if we compute the correlation coefficient of Weight and Age, we find $r = 0.8139$, indicating that as a bear gets older, its weight increases. The coefficient for Age is -1.372, however, suggesting a negative correlation. The variables Headlen and Length also experience this phenomenon. Again, this is due to the presence of multicollinearity.

In a situation like this, it may be best to construct separate models each with only one independent variable and compare R^2 for each model to see which variable is best in predicting weight. Note that there is no need to compare the adjusted R^2 since each model has only one variable. We summarize the results:

Age	Headlen	Headwth	Neck	Length	Chest
66.25%	78.17%	78.36%	94.26%	80.52%	98.31%

Based on these results, Chest is the best predictor of Weight. Here is the Excel output for this model:

1	SUMMARY OUTPUT						
2							
3	*Regression Statistics*						
4	Multiple R	0.9915					
5	R Square	0.9831					
6	Adjusted R Square	0.9803					
7	Standard Error	19.6208					
8	Observations	8					
9							
10	ANOVA						
11		*df*	*SS*	*MS*	*F*	*Significance F*	
12	Regression	1	134338.1341	134338.1341	348.9505	1.51905E-06	
13	Residual	6	2309.8659	384.9776			
14	Total	7	136648				
15							
16		*Coefficients*	*Standard Error*	*t Stat*	*P-value*	*Lower 95%*	*Upper 95%*
17	Intercept	-194.6104	25.9242	-7.5069	0.0003	-258.0447	-131.1761
18	Chest	11.4155	0.6111	18.6802	0.0000	9.9202	12.9109

The model would be Weight $= -194.6104 + 11.4155$(Chest). For example, if a bear has a chest measuring 40 inches around, the model predicts its average weight would be $-194.6104 + 11.4155(40) = 262$ lb. Based on the coefficient, for each additional inch in the chest measurement, the bear's weight increases by 11.4155 lb.

For cases involving a large number of independent variables, many statistical software packages include a program for performing **stepwise regression**, whereby different combinations are tried until the best model is obtained. (However, in some cases, the validity of such solutions may be questionable.)

When we discussed regression in Section 11–3, we listed four common errors that should be avoided when using regression equations to make predictions. These same errors should be avoided when using multiple regression equations. Be especially careful about concluding that a cause–effect relationship exists.

USING TECHNOLOGY

EXCEL: First enter the sample data in columns.

EXCEL (Prior to 2007): Select **Tools** from the main menu, then select **Data Analysis** and **Regression**. In the dialog box, enter the range of values for the dependent (y) variable, then enter the range of values for the independent (x) variables, which must be in adjacent columns. An example output is illustrated in this section.

EXCEL 2007: Click **Data** on the main menu, then **Data Analysis** from the **Analysis** tab. Proceed as above.

MINITAB 15: From the **Stat** menu, choose **Regression** and then **Regression** again. In the dialog box, choose the column with the dependent variable and the columns with the independent (**Predictor**) variables. Under **Options**, there is a checkbox for VIF.

STATDISK: Select **Analysis**, then **Multiple Regression**. Check the boxes with the columns containing both the independent variables and the dependent variable. (You may have up to 8 independent variables.) Enter the column in the box for the dependent variable. Click on **Evaluate**. Outputs include the multiple regression coefficients, the components of the regression ANOVA table (including the P-value), the multiple coefficient of determination R^2, and the adjusted R^2.

11-5 Exercises A: Basic Skills and Concepts

In Exercises 1–4, refer to the Excel display below and answer the given questions.

1. Construct the multiple regression equation that expresses earthquake magnitude in terms of latitude, longitude, and depth.

2. Identify
   a. the P-value corresponding to the overall significance of the multiple regression equation.
   b. the value of the multiple coefficient of determination R^2.
   c. the adjusted value of R^2.

EXCEL DISPLAY

	A	B	C	D	E	F	G
1	SUMMARY OUTPUT						
2							
3	Regression Statistics						
4	Multiple R	0.409795					
5	R Square	0.167932					
6	Adjusted R Square	0.145239					
7	Standard Error	1.067112					
8	Observations	114					
9							
10	ANOVA						
11		df	SS	MS	F	Significance F	
12	Regression	3	25.28057	8.426856	7.400232	0.000146209	
13	Residual	110	125.2601	1.138728			
14	Total	113	150.5407				
15							
16		Coefficients	Standard Err	t Stat	P-value	Lower 95%	Upper 95%
17	Intercept	-2.02693	1.26949	-1.59665	0.113212	-4.542766075	0.488899215
18	Lat	0.069004	0.017405	3.964659	0.000131	0.034511774	0.103496089
19	Long	0.008697	0.005985	1.453163	0.149025	-0.00316379	0.020558711
20	Depth	-0.02623	0.012521	-2.09508	0.038459	-0.051045566	-0.001418796

3. Is the multiple regression equation usable for predicting an earthquake's magnitude based on its recorded latitude, longitude, and depth? Why or why not?

4. Seismic activity has been detected as 48.2°N latitude and 124.99°W longitude, at a depth of 1 m.
   a. Find the predicted magnitude of the earthquake.
   b. The seismic activity in question actually had a magnitude of 0.9 ML. How accurate is the predicted weight from part (a)?

In Exercises 5–8, refer to the temperature and precipitation data in Data Set 4 of Appendix B. Let the dependent variable be AVTEMP, and let the independent variables be those given in the exercise. (The data sets are already stored on the data disk and STATDISK.) Use software such as Excel, Minitab, or STATDISK to answer these questions:

5. If AVPRECIP and AVSNOW are used to predict AVTEMP, why is there no presence of multicollinearity?

6. Construct a model in which AVPRECIP and AVSNOW are used to predict AVTEMP. If a city has average precipitation of 900 mm per year and average snowfall of 225 cm per year, what is the city's predicted average temperature, accurate to 1 decimal?

7. Analysis indicates that the data from Prince Rupert, B.C. (third from the bottom of the data) is outlier data. If we remove this set of data, what is the change in R^2 and the regression ANOVA P-value? Also, reconstruct the model in Question 6 and repredict the average temperature based on the values given in Question 6.

8. Without the outlier data from Prince Rupert, and based on adjusted R^2, regression ANOVA P-value, and variable significance based on the individual t tests, which of these three models is best in predicting average annual temperature: (#1) AVPRECIP and AVSNOW, (#2) AVPRECIP, or (#3) AVSNOW?

In Exercises 9–15, refer to the weights of garbage in Data Set 1 of Appendix B. Let the dependent variable be HHSIZE (household size), and let the independent variables be those given in the exercise. (The data sets are already stored on the data disk and STATDISK.) Use software such as Excel, Minitab, or STATDISK to answer these questions:

a. *Find the multiple regression equation that expresses the dependent variable HHSIZE in terms of the given independent variables.*

b. *Identify the values of the multiple coefficient of determination R^2, the adjusted R^2, and (if Excel is used) the P-value corresponding to the overall significance.*

c. *Does the multiple regression equation seem suitable for making predictions of household size based on the given independent variables?*

9. YARD

10. PLASTIC

11. PLASTIC and PAPER

12. METAL, PLASTIC, and FOOD

13. METAL and PLASTIC

14. PAPER and GLASS

15. METAL, PAPER, PLASTIC, GLASS, and FOOD

16. Refer to Data Set 5 in Appendix B and find the best multiple regression equation with number of Senate sittings as the dependent variable. (Don't include

the number of each Parliament as an independent variable.) Is this equation suitable for predicting the number of Senate sittings in a given Parliament? Why or why not? [*Hint:* To efficiently try out the regressions of the dependent variable with various combinations of independent variables, try sending each output to its own (suitably labelled) spreadsheet page. Where the differences between significance levels are very small, the real-world judgment of "best" regression may depend partly on non-mathematical factors, such as the ease of getting data for the different variables.]

11-5 Exercises B: Beyond the Basics

17. In some cases, the best-fitting multiple regression equation is of the form $\hat{y} = b_0 + b_1x + b_2x^2$. The graph of such an equation is a parabola. Using the data set listed in the margin, let $x_1 = x$, and, creating an x^2 row, let $x_2 = x^2$. Find the multiple regression equation for the parabola that best fits the given data. Based on the value of the multiple coefficient of determination, how well does this equation fit the data?

x	1	3	4	7	5
y	5	14	19	42	26

■ VOCABULARY LIST

adjusted coefficient of
   determination   655
bivariate data   600
centroid   612
coefficient of
   determination   639
confidence interval
   for μ_y 645
correlation   600
correlation coefficient   603
cross product   604
dependent variable   619
dummy variable   654
explained deviation   638
explained variation   639
independent variable   619
influential point   625
least-squares property   627

linear correlation
   coefficient   603
lurking variable   607
marginal change   625
mathematical model   628
multicollinearity   657
multiple coefficient of
   determination   655
multiple regression   652
multiple regression
   equation   652
outlier   625
Pearson's product moment
   correlation coefficient   603
predicted value   623
prediction interval   645
predictor variable   619
qualitative variable   654

regression ANOVA   641
regression equation   619
regression line   619
residual   626
residual plot   629
response variable   619
scatter diagram   601
scatterplot   601
standard error of
   estimate   645
stepwise regression   662
total deviation   637
total variation   639
unexplained
   deviation   638
unexplained
   variation   639
variable significance   657

■ REVIEW

In this chapter we presented basic methods for using paired data to investigate the relationship between two or more variables. We generally limited our discussion to linear relationships because consideration of nonlinear relationships requires more advanced mathematics.

- Section 11–2 used scatter diagrams and the linear correlation coefficient to decide whether there is a linear correlation between two variables.

- Section 11–3 presented methods for finding the equation of the regression line, which (by the least-squares criterion) best fits the paired data. When there is a significant linear correlation, the regression line can be used to predict the value of a variable, given some value of the other variable. We saw that regression equations exemplify the more general concept of a mathematical model, which is a function that can be used to describe a relationship between variables. The possibility of *nonlinear* relationships was discussed. Residual plots were discussed as a way to determine model sufficiency.

- Section 11–4 introduced the concept of total variation, with components of explained and unexplained variation. We defined the coefficient of determination r^2 to be the quotient obtained by dividing explained variation by total variation. We introduced regression ANOVA in order to determine model significance. We introduced t tests and confidence intervals for the model slope. We also developed methods for constructing prediction intervals, which are helpful in judging the accuracy of predicted values as well as confidence intervals for μ_y.

- In Section 11–5 we considered multiple regression, which allows us to investigate relationships among several variables, including dummy (qualitative) variables. We discussed procedures for obtaining a multiple regression equation, as well as the values of the multiple coefficient of determination R^2, the adjusted R^2, the regression ANOVA P-value, individual t tests, and the presence of multicollinearity to determine the "best" model.

REVIEW EXERCISES

In Exercises 1–4, use the data in the accompanying table. The data come from a study of ice cream consumption that spanned the springs and summers of three years. The ice cream consumption is in litres per capita per week, price of the ice cream is in dollars, family income of consumers is in dollars per week, and temperature is in degrees Celsius.

Consumption	0.219	0.212	0.223	0.241	0.231	0.195	0.186	0.164	0.153	0.145
Price	2.38	2.48	2.45	2.46	2.39	2.31	2.43	2.36	2.34	2.45
Income	540	548	562	554	527	540	568	548	527	548
Temperature	5	13	17	20	21	18	16	8	0	−4

Based on data from Kadiyala, *Econometrica*, Vol. 38.

1. a. Use a 0.05 significance level to test for a linear correlation between consumption and price.

b. Find the equation of the regression line that expresses consumption (y) in terms of price (x).

c. What is the best predicted consumption amount if the price is $2.43?

2. a. Use a 0.05 significance level to test for a linear correlation between consumption and income.

b. Find the equation of the regression line that expresses consumption (y) in terms of income (x).

c. What is the best predicted consumption amount if the income is $562?

3. a. Use a 0.05 significance level to test for a linear correlation between consumption and temperature.

b. Find the equation of the regression line that expresses consumption (y) in terms of temperature (x).

c. What is the best predicted consumption amount if the temperature is 0°C?

4. Use software such as Excel, Minitab, or STATDISK to find the multiple regression equation of the form $\hat{y} = b_0 + b_1x_1 + b_2x_2 + b_3x_3$, where the dependent variable y represents consumption, x_1 represents price, x_2 represents income, and x_3 represents temperature. Also identify the value of the multiple coefficient of determination R^2, the adjusted R^2, and the P-value representing the overall significance of the multiple regression equation. Can the regression equation be used to predict ice cream consumption? Are any of the equations from Exercises 1–3 better?

In Exercises 5–8, use the sample data in the accompanying table. The table includes the results for the top performers in the 1996 Molson Indy Toronto. Note that the data in the Final Position column are ordinal and thus should not be used in the regressions. The distance between the first and second final positions, for example, may be quite different from the distance between the second and third final positions. Intervals between starting positions, however, are assumed to correspond to consistent track distances. Times are given in seconds and speeds in miles per hour.

Final Position	Starting Position	Qualifying Time	Qualifying Speed	Points
1	3	58.471	109.838	20
2	2	58.145	110.455	17
3	8	59.016	108.825	14
4	5	58.519	109.748	12
5	13	59.189	108.507	10
6	12	59.175	108.533	8
7	14	59.220	108.450	6
8	11	59.120	108.634	5
9	18	59.680	107.615	4
10	4	58.490	109.803	3

Based on data from Molson Indy Toronto.

5. a. Use a 0.05 significance level to test for a linear correlation between qualifying speed and starting position.

b. Find the equation of the regression line that expresses starting position (y) in terms of qualifying speed (x).

c. What is the best predicted starting position (rounded to a whole number) for a driver whose qualifying speed is 109.066?

6. a. Use a 0.05 significance level to test for a linear correlation between qualifying speed and points.

b. Find the equation of the regression line that expresses points (y) in terms of qualifying speed (x).

c. What is the best predicted number of points (rounded to a whole number) for a driver whose qualifying speed is 109.066?

7. Use only the paired data for starting position and qualifying speed. For a driver whose qualifying speed is 109.066, find a 95% prediction interval estimate of the starting position.

8. Let y = points, x_1 = starting position, and x_2 = qualifying time. Use software such as Excel or STATDISK to find the multiple regression equation of the form $\hat{y} = b_0 + b_1 x_1 + b_2 x_2$. Also identify the value of the multiple coefficient of determination R^2, the adjusted R^2, and the P-value representing the overall significance of the multiple regression equation. Based on the results, should the multiple regression equation be used for making predictions? Why or why not?

CUMULATIVE REVIEW EXERCISES

1. In 1970, the mean time between eruptions of the Old Faithful geyser was 66 minutes. Refer to the intervals (in minutes) between eruptions for the more recent data listed in Data Set 16 of Appendix B.

a. Test the claim of Yellowstone National Park geologist Rick Hutchinson that eruptions now occur at intervals that are longer than in 1970.

b. Construct a 95% confidence interval for the mean time between eruptions.

2. The Marc Michael Advertising Company has prepared two different television commercials for Taylor's women's jeans. One commercial is humorous and the other is serious. A test screening involves a standard scale with higher scores indicating more favourable responses. The results are listed here. Based on the results, does one commercial seem to be better? Is the issue of correlation relevant to this situation?

Consumer	A	B	C	D	E	F	G	H
Humorous commercial	26	33	19	18	29	27	23	24
Serious commercial	21	30	14	14	24	22	21	19

3. In studying the effects of heredity and environment on intelligence, it has been helpful to analyze IQs of identical twins who were separated soon after birth.

Identical twins share identical genes inherited from the same fertilized egg. By studying identical twins raised apart, we can eliminate the variable of heredity and better isolate the effects of environment. The accompanying table shows the IQs of identical twins (older twins are x) raised apart. Use the sample data to determine whether there is a relationship between IQs of the twins. Write a summary statement about the effect of heredity and environment on intelligence. Note that your conclusions will be based on this relatively small sample of 12 pairs of identical twins.

x	107	96	103	90	96	113	86	99	109	105	96	89
y	111	97	116	107	99	111	85	108	102	105	100	93

Based on data from "IQs of Identical Twins Reared Apart," by Arthur Jensen, *Behavioral Genetics*.

TECHNOLOGY PROJECT

Use Excel or STATDISK to retrieve the bear data found in Data Set 3 of Appendix B. (Data Set 3 is stored on the data disk.) Using the complete data set, find the linear correlation coefficient and equation of the regression line when the dependent variable of weight is paired with (a) head length, (b) head width, (c) neck size, (d) length, and (e) chest size. Which of these regression equations is best for predicting the weight of a bear given its head length, head width, neck size, length, and chest size? Also, find the multiple regression equation obtained when the dependent variable of weight is expressed in terms of the independent variables of head length, head width, neck size, and chest size. Interpret the computer results. Is the resulting multiple regression equation suitable for making predictions of a bear's weight based on its head length, head width, neck size, and chest size? Why or why not?

When hiking (with a tape measure but no scale) on the Grizzly Lake trail in Yellowstone National Park, one of the authors once encountered a bear, which he wrestled to the ground and rendered temporarily unconscious. (*Editor's note:* The truth is that he saw a bear 250 m away.) Given all of the preceding results, what is the best predicted weight of the bear that was measured and found to have a 12.3 in. head length, a 4.9 in. head width, a 19.0 in. neck size, a 55.3 in. length, and a 31.4 in. chest size?

FROM DATA TO DECISION

How Much Hockey "Star Power" Should You Buy?

Hockey-team executives must regularly assess the value of buying (or trading away) expensive star players for their team. Do the expensive new players deliver when it comes to team wins? The accompanying table lists NHL teams' payrolls (in millions of dollars, assessed after the trading deadline for the 1995–96 season) and points for that season (as of March 29, 1996). Use the methods of this chapter to analyze the data. Assuming the 1996 data are representative of future years, determine whether there is a correlation between payroll and points. How much should a team's payroll be if its season ended with 75 points? What is the highest point total you could expect if payroll was capped at $15 million? Which was the better deal in 1996: Montreal's payroll or Toronto's? Why?

Team	Payroll	Points
Edmonton	11.24	66
Ottawa	11.58	34
N.Y. Islanders	13.02	48
San Jose	14.14	43
Anaheim	14.14	65
Calgary	14.36	73
Tampa Bay	14.83	77
Buffalo	17.19	63
Montreal	17.25	84
Washington	17.40	82
Los Angeles	17.51	61
Florida	17.69	85
Boston	20.12	79
Hartford	20.47	69
Dallas	20.77	61
Chicago	21.07	86
Colorado	21.25	96
Philadelphia	21.91	91
Winnipeg	22.78	69
Pittsburgh	22.83	94
Toronto	22.93	72
New Jersey	24.47	80
Vancouver	25.33	75
Detroit	28.42	119
St. Louis	30.45	76
N.Y. Rangers	33.84	90

Based on data from "Making a Point Is Expensive" in the *Toronto Star*, March 31, 1996.

■ COOPERATIVE GROUP ACTIVITIES

1. **Out-of-class activity:** Divide into groups of 3 or 4 people. Investigate the relationship between two variables by collecting your own paired sample data and using the methods of this chapter to determine whether there is a significant linear correlation. Also identify the regression equation and describe a procedure for predicting values of one of the variables when given values of the other variable. Suggested topics:

- Is there a relationship between taste and cost of different brands of chocolate chip cookies (or colas)? (Taste can be measured on some scale, such as 1 to 10.)

- Is there a relationship between salaries of professional baseball (or basketball, or football) players and their season achievements?

- Rates versus weights: Is there a relationship between car fuel consumption rates and car weights? If so, what is it?

- Is there a relationship between the lengths of men's (or women's) feet and their heights?

- Is there a relationship between student grade-point averages and the amount of television watched? If so, what is it?

- Is there a relationship between heights of fathers (or mothers) and heights of their first sons (or daughters)?

2. **In-class activity:** Divide into groups of 8 to 12 people. For each group member, *measure* his or her height and *measure* his or her arm span. For the arm span, the subject should stand with arms extended, like the wings on an airplane. It's easy to mark the height and arm span on a chalkboard, then measure the distances there. Using the paired sample data, is there a correlation between height and arm span? If so, find the regression equation with height expressed in terms of arm span. Can arm span be used as a reasonably good predictor of height?

3. **In-class activity:** Divide into groups of 8 to 12 people. For each group member, record the pulse rate by counting the number of heart beats in one minute. Also record height. Is there a relationship between pulse rate and height? If so, what is it?

4. **In-class activity:** Divide into groups of 8 to 12 people. For each group member, use a string and ruler to measure head circumference and forearm length. Is there a relationship between these two variables? If so, what is it?

INTERNET PROJECT

Linear Regression

The linear correlation coefficient is a tool that can be used to measure the strength of the linear relationship between two variables. We can evaluate the linear correlation coefficient for any two sets of paired data, and then we can proceed to analyze the resulting values. Does it make sense for the two variables to be linearly correlated? Could a high correlation be caused by a third variable that is correlated with each of the two original variables? The Internet Project for this chapter can be found at either of the following websites:

http://www.mathxl.com   or   http://www.mystatlab.com

Click on Internet Project, then on Chapter 11. There you will be told how to find several paired data sets in the fields of sports, medicine, and economics. You will then apply the correlation and regression methods of this chapter to answer some of the questions that arise in this type of analysis.

12

Nonparametric Statistics

12-1 Overview

Chapter objectives are identified. The general nature and advantages and disadvantages of nonparametric methods are presented, and the concept of ranks is introduced.

12-2 Sign Test

The sign test is used to test the claim that two sets of dependent data have the same median.

12-3 Wilcoxon Signed-Ranks Test

The Wilcoxon signed-ranks test serves two purposes: to test the claim of a specified median and to test the claim that two sets of dependent data come from identical populations.

12-4 Wilcoxon Rank-Sum Test for Two Independent Samples

The Wilcoxon rank-sum test is used to test the claim that two independent samples come from identical populations.

12-5 Tests for Multiple Samples

The Kruskal-Wallis test is used to test the claim that several independent samples come from identical populations. The Friedman test is the equivalent test for dependent samples.

12-6 Rank Correlation

The rank correlation coefficient is used to test for an association between two sets of paired data.

12-7 Runs Test for Randomness

The runs test is used to test for randomness in the way data are selected.

CHAPTER PROBLEM

Do short and long Parliaments occur in a nonrandom pattern?

Of the first 35 Parliaments since Confederation, most lasted for more than three years (typically four to five years). Yet nearly a quarter of the Parliaments (8 of them) were relatively brief, lasting only one or two years. Is there any evidence that the short and long Parliaments occur in a nonrandom pattern? If there is, would you want to run for an election if you knew a brief session was coming up next? Or, in fact, does the alternation between Parliament lengths appear to be random?

The sequence of typical-length Parliaments (T) and brief Parliaments (B) is listed in Figure 12-1. (The sequence is read from left to right, continuing on the next row.) In Section 12-7 we will consider a test for determining whether such a sequence is random.

Figure 12-1 Length of Parliaments in Session

```
T   B   T   T   T   T   T   T   T   T   B   T   T   T   B   T   T   T
T   T   T   T   B   T   B   T   B   T   B   T   B   T   T   T   T
```

12-1 Overview

The methods of inferential statistics presented in Chapters 6, 7, 8, 9, and 11 are called *parametric methods* because they are based on sampling from a population with specific parameters, such as the mean μ, standard deviation σ, or proportion p. Those parametric methods usually must conform to some fairly strict conditions, such as a requirement that the sample data come from a normally distributed population. In this chapter we introduce nonparametric methods, which do not have such strict requirements.

DEFINITIONS

Parametric tests require assumptions about the nature or shape of the populations involved; **nonparametric tests** do not require such assumptions. Consequently, nonparametric tests of hypotheses are often called **distribution-free tests**.

Although the term *nonparametric* strongly suggests that the test is not based on a parameter, there are some nonparametric tests that do depend on a parameter such as the median, but they don't require a particular distribution. Although *distribution-free* is a more accurate description, the term *nonparametric* is more commonly used. The following are major advantages and disadvantages of nonparametric methods.

Advantages of Nonparametric Methods

1. Nonparametric methods can be applied to a wide variety of situations because they do not have the more rigid requirements associated with parametric methods. In particular, nonparametric methods do not require normally distributed populations.

2. Unlike parametric methods, nonparametric methods can often be applied to nonnumerical data such as the genders of survey respondents, or to ordinal scale data such as survey ratings.

3. Nonparametric methods usually involve simpler computations than the corresponding parametric methods and are therefore easier to understand and apply.

Disadvantages of Nonparametric Methods

1. Nonparametric methods may appear to waste information in cases where exact numerical data are converted to a qualitative form. For example, in the nonparametric sign test (described in Section 12–2), weight losses by dieters are recorded simply as negative signs; the actual magnitudes of the weight losses are ignored.

2. Nonparametric tests are not as efficient as parametric tests, so with a nonparametric test we generally need stronger evidence (such as a larger sample or greater differences) before we reject a null hypothesis.

When the requirements of population distributions are satisfied, nonparametric tests are generally less efficient than their parametric counterparts, but the reduced efficiency can be compensated for by an increased sample size. For example, Section 12–6 will deal with a concept called *rank correlation*, which has an efficiency rating of 0.91 when compared to the linear correlation presented in Chapter 11. This means that all other things being equal, nonparametric rank correlation requires 100 sample observations to achieve the same results as 91 sample observations analyzed through parametric linear correlation, assuming the stricter requirements for using the parametric method are met.

Table 12-1 lists the nonparametric methods covered in this chapter, along with the corresponding parametric approach and **efficiency** rating. Table 12-1 shows that several nonparametric tests have efficiency ratings above 0.90, so the lower efficiency might not be a critical factor in choosing between parametric or nonparametric methods. More important is that we avoid using the parametric tests when their required assumptions are not satisfied.

Ranks

Sections 12–3 through 12–6 use methods based on ranks. First, we look briefly at the nature of ranks.

DEFINITION

Data are *sorted* when they are arranged according to some criterion, such as smallest to largest or best to worst. A **rank** is a number assigned to an individual sample item according to its order in the sorted list. The first item is assigned a rank of 1, the second item is assigned a rank of 2, and so on.

Table 12-1 Comparison of Parametric and Nonparametric Tests

Application	Parametric Test	Nonparametric Test	Efficiency of Nonparametric Test with Normal Population
Matched pairs of sample data	*t* test or *Z* test	Sign test	0.63
		Wilcoxon signed-ranks test	0.95
Two independent samples	*t* test or *Z* test	Wilcoxon rank-sum test	0.95
Several independent samples	Analysis of variance (*F* test)	Kruskal-Wallis test	0.95
		Friedman test	0.95
Correlation	Linear correlation	Rank correlation test	0.91
Randomness	No parametric test	Runs test	No basis for comparison

The numbers 5, 3, 40, 10, and 12 can be arranged from lowest to highest as 3, 5, 10, 12, and 40, and these numbers have ranks of 1, 2, 3, 4, and 5, respectively:

5	3	40	10	12	Original values
3	5	10	12	40	Values arranged in order
↑	↑	↑	↑	↑	
1	2	3	4	5	Ranks

Handling ties in ranks: If a tie in ranks occurs, the usual procedure is to find the mean of the ranks involved and then assign this mean rank to each of the tied items, as in the following example.

The numbers 3, 5, 5, 10, and 12 are given ranks of 1, 2.5, 2.5, 4, and 5, respectively. In this case, ranks 2 and 3 were tied, so we found the mean of 2 and 3 (which is 2.5) and assigned it to the values that created the tie:

2 and 3 are tied

In the above examples, assigning the ranks was easy to keep track of since there were only 5 values. Imagine if there were 20 or 30 and there are ties on top of that. You want to make sure you assign the correct rank to each value. Fortunately, there is a check that can be easily used.

Sum of Arithmetic Series

The sum of the numbers 1, 2, 3, . . . n is $\frac{n(n + 1)}{2}$. For example, the sum of 1 through 5 is 15. Since $n = 5$, sum $= (5)(6)/2 = 15$. In the second example above, $1 + 2.5 + 2.5 + 4 + 5 = 15$. Since this equals the sum of the arithmetic series, the ranks were assigned correctly.

12-2 Sign Test

In this section we introduce the *sign test* procedure, which is among the easiest for nonparametric tests.

> **DEFINITION**
>
> The **sign test** is a nonparametric (distribution-free) test that uses plus and minus signs to test different claims, including:
>
> 1. Claims involving matched pairs of sample data
> 2. Claims involving nominal data
> 3. Claims about the median of a single population

Assumptions

1. The sample data have been randomly selected.
2. There is no requirement that the sample data come from a population with a particular distribution, such as a normal distribution.

Basic Concept of the Sign Test: To prepare for the sign test, we convert the raw data into plus and minus signs. Then we test whether the frequencies of these plus and minus signs are significantly different. Figure 12-2 summarizes the sign test procedure, which will be illustrated with examples in this section.

For consistency and ease, we will stipulate the following notation.

NOTATION FOR THE SIGN TEST

x = the test statistic representing the number of times the *less frequent* sign occurs

n = the total number of positive and negative signs combined

TEST STATISTIC FOR THE SIGN TEST

For $n \le 25$: Use x (the number of times the less frequent sign occurs)

For $n > 25$: Use
$$z = \frac{(x + 0.5) - \left(\dfrac{n}{2}\right)}{\dfrac{\sqrt{n}}{2}}$$

(based on a normal approximation).

Figure 12-2
Sign Test Procedure

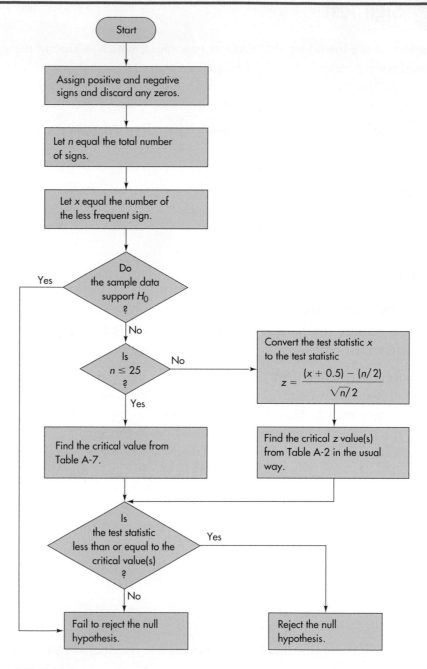

Critical Values

1. For $n \le 25$: Find critical x values in Table A-7.

2. For $n > 25$: Find critical z values in Table A-2.

Caution: When applying the sign test in a one-tailed test, be very careful to avoid making the wrong conclusion when one sign occurs significantly more often than the other, but the sample data are *consistent with* the null hypothesis. If the sense of the data is consistent with (instead of conflicting with) the null hypothesis, then fail to reject the null hypothesis and don't proceed with the sign test. Figure 12-2 summarizes the procedure for the sign test and includes this check: Do the sample data "support" (in the sense of being consistent with) H_0? If the answer is yes, fail to reject the null hypothesis. *It is always important to think about the data and to avoid relying on blind calculations or computer results.*

Claims Involving Matched Pairs of Sample Data

When using the sign test with data that are matched by pairs, we convert the raw data to plus and minus signs as follows:

1. We subtract each value of the second variable from the corresponding value of the first variable.
2. We record only the sign of the difference found in Step 1. We *exclude ties*: That is, we exclude from further analysis any matched pairs in which both values are equal. (For other ways to handle ties, see Exercise 14.)

The key concept underlying the sign test is this:

If the two sets of data have equal medians, the number of positive signs should be approximately equal to the number of negative signs.

EXAMPLE

Table 12-2 consists of sample data obtained when 14 subjects are tested for reaction times with their left and right hands. (Only right-handed subjects were used.) Use a 0.05 significance level to test the claim of no difference between the right- and left-hand reaction times.

SOLUTION

Here's the basic idea: If people generally have the same reaction times with their right and left hands, the numbers of positive and negative signs should be approximately equal—but in Table 12-2 we have 13 negative signs and 1 positive sign. Are the numbers of positive

Table 12-2 Reaction Times (in thousandths of a second) of 14 Subjects														
Right hand	191	97	116	165	116	129	171	155	112	102	188	158	121	133
Left hand	224	171	191	207	196	165	177	165	140	188	155	219	177	174
Sign of difference (right hand – left hand)	−	−	−	−	−	−	−	−	−	−	+	−	−	−

and negative signs approximately equal, or are they significantly different? We follow the same basic steps for testing hypotheses as outlined in Figure 7-4.

Steps 1, 2, 3: The null hypothesis is the claim of no difference between the right- and left-hand reaction times, and the alternative hypothesis is the claim that there is a difference.

H_0: There is no difference.

(The median of the differences is equal to 0.)

H_1: There is a difference.

(The median of the differences is not equal to 0.)

Step 4: The significance level is $\alpha = 0.05$.

Step 5: We are using the nonparametric sign test.

Step 6: The test statistic x is the number of times the less frequent sign occurs. Table 12-2 includes differences with 13 negative signs and 1 positive sign. We let x equal the smaller of 13 and 1, so $x = 1$. Also, $n = 14$ (the total number of positive and negative signs combined). Our test is two-tailed with $\alpha = 0.05$, and reference to Table A-7 shows that the critical value is 2. (See Figure 12-2.)

Step 7: With a test statistic of $x = 1$ and a critical value of 2, we reject the null hypothesis of no difference. (See Note 2 included with Table A-7: "Reject the null hypothesis if the number of the less frequent sign (x) is less than or equal to the value in the table." Because $x = 1$ is less than or equal to 2, we reject the null hypothesis.)

Interpretation

There is sufficient evidence to warrant rejection of the claim that the median of the differences is equal to 0; that is, there is sufficient evidence to warrant rejection of the claim that there is no difference between the right- and left-hand reaction times. This is the same conclusion reached in Section 8–2 where we used a parametric t test, but sign test results do not always agree with parametric test results.

Claims Involving Nominal Data

Recall that nominal data consist of names, labels, or categories only. Although such a nominal data set limits the calculations that are possible, we can identify the *proportion* of the sample data that belong to a particular category, and we can test claims about the corresponding population proportion p. The following example uses nominal data consisting of genders (male/female). The sign test is used by representing women with $+$ signs and men with $-$ signs. (Those signs are chosen arbitrarily, honest.) Also note the procedure for handling cases in which $n > 25$.

EXAMPLE

The Gagné and Turner Advertising Company claims that its hiring practices are fair: "We do not discriminate on the basis of gender, and the fact that 40 of the last 50 new employees are men is just a fluke." The company acknowledges that applicants are about half men and half women, and all have met the basic job-qualification standards. Test the null hypothesis that men and women are hired equally by this company. Use a significance level of 0.05.

SOLUTION

Let p denote the population proportion of hired men. The claim of no discrimination implies that the proportions of hired men and women are both equal to 0.5, so that $p = 0.5$. The null and alternative hypotheses can therefore be stated as follows.

$H_0: p = 0.5$ (The proportion of hired men is equal to 0.5.)

$H_1: p \neq 0.5$

Denoting hired women by $+$ and hired men by $-$, we have 10 positive signs and 40 negative signs. Refer now to the flowchart in Figure 12-2. The test statistic x is the smaller of 10 and 40, so $x = 10$. This test involves two tails because a disproportionately low number of either gender will cause us to reject the claim of equality. The sample data are not consistent with the null hypothesis because 10 and 40 are not precisely equal. Continuing with the procedure in Figure 12-2, we note that the value of $n = 50$ is above 25, so the test statistic x is converted (using a correction for continuity) to the test statistic z as follows:

$$z = \frac{(x + 0.5) - \left(\dfrac{n}{2}\right)}{\dfrac{\sqrt{n}}{2}}$$

$$= \frac{(10 + 0.5) - \left(\dfrac{50}{2}\right)}{\dfrac{\sqrt{50}}{2}} = -4.10$$

With $\alpha = 0.05$ in a two-tailed test, the critical values are $z = \pm 1.96$. The test statistic $z = -4.10$ falls beyond these critical values (see Figure 12-3), so we reject the null hypothesis of equality.

Interpretation

There is sufficient sample evidence to warrant rejection of the claim that the hiring practices are fair. This company is in trouble.

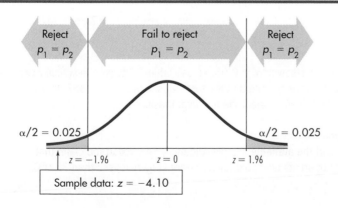

Figure 12-3
Testing the Claim That
Hiring Practices Are Fair

Claims About the Median of a Single Population

In the next example, we illustrate how to use the sign test for testing a claim about the median of a single population. Observe that in such applications, the negative and positive signs are based on the claimed value of the median.

EXAMPLE

In Chapter 7 we used sample data to test the claim that the mean body temperature of adults accepted for voluntary surgery is not 37.0°C. With the same data, use the sign test to test the claim that the median is less than 37.0°C. The data set used in Chapter 7 has 108 subjects—94 subjects with temperatures below 37.0°C, 12 subjects with temperatures above 37.0°C, and 2 temperatures equal to 37.0°C.

SOLUTION

The claim that the median is less than 37.0°C is the alternative hypothesis, while the null hypothesis is the claim that the median is at least 37.0°C:

$$H_0: \text{median is at least } 37.0°C \text{ (median} \geq 37.0°C)$$
$$H_1: \text{median is less than } 37.0°C \text{ (median} < 37.0°C)$$

Following the procedure outlined in Figure 12-2, we discard the 2 zeros, we use the negative sign ($-$) to denote each temperature that is below 37.0°C, and we use the positive sign ($+$) to denote each temperature that is above 37.0°C. We therefore have 94 negative signs and 12 positive signs, so $n = 106$ and $x = 12$ (the number of the less frequent sign). The sample data thus conflict with the null hypothesis, because fewer than half of the 106 temperatures are at least 37.0°C. We must now proceed to determine whether this conflict is significant. (If the sample data did not conflict with the null hypothesis, we could immediately terminate the test by concluding that we fail to reject the null hypothesis.) The value of n exceeds 25, so we convert the test statistic x to the test statistic z:

$$z = \frac{(x + 0.5) - \left(\dfrac{n}{2}\right)}{\dfrac{\sqrt{n}}{2}}$$

$$= \frac{(12 + 0.5) - \left(\dfrac{106}{2}\right)}{\dfrac{\sqrt{106}}{2}} = -7.867$$

In this one-tailed test with $\alpha = 0.05$, we use Table A-2 to get the critical z value of -1.645. From Figure 12-4 we can see that the test statistic of $z = -7.867$ does fall within the critical region. We therefore reject the null hypothesis.

Interpretation
On the basis of the available sample evidence, we support the claim that the median body temperature of adults accepted for voluntary surgery is less than 37.0°C.

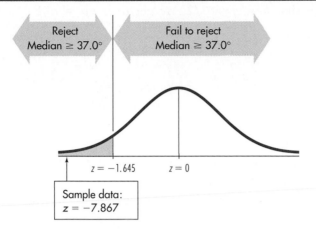

Figure 12-4
Testing the Claim
That the Median Is
Less Than 37.0°C

In this sign test of the claim that the median is below 37.0°C, we get a test statistic of $z = -7.867$, but a parametric test of the claim that $\mu < 37.0$°C results in a test statistic of $z = -10.12$. Because -7.867 isn't as extreme as -10.12, we see that the sign test isn't as sensitive as the parametric test. Both tests lead to rejection of the null hypothesis, but the sign test doesn't consider the sample data to be as extreme, partly because the sign test uses only information about the direction of the differences between pairs of data, ignoring the magnitudes of those differences. The next section introduces the Wilcoxon signed-ranks test, which largely overcomes that disadvantage.

Rationale for the Test Statistics Used with the Sign Test

Rationale for the test statistic used when $n \leq 25$: The critical values for the sign test, when $n \leq 25$, are found in Table A-7 in Appendix A. The entries in Table A-7 were found by using the binomial probability distribution (Section 4–3). The binomial probability distribution applies because the results fall into two categories (positive sign, negative sign) and we have a fixed number of independent cases (or pairs of values).

Rationale for the test statistic used when $n > 25$: Table A-7 applies only for n up to 25, so we need another procedure for finding critical values when $n > 25$. When $n > 25$, the test statistic z is based on a normal approximation to the binomial probability distribution with $p = q = 1/2$. Recall that in Section 5–6 we saw that the normal approximation to the binomial distribution is acceptable when both $np \geq 5$ and $nq \geq 5$. Recall also that in Section 4–4 we saw that $\mu = np$ and $\sigma = \sqrt{npq}$ for binomial experiments. Because this sign test assumes that $p = q = 1/2$, we meet the $np \geq 5$ and $nq \geq 5$ prerequisites whenever $n \geq 10$. Also, with the assumption that $p = q = 1/2$, we get $\mu = np = n/2$ and $\sigma = \sqrt{npq} = \sqrt{n/4} = \sqrt{n}/2$, so

$$z = \frac{x - \mu}{\sigma} \quad \text{becomes} \quad z = \frac{x - \left(\dfrac{n}{2}\right)}{\dfrac{\sqrt{n}}{2}}$$

Finally, we replace x by $x + 0.5$ as a correction for continuity. That is, the values of x are discrete, but because we are using a continuous probability distribution, a discrete value such as 10 is actually represented by the interval from 9.5 to 10.5. Because x represents the less frequent sign, we act conservatively by concerning ourselves only with $x + 0.5$; we thus get the test statistic z, as given in the equation and in Figure 12-2.

USING TECHNOLOGY

EXCEL: You can perform the sign test using the following Excel template. It relies on the fact that the sign test is really based on binomial probabilities: That is, what is the probability, given n positive or negative differences between paired samples and assuming the null hypothesis of no difference between the samples, of randomly getting x or fewer occurrences of the less frequent sign? For the one-tailed test, the P-value is taken directly from the binomial result; for a two-tailed test, it is the binomial result times 2. On your worksheet, type the indicated labels. Enter the values for x (the number of times the less frequent sign appears) and n (the number of positive and negative signs combined) and the given formulas for the P-values. If the displayed P-value is less than or equal to the significance level of α, we reject the null hypothesis; otherwise, we fail to reject the null hypothesis.

EXCEL DISPLAY
Sign Test

	A	B	C
1	x	1	
2	n	14	
3	P (one tail)	0.0009155	
4	P (two tails)	0.0018311	
5			
6			

Cell	Formula	Explanation
B3	=BINOMDIST(B1,B2,0.5, TRUE)	Calculate as a binomial test, for specified x and n, and $p = 0.5$
B4	=2*B3	P-value doubles for a two-tailed test

Use appropriate cell addresses for data.

MINITAB 15:   For a test for one median, enter the data in C1. From the **Stat** menu, choose **Nonparametrics** and then **1-Sample Sign**. Select C1 for it to appear in the **Variables** window. Check the **Test median** radio button and enter the value of the hypothesized median. From the pull-down menu, choose the alternative hypothesis you want.

MINITAB DISPLAY
Sign Test

↓	C1	C2	C3	C4	C5	C6	C7	C8
1	9.40							
2	7.44							
3	10.49							
4	12.55							
5	12.40							
6	13.47							
7	5.63							
8	9.53							
9								
10								
11								
12								
13								
14								

1-Sample Sign

Variables:
C1

○ Confidence interval
   Level: 95.0

● Test median: 8
   Alternative: greater than ▾

Select   Help   OK   Cancel

STATDISK: Select **Analysis** from the main menu bar, then select **Sign Test**. Select the option **Given Number of Signs** if you know the number of plus and minus signs, or select **Given Pairs of Values** if you prefer to enter matched pairs of data. After making the required entries in the dialog box, the displayed results will include the test statistic, critical value, and conclusion.

12-2 Exercises A: Basic Skills and Concepts

In Exercises 1–12, use the sign test.

1. Do strong El Niño phenomena increase temperatures during springtime? Using the data in the following table, test the claim that the progression of a strong El Niño, from the spring of onset to the next spring, has no effect on Canadian temperature departures from seasonal norms. Use a 0.05 level of significance.

El Niño Cycle	A	B	C	D	E	F	G
Spring of El Niño onset	0.5	0.2	−1.2	−1	0.6	1.3	−0.4
Second spring of El Niño	1.5	0.0	0.9	−0.5	1.0	0.1	3.1

Based on data from Environment Canada.

2. An environmental study compared the use of motorized vehicles in different countries over time. The accompanying table shows comparative data (in

billions of vehicle-kilometres driven) for two countries in selected years. Use a 0.025 significance level to test the claim that there are more vehicle-kilometres driven annually in Japan than in France. What might explain this result?

	A	B	C	D	E	F
Japan	226.0	286.3	389.1	447.0	471.1	520.7
France	208.0	260.7	328.1	328.1	376.0	414.0

Based on data from www.swishweb.com.

3. The Alumni Association of the University of British Columbia tries to track the whereabouts and careers of UBC graduates. In a sample of 88 graduates known to be living outside of Canada, 51 are living in the United States. At the 0.05 significance level, test the claim that most (more than half) of UBC graduates who live outside of Canada are living in the United States (based on data in *UBC Reports*, May 21, 1998.).

4. A college aptitude test was given to 100 randomly selected high-school seniors. After a period of intensive training, a similar test was given to the same students: 59 students received higher grades, 36 received lower grades, and 5 received the same grades. At the 0.05 level of significance, test the claim that the training is effective.

5. The Joliette Manufacturing Company claims that hiring is done without any gender bias. If 18 of the last 25 new employees are women, and job applicants are about half men and half women who are all qualified, is there sufficient evidence to charge gender bias? Use a 0.01 level of significance, because we don't want to make such a serious charge unless there is very strong evidence.

6. Refer to Data Set 16 in Appendix B. Test the claim that the intervals between eruptions of the Old Faithful geyser have a median greater than 77 min, which was the median about 20 years ago.

7. Using the weights of only the *brown* M&Ms listed in Data Set 11 of Appendix B, test the claim that the median is greater than 0.9085 g. (For the 1498 M&Ms to produce a total package weight of 1361 g, the mean weight must be at least 0.9085 g.) Use a 0.05 significance level. Based on the result, does it appear that the package is labelled correctly?

8. Refer to Data Set 6 in Appendix B. Use the paired data consisting of the return on invested capital (%) and return on equity (%) for the top 50 companies. If either value is not available (n.a.) for a company, omit that company from the analysis. Using a 0.05 level of significance, test the claim that for top companies, the median difference in these two types of financial return is 0. Based on these results for top companies, do return on invested capital and return on equity appear to be about the same?

9. Refer to Data Set 20 in Appendix B. If annual person-days lost due to work stoppages from 1976 to 1999 are analyzed, for six groups of transportation industries, the median annual number of person-days lost overall is 10,595. Find the median number of person-days lost due to work stoppages in just the truck-based transportation industry for 1976 to 1999. Test at the 0.05 level of significance, and assume that the data in Data Set 20 are representative of *all* years. Test the claim that the median number of person-days lost in the truck-based group is smaller than the median of 10,595 for all transportation groups combined.

10. The Life Trust Insurance Company funded a university study of drinking and driving. After 30 randomly selected drivers were tested for reaction times, they were given two drinks and tested again, with the result that 22 had slower reaction times, 6 had faster reaction times, and 2 received the same scores as before the drinks. At the 0.01 significance level, test the claim that the drinks had no effect on the reaction times. Based on these very limited results, does it appear that the insurance company is justified in charging higher rates for those who drink and drive?

11. Mental measurements of young children are often made by giving them blocks and telling them to build a tower as tall as possible. One experiment of block building was repeated a month later, with the times (in seconds) listed below. Use a 0.01 level of significance to test the claim that there is no difference between the two times.

First trial	30	19	19	23	29	178	42	20	12	39	14	81	17	31	52
Second trial	30	6	14	8	14	52	14	22	17	8	11	30	14	17	15

Based on data from "Tower Building," by Johnson and Courtney, *Child Development*, Vol. 3.

12. Refer to the number of Senate sittings for Parliaments in Data Set 5 of Appendix B. Use the data for Parliaments of at least three years' duration. Assuming past Parliaments are representative of Parliaments in general, test the claim that the median number of Senate sittings for Parliaments of at least three years' duration is less than 300.

12-2 Exercises B: Beyond the Basics

13. A sample of 10 randomly selected IQ scores is obtained from the population of professional athletes. We want to test the claim that the population is centred at the IQ score of 100.
   a. Is it possible to have sample data such that the sign test leads to rejection of the null hypothesis that the median is equal to 100 while the *t* test

conclusion (Chapter 4) is failure to reject the null hypothesis of $\mu = 100$? If so, construct such a data set.

b. Is it possible to have sample data such that the sign test conclusion is failure to reject the null hypothesis that the median is equal to 100 while the t test conclusion (Chapter 4) is rejection of the null hypothesis of $\mu = 100$? If so, construct such a data set.

14. In the sign test procedure described in this section, we excluded ties (represented by 0 instead of a sign of $+$ or $-$). A second approach is to treat half of the 0s as positive signs and half as negative signs. (If the number of 0s is odd, exclude one so that they can be divided equally.) With a third approach, in two-tailed tests make half of the 0s positive and half negative; in one-tailed tests make all 0s either positive or negative, whichever supports the null hypothesis. Assume that in using the sign test on a claim that the median score is at least 100, we get 60 scores below 100, 40 scores above 100, and 21 scores equal to 100. Identify the test statistic and conclusion for the three different ways of handling differences of 0. Assume a 0.05 significance level in all three cases.

15. Table A-7 lists critical values for limited choices of α. Use Table A-1 to add a new column in Table A-7 (down to $n = 15$) that represents a significance level of 0.03 in one tail or 0.06 in two tails. For any particular n we use $p = 0.5$, as the sign test requires the assumption that

$$P(\text{positive sign}) = P(\text{negative sign}) = 0.5$$

The probability of x or fewer like signs is the sum of the probabilities for values up to and including x.

12-3 Wilcoxon Signed-Ranks Test

In Section 12–2 we used the sign test to analyze sample data from a single population as well as those consisting of matched pairs. The sign test used only the signs of the differences and did not use their actual magnitudes (how large the numbers are). This section introduces the *Wilcoxon signed-ranks test*, which is also used in the above scenarios. By using ranks (as defined in Section 12–1), this test takes the magnitudes of the differences into account. Because the Wilcoxon signed-ranks test incorporates and uses more information than the sign test, it tends to yield conclusions that better reflect the true nature of the data.

DEFINITION

The **Wilcoxon signed-ranks test** is a nonparametric test that uses ranks of sample data from a single population to test a claim about a hypothesized median, or consisting of matched pairs to test for differences in the population distributions.

Wilcoxon Signed-Ranks Test for One Median

As with the Z or t test for one mean, the hypotheses can be set up as right-tailed, left-tailed, or two-tailed. We use the symbol m_0 to represent the hypothesized median:

Right-tailed	H_0: median $\leq m_0$	H_1: median $> m_0$
Left-tailed	H_0: median $\geq m_0$	H_1: median $< m_0$
Two-tailed	H_0: median $= m_0$	H_1: median $\neq m_0$

Assumptions

1. The sample data have been randomly selected.
2. The population has a distribution that is approximately *symmetric about the median*, meaning that the left half of its histogram is roughly a mirror image of its right half. (There is no requirement that the data have a normal distribution.)

Procedure for Finding the Value of the Test Statistic

STEP 1: For each value, find the difference d by subtracting the hypothesized median from the value. Keep the signs, but discard any values for which $d = 0$.

STEP 2: *Ignore the signs of the differences*, then sort the differences from lowest to highest, and replace the differences by the corresponding rank value (as described in Section 12–1). When differences have the same numerical value, assign to them the mean of the ranks involved in the tie.

STEP 3: Attach to each rank the sign of the difference from which it came. That is, insert those signs that were ignored in Step 2.

STEP 4: Find the sum of the absolute values of the negative ranks (called T_-). Also find the sum of the positive ranks (called T_+).

STEP 5: If $n \leq 30$, the value of the test statistic depends on the tail of the alternative hypothesis. For the right-tailed test, the test statistic $= T_-$; for the left-tailed test, the test statistic $= T_+$. For the two-tailed test, let T be the *smaller* of T_- and T_+. If $n > 30$, we can convert to z:

For $n > 30$:
$$z = \frac{T - \dfrac{n(n+1)}{4}}{\sqrt{\dfrac{n(n+1)(2n+1)}{24}}}$$

STEP 6: Let n be the number of pairs of data for which the difference d is not 0.

STEP 7: Determine the critical value based on the sample size.

STEP 8: When forming the conclusion, reject the null hypothesis if the sample data lead to a test statistic that is in the critical region—that is, the test statistic is less than or equal to the critical value(s). Otherwise, fail to reject the null hypothesis.

NOTATION

T = the smaller of the following two sums:

1. The sum of the absolute values of the negative ranks (T_-)
2. The sum of the positive ranks (T_+)

CRITICAL VALUES

If $n \leq 30$, the critical T value is found in Table A-8.

If $n > 30$, the critical z values are found in Table A-2.

EXAMPLE

In Chapter 7, we tested the claim that the mean axial loads of 12-oz aluminum cans is greater than 165 lbs. We reproduce the data here:

270   273   258   204   254   228   282

Test the claim that the median axial load is greater than 220 lbs at a 0.05 level of significance.

SOLUTION

The null and alternative hypotheses are:

$$H_0: \text{median} \leq 220$$
$$H_1: \text{median} > 220$$

If the null hypothesis is true, we should expect an equal number of values on either side of the hypothesis median of 220.

Table 12-3 Axial Loads of Seven 12-oz aluminum cans

Value	270	273	258	204	254	228	282
Difference d	50	53	38	−16	34	8	62
Ranks of differences	5	6	4	2	3	1	7
Signed ranks	5	6	4	−2	3	1	7

Step 1: We subtract the hypothesized median of 220 from each of the values so that $d =$ value $- 220$.

Step 2: Ignoring their signs, we compute the ranks of the differences.

Step 3: Attach to each rank the sign of the correspondent difference. If the median is actually greater than 220, we expect T_- to be appreciably less than T_+.

Step 4: From the signed ranks, $T_- = 2$. Since the sum of all the values is $(7)(8)/2 = 28$, $T_+ = 28 - 2 = 26$.

Step 5: Since this is a right-tailed test, the value of the test statistic is $T_- = 2$.

Step 6: There are no differences of zero; therefore, $n = 7$.

Step 7: Using Table A-8, we reject the null hypothesis if the value of the test statistic is less than or equal to 4.

Step 8: With the test statistic value of 2 being less than the critical value of 4, we reject the null hypothesis.

Interpretation

The median axial load of 12-oz aluminum cans is greater than 220 lbs.

Wilcoxon Signed-Ranks Test for Two Dependent Samples

The procedure is similar to the Wilcoxon signed-ranks test for one median except that the differences are computed by subtracting the second value from the first value. The null and alternative hypotheses for a two-tailed test are:

H_0: The two samples come from populations with the same distribution.

H_1: The two samples come from populations with different distributions.

EXAMPLE

Table 12-4 consists of sample data obtained when 14 subjects are tested for reaction times with their left and right hands. (Only right-handed subjects were used.) Use the Wilcoxon signed-ranks test to test the claim of no difference (i.e., difference has median 0) between reaction times with right and left hands. Use a significance level of $\alpha = 0.05$.

Table 12-4 Reaction Times (in thousandths of a second) of 14 Subjects

Right hand	191	97	116	165	116	129	171	155	112	102	188	158	121	133
Left hand	224	171	191	207	196	165	177	165	140	188	155	219	177	174
Differences d	-33	-74	-75	-42	-80	-36	-6	-10	-28	-86	$+33$	-61	-56	-41
Ranks of differences	4.5	11	12	8	13	6	1	2	3	14	4.5	10	9	7
Signed ranks	-4.5	-11	-12	-8	-13	-6	-1	-2	-3	-14	$+4.5$	-10	-9	-7

The null and alternative hypotheses are as follows:

H_0: There is no difference between right- and left-hand reaction times.

H_1: There is a difference between right- and left-hand reaction times.

The significance level is $\alpha = 0.05$. We are using the Wilcoxon signed-ranks test procedure, so the test statistic is calculated by using the eight-step procedure presented earlier in this section.

Step 1: In Table 12-4, the row of differences is obtained by computing $d =$ right hand $-$ left hand for each pair of data.

Step 2: Ignoring their signs, we rank the absolute differences from lowest to highest. Note that the differences of -33 and $+33$ are tied for ranks of 4 and 5, so we assign a rank of 4.5 to each of them.

Step 3: The bottom row of Table 12-4 is created by attaching to each rank the sign of the corresponding difference. If there is no difference between right- and left-hand reaction times, we expect the number of positive ranks to be approximately equal to the number of negative ranks.

Step 4: We now find the sum of the absolute values of the negative ranks, and we also find the sum of the positive ranks.

$$\text{Sum of positive ranks:} \quad T_+ = 4.5$$

Sum of absolute values of negative ranks: $T_- = 100.5$, which is

$$4.5 + 11 + 12 + 8 + 13 + 6 + 1 + 2 + 3 + 14 + 10 + 9 + 7$$

Note that the sum of all the values is $(14)(15)/2 = 105$. We could compute $T_- = 105 - 4.5 = 100.5$.

Step 5: Letting T be the smaller of the two sums found in Step 4, we find that $T = 4.5$.

Step 6: Letting n be the number of pairs of data for which the difference d is not 0, we have $n = 14$.

Step 7: Because $n = 14$, we have $n \leq 30$, so we use a test statistic of $T = 4.5$ (and we do not calculate a z test statistic). Also, because $n \leq 30$, we use Table A-8 to find the critical value of 21, given $\alpha = 0.05$ (two tails).

Step 8: The test statistic $T = 4.5$ is less than or equal to the critical value of 21, so we reject the null hypothesis.

Interpretation

It appears that there is a difference between right- and left-hand reaction times. Examining the sample data, we see that there are 13 negative signs and only 1 positive sign. Because negative signs result from lower values with the right hand, it appears that right-handed people have faster reaction times with their right hands.

Because of the faster reaction time with the right hand, the above test could have been conducted as a left-tailed test. The null and alternative hypotheses would be:

H_0: Right-hand reactions times are not less than left-hand reaction times.

H_1: Right-hand reaction times are less than left-hand reaction times.

Since this is a left-tailed test, the test statistic is $T_+ = 4.5$. Using Table A-8 and setting $\alpha = 0.05$, we reject the null hypothesis if the test statistic is less than or equal to 26. Since this is the case, we reject the null hypothesis and conclude that for right-handed people, reaction times with their right hands are less than those with their left hands.

If we use the sign test with the preceding example, we will arrive at the same conclusion. Although the sign test and the Wilcoxon signed-ranks test agree in this particular case, there are other cases in which not all tests agree.

Rationale: In the preceding example the unsigned ranks of 1 through 14 have a total of 105, so if there are no significant differences, each of the two signed-rank totals should be around $105 \div 2$, or 52.5. The table of critical values shows that at the 0.05 level of significance with 14 pairs of data, a 21–84 split represents a significant departure from the null hypothesis, and any split that is farther apart (such as 20–85 or 19–86 or, in our case, 4.5–100.5) will also represent a significant departure from the null hypothesis. Conversely, splits like 22–83 do not represent significant departures away from a 52.5–52.5 split, and they would not be a basis for rejecting the null hypothesis. The Wilcoxon signed-ranks test is based on the lower rank total, so instead of analyzing both numbers constituting the split, we consider only the lower number.

The sum $1 + 2 + 3 + \cdots + n$ of all the ranks is equal to $n(n + 1)/2$; if this is a rank sum to be divided equally between two categories (positive and negative), each of the two totals should be near $n(n + 1)/4$, which is half of $n(n + 1)/2$. Recognition of this principle helps us understand the test statistic used when $n > 30$. The denominator in that expression represents a standard deviation of T and is based on the principle that

$$1^2 + 2^2 + 3^3 + \cdots + n^2 = \frac{n(n + 1)(2n + 1)}{6}$$

The Wilcoxon signed-ranks test can only be used for dependent (matched) data. Section 12–4 will describe a rank-sum test that can be applied to two sets of independent data that are not paired.

USING TECHNOLOGY

STATDISK: Select **Analysis** from the main menu bar, then select **Wilcoxon Tests** and proceed to use the Wilcoxon signed-ranks test for dependent samples. The STATDISK display will include the test statistic, critical value, and conclusion.

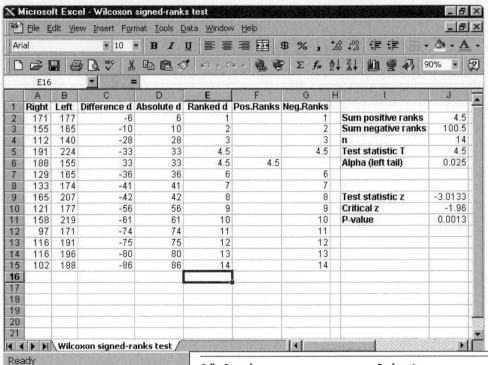

EXCEL DISPLAY
Wilcoxon Signed-Ranks Test

Cell	Formula	Explanation
C2	=A2-B2	The difference *d* between the paired scores
D2	=ABS(C2)	Ignores the signs of the differences
F2	=IF(C2>0,E2,"")	Records absolute ranks for the positive differences
G2	=IF(C2<0,E2, "")	Records absolute ranks for the negative differences
J2	=SUM(F2:F15)	Finds sum of recorded ranks for positive differences
J3	=SUM(G2:G15)	Finds sum of recorded ranks for negative differences
J4	=COUNT(F2:G15)	*n* (does not include cases of no difference)
J5	=MIN(J2:J3)	*T* (the smaller of the two sums for + and − ranks)
J9	=(J5-(J4*(J4+1)/4))/ SQRT(J4*(J4+1)*(2*J4+1)/24)	*z* (see boxed formula for test statistic, for $n > 30$)
J10	=NORMSINV(J6)	Lower critical value for *z*, for $n > 30$
J11	=NORMSDIST(J9)	*P*-value, for $n > 30$ (for two-tailed test, multiply by 2)

Use appropriate cell addresses for data.

Note: Ignore calculations for Cells J9:J11 if $n \leq 30$.

EXCEL: In Excel, the Wilcoxon signed-ranks test can be partly automated with the following templates. Set up the column headings as shown in the left part of the figure, placing the headings and data for the two matched variables in the two left columns. Enter the given formulas for the first values of the difference and absolute difference; copy and paste them to fill the columns. Then sort the four left columns in order of the absolute differences. To do so, highlight those columns, including the headings. Select **Sort** from the **Data** menu and, in the dialog box that appears, select "Absolute d" for the **Sort by** option. Then specify **Ascending** and **Header** row, and click on **OK**. Fill the "Ranked d" column with the ordered numbers from 1 to the number of data pairs. Inspect the "Absolute d" column for tied values. If there are any, manually replace their ranks in the "Ranked d" column with the mean of the ranks that first appear in that column. (For example, in the sorted display shown, there are two 33s in the "Absolute d" column. Originally, the first 33 had a rank of 4 and the second 33 had a rank of 5. Both ranks were manually replaced with 4.5.) Next, complete Columns F and G by entering and copying the appropriate formulas, which are shown. These distinguish between the positive and negative ranks.

Input the indicated labels in Column I and formulas in Column J to complete the calculations. For alpha, enter only the left-tailed portion. Note that if $n \leq 30$, you should look up the displayed T value in Appendix A-8 rather than use the z test statistic at the bottom of the template.

MINITAB 15: For the test for one median, enter the data in C1. From the **Stat** menu, choose **Nonparametrics** and then **1-Sample Wilcoxon**. Select C1 for it to appear in the **Variables** window. Check the **Test median** radio button and enter the value of the hypothesized median. From the pull-down menu, choose the alternative hypothesis you want.

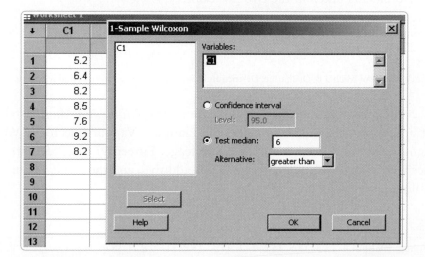

MINITAB DISPLAY
Wilcoxon Test for 1 Median

12-3 Exercises A: Basic Skills and Concepts

In Exercises 1–4, refer to the sample data for the given exercises in Section 12–2. Instead of the sign test, use the Wilcoxon signed-ranks test to test the claim that both samples come from populations having the same distribution. Use the significance level α that is given below.

1. Exercise 1; $\alpha = 0.05$
2. Exercise 2; $\alpha = 0.05$
3. Exercise 8; $\alpha = 0.01$
4. Exercise 11; $\alpha = 0.01$

In Exercises 5–8, use the Wilcoxon signed-ranks test.

5. A study was conducted to investigate the effectiveness of hypnotism in reducing pain. Results for randomly selected subjects are given below. The measurements are in centimetres on a pain scale. At the 0.01 significance level, test the claim that hypnotism has no effect.

Subject	A	B	C	D	E	F	G	H
Before hypnosis	6.6	6.5	9.0	10.3	11.3	8.1	6.3	11.6
After hypnosis	6.8	2.4	7.4	8.5	8.1	6.1	3.4	2.0

Based on "An Analysis of Factors That Contribute to the Efficacy of Hypnotic Analgesia," by Price and Barber, *Journal of Abnormal Psychology*, Vol. 96, No. 1.

6. In a study of the effect of posture on body temperature, data were collected for 11 subjects. The values given in the table are in degrees Celsius and represent the measured body temperatures of the 11 subjects first in a sitting position and then in a supine (lying face up) position. At the 0.01 level of significance, test the claim that there is no significant difference between the measurements taken in the two positions.

Sitting	37.61	37.07	37.2	37.23	37.38	37.28	37.12	37.55	37.04	37.78	37.59
Supine	37.19	36.91	36.7	36.80	36.96	37.11	36.82	37.36	36.92	37.22	37.25

Data courtesy of Michel B. Ducharme, Defence and Civil Institute of Environmental Medicine, Toronto.

7. According to a paid promotional article in *The Globe and Mail*, installation of an automotive accessory called the Platinum Vapour Injector (P.V.I.) can significantly increase a vehicle's gas mileage. Fifteen identical 5.0-litre vehicles were tested without, then with, this additive, and the results, in miles per gallon, are as shown.

Vehicle	A	B	C	D	E	F	G	H	I	J	K	L	M	N	O
Without P.V.I.	14.4	13.5	16.9	15.6	14.6	11.5	15.9	11.7	17.2	13.0	16.9	19.0	17.3	15.7	15.5
With P.V.I.	21.3	19.9	24.8	22.5	20.5	15.9	21.5	15.7	22.1	16.7	21.1	21.0	19.1	16.8	13.5

a. Assume that the tested cars are representative of cars in general, and use the Wilcoxon signed-ranks test at the 0.05 level of significance. Test the claim that the P.V.I. additive does affect vehicles' gas mileage.

b. Note that the vehicles tested in this example were not randomly selected, and in fact they all represent the same model of car under standardized conditions. Is the assumption made in part (a), that the test results can be potentially applied to all cars, a reasonable one? Explain why or why not.

8. A company offers a daily list of stock picks to paid subscribers. Its website, called *Track Record for TSE*, shows the prices of stocks it has recommended (1) on the day the "buy" recommendation was made, and (2) on a subsequent day within the next week or two. The accompanying table is a sample from this data, for November to December 2000. At the 0.05 significance level, do the data support the company's claim that, for stocks bought and sold at the prices cited, the buy and sell prices are different? If clients make their own decisions about when to sell a stock, is the displayed data sufficient to support a claim that the clients of this company made money? Explain.

Stock Pick Number	1	2	3	4	5	6	7	8	9	10	11	12
Entry price	4.87	22.85	14.05	13.80	22.70	10.00	25.45	18.90	20.90	4.50	31.30	14.00
Exit price	5.35	23.65	14.25	14.80	25.25	9.65	26.70	20.50	25.25	9.15	41.25	14.40

9. In an exercise in Chapter 7, the claim that chickens fed with enriched feed attain a mean weight greater than 1.77 kg was tested. In an experiment with an enriched feed mixture, nine chickens are born that attain the weights (in kg) given below.

    1.74   1.77   1.90   1.80   1.88   1.87   1.79   1.81   1.89

Test that the median weight is greater than 1.77 kg at a 0.05 level of significance.

12-3 Exercises B: Beyond the Basics

10. Assume that the Wilcoxon signed-ranks test is being used for a two-tailed hypothesis test with a significance level of 0.05.
    a. With $n = 10$ pairs of data, find the lowest and highest possible values of T.
    b. With $n = 50$ pairs of data, find the lowest and highest possible values of T.
    c. If there are $n = 100$ pairs of data with no differences of 0 and no tied ranks, find the critical value of T.

12-4 Wilcoxon Rank-Sum Test for Two Independent Samples

In this section we introduce the *Wilcoxon rank-sum test*, which is a nonparametric test that can be applied to two independent sets of sample data. Recall that two samples are independent if the sample values selected from one population are not related or somehow matched or paired with the sample values from the other population.

DEFINITION

The **Wilcoxon rank-sum test** is a nonparametric test that uses ranks of sample data from two independent populations. It is used to test the null hypothesis that the two independent samples come from populations with the same distribution. (That is, the two populations are identical.) The alternative hypothesis is that the two population distributions are different in some way.

H_0: The two samples come from populations with the same distribution. (That is, the two populations are identical.)

H_1: The two samples come from populations with different distributions. (That is, the two distributions are different in some way.)

The Wilcoxon rank-sum test is equivalent to the **Mann-Whitney U test** (see Exercise 10 below), which is included in some other textbooks and software packages. The basis for the procedure used in the Wilcoxon rank-sum test is this principle: If two samples are drawn from identical populations and the individual values are all ranked as one combined collection of values, then the high and low ranks should fall evenly between the two samples. If the low ranks are found predominantly in one sample and the high ranks are found predominantly in the other sample, we suspect that the two populations are not identical.

Assumptions

1. There are two independent samples that were randomly selected.
2. There is no requirement that the two populations have a normal distribution or any other particular distribution.

Note that unlike the corresponding hypothesis tests in Sections 8–3 and 8–5, the Wilcoxon rank-sum test does *not* require normally distributed populations. Also, the Wilcoxon rank-sum test can be used with data at the ordinal level of measurement, such as data consisting of ranks. In contrast, the parametric methods of Sections 8–3 and 8–5 cannot be used with data at the ordinal level of measurement.

In Table 12-1 we noted that the Wilcoxon rank-sum test has a 0.95 efficiency rating when compared with the parametric t test or Z test. Because this test has such a high efficiency rating and involves easier calculations, it is often preferred to the parametric tests presented in Sections 8–3 and 8–5, even when the requirement of normality is satisfied.

As with the Wilcoxon signed-ranks test, the test statistic and critical values depend on the sample sizes. If $n_1 \leq 10$ and $n_2 \leq 10$, the critical values are found in Table A-13. This is known as the small sample procedure. If $n_1 > 10$ or $n_2 > 10$, we can convert to Z and use Table A-2 to find the critical value. This is known as the large sample procedure.

Small Sample Procedure for Finding the Value of the Test Statistic

1. Temporarily combine the two samples into one big sample, sort and then replace each sample value with its rank. (The lowest value gets a rank of 1, the next lowest value gets a rank of 2, and so on. For more details about ranks, see Section 12–1.)

2. Choose n_1 according to your alternative hypothesis. Find the sum of the ranks for this sample. This is the value of the test statistic which we call T_1.

3. From Table A-13, T_L and T_U represent the lower and upper critical values, respectively. The decision to reject the null hypothesis is based on these criteria:

   Right-tailed test: Reject the null hypothesis if $T_1 \geq T_U$.

   Left-tailed test: Reject the null hypothesis if $T_1 \leq T_L$.

   Two-tailed test: Reject the null hypothesis if $T_1 \leq T_L$ or $T_1 \geq T_U$.

You will notice in Table A-13 that the critical values are split into two parts. If $\alpha = 0.05$, the critical values are from the lower half for one-tailed tests (either right- or left-tailed) while they are from the upper half for two-tailed tests.

EXAMPLE

A researcher held two focus groups, one of each gender, to watch a new romantic comedy movie. It asked the participants to rate the movie on a scale from 1 to 10, 10 being best. The researcher wanted to determine if females would like the movie more, that is, the median rating for females is higher than that of males. These were the results:

Male	4	7	6	4	3	6	5	4
Female	7	8	7	3	7	7	8	5

Test the hypothesis at a 0.05 level of significance.

SOLUTION

The null and alternative hypotheses are:

H_0: median (females) $\leq$ median (males)

H_1: median (females) $>$ median (males)

If the null hypothesis is true, we would expect the combined distribution of the male and female ratings to be roughly equivalent. This would be reflected in their respective rank sums.

Since this is set up as a right-tailed test, we choose females as the first sample. Thus T_1 is the rank sum of the females. To find this value, we combine the 8 values from each group and compute their ranks. The results are summarized in Table 12-5.

Table 12-5 **Computation of Ranks for Females and Males**

Observation	1	2	3	4	5	6	7	8	9	10	11	12	13	14	15	16
Value	3	3	4	4	4	5	5	6	6	7	7	7	7	7	8	8
Group	M	F	M	M	M	M	F	M	M	M	F	F	F	F	F	F
Rank	1.5	1.5	4	4	4	6.5	6.5	8.5	8.5	12	12	12	12	12	15.5	15.5

$T_1 = 1.5 + 6.5 + 12 + 12 + 12 + 12 + 15.5 + 15.5 = 87$. As a check, the rank sum for the males is $1.5 + 4 + 4 + 4 + 6.5 + 8.5 + 8.5 + 12 = 49$. The sum of both rank sums is $87 + 49 = 136$. The sum of the arithmetic series is $(16)(17)/2 = 136$. This ensures we did the arithmetic correctly.

Using Table A-13, $n_1 = 8$ and $n_2 = 8$. Using the lower half of the table, we find $T_U = 84$. We reject the null hypothesis if $T_1 \geq 84$. Since $T_1 = 87$, we reject the null hypothesis.

Interpretation
We conclude the median rating of the females is greater than that of the males. Females like the movie better than males.

Large Sample Procedure for Finding the Value of the Test Statistic

Steps 1 and 2 are the same for the large sample procedure. For step 3, calculate the value of the Z test statistic as shown in the Test Statistic box.

NOTATION FOR THE WILCOXON RANK-SUM TEST

n_1 = size of sample 1

n_2 = size of sample 2

R_1 = sum of ranks for sample 1

R_2 = sum of ranks for sample 2

R = same as R_1 (sum of ranks for sample 1)

μ_R = mean of the sample R values that is expected when the two populations are identical

σ_R = standard deviation of the sample R values that is expected when the two populations are identical

If you are testing the null hypothesis of identical populations and both sample sizes are greater than 10, then the sampling distribution of R is approximately normal with mean μ_R and standard deviation σ_R, and the test statistic is as follows.

$$z = \frac{R - \mu_R}{\sigma_R}$$

where $\quad \mu_R = \dfrac{n_1(n_1 + n_2 + 1)}{2}$

$\sigma_R = \sqrt{\dfrac{n_1 n_2(n_1 + n_2 + 1)}{12}}$

n_1 = size of the sample from which the rank sum R is found

n_2 = size of the other sample

R = sum of ranks of the sample with size n_1

Critical values: Critical values can be found in Table A-2 (because the test statistic is based on the normal distribution). The expression for μ_R is a variation of the following result of mathematical induction: The sum of the first n positive integers is given by $1 + 2 + 3 + \cdots + n = n(n + 1)/2$. The expression for σ_R is a variation of a result that states that the integers $1, 2, 3, \ldots, n$ have standard deviation $\sqrt{(n^2 - 1)/12}$.

EXAMPLE

Samples of M&M plain candies are randomly selected, and the red and yellow M&Ms are weighed, with the results listed in the margin (from Data Set 11 in Appendix B). Use a 0.05 level of significance to test the claim that weights of red M&Ms and yellow M&Ms have the same distribution.

Weights (in grams) of M&Ms			
Red		Yellow	
0.870	(2)	0.906	(19)
0.933	(35)	0.978	(45)
0.952	(42)	0.926	(34)
0.908	(21)	0.868	(1)
0.911	(24.5)	0.876	(5)
0.908	(21)	0.968	(44)
0.913	(27)	0.921	(30)
0.983	(46)	0.893	(15)
0.920	(29)	0.939	(38)
0.936	(37)	0.886	(10.5)
0.891	(13)	0.924	(32)
0.924	(32)	0.910	(23)
0.874	(4)	0.877	(6)
0.908	(21)	0.879	(7.5)
0.924	(32)	0.941	(40)
0.897	(16)	0.879	(7.5)
0.912	(26)	0.940	(39)
0.888	(12)	0.960	(43)
0.872	(3)	0.989	(47)
0.898	(17)	0.900	(18)
0.882	(9)	0.917	(28)
		0.911	(24.5)
		0.892	(14)
		0.886	(10.5)
		0.949	(41)
		0.934	(36)
$n_1 = 21$		$n_2 = 26$	
$R_1 = 469.5$		$R_2 = 658.5$	

SOLUTION

The null and alternative hypotheses are as follows:

H_0: Red and yellow M&M plain candies have weights with identical populations.

H_1: The two populations are not identical.

Rank all 47 weights combined, beginning with a rank of 1 (assigned to the lowest weight of 0.868 g). Ties in ranks are handled as described in Section 12–1: Find the mean of the ranks involved and assign this mean rank to each of the tied values. The ranks corresponding to the individual sample values are shown in parentheses in the table. R denotes the sum of the ranks for the sample we choose as sample 1. If we choose the red M&Ms, we get

$$R = 2 + 35 + 42 + \cdots + 9 = 469.5$$

Because there are 21 red M&Ms, we have $n_1 = 21$. Also, $n_2 = 26$, because there are 26 yellow M&Ms. We can now determine the values of μ_R, σ_R, and the test statistic z:

$$\mu_R = \frac{n_1(n_1 + n_2 + 1)}{2} = \frac{21(21 + 26 + 1)}{2} = 504$$

$$\sigma_R = \sqrt{\frac{n_1 n_2(n_1 + n_2 + 1)}{12}} = \sqrt{\frac{(21)(26)(21 + 26 + 1)}{12}} = 46.73$$

$$z = \frac{R - \mu_R}{\sigma_R} = \frac{469.5 - 504}{46.73} = -0.74$$

The test is two-tailed because a large positive z score would indicate that the higher ranks are found disproportionately in the first sample, and a large negative z score would indicate that the first sample had a disproportionate share of lower ranks. In either case, we would have strong evidence against the claim that the two samples come from identical populations.

The significance of the test statistic z can be treated in the same manner as in previous chapters. We are now testing (with $\alpha = 0.05$) the hypothesis that the two populations are the same, so we have a two-tailed test with critical z values of 1.96 and 21.96. The test statistic of $z = -0.74$ does not fall within the critical region, so we fail to reject the null hypothesis that red and yellow M&Ms have the same weights.

We can verify that if we interchange the two sets of weights and consider the sample of yellow M&Ms to be first, $R = 658.5$, $\mu_R = 624$, $\sigma_R = 46.73$, and $z = 0.74$, so the conclusion is the same.

EXAMPLE

A labour specialist has compiled Data Set 20 in Appendix B, in order to compare the annual amounts of work time lost (in person-days) in the rail and air transportation industries due to work stoppages. Because these data are not normally distributed, you suggest that she use the Wilcoxon rank-sum test. Based on annual totals for 15 different, randomly

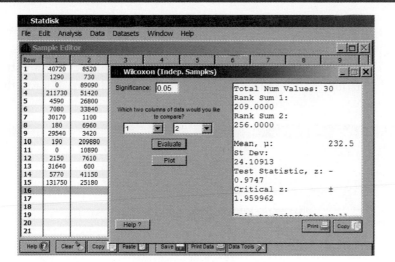

**Figure 12-5
Stoppages**

Rail	Air
40,720	8,520
1,290	730
0	89,090
211,730	51,420
4,590	26,800
7,080	33,840
30,170	1,100
180	6,960
29,540	3,420
190	209,880
0	10,890
2,150	7,610
31,640	600
5,770	41,150
137,150	25,180

selected years for each transportation group (see margin), test the claim that the person-days lost per year both come from the same distribution.

SOLUTION

The null and alternative hypotheses are as follows:

H_0: Numbers of person-days lost per year due to work stoppages, in the rail and air transportation industries, come from populations with the same distribution.

H_1: The two distributions differ in some way.

Instead of manually calculating the rank sums, we refer to the STATDISK display shown in Figure 12-5. The display shows that the first rank sum (for rail transportation) is 209.0, which equals R. The test statistic z is -0.97473, and (assuming $\alpha = 0.05$) the critical z values are ± 1.9600. The conclusion is that we fail to reject the null hypothesis that the distributions are the same.

Interpretation

The annual numbers of person-days lost due to work stoppages in the rail and air transportation industries show great variation, and have nonnormal distributions. That is why we avoided a parametric test. Based on the Wilcoxon rank-sum test, there is insufficient evidence to conclude that the data are distributed differently for these two groups.

USING TECHNOLOGY

STATDISK: Select **Analysis** from the main menu bar, then select **Wilcoxon Tests** and proceed to use the Wilcoxon rank-sum test for *independent* samples. The STATDISK display will include the rank sums, sample size, test statistic, critical value, and conclusion.

In Excel, the Wilcoxon ranked-sum test can be partly automated with the following template. Set up the column headings as shown in the left part of the figure. Fill in the leftmost column with the data from *both* samples. Then, in the second column, identify the sample

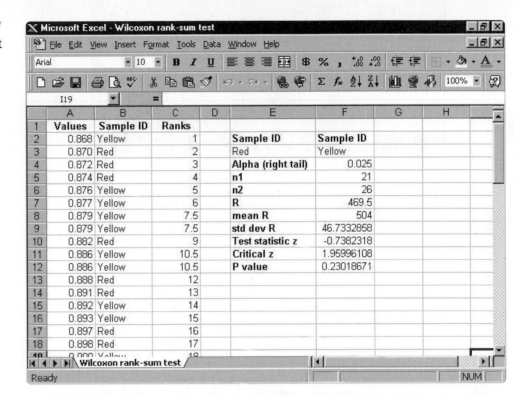

Cell	Formula	Explanation
F5	=DCOUNT(A1:C48,1,E2:E3)	n for the first sample, whose ID is specified in E3
F6	=DCOUNT(A1:C48,1,F2:F3)	n for the second sample, whose ID is specified in F3
F7	=DSUM(A1:C48,3,E2:E3)	Sum of the ranks for the first sample
F8	=F5*(F5+F6+1)/2	μ_R (see the formula box for the test statistic)
F9	=SQRT(F5*F6*(F5+F6+1)/12)	σ_R (see the formula box for the test statistic)
F10	=(F7-F8)/F9	Test statistic z (see the formula box)
F11	=NORMSINV(1-F4)	Upper critical value
F12	=1-NORMSDIST(ABS(F10))	P-value (for two-tailed test, multiply by 2)

Use appropriate cell addresses for data.

that each data value comes from. Once the data are entered, sort the two left columns in order of values. To do so, highlight those two columns, including the headings. Select **Sort** from the **Data** menu and, in the dialog box that appears, select "Values" for the **Sort** by option. Then specify **Ascending** and **Header** row, and click on **OK**. Now fill the "Ranks" column with the ordered numbers from 1 to the number of data values (in both samples combined). Inspect the "Values" column for tied values. If there are any, manually replace their ranks in the "Ranks" column with the mean of the ranks that first appear in that column. (For example, since 0.879 appears twice in the "Values" column, with the first 0.879 having a rank of 7 and the second having a rank of 8, both ranks have been manually replaced with 7.5.)

Also shown is the template for completing the calculations. For alpha, enter only the right-tailed portion. In Cells E2 and F2, enter the same heading that is in the middle column of your input block; use Cells E3 and F3 to enter the IDs you used, in the input block, for the first and second samples, respectively. All other values will be calculated, based on the indicated formulas.

MINITAB 15: Minitab does the Mann-Whitney test, which is equivalent to the Wilcoxon ranks-sum test. Enter one sample in C1, the other in C2. From the **Stat** menu, choose **Nonparametrics** and then **Mann-Whitney**. Select the appropriate columns for your samples. From the pull-down menu, choose the alternative hypothesis.

MINITAB DISPLAY
Mann-Whitney Test

↓	C1	C2
	group 1	group 2
1	5.2	7.2
2	6.4	8.4
3	8.2	5.5
4	8.5	5.2
5	7.6	4.1
6	9.2	8.2
7	8.2	7.2
8		
9		
10		
11		
12		
13		

Mann-Whitney

C1   group 1
C2   group 2

First Sample: 'group 1'

Second Sample: 'group 2'

Confidence level: 95.0

Alternative: not equal

Select

Help   OK   Cancel

In Exercises 1–8, use the Wilcoxon rank-sum test.

1. The example in this section tested the null hypothesis that red and yellow M&M plain candies have weights with identical populations. Refer to Data Set 11 in Appendix B and test the claim that red and brown M&M plain candies have weights with identical populations. Use a 0.05 level of significance.

2. Listed below are time intervals (in minutes) between eruptions of the Old Faithful geyser in Yellowstone National Park. Using a 0.05 significance level, test the claim that the times are the same for both years.

1951	74	60	74	42	74	52	65	68	62	66	62	60
1996	88	86	85	89	83	85	91	68	91	56	89	94

Based on data from geologist Rick Hutchinson and the National Park Service.

3. Provincial governments have been re-examining their delivery of services, such as health care. The accompanying table is based on one random sample of hospitals in New Brunswick and another in Nova Scotia, and shows the total number of beds in each hospital. At the 0.05 significance level, test the claim that the two provinces have the same distribution of hospital-bed numbers. If the distributions are different, do we have sufficient information to conclude that the numbers of hospital beds per person are different in the two provinces? Explain.

New Brunswick		Nova Scotia	
23	47	15	13
153	47	27	136
397	15	8	26
12	500	85	311
15	56	12	132
398		64	

Based on data from *Guide to Canadian Healthcare Facilities,*
1998–1999, Vol. 6.

4. An umbrella maker is deciding whether to concentrate on eastern or western Canada. To help decide, she randomly selects cities from both parts of the country, and obtains the corresponding annual precipitation data (in millimetres). At the 0.01 significance level, test the claim that, in general, eastern and western cities have the same amount of annual precipitation. Based on your result, can we conclude that one side of Canada would be a better market for umbrellas than the other?

Eastern Cities				Western Cities			
1169	1174	879	1097	398	647	402	266
1282	712	1109	1514	467	454	628	424
762	1444	769	908	526	2523	467	384
1491	909	946		1113			

5. The "bootstrap" (simulated) samples below are based on summary data for building statistics, in *The Durham Business News*, June 1998. They show the values of building permits, in thousand of dollars, in large and small communities in the region. At the 0.05 significance level, test the claim that both sizes of community are issuing the same dollar values of building permits. Is there evidence that one size of community is attracting larger-value projects than the other?

Large Communities				Small Communities			
118	118	118	57	29	203	29	203
111	118	123	123	203	29	52	77
57	111	118	118	203	203	203	29
111	111	111	111	203	203	77	77
57	57			203	52		

6. Sample data were collected in a study of calcium supplements and their effects on blood pressure. A placebo group and a calcium group began the study with measures of blood pressures. At the 0.05 significance level, test the claim that the two sample groups come from populations with the same blood pressure levels. (The data are based on "Blood Pressure and Metabolic Effects of Calcium Supplementation in Normotensive White and Black Men," by Lyle et al., *Journal of the American Medical Association*, Vol. 257, No. 13.)

Placebo Group				Calcium Group			
124.6	104.8	96.5	116.3	129.1	123.4	102.7	118.1
106.1	128.8	107.2	123.1	114.7	120.9	104.4	116.3
118.1	108.5	120.4	122.5	109.6	127.7	108.0	124.3
113.6				106.6	121.4	113.2	

7. The arrangement of test items was studied for its effect on anxiety. Sample results are as follows:

Easy to Difficult				Difficult to Easy			
24.64	39.29	16.32	32.83	33.62	34.02	26.63	30.26
28.02	33.31	20.60	21.13	35.91	26.68	29.49	35.32
26.69	28.90	26.43	24.23	27.24	32.34	29.34	33.53
7.10	32.86	21.06	28.89	27.62	42.91	30.20	32.54
28.71	31.73	30.02	21.96				
25.49	38.81	27.85	30.29				
30.72							

At the 0.05 level of significance, test the claim that the two samples come from populations with the same scores. (The data are based on "Item Arrangement, Cognitive Entry Characteristics, Sex and Test Anxiety as Predictors of Achievement Examination Performance," by Klimko, *Journal of Experimental Education*, Vol. 52, No. 4.)

8. Refer to Data Set 9 in Appendix B. Use a 0.01 significance level to test the claim that the population of cotinine levels is the same for these two groups: Nonsmokers who are not exposed to environmental tobacco smoke (labelled as NOETS), and nonsmokers who are exposed to tobacco smoke (labelled as ETS). What do the results suggest about "second-hand" smoke?

9. Two separate classes of 6 and 8 were taught the same subject using different pedagogies. The final marks of the students were as follows:

Class A	70	90	82	64	86	77	
Class B	86	78	90	82	65	87	80	88

No assumptions about the normality of the data were made. Is there any significant difference in the median grade between the two pedagogies? Test at a 0.05 level of significance.

Exercises B: Beyond the Basics

10. The Mann-Whitney U test is equivalent to the Wilcoxon rank-sum test for independent samples in the sense that it applies to the same situations and always leads to the same conclusions. In the Mann-Whitney U test we calculate

$$z = \frac{U - \frac{n_1 n_2}{2}}{\sqrt{\frac{n_1 n_2 (n_1 + n_2 + 1)}{12}}}$$

where

$$U = n_1 n_2 + \frac{n_1(n_1 + 1)}{2} - R$$

Show that if the expression for U is substituted into the preceding expression for z, we get the same test statistic (with opposite sign) used in the Wilcoxon rank-sum test for two independent samples.

11. Assume that we have two treatments (A and B) that produce measurable results, and we have only two observations for Treatment A and two observations for Treatment B. We cannot use the test statistic given in this section because both sample sizes do not exceed 10.

a. Complete the accompanying table by listing the five rows corresponding to the other five possible rankings, and enter the corresponding rank sums for treatment A.

b. List the possible values of R, along with their corresponding probabilities. [Assume that the rows of the table from part (a) are equally likely.]

c. Is it possible, at the 0.10 significance level, to reject the null hypothesis that there is no difference between treatments A and B? Explain.

Rank 1 2 3 4	Rank Sum for Treatment A
A A B B	3

12. When using the Wilcoxon rank-sum test, is the effect of an outlier large or small? Explain.

12-5 Tests for Multiple Samples

In this section we introduce two tests: the *Kruskal-Wallis test*, which is used to test the claim that several different independent samples come from identical populations and the *Friedman test*, which is the equivalent test for dependent samples. In Chapter 9, we used one-way and two-way analysis of variance (ANOVA) to test hypotheses that differences in means among several samples are significant, but that method requires that all of the involved populations have normal distributions. The Kruskal-Wallis and Friedman tests do not require normal distributions.

Kruskal-Wallis Test

DEFINITION

The **Kruskal-Wallis test** (also called the **H test**) is a nonparametric test that uses ranks of sample data from three or more independent populations. It is used to test the null hypothesis that the independent samples come from populations with the same distribution. The alternative hypothesis is that the population distributions are different in some way.

H_0: The samples come from populations with the same distribution.

H_1: The samples come from populations with different distributions.

Assumptions

1. We have at least three samples, all of which are random.

2. Each sample has at least five observations. (If samples have fewer than five observations, refer to special tables of critical values, such as *CRC Standard Probability and Statistics Tables and Formulae*, published by CRC Press.)

3. There is no requirement that the populations have a normal distribution or any other particular distribution.

In applying the Kruskal-Wallis test, we compute the **test statistic H, which has a distribution that can be approximated by the chi-square distribution as long as each sample has at least five observations.** When we use the chi-square distribution in this context, the number of degrees of freedom is $k - 1$, where k is the number of samples. (For a quick review of the key features of the chi-square distribution, see Section 6–5.)

Procedure for Finding the Value of the Test Statistic

1. Temporarily combine all samples into one big sample and assign a rank to each sample value. (Sort the values from lowest to highest, and in cases of ties, assign to each observation the mean of the ranks involved.)
2. For each sample, find the sum of the ranks and find the sample size.
3. Using the results from Step 2, calculate the test statistic H. The relevant notation and test statistic are as follows.

NOTATION FOR THE KRUSKAL-WALLIS TEST

N = total number of observations in all samples combined

k = number of samples

R_1 = sum of ranks for sample 1

n_1 = number of observations in sample 1

For sample 2, the sum of ranks is R_2 and the number of observations is n_2, and similar notation is used for the other samples.

TEST STATISTIC FOR THE KRUSKAL-WALLIS TEST

$$H = \frac{12}{N(N + 1)} \left(\frac{R_1^2}{n_1} + \frac{R_2^2}{n_2} + \cdots + \frac{R_k^2}{n_k} \right) - 3(N + 1)$$

where degrees of freedom = $k - 1$.

Critical Values

1. The test is *right-tailed*.
2. Use Table A-4 (because the test statistic H can be approximated by a chi-square distribution).
3. Degrees of freedom = $k - 1$.

The test statistic H is basically a measure of the variance of the rank sums $R_1, R_2, \ldots, R_k$. If the ranks are distributed evenly among the sample groups, then H should be a relatively small number. If the samples are very different, then the ranks will be excessively low in some groups and high in others, with the net effect that H will be large. Consequently, only large values of H lead to rejection of the null hypothesis that the samples come from identical populations. **The Kruskal-Wallis test is therefore a right-tailed test.**

EXAMPLE

In an example for the previous section, we illustrated how Data Set 20 in Appendix B could be used to compare the annual amounts of work time lost (in person-days) due to work stoppages, in two related industries. The data are not normally distributed. Suppose you need data compared for three industries. Based on annual totals for 15 different, randomly selected years for each of three transportation groups—water-based, truck, and bus/urban transport—use the Kruskal-Wallis test for testing the claim that the annual person-days lost due to work stoppages in the three industries come from the same population.

Table 12-6 **Annual Person-Days Lost in Three Transportation Industries Due to Work Stoppages**

Water		Truck		Bus & Urban	
0	(1.5)	30,850	(28)	71,630	(38)
56,830	(35)	10,740	(15)	28,150	(26)
20,160	(22)	3,680	(8)	84,640	(40)
15,010	(18)	15,450	(19)	65,430	(36)
32,500	(30)	98,740	(42)	37,570	(31)
10,360	(13)	820	(5)	25,400	(25)
19,620	(21)	19,400	(20)	42,460	(32)
0	(1.5)	100,390	(43)	43,050	(34)
9,540	(12)	8,860	(11)	31,070	(29)
13,450	(16)	129,920	(44)	223,720	(45)
8,080	(10)	1,700	(60)	6,000	(9)
210	(3)	29,600	(27)	83,910	(39)
85,550	(41)	550	(4)	42,820	(33)
1,850	(7)	21,620	(24)	70,990	(37)
10,510	(14)	14,100	(17)	21,490	(23)
$n_1 = 15$		$n_2 = 15$		$n_3 = 15$	
$R_1 = 245$		$R_2 = 313$		$R_3 = 477$	

SOLUTION

We will follow the hypothesis testing procedure summarized in Figure 7-4.

Steps 1, 2, and 3: The null and alternative hypotheses are:

H_0: The annual numbers of person-days lost per year due to work stoppages for the three industries come from the same population.

H_1: These numbers do not come from the same population.

Step 4: No significance level was specified. In the absence of any overriding circumstances, we use $\alpha = 0.05$.

Step 5: Because the three populations are not normally distributed, we use the Kruskal-Wallis test.

Step 6: In determining the value of the test statistic H, we must first rank all of the data. We begin with the lowest value of 0; because there are two zeros, we follow the rule for ties, and assign both of the zeros the rank 1.5, which is the mean of ranks 1 and 2. Ranks are shown in parentheses with the original data in Table 12-6. Next we find the sample size, n, and sum of ranks, R, for each sample; these values are shown at the bottom of Table 12-6. Because the total number of observations is 45, we have $N = 45$. To ensure that we calculated the rank sums correctly, note that $245 + 313 + 477 = 1035$. The sum of the arithmetic series is $(45)(46)/2 = 1035$. We can now evaluate the test statistic as follows:

$$H = \frac{12}{N(N + 1)}\left(\frac{R_1^2}{n_1} + \frac{R_2^2}{n_2} + \cdots + \frac{R_k^2}{n_k}\right) - 3(N + 1)$$

$$= \frac{12}{45(45 + 1)}\left(\frac{245^2}{15} + \frac{313^2}{15} + \frac{477^2}{15}\right) - 3(45 + 1) = 10.994$$

Because each sample has at least five observations, the distribution of H is approximately a chi-square distribution with $k - 1$ degrees of freedom. The number of samples is $k = 3$, so we have $3 - 1 = 2$ degrees of freedom. Refer to Table A-4 to find the critical value of 5.991, which corresponds to 2 degrees of freedom and a 0.05 significance level.

Step 7: The test statistic $H = 10.994$ is in the critical region bounded by 5.991, so we reject the null hypothesis that the populations are the same.

Interpretation

The annual numbers of person-days lost due to work stoppages in these industry groups do not appear to come from the same population.

Rationale: The test statistic H, as presented earlier, is the rank version of the test statistic F used in the analysis of variance discussed in Chapter 9. When we deal with ranks R instead of raw scores x, many components are predetermined. For example, the sum of all ranks can be expressed as $N(N + 1)/2$, where N is the total number of scores in all samples combined. The expression

$$H = \frac{12}{N(N + 1)} \, \Sigma n_i(\overline{R}_i - \overline{\overline{R}})^2$$

where

$$\overline{R}_i = \frac{R_i}{n_i} \qquad \overline{\overline{R}} = \frac{\Sigma R_i}{\Sigma n_i}$$

combines weighted variances of ranks to produce the test statistic H given here. This expression for H is algebraically equivalent to the expression for H given earlier as the test statistic. The earlier form of H (not the one given here) is easier to work with. In comparing the procedures of the parametric F test for analysis of variance and the nonparametric Kruskal-Wallis test, we see that in the absence of computer software, the Kruskal-Wallis test is much simpler to apply. We need not compute the sample variances and sample means. We do not require normal population distributions. Life becomes so much easier. However, the Kruskal-Wallis test is not as efficient as the F test, and it may require more dramatic differences for the null hypothesis to be rejected.

It should be noted that, unlike the situation with ANOVA, there is no follow-up test per se such as Tukey's method of simultaneous confidence intervals to determine which pairs are significantly different. If you reject the null hypothesis in the Kruskal-Wallis test, the next follow-up test would be the Wilcoxon rank-sum test. A judicious approach may be to compare the groups with the largest difference in their rank sums. In the above example, we could begin by comparing those in the water transportation industry to those in the bus and urban industry since the difference in the rank sums of these two industries is the largest.

Friedman Test

The **Friedman test** is similar to the Kruskal-Wallis test with the exception that it is used for dependent samples. The null and alternative hypotheses are constructed in the same manner and the critical value is found in same way using the chi-square distribution and $k - 1$ degrees of freedom.

Procedure for Finding the Value of the Test Statistic

1. Rank the data *within each block* from lowest to highest.
2. For each sample, find the sum of the ranks.
3. Using the results from Step 2, calculate the test statistic FR. The relevant notation and test statistic are as follows.

NOTATION FOR THE FRIEDMAN TEST

k = number of samples

b = number of blocks

R_1 = sum of ranks for sample 1

For sample 2, the sum of ranks is R_2 and similar notation is used for the other samples.

$$FR = \frac{12}{bk(k + 1)} (R_1^2 + R_2^2 + \cdots + R_k^2) - 3b(k + 1)$$

where degrees of freedom $= k - 1$.

EXAMPLE

In a mall-intercept survey, people were asked to rate 3 new logos a company was thinking of using on its product label from a scale from 1 to 10, 10 being best. For a sample of people, these were the results:

	Logo 1	Logo 2	Logo 3
Person 1	8	9	8
Person 2	8	8	7
Person 3	8	9	7
Person 4	7	8	7
Person 5	8	9	8

Is there any significant difference in the median ratings of the 3 logos? Test at a 0.05 level of significance.

SOLUTION

The null and alternative hypotheses are:

H_0: median(logo 1) = median(logo 2) = median(logo 3)

H_1: at least 2 of the logos have different median ratings

With $k = 3$, there are 2 degrees of freedom. Using Table A-4, the critical value at a 0.05 level of significance is 5.991. To compute the test statistic, we rate the logos for each person.

	Logo 1	Logo 2	Logo 3
Person 1	1.5	3	1.5
Person 2	2.5	2.5	1
Person 3	2	3	1
Person 4	1.5	3	1.5
Person 5	1.5	3	1.5
Rank sum	9	14.5	6.5

Checking the computation of the rank sums is slightly more difficult than with the Kruskal-Wallis test. Note that $9 + 14.5 + 6.5 = 30$. Since each person rates 3 logos, within each block, the sum of the arithmetic series is $(3)(4)/2 = 6$. With $b = 5$, the total of the ranks sums is $(6)(5) = 30$. We can now compute the test statistic.

$$FR = \frac{12}{bk(k + 1)} (R_1^2 + R_2^2 + \cdots + R_k^2) - 3b(k + 1)$$

$$= \frac{12}{(5)(3)(4)} (9^2 + 14.5^2 + 6.5^2) - (3)(5)(4) = 6.7$$

Since the test statistic of 6.7 is greater than the critical value of 5.991, we reject the null hypothesis.

Interpretation
We conclude that the median rating of at least 2 logos is significantly different.

As with the Kruskal-Wallis test, if we reject the null hypothesis in the Friedman test, there is no follow-up test per se. The best follow-up procedure would be to conduct the Wilcoxon signed-ranks test on pairs of data beginning with the pairs that have the largest difference in their rank sums. In the above example, we could begin by comparing Logo 2 to Logo 3.

Rationale for the test statistic: If each sample has the same average rank (call it R) then

$$\sum R = \frac{bk(k + 1)}{2}$$

since the rank sum within each block is $k(k + 1)/2$. Since there are k samples, then

$$R = \frac{bk(k + 1)}{2k} = \frac{b(k + 1)}{2}$$

$$R^2 = \frac{b^2(k + 1)^2}{4}$$

$$\sum R^2 = \frac{b^2(k + 1)^2}{4} + \frac{b^2(k + 1)^2}{4} + \cdots + \frac{b^2(k + 1)^2}{4} \ (k \text{ terms})$$

$$= \frac{kb^2(k + 1)^2}{4}$$

The test statistic becomes

$$FR = \frac{12}{bk(k + 1)} \left(\frac{kb^2(k + 1)^2}{4} \right) - 3b(k + 1)$$

$$= 3b(k + 1) - 3b(k + 1) = 0$$

which is the lowest possible test statistic value possible under the χ^2 distribution. Regardless of the degrees of freedom, the P-value $= P(\chi^2 > 0) = 1$.

STATDISK: Select **Analysis** from the main menu bar, then select **Kruskal-Wallis Test**. STATDISK will display the sum of the ranks for each sample, the **H test** statistic, the critical value, and the conclusion.

Kruskal-Wallis test: To enter the data, put the values in C1 and the sample to which they belong in C2. From the **Stat** menu, choose **Nonparametrics** and then **Kruskal-Wallis**. Choose C1 as the response and C2 as the factor.

Friedman test: To enter the data, put the values in C1, the sample they belong to in C2, and the block to which they belong in C3. From the **Stat** menu, choose **Nonparametrics** and then **Friedman**. Choose C1 as the response, C2 as the treatment, and C3 as the blocks.

MINITAB DISPLAY
Kruskal-Wallis Test

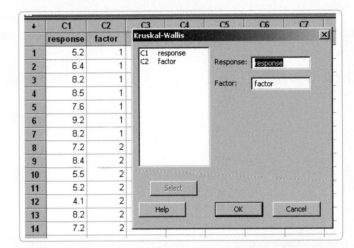

MINITAB DISPLAY
Friedman Test

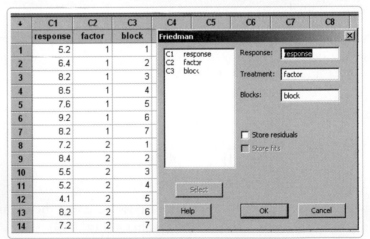

EXCEL: In Excel, use the following template. Set up the column headings as shown in the left part of the figure. Fill in the leftmost column with the data from all of the samples. Then, in the second column, identify the sample that each data value comes from. Once the data are entered, sort the two left columns in order of values.

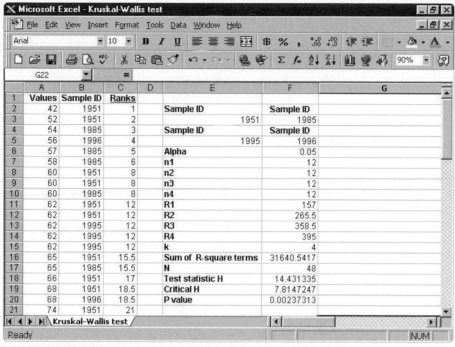

Cell	Formula	Explanation
F7	=DCOUNT(A1:C49,1,E2:E3)	n for the first sample, whose ID is specified in E3
F8	=DCOUNT(A1:C49,1,F2:F3)	n for the second sample, whose ID is specified in F3
F9	=DCOUNT(A1:C49,1,E4:E5)	n for the third sample, whose ID is specified in E5
F10	=DCOUNT(A1:C49,1,F4:F5)	n for the fourth sample, whose ID is specified in F5 (Leave F10 blank if there is no fourth sample.)
F11	=DSUM(A1:C49,3,E2:E3)	Sum of the ranks for the first sample
F12	=DSUM(A1:C49,3,F2:F3)	Sum of the ranks for the second sample
F13	=DSUM(A1:C49,3,E4:E5)	Sum of the ranks for the third sample
F14	=DSUM(A1:C49,3,F4:F5)	Sum of the ranks for the fourth sample (Leave F14 blank if there is no fourth sample.)
F16	=IF(F15=4,F11^2/F7+F12^2/F8+F13^2/F9+F14^2/F10,F11^2/F7+F12^2/F8+F13^2/F9)	Sum of each sample's R^2/n (an intermediate calculation within the boxed formula for the test statistic) (Based on 3 or 4 samples)
F17	=SUM(F7:F10)	Total number of observations in all samples combined
F18	=(12/(F17*(F17+1)))*F16-(3*(F17+1))	Test statistic H (see boxed formula for the test statistic)
F19	=CHIINV(F6,F15-1)	Critical value, based on the chi-square approximation
F20	=CHIDIST(F18,F15-1)	P-value, based on the chi-square approximation

Use appropriate cell addresses for data.

Note: The chi-square approximation assumes that each sample has at least 5 observations.

To do so, highlight those two columns, including the headings. Select **Sort** from the **Data** menu and, in the dialog box that appears, select "Values" for the **Sort by** option. Then specify **Ascending** and **Header row**, and click on **OK**. Now fill the "Ranks" column with the ordered numbers from 1 to the number of data values (in all samples combined). Inspect the "Values" column for tied values. If there are any, manually replace their ranks in the "Ranks" column with the mean of the ranks that first appear in that column. (For example, since 65 appears twice in the "Values" column, with the first 65 having a rank of 15 and the second having a rank of 16, both ranks have been manually replaced with 15.5.)

We show the template for completing the calculations. Enter the full (right-tailed) alpha manually. In Cells E2, F2, E4, and F4, enter the same heading that is in the middle column of your input block. Use Cells E3, F3, E5, and F5 to enter the IDs you used, in the input block, for the samples (such as 1951 and 1985 in the example). In Cell F15, indicate whether three or four samples were used; the template does not handle more than four samples. All other values will be calculated, based on the indicated formulas.

 ## 12-5 Exercises A: Basic Skills and Concepts

In Exercises 1–8, use the Kruskal-Wallis test.

1. The accompanying table lists the numbers of House sittings in each of 18 Parliaments, randomly selected from three different time periods. Parliaments lasting less than a year were not sampled. Using a significance level of $\alpha = 0.05$, test the claim that for the three periods sampled, the numbers of House sittings per Parliament are identical.

Time Period		
1	2	3
283	355	447
568	605	687
303	292	405
254	366	707
262	565	767
458	390	590

Based on data from *Duration of Sessions of Parliament,* Library of Parliament.

2. Has usage of the Internet increased? Very likely, more people are using the Internet than in the past, especially from home. But if a person has been using the Internet all along, are they are now spending more time on it? The table in the margin shows results from three independent surveys, conducted in 1997, 1998, and 1999; the specific values are the hours that respondents spend per week on the Internet. Use the 0.05 significance level to test the claim that the three samples come from identical populations.

3. Refer to Data Set 21 in Appendix B. A desirable trait for a top-rate diamond is to be flawless. Clarity is a measure of this characteristic—with 1 being the best. At the 0.05 level, test the claim that the prices of diamonds are not the same for each of the four clarity values listed (2, 3, 4, and 5). Would you conclude that clarity is a major determinant of price? If you need more information to answer the question, explain.

4. Flammability tests were conducted on children's sleepwear. The Vertical Semirestrained Test was used, in which pieces of fabric were burned under controlled conditions. After the burning stopped, the length of the charred portion was measured and recorded. Results are given on the following page for the same fabric tested at different laboratories (based on data from Minitab, Inc.). Because the same fabric was used, the different laboratories should have obtained the same results. Did they?

Hours per Week that Canadian Adults Spend on the Internet

Summer 1997	Summer 1998	Fall 1999
0.0	1.8	4.2
7.6	6.1	5.2
0.7	4.1	8.8
19.1	4.3	4.3
0.6	0.8	1.5
9.3	7.9	14.5
3.7	3.9	6.6
1.8	8.6	16.0
4.8	16.3	18.9
0.0	0.0	0.0
0.2	0.4	0.4
1.1	1.3	1.2
4.1	14.7	1.7
1.4	1.4	3.9
2.6	3.6	3.1
10.5	12.3	10.5
4.7	4.3	4.9
0.0	0.0	0.6
14.5	13.9	1.8

Laboratory

1	2	3	4	5
2.9	2.7	3.3	3.3	4.1
3.1	3.4	3.3	3.2	4.1
3.1	3.6	3.5	3.4	3.7
3.7	3.2	3.5	2.7	4.2
3.1	4.0	2.8	2.7	3.1
4.2	4.1	2.8	3.3	3.5
3.7	3.8	3.2	2.9	2.8
3.9	3.8	2.8	3.2	
3.1	4.3	3.8	2.9	
3.0	3.4	3.5		
2.9	3.3			

5. Refer to Data Set 11 in Appendix B. At the 0.05 significance level, test the claim that the weights of M&Ms are the same for each of the six different colour populations. If it is the intent of Mars, Inc., to make the candies so that the different colour populations are the same, do your results suggest that the company has a problem that requires corrective action?

6. Refer to Data Set 9 in Appendix B. Use a 0.01 significance level to test the claim that the populations of cotinine levels are identical for these three groups: nonsmokers who are not exposed to environmental tobacco smoke, nonsmokers who are exposed to tobacco smoke, and people who smoke. What do the results suggest about second-hand smoke?

7. The accompanying table lists the lengths, in minutes, and star ratings of televised movies rated 18+. Use a 0.05 significance level to test whether there is any difference in movie length based on star rating. Based on the results, does it appear that bad adult-rated movies are longer than good ones, or does it just seem that way?

(2.0–2.5 Stars) Fair	(3.0–3.5 Stars) Good	(4.0 Stars) Excellent
110	115	72
114	133	120
96	106	106
101	93	104
155	129	159

8. The City Resource Recovery Company (CRRC) collects the waste discarded by households in a region. Discarded waste must be separated into categories of metal, paper, plastic, and glass. In planning for the equipment needed to collect and process the garbage, CRRC refers to the data in Data Set 1 in Appendix B and uses Excel to obtain the results shown in the following display. At the 0.05 level of significance, test the claim that the four specific populations of garbage (metal, paper, plastic, glass) are the same. Based on the results, does it appear that these four categories require the same collection and processing resources? Does any single category seem to be a particularly large part of the waste management problem? (*Hint:* Look at the sums of the ranks. Also note that the *P*-value is in scientific notation.)

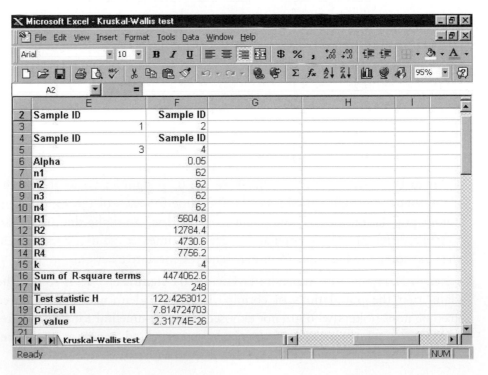

9. For a store with 3 locations, daily sales (in thousands) were taken from each location for a series of dates:

	Store 1	Store 2	Store 3
Aug. 1	8.2	7.8	8.9
Aug. 5	10.3	10.1	10.5
Aug. 9	6.7	7.3	7.8
Aug. 13	8.7	9.2	9.8
Aug. 17	11.3	10.9	11.6
Aug. 21	6.7	7.3	7.5
Aug. 25	12.3	12.7	13.1
Aug. 29	5.9	6.2	7.8

Is there any significant difference in median daily sales between the 3 locations? Test at a 0.05 level of significance.

Exercises B: Beyond the Basics

10. a. In general, how is the value of the test statistic H affected if a constant is added to (or subtracted from) each score?
    b. In general, how is the value of the test statistic H affected if each score is multiplied (or divided) by a positive constant?
    c. In general, how is the value of the test statistic H affected if a single sample value is changed to a value that causes it to become an outlier?

11. For three samples, each of size 5, what are the largest and smallest possible values of H?

12. In using the Kruskal-Wallis test, there is a correction factor that should be applied whenever there are many ties: Divide H by

$$1 - \frac{\Sigma T}{N^3 - N}$$

Here $T = t^3 - t$, where t is the number of observations that are tied for a group of tied scores. That is, find t for each group of tied scores, then compute the value of T, then add the T values to get ΣT. For the example presented in this section, use this procedure to find the corrected value of H. Does the corrected value of H differ substantially from the value of 14.431 that was found in this section?

13. Show that for the case of two samples, the Kruskal-Wallis test is equivalent to the Wilcoxon rank-sum test. This can be done by showing that for the case of two samples, the test statistic H equals the square of the test statistic z used in the Wilcoxon rank-sum test. Also note that with 1 degree of freedom, the critical values of χ^2 correspond to the square of the critical z score.

12-6 Rank Correlation

In this section we describe how the nonparametric method of rank correlation is used with paired data to test for an association between two variables. In Chapter 11 we used paired sample data to compute values for the linear correlation coefficient *r*, but in this section we use *ranks* as the basis for measuring the strength of the correlation between two variables.

DEFINITIONS

> The **rank correlation test** (or **Spearman's rank correlation test**) is a nonparametric test that uses ranks of sample data consisting of matched pairs. It is used to test for an association between two variables, so the null and alternative hypotheses are as follows (where ρ_s denotes the **rank correlation coefficient**—the correlation coefficient of the ranks—for the entire population.)
>
> H_0: $\rho_s = 0$ (There is no correlation between the two variables.)
> H_1: $\rho_s \neq 0$ (There is a correlation between the two variables.)

Assumptions

1. The sample data have been randomly selected.

2. Unlike the parametric methods of Section 11–2, there is no requirement that the sample pairs of data have a bivariate normal distribution. There is no requirement of a normal distribution for any population.

Advantages: Rank correlation has some distinct advantages over the parametric methods discussed in Chapter 11:

1. The nonparametric method of rank correlation can be used in a wider variety of circumstances than the parametric method of linear correlation. With rank correlation, we can analyze paired data that can be ranked but not measured. For example, if two judges rank 30 different gymnasts, we can use those ranks to test for a correlation between the two judges, but we cannot test for a correlation using linear correlation.

   Even when measured data are available, we can sometimes get a clearer result by converting the data into ranks. For example, the severity of industrial accidents is often reported in number of days of lost work time. But these exact numbers may cluster near arbitrary values, such as cutoffs for time-off support by Workers' Compensation. If you are comparing the magnitudes of energy sources and the severity of related accidents, converting the data to ranks may provide clearer results.

2. Rank correlation can be used to detect some (not all) relationships that are not linear. (An example will be given later in this section.)

3. The computations for rank correlation are much simpler than those for linear correlation, as can be readily seen by comparing the formulas used to compute these statistics. If calculators or computers are not available, the rank correlation coefficient is easier to compute.

Disadvantage: A disadvantage of rank correlation is its efficiency rating of 0.91, as described in Section 12–1. This efficiency rating indicates that with all other circumstances being equal, the nonparametric approach of rank correlation requires 100 pairs of sample data to achieve the same results as only 91 pairs of sample observations analyzed through the parametric approach, assuming the stricter requirements of the parametric approach are met.

We will use the following notation, which closely parallels the notation used in Chapter 11 for linear correlation. (Recall from Chapter 11 that r denotes the linear correlation coefficient for sample paired data, ρ denotes the linear correlation coefficient for all paired data in a population, and n denotes the number of pairs of data.)

NOTATION

r_s = rank correlation coefficient for sample paired data (r_s is a sample statistic)

ρ_s = rank correlation coefficient for all the population data (ρ_s is a population parameter)

n = number of pairs of data

d = difference between ranks for the two observations within a pair

We use the notation r_s for the rank correlation coefficient so that we don't confuse it with the linear correlation coefficient r. The subscript "s" has nothing to do with standard deviation; it is used in honour of Charles Spearman (1863–1945), who originated the rank correlation approach. In fact, r_s is often called **Spearman's rank correlation coefficient.**

TEST STATISTIC FOR THE RANK CORRELATION COEFFICIENT

$$r_s = 1 - \frac{6\Sigma d^2}{n(n^2 - 1)}$$

where each value of d is a difference between the ranks for a pair of sample data.

The procedure for using the rank correlation procedure is summarized in Figure 12-6. That figure includes the following test statistic.

Figure 12-6
Rank Correlation for
Testing $H_0: \rho_s = 0$

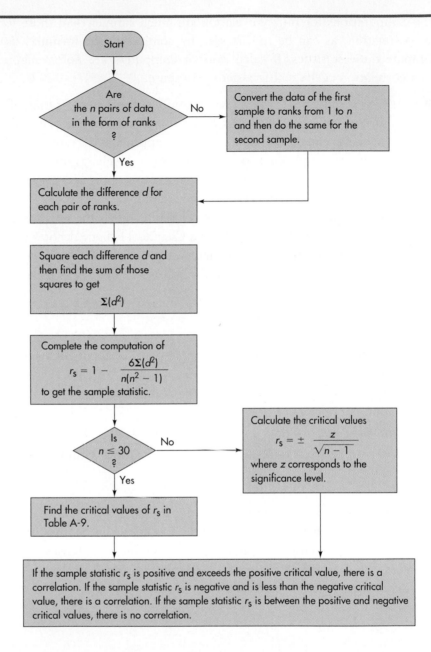

Critical Values: If $n \le 30$, critical values are found in Table A-9.
If $n > 30$, critical values are found by using Formula 12-1.

Formula 12-1 $\qquad\qquad\qquad r_s = \dfrac{\pm z}{\sqrt{n-1}} \qquad$ (critical values when $n > 30$)

where the value of z corresponds to the significance level.

Handling Ties: We handle ties in ranks of the original sample values as in the preceding sections of this chapter: Find the mean of the ranks involved in the tie, and then assign the mean rank to each of the tied items. This test statistic yields the exact value of r_s only if there are no ties. With a relatively small number of ties, the test statistic is a good approximation of r_s. (When ties occur, we can get an exact value of r_s by ranking the data and using Formula 11-1 for the linear correlation coefficient; after finding the value of r_s, we can proceed with the methods of this section.)

EXAMPLE

Every year, *Maclean's* magazine ranks Canadian universities by a number of criteria. The universities are grouped into categories such as primarily undergraduate, comprehensive, and so on, and ranked within these categories. Table 12-7 is based on one year's results for the 13 comprehensive universities. Is there a correlation between the universities' rankings for numbers of student awards and their rankings for number of faculty awards? The linear correlation coefficient r (Section 11–2) should not be used because it requires normal distributions, but the data consist of ranks that are not normally distributed. Instead, use the rank correlation coefficient to test the claim that there is a relationship between the rankings for student awards and faculty awards (that is, $\rho_s \neq 0$). Use a significance level of $\alpha = 0.05$.

Table 12-7 Rankings of Comprehensive Universities

University	Simon Fraser	Guelph	Victoria	Waterloo	York	Memorial	Carleton
Student awards	3	2	4	1	11	12	6
Faculty awards	1	9	5	4	2	12	7
d	2	−7	−1	−3	9	0	−1
d^2	4	49	1	9	81	0	1

University	Windsor	New Brunswick	Regina	Concordia	UQAM	Trois-Rivières	
Student awards	13	5	10	7	8	9	
Faculty awards	6	13	11	8	3	10	
d	7	−8	−1	−1	5	−1	
d^2	49	64	1	1	25	1	→Total = 286

SOLUTION

We follow the same basic steps for testing hypotheses as outlined in Figure 7-4.

Step 1: The claim of a correlation is expressed symbolically as $\rho_s \neq 0$.
Step 2: The negation of the claim in Step 1 is $\rho_s = 0$.

Step 3: Because the null hypothesis must contain the condition of equality, we have

$$H_0: \rho_s = 0$$
$$H_1: \rho_s \neq 0 \quad \text{(original claim)}$$

Step 4: The significance level is $\alpha = 0.05$.

Step 5: As we noted, we cannot use the linear correlation approach of Section 11–2 because ranks do not satisfy the requirement of a normal distribution. We will use the rank correlation approach instead.

Step 6: We now find the value of the test statistic. Table 12-7 shows the calculation of the differences d and their squares d^2, which results in a value of $\Sigma d^2 = 286$. With $n = 13$ (for 13 pairs of data) and $\Sigma d^2 = 286$, we can find the value of the test statistic r_s as follows:

$$r_s = 1 - \frac{6\Sigma d^2}{n(n^2 - 1)} = 1 - \frac{6(286)}{13(13^2 - 1)}$$

$$= 1 - \frac{1716}{2184} = 0.214$$

Now refer to Table A-9 to determine that the critical values are 60.566 (based on $\alpha = 0.05$ and $n = 13$). Because the test statistic $r_s = 0.214$ does not exceed the critical value of 0.566, we fail to reject the null hypothesis.

Interpretation

When the universities' rankings for numbers of student awards and faculty awards are compared, there is not sufficient evidence to support the claim of a correlation. It appears that different unrelated university features are associated with student and faculty awards.

EXAMPLE

Suppose that there were 40 comprehensive universities in Canada, and that *Maclean's* ranked them with respect to numbers of student awards and the numbers of faculty awards. If the test statistic r_s is found to be 0.291, and if the significance level is $\alpha = 0.05$, what do you conclude about the correlation between the two rankings?

SOLUTION

Because there are 40 pairs of data, we have $n = 40$. Because n exceeds 30, we find the critical values from Formula 12-1 instead of Table A-9. With $\alpha = 0.05$ in two tails, we let $z = 1.96$ to get

$$r_s = \frac{\pm 1.96}{\sqrt{40 - 1}} = \pm 0.314$$

The test statistic of $r_s = 0.291$ does not exceed the critical value of 0.314, so we fail to reject the null hypothesis. There is not sufficient evidence to support the claim of a correlation between the rankings for student and faculty awards.

The next example is intended to illustrate the principle that rank correlation can sometimes be used to detect relationships that are not linear.

EXAMPLE

Ten students study for a test; the following table lists the number of hours studied (x) and the corresponding number of correct answers (y).

Hours studied (x)	5	9	17	1	2	21	3	29	7	100
Correct answers (y)	6	16	18	1	3	21	7	20	15	22

At the 0.05 level of significance, test the claim that there is a relationship between hours studied and the number of correct answers.

SOLUTION

In part because of the x value of 100 hours (an outlier), the distribution requirements are not met for using the linear regression methods of Chapter 11. (Examine Figure 12-7.) We could choose to discard the outlier, but this wastes a potentially useful data pair. Instead we will use the Spearman's rank correlation approach. Testing the null hypothesis of null rank correlation, we have:

$$H_0: \rho_s = 0$$
$$H_1: \rho_s \neq 0$$

Refer to Figure 12-6, which we follow in this solution. Following the procedures of Section 12–1, we convert the original values into ranks as shown in the table.

Ranks for hours studied	4	6	7	1	2	8	3	9	5	10
Ranks for correct answers	3	6	7	1	2	9	4	8	5	10
d	1	0	0	0	0	−1	−1	1	0	0
d^2	1	0	0	0	0	1	1	1	0	0 → Total = 4

After expressing all data as ranks, we calculate the differences, d, and then square them. The sum of the d^2 values is 4. We now calculate

$$r_s = 1 - \frac{6\sum d^2}{n(n^2 - 1)} = 1 - \frac{6(4)}{10(10^2 - 1)}$$

$$= 1 - \frac{24}{990} = 0.976$$

Proceeding with Figure 12-6, we have $n = 10$, so we answer yes when asked if $n \leq 30$. We use Table A-9 to get the critical values of 60.648. Finally, the sample statistic of 0.976 exceeds 0.648, so we reject the null hypothesis of no significant correlation.

Interpretation
More hours of study appear to be associated with higher grades. (You didn't really think we would suggest otherwise, did you?)

Figure 12-7
Scatter Diagram for Hours Studied and Correct Answers

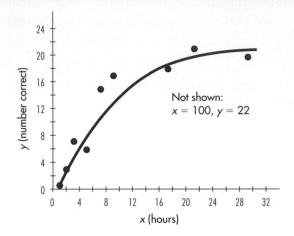

In the above example, if we compute the linear correlation coefficient r (using Formula 11-1) for the original data, we get $r = 0.629$, which leads to the conclusion that there is not enough evidence to support the claim of a significant linear correlation at the 0.05 level of significance. If we examine the scatter diagram in Figure 12-7, we can see that there does seem to be a relationship, but it's not linear. This last example illustrates these two advantages of the nonparametric approach over the parametric approach: (1) With rank correlation, we can sometimes detect relationships that are not linear, and (2) Spearman's rank correlation coefficient r_s is less sensitive to an outlier, such as the 100 hours in the preceding data.

USING TECHNOLOGY

STATDISK: Select **Analysis** from the main menu bar, then select **Rank Correlation**. Next, click **Evaluate**.

EXCEL: Create a data range in which each sample value has been replaced by its corresponding rank, and these ranks are in neighbouring columns; then, using the data range as input, calculate the linear correlation coefficient with the same Excel procedures used in Section 11–2.

If the data are not already ranked, align the data in two neighbouring columns and add the headings as shown in the following figure. The original data are in the columns "Variable 1" and "Variable 2." Sort the table in order of the "Variable 1" column. Now fill the "Ranks 1" column with the ordered numbers from 1 to the number of data pairs. Inspect the "Variable 1" column for tied values. If there are any, then manually replace their ranks in the "Ranks 1" column with the mean of the ranks that first appear there. (If, for example, the two smallest values in the "Variable 1" column were both 1, then they would both be assigned the rank 1.5, which is the mean of the original ranks 1 and 2.) Now re-sort the full data range, including the "Ranks 1" column, based on the "Variable 2" column, and repeat the above procedures to fill in the "Ranks 2" column. When this is complete, the "Ranks 1" and "Ranks 2" columns can be the input range for the Excel procedures for calculating the linear correlation coefficient r.

	A	B	C	D	E	F	G	H	I
1	Variable 1	Variable 2	Ranks 1	Ranks 2					
2	1	1	1	1					
3	2	3	2	2					
4	5	6	4	3					
5	3	7	3	4					
6	7	15	5	5					
7	9	16	6	6					
8	17	18	7	7					
9	29	20	9	8					
10	21	21	8	9					
11	100	22	10	10					
12									

EXCEL DISPLAY
Rank Correlation

MINITAB 15: In the spreadsheet part, create a data range in which each sample value has been replaced by its corresponding rank, and these ranks are in neighbouring columns. From the **Stat** menu, choose **Basic Statistics** and then **Correlation**. Select the variables and click **OK**.

12-6 Exercises A: Basic Skills and Concepts

1. For each of the following samples of paired ranks, sketch a scatter diagram, find the value of r_s, and state whether there appears to be a correlation between x and y.

   a.

x	1	3	5	4	2
y	1	3	5	4	2

   b.

x	1	2	3	4	5
y	5	4	3	2	1

   c.

x	1	2	3	4	5
y	2	5	3	1	4

2. Find the critical value(s) for r_s by using either Table A-9 or Formula 12-1, as appropriate. Assume two-tailed cases, where α represents the level of significance and n represents the number of pairs of data.

   a. $n = 20$, $\alpha = 0.05$ **b.** $n = 50$, $\alpha = 0.05$
   c. $n = 40$, $\alpha = 0.02$ **d.** $n = 15$, $\alpha = 0.01$
   e. $n = 82$, $\alpha = 0.04$

In Exercises 3–18, use the rank correlation coefficient to test the claim of no correlation between the two variables. Use a significance level of $\alpha = 0.05$.

Job	Salary Rank	Stress Rank
Stockbroker	2	2
Zoologist	6	7
Electrical engineer	3	6
School principal	5	4
Hotel manager	7	5
Bank officer	10	8
Occ. safety inspector	9	9
Home economist	8	10
Psychologist	4	3
Airline pilot	1	1

Based on data from *The Jobs Rated Almanac.*

3. The accompanying table lists salary rankings and stress rankings for randomly selected jobs. Does it appear that salary increases as stress increases?

4. Exercise 3 includes paired salary and stress level ranks for 10 randomly selected jobs. The physical demands of the jobs were also ranked; the salary and physical demand ranks are given below. Does there appear to be a relationship between the salary of a job and its physical demands?

Salary	2	6	3	5	7	10	9	8	4	1
Physical demand	5	2	3	8	10	9	1	7	6	4

Based on data from *The Jobs Rated Almanac.*

5. Ten jobs were randomly selected and ranked according to stress level and physical demand, with the results given below. Does there appear to be a relationship between the stress levels of jobs and their physical demands?

Stress level	2	7	6	4	5	8	9	10	3	1
Physical demand	5	2	3	8	10	9	1	7	6	4

Based on data from *The Jobs Rated Almanac.*

6. Most truck industry carriers operate within a single province or region. About 1% operate across multiple jurisdictions, including nationally. Comparing years ranked by total carrier activity (from the period 1984–94) and ranked numbers of multi-jurisdiction carriers per year, assess whether the number of these carriers and the year sequence are related.

Years	7	2	9	11	1	10	3	5	4	8	6
Numbers of carriers	7	2	9.5	11	1	8	3	9.5	5.5	4	5.5

Based on data from Transport Canada.

7. In studying the effects of heredity and environment on intelligence, scientists have learned much by analyzing the IQ scores of identical twins who were separated soon after birth. Identical twins have identical genes, which they inherited from the same fertilized egg. By studying identical twins raised apart, we can eliminate the variable of heredity and better isolate the effects of environment. Following are the IQ scores of identical twins (first-born twins are x) raised apart.

x	107	96	103	90	96	113	86	99	109	105	96	89
y	111	97	116	107	99	111	85	108	102	105	100	93

Based on data from "IQ's of Identical Twins Reared Apart," by Arthur Jensen, *Behavioral Genetics.*

8. It's a baseball cliché that "pitching wins games." Test that claim based on the accompanying data. For eight randomly selected American League teams for the 2000 season, the following data compares the Earned Run Average (ERA) for the teams' pitchers (lower is considered better) and total team wins (based on data from CNN/Sports Illustrated).

Team ERA	4.53	4.58	4.84	5.02	5.16	5.17	5.48	5.52
Team wins	91	91	90	82	69	83	77	71

9. Researchers have studied bears by anesthetizing them in order to obtain vital measurements. The data in the table represent the first eight male bears in Data Set 3 from Appendix B.

x Length (in.)	53.0	67.5	72.0	72.0	73.5	68.5	73.0	37.0
y Weight (lb)	80	344	416	348	262	360	332	34

10. An amateur biologist has read that the lifespans of mammals are related to their metabolic rates. The relation, if any, is not linear, so she will test the claim based on the rank correlation coefficient. She begins by randomly selecting six species, and then recording their maximum lifespans and their typical metabolic rates (see Data Set 19 in Appendix B for more details). Is there a significant correlation between these variables?

Metabolic rate	0.6	0.9	1.3	1.4	11	36
Maximum life	70	45	115	20	3.5	0.3

11. The accompanying table lists educational requirements (in years) and typical salary (in thousands of dollars) for a sample of professions. Based on the result, can you expect to earn more if you choose a profession that requires more years of education? How do the results change if the salaries are entered as 61,100, 36,000, . . . , 14,800?

Years of education	16	15	14	14	13	13	12	11
Income	61.1	36.0	62.6	32.9	27.2	36.9	36.7	29.6
Years of education	13	11	10	12	11	12	10	10
Income	37.6	28.4	26.5	22.2	23.1	24.8	21.3	14.8

Based on data from Statistics Canada.

12. Refer to Data Set 21 in Appendix B and use the paired data for the price and weight (carats) of diamonds. Is there a correlation, suggesting that heavier diamonds are generally of greater value?

13. Refer to Data Set 16 in Appendix B and use the paired data for durations and intervals after eruptions of the Old Faithful geyser. Is there a correlation, suggesting that the interval after an eruption is related to the duration of the eruption?

14. Refer to Data Set 16 in Appendix B and use the paired data for intervals after eruptions and heights of eruptions of the Old Faithful geyser. Is there a correlation, suggesting that the interval after an eruption is related to the height of the eruption?

15. Refer to Data Set 2 in Appendix B and use the paired data consisting of the first and second columns of body temperatures for subjects who are initially sitting. Based on the result, does there appear to be a correlation between temperatures in the initially sitting postures and supine postures? If so, could researchers reduce their laboratory expenses by measuring only one of these two variables?

16. Refer to Data Set 2 in Appendix B and use the paired data consisting of the first and second columns of body temperatures for subjects who are initially standing. Based on the result, does there appear to be a correlation between temperatures in the initially standing postures and supine postures? If so, could researchers reduce their laboratory expenses by measuring only one of these two variables?

17. Refer to Data Set 7 in Appendix B, which looks at temperature and precipitation departures from seasonal norms. Pick out the data pairs corresponding to years of strong El Niños. Compare (a) the temperature departure for the first spring of El Niño and (b) the temperature departure for the second spring of El Niño. Based on the result, does there appear to be a correlation between temperatures in the first spring and those in the second spring of El Niño phenomena? If so, could researchers use this information to help predict the temperature departures for the second spring of El Niño?

18. Refer to Data Set 7 in Appendix B, which looks at temperature and precipitation departures from seasonal norms. Pick out the data pairs corresponding to years of strong El Niños. Compare (a) the precipitation departure for the first spring of El Niño and (b) the precipitation departure for the second spring of El Niño. Based on the result, does there appear to be a correlation between precipitation in the first spring and that in the second spring of El Niño phenomena? If so, could researchers use this information to help predict the precipitation departures for the second spring of El Niño?

12-6 Exercises B: Beyond the Basics

19. One alternative to using Table A-9 involves an approximation of critical values for r_s given as

$$r_s = \pm\sqrt{\frac{t^2}{t^2 + n - 2}}$$

Here t is the t score from Table A-3 corresponding to the significance level and $n - 2$ degrees of freedom. Apply this approximation to find critical values of r_s for the following cases.

a. $n = 8$, $\alpha = 0.05$ b. $n = 15$, $\alpha = 0.05$

c. $n = 30$, $\alpha = 0.05$ d. $n = 30$, $\alpha = 0.01$

e. $n = 8$, $\alpha = 0.01$

20. a. How is r_s affected if the scale for one of the variables is changed from feet to inches?

b. How is r_s affected if one variable is ranked from low to high while the other variable is ranked from high to low?

c. How is r_s affected if the two variables are interchanged?

d. One researcher ranks both variables from low to high, while another researcher ranks both variables from high to low. How will their values of r_s compare?

e. How is r_s affected if each value of x is replaced by $\log x$?

12-7 Runs Test for Randomness

We have seen many statistical procedures used in this book that require the random selection of data. In this section we describe the runs test, which is one method used to determine whether a sequence of sample data have the desired characteristic of randomness.

DEFINITIONS

A **run** is a sequence of data having the same characteristic; the sequence is preceded and followed by data with a different characteristic or by no data at all.

The **runs test** uses the number of runs in a sequence of sample data to test for randomness in the order of the data.

Assumptions

1. The sample data are arranged according to some ordering scheme such as the order in which the samples were obtained.

2. Each data value can be categorized into one of two separate categories.

3. The runs test for randomness is based on the order in which the data occur; it is not based on the frequency of the data. (For example, a sequence of 3 men and 20 women might appear to be random by this test; whether 3 men and 20 women constitute a biased sample is not addressed by the runs test.)

n_1 = number of elements in the sequence that have one particular characteristic. (The characteristic chosen for n_1 is arbitrary.)

n_2 = number of elements in the sequence that have the other characteristic

G = number of runs

EXAMPLE

Manufacturers of cola conduct market research to determine consumer preferences. When 10 consumers are asked whether they prefer diet cola or regular cola, the responses of the first 10 subjects are listed in the order in which they were obtained. We use D to denote a consumer who prefers *diet* cola and R to denote a consumer who prefers *regular* cola. Find the number of runs G, and also identify the values of n_1 and n_2.

D D D D	R R	D D D	R
1st run	2nd run	3rd run	4th run

SOLUTION

There are exactly four runs, as shown, so $G = 4$. Letting n_1 represent the number of diet (D) colas, we have $n_1 = 7$ because there are seven diet colas present. It follows that the number of regular (R) colas is described by $n_2 = 3$.

Fundamental Principle of the Runs Test

The fundamental principle of the runs test can be briefly stated as follows: **Reject randomness if the number of runs is very low or very high.**

- Example: DDDDDRRRRR is not random because it has only 2 runs, so the number of runs is very low.

- Example: DRDRDRDRDR is not random because there are 10 runs, which is very high (relative to a 10-value data set).

We reject randomness if the number of runs G is too small or too large, but how do we determine exactly which values of G are too small or too large? We use the following criterion:

5% CUTOFF CRITERION

Reject randomness of the data if the number of runs G is so small or so large that in repeated samplings, a value at least as extreme as G will occur 5% of the time or less.

Although this criterion wins no prizes for simplicity, it's quite easy to apply if we use Table A-10. Table A-10 identifies those values of G that are so small or so large that they belong to the category of exceptional sequences that happen 5% of the time or less. Using Table A-10, the 5% criterion can be simply restated as follows:

SIMPLIFIED 5% CUTOFF CRITERION

Reject randomness if the number of runs G is
- less than or equal to the smaller Table A-10 entry or
- greater than or equal to the larger entry in that table

For example, the sequence

D D D D R R R R R

results in these values: $n_1 = 4$, $n_2 = 5$, and $G = 2$ runs. With $n_1 = 4$ and $n_2 = 5$, Table A-10 indicates that we should reject randomness if the number of runs is 2 or less or 9 or greater. With $G = 2$ runs, we therefore reject randomness according to the simplified 5% cutoff criterion based on Table A-10.

Rationale: How do we find the cutoff values separating random sequences from those that are not? That is, how do we find the critical values displayed in Table A-10? Let's reconsider the sequence DDDDRRRRR (from the previous paragraph), for which $G = 2$, $n_1 = 4$ and $n_2 = 5$. If we refer to Table A-10, we find that for $n_1 = 4$ and $n_2 = 5$, the cutoff values are 2 and 9. If we had an abundance of time and patience, we could list all possible sequences of 4 diet colas and 5 regular colas. Examination of that list would reveal these facts:

- There are 126 different possible sequences of 4 diet and 5 regular colas.
- Among the 126 different possible cases, 2 cases have 2 runs, 7 cases have 3 runs, and so on, as summarized in Table 12-8.

Based on the 5% cutoff criterion, and based on the 126 cases summarized in Table 12-8, we should reject randomness if the number of runs G is 2 or 9 because, with randomly selected data, we will get 2 runs or 9 runs only 2.38% of the time. (With only 2 cases having 2 runs and with only 1 case having 9 runs, the number of runs G is excessively low or high in 3 cases, which is 3/126 or

Table 12-8 Frequency Table of the Number of Runs

Number of runs	2	3	4	5	6	7	8	9
Frequency in 126 cases	2	7	24	30	36	18	8	1

Country	Relative Currency Valuation
U.S.	H
Japan	L
U.K.	H
China	L
Germany	L
South Korea	L
Mexico	L
Belgium	L
France	L
Italy	L

2.38% of the total number of cases.) It is easy to get 3, 4, 5, 6, 7, or 8 runs, because these values occur more than 95% of the time; but it is unusual to get 2 or 9 runs because these values occur only 2.38% of the time. Table A-10 is constructed by this same principle: It displays the G values, for each case, such that there is a 5% chance or less of such a G value occurring if the sequence of data is random.

EXAMPLE

Canada is a trading nation that depends on exports to other countries. A factor that influences trade is the foreign exchange value of the Canadian dollar in relation to the currencies of our trading partners. An ordered list of the top 10 countries for Canadian exports is given in the margin. Next to each country is indicated whether, in the international currency markets, its currency trades at a higher (H) or lower (L) value than the Canadian dollar. Use the (simplified) 5% cutoff criterion to test a claim that the sequence of higher and lower currency valuations is random. (Note that we are not testing for a *bias* in favour of one classification or the other. There are 2 H currencies and 8 L currencies; we are investigating only the randomness in the way they appear in the given sequence.)

SOLUTION

We must first find the values of n_1, n_2, and the number of runs G:

n_1 = number of currencies valued higher than the Canadian dollar = 2

n_2 = number of currencies valued less than the Canadian dollar = 8

G = number of runs = 4

We refer to Table A-10 to find the cutoff values of 2 and 9. Because $G = 7$ is not less than or equal to 2, nor is it greater than or equal to 9, we do not reject randomness.

Interpretation
The sequence in which higher-valued and lower-valued currencies appear in the ordered list of Canada's export partners appears to be random.

Large Sample Cases

Table A-10 applies when the following three conditions are all met:

1. We are using 5% as the cutoff for sequences that have too few or too many runs.

2. $n_1 \leq 20$.

3. $n_2 \leq 20$.

If we wish to use the runs test for randomness but $n_1 > 20$ or $n_2 > 20$ or $\alpha \neq 0.05$, we use the property that the number of runs G has a distribution that is approximately normal with mean μ_G and standard deviation σ_G described as follows:

Formula 12-2
$$\mu_G = \frac{2n_1 n_2}{n_1 + n_2} + 1$$

Formula 12-3
$$\sigma_G = \sqrt{\frac{(2n_1 n_2)(2n_1 n_2 - n_1 - n_2)}{(n_1 + n_2)^2 (n_1 + n_2 - 1)}}$$

After finding the values of μ_G and σ_G, the test statistic can be computed as $z = (G - \mu_G)/\sigma_G$. The normal approximation (with test statistic z) is quite good. If the entire table of critical values (Table A-10) had been computed using this normal approximation, no critical value would be off by more than one unit.

TEST STATISTIC FOR THE RUNS TEST FOR RANDOMNESS

If $\alpha = 0.05$ and $n_1 \leq 20$ and $n_2 \leq 20$, the test statistic is G.

If $\alpha \neq 0.05$ or $n_1 > 20$ or $n_2 > 20$, the test statistic is

$$z = \frac{G - \mu_G}{\sigma_G} \qquad \text{(where } \mu_G \text{ and } \sigma_G \text{ are from Formulas 12-2 and 12-3)}$$

Critical Values

1. If the test statistic is G, critical values are found in Table A-10.

2. If the test statistic is z, critical values are found in Table A-2 by using the same procedures introduced in Chapter 6.

Figure 12-8 summarizes the procedures for the runs test for randomness and includes cases in which the test statistic is G as well as cases in which the test statistic is z.

EXAMPLE

Refer to the sequence of typical length (T) and brief (B) Parliaments in Figure 12-1. Is there any evidence that these T and B Parliaments occur in a nonrandom pattern? Use a 0.05 significance level in testing the sequence for randomness.

SOLUTION

The null and alternative hypotheses are as follows:

H_0: The sequence is random.

H_1: The sequence is not random.

Figure 12-8
Runs Test for Randomness

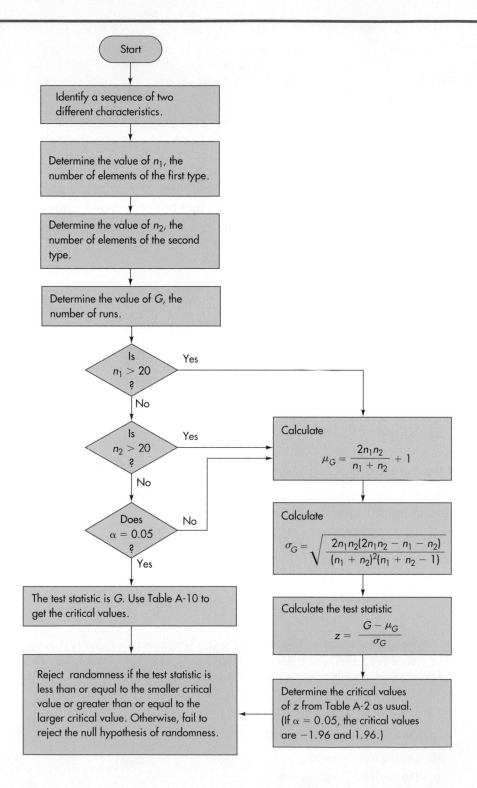

The significance level is $\alpha = 0.05$; we are using the runs test for randomness. The test statistic is obtained by first finding the number of Ts, the number of Bs, and the number of runs. After analyzing the sequence, we find that

$$n_1 = \text{number of Ts} = 27$$
$$n_2 = \text{number of Bs} = 8$$
$$G = \text{number of runs} = 17$$

As we follow Figure 12-8, we answer yes to "Is $n_1 > 20$?" We therefore need to evaluate μ_G and σ_G before we can determine the test statistic. Using Formulas 12-2 and 12-3 we get

$$\mu_G = \frac{2n_1 n_2}{n_1 + n_2} + 1 = \frac{2(27)(8)}{27 + 8} + 1 = 13.343$$

$$\sigma_G = \sqrt{\frac{(2n_1 n_2)(2n_1 n_2 - n_1 - n_2)}{(n_1 + n_2)^2(n_1 + n_2 - 1)}}$$

$$= \sqrt{\frac{[2(27)(8)]\,[2(27)(8) - 27 - 8]}{(27 + 8)^2(27 + 8 - 1)}} = 2.029$$

We can now find the test statistic:

$$z = \frac{G - \mu_G}{\sigma_G} = \frac{17 - 13.343}{2.029} = 1.802$$

Because the significance level is $\alpha = 0.05$ and we have a two-tailed test, the critical values are $z = -1.96$ and $z = 1.96$. The test statistic of $z = 1.802$ does not fall within the critical region, so we do not reject the null hypothesis of randomness.

Interpretation
The given sequence appears to be random. No matter what pattern has come before, the pattern provides no basis to guess whether the next Parliament is likely to be brief or long.

Alternative Application: Test for Randomness Above or Below the Mean or Median

In each of the preceding examples, the data clearly fit into two categories, but we can also test for randomness in the way numerical data fluctuate above or below a mean or median. To test for randomness above and below the median, for example, we use the sample data to find the value of the median, then replace each individual score with the letter A if it is *above* the median, and replace it with B if it is *below* the median. Delete any values that are equal to the median. It is helpful to write the As and Bs directly above the numbers they represent because this makes checking easier and also reduces the chance of having the wrong number of letters.

After finding the sequence of A and B letters, we can apply the runs test as described earlier. Economists use this test to help identify trends or cycles. An upward economic trend would contain a predominance of Bs at the beginning and As at the end, so the number of runs would be small. A downward trend would have As dominating at the beginning and Bs at the end, with a low number of runs. A cyclical pattern would yield a sequence that systematically changes, so the number of runs would tend to be large.

USING TECHNOLOGY

STATDISK: STATDISK will perform the runs test, but, because of the nature of the data, you must first determine the values of n_1, n_2, and the number of runs G. Select **Analysis** from the main menu bar, then select **Runs Test**. The STATDISK display will include the upper and lower critical values, the mean and standard deviation for the number of runs expected with a random sequence, and the conclusion.

EXCEL: To create a template in Excel, enter the right-tailed portion of alpha, n_1, n_2, and G. The formulas shown will calculate the other values. But for a two-tailed test, multiply the displayed P-value by 2. (Note that if n_1 and n_2 are both ≤ 20 and two-tailed $\alpha = 0.05$, the G cutoff returned by the template, which is based on a normal approximation, may differ slightly from the cutoff given in Table A-10.)

EXCEL DISPLAY
Runs Test for Randomness

	A	B	C	D	E	F	G	H	
2	Alpha (right tail)	0.025							
3	n1	27							
4	n2	8							
5	G	17							
6	Mean G	13.342857							
7	Std dev G	2.0292223							
8	Test statistic z	1.8022387							
9	Critical z	1.9599611							
10	P value	0.0357539							
11									
12									
13									
14									
15									

Cell	Formula	Explanation
B6	=(2*B3*B4)/(B3+B4)+1	Formula 12-2
B7	=SQRT((((2*B3*B4)*(2*B3*B4-B3-B4)) /((B3+B4)^2*(B3+B4-1))))	Formula 12-3
B8	=(B5-B6)/B7	Test statistic z (see boxed formula)
B9	=NORMSINV(1-B2)	Upper critical value
B10	=1-NORMSDIST(ABS(B8))	P-value (for two-tailed test, multiply by 2)

Use appropriate cell addresses for data.

MINITAB 15: Enter the data in numeric form in C1. From the **Stat** menu, choose **Nonparametrics** and then **Runs Test.** Select C1 into the **Variables** window. Choose the radio button above and below the mean unless you have a specific value in mind.

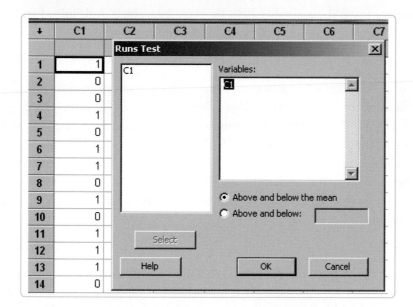

12-7 Exercises A: Basic Skills and Concepts

In Exercises 1–4, use the given sequence to determine the values of n_1, n_2, the number of runs G, and the 5% cutoff values from Table A-10.

1. A A A B B B B B A A A A A A

2. T F T F T F T F T F T T T T T T T T T T

3. O O O O O O O O O O E E O O E E

4. M F F F F M F F M M M F F F F F

In Exercises 5–12, use the runs test of this section to determine whether the given sequence is random. Use a significance level of $\alpha = 0.05$. (All data are listed in order by row.)

5. In conducting research for this book, one of the authors recorded the outcomes of a roulette wheel in the Stardust Casino. (Yes, it was hard work, but somebody had to do it.) Test for randomness of odd (O) and even (E) numbers

for the results given in the following sequence. What would a lack of random-ness mean to the author? To the casino?

O O E E E E O O E O E O O O O O O E O E

6. The Niko Music Company uses a machine to produce compact discs that must meet certain specifications. When a sample of discs is selected in sequence from the assembly line, they are examined and judged to be defec-tive (D) or acceptable (A), with the following results.

A A A A A A A A A D D A A A A A A A A A D D

7. Natural Resources Canada reports regularly on seismic events (earthquakes) in Canada. In just the three days—January 15 to 17, 2001—there were 24 such events. Earthquake magnitudes are recorded in a variety of scales, of which Canada primarily uses ML (Magnitude Local = Richter Scale) and MN (Magnitude Nuttli). For that January sequence of 24 seismic events, the sequence in which the two reporting scales (L for ML, N for MN) were used is shown below. Test the claim that the order in which the two scales appear is random.

N N N N N L N L L L N L N N L N L L N L L L

8. Test the claim that the sequence of World Series wins by American League and National League teams is random. Given below are recent results with American and National league teams represented by A and N, respectively. What does the result suggest about the abilities of the two leagues?

A N A A A N N A A N N N N A A A N A N A N A N A A A N A

9. Since 1935, there have been 11 changes of prime minister in Canada, includ-ing Pierre Trudeau's re-election in 1980 after a period in Opposition. The political party of each new prime minister is indicated by L for Liberal and P for Progressive Conservative. Does it appear that the Liberal and Progressive Conservative parties became the party of new prime ministers in a random sequence?

L L P L L P L L P P L

10. As an aid to its farming industries, Alberta tracks the dates of last frost across the province. At some sites, the last frost occurs in mid-May, while others reach this event a month or more later. Tracking this information in five-day intervals, the accompanying list shows the sequence in which new sites are added to the last-frost list. (For example, two sites had their last frost within the first interval, 12 sites had their last frost in the next interval, and so on.) Find the median of the values, then replace each value by A if it is above the median and by B if it is below the median. Omit from the analysis any values that are exactly at the median. Then apply the runs test to the resulting

sequence of As and Bs. What does the result suggest about the pattern in which sites across Alberta are added to the last-frost list? If you knew more about the latitude of each site in relation to its last frost date, might this cause you to reassess your conclusion from the runs test?

2   12   17   15   8   10   3   0   0   0   1

Based on "Freezing Dates for Alberta Locations," by Peter Dzikowski,
Conservation and Development Branch of Alberta Agriculture.

11. Business trends and economics applications are often analyzed with the runs test. The accompanying list shows, in order by row, the net gain or loss in the S&P/TSX Composite Index (formerly the TSE 300 Composite Index) over each month in a recent sequence of months. Replace each value by G if it is a gain (above zero) or L if it is a loss (below 0). Then apply the runs test to the resulting sequence of Gs and Ls. What does the result suggest about the pattern of net monthly gains and losses in the index? Given the pattern you have identified, is it possible to make money?

69.07	48.26	−307.62	135.75	405.69	136.26	181.24
−176.18	428.46	−209.30	−379.16	218.38	289.38	428.31
294.61	−151.52	−92.01	−322.26	−608.80		

12. Listed below are the bonus numbers selected in consecutive Lotto 6/49 drawings. Test for randomness of odd (O) and even (E) numbers. What would lack of randomness mean for those who play the lottery?

30   27   25   8   26   45   31   19   31   11   23   9   12   46   5   18
22   18   49   49   28   14   25   41   17   3   20   10   36   37   45   37

13. The Restaurant Market Research Company hires Diana Kriplani to survey adults about their dining habits. A check of her first 40 results shows that their genders are represented in the following sequence. Use the runs test to determine whether the subjects are randomly selected with respect to gender.

F F F F   F M M M M M   F F   F F M M M M M F F
F F F M   M M M M M M   M M M M M F F F F   F F F

14. Many mathematicians agree that the decimal digits of π cannot be distinguished from any random sequence. Given below are the first 100 decimal places of π. Test for randomness of odd (O) and even (E) digits.

1   4   1   5   9   2   6   5   3   5   8   9   7   9   3   2   3   8   4   6
2   6   4   3   3   8   3   2   7   9   5   0   2   8   8   4   1   9   7   1
6   9   3   9   9   3   7   5   1   0   5   8   2   0   9   7   4   9   4   4
5   9   2   3   0   7   8   1   6   4   0   6   2   8   6   2   0   8   9   9
8   6   2   8   0   3   4   8   2   5   3   4   2   1   1   7   0   6   7   9

15. Use the 100 decimal digits for π given in Exercise 14. Test for randomness above (A) and below (B) the value of 4.5.

16. Test the claim that the sequence of World Series wins by American League and National League teams is random. Given below are recent results, with American and National League teams represented by A and N, respectively.

```
A N A N N N A A A A N A A A A N A N N A
A N N A A A A N A N N A A A A A N A N A
N A N A A A A A A A N N A N A N N A A N
N N A N A N A N A A A N N A A N N N N A
A A N A N A N A A A N A
```

12-7 Exercises B: Beyond the Basics

17. Using the elements A, A, B, B, what is the minimum number of possible runs that can be arranged? What is the maximum number of runs? Now refer to Table A-10 to find the 5% cutoff G values for $n_1 = n_2 = 2$. What do you conclude about this case?

18. a. Using all of the elements A, A, A, B, B, B, B, B, B, list the 84 different possible sequences.
    b. Find the number of runs for each of the 84 sequences.
    c. Use the results from parts (a) and (b) to find your own 5% cutoff values for G.
    d. Compare your results to those given in Table A-10.

■ VOCABULARY LIST

distribution-free tests 674	rank 675	Spearman's rank correlation
efficiency 675	rank correlation	coefficient 723
Friedman test 713	coefficient 722	Spearman's rank
H test 709	rank correlation	correlation test 722
Kruskal-Wallis test 709	test 722	Wilcoxon rank-sum
Mann-Whitney U test 698	run 733	test 698
nonparametric tests 674	runs test 733	Wilcoxon signed-ranks
parametric tests 674	sign test 677	test 688

■ REVIEW

In this chapter we examined seven different nonparametric methods for analyzing sample data. Nonparametric tests are also called distribution-free tests because they do not require that the populations have a particular distribution, such as a normal distribution. However, if the population distribution is normal,

Table 12-9 Summary of Nonparametric Tests

Nonparametric Test	Function	Parametric Test
Sign test (Section 12-2)	Test for claimed value of average with one sample	*t* test or *Z* test (Sections 7-2, 7-3, 7-4)
	Test for difference between two dependent samples	*t* test or *Z* test (Section 8-2)
Wilcoxon signed-ranks test (Section 12-3)	Test for difference between two dependent samples	*t* test or *Z* test (Section 8-2)
Wilcoxon rank-sum test (Section 12-4)	Test for difference between two independent samples	*t* test or *Z* test (Sections 8-3, 8-5)
Kruskal-Wallis test (Section 12-5)	Test for more than two independent samples coming from identical populations	Analysis of variance (Section 9-2)
Friedman test (Section 12-5)	Test for more than two dependent samples coming from identical populations	Analysis of variance (Section 9-3)
Rank correlation (Section 12-6)	Test for relationship between two variables	Linear correlation (Section 11-2)
Runs test (Section 12-7)	Test for randomness of sample data	(No parametric test)

then nonparametric tests are not as efficient as parametric tests, so we would need stronger evidence before we reject a null hypothesis.

Table 12.9 lists the nonparametric tests presented in this chapter, along with their functions. The table also lists the corresponding parametric tests.

REVIEW EXERCISES

In Exercises 1–12, use a 0.05 significance level with the indicated test. If no particular test is specified, use the appropriate nonparametric test from this chapter.

1. The accompanying table shows samples of data about Canadian prime ministers, popes, and British monarchs. The values are the numbers of years the subjects lived after they were first elected or were crowned. Test the claim that the three populations are not all the same.

Prime ministers	17	23	22	23	29	40	9	19	25
Popes	3	18	25	6	15	11	19		
Monarchs	17	13	12	33	10	7	25		

2. A number of companies sell both cars and trucks in Canada. Do a company's car sales go up (or down) in tandem with its truck sales? The accompanying table shows the annual percent changes in car and truck sales for a selection

of companies that offer both. Use the Wilcoxon signed-ranks test to test the hypothesis that there is no difference in a company's percent change in sales for cars and trucks. By its use of ranks, the Wilcoxon test is not sensitive to the magnitude of differences where they occur. Inspecting these data, would you be inclined to predict changes in car sales based on changes in truck sales?

Car change	34.3	9	−3.7	−31.9	79.6	27.8	6.4	2.7	−20.8
Truck change	24.1	73	13.3	−8.6	−33.3	0	33.9	59.3	−19.4

Based on data in *The Globe and Mail*.

3. A Regina manufacturing company uses a machine to produce surgical knives that must meet certain specifications. A sample of knives is randomly selected and each knife is judged to be defective (D) or acceptable (A), with the results given below. Use the runs test for randomness to determine whether the sequence of defective and acceptable knives is random.

D A A A A D A A A A D A A A A
D A A A A D A A A A D A A A A

4. The accompanying table shows a sample from poll results by *The Globe and Mail Report on Business* and *Angus Reid*, which compare CEO evaluations of the 25 top-performing firms over two years. For the 2000 results, each company is given a score that reflects the respondents' choices. For 1999, the companies' scores have been converted to ranks (where 1 is best). Test the claim that there is a correlation between a company's scores in the two years.

2000 score	58	42	61	346	70	85	41	903	95
1999 rank	18	21	15	7	10	31	61	3	14

5. Internet search engines often return large lists of results for the search lists that are entered. Some guides suggest refining the search by adding more words to the search list. To test whether this suggestion is effective, the accompanying table shows the numbers of results returned by Yahoo Canada for each of eight randomly generated, two-word search lists. The bottom row shows the numbers of results returned when each two-word search list is supplemented by a randomly generated third word. Use the sign test to test the claim that there is no difference in the numbers of results returned before and after adding the additional word to the search list. Is simply adding more words to a search list an effective strategy to reduce the numbers of results?

Trial	A	B	C	D	E	F	G	H
Two words	3589	1,071,575	387,835	88,535	71,640	1369	236,140	208,375
Three words	3610	793,961	695,933	72,468	66,339	1861	166,103	315,455

6. The Medassist Pharmaceutical Company conducted a study to determine whether a drug affects eye movements. A standardized scale was developed, and the drug was administered to one group, while a control group was given a placebo that produces no effects. The eye movement ratings of subjects are listed below. Test the claim that the drug has no effect on eye movements. Use the Wilcoxon rank-sum test.

Medicated group	Control group
652 512 711 621 508 603 787	674 676 821 830 565 821 837 652 549
747 516 624 627 777 729	668 772 563 703 789 800 711 598

7. To test the effectiveness of the use of technology in a statistics course, four different sections were taught with different combinations of calculators and computer software, with final course averages listed below for randomly selected students from each section. Test the claim that the samples come from populations that are not all the same.

Calculators Only	Computers Only	Calculators and Computers	No Calculators and No Computers
74 85 91 62	82 87 60 71	78 89 82 64	66 78 83 55
73 87 66 80	77 63 84 70	77 91 73 85	72 57 81 65

8. A statistics quiz consists of true-false questions with the answers listed as shown.
   a. Examine the sequence and try to identify a pattern showing that the answers are not random.
   b. What do you conclude from using the runs test for randomness?
   c. How do you explain the discrepancy between parts (a) and (b)?

   T F T T T T F T T T T T F T T T T T F T T T T T F T T T

9. A consumer investigator obtained prices from mail-order companies and computer stores. The accompanying lists show the prices (in dollars) quoted for cartons of floppy disks from various manufacturers. Use the Wilcoxon rank-sum test to test the claim that there is no difference between mail order and store prices.

Mail Order				Computer Store			
23.00	26.00	27.99	31.50	30.99	33.98	37.75	38.99
32.75	27.00	27.98	24.50	35.79	33.99	34.79	32.99
24.75	28.15	29.99	29.99	29.99	33.00	32.00	

10. A study was conducted to investigate some effects of physical training. Sample data (weights in kilograms) are listed in the margin. (See "Effect of Endurance Training on Possible Determinants of VO_2 During Heavy Exercise," by Casaburi et al., *Journal of Applied Physiology*, Vol. 62, No. 1.) Use the sign test to test the claim that the training has no effect on weight.

Pre-training	Post-training
99	94
57	57
62	62
69	69
74	66
77	76
59	58
92	88
70	70
85	84

11. Do Exercise 10 using the Wilcoxon signed-ranks test.

12. Two measures of health care are the mean number of people per physician, and the percentage of the population who reside fewer than 5 km from a physician. In the table below, the 10 provinces are ranked based on these measures. For "People/physician," a rank of 1 signifies the fewest people per physician; for "% < 5 km," a rank of 1 signifies the largest proportion of people who reside less than 5 km from a physician. Is there a correlation between these two measures of health care?

Province	BC	AB	SK	MB	ON	PQ	NB	NS	PE	NL
People/physician	1	7	8	5	4	3	9	2	10	6
% <5 km	1	4	8	5	3	2	10	7	9	6

Based on data from Statistics Canada.

13. A focus group of people aged 18–24 were asked to rate 4 new video games on a scale from 1 to 10, 10 being best. These were the results:

	Game 1	Game 2	Game 3	Game 4
Person 1	5	7	7	4
Person 2	7	10	4	5
Person 3	7	8	7	7
Person 4	4	6	9	5
Person 5	6	8	6	8
Person 6	7	9	6	8
Person 7	7	9	4	5
Person 8	7	9	6	9

Is there any significant difference in the median ratings of the games? Test at a 0.05 level of significance.

14. A manufacturing firm wanted to determine if the median amount of waste per unit produced by a new type of lathe was significantly less than that of its current lathe. Samples from each lathe produced the following results (in grams):

New	13.6	18.7	18.7	13.5	12.3	17.1
Existing	18.3	18.2	21	17	18.9	21.2

Test at a 0.05 level of significance.

◼ CUMULATIVE REVIEW EXERCISES

1. The Broadbent Opinion Research Organization assigned a pollster to collect data from 30 randomly selected adults. As the data were submitted to the company, the genders of the interviewed subjects were noted. The sequence obtained is shown in the accompanying list.

a. At the 0.05 significance level, test the claim that the sequence is random.

b. At the 0.05 significance level, test the claim that the proportion of women is different from 0.5.

c. Use the sample data to construct a 95% confidence interval for the proportion of women.

d. What do the preceding results suggest? Is the sample biased against either gender? Was the sample obtained in a random sequence? If you are the manager, do you have any problems with these results?

M M F M M F M M F M F M M F M
F M F M M F M M F M M F M M M

2. The Kamloops Weight Clinic claims that everyone can lose weight after two weeks in its program. The following table lists the weights, in kilograms, of randomly selected clients before and after the program. Use the indicated test with a 0.05 level of significance to test the hypothesis that the program has no effect on people's weight.

a. Sign test

b. Wilcoxon signed-ranks test

c. t test for a claim about two dependent samples (Section 8–2)

d. How do the preceding results support the statement that nonparametric tests lack the sensitivity of parametric tests (so stronger evidence is required before a null hypothesis is rejected)?

Client	A	B	C	D	E	F	G	H	I	J
Weight before entering program	75	81	77	87	85	74	77	104	95	100
Weight after entering program	70	82	77	85	80	75	79	100	84	93

3. Refer to the annual numbers of person-days lost in the water-based transportation industry due to work stoppages in Data Set 20 in Appendix B. Find the following:

a. Mean

b. Median

c. Mode

d. Range

e. Standard Deviation

f. Any outliers

TECHNOLOGY PROJECT

In Section 12–4 we saw that the Wilcoxon rank-sum test can be used to test the null hypothesis that two independent samples come from the same distribution, and the alternative hypothesis is the claim that the two distributions differ in some way. Use

Excel, STATDISK, or any other statistical software package to run the Wilcoxon rank-sum test and determine whether it detects the difference in the two distributions. The two samples are from populations with the same mean of 100 and the same standard deviation of 15, but the distributions are different (normal, uniform).

Sample randomly selected from a *normally* distributed population with mean 100 and standard deviation 15:

82	92	84	107	104	130	84	125	110	96
98	83	87	99	108	97	84	122	106	114
78	117	105	100	94	114	135	99	109	79
106	83	72	105	108	102	100	106	109	119
95	124	95	86	113	83	93	95	99	122
77	106	87	94	92	95	112	95	101	65
69	108	125	105	132	107	101	94	122	80
101	111	108	107	96	134	105	51	85	141
94	102	74	84	101	99	109	83	116	119
118	114	129	98	101	117	92	96	83	114

Sample randomly selected from a *uniformly* distributed population with a minimum of 74 and a maximum of 126 (such a population has a mean of 100 and a standard deviation of 15):

111	104	96	109	93	78	88	85	88	91
92	108	124	114	82	110	100	104	84	113
111	116	95	121	82	120	110	115	93	124
118	94	102	106	104	126	114	123	115	82
84	100	118	76	82	110	81	123	108	96
107	91	89	122	92	100	122	108	81	107
89	123	81	111	118	97	94	86	109	74
87	113	97	78	115	95	117	77	125	109
77	82	111	80	117	108	107	102	94	93
114	94	112	94	116	111	88	101	81	110

FROM DATA TO DECISION

Is Hockey Attendance Random?

Owners of hockey franchises, and many others, are vitally interested in the attendance patterns at hockey games. If there is evidence of nonrandom trends in attendance, then future marketing plans, budgets, and scheduling may all be affected. The accompanying list shows the attendance figures for the Vancouver Canucks' 41 home games during the 1996–97 and 1997–98 seasons.

a. Consider the sequence from the first game in the 1996–97 season to the last game of the 1997–98 season. Use the runs test to test the sequence for randomness above and below the median, which is 17,507.

b. Consider these four time segments within the data: (**1A**) games 1 to 20 of the 1996–97 season; (**1B**) games 21 to 41 of the 1996/97 season; (**2A**) games 1 to 20 of the 1997–98 season; and (**2B**) games 21 to 41 of the 1997–98 season. Use the Kruskal-Wallis test to test the claim that the four time segments all had attendance numbers drawn from the same population.

c. Draw a run chart based on the full sequence of 82 games, combining the two 41-game seasons in order, with the horizontal scale plotting the game numbers from 1 to 82 and the vertical scale plotting the attendance number corresponding to each game number. (See Section 13–2 for more details on run charts.) Note any pattern that may suggest the attendance numbers are not randomly sequenced.

d. In the data, there is one attendance number that appears much more frequently than the others. Identify that number and speculate as to why it appears so often. Does the recurrence of that number have an effect on the tests performed in parts (a), (b), and (c)? If it does, how? If not, do you think if there were *more* recurrences of that value, it could have an effect on the tests? Explain your answer.

Game	1996–97 Season Attendance	1997–98 Season Attendance	Game	1996–97 Season Attendance	1997–98 Season Attendance
1	17,501	10,500	22	18,422	17,003
2	14,109	17,312	23	15,017	17,679
3	15,691	18,422	24	16,010	15,822
4	15,142	14,535	25	17,331	17,808
5	17,807	16,304	26	17,243	18,422
6	18,422	16,880	27	18,422	17,421
7	17,546	17,437	28	18,422	18,422
8	18,422	18,422	29	18,422	18,422
9	16,202	18,422	30	17,595	18,422
10	17,411	14,740	31	17,492	15,613
11	18,422	16,454	32	17,513	15,370
12	16,605	18,422	33	18,422	16,452
13	18,422	18,422	34	17,535	18,422
14	16,313	15,860	35	16,740	17,699
15	18,422	18,422	36	15,076	17,478
16	17,854	15,401	37	16,050	15,662
17	18,422	17,128	38	18,028	17,794
18	16,125	18,422	39	18,422	15,239
19	18,422	18,422	40	16,809	16,602
20	17,378	18,422	41	18,422	17,787
21	18,105	14,549			

COOPERATIVE GROUP ACTIVITIES

1. **In-class activity:** Use the existing class seating arrangement and apply the runs test to determine whether the students are arranged randomly according to gender. After recording the seating arrangement, analysis can be done in subgroups of 3 or 4 students.

2. **In-class activity:** Divide into groups of 8 to 12 people. For each group member, *measure* his or her height and *measure* his or her arm span. For the arm span, the subject should stand with arms extended, like the wings on an airplane. It's easy to mark the height and arm span on a chalkboard, then measure the distances there. Divide the following tasks among subgroups of 3 or 4 people.

   a. Use rank correlation with the paired sample data to determine whether there is a correlation between height and arm span.

   b. Use the sign test to test for a difference between the two variables.

   c. Use the Wilcoxon signed-ranks test to test for a difference between the two variables.

3. **In-class activity:** Do Activity 2 using pulse rate instead of arm span. Measure pulse rates by counting the number of heartbeats in one minute.

4. **In-class activity:** Divide into groups of about 10 or 12 students and use the same reaction timer included with Chapter 5. Each group member should be tested for right-hand reaction time and left-hand reaction time. Analyze the results using methods from this chapter. State the methods used and the conclusions reached.

5. **Out-of-class activity:** See the preceding From Data to Decision project, which involves analysis of attendance data for all home games played by a hockey team over two seasons. Suppose you are helping to write a video game that incorporates simulated attendance data over two fictional seasons (to simulate crowd-noise effects and so on). Design a procedure that could produce 82 random numbers, such that the sequence will have the same minimum, maximum, and median values as the actual attendances in the preceding project. Once you have generated the 82 numbers, test your results by using the techniques suggested in parts (a), (b), and (c) of the From Data to Decision project. How do your results compare to those obtained in that project? Write a report that clearly describes the process you designed. Include your analyses and conclusions.

Nonparametric Tests

This chapter introduced several different nonparametric or distribution-free tests. Such tests allow you to test hypotheses without making assumptions about the distributions underlying the associated data sets. The Internet Project for this chapter is found at either of the following websites:

http://www.mathxl.com   or   http://www.mystatlab.com

Click on Internet Project, then on Chapter 12. There you will be directed to revisit some of the hypotheses tested in earlier Internet Projects, but this time you will apply an appropriate nonparametric test. In the second part of the project, you will be asked to apply the runs test to determine randomness of a sequence.

APPENDIX A: TABLES ... 756

Table A-1 Binomial Probabilities ... 756

Table A-2 Standard Normal (z) Distribution ... 759

Table A-3 t Distribution ... 760

Table A-4 Chi-Square (χ^2) Distribution ... 761

Table A-5 F Distribution ... 763

Table A-6 Critical Values of the Pearson Correlation Coefficient r ... 769

Table A-7 Critical Values for the Sign Test ... 770

Table A-8 Critical Values of T for the Wilcoxon Signed-Ranks Test ... 771

Table A-9 Critical Values of Spearman's Rank Correlation Coefficient r_s ... 772

Table A-10 Critical Values for Number of Runs G ... 773

Table A-11 Percentage Points of the Studentized Range ... 774

Table A-12 Table of Critical Values for Lilliefors Normality Test ... 777

Table A-13 Wilcoxon Rank Sum Table ... 778

APPENDIX B: DATA SETS ... 779

Data Set 1: Weights of Household Garbage for One Week ... 779

Data Set 2: Body Temperatures of Healthy Adults in Varied Postures ... 781

Data Set 3: Bears ... 782

Data Set 4: Temperature and Precipitation for Canadian Cities ... 784

Data Set 5: Sitting Days and Durations of Sessions of Parliament ... 785

Data Set 6: The Financial Post Top 100 Companies for 1997 ... 786

Data Set 7: Temperature and Precipitation Departures from Seasonal Means ... 789

Data Set 8: Survey of 100 Statistics Students ... 790

Data Set 9: Passive and Active Smoke ... 793

Data Set 10: Annual Rate of Return of 40 Mutual Funds ... 794

Data Set 11: Weights of a Sample of M&M Plain Candies ... 795

Data Set 12: Lotto 6/49 Results ... 796

Data Set 13: Weights of Quarters ... 797

Data Set 14: Labour Force 15 Years and Over by Occupation, by Province ... 798

Data Set 15: Axial Loads of Aluminum Cans ... 800

Data Set 16: Old Faithful Geyser ... 801

Data Set 17: Sample Dates of Last Spring Frost at Two Groups of Alberta Locations ... 802

Data Set 18: Body Temperatures of Adults Accepted for Voluntary Surgical Procedures ... 803

Data Set 19: Lifespans of Animals ... 805

Data Set 20: Person-Days Lost in Labour Stoppages in the Transportation Industry ... 806

Data Set 21: Diamonds ... 807

Data Set 22: Canadian Exports and Imports by World Area ... 808

APPENDIX C: BIBLIOGRAPHY ... 809

APPENDIX D: ANSWERS TO ODD-NUMBERED EXERCISES (AND ALL REVIEW EXERCISES AND CUMULATIVE REVIEW EXERCISES) ... 810

APPENDIX A: Tables

Table A-1 Binomial Probabilities

		p													
n	*x*	.01	.05	.10	.20	.30	.40	.50	.60	.70	.80	.90	.95	.99	*x*
2	0	980	902	810	640	490	360	250	160	090	040	010	002	0+	0
	1	020	095	180	320	420	480	500	480	420	320	180	095	020	1
	2	0+	002	010	040	090	160	250	360	490	640	810	902	980	2
3	0	970	857	729	512	343	216	125	064	027	008	001	0+	0+	0
	1	029	135	243	384	441	432	375	288	189	096	027	007	0+	1
	2	0+	007	027	096	189	288	375	432	441	384	243	135	029	2
	3	0+	0+	001	008	027	064	125	216	343	512	729	857	970	3
4	0	961	815	656	410	240	130	062	026	008	002	0+	0+	0+	0
	1	039	171	292	410	412	346	250	154	076	026	004	0+	0+	1
	2	001	014	049	154	265	346	375	346	265	154	049	014	001	2
	3	0+	0+	004	026	076	154	250	346	412	410	292	171	039	3
	4	0+	0+	0+	002	008	026	062	130	240	410	656	815	961	4
5	0	951	774	590	328	168	078	031	010	002	0+	0+	0+	0+	0
	1	048	204	328	410	360	259	156	077	028	006	0+	0+	0+	1
	2	001	021	073	205	309	346	312	230	132	051	008	001	0+	2
	3	0+	001	008	051	132	230	312	346	309	205	073	021	001	3
	4	0+	0+	0+	006	028	077	156	259	360	410	328	204	048	4
	5	0+	0+	0+	0+	002	010	031	078	168	328	590	774	951	5
6	0	941	735	531	262	118	047	016	004	001	0+	0+	0+	0+	0
	1	057	232	354	393	303	187	094	037	010	002	0+	0+	0+	1
	2	001	031	098	246	324	311	234	138	060	015	001	0+	0+	2
	3	0+	002	015	082	185	276	312	276	185	082	015	002	0+	3
	4	0+	0+	001	015	060	138	234	311	324	246	098	031	001	4
	5	0+	0+	0+	002	010	037	094	187	303	393	354	232	057	5
	6	0+	0+	0+	0+	001	004	016	047	118	262	531	735	941	6
7	0	932	698	478	210	082	028	008	002	0+	0+	0+	0+	0+	0
	1	066	257	372	367	247	131	055	017	004	0+	0+	0+	0+	1
	2	002	041	124	275	318	261	164	077	025	004	0+	0+	0+	2
	3	0+	004	023	115	227	290	273	194	097	029	003	0+	0+	3
	4	0+	0+	003	029	097	194	273	290	227	115	023	004	0+	4
	5	0+	0+	0+	004	025	077	164	261	318	275	124	041	002	5
	6	0+	0+	0+	0+	004	017	055	131	247	367	372	257	066	6
	7	0+	0+	0+	0+	0+	002	008	028	082	210	478	698	932	7
8	0	923	663	430	168	058	017	004	001	0+	0+	0+	0+	0+	0
	1	075	279	383	336	198	090	031	008	001	0+	0+	0+	0+	1
	2	003	051	149	294	296	209	109	041	010	001	0+	0+	0+	2
	3	0+	005	033	147	254	279	219	124	047	009	0+	0+	0+	3
	4	0+	0+	005	046	136	232	273	232	136	046	005	0+	0+	4
	5	0+	0+	0+	009	047	124	219	279	254	147	033	005	0+	5
	6	0+	0+	0+	001	010	041	109	209	296	294	149	051	003	6
	7	0+	0+	0+	0+	001	008	031	090	198	336	383	279	075	7
	8	0+	0+	0+	0+	0+	001	004	017	058	168	430	663	923	8

NOTE: 0+ represents a positive probability less than 0.0005.

(continued)

Table A-1 Binomial Probabilities

n	x	.01	.05	.10	.20	.30	.40	.50	.60	.70	.80	.90	.95	.99	x
9	0	914	630	387	134	040	010	002	0+	0+	0+	0+	0+	0+	0
	1	083	299	387	302	156	060	018	004	0+	0+	0+	0+	0+	1
	2	003	063	172	302	267	161	070	021	004	0+	0+	0+	0+	2
	3	0+	008	045	176	267	251	164	074	021	003	0+	0+	0+	3
	4	0+	001	007	066	172	251	246	167	074	017	001	0+	0+	4
	5	0+	0+	001	017	074	167	246	251	172	066	007	001	0+	5
	6	0+	0+	0+	003	021	074	164	251	267	176	045	008	0+	6
	7	0+	0+	0+	0+	004	021	070	161	267	302	172	063	003	7
	8	0+	0+	0+	0+	0+	004	018	060	156	302	387	299	083	8
	9	0+	0+	0+	0+	0+	0+	002	010	040	134	387	630	914	9
10	0	904	599	349	107	028	006	001	0+	0+	0+	0+	0+	0+	0
	1	091	315	387	268	121	040	010	002	0+	0+	0+	0+	0+	1
	2	004	075	194	302	233	121	044	011	001	0+	0+	0+	0+	2
	3	0+	010	057	201	267	215	117	042	009	001	0+	0+	0+	3
	4	0+	001	011	088	200	251	205	111	037	006	0+	0+	0+	4
	5	0+	0+	001	026	103	201	246	201	103	026	001	0+	0+	5
	6	0+	0+	0+	006	037	111	205	251	200	088	011	001	0+	6
	7	0+	0+	0+	001	009	042	117	215	267	201	057	010	0+	7
	8	0+	0+	0+	0+	001	011	044	121	233	302	194	075	004	8
	9	0+	0+	0+	0+	0+	002	010	040	121	268	387	315	091	9
	10	0+	0+	0+	0+	0+	0+	001	006	028	107	349	599	904	10
11	0	895	569	314	086	020	004	0+	0+	0+	0+	0+	0+	0+	0
	1	099	329	384	236	093	027	005	001	0+	0+	0+	0+	0+	1
	2	005	087	213	295	200	089	027	005	001	0+	0+	0+	0+	2
	3	0+	014	071	221	257	177	081	023	004	0+	0+	0+	0+	3
	4	0+	001	016	111	220	236	161	070	017	002	0+	0+	0+	4
	5	0+	0+	002	039	132	221	226	147	057	010	0+	0+	0+	5
	6	0+	0+	0+	010	057	147	226	221	132	039	002	0+	0+	6
	7	0+	0+	0+	002	017	070	161	236	220	111	016	001	0+	7
	8	0+	0+	0+	0+	004	023	081	177	257	221	071	014	0+	8
	9	0+	0+	0+	0+	001	005	027	089	200	295	213	087	005	9
	10	0+	0+	0+	0+	0+	001	005	027	093	236	384	329	099	10
	11	0+	0+	0+	0+	0+	0+	0+	004	020	086	314	569	895	11
12	0	886	540	282	069	014	002	0+	0+	0+	0+	0+	0+	0+	0
	1	107	341	377	206	071	017	003	0+	0+	0+	0+	0+	0+	1
	2	006	099	230	283	168	064	016	002	0+	0+	0+	0+	0+	2
	3	0+	017	085	236	240	142	054	012	001	0+	0+	0+	0+	3
	4	0+	002	021	133	231	213	121	042	008	001	0+	0+	0+	4
	5	0+	0+	004	053	158	227	193	101	029	003	0+	0+	0+	5
	6	0+	0+	0+	016	079	177	226	177	079	016	0+	0+	0+	6
	7	0+	0+	0+	003	029	101	193	227	158	053	004	0+	0+	7
	8	0+	0+	0+	001	008	042	121	213	231	133	021	002	0+	8
	9	0+	0+	0+	0+	001	012	054	142	240	236	085	017	0+	9
	10	0+	0+	0+	0+	0+	002	016	064	168	283	230	099	006	10
	11	0+	0+	0+	0+	0+	0+	003	017	071	206	377	341	107	11
	12	0+	0+	0+	0+	0+	0+	0+	002	014	069	282	540	886	12

NOTE: 0+ represents a positive probability less than 0.0005.

(continued)

Table A-1 Binomial Probabilities

n	x	.01	.05	.10	.20	.30	.40	.50	.60	.70	.80	.90	.95	.99	x
13	0	878	513	254	055	010	001	0+	0+	0+	0+	0+	0+	0+	0
	1	115	351	367	179	054	011	002	0+	0+	0+	0+	0+	0+	1
	2	007	111	245	268	139	045	010	001	0+	0+	0+	0+	0+	2
	3	0+	021	100	246	218	111	035	006	001	0+	0+	0+	0+	3
	4	0+	003	028	154	234	184	087	024	003	0+	0+	0+	0+	4
	5	0+	0+	006	069	180	221	157	066	014	001	0+	0+	0+	5
	6	0+	0+	001	023	103	197	209	131	044	006	0+	0+	0+	6
	7	0+	0+	0+	006	044	131	209	197	103	023	001	0+	0+	7
	8	0+	0+	0+	001	014	066	157	221	180	069	006	0+	0+	8
	9	0+	0+	0+	0+	003	024	087	184	234	154	028	003	0+	9
	10	0+	0+	0+	0+	001	006	035	111	218	246	100	021	0+	10
	11	0+	0+	0+	0+	0+	001	010	045	139	268	245	111	007	11
	12	0+	0+	0+	0+	0+	0+	002	011	054	179	367	351	115	12
	13	0+	0+	0+	0+	0+	0+	0+	001	010	055	254	513	878	13
14	0	869	488	229	044	007	001	0+	0+	0+	0+	0+	0+	0+	0
	1	123	359	356	154	041	007	001	0+	0+	0+	0+	0+	0+	1
	2	008	123	257	250	113	032	006	001	0+	0+	0+	0+	0+	2
	3	0+	026	114	250	194	085	022	003	0+	0+	0+	0+	0+	3
	4	0+	004	035	172	229	155	061	014	001	0+	0+	0+	0+	4
	5	0+	0+	008	086	196	207	122	041	007	0+	0+	0+	0+	5
	6	0+	0+	001	032	126	207	183	092	023	002	0+	0+	0+	6
	7	0+	0+	0+	009	062	157	209	157	062	009	0+	0+	0+	7
	8	0+	0+	0+	002	023	092	183	207	126	032	001	0+	0+	8
	9	0+	0+	0+	0+	007	041	122	207	196	086	008	0+	0+	9
	10	0+	0+	0+	0+	001	014	061	155	229	172	035	004	0+	10
	11	0+	0+	0+	0+	0+	003	022	085	194	250	114	026	0+	11
	12	0+	0+	0+	0+	0+	001	006	032	113	250	257	123	008	12
	13	0+	0+	0+	0+	0+	0+	001	007	041	154	356	359	123	13
	14	0+	0+	0+	0+	0+	0+	0+	001	007	044	229	488	869	14
15	0	860	463	206	035	005	0+	0+	0+	0+	0+	0+	0+	0+	0
	1	130	366	343	132	031	005	0+	0+	0+	0+	0+	0+	0+	1
	2	009	135	267	231	092	022	003	0+	0+	0+	0+	0+	0+	2
	3	0+	031	129	250	170	063	014	002	0+	0+	0+	0+	0+	3
	4	0+	005	043	188	219	127	042	007	001	0+	0+	0+	0+	4
	5	0+	001	010	103	206	186	092	024	003	0+	0+	0+	0+	5
	6	0+	0+	002	043	147	207	153	061	012	001	0+	0+	0+	6
	7	0+	0+	0+	014	081	177	196	118	035	003	0+	0+	0+	7
	8	0+	0+	0+	003	035	118	196	177	081	014	0+	0+	0+	8
	9	0+	0+	0+	001	012	061	153	207	147	043	002	0+	0+	9
	10	0+	0+	0+	0+	003	024	092	186	206	103	010	001	0+	10
	11	0+	0+	0+	0+	001	007	042	127	219	188	043	005	0+	11
	12	0+	0+	0+	0+	0+	002	014	063	170	250	129	031	0+	12
	13	0+	0+	0+	0+	0+	0+	003	022	092	231	267	135	009	13
	14	0+	0+	0+	0+	0+	0+	0+	005	031	132	343	366	130	14
	15	0+	0+	0+	0+	0+	0+	0+	0+	005	035	206	463	860	15

NOTE: 0+ represents a positive probability less than 0.0005.

From Frederick C. Mosteller, Robert E. K. Rourke, and George B. Thomas, Jr., *Probability with Statistical Applications*, 2nd ed., © 1970 Addison-Wesley Publishing Co., Reading, MA. Reprinted with permission.

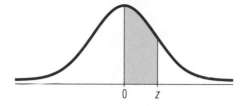

Table A-2 Standard Normal (z) Distribution

z	.00	.01	.02	.03	.04	.05	.06	.07	.08	.09
0.0	.0000	.0040	.0080	.0120	.0160	.0199	.0239	.0279	.0319	.0359
0.1	.0398	.0438	.0478	.0517	.0557	.0596	.0636	.0675	.0714	.0753
0.2	.0793	.0832	.0871	.0910	.0948	.0987	.1026	.1064	.1103	.1141
0.3	.1179	.1217	.1255	.1293	.1331	.1368	.1406	.1443	.1480	.1517
0.4	.1554	.1591	.1628	.1664	.1700	.1736	.1772	.1808	.1844	.1879
0.5	.1915	.1950	.1985	.2019	.2054	.2088	.2123	.2157	.2190	.2224
0.6	.2257	.2291	.2324	.2357	.2389	.2422	.2454	.2486	.2517	.2549
0.7	.2580	.2611	.2642	.2673	.2704	.2734	.2764	.2794	.2823	.2852
0.8	.2881	.2910	.2939	.2967	.2995	.3023	.3051	.3078	.3106	.3133
0.9	.3159	.3186	.3212	.3238	.3264	.3289	.3315	.3340	.3365	.3389
1.0	.3413	.3438	.3461	.3485	.3508	.3531	.3554	.3577	.3599	.3621
1.1	.3643	.3665	.3686	.3708	.3729	.3749	.3770	.3790	.3810	.3830
1.2	.3849	.3869	.3888	.3907	.3925	.3944	.3962	.3980	.3997	.4015
1.3	.4032	.4049	.4066	.4082	.4099	.4115	.4131	.4147	.4162	.4177
1.4	.4192	.4207	.4222	.4236	.4251	.4265	.4279	.4292	.4306	.4319
1.5	.4332	.4345	.4357	.4370	.4382	.4394	.4406	.4418	.4429	.4441
1.6	.4452	.4463	.4474	.4484	.4495 *	.4505	.4515	.4525	.4535	.4545
1.7	.4554	.4564	.4573	.4582	.4591 ↑	.4599	.4608	.4616	.4625	.4633
1.8	.4641	.4649	.4656	.4664	.4671	.4678	.4686	.4693	.4699	.4706
1.9	.4713	.4719	.4726	.4732	.4738	.4744	.4750	.4756	.4761	.4767
2.0	.4772	.4778	.4783	.4788	.4793	.4798	.4803	.4808	.4812	.4817
2.1	.4821	.4826	.4830	.4834	.4838	.4842	.4846	.4850	.4854	.4857
2.2	.4861	.4864	.4868	.4871	.4875	.4878	.4881	.4884	.4887	.4890
2.3	.4893	.4896	.4898	.4901	.4904	.4906	.4909	.4911	.4913	.4916
2.4	.4918	.4920	.4922	.4925	.4927	.4929	.4931	.4932	.4934	.4936
2.5	.4938	.4940	.4941	.4943	.4945	.4946	.4948	.4949 *	.4951	.4952
2.6	.4953	.4955	.4956	.4957	.4959	.4960	.4961	.4962 ↑	.4963	.4964
2.7	.4965	.4966	.4967	.4968	.4969	.4970	.4971	.4972	.4973	.4974
2.8	.4974	.4975	.4976	.4977	.4977	.4978	.4979	.4979	.4980	.4981
2.9	.4981	.4982	.4982	.4983	.4984	.4984	.4985	.4985	.4986	.4986
3.0	.4987	.4987	.4987	.4988	.4988	.4989	.4989	.4989	.4990	.4990
3.10 and higher	.4999									

NOTE: For values of z above 3.09, use 0.4999 for the area.

*Use these common values that result from interpolation:

z score	Area
1.645	0.4500 ←
2.575	0.4950 ←

From Frederick C. Mosteller and Robert E. K. Rourke, *Sturdy Statistics*, 1973, Addison-Wesley Publishing Co., Reading, MA. Reprinted with permission of Frederick Mosteller.

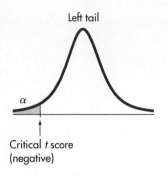

Left tail

α

Critical *t* score
(negative)

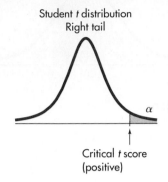

Student *t* distribution
Right tail

α

Critical *t* score
(positive)

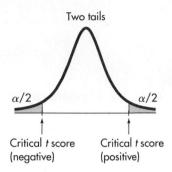

Two tails

α/2   α/2

Critical *t* score   Critical *t* score
(negative)   (positive)

Table A-3 *t* Distribution

	α					
Degrees of freedom	.005 (one tail) .01 (two tails)	.01 (one tail) .02 (two tails)	.025 (one tail) .05 (two tails)	.05 (one tail) .10 (two tails)	.10 (one tail) .20 (two tails)	.25 (one tail) .50 (two tails)
1	63.657	31.821	12.706	6.314	3.078	1.000
2	9.925	6.965	4.303	2.920	1.886	.816
3	5.841	4.541	3.182	2.353	1.638	.765
4	4.604	3.747	2.776	2.132	1.533	.741
5	4.032	3.365	2.571	2.015	1.476	.727
6	3.707	3.143	2.447	1.943	1.440	.718
7	3.500	2.998	2.365	1.895	1.415	.711
8	3.355	2.896	2.306	1.860	1.397	.706
9	3.250	2.821	2.262	1.833	1.383	.703
10	3.169	2.764	2.228	1.812	1.372	.700
11	3.106	2.718	2.201	1.796	1.363	.697
12	3.054	2.681	2.179	1.782	1.356	.696
13	3.012	2.650	2.160	1.771	1.350	.694
14	2.977	2.625	2.145	1.761	1.345	.692
15	2.947	2.602	2.132	1.753	1.341	.691
16	2.921	2.584	2.120	1.746	1.337	.690
17	2.898	2.567	2.110	1.740	1.333	.689
18	2.878	2.552	2.101	1.734	1.330	.688
19	2.861	2.540	2.093	1.729	1.328	.688
20	2.845	2.528	2.086	1.725	1.325	.687
21	2.831	2.518	2.080	1.721	1.323	.686
22	2.819	2.508	2.074	1.717	1.321	.686
23	2.807	2.500	2.069	1.714	1.320	.685
24	2.797	2.492	2.064	1.711	1.318	.685
25	2.787	2.485	2.060	1.708	1.316	.684
26	2.779	2.479	2.056	1.706	1.315	.684
27	2.771	2.473	2.052	1.703	1.314	.684
28	2.763	2.467	2.048	1.701	1.313	.683
29	2.756	2.462	2.045	1.699	1.311	.683
Large (z)	2.575	2.327	1.960	1.645	1.282	.675

Table A-4 Chi-Square (χ^2) Distribution

Degrees of freedom	Area to the Right of the Critical Value									
	0.995	0.99	0.975	0.95	0.90	0.10	0.05	0.025	0.01	0.005
1	—	—	0.001	0.004	0.016	2.706	3.841	5.024	6.635	7.879
2	0.010	0.020	0.051	0.103	0.211	4.605	5.991	7.378	9.210	10.597
3	0.072	0.115	0.216	0.352	0.584	6.251	7.815	9.348	11.345	12.838
4	0.207	0.297	0.484	0.711	1.064	7.779	9.488	11.143	13.277	14.860
5	0.412	0.554	0.831	1.145	1.610	9.236	11.071	12.833	15.086	16.750
6	0.676	0.872	1.237	1.635	2.204	10.645	12.592	14.449	16.812	18.548
7	0.989	1.239	1.690	2.167	2.833	12.017	14.067	16.013	18.475	20.278
8	1.344	1.646	2.180	2.733	3.490	13.362	15.507	17.535	20.090	21.955
9	1.735	2.088	2.700	3.325	4.168	14.684	16.919	19.023	21.666	23.589
10	2.156	2.558	3.247	3.940	4.865	15.987	18.307	20.483	23.209	25.188
11	2.603	3.053	3.816	4.575	5.578	17.275	19.675	21.920	24.725	26.757
12	3.074	3.571	4.404	5.226	6.304	18.549	21.026	23.337	26.217	28.299
13	3.565	4.107	5.009	5.892	7.042	19.812	22.362	24.736	27.688	29.819
14	4.075	4.660	5.629	6.571	7.790	21.064	23.685	26.119	29.141	31.319
15	4.601	5.229	6.262	7.261	8.547	22.307	24.996	27.488	30.578	32.801
16	5.142	5.812	6.908	7.962	9.312	23.542	26.296	28.845	32.000	34.267
17	5.697	6.408	7.564	8.672	10.085	24.769	27.587	30.191	33.409	35.718
18	6.265	7.015	8.231	9.390	10.865	25.989	28.869	31.526	34.805	37.156
19	6.844	7.633	8.907	10.117	11.651	27.204	30.144	32.852	36.191	38.582
20	7.434	8.260	9.591	10.851	12.443	28.412	31.410	34.170	37.566	39.997
21	8.034	8.897	10.283	11.591	13.240	29.615	32.671	35.479	38.932	41.401
22	8.643	9.542	10.982	12.338	14.042	30.813	33.924	36.781	40.289	42.796
23	9.260	10.196	11.689	13.091	14.848	32.007	35.172	38.076	41.638	44.181
24	9.886	10.856	12.401	13.848	15.659	33.196	36.415	39.364	42.980	45.559
25	10.520	11.524	13.120	14.611	16.473	34.382	37.652	40.646	44.314	46.928
26	11.160	12.198	13.844	15.379	17.292	35.563	38.885	41.923	45.642	48.290
27	11.808	12.879	14.573	16.151	18.114	36.741	40.113	43.194	46.963	49.645
28	12.461	13.565	15.308	16.928	18.939	37.916	41.337	44.461	48.278	50.993
29	13.121	14.257	16.047	17.708	19.768	39.087	42.557	45.722	49.588	52.336
30	13.787	14.954	16.791	18.493	20.599	40.256	43.773	46.979	50.892	53.672
40	20.707	22.164	24.433	26.509	29.051	51.805	55.758	59.342	63.691	66.766
50	27.991	29.707	32.357	34.764	37.689	63.167	67.505	71.420	76.154	79.490
60	35.534	37.485	40.482	43.188	46.459	74.397	79.082	83.298	88.379	91.952
70	43.275	45.442	48.758	51.739	55.329	85.527	90.531	95.023	100.425	104.215
80	51.172	53.540	57.153	60.391	64.278	96.578	101.879	106.629	112.329	116.321
90	59.196	61.754	65.647	69.126	73.291	107.565	113.145	118.136	124.116	128.299
100	67.328	70.065	74.222	77.929	82.358	118.498	124.342	129.561	135.807	140.169

From Donald B. Owen, *Handbook of Statistical Tables*, © 1962 Addison-Wesley Publishing Co., Reading, MA. Reprinted with permission of the publisher. *(continued)*

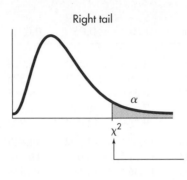

Right tail

To find this value, use the column with the area α given at the top of the table.

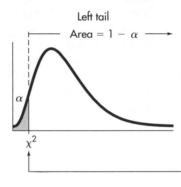

Left tail

Area = $1 - \alpha$

To find this value, determine the area of the region to the right of this boundary (the unshaded area) and use the column with this value at the top. If the left tail has area α, use the column with the value of $1 - \alpha$ at the top of the table.

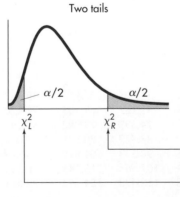

Two tails

$\alpha/2$ $\alpha/2$

χ_L^2 χ_R^2

To find this value, use the column with area $\alpha/2$ at the top of the table.

To find this value, use the column with area $1 - \alpha/2$ at the top of the table.

Table A-5 F Distribution ($\alpha = 0.01$ in the right tail)

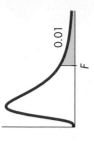

0.01

F

Denominator degrees of freedom (df_2)	\	Numerator degrees of freedom (df_1)								
		1	2	3	4	5	6	7	8	9
1		4052.2	4999.5	5403.4	5624.6	5763.6	5859.0	5928.4	5981.1	6022.5
2		98.503	99.000	99.166	99.249	99.299	99.333	99.356	99.374	99.388
3		34.116	30.817	29.457	28.710	28.237	27.911	27.672	27.489	27.345
4		21.198	18.000	16.694	15.977	15.522	15.207	14.976	14.799	14.659
5		16.258	13.274	12.060	11.392	10.967	10.672	10.456	10.289	10.158
6		13.745	10.925	9.7795	9.1483	8.7459	8.4661	8.2600	8.1017	7.9761
7		12.246	9.5466	8.4513	7.8466	7.4604	7.1914	6.9928	6.8400	6.7188
8		11.259	8.6491	7.5910	7.0061	6.6318	6.3707	6.1776	6.0289	5.9106
9		10.561	8.0215	6.9919	6.4221	6.0569	5.8018	5.6129	5.4671	5.3511
10		10.044	7.5594	6.5523	5.9943	5.6363	5.3858	5.2001	5.0567	4.9424
11		9.6460	7.2057	6.2167	5.6683	5.3160	5.0692	4.8861	4.7445	4.6315
12		9.3302	6.9266	5.9525	5.4120	5.0643	4.8206	4.6395	4.4994	4.3875
13		9.0738	6.7010	5.7394	5.2053	4.8616	4.6204	4.4410	4.3021	4.1911
14		8.8616	6.5149	5.5639	5.0354	4.6950	4.4558	4.2779	4.1399	4.0297
15		8.6831	6.3589	5.4170	4.8932	4.5556	4.3183	4.1415	4.0045	3.8948
16		8.5310	6.2262	5.2922	4.7726	4.4374	4.2016	4.0259	3.8896	3.7804
17		8.3997	6.1121	5.1850	4.6690	4.3359	4.1015	3.9267	3.7910	3.6822
18		8.2854	6.0129	5.0919	4.5790	4.2479	4.0146	3.8406	3.7054	3.5971
19		8.1849	5.9259	5.0103	4.5003	4.1708	3.9386	3.7653	3.6305	3.5225
20		8.0960	5.8489	4.9382	4.4307	4.1027	3.8714	3.6987	3.5644	3.4567
21		8.0166	5.7804	4.8740	4.3688	4.0421	3.8117	3.6396	3.5056	3.3981
22		7.9454	5.7190	4.8166	4.3134	3.9880	3.7583	3.5867	3.4530	3.3458
23		7.8811	5.6637	4.7649	4.2636	3.9392	3.7102	3.5390	3.4057	3.2986
24		7.8229	5.6136	4.7181	4.2184	3.8951	3.6667	3.4959	3.3629	3.2560
25		7.7698	5.5680	4.6755	4.1774	3.8550	3.6272	3.4568	3.3239	3.2172
26		7.7213	5.5263	4.6366	4.1400	3.8183	3.5911	3.4210	3.2884	3.1818
27		7.6767	5.4881	4.6009	4.1056	3.7848	3.5580	3.3882	3.2558	3.1494
28		7.6356	5.4529	4.5681	4.0740	3.7539	3.5276	3.3581	3.2259	3.1195
29		7.5977	5.4204	4.5378	4.0449	3.7254	3.4995	3.3303	3.1982	3.0920
30		7.5625	5.3903	4.5097	4.0179	3.6990	3.4735	3.3045	3.1726	3.0665
40		7.3141	5.1785	4.3126	3.8283	3.5138	3.2910	3.1238	2.9930	2.8876
60		7.0771	4.9774	4.1259	3.6490	3.3389	3.1187	2.9530	2.8233	2.7185
120		6.8509	4.7865	3.9491	3.4795	3.1735	2.9559	2.7918	2.6629	2.5586
∞		6.6349	4.6052	3.7816	3.3192	3.0173	2.8020	2.6393	2.5113	2.4073

(continued)

Table A-5 F Distribution ($\alpha = 0.01$ in the right tail)

Numerator degrees of freedom (df_1)

df_2	10	12	15	20	24	30	40	60	120	∞
1	6055.8	6106.3	6157.3	6208.7	6234.6	6260.6	6286.8	6313.0	6339.4	6365.9
2	99.399	99.416	99.433	99.449	99.458	99.466	99.474	99.482	99.491	99.499
3	27.229	27.052	26.872	26.690	26.598	26.505	26.411	26.316	26.221	26.125
4	14.546	14.374	14.198	14.020	13.929	13.838	13.745	13.652	13.558	13.463
5	10.051	9.8883	9.7222	9.5526	9.4665	9.3793	9.2912	9.2020	9.1118	9.0204
6	7.8741	7.7183	7.5590	7.3958	7.3127	7.2285	7.1432	7.0567	6.9690	6.8800
7	6.6201	6.4691	6.3143	6.1554	6.0743	5.9920	5.9084	5.8236	5.7373	5.6495
8	5.8143	5.6667	5.5151	5.3591	5.2793	5.1981	5.1156	5.0316	4.9461	4.8588
9	5.2565	5.1114	4.9621	4.8080	4.7290	4.6486	4.5666	4.4831	4.3978	4.3105
10	4.8491	4.7059	4.5581	4.4054	4.3269	4.2469	4.1653	4.0819	3.9965	3.9090
11	4.5393	4.3974	4.2509	4.0990	4.0209	3.9411	3.8596	3.7761	3.6904	3.6024
12	4.2961	4.1553	4.0096	3.8584	3.7805	3.7008	3.6192	3.5355	3.4494	3.3608
13	4.1003	3.9603	3.8154	3.6646	3.5868	3.5070	3.4253	3.3413	3.2548	3.1654
14	3.9394	3.8001	3.6557	3.5052	3.4274	3.3476	3.2656	3.1813	3.0942	3.0040
15	3.8049	3.6662	3.5222	3.3719	3.2940	3.2141	3.1319	3.0471	2.9595	2.8684
16	3.6909	3.5527	3.4089	3.2587	3.1808	3.1007	3.0182	2.9330	2.8447	2.7528
17	3.5931	3.4552	3.3117	3.1615	3.0835	3.0032	2.9205	2.8348	2.7459	2.6530
18	3.5082	3.3706	3.2273	3.0771	2.9990	2.9185	2.8354	2.7493	2.6597	2.5660
19	3.4338	3.2965	3.1533	3.0031	2.9249	2.8442	2.7608	2.6742	2.5839	2.4893
20	3.3682	3.2311	3.0880	2.9377	2.8594	2.7785	2.6947	2.6077	2.5168	2.4212
21	3.3098	3.1730	3.0300	2.8796	2.8010	2.7200	2.6359	2.5484	2.4568	2.3603
22	3.2576	3.1209	2.9779	2.8274	2.7488	2.6675	2.5831	2.4951	2.4029	2.3055
23	3.2106	3.0740	2.9311	2.7805	2.7017	2.6202	2.5355	2.4471	2.3542	2.2558
24	3.1681	3.0316	2.8887	2.7380	2.6591	2.5773	2.4923	2.4035	2.3100	2.2107
25	3.1294	2.9931	2.8502	2.6993	2.6203	2.5383	2.4530	2.3637	2.2696	2.1694
26	3.0941	2.9578	2.8150	2.6640	2.5848	2.5026	2.4170	2.3273	2.2325	2.1315
27	3.0618	2.9256	2.7827	2.6316	2.5522	2.4699	2.3840	2.2938	2.1985	2.0965
28	3.0320	2.8959	2.7530	2.6017	2.5223	2.4397	2.3535	2.2629	2.1670	2.0642
29	3.0045	2.8685	2.7256	2.5742	2.4946	2.4118	2.3253	2.2344	2.1379	2.0342
30	2.9791	2.8431	2.7002	2.5487	2.4689	2.3860	2.2992	2.2079	2.1108	2.0062
40	2.8005	2.6648	2.5216	2.3689	2.2880	2.2034	2.1142	2.0194	1.9172	1.8047
60	2.6318	2.4961	2.3523	2.1978	2.1154	2.0285	1.9360	1.8363	1.7263	1.6006
120	2.4721	2.3363	2.1915	2.0346	1.9500	1.8600	1.7628	1.6557	1.5330	1.3805
∞	2.3209	2.1847	2.0385	1.8783	1.7908	1.6964	1.5923	1.4730	1.3246	1.0000

Denominator degrees of freedom (df_2)

From Maxine Merrington and Catherine M. Thompson, "Tables of Percentage Points of the Inverted Beta (F) Distribution," *Biometrika 33* (1943): 80–84. Reproduced with permission of the Biometrika Trustees.

(continued)

Table A-5 F Distribution ($\alpha = 0.025$ in the right tail)

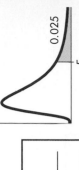

	Numerator degrees of freedom (df_1)								
Denominator degrees of freedom (df_2)	1	2	3	4	5	6	7	8	9
1	647.79	799.50	864.16	899.58	921.85	937.11	948.22	956.66	963.28
2	38.506	39.000	39.165	39.248	39.298	39.331	39.335	39.373	39.387
3	17.443	16.044	15.439	15.101	14.885	14.735	14.624	14.540	14.473
4	12.218	10.649	9.9792	9.6045	9.3645	9.1973	9.0741	8.9796	8.9047
5	10.007	8.4336	7.7636	7.3879	7.1464	6.9777	6.8531	6.7572	6.6811
6	8.8131	7.2599	6.5988	6.2272	5.9876	5.8198	5.6955	5.5996	5.5234
7	8.0727	6.5415	5.8898	5.5226	5.2852	5.1186	4.9949	4.8993	4.8232
8	7.5709	6.0595	5.4160	5.0526	4.8173	4.6517	4.5286	4.4333	4.3572
9	7.2093	5.7147	5.0781	4.7181	4.4844	4.3197	4.1970	4.1020	4.0260
10	6.9367	5.4564	4.8256	4.4683	4.2361	4.0721	3.9498	3.8549	3.7790
11	6.7241	5.2559	4.6300	4.2751	4.0440	3.8807	3.7586	3.6638	3.5879
12	6.5538	5.0959	4.4742	4.1212	3.8911	3.7283	3.6065	3.5118	3.4358
13	6.4143	4.9653	4.3472	3.9959	3.7667	3.6043	3.4827	3.3880	3.3120
14	6.2979	4.8567	4.2417	3.8919	3.6634	3.5014	3.3799	3.2853	3.2093
15	6.1995	4.7650	4.1528	3.8043	3.5764	3.4147	3.2934	3.1987	3.1227
16	6.1151	4.6867	4.0768	3.7294	3.5021	3.3406	3.2194	3.1248	3.0488
17	6.0420	4.6189	4.0112	3.6648	3.4379	3.2767	3.1556	3.0610	2.9849
18	5.9781	4.5597	3.9539	3.6083	3.3820	3.2209	3.0999	3.0053	2.9291
19	5.9216	4.5075	3.9034	3.5587	3.3327	3.1718	3.0509	2.9563	2.8801
20	5.8715	4.4613	3.8587	3.5147	3.2891	3.1283	3.0074	2.9128	2.8365
21	5.8266	4.4199	3.8188	3.4754	3.2501	3.0895	2.9686	2.8740	2.7977
22	5.7863	4.3828	3.7829	3.4401	3.2151	3.0546	2.9338	2.8392	2.7628
23	5.7498	4.3492	3.7505	3.4083	3.1835	3.0232	2.9023	2.8077	2.7313
24	5.7166	4.3187	3.7211	3.3794	3.1548	2.9946	2.8738	2.7791	2.7027
25	5.6864	4.2909	3.6943	3.3530	3.1287	2.9685	2.8478	2.7531	2.6766
26	5.6586	4.2655	3.6697	3.3289	3.1048	2.9447	2.8240	2.7293	2.6528
27	5.6331	4.2421	3.6472	3.3067	3.0828	2.9228	2.8021	2.7074	2.6309
28	5.6096	4.2205	3.6264	3.2863	3.0626	2.9027	2.7820	2.6872	2.6106
29	5.5878	4.2006	3.6072	3.2674	3.0438	2.8840	2.7633	2.6686	2.5919
30	5.5675	4.1821	3.5894	3.2499	3.0265	2.8667	2.7460	2.6513	2.5746
40	5.4239	4.0510	3.4633	3.1261	2.9037	2.7444	2.6238	2.5289	2.4519
60	5.2856	3.9253	3.3425	3.0077	2.7863	2.6274	2.5068	2.4117	2.3344
120	5.1523	3.8046	3.2269	2.8943	2.6740	2.5154	2.3948	2.2994	2.2217
∞	5.0239	3.6889	3.1161	2.7858	2.5665	2.4082	2.2875	2.1918	2.1136

(continued)

Table A-5 F Distribution ($\alpha = 0.025$ in the right tail)

Numerator degrees of freedom (df_1)

df_2	10	12	15	20	24	30	40	60	120	∞
1	968.63	976.71	984.87	993.10	997.25	1001.4	1005.6	1009.8	1014.0	1018.3
2	39.398	39.415	39.431	39.448	39.456	39.465	39.473	39.481	39.490	39.498
3	14.419	14.337	14.253	14.167	14.124	14.081	14.037	13.992	13.947	13.902
4	8.8439	8.7512	8.6565	8.5599	8.5109	8.4613	8.4111	8.3604	8.3092	8.2573
5	6.6192	6.5245	6.4277	6.3286	6.2780	6.2269	6.1750	6.1225	6.0693	6.0153
6	5.4613	5.3662	5.2687	5.1684	5.1172	5.0652	5.0125	4.9589	4.9044	4.8491
7	4.7611	4.6658	4.5678	4.4667	4.4150	4.3624	4.3089	4.2544	4.1989	4.1423
8	4.2951	4.1997	4.1012	3.9995	3.9472	3.8940	3.8398	3.7844	3.7279	3.6702
9	3.9639	3.8682	3.7694	3.6669	3.6142	3.5604	3.5055	3.4493	3.3918	3.3329
10	3.7168	3.6209	3.5217	3.4185	3.3654	3.3110	3.2554	3.1984	3.1399	3.0798
11	3.5257	3.4296	3.3299	3.2261	3.1725	3.1176	3.0613	3.0035	2.9441	2.8828
12	3.3736	3.2773	3.1772	3.0728	3.0187	2.9633	2.9063	2.8478	2.7874	2.7249
13	3.2497	3.1532	3.0527	2.9477	2.8932	2.8372	2.7797	2.7204	2.6590	2.5955
14	3.1469	3.0502	2.9493	2.8437	2.7888	2.7324	2.6742	2.6142	2.5519	2.4872
15	3.0602	2.9633	2.8621	2.7559	2.7006	2.6437	2.5850	2.5242	2.4611	2.3953
16	2.9862	2.8890	2.7875	2.6808	2.6252	2.5678	2.5085	2.4471	2.3831	2.3163
17	2.9222	2.8249	2.7230	2.6158	2.5598	2.5020	2.4422	2.3801	2.3153	2.2474
18	2.8664	2.7689	2.6667	2.5590	2.5027	2.4445	2.3842	2.3214	2.2558	2.1869
19	2.8172	2.7196	2.6171	2.5089	2.4523	2.3937	2.3329	2.2696	2.2032	2.1333
20	2.7737	2.6758	2.5731	2.4645	2.4076	2.3486	2.2873	2.2234	2.1562	2.0853
21	2.7348	2.6368	2.5338	2.4247	2.3675	2.3082	2.2465	2.1819	2.1141	2.0422
22	2.6998	2.6017	2.4984	2.3890	2.3315	2.2718	2.2097	2.1446	2.0760	2.0032
23	2.6682	2.5699	2.4665	2.3567	2.2989	2.2389	2.1763	2.1107	2.0415	1.9677
24	2.6396	2.5411	2.4374	2.3273	2.2693	2.2090	2.1460	2.0799	2.0099	1.9353
25	2.6135	2.5149	2.4110	2.3005	2.2422	2.1816	2.1183	2.0516	1.9811	1.9055
26	2.5896	2.4908	2.3867	2.2759	2.2174	2.1565	2.0928	2.0257	1.9545	1.8781
27	2.5676	2.4688	2.3644	2.2533	2.1946	2.1334	2.0693	2.0018	1.9299	1.8527
28	2.5473	2.4484	2.3438	2.2324	2.1735	2.1121	2.0477	1.9797	1.9072	1.8291
29	2.5286	2.4295	2.3248	2.2131	2.1540	2.0923	2.0276	1.9591	1.8861	1.8072
30	2.5112	2.4120	2.3072	2.1952	2.1359	2.0739	2.0089	1.9400	1.8664	1.7867
40	2.3882	2.2882	2.1819	2.0677	2.0069	1.9429	1.8752	1.8028	1.7242	1.6371
60	2.2702	2.1692	2.0613	1.9445	1.8817	1.8152	1.7440	1.6668	1.5810	1.4821
120	2.1570	2.0548	1.9450	1.8249	1.7597	1.6899	1.6141	1.5299	1.4327	1.3104
∞	2.0483	1.9447	1.8326	1.7085	1.6402	1.5660	1.4835	1.3883	1.2684	1.0000

Denominator degrees of freedom (df_2)

From Maxine Merrington and Catherine M. Thompson, "Tables of Percentage Points of the Inverted Beta (F) Distribution," *Biometrika 33* (1943): 80–84. Reproduced with permission of the Biometrika Trustees.

(continued)

Table A-5 F Distribution ($\alpha = 0.05$ in the right tail)

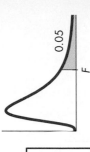

	Numerator degrees of freedom (df$_1$)								
Denominator degrees of freedom (df$_2$)	**1**	**2**	**3**	**4**	**5**	**6**	**7**	**8**	**9**
1	161.45	199.50	215.71	224.58	230.16	233.99	236.77	238.88	240.54
2	18.513	19.000	19.164	19.247	19.296	19.330	19.353	19.371	19.385
3	10.128	9.5521	9.2766	9.1172	9.0135	8.9406	8.8867	8.8452	8.8123
4	7.7086	6.9443	6.5914	6.3882	6.2561	6.1631	6.0942	6.0410	5.9988
5	6.6079	5.7861	5.4095	5.1922	5.0503	4.9503	4.8759	4.8183	4.7725
6	5.9874	5.1433	4.7571	4.5337	4.3874	4.2839	4.2067	4.1468	4.0990
7	5.5914	4.7374	4.3468	4.1203	3.9715	3.8660	3.7870	3.7257	3.6767
8	5.3177	4.4590	4.0662	3.8379	3.6875	3.5806	3.5005	3.4381	3.3881
9	5.1174	4.2565	3.8625	3.6331	3.4817	3.3738	3.2927	3.2296	3.1789
10	4.9646	4.1028	3.7083	3.4780	3.3258	3.2172	3.1355	3.0717	3.0204
11	4.8443	3.9823	3.5874	3.3567	3.2039	3.0946	3.0123	2.9480	2.8962
12	4.7472	3.8853	3.4903	3.2592	3.1059	2.9961	2.9134	2.8486	2.7964
13	4.6672	3.8056	3.4105	3.1791	3.0254	2.9153	2.8321	2.7669	2.7144
14	4.6001	3.7389	3.3439	3.1122	2.9582	2.8477	2.7642	2.6987	2.6458
15	4.5431	3.6823	3.2874	3.0556	2.9013	2.7905	2.7066	2.6408	2.5876
16	4.4940	3.6337	3.2389	3.0069	2.8524	2.7413	2.6572	2.5911	2.5377
17	4.4513	3.5915	3.1968	2.9647	2.8100	2.6987	2.6143	2.5480	2.4943
18	4.4139	3.5546	3.1599	2.9277	2.7729	2.6613	2.5767	2.5102	2.4563
19	4.3807	3.5219	3.1274	2.8951	2.7401	2.6283	2.5435	2.4768	2.4227
20	4.3512	3.4928	3.0984	2.8661	2.7109	2.5990	2.5140	2.4471	2.3928
21	4.3248	3.4668	3.0725	2.8401	2.6848	2.5727	2.4876	2.4205	2.3660
22	4.3009	3.4434	3.0491	2.8167	2.6613	2.5491	2.4638	2.3965	2.3419
23	4.2793	3.4221	3.0280	2.7955	2.6400	2.5277	2.4422	2.3748	2.3201
24	4.2597	3.4028	3.0088	2.7763	2.6207	2.5082	2.4226	2.3551	2.3002
25	4.2417	3.3852	2.9912	2.7587	2.6030	2.4904	2.4047	2.3371	2.2821
26	4.2252	3.3690	2.9752	2.7426	2.5868	2.4741	2.3883	2.3205	2.2655
27	4.2100	3.3541	2.9604	2.7278	2.5719	2.4591	2.3732	2.3053	2.2501
28	4.1960	3.3404	2.9467	2.7141	2.5581	2.4453	2.3593	2.2913	2.2360
29	4.1830	3.3277	2.9340	2.7014	2.5454	2.4324	2.3463	2.2783	2.2229
30	4.1709	3.3158	2.9223	2.6896	2.5336	2.4205	2.3343	2.2662	2.2107
40	4.0847	3.2317	2.8387	2.6060	2.4495	2.3359	2.2490	2.1802	2.1240
60	4.0012	3.1504	2.7581	2.5252	2.3683	2.2541	2.1665	2.0970	2.0401
120	3.9201	3.0718	2.6802	2.4472	2.2899	2.1750	2.0868	2.0164	1.9588
∞	3.8415	2.9957	2.6049	2.3719	2.2141	2.0986	2.0096	1.9384	1.8799

(continued)

Table A-5 F Distribution ($\alpha = 0.05$ in the right tail)

Numerator degrees of freedom (df_1)

Denominator degrees of freedom (df_2)	10	12	15	20	24	30	40	60	120	∞
1	241.88	243.91	245.95	248.01	249.05	250.10	251.14	252.20	253.25	254.31
2	19.396	19.413	19.429	19.446	19.454	19.462	19.471	19.479	19.487	19.496
3	8.7855	8.7446	8.7029	8.6602	8.6385	8.6166	8.5944	8.5720	8.5494	8.5264
4	5.9644	5.9117	5.8578	5.8025	5.7744	5.7459	5.7170	5.6877	5.6581	5.6281
5	4.7351	4.6777	4.6188	4.5581	4.5272	4.4957	4.4638	4.4314	4.3985	4.3650
6	4.0600	3.9999	3.9381	3.8742	3.8415	3.8082	3.7743	3.7398	3.7047	3.6689
7	3.6365	3.5747	3.5107	3.4445	3.4105	3.3758	3.3404	3.3043	3.2674	3.2298
8	3.3472	3.2839	3.2184	3.1503	3.1152	3.0794	3.0428	3.0053	2.9669	2.9276
9	3.1373	3.0729	3.0061	2.9365	2.9005	2.8637	2.8259	2.7872	2.7475	2.7067
10	2.9782	2.9130	2.8450	2.7740	2.7372	2.6996	2.6609	2.6211	2.5801	2.5379
11	2.8536	2.7876	2.7186	2.6464	2.6090	2.5705	2.5309	2.4901	2.4480	2.4045
12	2.7534	2.6866	2.6169	2.5436	2.5055	2.4663	2.4259	2.3842	2.3410	2.2962
13	2.6710	2.6037	2.5331	2.4589	2.4202	2.3803	2.3392	2.2966	2.2524	2.2064
14	2.6022	2.5342	2.4630	2.3879	2.3487	2.3082	2.2664	2.2229	2.1778	2.1307
15	2.5437	2.4753	2.4034	2.3275	2.2878	2.2468	2.2043	2.1601	2.1141	2.0658
16	2.4935	2.4247	2.3522	2.2756	2.2354	2.1938	2.1507	2.1058	2.0589	2.0096
17	2.4499	2.3807	2.3077	2.2304	2.1898	2.1477	2.1040	2.0584	2.0107	1.9604
18	2.4117	2.3421	2.2686	2.1906	2.1497	2.1071	2.0629	2.0166	1.9681	1.9168
19	2.3779	2.3080	2.2341	2.1555	2.1141	2.0712	2.0264	1.9795	1.9302	1.8780
20	2.3479	2.2776	2.2033	2.1242	2.0825	2.0391	1.9938	1.9464	1.8963	1.8432
21	2.3210	2.2504	2.1757	2.0960	2.0540	2.0102	1.9645	1.9165	1.8657	1.8117
22	2.2967	2.2258	2.1508	2.0707	2.0283	1.9842	1.9380	1.8894	1.8380	1.7831
23	2.2747	2.2036	2.1282	2.0476	2.0050	1.9605	1.9139	1.8648	1.8128	1.7570
24	2.2547	2.1834	2.1077	2.0267	1.9838	1.9390	1.8920	1.8424	1.7896	1.7330
25	2.2365	2.1649	2.0889	2.0075	1.9643	1.9192	1.8718	1.8217	1.7684	1.7110
26	2.2197	2.1479	2.0716	1.9898	1.9464	1.9010	1.8533	1.8027	1.7488	1.6906
27	2.2043	2.1323	2.0558	1.9736	1.9299	1.8842	1.8361	1.7851	1.7306	1.6717
28	2.1900	2.1179	2.0411	1.9586	1.9147	1.8687	1.8203	1.7689	1.7138	1.6541
29	2.1768	2.1045	2.0275	1.9446	1.9005	1.8543	1.8055	1.7537	1.6981	1.6376
30	2.1646	2.0921	2.0148	1.9317	1.8874	1.8409	1.7918	1.7396	1.6835	1.6223
40	2.0772	2.0035	1.9245	1.8389	1.7929	1.7444	1.6928	1.6373	1.5766	1.5089
60	1.9926	1.9174	1.8364	1.7480	1.7001	1.6491	1.5943	1.5343	1.4673	1.3893
120	1.9105	1.8337	1.7505	1.6587	1.6084	1.5543	1.4952	1.4290	1.3519	1.2539
∞	1.8307	1.7522	1.6664	1.5705	1.5173	1.4591	1.3940	1.3180	1.2214	1.0000

From Maxine Merrington and Catherine M. Thompson, "Tables of Percentage Points of the Inverted Beta (F) Distribution," *Biometrika* 33 (1943): 80–84. Reproduced with permission of the Biometrika Trustees.

Table A-6 Critical Values of the Pearson Correlation Coefficient r

n	$\alpha = .05$	$\alpha = .01$
4	.950	.999
5	.878	.959
6	.811	.917
7	.754	.875
8	.707	.834
9	.666	.798
10	.632	.765
11	.602	.735
12	.576	.708
13	.553	.684
14	.532	.661
15	.514	.641
16	.497	.623
17	.482	.606
18	.468	.590
19	.456	.575
20	.444	.561
25	.396	.505
30	.361	.463
35	.335	.430
40	.312	.402
45	.294	.378
50	.279	.361
60	.254	.330
70	.236	.305
80	.220	.286
90	.207	.269
100	.196	.256

NOTE: To test H_0: $\rho = 0$, against H_1: $\rho \neq 0$, reject H_0 if the absolute value of r is greater than the critical value in the table.

Table A-7 Critical Values for the Sign Test

	α			
n	.005 (one tail) .01 (two tails)	.01 (one tail) .02 (two tails)	.025 (one tail) .05 (two tails)	.05 (one tail) .10 (two tails)
1	*	*	*	*
2	*	*	*	*
3	*	*	*	*
4	*	*	*	*
5	*	*	*	0
6	*	*	0	0
7	*	0	0	0
8	0	0	0	1
9	0	0	1	1
10	0	0	1	1
11	0	1	1	2
12	1	1	2	2
13	1	1	2	3
14	1	2	2	3
15	2	2	3	3
16	2	2	3	4
17	2	3	4	4
18	3	3	4	5
19	3	4	4	5
20	3	4	5	5
21	4	4	5	6
22	4	5	5	6
23	4	5	6	7
24	5	5	6	7
25	5	6	7	7

NOTES:

1. *indicates that it is not possible to get a value in the critical region.
2. Reject the null hypothesis if the number of the less frequent sign (x) is less than or equal to the value in the table.
3. For values of n greater than 25, a normal approximation is used with

$$z = \frac{(x + 0.5) - \left(\frac{n}{2}\right)}{\frac{\sqrt{n}}{2}}$$

Table A-8 Critical Values of T for the Wilcoxon Signed-Ranks Test

	α			
n	.005 (one tail) .01 (two tails)	.01 (one tail) .02 (two tails)	.025 (one tail) .05 (two tails)	.05 (one tail) .10 (two tails)
5	*	*	*	1
6	*	*	1	2
7	*	0	2	4
8	0	2	4	6
9	2	3	6	8
10	3	5	8	11
11	5	7	11	14
12	7	10	14	17
13	10	13	17	21
14	13	16	21	26
15	16	20	25	30
16	19	24	30	36
17	23	28	35	41
18	28	33	40	47
19	32	38	46	54
20	37	43	52	60
21	43	49	59	68
22	49	56	66	75
23	55	62	73	83
24	61	69	81	92
25	68	77	90	101
26	76	85	98	110
27	84	93	107	120
28	92	102	117	130
29	100	111	127	141
30	109	120	137	152

NOTES:

1. *indicates that it is not possible to get a value in the critical region.
2. Reject the null hypothesis if the test statistic T is less than or equal to the critical value found in this table. Fail to reject the null hypothesis if the test statistic T is greater than the critical value found in this table.

From *Some Rapid Approximate Statistical Procedures,* Copyright © 1949, 1964 Lederle Laboratories Division of American Cyanamid Company. Reprinted with the permission of the American Cyanamid Company.

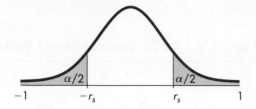

Table A-9 Critical Values of Spearman's Rank Correlation Coefficient r_s

n	$\alpha = 0.10$	$\alpha = 0.05$	$\alpha = 0.02$	$\alpha = 0.01$
5	.900	—	—	—
6	.829	.886	.943	—
7	.714	.786	.893	—
8	.643	.738	.833	.881
9	.600	.683	.783	.833
10	.564	.648	.745	.794
11	.523	.623	.736	.818
12	.497	.591	.703	.780
13	.475	.566	.673	.745
14	.457	.545	.646	.716
15	.441	.525	.623	.689
16	.425	.507	.601	.666
17	.412	.490	.582	.645
18	.399	.476	.564	.625
19	.388	.462	.549	.608
20	.377	.450	.534	.591
21	.368	.438	.521	.576
22	.359	.428	.508	.562
23	.351	.418	.496	.549
24	.343	.409	.485	.537
25	.336	.400	.475	.526
26	.329	.392	.465	.515
27	.323	.385	.456	.505
28	.317	.377	.448	.496
29	.311	.370	.440	.487
30	.305	.364	.432	.478

NOTE: For $n > 30$ use $r_s = \pm\, z/\sqrt{n-1}$, where z corresponds to the level of significance. For example, if $\alpha = 0.05$, then $z = 1.96$.

To test H_0: $\rho_s = 0$
against H_1: $\rho_s \neq 0$

From "Distribution of sums of squares of rank differences to small numbers of individuals," *The Annals of Mathematical Statistics*, Vol. 9, No. 2. Reprinted with permission of the Institute of Mathematical Statistics.

Table A-10 Critical Values for Number of Runs G

Value of n_1	Value of n_2																		
	2	3	4	5	6	7	8	9	10	11	12	13	14	15	16	17	18	19	20
2	1	1	1	1	1	1	1	1	1	1	2	2	2	2	2	2	2	2	2
	6	6	6	6	6	6	6	6	6	6	6	6	6	6	6	6	6	6	6
3	1	1	1	1	2	2	2	2	2	2	2	2	2	3	3	3	3	3	3
	6	8	8	8	8	8	8	8	8	8	8	8	8	8	8	8	8	8	8
4	1	1	1	2	2	2	3	3	3	3	3	3	3	3	4	4	4	4	4
	6	8	9	9	9	10	10	10	10	10	10	10	10	10	10	10	10	10	10
5	1	1	2	2	3	3	3	3	3	4	4	4	4	4	4	4	5	5	5
	6	8	9	10	10	11	11	12	12	12	12	12	12	12	12	12	12	12	12
6	1	2	2	3	3	3	3	4	4	4	4	5	5	5	5	5	5	6	6
	6	8	9	10	11	12	12	13	13	13	13	14	14	14	14	14	14	14	14
7	1	2	2	3	3	3	4	4	5	5	5	5	5	6	6	6	6	6	6
	6	8	10	11	12	13	13	14	14	14	14	15	15	15	16	16	16	16	16
8	1	2	3	3	3	4	4	5	5	5	6	6	6	6	6	7	7	7	7
	6	8	10	11	12	13	14	14	15	15	16	16	16	16	17	17	17	17	17
9	1	2	3	3	4	4	5	5	5	6	6	7	7	7	7	8	8	8	8
	6	8	10	12	13	14	14	15	16	16	16	17	17	18	18	18	18	18	18
10	1	2	3	3	4	5	5	5	6	6	7	7	7	7	8	8	8	8	9
	6	8	10	12	13	14	15	16	16	17	17	18	18	18	19	19	19	20	20
11	1	2	3	4	4	5	5	6	6	7	7	7	8	8	8	9	9	9	9
	6	8	10	12	13	14	15	16	17	17	18	19	19	19	20	20	20	21	21
12	2	2	3	4	4	5	6	6	7	7	7	8	8	8	9	9	9	10	10
	6	8	10	12	13	14	16	16	17	18	19	19	20	20	21	21	21	22	22
13	2	2	3	4	5	5	6	6	7	7	8	8	9	9	9	10	10	10	10
	6	8	10	12	14	15	16	17	18	19	19	20	20	21	21	22	22	23	23
14	2	2	3	4	5	5	6	7	7	8	8	9	9	9	10	10	10	11	11
	6	8	10	12	14	15	16	17	18	19	20	20	21	22	22	23	23	23	24
15	2	3	3	4	5	6	6	7	7	8	8	9	9	10	10	11	11	11	12
	6	8	10	12	14	15	16	18	18	19	20	21	22	22	23	23	24	24	25
16	2	3	4	4	5	6	6	7	8	8	9	9	10	10	11	11	11	12	12
	6	8	10	12	14	16	17	18	19	20	21	21	22	23	23	24	25	25	25
17	2	3	4	4	5	6	7	7	8	9	9	10	10	11	11	11	12	12	13
	6	8	10	12	14	16	17	18	19	20	21	22	23	23	24	25	25	26	26
18	2	3	4	5	5	6	7	8	8	9	9	10	10	11	11	12	12	13	13
	6	8	10	12	14	16	17	18	19	20	21	22	23	24	25	25	26	26	27
19	2	3	4	5	6	6	7	8	8	9	10	10	11	11	12	12	13	13	13
	6	8	10	12	14	16	17	18	20	21	22	23	23	24	25	26	26	27	27
20	2	3	4	5	6	6	7	8	9	9	10	10	11	12	12	13	13	13	14
	6	8	10	12	14	16	17	18	20	21	22	23	24	25	25	26	27	27	28

NOTE:
1. The entries in this table are the critical G values, assuming a two-tailed test with a significance level of $\alpha = 0.05$.
2. The null hypothesis of randomness is rejected if the total number of runs G is less than or equal to the smaller entry or greater than or equal to the larger entry.

From "Tables for testing randomness of groupings in a sequence of alternatives," *The Annals of Mathematical Statistics*, Vol. 14, No. 1. Reprinted with permission of the Institute of Mathematical Statistics.

Table A-11 Percentage Points of the Studentized Range

$\alpha = 0.10$

df/r	2	3	4	5	6	7	8	9	10	11	12	13	14	15	16	17	18	19	20	30	40	60	80	100
1	8.929	13.44	16.36	18.49	20.15	21.51	22.64	23.62	24.48	25.24	25.92	26.54	27.1	27.62	28.1	28.54	28.96	29.35	29.71	32.5	34.38	36.91	38.62	39.91
2	4.129	5.733	6.773	7.538	8.139	8.633	9.049	9.409	9.725	10.01	10.26	10.49	10.7	10.89	11.07	11.24	11.39	11.54	11.68	12.73	13.44	14.4	15.04	15.54
3	3.328	4.467	5.199	5.738	6.162	6.511	6.806	7.062	7.287	7.487	7.667	7.831	7.982	8.12	8.248	8.368	8.479	8.584	8.683	9.44	9.954	10.65	11.12	11.48
4	3.015	3.976	4.586	5.035	5.388	5.679	5.926	6.139	6.327	6.494	6.645	6.783	6.909	7.025	7.132	7.233	7.326	7.414	7.497	8.135	8.569	9.156	9.558	9.861
5	2.85	3.717	4.264	4.664	4.979	5.238	5.458	5.648	5.816	5.965	6.1	6.223	6.336	6.439	6.536	6.626	6.71	6.788	6.863	7.435	7.824	8.353	8.714	8.987
6	2.748	3.558	4.065	4.435	4.726	4.966	5.168	5.344	5.499	5.637	5.762	5.875	5.979	6.075	6.164	6.247	6.325	6.398	6.466	6.996	7.358	7.848	8.184	8.438
7	2.679	3.451	3.931	4.28	4.555	4.78	4.971	5.137	5.283	5.413	5.53	5.637	5.735	5.826	5.91	5.988	6.061	6.13	6.195	6.695	7.036	7.5	7.818	8.059
8	2.63	3.374	3.834	4.169	4.431	4.646	4.829	4.987	5.126	5.25	5.362	5.464	5.558	5.644	5.724	5.799	5.869	5.935	5.997	6.475	6.801	7.245	7.55	7.78
9	2.592	3.316	3.761	4.084	4.337	4.545	4.721	4.873	5.007	5.126	5.234	5.333	5.423	5.506	5.583	5.655	5.722	5.786	5.845	6.306	6.621	7.049	7.343	7.566
10	2.563	3.27	3.704	4.018	4.264	4.465	4.636	4.783	4.913	5.029	5.134	5.229	5.315	5.397	5.472	5.542	5.607	5.668	5.726	6.173	6.478	6.894	7.18	7.396
11	2.54	3.234	3.658	3.965	4.205	4.401	4.567	4.711	4.838	4.951	5.053	5.145	5.231	5.309	5.382	5.45	5.514	5.573	5.63	6.065	6.363	6.768	7.046	7.257
12	2.521	3.204	3.621	3.921	4.156	4.349	4.511	4.652	4.776	4.886	4.986	5.076	5.16	5.236	5.308	5.374	5.436	5.495	5.55	5.975	6.267	6.663	6.936	7.142
13	2.504	3.179	3.589	3.885	4.116	4.304	4.464	4.602	4.724	4.832	4.93	5.019	5.1	5.175	5.245	5.31	5.371	5.429	5.483	5.9	6.186	6.575	6.842	7.045
14	2.491	3.158	3.563	3.854	4.081	4.267	4.424	4.56	4.679	4.786	4.882	4.969	5.05	5.124	5.192	5.256	5.316	5.372	5.426	5.836	6.116	6.499	6.762	6.961
15	2.479	3.14	3.54	3.828	4.052	4.235	4.39	4.524	4.641	4.746	4.841	4.927	5.006	5.079	5.146	5.209	5.268	5.324	5.376	5.78	6.056	6.433	6.692	6.888
16	2.469	3.124	3.52	3.804	4.026	4.207	4.36	4.492	4.608	4.712	4.805	4.89	4.968	5.04	5.106	5.169	5.227	5.282	5.333	5.731	6.004	6.376	6.631	6.825
17	2.46	3.11	3.503	3.784	4.003	4.182	4.334	4.464	4.579	4.681	4.774	4.857	4.934	5.005	5.071	5.133	5.19	5.244	5.295	5.688	5.958	6.325	6.577	6.769
18	2.452	3.098	3.487	3.766	3.984	4.161	4.31	4.44	4.553	4.654	4.746	4.829	4.905	4.975	5.04	5.101	5.158	5.211	5.262	5.65	5.917	6.28	6.529	6.718
19	2.445	3.087	3.474	3.751	3.966	4.142	4.29	4.418	4.53	4.63	4.721	4.803	4.878	4.948	5.012	5.072	5.129	5.182	5.232	5.616	5.88	6.239	6.486	6.673
20	2.439	3.077	3.462	3.736	3.95	4.124	4.271	4.398	4.51	4.609	4.699	4.78	4.855	4.923	4.987	5.047	5.103	5.155	5.205	5.586	5.847	6.202	6.447	6.633
24	2.42	3.047	3.423	3.692	3.9	4.07	4.213	4.336	4.445	4.541	4.628	4.707	4.78	4.847	4.909	4.966	5.02	5.071	5.119	5.489	5.741	6.086	6.323	6.503
30	2.4	3.017	3.386	3.648	3.851	4.016	4.155	4.275	4.381	4.474	4.559	4.635	4.706	4.77	4.83	4.886	4.939	4.988	5.034	5.391	5.636	5.969	6.198	6.372
40	2.381	2.988	3.348	3.605	3.802	3.963	4.099	4.215	4.317	4.408	4.49	4.564	4.632	4.694	4.752	4.806	4.857	4.904	4.949	5.294	5.529	5.85	6.071	6.238
60	2.363	2.959	3.312	3.562	3.755	3.911	4.042	4.155	4.254	4.342	4.421	4.493	4.558	4.619	4.675	4.727	4.775	4.821	4.864	5.196	5.422	5.73	5.941	6.101
120	2.344	2.93	3.276	3.52	3.707	3.859	3.986	4.096	4.191	4.276	4.353	4.422	4.485	4.543	4.597	4.647	4.694	4.738	4.779	5.097	5.313	5.606	5.808	5.96
∞	2.326	2.902	3.24	3.478	3.661	3.808	3.931	4.037	4.129	4.211	4.285	4.351	4.412	4.468	4.519	4.568	4.612	4.654	4.694	4.997	5.202	5.48	5.669	5.812

(continued)

Table A-11 Percentage Points of the Studentized Range

$\alpha = 0.05$

df/r	2	3	4	5	6	7	8	9	10	11	12	13	14	15	16	17	18	19	20	30	40	60	80	100
1	17.97	26.98	32.82	37.08	40.41	43.12	45.4	47.36	49.07	50.59	51.96	53.2	54.33	55.36	56.32	57.22	58.04	58.83	59.56	65.15	68.92	73.97	77.4	79.98
2	6.085	8.331	9.799	10.88	11.73	12.43	13.03	13.54	13.99	14.4	14.76	15.09	15.39	15.65	15.92	16.14	16.38	16.57	16.78	18.27	19.28	20.66	21.59	22.29
3	4.501	5.91	6.825	7.502	8.037	8.478	8.852	9.177	9.462	9.717	9.946	10.15	10.35	10.52	10.69	10.84	10.98	11.11	11.24	12.21	12.86	13.76	14.36	14.82
4	3.926	5.04	5.757	6.287	6.706	7.053	7.347	7.602	7.826	8.027	8.208	8.373	8.524	8.664	8.793	8.914	9.027	9.133	9.233	10	10.53	11.24	11.73	12.1
5	3.635	4.602	5.218	5.673	6.033	6.33	6.582	6.801	6.995	7.167	7.323	7.466	7.596	7.716	7.828	7.932	8.03	8.122	8.208	8.875	9.33	9.949	10.37	10.69
6	3.46	4.339	4.896	5.305	5.628	5.895	6.122	6.319	6.493	6.649	6.789	6.917	7.034	7.143	7.244	7.338	7.426	7.508	7.586	8.189	8.601	9.162	9.547	9.839
7	3.344	4.165	4.681	5.06	5.359	5.606	5.815	5.997	6.158	6.302	6.431	6.55	6.658	6.759	6.852	6.939	7.02	7.097	7.169	7.727	8.11	8.631	8.989	9.26
8	3.261	4.041	4.529	4.886	5.167	5.399	5.596	5.767	5.918	6.053	6.175	6.287	6.389	6.483	6.571	6.653	6.729	6.801	6.869	7.395	7.756	8.248	8.586	8.843
9	3.199	3.948	4.415	4.755	5.024	5.244	5.432	5.595	5.738	5.867	5.983	6.089	6.186	6.276	6.359	6.437	6.51	6.579	6.643	7.144	7.488	7.958	8.281	8.526
10	3.151	3.877	4.327	4.654	4.912	5.124	5.304	5.46	5.598	5.722	5.833	5.935	6.028	6.114	6.194	6.269	6.339	6.405	6.467	6.948	7.278	7.73	8.041	8.276
11	3.113	3.82	4.256	4.574	4.823	5.028	5.202	5.353	5.486	5.605	5.713	5.811	5.901	5.984	6.062	6.134	6.202	6.265	6.325	6.79	7.109	7.546	7.847	8.075
12	3.081	3.773	4.199	4.508	4.75	4.95	5.119	5.265	5.395	5.51	5.615	5.71	5.797	5.878	5.953	6.023	6.089	6.151	6.209	6.66	6.97	7.394	7.687	7.908
13	3.055	3.734	4.151	4.453	4.69	4.884	5.049	5.192	5.318	5.431	5.533	5.625	5.711	5.789	5.862	5.931	5.995	6.055	6.112	6.551	6.853	7.267	7.552	7.769
14	3.033	3.701	4.111	4.407	4.639	4.829	4.99	5.13	5.253	5.364	5.463	5.554	5.637	5.714	5.785	5.852	5.915	5.973	6.029	6.459	6.754	7.159	7.437	7.649
15	3.014	3.673	4.076	4.367	4.595	4.782	4.94	5.077	5.198	5.306	5.403	5.492	5.574	5.649	5.719	5.785	5.846	5.904	5.958	6.379	6.669	7.065	7.338	7.546
16	2.998	3.649	4.046	4.333	4.557	4.741	4.896	5.031	5.15	5.256	5.352	5.439	5.519	5.593	5.662	5.726	5.785	5.843	5.896	6.31	6.594	6.983	7.252	7.456
17	2.984	3.628	4.02	4.303	4.524	4.705	4.858	4.991	5.108	5.212	5.306	5.392	5.471	5.544	5.612	5.675	5.734	5.79	5.842	6.249	6.529	6.912	7.176	7.377
18	2.971	3.609	3.997	4.276	4.494	4.673	4.824	4.955	5.071	5.173	5.266	5.351	5.429	5.501	5.567	5.629	5.688	5.743	5.794	6.195	6.471	6.848	7.108	7.307
19	2.96	3.593	3.977	4.253	4.468	4.645	4.794	4.924	5.037	5.139	5.231	5.314	5.391	5.462	5.528	5.589	5.647	5.701	5.752	6.147	6.419	6.791	7.048	7.244
20	2.95	3.578	3.958	4.232	4.445	4.62	4.768	4.895	5.008	5.108	5.199	5.282	5.357	5.427	5.492	5.553	5.61	5.663	5.714	6.104	6.372	6.74	6.994	7.187
24	2.919	3.532	3.901	4.166	4.373	4.541	4.684	4.807	4.915	5.012	5.099	5.179	5.251	5.319	5.381	5.439	5.494	5.545	5.594	5.968	6.226	6.578	6.822	7.007
30	2.888	3.486	3.845	4.102	4.301	4.464	4.601	4.72	4.824	4.917	5.001	5.077	5.147	5.211	5.271	5.327	5.379	5.429	5.475	5.833	6.08	6.417	6.65	6.827
40	2.858	3.442	3.791	4.039	4.232	4.388	4.521	4.634	4.735	4.824	4.904	4.977	5.044	5.106	5.163	5.216	5.266	5.313	5.358	5.7	5.934	6.255	6.477	6.645
60	2.829	3.399	3.737	3.977	4.163	4.314	4.441	4.55	4.646	4.732	4.808	4.878	4.942	5.001	5.056	5.107	5.154	5.199	5.241	5.566	5.789	6.093	6.302	6.462
120	2.8	3.356	3.685	3.917	4.096	4.241	4.363	4.468	4.56	4.641	4.714	4.781	4.842	4.898	4.95	4.998	5.043	5.086	5.126	5.434	5.644	5.929	6.126	6.275
∞	2.772	3.314	3.633	3.858	4.03	4.17	4.286	4.387	4.474	4.552	4.622	4.685	4.743	4.796	4.845	4.891	4.934	4.974	5.012	5.301	5.498	5.764	5.947	6.085

(continued)

Table A-11 Percentage Points of the Studentized Range

$\alpha = 0.01$

df/r	2	3	4	5	6	7	8	9	10	11	12	13	14	15	16	17	18	19	20	30	40	60	80	100
1	90.02	135	164.3	185.6	202.2	215.8	227.2	237	245.6	253.2	260	266.2	271.8	277	281.8	286.3	290.4	294.3	298	326	344.8	370.1	387.3	400.1
2	14.04	19.02	22.29	24.72	26.63	28.2	29.53	30.68	31.69	32.59	33.4	34.13	34.81	35.43	36	36.53	37.03	37.5	37.95	41.32	43.61	46.7	48.8	50.38
3	8.26	10.62	12.17	13.32	14.24	15	15.65	16.21	16.69	17.13	17.53	17.89	18.22	18.52	18.81	19.07	19.32	19.55	19.77	21.44	22.59	24.13	25.19	25.99
4	6.511	8.12	9.173	9.958	10.58	11.1	11.54	11.92	12.26	12.57	12.84	13.09	13.32	13.53	13.72	13.91	14.08	14.24	14.39	15.57	16.38	17.46	18.2	18.77
5	5.702	6.976	7.804	8.421	8.913	9.321	9.669	9.971	10.24	10.48	10.7	10.89	11.08	11.24	11.4	11.55	11.68	11.81	11.93	12.87	13.51	14.39	14.99	15.45
6	5.243	6.331	7.033	7.556	7.972	8.318	8.612	8.869	9.097	9.3	9.485	9.653	9.808	9.951	10.08	10.21	10.32	10.43	10.54	11.34	11.89	12.65	13.17	13.55
7	4.949	5.919	6.542	7.005	7.373	7.678	7.939	8.166	8.367	8.548	8.711	8.86	8.997	9.124	9.242	9.353	9.456	9.553	9.645	10.36	10.85	11.52	11.98	12.34
8	4.745	5.635	6.204	6.625	6.959	7.237	7.474	7.68	7.863	8.027	8.176	8.311	8.436	8.552	8.659	8.76	8.854	8.943	9.027	9.677	10.13	10.74	11.17	11.49
9	4.596	5.428	5.957	6.347	6.657	6.915	7.134	7.325	7.494	7.646	7.784	7.91	8.025	8.132	8.232	8.325	8.412	8.495	8.573	9.177	9.594	10.17	10.56	10.86
10	4.482	5.27	5.769	6.136	6.428	6.669	6.875	7.054	7.213	7.356	7.485	7.603	7.712	7.812	7.906	7.993	8.075	8.153	8.226	8.794	9.186	9.726	10.1	10.38
11	4.392	5.146	5.621	5.97	6.247	6.476	6.671	6.841	6.992	7.127	7.25	7.362	7.464	7.56	7.648	7.731	7.809	7.883	7.952	8.491	8.864	9.377	9.732	10
12	4.32	5.046	5.502	5.836	6.101	6.32	6.507	6.67	6.814	6.943	7.06	7.166	7.265	7.356	7.441	7.52	7.594	7.664	7.73	8.246	8.602	9.093	9.433	9.693
13	4.26	4.964	5.404	5.726	5.981	6.192	6.372	6.528	6.666	6.791	6.903	7.006	7.1	7.188	7.269	7.345	7.417	7.484	7.548	8.043	8.386	8.859	9.186	9.436
14	4.21	4.895	5.322	5.634	5.881	6.085	6.258	6.409	6.543	6.663	6.772	6.871	6.962	7.047	7.125	7.199	7.268	7.333	7.394	7.873	8.204	8.661	8.978	9.219
15	4.167	4.836	5.252	5.556	5.796	5.994	6.162	6.309	6.438	6.555	6.66	6.756	6.845	6.927	7.003	7.074	7.141	7.204	7.264	7.727	8.049	8.492	8.8	9.034
16	4.13	4.786	5.192	5.489	5.722	5.915	6.079	6.222	6.348	6.461	6.564	6.658	6.744	6.823	6.897	6.967	7.032	7.093	7.151	7.602	7.915	8.346	8.646	8.874
17	4.099	4.742	5.14	5.43	5.659	5.847	6.007	6.147	6.27	6.38	6.48	6.572	6.656	6.733	6.806	6.873	6.937	6.997	7.053	7.493	7.798	8.219	8.511	8.734
18	4.071	4.703	5.094	5.379	5.603	5.787	5.944	6.081	6.201	6.309	6.407	6.496	6.579	6.655	6.725	6.791	6.854	6.912	6.967	7.397	7.696	8.107	8.393	8.611
19	4.045	4.669	5.054	5.334	5.553	5.735	5.889	6.022	6.141	6.246	6.342	6.43	6.51	6.585	6.654	6.719	6.78	6.837	6.891	7.312	7.605	8.008	8.288	8.501
20	4.024	4.639	5.018	5.293	5.51	5.688	5.839	5.97	6.086	6.19	6.285	6.37	6.449	6.523	6.591	6.654	6.714	6.77	6.823	7.237	7.523	7.919	8.194	8.404
24	3.955	4.546	4.907	5.168	5.373	5.542	5.685	5.809	5.919	6.017	6.105	6.186	6.261	6.33	6.394	6.453	6.51	6.562	6.612	7.001	7.27	7.641	7.9	8.097
30	3.889	4.455	4.799	5.048	5.242	5.401	5.536	5.653	5.756	5.848	5.932	6.008	6.078	6.142	6.202	6.258	6.311	6.361	6.407	6.771	7.023	7.37	7.611	7.796
40	3.825	4.367	4.695	4.931	5.114	5.265	5.392	5.502	5.599	5.685	5.764	5.835	5.9	5.961	6.017	6.069	6.118	6.165	6.208	6.547	6.781	7.104	7.328	7.499
60	3.762	4.282	4.594	4.818	4.991	5.133	5.253	5.356	5.447	5.528	5.601	5.667	5.728	5.784	5.837	5.886	5.931	5.974	6.015	6.329	6.546	6.843	7.049	7.207
120	3.702	4.2	4.497	4.708	4.872	5.005	5.118	5.214	5.299	5.375	5.443	5.505	5.561	5.614	5.662	5.708	5.75	5.79	5.827	6.117	6.316	6.588	6.776	6.919
∞	3.643	4.12	4.403	4.603	4.757	4.882	4.987	5.078	5.157	5.227	5.29	5.348	5.4	5.448	5.493	5.535	5.574	5.611	5.645	5.911	6.092	6.338	6.507	6.636

From Gleason, J.R., 1998, *Quantiles of the Studentized Range in Distribution*, Stata Technical Bulletin, 46:6–10. Reproduced with permission of StataCorp LP, College Station, Texas.

Table A-12 Table of Critical Values for Lilliefors Normality Test

			$p =$		
n	0.80	0.85	0.90	0.95	0.99
4	0.300	0.319	0.352	0.381	0.417
5	0.285	0.299	0.315	0.337	0.405
6	0.265	0.277	0.294	0.319	0.364
7	0.247	0.258	0.276	0.300	0.348
8	0.233	0.244	0.261	0.285	0.331
9	0.223	0.233	0.249	0.271	0.311
10	0.215	0.224	0.239	0.258	0.294
11	0.206	0.217	0.230	0.249	0.284
12	0.199	0.212	0.223	0.242	0.275
13	0.190	0.202	0.214	0.234	0.268
14	0.183	0.194	0.207	0.227	0.261
15	0.177	0.187	0.201	0.220	0.257
16	0.173	0.182	0.195	0.213	0.250
17	0.169	0.177	0.189	0.206	0.245
18	0.166	0.173	0.184	0.200	0.239
19	0.163	0.169	0.179	0.195	0.235
20	0.160	0.166	0.174	0.190	0.231
25	0.142	0.147	0.158	0.173	0.200
30	0.131	0.136	0.144	0.161	0.187
Over 30	$0.736/\sqrt{n}$	$0.768/\sqrt{n}$	$0.805/\sqrt{n}$	$0.886/\sqrt{n}$	$1.031/\sqrt{n}$

Reprinted with permission from *The Journal of the American Statistical Association*. Copyright 1967 by the American Statistical Association. All Rights Reserved.

Table A-13 Wilcoxon Rank Sum Table

$\alpha = 0.025$ one-sided; $\alpha = 0.05$ two-sided

n2 \ n1	3 T_L	3 T_U	4 T_L	4 T_U	5 T_L	5 T_U	6 T_L	6 T_U	7 T_L	7 T_U	8 T_L	8 T_U	9 T_L	9 T_U	10 T_L	10 T_U
3	5	16	6	18	6	21	7	23	7	26	8	28	8	31	9	33
4	6	18	11	25	12	28	12	32	13	35	14	38	15	41	16	44
5	6	21	12	28	18	37	19	41	20	45	21	49	22	53	24	56
6	7	23	12	32	19	41	26	52	28	56	29	61	31	65	32	70
7	7	26	13	35	20	45	28	56	37	68	39	73	41	78	43	83
8	8	28	14	38	21	49	29	61	39	73	49	87	51	93	54	98
9	8	31	15	41	22	53	31	65	41	78	51	93	63	108	66	114
10	9	33	16	44	24	56	32	70	43	83	54	98	66	114	79	131

$\alpha = 0.05$ one-sided; $\alpha = 0.10$ two-sided

n2 \ n1	3 T_L	3 T_U	4 T_L	4 T_U	5 T_L	5 T_U	6 T_L	6 T_U	7 T_L	7 T_U	8 T_L	8 T_U	9 T_L	9 T_U	10 T_L	10 T_U
3	6	15	7	17	7	20	8	22	9	24	9	27	10	29	11	31
4	7	17	12	24	13	27	14	30	15	33	16	36	17	39	18	42
5	7	20	13	27	19	36	20	40	22	43	24	46	25	50	26	54
6	8	22	14	30	20	40	28	50	30	54	32	58	33	63	35	67
7	9	24	15	33	22	43	30	54	39	66	41	71	43	76	46	80
8	9	27	16	36	24	46	32	58	41	71	52	84	54	90	57	95
9	10	29	17	39	25	50	33	63	43	76	54	90	66	105	69	111
10	11	31	18	42	26	54	35	67	46	80	57	95	69	111	83	127

From *Some Rapid Approximate Statistical Procedures,* Copyright © 1949, 1964 Lederle Laboratories Division of American Cyanamid Company. Reprinted with permission from American Cyanamid Company.

APPENDIX B: Data Sets

Data Set 1: Weights of Household Garbage for One Week (in pounds)

HOUSEHOLD = household number
HHSIZE = household size
METAL = weight of discarded metal
PAPER = weight of discarded paper goods
PLASTIC = weight of discarded plastic goods
GLASS = weight of discarded glass products
FOOD = weight of discarded food items
YARD = weight of discarded yard waste
TEXTILE = weight of discarded textile goods
OTHER = weight of discarded goods not included in the above categories
TOTAL = total weight of discarded materials

HOUSEHOLD	HHSIZE	METAL	PAPER	PLASTIC	GLASS	FOOD	YARD	TEXTILE	OTHER	TOTAL
1	2	1.09	2.41	0.27	0.86	1.04	0.38	0.05	4.66	10.76
2	3	1.04	7.57	1.41	3.46	3.68	0.00	0.46	2.34	19.96
3	3	2.57	9.55	2.19	4.52	4.43	0.24	0.50	3.60	27.60
4	6	3.02	8.82	2.83	4.92	2.98	0.63	2.26	12.65	38.11
5	4	1.50	8.72	2.19	6.31	6.30	0.15	0.55	2.18	27.90
6	2	2.10	6.96	1.81	2.49	1.46	4.58	0.36	2.14	21.90
7	1	1.93	6.83	0.85	0.51	8.82	0.07	0.60	2.22	21.83
8	5	3.57	11.42	3.05	5.81	9.62	4.76	0.21	10.83	49.27
9	6	2.32	16.08	3.42	1.96	4.41	0.13	0.81	4.14	33.27
10	4	1.89	6.38	2.10	17.67	2.73	3.86	0.66	0.25	35.54
11	4	3.26	13.05	2.93	3.21	9.31	0.70	0.37	11.61	44.44
12	7	3.99	11.36	2.44	4.94	3.59	13.45	4.25	1.15	45.17
13	3	2.04	15.09	2.17	3.10	5.36	0.74	0.42	4.15	33.07
14	5	0.99	2.80	1.41	1.39	1.47	0.82	0.44	1.03	10.35
15	6	2.96	6.44	2.00	5.21	7.06	6.14	0.20	14.43	44.44
16	2	1.50	5.86	0.93	2.03	2.52	1.37	0.27	9.65	24.13
17	4	2.43	11.08	2.97	1.74	1.75	14.70	0.39	2.54	37.60
18	4	2.97	12.43	2.04	3.99	5.64	0.22	2.47	9.20	38.96
19	3	1.42	6.05	0.65	6.26	1.93	0.00	0.86	0.00	17.17
20	3	3.60	13.61	2.13	3.52	6.46	0.00	0.96	1.32	31.60
21	2	4.48	6.98	0.63	2.01	6.72	2.00	0.11	0.18	23.11
22	2	1.36	14.33	1.53	2.21	5.76	0.58	0.17	1.62	27.56
23	4	2.11	13.31	4.69	0.25	9.72	0.02	0.46	0.40	30.96
24	1	0.41	3.27	0.15	0.09	0.16	0.00	0.00	0.00	4.08
25	4	2.02	6.67	1.45	6.85	5.52	0.00	0.68	0.03	23.22
26	6	3.27	17.65	2.68	2.33	11.92	0.83	0.28	4.03	42.99
27	11	4.95	12.73	3.53	5.45	4.68	0.00	0.67	19.89	51.90

(continued)

HOUSEHOLD	HHSIZE	METAL	PAPER	PLASTIC	GLASS	FOOD	YARD	TEXTILE	OTHER	TOTAL
28	3	1.00	9.83	1.49	2.04	4.76	0.42	0.54	0.12	20.20
29	4	1.55	16.39	2.31	4.98	7.85	2.04	0.20	1.48	36.80
30	3	1.41	6.33	0.92	3.54	2.90	3.85	0.03	0.04	19.02
31	2	1.05	9.19	0.89	1.06	2.87	0.33	0.01	0.03	15.43
32	2	1.31	9.41	0.80	2.70	5.09	0.64	0.05	0.71	20.71
33	2	2.50	9.45	0.72	1.14	3.17	0.00	0.02	0.01	17.01
34	4	2.35	12.32	2.66	12.24	2.40	7.87	4.73	0.78	45.35
35	6	3.69	20.12	4.37	5.67	13.20	0.00	1.15	1.17	49.37
36	2	3.61	7.72	0.92	2.43	2.07	0.68	0.63	0.00	18.06
37	2	1.49	6.16	1.40	4.02	4.00	0.30	0.04	0.00	17.41
38	2	1.36	7.98	1.45	6.45	4.27	0.02	0.12	2.02	23.67
39	2	1.73	9.64	1.68	1.89	1.87	0.01	1.73	0.58	19.13
40	2	0.94	8.08	1.53	1.78	8.13	0.36	0.12	0.05	20.99
41	3	1.33	10.99	1.44	2.93	3.51	0.00	0.39	0.59	21.18
42	3	2.62	13.11	1.44	1.82	4.21	4.73	0.64	0.49	29.06
43	2	1.25	3.26	1.36	2.89	3.34	2.69	0.00	0.16	14.95
44	2	0.26	1.65	0.38	0.99	0.77	0.34	0.04	0.00	4.43
45	3	4.41	10.00	1.74	1.93	1.14	0.92	0.08	4.60	24.82
46	6	3.22	8.96	2.35	3.61	1.45	0.00	0.09	1.12	20.80
47	4	1.86	9.46	2.30	2.53	6.54	0.00	0.65	2.45	25.79
48	4	1.76	5.88	1.14	3.76	0.92	1.12	0.00	0.04	14.62
49	3	2.83	8.26	2.88	1.32	5.14	5.60	0.35	2.03	28.41
50	3	2.74	12.45	2.13	2.64	4.59	1.07	0.41	1.14	27.17
51	10	4.63	10.58	5.28	12.33	2.94	0.12	2.94	15.65	54.47
52	3	1.70	5.87	1.48	1.79	1.42	0.00	0.27	0.59	13.12
53	6	3.29	8.78	3.36	3.99	10.44	0.90	1.71	13.30	45.77
54	5	1.22	11.03	2.83	4.44	3.00	4.30	1.95	6.02	34.79
55	4	3.20	12.29	2.87	9.25	5.91	1.32	1.87	0.55	37.26
56	7	3.09	20.58	2.96	4.02	16.81	0.47	1.52	2.13	51.58
57	5	2.58	12.56	1.61	1.38	5.01	0.00	0.21	1.46	24.81
58	4	1.67	9.92	1.58	1.59	9.96	0.13	0.20	1.13	26.18
59	2	0.85	3.45	1.15	0.85	3.89	0.00	0.02	1.04	11.25
60	4	1.52	9.09	1.28	8.87	4.83	0.00	0.95	1.61	28.15
61	2	1.37	3.69	0.58	3.64	1.78	0.08	0.00	0.00	11.14
62	2	1.32	2.61	0.74	3.03	3.37	0.17	0.00	0.46	11.70

Data provided by Masakazu Tani, the Garbage Project, University of Arizona.

Data Set 2: Body Temperatures of Healthy Adults in Varied Postures (in degrees Celsius)

SIT1 = subject's initial temperature while sitting
SUPINE1 = subject's temperature after lying down
SIT2 = subject's temperature after sitting again

STAND1 = subject's initial temperature while standing
SUPINE = subject's temperature after lying down
STAND2 = subject's temperature after standing again

INITIALLY SITTING

SUBJECT	SIT1	SUPINE1	SIT2
1	37.61	37.19	37.51
2	37.07	36.91	37.15
3	37.20	36.70	37.24
4	37.23	36.80	37.34
5	37.38	36.96	37.64
6	37.28	37.11	37.31
7	37.12	36.82	37.53
8	37.55	37.36	37.41
9	37.04	36.92	37.60
10	37.78	37.22	37.85
11	37.59	37.25	37.63

INITIALLY STANDING

SUBJECT	STAND1	SUPINE2	STAND2
1	37.63	37.22	37.80
2	37.17	36.95	37.35
3	36.93	36.47	37.06
4	37.35	36.85	37.63
5	37.73	36.91	37.74
6	37.39	36.99	37.70
7	37.57	36.73	37.99
8	37.89	37.32	37.89
9	37.50	36.96	37.79
10	37.98	37.30	37.96
11	37.36	36.98	37.60

Data provided by Michel B. Ducharme, Defence and Civil Institute of Environmental Medicine, Toronto.

Data Set 3: Bears (wild bears anesthetized)

BEARAGE = age in months
MONTH = month of measurement (1 = January, 2 = February, etc.)
BEARSEX = male (1) or female (2)
HEADLEN = length of head in inches
HEADWTH = width of head in inches
NECK = distance around neck in inches
LENGTH = length of body in inches
CHEST = distance around chest in inches
WEIGHT = weight in pounds

BEARAGE	MONTH	BEARSEX	HEADLEN	HEADWTH	NECK	LENGTH	CHEST	WEIGHT
19	7	1	11.0	5.5	16.0	53.0	26.0	80
55	7	1	16.5	9.0	28.0	67.5	45.0	344
81	9	1	15.5	8.0	31.0	72.0	54.0	416
115	7	1	17.0	10.0	31.5	72.0	49.0	348
104	8	2	15.5	6.5	22.0	62.0	35.0	166
100	4	2	13.0	7.0	21.0	70.0	41.0	220
56	7	1	15.0	7.5	26.5	73.5	41.0	262
51	4	1	13.5	8.0	27.0	68.5	49.0	360
57	9	2	13.5	7.0	20.0	64.0	38.0	204
53	5	2	12.5	6.0	18.0	58.0	31.0	144
68	8	1	16.0	9.0	29.0	73.0	44.0	332
8	8	1	9.0	4.5	13.0	37.0	19.0	34
44	8	2	12.5	4.5	10.5	63.0	32.0	140
32	8	1	14.0	5.0	21.5	67.0	37.0	180
20	8	2	11.5	5.0	17.5	52.0	29.0	105
32	8	1	13.0	8.0	21.5	59.0	33.0	166
45	9	1	13.5	7.0	24.0	64.0	39.0	204
9	9	2	9.0	4.5	12.0	36.0	19.0	26
21	9	1	13.0	6.0	19.0	59.0	30.0	120
177	9	1	16.0	9.5	30.0	72.0	48.0	436
57	9	2	12.5	5.0	19.0	57.5	32.0	125
81	9	2	13.0	5.0	20.0	61.0	33.0	132
21	9	1	13.0	5.0	17.0	54.0	28.0	90
9	9	1	10.0	4.0	13.0	40.0	23.0	40
45	9	1	16.0	6.0	24.0	63.0	42.0	220
9	9	1	10.0	4.0	13.5	43.0	23.0	46
33	9	1	13.5	6.0	22.0	66.5	34.0	154
57	9	2	13.0	5.5	17.5	60.5	31.0	116
45	9	2	13.0	6.5	21.0	60.0	34.5	182
21	9	1	14.5	5.5	20.0	61.0	34.0	150
10	10	1	9.5	4.5	16.0	40.0	26.0	65
82	10	2	13.5	6.5	28.0	64.0	48.0	356
70	10	2	14.5	6.5	26.0	65.0	48.0	316

(continued)

BEARAGE	MONTH	BEARSEX	HEADLEN	HEADWTH	NECK	LENGTH	CHEST	WEIGHT
10	10	1	11.0	5.0	17.0	49.0	29.0	94
10	10	1	11.5	5.0	17.0	47.0	29.5	86
34	10	1	13.0	7.0	21.0	59.0	35.0	150
34	10	1	16.5	6.5	27.0	72.0	44.5	270
34	10	1	14.0	5.5	24.0	65.0	39.0	202
58	10	2	13.5	6.5	21.5	63.0	40.0	202
58	10	1	15.5	7.0	28.0	70.5	50.0	365
11	11	1	11.5	6.0	16.5	48.0	31.0	79
23	11	1	12.0	6.5	19.0	50.0	38.0	148
70	10	1	15.5	7.0	28.0	76.5	55.0	446
11	11	2	9.0	5.0	15.0	46.0	27.0	62
83	11	2	14.5	7.0	23.0	61.5	44.0	236
35	11	1	13.5	8.5	23.0	63.5	44.0	212
16	4	1	10.0	4.0	15.5	48.0	26.0	60
16	4	1	10.0	5.0	15.0	41.0	26.0	64
17	5	1	11.5	5.0	17.0	53.0	30.5	114
17	5	2	11.5	5.0	15.0	52.5	28.0	76
17	5	2	11.0	4.5	13.0	46.0	23.0	48
8	8	2	10.0	4.5	10.0	43.5	24.0	29
83	11	1	15.5	8.0	30.5	75.0	54.0	514
18	6	1	12.5	8.5	18.0	57.3	32.8	140

Data provided by Gary Alt and Minitab, Inc.

Data Set 4: Temperature and Precipitation for Canadian Cities

AVTEMP = average annual temperature in degrees Celsius
AVPRECIP = average annual precipitation in millimetres
AVSNOW = average annual snowfall in centimetres

CITY	AVTEMP	AVPRECIP	AVSNOW
St. John's, NF	4.8	1514	359
Charlottetown, PE	5.4	1169	331
Halifax, NS	6.1	1491	271
Sydney, NS	5.7	1400	318
Yarmouth, NS	6.9	1282	208
Chatham, NB	4.8	1097	333
Fredericton, NB	5.4	1109	290
Saint John, NB	5.0	1444	293
Arvida, PQ	3.0	908	271
Montreal, PQ	6.2	946	235
Pointe-au-Père, PQ	3.1	851	293
Quebec City, PQ	4.1	1174	343
Schefferville, PQ	−4.8	769	387
Sherbrooke, PQ	4.0	1075	323
Kapuskasing, ON	0.5	858	320
London, ON	7.3	909	209
Ottawa, ON	5.7	879	227
Thunder Bay, ON	2.3	712	213
Toronto, ON	7.3	762	131
Churchill, MB	−7.2	402	196
The Pas, MB	−0.6	454	170
Winnipeg, MB	2.2	526	126
Prince Albert, SK	0.1	398	122
Regina, SK	2.2	384	116
Beaverlodge, AB	1.6	467	195
Calgary, AB	3.4	424	153
Edmonton, AB	1.6	467	138
Kamloops, BC	8.3	266	92
Prince George, BC	3.3	628	242
Prince Rupert, BC	6.7	2523	152
Vancouver, BC	9.8	1113	60
Victoria, BC	10.0	647	32

Based on data from Environment Canada.

Data Set 5: Sitting Days and Durations of Sessions of Parliament

YEARS = approximate duration of Parliament
DAYS = number of days in sesssion
HOUSE = number of House sittings
SENATE = number of Senate sittings

PARLIAMENT	YEARS	DAYS	HOUSE	SENATE
1	5.0	479	283	255
2	1.0	177	70	58
3	5.0	362	262	240
4	4.0	379	254	231
5	4.0	472	324	257
6	4.0	375	256	207
7	5.0	695	458	363
8	5.0	593	402	302
9	4.0	578	394	261
10	4.0	712	460	286
11	2.0	550	303	187
12	5.0	934	568	357
13	4.0	515	355	241
14	4.0	550	366	217
15	1.0	177	111	42
16	4.0	493	292	158
17	5.0	837	507	283
18	4.0	533	390	201
19	5.0	1750	565	241
20	4.0	748	487	283
21	4.5	1154	527	339
22	4.0	787	507	285
23	1.0	111	78	45
24	4.5	919	605	352
25	1.0	132	72	43
26	3.0	788	418	207
27	2.0	828	405	179
28	5.0	1451	687	366
29	1.5	491	256	141
30	4.5	1640	767	478
31	0.7	67	49	31
32	4.0	1541	707	383
33	4.0	1423	697	353
34	5.0	1699	590	316
35	3.0	1189	447	229

Based on data from *Duration of Sessions of Parliament*, Library of Parliament.

Data Set 6: The *Financial Post* Top 100 Companies for 1997

RANK = rating by revenue for 1997
CODE = industry code (see table after data)
REVENUES = 1997 revenues in $1,000s
EMPLOY = number of employees
CAPITAL = return on invested capital (as a percent)
EQUITY = return on equity (as a percent)
FOREIGN = registered foreign ownership (as a percent) if over 10%

RANK	COMPANY	CODE	REVENUES	EMPLOY	CAPITAL	EQUITY	FOREIGN
1	General Motors of Canada	s	34,249,489	29,000	n.a.	n.a.	100
2	BCE	aa	33,191,000	122,000	15.61	14.38	11
3	Ford Motor Co. of Canada	s	27,911,591	24,402	26.49	30.13	100
4	The Seagram Co.	f	17,160,728	30,000	6.47	5.37	
5	Chrysler Canada	s	16,688,000	16,000	6.33	6.09	100
6	TransCanada PipeLines	m	14,242,800	3,042	11.42	11.62	15
7	George Weston Ltd.	f	13,921,000	83,000	17.64	14.48	
8	The Thomson Corp.	w	12,137,404	50,000	8.98	11.82	
9	Onex Corp.	f	11,212,384	43,000	8.39	6.38	
10	Alcan Aluminium	n	10,768,034	33,000	10.10	9.48	39
11	Imasco	f	10,008,000	27,000	2.90	21.97	48
12	Canadian Pacific	f	9,560,000	33,600	15.76	17.56	50
13	Imperial Oil	t	9,512,000	7,096	20.55	18.93	82
14	Power Corp. of Canada	f	8,615,000	15,505	n.a.	13.24	
15	Bombardier	ht	8,508,900	47,000	17.55	17.09	n.a.
16	Magna International	s	7,691,800	36,000	24.61	23.50	
17	IBM Canada	it	7,400,000	15,383	n.a.	n.a.	100
18	Westcoast Energy	m	7,312,000	5,932	9.88	10.64	26
19	Quebecor	w	7,013,346	37,000	12.86	12.38	
20	The Oshawa Group	l	6,813,100	16,050	8.68	5.69	
21	Hudson's Bay Co.	o	6,446,652	57,000	n.a.	n.a.	
22	Noranda	r	6,407,000	21,667	4.66	5.22	
23	Amoco Canada Petroleum Co.	t	6,176,701	2,200	n.a.	n.a.	100
24	Petro-Canada	t	6,017,000	5,749	10.91	8.04	17
25	Provigo	l	5,956,200	22,000	25.70	30.49	
26	EdperBrascan Corp.	f	5,886,000	50,000	10.89	18.76	
27	Air Canada	cc	5,572,000	21,215	14.66	35.29	16
28	Shell Canada	t	5,445,000	3,593	18.08	14.84	80
29	NOVA Corp.	m	4,840,000	5,500	10.30	8.20	15
30	Canada Safeway	l	4,719,500	31,000	n.a.	n.a.	100
31	Sears Canada	o	4,583,479	38,545	16.01	11.70	55
32	MacMillan Bloedel	v	4,521,000	10,592	n.a.	n.a.	11
33	Canadian National Railway	cc	4,352,000	24,081	15.71	13.90	64

(continued)

RANK	COMPANY	CODE	REVENUES	EMPLOY	CAPITAL	EQUITY	FOREIGN
34	Saskatchewan Wheat Pool	a	4,229,325	3,873	13.47	9.71	
35	Canadian Ultramar Co.	t	4,175,400	1,176	n.a.	n.a.	100
36	McCain Food	k	4,150,639	16,000	n.a.	n.a.	
37	Laidlaw	cc	4,147,513	79,500	11.84	24.64	15
38	Canadian Tire Corp.	o	4,057,197	4,950	15.02	11.44	
39	Jim Pattison Group	f	4,000,000	18,000	n.a.	n.a.	
40	Honda Canada	s	3,794,562	2,300	n.a.	n.a.	100
41	Abitibi-Consolidated	v	3,747,000	14,000	n.a.	n.a.	25
42	Maple Leaf Foods	k	3,678,419	11,200	13.44	14.74	
43	Moore Corp.	y	3,642,902	20,084	6.84	3.99	40
44	Mitsui & Co. (Canada)	q	3,581,714	100	10.26	8.10	100
45	Métro-Richelieu	l	3,432,300	6,500	30.00	24.67	10
46	Mobil Oil Canada	t	3,307,485	676	8.39	58.24	100
47	Inco	r	3,277,348	14,278	2.66	1.24	55
48	Potash Corp. of Saskatchewan	e	3,220,481	5,751	13.93	16.09	85
49	Empire Co.	l	3,149,773	17,000	11.87	11.91	
50	Stelco	z	3,149,000	11,706	12.94	10.14	
51	Hollinger	w	3,116,869	18,729	17.15	n.a.	
52	Cargill	a	3,088,000	3,000	n.a.	n.a.	100
53	Canadian Airlines Corp.	cc	3,075,500	15,706	7.75	n.a.	21
54	Dofasco	z	3,070,400	7,070	11.88	10.75	
55	Anglo-Canadian Telephone Co.	aa	3,058,100	n.a.	16.86	11.31	100
56	Charlwood Pacific Group	y	2,886,000	6,000	n.a.	n.a.	
57	Rogers Communications	d	2,695,322	10,300	n.a.	n.a.	
58	Agrium	e	2,683,216	4,400	28.53	27.47	30
59	Price Costco Canada	o	2,653,100	6,065	n.a.	n.a.	100
60	Toyota Canada	dd	2,639,067	462	n.a.	n.a.	100
61	General Electric Canada	n	2,537,463	6,469	19.85	24.27	100
62	IPL Energy	m	2,520,000	5,000	12.88	14.04	
63	Great Atlantic & Pacific Co. of Canada	l	2,457,966	19,700	n.a.	n.a.	100
64	Philip Services Corp.	y	2,424,338	12,500	n.a.	n.a.	80
65	Federated Co-operatives	dd	2,411,466	2,352	34.65	29.94	
66	SHL Systemhouse Co.	it	2,400,000	9,400	n.a.	n.a.	100
67	Westburne	dd	2,334,385	5,071	23.86	16.23	16
68	Finning International	dd	2,327,064	5,494	15.13	16.19	
69	Medis Health & Pharmaceutical Services	dd	2,312,226	1,800	n.a.	n.a.	100
70	United Dominion Industries	n	2,291,069	11,000	15.78	15.99	35
71	Dow Chemical Canada	e	2,271,000	n.a.	n.a.	n.a.	100
72	Cascades	v	2,208,257	12,000	7.42	9.61	
73	Kraft Canada	k	2,169,000	3,900	n.a.	n.a.	100
74	Alberta Wheat Pool	a	2,051,654	1,433	6.97	5.61	
75	ATCO	f	2,045,100	4,500	13.36	13.75	
76	James Richardson & Sons	f	2,043,395	1,335	10.50	7.72	

(continued)

RANK	COMPANY	CODE	REVENUES	EMPLOY	CAPITAL	EQUITY	FOREIGN
77	TELUS Corp.	aa	2,019,823	8,972	15.99	13.06	14
78	DuPont Canada	n	1,997,810	3,372	28.45	20.57	77
79	Avenor	v	1,991,700	6,760	n.a.	n.a.	
80	Chevron Canada Resources	t	1,972,871	760	6.87	n.a.	100
81	Domtar	v	1,938,000	4,700	4.04	1.91	
82	Parmalat Canada	k	1,893,000	4,600	n.a.	n.a.	100
83	West Fraser Timber Co.	v	1,869,761	6,500	9.72	8.20	49
84	United Grain Growers	a	1,866,373	1,544	10.00	6.14	47
85	Coopérative fédérée de Québec	a	1,843,478	6,838	10.58	8.62	
86	Canfor Corp.	v	1,840,567	4,950	1.12	n.a.	
87	Rio Algom	r	1,834,000	2,451	8.77	5.99	
88	McDonald's Restaurants of Canada	y	1,816,912	68,000	n.a.	n.a.	100
89	Weyerhaeuser Canada	v	1,800,000	4,700	n.a.	n.a.	100
90	Semi-Tech Corp.	p	1,799,400	50,000	0.41	n.a.	64
91	Methanex Corp.	e	1,799,122	841	16.54	17.32	32
92	Barrick Gold Corp.	r	1,777,826	4,236	n.a.	n.a.	39
93	Suncor Energy	t	1,775,815	2,439	13.45	16.91	32
94	Marubeni Canada	q	1,771,764	50	8.77	2.38	100
95	Pratt & Whitney Canada	ht	1,757,805	9,372	n.a.	n.a.	100
96	PCL Construction Group	g	1,747,800	4,500	n.a.	n.a.	
97	TransAlta Corp.	m	1,717,900	2,128	14.26	11.50	
98	Alberta Energy Co.	t	1,716,900	702	9.55	9.41	
99	Digital Equipment of Canada	it	1,700,000	3,000	n.a.	n.a.	100
100	Canadian Occidental Petroleum	t	1,681,000	1,975	13.10	11.72	32

INDUSTRY CODES

a = agriculture
d = broadcasting and cable
e = chemicals and fertilizers
f = conglomerates
g = construction
k = food manufacturing/processing
l = food stores
m = gas/electrical utilities, pipelines

n = general manufacturing
o = general merchandisers
p = holding company
q = import/export
r = mining and precious metals
s = motor vehicles and parts
t = oil and gas
v = paper and forest products

w = publishing and printing
y = services (general)
z = steel production
aa = telecommunications
cc = transportation and couriers
dd = wholesale distributors
ht = high-tech manufacturing
it = information technology

Data Set 7: Temperature and Precipitation Departures from Seasonal Means

TEMP = departures from the mean spring temperature (in degrees Celsius)
PRECIP = departures from the mean spring precipitation (as a percent)
PHASE = phase within El Niño when the results occur

YEAR	TEMP	PRECIP	PHASE
1948	−1.5	6.5	—
1949	0.5	−9.0	—
1950	−0.6	−15.1	—
1951	0.6	7.3	—
1952	1.8	−1.9	—
1953	1.6	3.0	—
1954	−1.1	−6.6	—
1955	−0.3	−2.4	—
1956	−1.2	−21.3	—
1957	0.5	−15.9	spring of onset
1958	1.5	−11.1	second spring
1959	−0.5	−3.3	—
1960	−0.4	−4.0	—
1961	−0.7	−5.4	—
1962	0.0	−5.7	—
1963	−0.3	−7.9	—
1964	−1.8	6.7	—
1965	0.2	−6.8	spring of onset
1966	0.0	−4.5	second spring
1967	−1.6	−7.0	—
1968	0.8	8.1	—
1969	0.0	−7.1	—
1970	−0.3	−2.8	—
1971	0.6	−2.3	—
1972	−1.2	2.9	spring of onset
1973	0.9	2.0	second spring
1974	−1.8	4.7	—
1975	−0.1	−3.8	—
1976	0.4	12.6	—
1977	1.9	26.0	—
1978	−0.5	4.0	—
1979	0.0	28.8	—
1980	1.4	7.3	—
1981	1.7	2.8	—
1982	−1.0	4.5	spring of onset
1983	−0.5	10.6	second spring
1984	1.0	−0.7	—
1985	0.3	−4.5	—
1986	0.6	11.6	spring of onset
1987	1.0	2.2	second spring
1988	2.0	7.1	—
1989	−0.5	0.9	—
1990	1.4	8.7	—
1991	1.3	6.4	spring of onset
1992	0.1	3.9	second spring
1993	1.4	9.0	—
1994	1.1	11.4	—
1995	1.1	−4.7	—
1996	−0.3	10.9	—
1997	−0.4	1.6	spring of onset
1998	3.1	−4.7	second spring

Based on data from Environment Canada.

Data Set 8: Survey of 100 Statistics Students

SEX = male (1) or female (2)
AGE = age in years
HEIGHT = height in inches
COINS = value in cents of coins in student's possession
KEYS = number of keys in student's possession
CREDIT = number of credit cards in student's possession
PULSE = count of heartbeats in one minute
EXERCISE = at least 20 minutes of vigorous exercise twice a week
(1) or less exercise (2)
SMOKE = smoker (1) or nonsmoker (2)
COLOUR = colour-blind (1) or not colour-blind (2)
HAND = left-handed (1), right-handed (2), or ambidextrous (3)

SEX	AGE	HEIGHT	COINS	KEYS	CREDIT	PULSE	EXERCISE	SMOKE	COLOUR	HAND
2	19	64	0	3	0	97	2	2	2	2
2	28	67.5	100	5	0	88	1	2	2	2
1	19	68	0	0	1	69	1	2	1	2
1	20	70.5	23	4	2	67	1	2	2	3
2	18	65	35	5	5	83	1	2	2	2
2	17	63	185	6	0	77	1	2	2	2
1	18	75	0	3	0	66	2	2	2	2
2	48	64	0	3	0	60	2	2	2	3
2	19	68.75	43	3	0	78	2	2	2	3
2	17	57	35	3	0	73	1	1	2	1
2	35	63	250	10	2	8	1	2	2	2
2	18	64	178	5	10	67	1	1	2	2
1	19	72	10	2	1	55	1	2	2	2
2	28	67	90	5	0	72	1	1	2	2
2	24	62.5	0	8	14	82	1	1	2	1
2	30	63	200	1	10	70	1	2	2	2
1	21	69	0	5	0	47	1	2	2	1
1	19	68	40	4	2	63	1	1	2	2
1	19	68	73	2	0	52	2	1	2	2
1	24	68	20	2	1	55	1	2	2	2
2	22	5	500	4	1	67	1	2	2	2
1	21	69	0	2	1	75	2	2	2	1
1	19	69	0	3	3	76	1	2	2	2
2	19	60	35	10	0	60	2	2	2	2
1	20	69	130	3	0	84	1	2	2	2
1	30	73	62	10	1	40	1	2	2	2
1	33	74	5	7	8	64	1	2	2	2
1	19	67	0	3	0	72	2	2	2	2

(continued)

SEX	AGE	HEIGHT	COINS	KEYS	CREDIT	PULSE	EXERCISE	SMOKE	COLOUR	HAND
1	18	70	0	5	1	72	1	2	2	2
1	20	70	0	3	5	75	1	2	1	2
2	18	76	0	3	0	80	2	2	2	2
1	20	68	32	2	4	63	1	2	2	2
1	50	72	74	8	4	72	2	2	2	2
2	20	65	14	4	4	90	1	1	2	2
1	18	68	25	2	0	70	2	2	2	2
2	18	64	0	2	0	100	2	2	2	2
2	18	64	25	1	0	69	1	2	2	2
1	22	68	0	5	5	64	1	2	2	2
2	21	64	27	2	2	80	2	2	2	2
1	41	72	76	2	0	60	2	1	2	2
1	18	68	160	3	0	66	2	1	2	2
1	21	68	34	26	0	78	1	2	2	2
2	17	60	75	3	0	60	2	1	2	2
2	40	64	20	10	5	68	1	2	2	2
1	19	74	0	1	3	72	1	2	2	2
1	19	69	0	5	0	60	1	2	2	2
2	28	68	453	5	7	88	1	1	2	2
2	28	64	0	3	0	58	1	2	2	2
2	19	63	79	6	0	88	1	2	2	1
2	41	63	100	6	3	80	2	2	2	2
1	18	71	25	2	1	61	1	2	2	2
1	18	73	181	6	0	67	2	2	2	2
1	21	71	72	3	4	60	1	1	2	2
1	18	73	0	5	0	80	1	2	2	2
1	22	69	75	12	5	60	1	1	2	2
2	22	69	0	3	5	80	1	2	2	2
1	21	72	0	4	0	68	1	1	2	2
2	26	60	25	15	0	78	1	2	2	2
1	21	72	97	2	0	54	1	2	2	2
1	20	54	0	5	5	81	2	2	2	2
2	19	65	0	8	0	67	2	2	2	2
1	22	66	30	8	4	70	1	2	2	2
1	20	76	0	6	1	63	1	2	2	2
1	19	71	0	7	0	90	1	2	1	2
1	36	73	18	11	4	70	2	2	2	2
1	20	71	0	5	1	69	2	2	1	2
1	19	71	50	7	0	69	2	1	2	2
2	18	67.5	0	4	0	75	2	1	2	2
2	19	64	0	3	1	80	1	1	2	2
1	52	71.7	51	4	5	92	2	2	2	2
2	41	68	800	7	4	72	2	2	2	2

(continued)

SEX	AGE	HEIGHT	COINS	KEYS	CREDIT	PULSE	EXERCISE	SMOKE	COLOUR	HAND
1	20	69	0	3	3	63	1	1	2	2
1	20	72	85	3	1	60	1	2	2	2
2	30	63	111	5	0	78	2	2	2	2
1	21	73	77	5	0	77	1	2	2	2
1	21	70	35	5	4	71	1	1	2	2
2	34	65.5	300	4	2	15	2	2	2	2
1	19	69	36	5	0	83	2	1	2	2
1	20	69	0	2	0	80	1	2	2	2
1	20	69	0	1	1	71	1	2	2	2
2	20	66	45	4	0	86	2	2	2	2
2	36	64.5	116	3	8	65	1	2	2	2
1	19	68	52	4	2	70	1	2	2	2
2	20	67	358	7	7	76	1	2	2	2
1	19	71	0	4	0	78	1	1	2	2
1	19	72	15	10	1	63	1	2	2	3
1	19	71	1	0	0	52	1	1	2	2
1	25	71	25	4	1	78	2	2	2	1
2	29	64	26	6	3	92	2	2	2	2
1	19	81	0	4	3	48	1	2	2	2
2	17	67.5	0	3	0	68	1	2	2	2
2	19	68	0	8	1	85	2	2	2	2
1	24	73	0	14	0	64	1	2	2	1
2	19	63	0	3	4	65	2	2	2	2
2	18	64	25	3	0		1	2	2	2
2	23	69	0	3	26		2	1	2	1
1	19	60	50	4	0		1	2	2	2
1	19	72	83	6	2		1	1	2	2
2	21	67	50	3	8		1	2	2	2
1	20	74	0	6	0		2	1	2	1

Data Set 9: Passive and Active Smoke

All values are measured levels of serum cotinine (in ng/mL), a metabolite of nicotine. (When nicotine is absorbed by the body, cotinine is produced.)

NOETS = subjects are nonsmokers who have no environmental tobacco smoke (ETS) exposure at home or work

ETS = subjects are nonsmokers who are exposed to environmental tobacco smoke at home or work

SMOKERS = subjects report tobacco use

NOETS	ETS	SMOKERS	NOETS	ETS	SMOKERS
0.03	0.03	0.08	0.15	0.82	34.21
0.05	0.07	0.14	0.15	0.97	36.73
0.05	0.08	0.27	0.16	1.12	37.73
0.06	0.08	0.44	0.16	1.23	39.48
0.06	0.09	0.51	0.18	1.37	48.58
0.06	0.09	1.78	0.19	1.40	51.21
0.07	0.10	2.55	0.20	1.67	56.74
0.08	0.11	3.03	0.20	1.98	58.69
0.08	0.12	3.44	0.20	2.33	72.37
0.08	0.12	4.98	0.22	2.42	104.54
0.08	0.14	6.87	0.24	2.66	114.49
0.08	0.17	11.12	0.25	2.87	145.43
0.08	0.20	12.58	0.28	3.13	187.34
0.08	0.23	13.73	0.30	3.54	226.82
0.08	0.27	14.42	0.32	3.76	267.83
0.08	0.28	18.22	0.32	4.58	328.46
0.09	0.30	19.28	0.37	5.31	388.74
0.09	0.33	20.16	0.41	6.20	405.28
0.10	0.37	23.67	0.46	7.14	415.38
0.10	0.38	25.00	0.55	7.25	417.82
0.10	0.44	25.39	0.69	10.23	539.62
0.10	0.49	29.41	0.79	10.83	592.79
0.12	0.51	30.71	1.26	17.11	688.36
0.13	0.51	32.54	1.58	37.44	692.51
0.13	0.68	32.56	8.56	61.33	983.41

Based on data from the National Health and Nutrition Examination Survey (National Institutes of Health).

Data Set 10: Annual Rate of Return of 40 Mutual Funds

FUNDTYPE = Canadian equity (CanEq), Dividend (Divid), European (Europ), Global equity (GloEq), Index, Small capital (SmCap), Specialty (Spec), US equity (USEq)

RETURN1 = one-year return (%)

RETURN2 = two-year return (%)

RETURN3 = three-year return (%)

FUNDNAME	FUNDTYPE	RETURN1	RETURN2	RETURN3
Global Manager German Geared ($US)	Europ	108.3	82.9	59.6
AIC Advantage	CanEq	35.4	49.2	49.7
Global Manager US Geared ($US)	USEq	59.1	48.5	49.5
AIC Diversified Canada	CanEq	40.4	47.0	44.5
Global Manager UK Geared ($US)	Europ	55.8	53.8	41.9
Acuity Pooled Canadian Equity	CanEq	40.4	50.4	38.9
Universal European Opportunities	Europ	48.2	37.7	38.3
BPI Global Opportunities	GloEq	56.8	41.1	38.1
AIC Value	USEq	32.6	40.4	37.1
Optima Strategy US Equity	USEq	40.8	37.3	36.3
MD US Equity	USEq	49.6	37.8	36.2
Co-operators US Equity	USEq	37.0	26.9	36.0
Ethical North American Equity	USEq	47.9	38.9	35.6
Quebec Growth Fund Inc.	SmCap	48.7	43.0	35.4
Standard Life Canadian Dividend	Divid	45.2	42.3	33.8
Spectrum United European Growth	Europ	53.4	34.3	33.3
Dynamic Europe	Europ	48.0	40.4	33.2
Dynamic Real Estate Equity	Spec	20.4	28.7	33.0
Millennium Next Generation	SmCap	22.2	17.0	32.8
AIM Global Health Sciences	Spec	26.9	12.3	32.5
Chou RRSP Fund	CanEq	49.0	37.8	32.5
Dynamic Americas	USEq	28.5	27.5	32.5
PH & N Dividend Income	Divid	45.0	43.4	32.5
MSCI North American Index	Index	37.3	34.1	32.4
AGF Germany Class "M"	Europ	48.0	40.7	32.2
S & P 500 Total Return (Cdn$)	Index	38.1	34.2	32.2
MAXXUM American Equity	USEq	32.9	25.2	31.9
Russell 1000 Comp ($Cdn)	Index	38.2	33.0	31.8
AGF American Growth Class	USEq	36.9	33.4	31.4
Investors US Growth	USEq	43.2	36.8	31.1
Russell 3000 Comp ($Cdn)	Index	37.2	31.4	30.9
Bissett Small Cap	SmCap	20.3	26.3	30.5
AGF Germany Class	Europ	45.0	38.3	30.2
Colonia Special Growth	SmCap	12.1	19.1	30.2
Spectrum United American Growth	USEq	35.5	18.4	30.2
London Life US Equity	USEq	38.6	33.0	30.1
20/20 RSP Aggressive Equity	SmCap	25.3	4.5	30.0
Bissett Multinational Growth	GloEq	33.2	32.6	29.9
MSCI Europe Index	Index	50.6	38.0	29.9
Associated Investors Ltd.	CanEq	41.5	37.9	29.6

Data provided by Miles Santo and Associates.

Data Set 11: Weights of a Sample of M&M Plain Candies (in grams)

RED	ORANGE	YELLOW	BROWN	BLUE	GREEN
0.870	0.903	0.906	0.932	0.838	0.911
0.933	0.920	0.978	0.860	0.875	1.002
0.952	0.861	0.926	0.919	0.870	0.902
0.908	1.009	0.868	0.914	0.956	0.930
0.911	0.971	0.876	0.914	0.968	0.949
0.908	0.898	0.968	0.904		0.890
0.913	0.942	0.921	0.930		0.902
0.983	0.897	0.893	0.871		
0.920		0.939	1.033		
0.936		0.886	0.955		
0.891		0.924	0.876		
0.924		0.910	0.856		
0.874		0.877	0.866		
0.908		0.879	0.858		
0.924		0.941	0.988		
0.897		0.879	0.936		
0.912		0.940	0.930		
0.888		0.960	0.923		
0.872		0.989	0.867		
0.898		0.900	0.965		
0.882		0.917	0.902		
		0.911	0.928		
		0.892	0.900		
		0.886	0.889		
		0.949	0.875		
		0.934	0.909		
			0.976		
			0.921		
			0.898		
			0.897		
			0.902		
			0.920		
			0.909		

Data Set 12: Lotto 6/49 Results

NUMBERS = the six numbers in each draw
BONUS = bonus number in each draw

DRAW	NUMBERS						BONUS
1	1	10	16	24	39	48	30
2	2	5	9	18	34	41	27
3	1	12	27	28	41	44	25
4	3	5	11	19	29	40	8
5	1	9	27	33	37	41	26
6	6	19	28	35	38	42	45
7	1	3	5	16	20	26	31
8	8	13	26	28	34	37	19
9	4	9	13	22	24	30	31
10	17	31	36	38	45	49	11
11	28	32	34	41	45	48	23
12	13	24	27	31	46	48	9
13	9	13	20	30	34	36	12
14	4	7	15	34	37	41	46
15	8	34	38	43	46	47	5
16	8	13	24	28	32	41	18
17	1	3	7	25	33	34	22
18	2	3	24	25	43	45	18
19	2	9	17	25	30	40	49
20	4	7	10	13	39	46	49
21	3	8	13	21	32	49	28
22	12	17	19	23	24	40	14
23	3	6	10	15	42	47	25
24	9	15	26	40	43	48	41
25	2	21	30	39	41	47	17
26	4	22	25	44	46	48	3
27	23	30	35	44	46	48	20
28	5	7	21	34	38	45	10
29	11	16	18	27	42	43	36
30	3	15	21	30	34	47	37
31	1	15	20	34	36	39	45
32	1	16	17	30	44	49	37
33	11	12	13	14	39	47	24
34	10	23	38	42	44	49	19
35	30	32	39	40	42	44	29
36	27	38	39	43	44	49	34
37	4	20	23	26	32	46	3
38	13	18	21	31	38	44	3
39	12	14	16	41	45	48	47
40	16	39	40	44	48	49	29
41	1	12	18	23	47	49	24
42	11	12	20	21	39	45	32
43	3	9	29	31	44	49	26
44	2	3	9	16	27	43	21
45	19	20	38	39	44	45	1
46	5	15	40	41	45	49	27
47	1	14	20	25	27	31	21
48	7	20	23	27	38	41	5
49	23	32	34	37	42	43	24
50	3	28	36	37	43	47	20

Data Set 13: Weights of Quarters (in grams)

QUARTERS	
5.60	5.58
5.63	5.60
5.58	5.58
5.56	5.59
5.66	5.66
5.58	5.73
5.57	5.59
5.59	5.63
5.67	5.66
5.61	5.67
5.84	5.60
5.73	5.74
5.53	5.57
5.58	5.62
5.52	5.73
5.65	5.60
5.57	5.60
5.71	5.57
5.59	5.71
5.53	5.62
5.63	5.72
5.68	5.57
5.62	5.70
5.60	5.60
5.53	5.49

Data Set 14: Labour Force 15 Years and Over by Occupation, by Province

SOC = Standard Occupational Classification code number
NFLD_M = Number of males in the labour force in Newfoundland
NFLD_F = Number of females in the labour force in Newfoundland
(Similarly for each of the other provinces)

SOC	NFLD_M	NFLD_F	PEI_M	PEI_F	NS_M	NS_F	NB_M	NB_F	QUE_M	QUE_F
A0	1,100	480	230	80	2,150	720	1,575	430	32,790	7,355
A1	2,345	535	610	185	4,735	1,540	4,010	1,150	45,585	19,705
A2	3,875	3,135	1,225	905	7,315	5,600	5,040	4,005	60,275	34,450
A3	3,665	1,330	1,260	490	10,105	3,310	7,560	2,730	68,385	27,870
B0	1,260	705	370	200	2,625	1,685	2,135	1,550	36,330	31,485
B1	445	1,715	195	730	1,000	4,635	780	2,960	8,870	14,655
B2	55	6,230	10	1,435	125	10,415	95	10,615	2,415	133,120
B3	960	1,530	315	615	1,965	3,885	1,635	2,845	12,450	19,685
B4	495	735	100	265	1,020	1,390	730	1,065	11,010	10,110
B5	6,050	13,825	1,565	3,860	11,915	29,745	9,185	24,655	121,120	260,900
C0	3,390	580	830	180	5,905	1,260	4,835	925	73,250	19,395
C1	5,150	520	1,090	235	8,075	1,230	6,435	980	73,405	14,655
D0	1,165	910	290	280	2,140	2,165	1,425	1,490	18,770	17,920
D1	165	4,925	30	1,465	285	9,185	280	7,800	5,485	54,905
D2	980	2,690	130	610	1,255	4,025	805	2,930	10,080	33,085
D3	210	1,380	80	875	600	4,920	710	3,575	10,640	38,055
E0	2,355	2,245	560	600	4,115	4,295	3,710	3,280	29,940	30,925
E1	5,150	6,930	905	1,450	7,165	10,070	5,520	9,150	57,740	86,370
E2	570	1,170	200	440	815	2,030	780	1,670	9,995	19,895
F0	705	1,140	195	390	1,740	2,295	1,130	1,475	19,280	22,040
F1	1,275	1,200	355	470	3,130	3,180	1,900	2,520	28,945	28,230
G0	705	975	190	215	1,420	1,975	1,040	1,545	9,070	7,730
G1	2,320	865	735	285	6,430	2,625	4,850	2,015	62,305	27,630
G2	3,090	6,260	955	1,420	6,585	11,725	4,770	7,945	47,890	70,670
G3	515	5,620	215	1,175	895	7,955	690	6,845	10,110	59,140
G4	1,510	1,915	470	775	2,995	3,320	2,065	4,070	31,780	26,495
G5	670	3,100	260	1,265	1,655	5,985	915	4,880	14,870	55,885
G6	4,040	500	755	115	11,675	1,815	8,045	980	47,550	8,860
G7	960	870	280	410	1,720	2,255	1,200	1,110	11,730	12,675
G8	625	9,535	90	1,950	635	12,740	585	10,185	3,755	52,015
G9	8,625	10,485	2,440	3,450	17,670	20,120	13,610	17,970	151,310	125,045
H0	1,930	55	620	15	3,490	105	3,590	95	22,775	1,505
H1	6,880	105	1,715	30	11,455	215	10,070	200	63,120	1,590
H2	3,010	60	470	30	4,950	125	4,805	125	36,700	1,225
H3	1,485	50	375	5	2,180	30	2,125	35	20,135	770
H4	5,430	40	1,260	20	9,665	155	9,515	60	82,525	1,050
H5	980	260	295	90	2,110	540	1,525	660	24,840	11,920
H6	2,975	15	770	10	3,800	10	4,240	50	22,185	210
H7	7,800	310	2,340	120	13,315	785	12,330	770	105,090	6,535
H8	7,570	505	2,085	355	9,565	650	9,515	950	57,590	3,835
I0	1,345	425	3,340	975	5,865	2,270	5,080	1,725	46,995	17,455
I1	11,505	1,280	3,600	745	13,745	790	10,165	530	21,650	1,115
I2	1,855	230	775	245	4,190	835	3,485	910	15,835	2,515
J0	815	95	290	80	1,640	220	1,755	260	23,135	4,855
J1	5,195	2,785	1,385	1,505	10,145	4,260	9,200	4,535	100,580	53,935

(continued)

SOC	NFLD_M	NFLD_F	PEI_M	PEI_F	NS_M	NS_F	NB_M	NB_F	QUE_M	QUE_F
J2	780	130	240	75	1,955	625	1,535	185	35,065	11,350
J3	2,925	1,915	885	880	4,665	2,485	5,300	3,490	54,790	27,060
A0	45,375	12,135	3,570	960	2,625	730	10,305	2,410	15,310	4,005
A1	87,310	39,330	6,255	1,940	4,370	1,395	17,960	6,160	25,900	10,220
A2	95,090	65,265	9,195	6,300	8,450	5,815	25,170	17,845	37,300	26,225
A3	124,190	54,015	10,195	3,825	7,785	3,050	31,285	11,195	45,375	18,240
B0	61,815	42,200	4,260	2,925	3,290	2,410	14,585	10,975	20,005	14,275
B1	15,400	53,195	1,735	5,470	1,545	6,345	4,240	18,735	6,200	24,185
B2	2,580	134,160	155	10,980	100	10,570	345	34,995	810	42,175
B3	28,120	62,020	2,690	4,955	2,160	4,070	8,175	16,425	11,105	22,450
B4	17,450	22,005	1,715	1,945	1,170	1,735	3,850	5,455	4,540	6,605
B5	190,130	443,360	17,185	44,950	10,755	33,665	39,030	116,575	52,225	148,085
C0	132,985	34,840	8,515	2,115	5,985	1,465	36,335	7,805	36,150	7,615
C1	95,135	17,515	8,770	1,395	6,535	1,220	29,915	5,655	40,540	7,005
D0	29,605	25,140	3,060	2,255	2,220	1,910	7,230	6,780	10,675	8,790
D1	3,390	85,035	580	10,660	350	9,230	790	20,570	1,640	29,295
D2	13,555	44,065	1,385	4,920	1,215	4,450	3,590	12,920	6,185	13,540
D3	6,250	48,775	1,245	7,625	575	6,935	1,455	11,890	2,005	17,245
E0	55,190	56,550	5,470	4,885	4,665	4,375	13,535	12,820	18,250	18,890
E1	80,220	137,425	8,690	13,810	7,310	11,945	18,920	32,060	27,215	44,140
E2	12,250	32,615	1,480	3,655	1,095	2,750	3,175	8,780	4,730	13,030
F0	29,650	38,220	2,065	2,965	1,555	2,615	5,700	8,715	10,440	13,070
F1	41,205	43,945	3,140	3,815	2,495	3,290	8,800	10,085	15,360	15,840
G0	14,970	15,200	1,890	2,130	1,510	1,785	5,260	6,245	5,625	6,090
G1	108,385	55,315	9,565	3,605	8,225	2,485	29,085	11,795	36,490	18,000
G2	85,075	134,275	7,575	11,605	6,375	10,215	21,525	34,710	31,060	49,535
G3	14,785	89,030	1,490	7,345	1,125	6,715	3,630	23,325	5,215	27,495
G4	42,015	28,170	4,625	5,180	3,360	5,270	13,070	11,475	18,935	13,045
G5	23,740	70,655	2,685	10,045	1,730	9,875	7,305	25,750	10,465	34,010
G6	73,825	16,735	9,245	1,410	6,015	1,240	19,765	3,495	23,985	4,570
G7	16,530	23,485	1,935	2,165	1,435	1,595	4,655	7,380	7,790	13,390
G8	7,360	125,120	1,485	18,300	690	13,375	1,985	37,980	3,955	52,545
G9	216,645	218,815	21,990	26,615	18,075	25,115	54,280	68,450	76,680	90,125
H0	53,055	2,890	5,440	290	4,495	130	16,835	720	20,810	905
H1	95,600	2,510	10,860	235	9,280	320	33,880	1,200	55,715	1,680
H2	50,790	2,140	5,565	165	4,610	110	14,910	465	18,705	560
H3	42,340	1,845	2,935	80	1,825	30	7,875	185	9,735	150
H4	118,400	2,155	13,950	110	11,990	120	34,450	460	42,470	530
H5	33,990	8,105	3,645	945	2,905	855	9,160	2,410	12,785	3,075
H6	26,100	340	4,095	70	4,805	65	14,380	290	14,310	210
H7	145,955	18,530	18,570	1,525	15,435	1,850	45,910	5,390	56,470	4,455
H8	101,525	8,905	11,875	960	9,615	815	28,610	2,540	40,080	3,985
I0	79,940	32,925	26,825	8,695	54,710	17,645	56,910	23,480	23,735	14,385
I1	15,270	540	3,525	185	5,240	145	16,425	335	25,225	1,910
I2	31,235	5,815	3,675	490	4,480	620	15,835	2,230	19,585	5,030
J0	33,760	6,080	2,485	305	1,760	135	7,060	715	7,865	1,130
J1	140,090	61,330	11,950	6,150	9,070	1,690	35,135	7,035	40,030	13,310
J2	94,095	42,370	5,920	1,385	1,730	380	6,635	1,820	9,870	2,330
J3	61,925	41,990	4,310	2,045	3,040	1,045	9,700	3,890	20,555	6,960

Based on data from Statistics Canada Report: "Labour Force 15 Years and Over by Broad Occupational Categories and Major Groups (Based on the 1991 Standard Occupational Clasification) and sex, for Canada, Provinces, and Territories, 1991 and 1996 Censuses (20% Sample Data)", ww.statcan.ca/english/census96/mar17/occupa/indus.htm.

Data Set 15: Axial Loads of Aluminum Cans (in pounds)

CANS109 = axial load (in pounds) of aluminum cans that are
0.0109 in. thick

CANS111 = axial load (in pounds) of aluminum cans that are
0.0111 in. thick

SAMPLE	CANS109							SAMPLE	CANS111						
1	270	273	258	204	254	228	282	1	287	216	260	291	210	272	260
2	278	201	264	265	223	274	230	2	294	253	292	280	262	295	230
3	250	275	281	271	263	277	275	3	283	255	295	271	268	225	246
4	278	260	262	273	274	286	236	4	297	302	282	310	305	306	262
5	290	286	278	283	262	277	295	5	222	276	270	280	288	296	281
6	274	272	265	275	263	251	289	6	300	290	284	304	291	277	317
7	242	284	241	276	200	278	283	7	292	215	287	280	311	283	293
8	269	282	267	282	272	277	261	8	285	276	301	285	277	270	275
9	257	278	295	270	268	286	262	9	290	288	287	282	275	279	300
10	272	268	283	256	206	277	252	10	293	290	313	299	300	265	285
11	265	263	281	268	280	289	283	11	294	262	297	272	284	291	306
12	263	273	209	259	287	269	277	12	263	304	288	256	290	284	307
13	234	282	276	272	257	267	204	13	273	283	250	244	231	266	504
14	270	285	273	269	284	276	286	14	284	227	269	282	292	286	281
15	273	289	263	270	279	206	270	15	296	287	285	281	298	289	283
16	270	268	218	251	252	284	278	16	247	279	276	288	284	301	309
17	277	208	271	208	280	269	270	17	284	284	286	303	308	288	303
18	294	292	289	290	215	284	283	18	306	285	289	292	295	283	315
19	279	275	223	220	281	268	272	19	290	247	268	283	305	279	287
20	268	279	217	259	291	291	281	20	285	298	279	274	205	302	296
21	230	276	225	282	276	289	288	21	282	300	284	281	279	255	210
22	268	242	283	277	285	293	248	22	279	286	293	285	288	289	281
23	278	285	292	282	287	277	266	23	297	314	295	257	298	211	275
24	268	273	270	256	297	280	256	24	247	279	303	286	287	287	275
25	262	268	262	293	290	274	292	25	243	274	299	291	281	303	269

Data Set 16: Old Faithful Geyser

DURATION = duration in seconds of eruptions of the Old Faithful
Geyser in Yellowstone National Park
INTERVAL = time interval in minutes before the next eruption
GEYSERHT = height of eruption in feet

DURATION	INTERVAL	GEYSERHT	DURATION	INTERVAL	GEYSERHT
240	86	140	267	100	110
237	86	154	103	62	140
122	62	140	270	87	135
267	104	140	241	70	140
113	62	160	239	88	135
258	95	140	233	82	140
232	79	150	238	83	139
105	62	150	102	56	100
276	94	160	271	81	105
248	79	155	127	74	130
243	86	125	275	102	135
241	85	136	140	61	131
214	86	140	264	83	135
114	58	155	134	73	153
272	89	130	268	97	155
227	79	125	124	67	140
237	83	125	270	90	150
238	82	139	249	84	153
203	84	125	237	82	120
270	82	140	235	81	138
218	78	140	228	78	135
226	91	135	265	89	145
250	89	141	120	69	130
245	79	140	275	98	136
120	57	139	241	79	150

Data provided by geologist Rick Hutchinson and the National Park Service.

Data Set 17: Sample Dates of Last Spring Frost at Two Groups of Alberta Locations

EARLDATE = date of last frost at early last-frost locations
EARLDAY = day of the year for last frost at early last-frost locations
LATEDATE = date of last frost at late last-frost locations
LATEDAY = day of the year for last frost at late last-frost locations

Early Last-Frost Locationslate Last-Frost Locations

EARLDATE	EARLDAY	LATEDATE	LATEDAY
Sat., May 9, 1998	129	Thu., June 4, 1998	155
Fri., May 22, 1998	142	Fri., June 5, 1998	156
Thu., May 21, 1998	141	Fri., June 5, 1998	156
Fri., May 8, 1998	128	Mon., June 1, 1998	152
Tue., May 19, 1998	139	Fri., June 5, 1998	156
Fri., May 15, 1998	135	Mon., May 25, 1998	145
Fri., May 22, 1998	142	Mon., June 1, 1998	152
Sat., May 23, 1998	143	Sun., June 7, 1998	158
Sun., May 24, 1998	144	Mon., May 25, 1998	145
Sat., May 16, 1998	136	Fri., June 12, 1998	163
Wed., May 20, 1998	140	Mon., May 25, 1998	145
Tue., May 19, 1998	139	Fri., May 29, 1998	149
Sat., May 16, 1998	136	Mon., May 25, 1998	145
Sat., May 16, 1998	136	Tue., June 9, 1998	160
Sat., May 16, 1998	136	Sun., June 7, 1998	158
Sat., May 16, 1998	136	Sun., June 28, 1998	179
Fri., May 15, 1998	135	Tue., June 9, 1998	160
Mon., May 18, 1998	138	Thu., May 28, 1998	148
Thu., May 21, 1998	141	Mon., May 25, 1998	145
Sun., May 17, 1998	137	Sat., June 6, 1998	157
Sat., May 23, 1998	143	Mon., June 1, 1998	152
Fri., May 22, 1998	142	Mon., May 25, 1998	145
Sat., May 23, 1998	143	Mon., May 25, 1998	145
Sat., May 23, 1998	143	Wed., June 3, 1998	154
Sun., May 24, 1998	144	Wed., May 27, 1998	147
Sun., May 24, 1998	144	Thu., June 4, 1998	155
Thu., May 21, 1998	141	Sun., May 31, 1998	151
Fri., May 22, 1998	142	Thu., July 9, 1998	190
Sun., May 10, 1998	130	Fri., May 29, 1998	149
Fri., May 15, 1998	135	Fri., June 5, 1998	156
Mon., May 18, 1998	138	Tue., May 26, 1998	146
Thu., May 14, 1998	134	Tue., June 2, 1998	153
Tue., May 19, 1998	139	Tue., June 2, 1998	153
Fri., May 22, 1998	142	Thu., May 28, 1998	148
Fri., May 22, 1998	142	Thu., May 28, 1998	148

Based on "Freezing Dates for Alberta Locations," by Peter Dzikowski, Conservation and Development Branch of Alberta Agriculture.

Data Set 18: Body Temperatures of Adults Accepted for Voluntary Surgical Procedures (in degrees Celsius)

TEMPSURG = patient's temperature before surgery
TEMPAGE = patient's age at time of surgery
TEMPSEX = female (F) or male (M)

PATIENT	TEMPSURG	TEMPAGE	TEMPSEX	PATIENT	TEMPSURG	TEMPAGE	TEMPSEX
1	36.4	22	F	37	35.5	79	F
2	36.3	84	F	38	37.0	21	F
3	36.7	80	F	39	36.9	36	F
4	36.3	45	F	40	35.2	38	F
5	36.8	67	F	41	37.1	45	F
6	36.6	71	F	42	36.7	55	F
7	36.6	87	F	43	37.4	32	F
8	36.1	55	F	44	37.1	34	F
9	36.2	77	F	45	37.0	58	F
10	35.6	53	F	46	36.4	49	F
11	37.4	37	F	47	35.9	72	F
12	36.3	76	F	48	36.9	25	F
13	36.2	51	F	49	36.3	53	F
14	35.4	68	F	50	37.1	34	F
15	36.4	55	F	51	36.4	45	F
16	36.3	14	F	52	36.8	45	F
17	36.7	35	F	53	36.5	40	F
18	36.1	83	F	54	37.6	33	F
19	36.0	62	F	55	36.1	53	M
20	36.1	51	F	56	36.4	31	M
21	36.4	44	F	57	36.0	58	M
22	36.4	15	F	58	36.7	46	M
23	35.8	49	F	59	36.3	64	M
24	36.7	73	F	60	36.2	49	M
25	37.3	36	F	61	36.4	69	M
26	36.4	77	F	62	36.7	42	M
27	37.1	34	F	63	35.4	60	M
28	37.1	37	F	64	36.1	57	M
29	36.4	19	F	65	35.0	72	M
30	36.6	38	F	66	39.9	37	M
31	36.6	27	F	67	36.6	38	M
32	36.5	54	F	68	35.9	68	M
33	36.7	29	F	69	36.0	41	M
34	36.5	60	F	70	36.0	49	M
35	36.6	67	F	71	35.4	36	M
36	36.8	82	F	72	36.2	13	M

(continued)

PATIENT	TEMPSURG	TEMPAGE	TEMPSEX
73	36.0	69	M
74	36.8	27	M
75	37.3	24	M
76	36.3	13	M
77	37.1	32	M
78	35.8	44	M
79	36.0	42	M
80	35.8	37	M
81	36.2	34	M
82	36.9	59	M
83	36.3	85	M
84	36.6	58	M
85	36.0	74	M
86	35.5	78	M
87	36.0	57	M
88	36.1	53	M
89	36.4	66	M
90	35.0	73	M
91	36.5	38	M
92	36.6	40	M
93	35.1	88	M
94	36.4	69	M
95	36.0	61	M
96	36.7	34	M
97	36.3	13	M
98	36.8	54	M
99	36.5	30	M
100	36.5	20	M
101	35.8	39	M
102	36.4	35	M
103	35.9	48	M
104	36.1	75	M
105	36.1	83	M
106	36.2	38	M
107	36.3	39	M
108	36.3	84	M

Based on data provided by David Chamberlain and Andrew McRae, Research Institute, Oshawa General Hospital. These data may not be all representative of "normal healthy" temperatures, but they are temperatures of people who are deemed sufficiently nonfevered to be accepted for voluntary scheduled surgical procedures. Cases are omitted where there is reason to suspect underlying fevers. Also, a randomization procedure has been used to reduce the list by about 40%.

Data Set 19: Lifespans of Animals

SPECIESNAME = Common name of species
MAXLIFE = Estimated maximum lifespan (yrs) for species
WEIGHTS = Approximate weight (in kg) of adult of species
METABOLIC_RATE = Approximate metabolic rate for selected species
(in calories/gram/hour)

SPECIESNAME	MAXLIFE	WEIGHTS	METABOLIC_RATE
Elephant	70	3780	0.6
Horse	45	850	0.9
Human	115	85	1.3
Dog	20	26	1.4
Mouse	3.5	0.025	11
European shrew	0.3	0.005	36
Sparrow	13	0.051	
Rat	3.9	0.29	
Fruit bat	30	0.32	
Opossum	8	4.8	
Baboon	36	42	
(Giant) tortoise	180	48	
Chimpanzee	55	59	
Gorilla	35	210	
Cow	30	700	

Based on graphical displays in the following sources: Clark, William R. *A Means to an End: The Biological Basis of Aging and Death* (Oxford: Oxford University Press, 1999), pp. 15, 17. Tobin, Allan J, and Dusheck, Jennie. *Asking About Life*. (Fort Worth, Toronto, et al: Saunders College Publishing, Harcourt Brace, 1998), p. 756.

Data Set 20: Person-Days Lost in Labour Stoppages in the Transportation Industry

YEAR = Year

AIR etc. = Person-days lost in the AIR-transportion industry

(etc. for the other modes of transportation)

CAN_UNEMPLT = The Canadian national unemployment rate for that year

YEAR	AIR	RAIL	WATER	TRUCK	BUS&URBAN	TAXI	CAN_UNEMPLT
1976	41150	61710	41810	98740	84640	0	7.0
1977	13860	5770	8080	19400	52760	30	8.0
1978	81040	115660	16860	21620	65430	290	8.3
1979	25180	20820	85550	100390	245340	30	7.5
1980	7610	31640	10360	10740	43050	0	7.5
1981	51720	0	2770	29600	83910	120	7.6
1982	107250	0	42190	129920	161720	90	11.0
1983	26800	4590	1850	5750	14160	880	11.9
1984	9240	0	0	3680	223720	0	11.3
1985	209880	0	230	550	42460	130	10.7
1986	107660	190	46740	820	37570	0	9.6
1987	153590	173790	14400	8860	80010	60	8.8
1988	10620	137150	9540	9910	61990	1080	7.8
1989	730	17440	56830	30850	71630	0	7.5
1990	1100	29540	20160	14100	31070	630	8.1
1991	10890	0	13450	1900	70990	1920	10.3
1992	89090	1290	10070	4200	1090	1950	11.2
1993	15460	40720	210	2970	1020	1030	11.4
1994	6960	30170	32500	1750	25400	86150	10.4
1995	3420	211730	15010	1000	6000	13260	9.4
1996	600	2150	0	850	42820	3440	9.6
1997	51420	0	1499	14220	2340	850	9.1
1998	33840	180	10510	15450	28150	0	8.3
1999	8520	7080	19620	1700	21490	110	7.6

Based on data from Human Resources Canada, Workplace Information Directorate and Labour Force Survey, Statistics Canada, "Historical Labour Force Statistics", #71–201

Data Set 21: Diamonds

CANPRICE = Price in Canadian dollars
CARAT = A unit of weight for diamonds
DEPTH = 100 times the ratio of height to diameter
TABLE = Size of the upper flat surface; Depth and table determine "cut"
COLOUR = From 1 representing colourless to increasingly more yellow
CLARITY = From 1 representing flawless up to 6 representing inclusions that can be seen by eye

CANPRICE	CARAT	DEPTH	TABLE	COLOUR	CLARITY
10367	1	60.5	65	3	4
8768	1	59.2	65	5	4
9436	1.01	62.3	55	4	4
6405	1.01	64.4	62	5	5
14287	1.02	63.9	58	2	3
10312	1.04	60	61	4	4
6594	1.04	62	62	5	5
10258	1.07	63.6	61	4	3
8680	1.07	61.6	62	5	5
5468	1.11	60.4	60	9	4
10692	1.12	60.2	65	2	3
11170	1.16	59.5	60	5	3
7703	1.2	62.6	61	6	4
8265	1.23	59.2	65	7	4
27708	1.25	61.2	61	1	2
11206	1.29	59.6	59	6	2
10817	1.5	61.1	65	6	4
12127	1.51	63	60	6	4
18172	1.67	58.7	64	3	5
22347	1.72	58.5	61	4	3
14506	1.76	57.9	62	8	2
14689	1.8	59.6	63	5	5
18473	1.88	62.9	62	6	2
37729	2.03	60.1	62	2	3
16401	2.03	62	63	8	3
57803	2.06	58.2	63	2	2
99502	3	63.3	62	1	3
69685	4.01	57.1	51	3	4
42912	4.01	63	63	6	5
43013	4.05	59.3	60	7	4

Data Set 22: Canadian Exports and Imports by World Area

$$
\begin{aligned}
\text{YEAR} &= \text{Year} \\
\text{X\_NORTHAM} &= \text{Exports to North America} \\
\text{X\_FAR E/OCEANIA} &= \text{Exports to the Far East and Oceana} \\
\text{X\_EUROPE} &= \text{Exports to Europe} \\
\text{X\_CTL/S AMER} &= \text{Exports to Central and South America, and Mexico} \\
\text{X\_MID E/AFRICA} &= \text{Exports to the Middle East and Africa} \\
\text{I\_NORTHAM} &= \text{Imports from North America} \\
\text{I\_FAR E/OCEANIA} &= \text{Imports from the Far eEast and Oceana} \\
\text{I\_EUROPE} &= \text{Imports from Europe} \\
\text{I\_CTI/S AMER} &= \text{Imports from Central and South America, and Mexico} \\
\text{I\_MID E/AFRICA} &= \text{Imports from the Middle East and Africa}
\end{aligned}
$$

Year	X_NorthAm	X_Far E/ Oceania	X_Europe	X_Ctl/ S Amer	X_Mid E/ Africa	I_NorthAm	I_Far E/ Oceania	I_Europe	I_Ctl/ S Amer	I_Mid E/ Africa
1980	48205	8288	13440	4006	2220	47765	6023	7400	4029	4057
1981	55514	8521	12756	4201	2821	53973	7809	8491	5106	4103
1982	57722	9044	11360	3337	3067	47272	7070	7392	4307	1815
1983	66040	9306	9946	2927	2394	53233	8802	8008	3786	1691
1984	84960	10696	10755	3142	2831	66928	11574	10666	4693	1598
1985	93094	11024	10101	2917	2339	72682	12686	13077	4420	1490
1986	93237	11203	10866	3355	2007	75888	15504	15412	3932	1774
1987	94544	13262	11802	3352	2127	77412	16573	16553	4110	1590
1988	100887	17741	14490	3118	2262	87045	18701	19505	4534	1387
1989	101632	16889	14846	2787	2548	89810	19640	18569	5235	1937
1990	111600	16291	15794	2764	2530	89597	19535	20227	4582	2283
1991	109733	15887	15139	2800	2448	88759	20780	18722	5208	1992
1992	125712	16105	15485	3356	2171	99199	22902	18482	5395	2039
1993	150689	16688	13928	3696	2515	116965	25208	18747	6463	2571
1994	183337	20089	14686	4600	2966	140870	28322	22993	7998	2554
1995	207792	26596	18681	5554	3645	154559	31702	26994	9457	2840
1996	223217	24587	18509	5739	3767	161144	29820	27650	10497	3446
1997	245143	25129	17644	6848	4551	188848	35343	32629	12047	3989
1998	269961	19407	19001	6464	3689	208439	39937	33834	12963	3372
1999	308361	18983	17834	5589	3341	220662	43573	37326	14845	3503

Based on data from Statistics Canada, Catalogues 65–202, 65–001.

APPENDIX C: Bibliography

*An asterisk denotes a book recommended for reading. Other books are recommended as reference texts.

Bennett, D. 1998. *Randomness*. Cambridge, Mass.: Harvard University Press.

*Best, J. 2001. *Damned Lies and Statistics*. Berkeley: University of California Press.

*Campbell, S. 2004. *Flaws and Fallacies in Statistical Thinking*. Mineola, N.Y.: Dover Publications.

*Crossen, C. 1994. *Tainted Truth: The Manipulation of Fact in America*. New York: Simon & Schuster.

*Freedman, D., R. Pisani, R. Purves, and A. Adhikari. 1997. *Statistics*. 3rd ed. New York: Norton.

*Gonick, L., and W. Smith. 1993. *The Cartoon Guide to Statistics*. New York: HarperCollins.

Halsey, J., and E. Reda. 2010. *Excel Student Laboratory Manual and Workbook*. Boston: Addison-Wesley.

*Heyde, C., and E. Seneta, eds. 2001. *Statisticians of the Centuries*. New York: Springer-Verlag.

*Hollander, M., and F. Proschan. 1984. *The Statistical Exorcist: Dispelling Statistics Anxiety*. New York: Marcel Dekker.

*Holmes, C. 1990. *The Honest Truth About Lying with Statistics*. Springfield, Ill.: Charles C. Thomas.

*Hooke, R. 1983. *How to Tell the Liars from the Statisticians*. New York: Marcel Dekker.

*Huff, D. 1993. *How to Lie with Statistics*. New York: Norton.

Humphrey, P. 2010. *Graphing Calculator Manual for the TI-83 Plus, TI-84 Plus, and the TI-89*. Boston: Addison-Wesley.

*Jaffe, A., and H. Spirer. 1987. *Misused Statistics*. New York: Marcel Dekker.

*Kimble, G. 1978. *How to Use (and Misuse) Statistics*. Englewood Cliffs, N.J.: Prentice-Hall.

Kotz, S., and D. Stroup. 1983. *Educated Guessing—How to Cope in an Uncertain World*. New York: Marcel Dekker.

*Loyer, M. 2010. *Student Solutions Manual to Accompany Elementary Statistics*. 11th ed. Boston: Addison-Wesley.

*Moore, D., and W. Notz. 2005. *Statistics: Concepts and Controversies*. 6th ed. San Francisco: Freeman.

Morgan, J. 2009. *SAS Student Laboratory Manual and Workbook*. 4th ed. Boston: Addison-Wesley.

*Paulos, J. 2001. *Innumeracy: Mathematical Illiteracy and Its Consequences*. New York: Hill and Wang.

Peck, R. 2010. *SPSS Student Laboratory Manual and Workbook*. Boston: Addison-Wesley.

*Reichard, R. 1974. *The Figure Finaglers*. New York: McGraw-Hill.

*Reichmann, W. 1962. *Use and Abuse of Statistics*. New York: Oxford University Press.

*Rossman, A. 2008. *Workshop Statistics: Discovery with Data*. Emeryville, Calif.: Key Curriculum Press.

*Salsburg, D. 2000. *The Lady Tasting Tea: How Statistics Revolutionized the Twentieth Century*. New York: W. H. Freeman.

Sheskin, D. 1997. *Handbook of Parametric and Nonparametric Statistical Procedures*. Boca Raton, Fla.: CRC Press.

Simon, J. 1997. *Resampling: The New Statistics*. 2nd ed. Arlington, VA.: Resampling Stats.

*Stigler, S. 1986. *The History of Statistics*. Cambridge, Mass.: Harvard University Press.

*Tanur, J., et al. 2006. *Statistics: A Guide to the Unknown*. 4th ed. Pacific Grove, Calif.: Brooks/Cole.

Triola, M. 2010. *Minitab Student Laboratory Manual and Workbook*. 11th ed. Boston: Addison-Wesley.

Triola, M. 2010. *STATDISK 11 Student Laboratory Manual and Workbook*. 11th ed. Boston: Addison-Wesley.

Triola, M., and L. Franklin. 1994. *Business Statistics*. Boston: Addison-Wesley.

Triola, M., and M. Triola. 2006. *Biostatistics for the Biological and Health Sciences*. Boston: Addison-Wesley.

*Tufte, E. 2001. *The Visual Display of Quantitative Information*. 2nd ed. Cheshire, Conn.: Graphics Press.

Tukey, J. 1977. *Exploratory Data Analysis*. Boston: Addison-Wesley.

Zwillinger, D., and S. Kokoska. 2000. *CRC Standard Probability and Statistics Tables and Formulae*. Boca Raton, Fla.: CRC Press.

APPENDIX D: Answers to Odd-Numbered Exercises (and ALL Review Exercises and Cumulative Review Exercises)

Section 1-2

1. Continuous
3. Discrete
5. Statistic
7. Parameter
9. Ordinal
11. Nominal
13. Interval
15. Nominal
17. Ordinal
19. The calculated result is meaningless. Although the different categories (political parties) have each been arbitrarily assigned a number, these numbers have no real computational significance.

Section 1-3

1. People with cell phones and unlisted numbers and people without telephones are excluded.
3. A study sponsored by the fruit industry is much more likely to reach conclusions favourable to that industry.
5. Because the respondents are self-selected, the survey results are not likely to be valid at all.
7. 62% of 8% of 1875 is only 93.
9. Mothers who eat lobsters tend to be wealthier and therefore healthier.
11. A maker of shoe polish has an obvious interest in the importance of the product and there are many ways in which this could affect the survey results.
13. The mean of 69.5 for all males is measured from birth, whereas males don't become conductors until they have already survived for about 30 years. When this is taken into account, the mean of 73.4 years is not significant.
15. The wording of the question is biased and tends to encourage negative responses. The sample size of 20 is too small. Survey respondents are self-selected instead of being selected by the newspaper. If 20 readers respond, the percentages should be multiples of 5; 87% and 13% are not possible results.

Section 1-4

1. Observational study
3. Experiment

5. Convenience
7. Stratified
9. Random
11. Stratified
13. Cluster
15. Random
17. Answers can vary. For example, you may design a procedure based on randomly generating students' ID numbers.
19. a. An advantage of open questions is that they provide the subject and the interviewer with a much wider variety of responses; a disadvantage is that open questions can be very difficult to analyze.
    b. An advantage of closed questions is that they reduce the chance of misinterpreting the topic; a disadvantage is that closed questions prevent the inclusion of valid responses the pollster might not have considered.
    c. Closed questions are easier to analyse with formal statistical procedures.
21. a. Not always. No.
    b. Not always. No.

Chapter 1 Review Exercises

1. a. Continuous
   b. Ratio
   c. Stratified
   d. Observational study
   e. The products that use the batteries may be damaged.
2. a. Ratio
   b. Ordinal
   c. Ordinal
   d. Interval
   e. Nominal
3. Because it is a mail survey, the respondents are self-selected and will likely include those with strong opinions about the issue. Self-selected respondents do not necessarily represent the views of everyone who invests.
4. a. Design the experiment so that the subjects don't know whether they are using Statistiszene or a placebo, and design it so that those who observe and evaluate the subjects do not know which subjects are using Statistiszene and which are using a placebo.

b. Blinding will help to distinguish between the effectiveness of Statistiszene and the placebo effect, whereby subjects and evaluators tend to believe that improvements are occurring just because some given treatment is given.

c. Subjects are put into different groups through a process of *random selection*.

d. Subjects are carefully chosen for the different groups so that those groups are made to be similar in ways that are important for the study.

e. Replication is used when the experiment is repeated. The sample size should be large enough so that we can see the true nature of any effects. It is important so that we are not misled by erratic behaviour of samples that are too small.

5. a. Systematic
   b. Random
   c. Cluster
   d. Stratified
   e. Convenience

6. Respondents often tend to round off to a nice even number like 50.

7. The sample could be biased by excluding those who work, those who don't eat at school, those who commute, and so on.

8. The figure is very precise, but it is probably not very accurate. The use of such a precise number may incorrectly suggest that it is also accurate.

Chapter 1 Cumulative Review Exercises

1. The second version of the question is substantially less confusing because it doesn't include a double negative. One possibility for a better question:

   "Which of the following two statements do you agree with more?
   • The Nazi extermination of Jews never happened.
   • The Nazi extermination of Jews definitely happened."

2. Answer varies.

Section 2-2

1. Class width: 6. Class midpoints: 2.5, 8.5, 14.5, 20.5, 26.5. Class boundaries: 20.5, 5.5, 11.5, 17.5, 23.5, 29.5.

3. Class width: 2.0. Class midpoints: 0.995, 2.995, 4.995, 6.995, 8.995.
   Class boundaries: −.005, 1.995, 3.995, 5.995, 7.995, 9.995

5.

Absences	Frequency	Relative Frequency	Cumulative Relative Frequency
0–5	39	0.195	0.195
6–11	41	0.205	0.400
12–17	38	0.190	0.590
18–23	40	0.200	0.790
24–29	42	0.210	1.000

7.

Weight (kg)	Frequency	Relative Frequency	Cumulative Relative Frequency
0.00–1.99	20	0.133	0.133
2.00–3.99	32	0.213	0.349
4.00–5.99	49	0.327	0.673
6.00–7.99	31	0.207	0.880
8.00–9.99	18	0.120	1.000

9. a. 0.4
   b. 0.41
   c. 0.39

11. a. 0.673
    b. 0.867
    c. 0.533

13. In Exercise 1, the numbers of absences are (approximately) evenly spread over the five classes, but the absences in Exercise 2 have frequencies that start relatively low, increase to a maximum in the middle class, then decrease to the last class.

15. Possible solutions include 0.25–0.74, or 0.26–0.75.

17.

Size of Group	Frequency
0–999	17
1000–1999	9
2000–2999	5
3000–3999	5
4000–4999	1
5000–5999	4
6000–6999	2
7000–7999	2
8000–8999	1
9000–9999	0
10000–10999	0
11000–11999	1

19.

Unemployment Rate	Frequency
7.0–7.49	1
7.5–7.99	6
8.0–8.49	4
8.5–8.99	1
9.0–9.49	2
9.5–9.99	2
10.0–10.49	2
10.5–10.99	1
11.0–11.49	4
11.5–11.99	1

These data do not appear random with respect to time. For example, the eight unemployment rates with values greater than 10% appear to occur in clusters—in sequences of several years in a row.

(1) The first two classes contain over half the total frequency. That is, most SOC occupational groups in Newfoundland contain fewer than 2000 males.

(2) Although the first two classes contain over half the total frequency, the remaining frequencies are tretched across eight additional classes. (*We say that such distributions are highly skewed.*)

21. Relative frequencies for precipitation: 0.250, 0.125, 0.219, 0.188, 0.063, 0.125, 0.00, 0.00, 0.00, 0.031. Relative frequencies for snowfall: 0.063, 0.031, 0.156, 0.094, 0.125, 0.125, 0.063, 0.125, 0.156, 0.063. The distributions are quite dissimilar; the relative amounts of precipitation and snowfall do not seem to correspond.

23.

Axial Loads Without Outlier	
Axial Load	Frequency
200–219	6
220–239	5
240–259	12
260–279	36
280–299	87
300–319	28

Axial Loads With Outlier	
Axial Load	Frequency
200–219	6
220–239	5
240–259	12
260–279	36
280–299	87
300–319	28
320–339	0
340–359	0
360–379	0
380–399	0
400–419	0
420–439	0
440–459	0
460–479	0
480–499	0
500–519	1

Including the outlier dramatically changes the frequency table. Ten additional classes have to be added; yet all but one of the extra classes is empty.

25. Yes. Here is just one possible example of a dramatic change in apparent distribution due to changing the number of classes in a frequency table.

Illustrative Data Set

0	12	17
1	12	17
3	13	17
3	15	26
3	16	27
4	16	27
5	16	27
5	16	27
5	16	28
5	17	28
5	17	29

Distribution With Five Classes	
Classes	Frequency
0–5	11
6–11	0
12–17	14
18–23	0
24–29	8

Distribution With Six Classes	
Classes	Frequency
0–4	6
5–9	5
10–14	3
15–19	11
20–24	0
25–29	8

Section 2–3

1. Only one eruption out of 200 took as long as 109 min from the time of the preceding eruption, so allocating 100 min would please almost all bus trip participants.

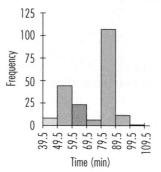

3.

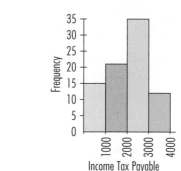

5. 570   571   577   581   583   583   584   589
   594   595   596   596   597   598   602   603

7.

3	67
4	00133
4	667889
5	022334
5	778999
6	0011123333444
6	556778
7	00222233
7	56

9. Networking appears to dominate as the most effective approach in getting a job.

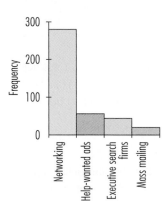

11.

13. The distribution of job-group-sizes appears to be similar for male and female Newfoundlanders. However, the number of job groups with about 500 members appears to be greater for women. The graph does not tell us if the groups with the higher (or lower) frequencies for men and women are actually the same SOC job categories in both cases.

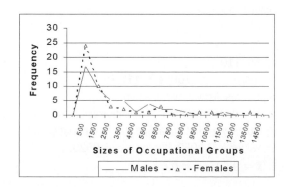

15.

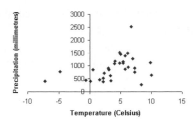

17. Skewed

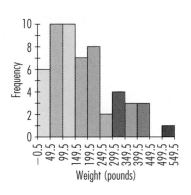

19. Bell-shaped

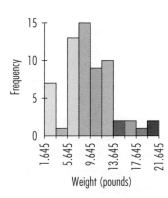

21. a. The two figures are given below. The histogram for π represents a more even distribution.

b. The number π is irrational (i.e., it cannot be represented as the quotient of two whole numbers). The decimal representation of the rational number 22/7 is obtained by dividing 7 into 22; because there are a maximum of 6 possible remainder digits at each step (1,2,3,4,5,6) that keep generating digits in the quotient, there are a maximum of six possible quotient digits that will repeat forever in a regular cycle.

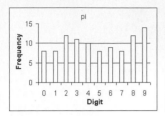 

23. The heights of the bars will be approximately halved. Also, depending on the nature of the data, the general shape may be changed.

Section 2-4

1. $\bar{x} = 16.2$; median $= 16$; mode $= 16$; midrange $= 16$

3. $\bar{x} = 6.14$; median $= 5.62$; mode: none; midrange $= 6.63$

5. Regular Coke: $\bar{x} = 0.37232$; median $= 0.37215$; (no mode); midrange $= 0.37270$.
Diet Coke: $\bar{x} = 0.35603$; median $= 0.35690$; (no mode); midrange $= 0.35575$.
Regular Coke does appear to weigh a small amount more than Diet Coke. Three measures of centre, the mean, median, and midrange, suggest the same result. It is possible that the sugar in Regular Coke weighs more than the artificial sweetener in Diet Coke.

7. 4000 BCE: $\bar{x} = 128.7$; median $= 128.5$; mode $= 131$; midrange $= 128.5$.
150 CE: $\bar{x} = 133.3$; median $= 133.5$; mode $= 126$; midrange $= 133.5$.
Yes. Three measures of centre, the mean, median, and midrange, suggest that head sizes increased over time. The modes appear to contradict this finding, but they are based on just two data values for each time period—and could just be coincidences.

9. $\bar{x} = 37.00$; median $= 36.96$; mode $= 36.91, 36.96, 37.22$; midrange $= 36.92$.

11. $\bar{x} = 182.9$; median $= 150.0$; mode $= 140, 150, 166, 202, 204, 220$; midrange $= 270.0$.

13. 74.4 min

15. 380. This year's level of 492 Mt is much higher than the average.

17. 82.0

19. a. $\bar{x} = 226,500$; median $= 175,600$; mode: none; midrange $= 251,450$.
b. Each result is increased by k.

c. Each result is multiplied by k.
d. They're not equal.

21. 1.092

23. a. 58,701
b. 60,067
c. 61,433
In this case, the open-ended class doesn't have too much effect on the mean. The mean is likely to be around 60,000, give or take about 1000.

25. a. 182.9 lb
b. 171.0 lb
c. 159.2 lb
The results differ by substantial amounts, suggesting that the mean of the original set of weights is strongly affected by extreme scores.

Section 2-5

1. range $= 4$; $s^2 = 1.64$; $s = 1.28$

3. range $= 3.33$; $s^2 = 1.223$; $s = 1.106$

5. Regular Coke: Range $= 0.0044$; $s^2 = 0.0000025337$; $s = 0.0015917$.
Diet Coke: Range $= 0.0063$; $s^2 = 0.0000063507$; $s = 0.0025201$.
By all these measures, there seems to be slightly more variation in the weight of Diet Coke than in the weight of Regular Coke.

7. 4000 BCE: range $= 19.0$; $s^2 = 21.5$; $s = 4.6$
150 CE: range $= 15.0$; $s^2 = 25.2$; $s = 5.0$

9. 0.233

11. 121.8

13. 14.7

15. 53.09 Mt

17. The population of batteries with $\sigma = 1$ month are much more consistent, so they have a smaller chance of failing much earlier than expected.

19. Dollars. Dollars.[2.]

21. a. 68%
b. 95%
c. 50, 110

23. a. range $= 259,700$; $s = 112,707$
b. The results will be the same.
c. The range and standard deviation will be multiplied by k.
d. No, they are not equal. The standard deviation of the log x values is 0.2062, but log $s = 5.0520$.
e. $\bar{x} = 97.45$; $s = 0.63$

25. Section 1: range = 19.0; s = 5.7
Section 2: range = 17.0; s = 6.7
The ranges suggest that Section 2 has less variation, but the standard deviations suggest that Section 1 has less variation.

27. I = 20.80; there is not significant skewness.

Section 2-6

1. a. −1.8
   b. 1.0
   c. 0.2
3. 2.77
5. 2.56; unusual
7. −2.43; no
9. 2.00; 2.50; 250 is better
11. 1.50; 1.54; 1.47; 398 is highest
13. 17
15. 60
17. 279
19. 276.5
21. 282
23. 230
25. 13
27. 58
29. 30850
31. 1825
33. 98740
35. 9385
37. a. 20
   b. 272
   c. 59
   d. Yes; yes
   e. No
39. a. Uniform
   b. Bell-shaped
   c. The shape of the distribution remains the same.
   d. There is no effect on the distribution of z scores.

Section 2-7

1.

3.

Actors:

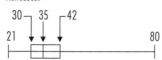

Actresses:

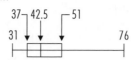

Oscar-winning actresses tend to be younger than actors.

5.

Male:

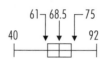

Female:

The two groups do not appear to be different.

7.

Red:

Yellow:

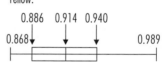

Red and yellow M&Ms appear to have weights that are about the same.

9.

11. Both distributions are skewed.

Chapter 2 Review Exercises

1.

Volume	Frequency
14–42	21
43–71	6
72–100	4
101–129	8
130–158	2
159–187	3
188–216	7
217–245	3
246–274	4
275–303	2

2.

Volume	Relative Frequency
14–42	0.350
43–71	0.100
72–100	0.067
101–129	0.133
130–158	0.033
159–187	0.050
188–216	0.117
217–245	0.050
246–274	0.067
275–303	0.033

3.

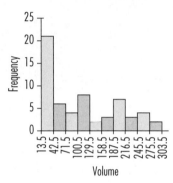

4. a. 25.5
   b. 72.5
   c. P_{62}

5. $s \approx 72.25$

6. $\bar{x} = 112.6$; $s = 85.5$

7.

0–4	*45788*4*4*2
5–9	78**5*3*
10–14	*68*56**
15–19	***1*2
20–24	3*2***
25–29	4*3***2
30–34	3****

8.

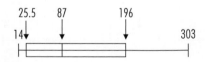

9. a. 124.4
   b. 121.0
   c. 135
   d. 148.0
   e. 158.0
   f. 48.6
   g. 2364.5

10. a. 23.08
   b. 23
   c. 23
   d. 30
   e. 30
   f. 5.77
   g. 33.28
   h. 19
   i. 19
   j. 24

11. a. No, the z score is 1.5, so 260 is within 2 standard deviations of the mean.
   b. -0.38
   c. About 95% of the scores should fall between 120 and 280.
   d. 220
   e. 40

12. 5.15; 1.67; 8 years is not unusual because it is within 2 standard deviations of the mean.

13.

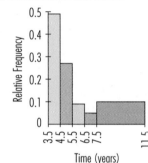

14. The score of 450 is better because its z score of -0.63 is greater than the other z score of -0.75.

15. The mean appears to be greater in 150 CE, so there does appear to be an upward shift in the maximum skull breadth.

16.

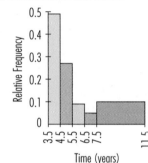

Chapter 2 Cumulative Review Exercises

1. a. $\bar{x}$ = 2.40; median = 2.20; mode: 0.0, 2.0, 2.1, 2.4, 4.4; midrange = 2.25

  b. s = 1.29; s^2 = 1.67; range = 4.50

  c. Continuous

  d. Ratio

2. a. Mode, because the other measures of central tendencies require calculations that cannot (or should not) be done with data at the nominal level of measurement.

  b. Convenience

  c. Cluster

3. No, the values should be weighted according to the corresponding number of cars sold. The weighted mean is $29,411.76.

Section 3-2

1. $-0.2, 3/2, \sqrt{2}$

3. 1/2

5. 0.210

7. 0.25; no

9. 0.320

11. 0.450. Not unusual.

13. a. 1/365. Unusual.

  b. 6/73 (reduced from 30/365). Not unusual.

15. 0.130. Not unusual!

17. 0.340; Not unusual; yes

19. 0.0896. Not unusual; yes.

21. 0.200. Not unusual.

23. a. bb, bg, gb, gg

  b. 1/4

  c. 1/2

25. b. 1/8

  c. 1/8

  d. 1/2

27. 4:1

29. a. 37:1

  b. 35:1

  c. The casino pays off at 35:1 so that it will make a profit, instead of paying off at 37:1, which would not yield a profit.

31. 3/13

33. 0.600

35. a. 4/1461

  b. 400/146,097

Section 3-3

1. a. No

  b. No

  c. Yes

3. a. 3/5

  b. 0.72

5. a. 4/13 (reduced from 16/52)

  b. 2/13 (reduced from 8/52)

7. 0.698

9. 0.290

11. 0.580

13. 0.62

15. 0.740

17. 0.54

19. 0.23

21. 0.15

23. 0.17

25. a. 17/60

  b. $P(A \text{ or } B)$ = 0.9

  c. $P(A \text{ or } B) < 0.9$

27. $P(A \text{ or } B) = P(A) + P(B) - 2P(A \text{ and } B)$

29. a.

	Inner City	Suburb	Outside	Total
Yes	8.5	35.8	0.9	45.2
No	1.8	27.6	2.9	32.3
Undecided	2.3	15.9	4.3	22.5
Total	12.6	79.3	8.1	100

  b. P(city or not yes) = P(city) + P(not yes) − P(city and not yes) = 99.1%

  c. P(city or not yes) = 100% − P(outside and yes) = 100% − 0.9% = 99.1%

Section 3-4

1. a. Dependent

  b. Independent

  c. Dependent

3. a.

	Import	Not Import	Total
Over 5	9	19	28
Not over 5	34	38	72
Total	43	57	100

  b. 0.2093

  c. 0.4722

  d. 0.6667

  e. LS = 0.09; RS = 0.1204; dependent

5.

	Income <100K	Income ≥100K	Total
Charity < 200	50	13	63
Charity ≥ 200	22	15	37
Total	72	28	100

   b. 0.6944

   c. 0.4643

   d. P(charity < 200) = 0.63 not equal either conditional probability; dependent

7. 0.16

9. a.

	Exercise	Not Exercise	Total
Age ≥80	14.7764	15.9176	30.694
Age <80	6.4236	62.8824	69.306
Total	21.2	78.8	100

   b. 0.4814

   c. 0.9073

   d. LS = 0.147764; RS = 0.065; dependent

11. a.

	< 25K	25K to 50K	50K+	Total
Male	60	185	155	400
Female	150	387	63	600
Total	210	572	218	1000

   b. 0.3875

   c. 0.6867

   d. LS = 0.155; RS = 0.0872; dependent

13. a. 0.5875

   b. 0.0107

   c. 0.1739

   d. 0.3125

15. 0.1

17. 0.55

19. 0.1

Section 3–5

1. a. 0.17361

   b. 0.15152

3. 0.00000369

5. 1/64

7. 0.0643; not common.

9. 0.0000416

11. 0.00000410; fraud

13. 0.3585; not strong evidence

15. a. 0.580495

   b. P(B | high-end) = 0.3594

   c. P(C | not high-end) = 0.6177

17. a. 0.439608

   b. 0.8692

19. 0.0192

Section 3–6

1. 720

3. 970,200

5. 720

7. 15

9. 79,833,600

11. 3,838,380

13. 1

15. $n!$

17. a. 10,000

   b. 13.9 hours

19. 1/1,000,000,000

21. 43,758

23. 5040

25. a. 3,838,380

   b. 1/3,838,380

   c. 3,838,379:1

27. 14,348,907

29. a. 230,300

   b. 1/230,300

31. 259,459,200

33. a. 17,576,000

   b. 1/17,576,000

35. 1/76,904,685; yes, if all other factors are equal.

37. 1/479,001,600; yes

39. 1,000,000; 151,200; 786,240

41. 2,095,681,645,538 (about 2 trillion)

43. a. Calculator: 3.0414093×10^{64}; approximation: 3.0363452×10^{64}

   b. 615

Chapter 3 Review Exercises

1. 0.35

2. 0.57

3. 0.0896

4. 0.18

5. 0.6

6. 0.27

7. 0.55

8. 0.51

9. 0.358

10. 0.0483; should not speed.

11. a. 1/56
    b. 336
12. 0.513
13. a. 9/19
    b. 10:9
    c. $5
14. 1/120
15. 0.000000531; no.
16. 1/12,870

Chapter 3 Cumulative Review Exercises

1. a. 67.5
   b. 13.09
   c. 0.683
   d. 0.049
   e. 0.0000011
2. a. 1/4
   b. 3/4
   c. 1/16
3. a. 151.3
   b. 90.5
   c. 130.0
   d. 337
   e. 0.667
   f. 0.424

Section 4-2

1. Continuous
3. Discrete
5. Probability distribution with $\mu = 1.799$, $\sigma^2 = 0.613$, $\sigma = 0.782$
7. Not a probability distribution because $\Sigma P(x) \neq 1$.
9. Probability distribution with $\mu = 1.3$, $s^2 = 0.71$, $\sigma = 0.84$
11. Probability distribution with $\mu = 0.5001$, $\sigma^2 = 0.4504$, $\sigma = 0.6711$
13. −26¢; 5.26¢
15. $2.00
17. $\mu = 1.5$, $\sigma^2 = 0.8$, $\sigma = 0.9$; minimum $= -0.3$ and maximum $= 3.3$, but reality dictates that the minimum and maximum are 0 and 3.
19. $\mu = 0.4$, $\sigma^2 = 0.3$, $\sigma = 0.5$
21. a. Yes
    b. No, $\Sigma P(x) > 1$
    c. Yes
    d. Yes

23. a. $\mu = 4.5$, $\sigma = 2.9$
    b. $\mu = 0$, $\sigma = 1$
    c. Yes
25. Same as the table given in Exercise 6; $\mu = 2.0$, $\sigma = 1.0$

Section 4-3

1. Not binomial; more than two outcomes
3. Binomial
5. Not binomial; more than two outcomes
7. Not binomial; trials are not independent
9. 0.243
11. 0.075
13. 0.141
15. 0.238
17. 0.032816
19. 0.993785
21. a. 0.205
    b. 0.828
    c. 0.120
23. 0.250
25. 0.007; study
27. 0.0095
29. 0.392 suggests that the defect rate could easily be 10%; there isn't sufficient evidence to conclude that the newly instituted measures were effective in lowering the defect rate.
31. 0.297; the results could occur by chance.
33. 0.999992
35. 0.0467

Section 4-4

1. $\mu = 32.0$, $\sigma^2 = 16.0$, $\sigma = 4.0$
3. $\mu = 267.0$, $\sigma^2 = 200.3$, $\sigma = 14.2$
5. $\mu = 12.5$, $\sigma^2 = 6.3$, $\sigma = 2.5$ Yes, because 20 randomly guessed correct answers is more than two standard deviations above the mean.
7. $\mu = 13.2$, $\sigma = 3.6$ Yes, because zero sevens in 500 trials is more than two standard deviations below the mean.
9. a. $\mu = 1140.0$, $\sigma = 7.5$
   b. Yes
11. a. $\mu = 153$, $\sigma = 11.27$
    b. No; the result is within two standard deviations.
13. a. $\mu = 1160$, $\sigma = 28.7$
    b. Yes, the result is higher than usual (not within two standard deviations); perhaps it was an exciting game!
15. a. $\mu = 1365$, $\sigma = 31.5$
    b. Yes, the result is unusual (not within two standard deviations).

17. a. $\mu = 44.6, \sigma = 6.1$
   b. 32.4, 56.9

Section 4-5

1. 0.180
3. 0.153
5. a. 0.039
   b. 0.127
   c. 0.181
7. a. 0.819
   b. 0.164
   c. 0.0164
9. 0.594
11. a. 0.497
   b. 0.348
   c. 0.122
   d. 0.0284
   e. 0.00497
   The expected frequencies of 139, 97, 34, 8, and 1.4 compare reasonably well to the actual frequencies of 144, 91, 32, 11, and 2. The Poisson distribution does provide good results.
13. Table: 0.130; Poisson formula: 0.129

Section 4-6

1. a. 1/13,983,816
   b. 258/13,983,816
   c. 246,820/13,983,816
   d. 6,096,454/13,983,816
3. 0.787
5. a. 0.2628
   b. 0.2626

Chapter 4 Review Exercises

1. a. A random variable is a variable that has a single numerical value (determined by chance) for each outcome of an experiment.
   b. A probability distribution gives the probability for each value of the random variable.
   c. Yes, because each probability value is between 0 and 1 and the sum of the probabilities is 1.
   d. $\mu = 3.4$
   e. $\sigma = 0.7$
2. a. 7.5
   b. 7.5

c. 2.5
d. Yes, using the range rule of thumb, the number is usually between 2.5 and 12.5, so 15 is unusually high.
e. 0.123
3. a. 0.216
   b. 0.798
   c. $\mu = 5.8, \sigma = 1.56$
4. a. 0.394
   b. 0.00347
   c. 0.000142
   d. 0.0375
5. a. 0.85
   b. 0.84
   c. Yes, because 4 is more than two standard deviations above the mean.
6. a. $\mu = 0.230$
   b. 0.0210

	r.f.
4	0.146
5	0.247
6	0.225
7	0.382

7. 0.877

Chapter 4 Cumulative Review Exercises

1. a.
   b. Yes, each cumulative frequency is a value between 0 and 1, and the sum of the frequencies is 1.
   c. 0.854
   d. 0.143
   e. 5.8
   f. 1.1
   g. 5.8 games; about 175,000 hot dogs
2. a. $\bar{x} = 9.7, s = 2.9$
   b. 0.4 isn't close to 0.0278
   c. 0.431
   d. Claim that the sample of 20 results is too small to obtain meaningful results.

Section 5-1

1. 0.6
3. 0.4
5. $\mu = 2.5; \sigma = 1.4434$
7. 0.3333

9. $P(\mu - 2\sigma) < X < \mu + 2\sigma) = 4\sigma/(b - a)$
Since $\sigma = (b - a)/\sqrt{12}$, we get $P(\mu - 2\sigma < X < \mu + 2\sigma) = 4/\sqrt{12} > 1$.

Since the probability cannot exceed 100%, we can deduce that 100% of a uniform distribution lies within 2 standard deviations of the mean, regardless of the values of a and b.

Section 5–2

1. 0.0987
3. 0.3133
5. 0.4987
7. 0.4901
9. 0.0049
11. 0.0183
13. 0.0863
15. 0.1203
17. 0.5319
19. 0.9890
21. 0.9545
23. 0.8412
25. 0.0099
27. 0.9759
29. 1.28°
31. −0.67°
33. 1.75°
35. −2.05°
37. a. 0.92
    b. 0.41
    c. 0.72
    d. −0.68
    e. −0.23
39. $\mu = 24$; $\sigma = 0.02$

Section 5–3

1. 0.1217
3. 0.9906
5. 0.2364
7. 0.1379
9. 0.0244
11. 0.0038; either a very rare event has occurred or the husband is not the father.

13. 0.9554
15. a. 0.0179
    b. 1343
17. 0.1706. This answer assumes a normal distribution, but the expenditure distribution is more likely to be bimodal: Many households will spend nothing on postsecondary textbooks (because no one is attending postsecondary institutions); and the minority who are buying textbooks will spend a great deal.
19. 0.0013. No, it is rare for a person in this age group to have a seriously elevated serum cholesterol level.
21. a. Normal distribution
    b. $\bar{x} = 0.9147, s = 0.0369$
    c. 0.0104
23. a. Normal distribution
    b. $\bar{x} = 35.66, s = 9.35$
    c. 0.271

Section 5–4

1. 66.9 in.
3. 61.5 in.
5. a. 9.1 years
    b. 0.0018
    c. 5.6 years.
7. 3.5 kg
9. 242 days
11. 135. Perhaps the employees of a "think tank" company could be held to this standard.
13. a. 0.9656
    b. 658.9
15. 3497.4
17. 11.9
19. 131.6; 782

Section 5-5

1. a. 0.1480
   b. 0.4896
3. a. 0.3783
   b. 0.0146
5. 0.1251
7. a. 0.0274
   b. 0.12%; level is acceptable
9. a. 0.3446
   b. 0.0838
   c. Because the original population has a normal distribution, samples of *any* size will yield sample means that are normally distributed.
11. 0.0668
13. 0.9345
15. 0.0392
17. a. 0.2981
    b. 0.0038
    c. Yes, because it is highly unlikely (with a probability of only 0.0038) that the mean will be that low because of chance.
19. a. 0.5675
    b. 0.9999
    c. Yes
21. a. 8; 5.385
    b. 2,3; 2,6; 2,8; 2,11; 2,18; 3,6; 3,8; 3,11; 3,18; 6,8; 6,11; 6,18; 8,11; 8,18; 11,18
    c. 2.5; 4.0; 5.0; 6.5; 10.0; 4.5; 5;5; 7.0; 10.5; 7.0; 8.5; 12.0; 9.5; 13.0; 14.5
    d. 8; 3.406
    e. $\mu_{\bar{x}} = 8 = \mu$;
       $\sigma_{\bar{x}} = 3.406 = \sigma/\sqrt{n} \cdot \sqrt{[(N-n)/(N-1)]} = 3.80777 \cdot 0.894427$
23. They are all near 0.975.

Section 5-6

1. The area to the right of 35.5
3. The area to the left of 41.5
5. The area to the left of 72.5
7. The area between 124.5 and 150.5
9. a. 0.183
   b. 0.1817
11. a. 0.996
    b. Normal approximation is not suitable.
13. 0.1841

15. 0.1020
17. 0.1159
19. 0.0668; not *very* unusual
21. 0.4207
23. 0.9821; possibly yes, depending on desired significance level
25. 0.1922; not unusual. The claim is accurate.
27. 0.7734
29. 0.2946
31. a. $0.4129 - 0.3264 = 0.0865$
    b. $0.3192 - 0.2643 = 0.0549$
    c. $0.0256 - 0.0228 = 0.0028$
    As *n* gets larger, the difference becomes smaller.

Section 5-7

1. 0.1534
3. 0.6065
5. 0.2865
7. 0.6471
9. 0.2564; not realistic

Chapter 5 Review Exercises

1. 60.83 in., 75.17 in.
2. 8.4%
3. 40.6%
4. 0.9381
5. 0.5198
6. a. 0.3944
   b. 0.3085
   c. 0.8599
   d. 0.6247
   e. 0.2426
7. 79 (rounded)
8. 671
9. 0.4168
10. a. 0.9115
    b. 0.8212
    c. 44,375 km
11. 0.5714
12. 0.7857
13. 0.4286
14. 0.2636
15. 0.3679
16. 0.3442
17. 0.7981

Chapter 5 Cumulative Review Exercises

1. a. 0.0005
   b. 0.243
   c. The requirement that $np \geq 5$ is not satisfied, indicating that the normal approximation would result in errors that are too large.
   d. 8.1
   e. 2.73
   f. No, 10 is within two standard deviations of the mean and is within the range of values that could easily occur by chance.
2. a. 200.46 ms
   b. 200.45 ms
   c. 200.5 ms
   d. 0.29 ms
   e. 0.14
   f. 7.5%
   g. 3.14% (using $\bar{x} = 200.46$ and $s = 0.29$) h. Yes

Section 6-2

1. 2.575
3. 2.33
5. a. 0.7 in.
   b. 63.5 in. $< \mu <$ 64.9 in.
7. a. 1.9
   b. 75.7 $< \mu <$ 79.5
9. 133.3 mm $< \mu <$ 135.7 mm; we are 95% confident that the interval from 133.3 mm to 135.7 mm actually does contain the true population mean.
11. 0.9499 mm $< \mu <$ 0.9513 mm. We are 90% confident that the mean length is very close to the specification (0.950 mm). With regard to the range specification (0.945 to 0.955), some measure of variation for the sample would be needed.
13. 543
15. 5.600 g $< \mu <$ 5.644 g; no, the quarters become lighter as they wear down.
17. 432
19. 3453.98$< \mu <$ 3476.94. No, the value claimed for the mean is outside the confidence interval.
21. 236
23. $\sigma = 3$, n = 782
25. $n = 62, \bar{x} = 9.428, s = 4.168$; 8.502 lb $< \mu <$ 10.354 lb
27. $n = 33$, $\bar{x} = 0.9128$, $s = 0.0395$; 0.8979 $< \mu <$ 0.9277; no, the sample is small.
29. a. 129.1 $< \mu <$ 134.9
    b. 147

31. (No confidence level is stated; the commonly used 95% confidence intervals are given here.) Men: 1.7324 $< \mu <$ 1.7616 Women: 1.5978 $< \mu <$ 1.6222. No. Because their respective confidence intervals do not overlap, it is unlikely that the populations of men's and women's heights have the same mean; so it is reasonable to conclude that the populations are different.

Section 6-3

1. 3.250
3. 2.528
5. a. 1.86 in.
   b. 62.3 in. $< \mu <$ 66.1 in.
7. a. 6.2
   b. 71.4 $< \mu <$ 83.8
9. $1066 $< \mu <$ $2506
11. 1.79 h $< \mu <$ 3.01 h. No, based on the sample, I am 47.5% confident that for the population, the mean hours required for paperwork is actually in the range from above 2.4h to 3.01h—which would leave less than 5.6 hours available for other work.
13. 3.84 $< \mu <$ 4.04
15. 43.13 h $< \mu <$ 45.37 h
17. $\bar{x} = 7.96, s = 1.60$; 7.11 $< \mu <$ 8.81
19. $n = 100, \bar{x} = 0.9147$ g, $s = 0.0369$ g; 0.9075 g $< \mu <$ 0.9219 g; there is reasonable agreement with the result from Exercise 18.
21. The confidence interval limits are closer than they should be.
23. 90% (approx.)

Section 6-4

1. 0.0300
3. 0.0174
5. 0.720 $< p <$ 0.780
7. 0.385 $< p <$ 0.417
9. 2401
11. 664
13. a. 64.0%
    b. 61.3% $< p <$ 66.7%
15. a. 0.134% $< p <$ 6.20% using x = 7, n = 221.
    b. It has not been shown that Ziac is causing dizziness as an adverse reaction.
17. 72.8% $< p <$ 92.0%
19. 9.53% $< p <$ 18.5%; yes
21. 2944
23. 82.8% $< p <$ 85.2%; yes

25. a. $0.00916 < p < 0.0233$
  b. 4229
27. $0.424 < p < 0.764$; yes, since 0.50 is within the interval
29. 1853
31. 89%
33. $p > 0.818$; 81.8%

Section 6–5

1. 13.120, 40.646
3. 43.188, 79.082
5. 1.79 in. $< \sigma < 4.75$ in.
7. $11.0 < \sigma < 20.4$
9. $1.65 < \sigma < 4.38$
11. $1.48 < \sigma < 5.47$; wider
13. 38.2 mL $< \sigma < 95.7$ mL; no, the fluctuation appears to be too high.
15. 3.0 mm $< \sigma < 5.0$ mm
17. a. 0.33 min $< \sigma < 0.87$ min
  b. 1.25 min $< \sigma < 3.33$ min
  c. The variation appears to be significantly lower with a single line. The single line appears to be better.
19. a. $s \approx 81.75$
  b. $66.22 < s < 130.65$
  c. Yes (Exclude the data for Victoria)
21. a. 98%
  b. 27.0

Chapter 6 Review Exercises

1. a. $22039.6 < \mu < 53991.8$
  b. We can see (from the wide difference between mean and median) that the distribution is far from normal; but the methods in Section 6–5 have a strong requirement of normality.
2. 2944
3. 5.47 yr $< \mu < 8.55$ yr
4. 2.92 yr $< s < 5.20$ yr
5. $0.509 < p < 0.571$; 3.1 percentage points
6. 221
7. $n = 16, \bar{x} = 72.6625, s = 2.5809236$; 71.29 cm $< \mu$ < 74.04 cm
8. $36.5 < \mu < 44.9$
9. 110
10. $21.5\% < p < 26.5\%$

Chapter 6 Cumulative Review Exercises

1. a. 70.20 in.
  b. 70.45 in.

c. 70.4 in.
d. 69.65 in.
e. 7.5 in.
f. 4.26 in.^{2}
g. 2.07 in.
h. 68.95 in.
i. 70.45 in.
j. 71.75 in.
k. ratio
l.

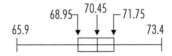

m. Answer varies, depending on the choices for the number of classes and the starting point. The histogram is approximately bell-shaped.
n. 69.02 in. $< \mu < 71.38$ in.
o. 1.49 in. $< \sigma < 3.25$ in.
p. 707
q. No, as neither the mean nor the variance are in the confidence intervals.
2. a. 0.0089
  b. $0.260 < p < 0.390$
  c. Because the confidence interval limits do not contain 0.25, it is unlikely that the expert is correct.

Section 7–2

1. Reject the claim that the coin is fair.
3. Do not accept the claim that women who eat blue M&M candies have a better chance of having a boy.
5. a. $\mu \neq 650$ mg
  b. $H_0: \mu = 650$ mg
  c. $H_1: \mu \neq 650$ mg
  d. Two-tailed
  e. The error of rejecting the claim that the mean equals 650 mg when it really does equal 650 mg.
  f. The error of failing to reject the claim that the mean is equal to 650 mg when it is really different from 650 mg.
  g. There is sufficient evidence to support the claim that the mean is different from 650 mg.
  h. There is not sufficient evidence to support the claim that the mean is different from 650 mg.
7. a. $p > 0.5$
  b. $H_0: p \leq 0.5$
  c. $H_1: p > 0.5$
  d. Right-tailed

e. The error of rejecting the claim that she is favoured by at most 1/2 of the voters when she really is favoured by at most 1/2 of them.

f. The error of failing to reject the claim that she is favoured by at most 1/2 of the voters when she is really favoured by more than 1/2 of them.

g. There is sufficient evidence to support the claim that she is favoured by more than 1/2 of the voters.

h. There is not sufficient evidence to support the claim that she is favoured by more than 1/2 of the voters.

9. a. $p > 0.02$

b. $H_0: p \leq 0.02$

c. $H_1: p > 0.02$

d. Right-tailed

e. The error of rejecting the hypothesis that the proportion is at most 0.02 when it really is at most 0.02.

f. The error of failing to reject the hypothesis that the proportion is at most 0.02 when it really is greater than 0.02.

g. There is sufficient evidence to support the claim that the proportion is greater than 0.02.

h. There is not sufficient evidence to support the claim that the proportion is more than 0.02.

11. a. $p < 0.1$

b. $H_0: p > 0.1$

c. $H_1: p < 0.1$

d. Left-tailed

e. The error of rejecting the hypothesis that the proportion is at least 0.1 when it really is at least 0.1.

f. The error of failing to reject the hypothesis that the proportion is at least 0.1 when it is really less than 0.1.

g. There is sufficient evidence to support the claim that the proportion is less than 0.1.

h. There is not sufficient evidence to support the claim that the proportion is less than 0.1.

13. 1.645

15. −2.33

17. ±1.645

19. −1.96

21. 14.61

23. −38.82

25. There must be a fixed value of μ so that a specific normal distribution can be used for calculation of the test statistic.

27. a. 0.0294

b. 0.9706; the probability that the null hypothesis will correctly be rejected if the alternate hypothesis is true.

Section 7–3

1. Test statistic: $z = 2.00$. Critical values: $z = \pm 1.96$. Reject $H_0: \mu = 75$. There is sufficient evidence to warrant rejection of the claim that the mean equals 75. P-value: 0.0456.

3. Test statistic: $z = -0.50$. Critical value: $z = -2.05$. Fail to reject $H_0: \mu \geq 2.50$. There is not sufficient evidence to support the claim that the mean is less than 2.50. P-value: 0.3085.

5. Test statistic: $z = 1.87$. Critical values: $z = \pm 1.96$. Fail to reject $H_0: \mu = 92.14$ in. There is not sufficient evidence to support the claim that the mean is different from 92.14 in. These balls do not appear to be "juiced." P-value: 0.0614.

7. Test statistic: $z = 14.61$. Critical value: $z = 2.33$. Reject $H_0: \mu \leq 0.21$. There is sufficient evidence to support the claim that the mean is greater than 0.21. Compulsive buyers do seem to get substantially higher scores than the general population. P-value: 0.0001.

9. Test statistic: $z = -1.30$. Critical value: $z = -1.645$ (assuming a significance level of 0.05). Fail to reject $H_0: \mu \geq 10\%$. There is not sufficient evidence to support the claim that Canadian equity funds fail to meet your investment criterion for foreign equity percentage. If the conclusion is a type II sampling error, it is possible that you would invest in a funds group that does not in fact meet your investment criterion. P-value: 0.0968.

11. Test statistic: $z = 3.36$. Critical value: $z = \pm 2.575$. Reject $H_0: \mu = 0.075$. There is sufficient evidence to warrant rejection of the claim that the mean accident rate for young adult males undergoing voluntary treatment for substance abuse equals the general driving population's mean accident rate of 0.075. The Ministry of Transportation has grounds to encourage these at-risk drivers who are seeking help. P-value: 0.0002.

13. Test statistic: $z = -1.31$. Critical value: $z = -1.645$. Fail to reject $H_0: \mu \geq 7.5$ yr. There is not sufficient evidence to support the manager's claim that the mean is less than 7.5 yr. P-value: 0.0951.

15. Test statistic: $z = -4.22$. Critical values: $z = \pm 2.575$. There is sufficient evidence to warrant rejection of the claim that the mean is equal to 600 mg. Don't buy this cold medicine because it is not as described on the label. P-value: 0.0002.

17. Test statistic: $z = -4.16$. Critical value: $z = -2.575$. Reject H_0: $\mu \geq 12$ mo. The sample data support the claim that the mean shelf life is less than 12 months. *P*-value: 0.0001.

19. Test statistic: $z = -8.63$. Critical value: $z = 1.645$. Fail to reject H_0: $\mu \leq 420$ hr. There is not sufficient evidence to support the claim of improved reliability. (In fact, it appears that quality has deteriorated by a significant amount.) *P*-value: 0.9999.

21. Test statistic: $z = -4.78$. Critical value: $z = -2.33$. Reject H_0: $\mu \geq 35$ lb. There is sufficient evidence to support the claim that the mean is less than 35 lb. *P*-value: 0.0001.

23. Test statistic: $z = 0.7135$. Critical value: $z = \pm1.96$. Fail to reject H_0: $\mu = 4.5$. There is not sufficient evidence to warrant rejection of the claim that the mean of the last digits of the numbers of sitting days is 4.5. In this respect, the numbers give no evidence of inaccurate reporting. *P*-value: 0.4755.

25. 11.0

27. 0.209

29. $\mu = 25.702$

31. 0.0797

Section 7-4

1. a. ±2.447
   b. -1.796
   c. 2.896

3. Test statistic: $t = -1.991$. Critical values: $t = \pm2.201$. Fail to reject H_0: $\mu = 64.8$. There is not sufficient evidence to support the claim that the mean is not equal to 64.8. From Table A-3, we get $0.05 < P\text{-value} < 0.10$; which leads to the same conclusion.

5. Test statistic: $t = 1.489$. Critical values: $t = \pm3.250$. *P*-value: 0.171. Fail to reject H_0: $\mu = \$1750$ (based on either the test statistic or the *P*-value > 0.01). There is not sufficient evidence to warrant rejection of the claim that the prices quoted in the paper have the same mean as the industry average.

7. Test statistic: $t = 0.626$. Critical value: $t = 2.132$. Fail to reject H_0: $\mu \leq 377$ yd. There is not sufficient evidence to support the scout's claim that the mean annual punt-return yardage for his recruits is greater than the mean for top-level punt returners.

9. Test statistic: $t = 4.662$. Critical values: $t = \pm2.228$. Reject H_0: $\mu = 37.0°$C. There is sufficient evidence to warrant rejection of the claim that the mean is equal to $37.0°$C.

11. Test statistic: $t = 3.344$. Critical value: $t = 2.718$. Reject H_0: $\mu \leq \$1800$. There is sufficient evidence to support the claim that the mean exceeds $1800.

13. Test statistic: $z = 1.00$. Critical value: $z = 1.96$. Fail to reject H_0: $\mu \leq \$2000$. There is not sufficient evidence to support the claim that the mean is greater than $2000. The new monitoring system will not be implemented.

15. Test statistic: $t = -3.408$. Critical value: $t = -2.602$. Reject H_0: $\mu \geq 10.00$. There is sufficient evidence to support the claim that the mean rating is less than 10.00.

17. Test statistic: $t = -6.158$. Critical value: $t = -2.764$. Reject H_0: $\mu \geq 2.4$ h. There is sufficient evidence to support the claim that the mean flight time is less than 2.4 h.

19. Test statistic: $t = -0.277$. Critical value: $t = -2.132$. Fail to reject H_0: $\mu \geq 0.9085$. There is not sufficient evidence to warrant rejection of the claim that the mean is at least 0.9085 g. We cannot conclude that the package contents disagree with the claimed weight printed on the label.

21. $n = 7, \bar{x} = 10601.2857, s = 1026.7175$. Test statistic: $t = -1.027$. Critical values: $t = \pm2.447$. Fail to reject H_0: $\mu = 11{,}000$ kWh. There is not sufficient evidence to warrant rejection of the claim that the mean is equal to 11,000 kWh.

23. $n = 15, \bar{x} = 69.5, s = 19.3$. Test statistic: $t = 1.264$ (1.271 is more accurate). Critical value: $t = 1.761$ (assuming that $\alpha = 0.05$). Fail to reject H_0: $\mu \leq 63.2$ s. There is not sufficient evidence to support the claim that the mean is greater than 63.2 s. Her new show does not appear to be significantly better.

25. Critical t changes to 2.821; *P*-value changes to 0.0855.

27. With $z = 1.645$, the table and the approximation both result in $t = 1.833$.

Section 7-5

1. Test statistic: $z = 2.19$. Critical values: $z = \pm1.96$. Reject H_0: $p = 0.01$. There is sufficient evidence to warrant rejection of the claim that with scanners, 1% of sales are overcharges. Based on these results, scanners appear to increase overcharges instead of helping to avoid them. *P*-value: 0.0286.

3. Test statistic: $z = 1.25$. Critical value: $z = 1.645$. Fail to reject H_0: $p \leq 0.04$. There is not sufficient evidence to warrant rejection of the claim that production is within control. No corrective actions appear to be necessary. *P*-value: 0.1056.

5. Test statistic: $z = 1.33$. Critical value: $z = 1.645$. Fail to reject H_0: $p \leq 0.5$. There is not sufficient evidence to support the claim that most students don't know what the Holocaust is. P-value: 0.0918.

7. Test statistic: $z = 1.42$. Critical value: $z = 1.645$. Fail to reject H_0: $p \leq 0.054$. There is not sufficient evidence to support the claim that the the the proportion of civilian doctors leaving the country who are from Saskatchewan has increased from its previous rate of 5.4%. P-value: 0.0778.

9. Test statistic: $z = -1.758$. Critical value: $z = -1.645$. Reject H_0: $p \geq 0.458$. There is sufficient evidence to support the claim that the proportion of air-strike incidents involving gulls has decreased. P-value: 0.0394.

11. Test statistic: $z = 5.96$. Critical values: $z = \pm 2.33$. Reject H_0: $p = 0.10$. There is sufficient evidence to warrant rejection of the claim that 10% of drivers use telephones. P-value: 0.0002.

13. Test statistic: $z = 0.99$. Critical values: $z = \pm 1.645$. Fail to reject H_0: $p = 0.08$. There is not sufficient evidence to warrant rejection of the claim that 8% of Seldane users experience drowsiness. P-value: 0.3222.

15. Test statistic: $z = -4.37$. Critical value: $z = -2.33$. Reject H_0: $p \geq 0.10$. There is sufficient evidence to support the claim that fewer than 10% of medical students prefer pediatrics. P-value: 0.0001.

17. Test statistic: $z = 0.83$. Critical value: $z = 1.28$. Fail to reject H_0: $p \leq 0.5$. There is not sufficient evidence to support the claim that the majority are smoking a year after therapy. Although many of the smokers did not stop with the nicotine patch therapy, there are many who did stop smoking, so the therapy is effective for them. P-value: 0.2033.

19. Test statistic: $z = -14.00$. Critical values: $z = \pm 2.575$. Reject H_0: $p = 0.5$. There is sufficient evidence to reject the claim that half of all adults eat their fruitcakes. Fruitcake producers should consider changes. P-value: 0.0002.

21. a. $n \cdot p = 4.884$; so the condition of $np \geq 5$ and $nq \geq 5$ is not met.
   b. Using binomial probability calculations, given $n = 44$, and $p = 11.1\%$, $P(5) = 0.1860$.
   c. We do not reject the null hypothesis that the proportion of cold/flu sufferers in Ottawa is less than or equal to 11.1%, because, under that hypothesis it would not be

at all unusual (probability = 0.1860) to find 5 cold/flu sufferers among 44 people tested.

23. 47% is not a possible result because, with new businesses, the only possible success rates are 0%, 5%, 10%, . . . , 100%.

Section 7–6

1. a. 1.735, 23.589
   b. 45.642
   c. 10.851

3. Test statistic: $x^2 = 114.586$. Critical values: $x^2 = 57.153$, 106.629. Reject H_0: $\sigma^2 = 43.7$ ft. There is sufficient evidence to support the claim that the standard deviation is different from 43.7 ft. Because the standard deviation seems to be larger than it was, the new production method results in less consistency and is therefore worse.

5. Test statistic: $x^2 = 9.016$. Critical value: $x^2 = 13.848$. Reject H_0: $\sigma \geq 6.2$ min. There is sufficient evidence to support the claim that there is lower variation with a single line. Although the waiting times are more consistent (so perceived by customers as fairer), the wait is not necessarily shorter.

7. Failing to meet the requirement means the standard deviation is greater than the allowed value. Test statistic: $x^2 = 22.036$. Critical value: $x^2 = 30.144$. Fail to reject H_0: $\sigma \leq 2.6$. There is not sufficient evidence to support the claim that the standard deviation for the X-ray tubes' peak-value voltages is greater than the permitted value of 2.6 kV.

9. Test statistic: $x^2 = 44.800$. Critical value: $x^2 = 51.739$. Reject H_0: $\sigma \geq 0.15$ oz. There is sufficient evidence to support the claim that the new machine fills bottles with lower variation. Based on the sample variation, the new machine should be purchased.

11. Test statistic: $x^2 = 69.135$. Critical value: $x^2 = 67.505$ (approximately). Reject H_0: $\sigma \leq 19.7$. There is sufficient evidence to support the claim that women have more variation.

13. $s = 2.340$. Test statistic: $x^2 = 15.628$. Critical values: $x^2 = 12.401$, 39.364 (assuming $\alpha = 0.05$). Fail to reject H_0: $\sigma = 2.9$ in. There is not sufficient evidence to support the claim that the standard deviation is different from 2.9 in.

15. a. $0.01 < P\text{-value} < 0.02$
   b. $P\text{-value} < 0.005$
   c. $0.005 < P\text{-value} < 0.01$

17. a. Estimated values: 74.216, 129.565; Table A-4 values: 74.222, 129.561

b. $x^2 = 117.093, 184.690$

19. Yes, this can happen. Try changing the first value in Example 13 to 48 (a rather moderate example of an outlier); x^2 jumps almost fourfold in value, and this reverses the conclusion of the hypothesis test.

Chapter 7 Review Exercises

1. a. $t = -1.833$

b. $z = \pm 2.575$

c. $x^2 = 8.907, 32.852$

2. a. $z = \pm 1.645$

b. $x^2 = 19.023$

c. $z = -2.33$

3. a. $\mu = \$90,000$

b. Two-tailed

c. Rejecting the claim that the mean equals $90,000 when it actually does equal that amount.

d. Failing to reject the claim that the mean equals $90,000 when it is actually different from that amount.

e. 0.05

4. a. $\sigma \geq 15$ s

b. Left-tailed

c. Rejecting the claim that the standard deviation is at least 15 s when it actually is at least 15 s.

d. Failing to reject the claim that the standard deviation is at least 15 s when it actually is less than 15 s.

e. 0.01

5. Not having one's teeth checked during the survey period means not having them checked annually. Test statistic: $z = 3.11$. Critical value: $z = 2.33$. Reject H_0: $p \leq 0.25$. There is sufficient evidence to support the claim that more than a quarter of people with dental insurance do not have their teeth checked annually.

6. Test statistic: $z = -2.73$. Critical value: $z = -2.33$. Reject H_0: $\mu \geq 5.00$. There is sufficient evidence to support the claim that the mean radiation dosage is below 5.00 mR.

7. Test statistic: $x^2 = 20.429$. The critical value is between 14.954 and 22.164, but it should be close to 18.559. Fail to reject H_0: $\sigma \geq 2.50$ mR. There is not sufficient evidence to support the claim that the standard deviation is less than 2.50 mR. If the standard deviation is too high, there may be some machines that emit dangerously high levels of radiation.

8. Test statistic: $z = 0.78$. Critical value: $z = 2.33$. Fail to reject H_0: $p \leq 0.267$. There is not sufficient evidence to support the claim that the proportion of farms with gross receipts over $100,000 is greater than it was in the 1991 census.

9. Test statistic: $t = -2.758$. Critical value: $t = -2.500$. Reject H_0: $\mu \geq 355$ mL. There is sufficient evidence to support the claim that the mean contents of the company's cola cans is less than the amount printed on the cans. If this is intentional, it would be cheating customers.

10. $\bar{x} = 23.29, s = 4.53$. Test statistic: $t = 3.240$. Critical values: $t = \pm 2.861$. Reject H_0: $\mu = 20$ mg. There is sufficient evidence to warrant rejection of the claim that the mean equals 20 mg. These pills are unacceptable because their contents do not correspond to the label.

11. Test statistic: $x^2 = 11.402$. Critical values: $x^2 = 2.180, 17.535$. Fail to reject H_0: $\sigma = 188.3$ days. There is not sufficient evidence to support the claim that the standard deviation is different from 188.3 days.

Chapter 7 Cumulative Review Exercises

1. a. 0.2358

b. 0.0020

2. a. $\bar{x} = 124.23, s = 22.52$

b. Test statistic: $t = 1.675$. Critical values: $t = \pm 2.132$. Fail to reject H_0: $\mu = 114.8$. There is not sufficient evidence to warrant rejection of the claim that the mean is equal to 114.8.

c. $112.23 < \mu < 136.23$; yes

d. Test statistic: $x^2 = 44.339$. Critical values: $x^2 = 6.262, 27.488$. Reject H_0: $\sigma = 13.1$. There is sufficient evidence to warrant rejection of the claim that the standard deviation is equal to 13.1.

e. Although the mean does not appear to change by a significant amount, the variation among the values appears to be significantly larger.

3. a. 6.3

b. 2.2

c. 0.0024 (or 0.0019 if unrounded statistics are used)

d. Based on the low probability value in part (c), reject H_0: $p = 0.25$. There is sufficient evidence to reject the claim that the subject made random guesses.

e. 423

4. a.

35.	24
35.	5689
36.	0111223333344444444
36.	5556666667777788899
37.	0011111344
37.	6

b. The distribution appears to be approximately normal.

c. A histogram could also be used to test for a normal distribution.

Section 8-2

1. a. $\bar{d} = 2.8$

b. $s_d = 3.6$

c. $t = 1.757$

d. $t = \pm 2.776$

3. $21.6 < \mu_d < 7.2$

5. a. Test statistic: $t = -1.496$. Critical values: $t = \pm 2.447$. Fail to reject H_0: $\mu_d = 0$. There is not sufficient evidence to support the claim that the progression of an El Niño from the spring of onset to the next spring affects temperature departures from spring seasonal norms.

b. $-2.30 < \mu_d < 0.55$. We have 95% confidence that the interval from -2.30 to 0.55 really does contain the true population mean difference.

7. a. If sales are growing, the mean difference (1987 minus 1997) will be negative. Test statistic: $t = -1.774$. Critical value: $t = -1.833$. Fail to reject H_0: $\mu_d < 0$. There is not sufficient evidence to support the claim that the fastest-growing companies do grow in sales after 10 years.

b. $-22.18 < \mu_d < 2.68$

9. a. Test statistic: $t = 0.384$. Critical values: $t = \pm 2.306$. Fail to reject H_0: $\mu_d = 0$. There is not sufficient evidence to reject the claim that the starting salaries of university graduates in Ontario and Quebec are equal. In both provinces, graduates with comparable degrees appear to earn about the same starting income.

b. $-1223.65 < \mu_d < 1712.53$

11. Test statistic: $t = 4.375$. Critical value: $t = 3.143$. Reject H_0: $\mu_d \le 0$. There is enough evidence to support the claim that flights scheduled one day in advance cost more than flights scheduled 30 days in advance. The best strategy would be to book a flight sooner rather than later.

13. a. $0.402 < \mu_d < 0.657$

b. Test statistic: $t = 9.247$. Critical values: $t = \pm 2.228$. Reject H_0: $\mu_d = 0$. There is sufficient evidence to support the claim that temperatures are different in the two postures. The differences in body temperatures for the two postures are not large, but they are significant.

15. Yes, the value of 64 ($64 million) for 1997 is far removed from all the other data. If the data pair that includes 64 is excluded from the analysis, the t test results are reversed: Test statistic t changes from -1.774 to -2.476; the critical value changes from -1.833 to -1.860. As a result, we now reject H_0: $\mu_d < 0$. The confidence interval changes dramatically, to $-8.692 < \mu_d < -0.308$. Yes, as shown here.

17. $\alpha = 0.025$

Section 8-3

1. Test statistic: $z = 0.79$. Critical values: $z = \pm 1.96$. Fail to reject H_0: $\mu_1 = \mu_2$. There is not sufficient evidence to warrant rejection of the claim that the two samples come from populations with the same mean.

3. $-3 < \mu_1 - \mu_2 < 7$; yes. The control group and experimental group do not appear to be significantly different.

5. a. Test statistic: $z = -0.47$. Critical values: $z = \pm 2.575$. Fail to reject H_0: $\mu_1 = \mu_2$. There is not enough evidence to reject the claim that there is no difference in foreign content between Canadian equity funds and Canadian balanced equity funds.

b. $-5.74 < \mu_1 - \mu_2 < 3.78$

7. a. Test statistic: $z = 2.06$. Critical values: $z = \pm 2.575$. Fail to reject H_0: $\mu_1 = \mu_2$. There is not sufficient evidence to warrant rejection of the claim that the two populations have the same mean. Advertise, because there does not appear to be a difference.

b. $-3.5 < \mu_1 - \mu_2 < 31.5$

9. a. H_0: $\mu_1 \le \mu_2$ and H_1: $\mu_1 > \mu_2$. Test statistic: $z = 0.28$. Critical value: $z = 1.645$. Fail to reject H_0: $\mu_1 \le \mu_2$. There is not sufficient evidence to support the claim that faculty whose last names begin with A to R earn more than faculty whose last names begin with S to Z.

b. $-8.31 < \mu_1 - \mu_2 < 11.11$. Yes. This indicates that there is not a significant difference between the two means.

11. a. Test statistic: $z = -2.54$. Critical value: $z = -2.05$. Reject H_0: $\mu_1 \ge \mu_2$. There is sufficient evidence to support the claim that nonsmokers with exposure to environmental tobacco smoke have higher levels of serum cotinine than nonsmokers without such exposure.

b. $-6.674 < \mu_1 - \mu_2 < -0.711$. No. Yes, exposed non-smokers appear to have significantly raised levels of serum cotinine.

13. Test statistic: $z = -0.954$. Critical value: $z = -1.645$. Fail to reject H_0: $\mu_1 \geq \mu_2$. There is not sufficient evidence to support the claim that the population of smaller communities has a higher mean value for building permits issued.

15. a. Test statistic: $z = -3.95$. Critical values: $z = \pm 1.96$. Reject H_0: $\mu_1 = \mu_2$. There is sufficient evidence to reject the claim that male and female statistics students have the same mean pulse rate.

   b. P-value (based on tables) $= 0.0002 < \alpha = 0.05$; reject H_0: $\mu_1 = \mu_2$.

   c. $-13.2 < \mu_1 - \mu_2 < -4.4$; reject H_0: $\mu_1 = \mu_2$.

   d. The results all agree.

17. P-value $= 0.0278 > \alpha = 0.02$. Fail to reject H_0: $\mu_1 = \mu_2$. There is not sufficient evidence to support the claim that the treatment group comes from a population with a different mean from the mean for the placebo population. (Note: The critical values shown in the display do not apply, since they are based on $\alpha = 0.05$.)

19. The boxplot for the 0.0111 in. cans uses a 5-number summary of 205, 275, 285, 294, 317. Before the deletion of the outlier, the 5-number summary was 205, 275, 285, 295, 504, so the only major change is shortening of the right whisker. In the hypothesis test, the test statistic changes from -5.48 to -5.66, so the conclusions will be the same. After deleting the outlier of 504, the confidence interval limits change from $(-21.6, -7.8)$ to $(-19.5, -7.3)$, so those changes are not substantial.

21. a. 50/3

   b. 2/3

   c. 52/3

   d. The range of the x-y values equals the range of the x values plus the range of the y values.

Section 8-4

1. Test statistic: $F = 2.2222$. Critical value: $F = 2.2090$. Reject H_0: $\sigma_1^2 = \sigma_2^2$. There is sufficient evidence to support the claim that the two population variances are different.

3. Test statistic: $F = 1.2949$. Critical value: $F = 1.6928$ (approximately). Fail to reject H_0: $\sigma_1 = \sigma_2$. There is not sufficient evidence to support the claim that the samples come from populations with different standard deviations.

5. Although these samples are not truly independent—they involve the same individuals—we do not possess the "before" and "after" data pairs for each specific individual that methods based on *dependent* samples would require. Test statistic: $F = 1.7831$. Critical value: $F = 1.3272$. (If using table, can use $df_1 = df_2 = 120$. $F = 1.3519$.) Reject H_0: $\sigma_1 \leq \sigma_2$. There is evidence to support the claim that the individuals who have received treatment represent a population with a standard deviation for an accident rate which is greater than the standard deviation for the accident rate among those who have not yet received treatment. (Due to the dependency in that data, apply this conclusion only for individuals who do go on to get treatment at some point.)

7. Test statistic: $F = 1.1142$. Critical value: $F = 1.9508$. (If using table, can use $df_1 = 40$ and $df_2 = 30$, $F = 2.0089$.) Fail to reject H_0: $\sigma_1 = \sigma_2$. There is not sufficient evidence to reject the claim that the standard deviation for shelf life is the same at both plants.

9. Test statistic: $F = 1.1800$. Critical value: $F = 5.4613$. Fail to reject H_0: $\sigma_1 = \sigma_2$. There is not sufficient evidence to warrant rejection of the claim that both groups of flight times come from populations with the same variance.

11. Test statistic: $F = 2.234$. Critical value: $F = 3.912$. (If using table, can use $df_1 = 10$; $df_2 = 9$. $F = 3.9639$.) Fail to reject H_0: $\sigma_1 = \sigma_2$. There is not sufficient evidence to reject the claim that both groups of ages come from populations with the same variance.

13. a. Reject H_0: $\sigma_1^2 = \sigma_2^2$.

   b. Fail to reject H_0: $\sigma_1^2 = \sigma_2^2$.

   c. Reject H_0: $\sigma_1^2 = \sigma_2^2$.

15. a. $F_L = 0.2484$, $F_R = 4.0260$

   b. $F_L = 0.2315$, $F_R = 5.5234$

   c. $F_L = 0.1810$, $F_R = 4.3197$

   d. $F_L = 0.3071$, $F_R = 4.7290$

   e. $F_L = 0.2115$, $F_R = 3.2560$

17. a. It doesn't change.

   b. It doesn't change.

   c. It doesn't change.

Section 8-5

1. F-test results: Test statistic is $F = 1.5625$. Critical value: $F = 2.8801$. Fail to reject $\sigma_1^2 = \sigma_2^2$. Test of means: Test statistic is $t = -0.990$. Critical values: $t = \pm 2.048$. Fail to reject

H_0: $\mu_1 = \mu_2$. There is not sufficient evidence to warrant rejection of the claim that the two populations have equal means.

3. $-15 < \mu_1 - \mu_2 < 5$

5. **a.** $0.1748 < \mu_1 - \mu_2 < 0.4373$; no. There is sufficient evidence to reject the equality of mean flight times in both directions.

   b. Test statistic: $t = 4.945$. Critical values: $t = 1.746$. P-value $= 0.00007338$. Reject H_0: $\mu_1 \leq \mu_2$. There is sufficient evidence to support the claim that flights from Vancouver to San Francisco take longer.

7. **a.** F-test results: Test statistic: $F = 1.1276$. Using a computer: Critical value: $F = 2.1952$. Fail to reject H_0: $\sigma_1^2 = \sigma_2^2$. Test of means: Test statistic: $t = 0.132$. Critical values: $t = \pm 1.975$. Fail to reject H_0: $\mu_1 = \mu_2$. There is not sufficient evidence to reject the claim that the past clients of both clinics have the same mean accident rate. The competing clinic is not justified in making its boast.

   b. $-0.097 < \mu_1 - \mu_2 < 0.111$

9. Test statistic: $t = 1.243$. Critical value: $t = 1.701$. Fail to reject H_0: $\mu_1 \leq \mu_2$. There is not sufficient evidence to support the claim that less flawed ("clearer") diamonds sell at a higher price than more flawed diamonds. Inspection of the raw data shows that, for the sampled high-quality diamonds, clarity never exceeds 5 (higher numbers indicate more visible flaws). To be sure that clarity is unrelated to price, we would need to also sample diamonds where clarity exceeds 5.

11. F-test results: Test statistic is $F = 2.9459$. Critical value: $F = 3.0074$. Fail to reject H_0: $\sigma_1^2 = \sigma_2^2$. Test of means: Test statistic is $t = -1.099$. Critical values: $t = \pm 2.052$. Fail to reject H_0: $\mu_1 = \mu_2$. There is not sufficient evidence to warrant rejection of the claim that red and orange M&Ms have the same mean weight. No corrective action is necessary.

13. F-test results: Test statistic: $F = 5.2849$. Critical value: $F = 2.7689$. Reject H_0: $\sigma_1^2 = \sigma_2^2$. Test of means: Test statistic: $t = -2.295$. Critical values: $t = \pm 2.179$ (based on the approximate method for finding df). Reject H_0: $\mu_1 = \mu_2$. There is sufficient evidence to reject the claim that the East and West have the same mean annual precipitation.

15. Test statistic: $t = 1.301$. Critical values: $t = \pm 2.086$. Fail to reject H_0: $\mu_1 = \mu_2$. There is not sufficient evidence to reject the claim that the ages of CEOs from top growth companies in Western Canada and in Quebec are the same.

17. F-test results: The test statistic doesn't exist, because it would be calculated with division by 0. However, the standard deviations do not appear to be equal, so reject H_0: $\sigma_1^2 = \sigma_2^2$. Test of means: Test statistic is $t = 15.322$. Critical values: $t = \pm 2.080$. Reject H_0: $\mu_1 = \mu_2$. There is sufficient evidence to warrant rejection of the claim that the two groups come from populations with the same mean. (There is a serious question about the requirement that the samples come from normally distributed populations, because the 22 scores in the second sample appear to be the same.)

19. The number of degrees of freedom changes from 9 to 10. The test statistic doesn't change, the critical value changes from $t = 1.833$ to $t = 1.812$, but the conclusions remain the same. The confidence interval limits change from $(-16.6, 192.6)$ to $(-15.1, 191.1)$.

Section 8-6

1. **a.** 8/15
   b. -0.58
   c. ± 1.96
   d. 0.5620

3. 49

5. Test statistic: $z = 2.17$. Critical value: $z = 1.645$. Reject H_0: $p_1 \leq p_2$. There is sufficient evidence to support the claim that Facility A has the greater proportion of product that contains below half the allowable level of foreign matter. We cannot conclude from this that Facility A necessarily has a more consistent level of quality than Facility B. For example, of the 70% of Facility A's output which is *not* below half the allowable level of foreign matter, we have no indication of *how much* extra impurity these samples contain.

7. Test statistic: $z = -2.85$. Critical value: $z = -2.33$. Reject H_0: $p_1 \geq p_2$. There is sufficient evidence to support the claim that fewer people, proportionally, support free trade in 2001 than in 1999. Yes. If, for example, n was 500 in 1999, the test statistic would change to $z = -2.31$, resulting in non-rejection of H_0: $p_1 \geq p_2$.

9. **a.** Test statistic: $z = -0.41$. Critical values: $z = \pm 2.575$. Fail to reject H_0: $p_1 = p_2$. There is not sufficient evidence to reject the claim that both industry groups have the same proportion of companies with a foreign ownership level of at least 10%.
   b. $-0.40 < p_1 - p_2 < 0.29$

11. Test statistic: $z = 3.86$. Critical value: $z = 2.33$. Reject H_0: $p_1 \leq p_2$. There is sufficient evidence to support the claim that arson is committed by a proportion of drinkers that is greater than the proportion of drinkers convicted of fraud. It does seem reasonable that drinking would have an effect on the type of crime.

13. Test statistic: $z = 4.47$. Critical value: $z = \pm 2.575$. Reject H_0: $p_1 = p_2$. There is sufficient evidence to reject the claim that the central-city refusal rate is the same as the refusal rate in other areas.

15. $-0.052 < p_1 - p_2 < 0.184$. Yes. There is no significant difference between the proportions of home wins for basketball and for football. To decide whether there is a home advantage at all, we would have to see whether each sport's individual confidence intervals for p at home include the value 0.50.

17. Test statistic: $z = 0.84$. Critical values: $z = \pm 1.96$. Fail to reject H_0: $p_1 - p_2 = 0.1$. There is not sufficient evidence to warrant rejection of the claim that when women are exposed to glycol ethers, their percentage of miscarriages is 10 percentage points more than the percentage for women not exposed to glycol ethers.

19. 2135

Chapter 8 Review Exercises

1. Test statistic: $z = 2.41$. Critical value: $z = 1.645$. Reject H_0: $p_1 \leq p_2$. There is sufficient evidence to support the claim that the percentage of people who stop after first seeing someone else being helped is greater than the percentage of people who stop without first seeing someone else being helped.

2. a. Test statistic: $t = -3.847$. Critical value: $t = -1.796$. Reject H_0: $\mu_d \geq 0$. There is sufficient evidence to support the claim that the Dozenol tablets are more soluble after the storage period.
   b. $12.4 < \mu_d < 45.6$

3. a. Test statistic: $z = 4.48$. Critical value: $z = 1.645$ (assuming a 0.05 significance level). Reject H_0: $p_1 \leq p_2$. There is sufficient evidence to support the claim that dyspepsia occurs at a higher rate among Viagra users.
   b. $0.0277 < p_1 - p_2 < 0.0699$; no; they are significantly different.

4. Test statistic: $t = -3.356$. Critical value: $t = \pm 3.106$ (based on the approximate method for finding df). Reject H_0: $\mu_1 = \mu_2$. There is sufficient evidence to reject the claim that the mean price for both types of flight is the same.

5. Test statistic: $F = 5.4505$. Critical values: $F = 4.2509$. Reject H_0: $\sigma_1^2 = \sigma_2^2$. The samples provide evidence to reject the claim of equal variances in the cost data for the two types of flights.

6. Test statistic: $z = 3.67$. Critical value: $z = 1.645$. Reject H_0: $p_1 \leq p_2$. There is sufficient evidence to support the claim that the question was answered correctly by a greater proportion of prepared students.

7. Test statistic: $t = 1.185$. Critical values: $t = \pm 2.262$. Fail to reject H_0: $\mu_d = 0$. There is not sufficient evidence to warrant rejection of the claim that the position has no effect.

8. Test statistic: $z = 0.96$. Critical value: $z = 2.33$. Fail to reject H_0: $\mu_1 \leq \mu_2$. There is not sufficient evidence to support the claim that the 10-day program results in scores with a lower mean.

9. F-test results: Test statistic is $F = 12.0013$. Critical value: $F = 2.5598$. Reject H_0: $\sigma_1^2 = \sigma_2^2$. Test of means: Test statistic is $t = 2.444$. Critical values: $t = \pm 2.110$. Reject H_0: $\mu_1 = \mu_2$. There is sufficient evidence to warrant rejection of the claim that both systems have the same mean.

10. a. F-test results: Test statistic is $F = 2.3623$. Critical value: $F = 3.4737$. Fail to reject H_0: $\sigma_1^2 = \sigma_2^2$. Test of means: Test statistic: $t = -1.335$. Critical values: $t = \pm 2.074$. Fail to reject H_0: $\mu_1 = \mu_2$. There is not sufficient evidence to reject the claim that men and women in these circumstances have the same mean temperature.
    b. $-0.659 < \mu_1 - \mu_2 < 0.143$

Chapter 8 Cumulative Review Exercises

1. a. 0.0707
   b. 0.369
   c. 0.104
   d. 0.054
   e. Test statistic: $z = -2.52$. Critical value: $z = -1.645$. Reject H_0: $p_1 \geq p_2$. There is sufficient evidence to support the claim that women are ticketed at a lower rate. We cannot conclude that men generally speed more, only that they are ticketed more. Perhaps men drive more or perhaps they are more likely to get a ticket when speeding.

2. a. Test statistic: $z = 1.95$. Critical values: $z = \pm 1.96$. Fail to reject H_0: $\mu = 0$. There is not sufficient evidence to warrant rejection of the claim that the sample comes from a population with a mean equal to 0.

b. Test statistic: $z = -1.87$. Critical values: $z = \pm1.96$. Fail to reject H_0: $\mu = 0$. There is not sufficient evidence to warrant rejection of the claim that the sample comes from a population with a mean equal to 0.

c. Test statistic is $z = 2.68$. Critical values: $z = \pm1.96$. Reject H_0: $\mu_1 = \mu_2$. There is sufficient evidence to warrant rejection of the claim that both shifts manufacture scales with the same mean error.

d. $0.0\text{ g} < \mu < 2.4\text{ g}$

e. $-2.9\text{ g} < \mu < 0.1\text{ g}$

f. $0.7\text{ g} < \mu_1 - \mu_2 < 4.5\text{ g}$

g. 234

Section 9-2

1. a. 6

b. $F = 3.33$

c. $F_c = 2.6207$

d. 0.032

e. There is sufficient evidence to reject the null hypothesis that the means are equal.

3. a.

Source	df	SS	MS	F
Factor	2	11.6238	5.8119	34.6826
Error	30	5.0272	0.1676	
Total	32	16.6510		

b. 3.3158

c. Reject the null hypothesis and conclude there is a significant difference in average body temperatures of at least two of the groups.

d. sitting/standing

5. Test statistic: $F = 1.4468$. Critical value: $F = 3.4028$. Fail to reject the null hypothesis of equal means. Although conglomerates appear to have a higher mean revenue, there is insufficient evidence to indicate that apparent differences in mean revenue are significant.

7. Test statistic: $F = 2.1188$. Critical value: $F = 3.4028$. Fail to reject the null hypothesis of equal means. Although conglomerates appear to have a greater mean number of employees, there is insufficient evidence to indicate that apparent differences in mean numbers of employees are significant.

9. Test statistic: $F = 2.9493$. Critical value: $F = 2.6060$. Reject the claim of equal means. The different laboratories do not appear to have the same mean.

11. Test statistic: $F = 4.3465$. Critical value: $F = 3.5546$. Reject the claim of equal means. The mean number of sittings appears to be different for different eras of Parliament.

13. Test statistic: $F = 0.0221$. Critical value: $F = 2.9752$. Fail to reject the claim that the mean weight (in carats) is the same for all four clarity levels. If you must compromise on clarity, it appears that your selection of carat levels is not affected.

15. a. 0.300300

b. 0.167331

c. $F = 1.794647$

d. 3.3158 (assuming $\alpha = 0.05$)

e. Fail to reject the claim of equal means for the three groups. The three groups now appear to have the same mean body temperature. By adding 1° to each temperature for the first sample, the variance between the samples decreased. The variance within samples did not change. The F test statistic decreased to reflect the decreased disparity among the three samples.

17. In each case, the test statistic F is not affected.

19. a. Test statistic: $t = -1.506$. Critical values: $t = \pm2.120$. Fail to reject the null hypothesis that the two industry groups have the same mean revenue.

b. Test statistic: $F = 2.2693$. Critical value: $F = 4.4940$. Fail to reject the null hypothesis that the two industry groups have the same mean revenue.

c. Test statistics: $t^2 = 2.2680 \cong F = 2.2693$. Critical values: $t^2 = 4.4944 \cong F = 4.4940$.

Section 9-3

1. a. 1263.396

b. 694.938

c. 349.729

d. 14.063

3. Test statistic: $F = 0.5033$. Critical value: $F = 4.0662$, assuming that the significance level is $\alpha = 0.05$. It appears that star rating does not have an effect on movie length.

5. Test statistic: $F = 0$. Critical value: $F = 10.1280$. It appears that CTR rating does not have an effect on movie length.

7. Test statistic: $F = 2.4740$. Critical value: $F = 4.7571$. Fail to reject the null hypothesis of no effects from the operators. It appears that the operators do not affect the numbers of support beams manufactured.

9. (The given results use the first nine values for each cell, but the pulse rate of 8 was excluded as being not feasible.) Test statistic: $F = 0.4200$. Critical value: $F = 4.1709$ (approximately). There does not appear to be an interaction between gender and smoking.

11. Test statistic: $F = 0.1814$. Critical value: $F = 4.1709$ (approximately, assuming $\alpha = 0.05$). Smoking does not appear to have an effect on pulse rates. (This is not surprising, because college students haven't been smoking very long.)
13. Nothing changes.
15. Answers vary.

Chapter 9 Review Exercises

1. Test statistic: $F = 46.90$. Critical value: $F = 3.0725$. Reject the null hypothesis of equal means.
2. Test statistic: $F = 3.0693$. Critical value: $F = 3.0725$. Fail to reject the claim (H_0) that the mean food costs are the same in the three cities.
3. Test statistic: $F = 0.1105$. Critical value: $F = 5.1432$. Fuel consumption does not appear to be affected by an interaction between transmission type and engine size.
4. Test statistic: $F = 1.5430$. Critical value: $F = 5.1432$. Fuel consumption does not appear to be affected by the engine size.
5. Test statistic: $F = 2.9589$. Critical value: $F = 5.9874$. The type of transmission does not appear to have an effect on fuel consumption.
6. Test statistic: $F = 24.2948$. Critical value: $F = 3.6667$. Reject the claim (H_0) that the interaction between the two variables does not affect brake performance. It appears that the interaction between snowmobile brand and speed does affect the mean brake performance of the vehicles. The P-value is 0.00000000438423.

Chapter 9 Cumulative Review Exercises

1. a.

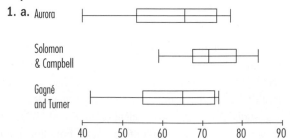

Aurora Advertising and Gagné and Turner Advertising appear to be very similar, but Solomon & Campbell Advertising seems to be somewhat higher.

b. Aurora: $\bar{x} = 62.8$, $s = 12.9$; Solomon & Campbell: $\bar{x} = 72.3$, $s = 8.2$; Gagné and Turner: $\bar{x} = 62.8$, $s = 11.6$.
c. F test results: Test statistic is $F = 2.4592$. Critical value: $F = 4.9949$. Fail to reject H_0: $\sigma_1^2 = \sigma_2^2$. Test of means: Test statistic is $t = -1.760$. Critical values: $t = \pm 2.145$. Fail to reject H_0: $\mu_1 = \mu_2$. There is not sufficient evidence to warrant rejection of the claim that the population of Aurora scores has a mean equal to the population mean for Solomon & Campbell scores.
d. Aurora: $52.0 < \mu < 73.5$; Solomon & Campbell: $65.4 < \mu < 79.1$; Gagné and Turner: $53.1 < \mu < 72.4$. Solomon & Campbell seems slightly higher.
e. Test statistic: $F = 1.9677$. Critical value: $F = 3.4668$. Fail to reject the null hypothesis of equal means. There is not sufficient evidence to warrant rejection of the claim that the three populations have the same mean reaction score. No single company stands out as being significantly better than the others.
2. a. 0.3300
   b. 0.0384
   c. 1/8
3. a. 2700.1
   b. 2602.8
   c.

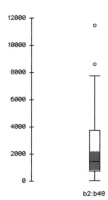

d. 11505 is clearly an outlier ($z = 3.38$).
   Although the boxplot also highlights the value 8625, because $z = 2.37$, the value appears consistent with the right-skewed data-distribution pattern (see the histogram for part (e).

e.

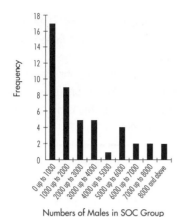

Numbers of Males in SOC Group

f. No, the populations are not normally distributed.

g. $P(B5) = 0.0477$

Section 10-2

1. Test statistic: $x^2 = 156.500$. Critical value: $x^2 = 21.666$. Reject H_0: The last digits occur with the same frequency. There is sufficient evidence to warrant rejection of the claim that the last digits occur with equal frequency. It appears that the students reported their weights.

3. **a.** df = 37, so $x^2 = 51.805$ (approximately)
   b. $0.10 < P\text{-value} < 0.90$
   c. There is not sufficient evidence to warrant rejection of the claim that the roulette slots are equally likely.

5. Test statistic: $x^2 = 5.185$. Critical value: $x^2 = 7.815$. There is not sufficient evidence to reject the claim that the work stoppages occur in the proportions indicated.

7. Test statistic: $x^2 = 23.431$. Critical value: $x^2 = 9.488$. Reject H_0: Accidents are distributed with the given percentages. There is sufficient evidence to warrant rejection of the claim that accidents are distributed with the given percentages. Rejection of the claim does not provide help in correcting the problem.

9. Test statistic: $x^2 = 4.200$. Critical value: $x^2 = 16.919$. Fail to reject H_0: The digits are uniformly distributed. There is not sufficient evidence to warrant rejection of the claim that the digits are uniformly distributed.

11. Test statistic: $x^2 = 8.381$. Critical value: $x^2 = 12.592$. Fail to reject H_0: The distribution of discounts is consistent with Zellers promotional claims. The data do not provide sufficient evidence to warrant rejection of the claim that discounts based on using Zellers' "Scratch and Save" cards are distributed in the proportions indicated.

13. Test statistic: $x^2 = 11.800$. Critical value: $x^2 = 9.488$ (assuming $\alpha = 0.05$). Reject H_0: The time intervals are uniformly distributed among the five categories. There is sufficient evidence to warrant rejection of the claim that the time intervals are uniformly distributed among the five categories.

15. Test statistic: $x^2 = 10.400$. Critical value: $x^2 = 7.815$. Reject H_0: The returns were evenly distributed among the four categories. There is sufficient evidence to warrant rejection of the claim that the returns from one-year funds were evenly distributed among the four categories from poor to excellent. There seems to have been a disproportionate number of good returns for mutual funds that year.

17. The test statistic changes from 13.325 to 373.059, so the outlier can have a dramatic effect on the value of x^2.

19. **a.**
$$x^2 = \frac{\left(f_1 - \dfrac{f_1 + f_2}{2}\right)^2}{\dfrac{f_1 + f_2}{2}} + \frac{\left(f_2 - \dfrac{f_1 + f_2}{2}\right)^2}{\dfrac{f_1 + f_2}{2}}$$
$$= \frac{(f_1 + f_2)^2}{f_1 + f_2}$$

b.
$$x^2 = \frac{\left(\dfrac{f_1}{f_1 + f_2} - 0.5\right)^2}{\dfrac{1/4}{f_1 + f_2}} = \frac{\left(f_1 + f_2\right)^2}{f_1 + f_2}$$

Critical values: The x^2 critical value is 3.841 and it is approximately equal to the square of $z = 1.96$.

21. Combine the last three cells. Test statistic: $x^2 = 2.059$. Critical value: $x^2 = 5.991$. Fail to reject H_0: The frequencies fit a Poisson distribution. There is not sufficient evidence to warrant rejection of the claim that the frequencies fit a Poisson distribution.

Section 10-3

1. Test statistic: $x^2 = 51.458$. Critical value: $x^2 = 6.635$. Reject the claim (H_0) that the proportions of agree/disagree responses are the same for the subjects interviewed by men and the subjects interviewed by women.

3. Test statistic: $x^2 = 1.174$. Critical value: $x^2 = 3.841$. Fail to reject the claim that the treatment is independent of the reaction. There is not enough evidence to be concerned.

5. Test statistic: $x^2 = 5.091$. Critical value: $x^2 = 7.815$ (assuming $\alpha = 0.05$). Fail to reject the claim that the car model and final position are independent. If all the same ratios held up in a sample five times larger, the test statistic x^2 would equal 25.455, and the P-value would be much less than the P-value for the table in the original problem.

7. Test statistic: $x^2 = 8.642$. Critical value: $x^2 = 9.210$. There is not sufficient evidence to warrant rejection of the claim that the distribution of planes flying and planes ordered is independent of the airline company.

9. Test statistic: $x^2 = 49.624$. Critical value: $x^2 = 3.841$. Reject the claim (H_0) that the patient's age and the sport where the injury occurred are independent variables. The data suggest that these two variables are, in fact, related. This test result does not establish a *causal* relationship between age and risk of injury. It may be that, proportionally, fewer children under 10 engage in basketball as their sport, compared with older people; perhaps that explains why fewer of the young people's sports injuries occur while playing basketball.

11. Test statistic: $x^2 = 2.327$. Critical value: $x^2 = 3.841$. There is not sufficient evidence to warrant rejection of the claim that the distribution of weights for offensive backs is independent of whether the player is a quarterback.

13. Test statistic: $x^2 = 49.731$. Critical value: $x^2 = 11.071$ (assuming $\alpha = 0.05$). Reject the claim that the type of crime is independent of whether the criminal drinks. Fraud appears to be unique in the sense that it is associated with drinkers at a much lower rate than the other crimes.

15. Test statistic: $x^2 = 0.122$. Critical value: $x^2 = 3.841$ (assuming $\alpha = 0.05$). Fail to reject the claim that gender of statistics students is independent of whether they smoke.

17. With Yates' correction: $x^2 = 1.205$. Without Yates' correction: $x^2 = 1.505$. In general, the test statistic decreases with Yates' correction.

Section 10-4

1. Test statistic $= 0.4109$. Critical value $= 0.285$. Reject H_0 and conclude the data is not normally distributed.

3. Test statistic $= 1.4256$. Critical value $= 11.071$. Fail to reject H_0 and conclude the data is normally distributed.

5. The original formula for the test statistic is $\chi^2 = \Sigma(O - E)^2/E$. If the number of values in each category is multiplied by N, the expected number of observations is also multiplied by N. The formula for the new test statistic is $\chi^2 = \Sigma(NO - NE)^2/NE = \Sigma N^2(O - E)^2/NE = \Sigma N(O - E)^2/E = N\Sigma(O - E)^2/E$. Thus, the original test statistic is multiplied by a factor of N.

Chapter 10 Review Exercises

1. Test statistic: $x^2 = 24.621$. Critical value: $x^2 = 5.991$. Reject the claim (H_0) that shelf life is independent of the additive formula that is used. The data suggest that the revised additives are associated with a changed distribution of product shelf life.

2. Test statistic: $x^2 = 1.714$. Critical value: $x^2 = 5.991$. Fail to reject the claim (H_0) that there is no difference in the proportion of springs with above-normal precipitation for the three categories of springs.

3. Test statistic: $x^2 = 7.600$. Critical value: $x^2 = 14.067$. There is not sufficient evidence to warrant rejection of the claim that all these species strike with equal frequencies. These data do not, of themselves, support a prevention policy that is focused on particular species.

4. Test statistic: $x^2 = 5.252$. Critical value: $x^2 = 3.841$. Reject the claim (H_0) that the patients' temperatures and genders are independent. The data provide evidence that patients' temperatures and genders are related.

5. Test statistic: $x^2 = 7.607$. Critical value: $x^2 = 5.991$. Reject the claim that the occurrence of headaches is independent of the group. The Seldane users had a lower headache rate than the placebo group, so headaches don't seem to be a problem.

Chapter 10 Cumulative Review Exercises

1. For both data sets, critical value $= 0.381$; memory test scores test statistic $= 0.2257$; fail to reject H_0; reasoning test scores test statistic $= 0.1591$; fail to reject H_0. Conclude that both data sets are normally distributed.

2. (Use the test for two dependent means, as in Section 8-2.) With $\bar{d} = -10.25$ and $s_d = 1.5$, the test statistic is $t = -13.667$. The critical value is $t = -2.353$ (assuming $\alpha = 0.05$). Reject H_0: $\mu_d \geq 0$. There is sufficient evidence to support the claim that the scores are higher after the training session. It appears that the training session is effective in raising scores.

3. (Use a test of homogeneity in a contingency table, as in Section 10–3.) Test statistic: $x^2 = 0.055$. Critical value: $x^2 = 7.815$ (assuming $\alpha = 0.05$). Fail to reject the claim that men and women choose the different answers in the same proportions. The letter selections appear to be the same for men and women.

4. (Use a test for equality between two independent means, as in Section 8-5.) Preliminary F test: Test statistic is $F = 1.0344$. Critical value: $F = 15.439$ (assuming $\alpha = 0.05$). Fail to reject $\sigma_1^2 = \sigma_2^2$. Test of means: Test statistic is $t = -2.014$. Critical values: $t = \pm 2.447$ (assuming $\alpha = 0.05$). Fail to reject $H_0: \mu_1 = \mu_2$. There is not sufficient evidence to warrant rejection of the claim that men and women have the same mean score.

Section 11–2

1. a. Significant linear correlation; 98.4%
   b. Significant linear correlation; 11.1%
   c. No significant linear correlation; 20.8%
3. a. Significant linear correlation
   b. $n = 5$, $\Sigma x = 25$, $\Sigma x^2 = 163$, $(\Sigma x)^2 = 625$, $\Sigma xy = 489$, $r = 0.997$
5. Test statistic: $r = 0.993$. Critical values: $r = \pm 0.707$. Reject the claim of no significant linear correlation. There does appear to be a linear correlation between chest size and weight. The results do not change if the chest measurements are converted to feet.
7. Test statistic: $r = 0.841$. Critical values: $r = \pm 0.707$. Reject the claim of no significant linear correlation. There does appear to be a linear correlation between weight of discarded plastic and household size.
9. a. Test statistic: $r = 0.819$. Critical value: $r = \pm 0.602$. Reject the null hypothesis of no significant correlation. There does appear to be a significant correlation between species' incubation and fledgling periods.
   b. For samples created by randomly selecting one individual bird from each of the 11 species, the mean r approaches the r calculated in (a); since randomly selected higher or lower values for incubation or fledgling periods will tend to cancel out. For samples created by randomly selecting *any* 11 birds from among the 11 species (not necessarily one per species), the mean of the r's is greater than in (a); because the correlation is weak for a few of the species, and by chance, many samples will not include data from those species.

11. Test statistic: $r = 0.772$. Critical values: $r = \pm 0.497$. Reject the claim of no significant linear correlation. There does appear to be a linear correlation between salary and years of education.
13. a. Test statistic: $r = 0.870$. Critical values: $r = \pm 0.279$. Reject the claim of no significant linear correlation. There does appear to be a linear correlation between the interval after an eruption and the duration of the eruption.
   b. Test statistic: $r = -.00978$. Critical value: $r = \pm 0.279$. Do not reject the null hypothesis of no significant correlation. There does not appear to be a correlation between the interval after an eruption and the height of an eruption.
   c. For predicting the time interval after an eruption before the next eruption, eruption duration would be more relevant than height—because, of these two, only duration has a significant correlation to the time interval.
15. a. Test statistic: $r = 0.767$. Critical values: $r = \pm 0.361$. Significant linear correlation.
   b. Test statistic: $r = -0.441$. Critical values: $r = \pm 0.361$. Significant linear correlation.
   c. Part (a) because there is a higher correlation.
17. Test statistic: $r = 0.199$. Critical values: $r = \pm 0.279$ (approximately). Fail to reject the claim of no significant linear correlation. There does not appear to be a linear correlation between annual spring temperature departures and precipitation departures from seasonal norms. This does not directly contradict the results from Exercise 14. Exercise 14 compares temperature and precipitation data for individual cities; Exercise 17 aggregates (combines) data from all across Canada into single, annual values.
19. With a linear correlation coefficient very close to 0, there does not appear to be a correlation, but the conclusion suggests that there is a correlation.
21. Although there is no *linear* correlation, the variables may be related in some other *nonlinear* way.
23. In parts (a), (b), and (c), the value of r does not change.
25. a. ± 0.272
   b. ± 0.189
   c. -0.378
   d. 0.549
   e. 0.658

27. The apparent correlation between earthquake magnitude and latitude is conditional on an earthquake's being detected. Earthquakes *detected* at higher latitudes tend to have greater magnitude, but we do not know whether *all* earthquakes (even undetected ones) tend to have greater magnitudes in the North.

Section 11-3

1. $\hat{y} = 1 + 2x$

3. $\hat{y} = 0 + 3x$

5. $\hat{y} = -187 + 11.3x$; 399 lb.

7. $\hat{y} = 0.549 + 3.27x$; 1.4 persons

9. $\hat{y} = -56.4 + 4.63x$; 110.3 days.

11. $\hat{y} = -36.1 + 5.58x$; \$36,400 (36.4, in \$1000s). No; for the three samples with $x = 13$, the actual incomes range from \$27,220 to \$37,620.

13. a. $\hat{y} = 41.9 + 0.179x$; 79 min.
 b. Use $\hat{y} = 80.7$ min.
 c. Use part (a), since part (b) simply uses mean y for prediction, but the significant regression explains some of the variation of data from the mean.

15. a. $\hat{y} = -9175.3 + 18179x$; \$18,093
 b. $\hat{y} = 42684 - 4535.9x$; \$29,076
 c. Part (a) because the carat (weight) has a higher correlation with price.

17. $\hat{y} = 0.633 + 1.70x$; -0.90%

19. a. 10.00
 b. 2.50

21. It is not an outlier because the point lies relatively close to the other data in the sample. It is not an influential point because it does not appreciably change the position or slope of the regression line.

23. $\hat{y} = -182 + 0.000351x$; $\hat{y} = -182 + 0.351x$. The slope is multiplied by 1000 and the y-intercept doesn't change. If each y entry is divided by 1000, the slope and y-intercept are both divided by 1000.

25. The linear equation $\hat{y} = -49.9 + 27.2x$ is better because it has $r = 0.997$, which is higher than $r = 0.963$ for $\hat{y} = -103.2 + 134.9\,ln\,(x)$.

27. Linear equation: $\hat{y} = 58.38 - 1.88x$
 $r = 0.597$; not significant
 Exponential equation: $\hat{y} = 50.744e^{-0.1507x}$
 $r = 0.942$; significant
 The exponential equation; because the correlation using the transformed data has the higher r value, and tests as significant.

Section 11-4

1. 0.04; 4%

3. 0.051; 5.1%

5. a. 154.8
 b. 0
 c. 154.8
 d. 1
 e. 0

7. a. 13.788646
 b. 5.711257
 c. 19.5
 d. 0.70711
 e. 0.9756

9. a. 14
 b. Because $s_e = 0$, $E = 0$ and there is no "interval" estimate.

11. a. 4.64 persons
 b. $1.95 < y < 7.32$

13. $\$11,800 < y < \$49,900$ 15. $\$23,400 < y < \$55,100$

17. $-7.9004 < \beta_0 < 1.7990$

19. a. $(n-2)s_e^2$
 b. $\dfrac{r^2 \cdot (\text{unexplained variation})}{1 - r^2}$
 c. $r = -0.949$

21. $-68.77 < \text{intercept} < -3.35$; $2.95 < x \text{ coefficient} < 8.21$

Section 11-5

1. $\hat{y} = -2.03 + 0.0690x_1 + 0.00870x_2 - 0.0262x_3$

3. Of limited usefulness. With a P-value of 0.00015, the equation has overall significance. However, the low r suggests that actual y values may vary considerably from predicted values.

5. VIF $= 1/(1 - 0.1544) = 1.1826$; no multicollinearity

7. AVTEMP $= 2.9615 + 0.0095(900) - (0.0322)(225) = 4.2$. The average annual temperature is 4.2°C.

9. a. $\hat{y} = 3.56 + 0.0980x_1$
 b. 0.021; 0.005; 0.256
 c. No

11. a. $\hat{y} = 1.15 + 1.41x_1 - 0.0144x_2$
 b. 0.564; 0.549; 0.000 (Excel's P-value of 2.33×10^{-11} is a *very* small number.)
 c. Yes

13. a. $\hat{y} = 0.554 + 0.491x_1 + 1.08x_2$
 b. 0.613; 0.600; 0.000 (in Excel, 7×10^{-13})
 c. Yes

15. a. HHSIZE $= 0.5211 + 0.4955$(METAL) $- 0.0222$ (PAPER) $+ 1.0613$(PLASTIC) $+ 0.0861$(GLASS) $- 0.0108$(FOOD)

b. $R^2 = 63.36\%$, Adjusted $R^2 = 60.08\%$ and the regression ANOVA P-value $= 3.8 \times 10^{-11}$

c. The model is highly significant suggesting that the equation is suitable for making predictions of household size. However, only two of the variables, metal and plastic, are significant. It may be best to construct a model (as in question 13) with only these two variables.

17. $\hat{y} = 2.17 + 2.44x + 0.464x^2$. Because $R^2 = 1$, the parabola fits perfectly.

Chapter 11 Review Exercises

1. a. Test statistic: $r = 0.320$. Critical values: $r = \pm 0.632$. Fail to reject the claim of no significant correlation. There does not appear to be a linear correlation between consumption and price.

b. $\hat{y} = -0.258 + 0.189x$

c. 0.202

2. a. Test statistic: $r = 0.112$. Critical values: $r = \pm 0.632$. Fail to reject the claim of no significant linear correlation. There does not appear to be a linear correlation between consumption and income.

b. $\hat{y} = 0.0421 + 0.000283x$

c. 0.2014

3. a. Test statistic: $r = 0.784$. Critical values: $r = \pm 0.632$. Reject the claim of no significant correlation. There does appear to be a linear correlation between consumption and temperature.

b. $\hat{y} = 0.162 + 0.00305x$

c. 0.162

4. $\hat{y} = 0.0301 + 0.250x_1 - 0.000862x_2 + 0.00317x_3$; $R^2 = 0.736$; adjusted $R^2 = 0.605$; P-value $= 0.0358$. This equation is a good predictor of ice cream consumption; but accounting for adjusted R^2, the equation from Exercise 3 is better.

5. a. Test statistic: $r = -0.976$. Critical values: $r = \pm 0.632$. Reject the claim of no significant correlation. There does appear to be a linear correlation between starting position and qualifying speed.

b. $\hat{y} = 667.138 - 6.036x$

c. 9 (rounded from 8.8)

6. a. Test statistic: $r = 0.598$. Critical values: $r = \pm 0.632$. Fail to reject the claim of no significant linear correlation. There does not appear to be a linear correlation between qualifying speeds and points.

b. $\hat{y} = -421 + 3.95x$

c. 10

7. $5.99 < y < 11.71$

8. $\hat{y} = -245 - 1.05x_1 + 4.48x_2$; $R^2 = 0.402$; adjusted $R^2 = 0.231$; P-value $= 0.166$. The multiple regression is not usable for predicting points, because the P-value is 0.166 and the adjusted R^2 is very small.

Chapter 11 Cumulative Review Exercises

1. $n = 50, \bar{x} = 80.7, s = 12.0$.

a. Test statistic: $z = 8.66$. Critical value: $z = 1.645$ (assuming $\alpha = 0.05$). Reject $H_0: \mu \le 66$. There is sufficient evidence to support the claim that the mean interval is now longer than 66 min.

b. $77.3 < \mu < 84.0$

2. (Use the methods of Section 8–2 for two matched samples.) Test statistic: $t = 10.319$. Critical values: $t = \pm 2.365$ (assuming $\alpha = 0.05$). Reject $H_0: \mu_d = 0$. There does appear to be a significant difference between the scores on the humorous commercial and those on the serious commercial. The humorous commercial appears to have higher scores and is therefore better. The issue of correlation is not relevant here; the issue is whether either commercial gets significantly higher scores than the other, not whether the scores are related.

3. Test for a significant linear correlation between the two variables: Test statistic: $r = 0.702$. Critical values: $r = \pm 0.576$ (assuming $\alpha = 0.05$). Reject the claim of no significant linear correlation. There does appear to be a significant linear correlation between the IQs of identical twins who were raised apart. Because $r^2 = 0.493$, we can conclude that 49.3% of the total variation in the IQs of the younger twins can be explained by the variation in IQs of the older twins. This suggests that 49.3% of the variation in IQs can be explained by heredity, and the other 50.7% is attributable to other factors.

Section 12–2

1. The test statistic $x = 2$ is not less than or equal to the critical value of 0. Fail to reject the claim that the temperature departures from seasonal norms do not differ between the first and second springs of an El Niño.

3. Because n > 25, use z test. (Instead of a right-tailed test to determine whether $x = 51$ is large enough to be significant, use a left-tailed test to determine whether $x = 88 - 51 = 37$ is small enough to be significant.) Test statistic: $z = -1.386$. Critical value: $z = -1.645$. Fail to reject the null hypothesis. There is not sufficient evidence to support the claim that more than half the UBC graduates living outside of Canada are living in the United States.

5. The test statistic $x = 7$ is not less than or equal to the critical value of 5. Fail to reject the null hypothesis that the proportions of men and women are equal. There is not sufficient evidence to charge gender bias.

7. (Instead of a right-tailed test to determine whether $x = 18$ is large enough to be significant, use a left-tailed test to determine whether $x = 15$ is small enough to be significant.) Convert $x = 15$ to the test statistic $z = -0.35$. Critical value: $z = -1.645$. Fail to reject H_0. There is not sufficient evidence to support the claim that the median is greater than 0.9085 g. We cannot conclude that the package is labelled incorrectly, because the median could be equal to 0.9085 g.

9. Person-days lost for the truck-based group were less than 10595 thirteen times and greater than 10595 eleven times, so use $x = 11$ and $n = 24$. The test statistic x is not less than or equal to the critical value of 7. Fail to reject the null hypothesis that person-days lost for the truck group were at least equal to 10595. There is not sufficient evidence to support the claim that truck-based transportation loses less time due to work stoppages than the median for all transportation groups combined.

11. The test statistic $x = 2$ is not less than or equal to the critical value of 1. Fail to reject the null hypothesis of no difference. There is not sufficient evidence to warrant rejection of the claim of no difference between the two times.

13. a. Yes. Example data set:
    20 120 120 120 120 120 140 160 180 200
    Sign test: Test statistic: $x = 1$. Critical values: $x = 1$ (assuming significance level $\alpha = 0.05$). Reject null hypothesis of median = 100.
    t test: Test statistic: $t = 1.964$. Critical values: $t = \pm 2.262$ (assuming significance level $\alpha = 0.05$). Fail to reject null hypothesis of mean = 100.
    b. Yes. Example data set:
    80 90 160 160 160 160 160 160 180 200
    Sign test: Test statistic: $x = 2$. Critical values: $x = 1$

(assuming significance level $\alpha = 0.05$). Fail to reject null hypothesis of median = 100.
    t test: Test statistic: $t = 4.329$. Critical values: $t = \pm 2.262$ (assuming significance level $\alpha = 0.05$). Reject null hypothesis of mean = 100.

15. *, *, *, *, *, *, 0, 0, 0, 1, 1, 1, 2, 2, 3, 3

Section 12-3

1. Test statistic: $T = 6$. Critical value: $T = 2$. Fail to reject the null hypothesis that both samples come from the same population distribution.

3. For $n > 30$; $T = 326.5$. Test statistic: $z = -0.64$. Critical values: $z = \pm 2.575$. Fail to reject the null hypothesis that both samples come from the same population distribution.

5. Test statistic: $T = 1$. Critical value: $T = 0$. Fail to reject the null hypothesis that both samples come from the same population distribution. Based on the available results, hypnotism does not appear to be effective.

7. a. Test statistic: $T = 3.5$. Critical value: $T = 25$. Reject the null hypothesis of no difference. There is sufficient evidence to support the claim that P.V.I. does affect vehicles' gas mileage.
    b. No. The data provide no basis for assuming that gas mileage would improve for all cars that use P.V.I. Perhaps the experimental results are unique to 5.0 litre vehicles, or to the one model that was tested. Or perhaps the standardized test conditions are not representative of actual urban or rural driving conditions.

9. Critical value = 6; T- = 2.5; reject H_0 and conclude the median weight of chickens fed with enriched feed is greater than 1.77 kg.

Section 12-4

1. $\mu_R = 577.5$, $\sigma_R = 56.358$, $R = 569$. Test statistic: $z = -0.15$. Critical values: $z = \pm 1.96$. Fail to reject the null hypothesis that red and brown M&Ms have weights with identical populations. They appear to be the same.

3. $\mu_R = 126.5$, $\sigma_R = 15.229$, $R = 139.5$, Test statistic: $z = 0.85$. Critical values: $z = \pm 1.96$. Fail to reject the null hypothesis that the two provinces have the same distribution of hospital bed numbers. (Regardless of these test results, we would need data about the two provinces' populations to determine whether the numbers of hospital beds per person differed between the provinces.)

5. $\mu_R = 333$, $\sigma_R = 31.607$, $R = 321$. Test statistic: $z = -0.38$. Critical values: $z = \pm 1.96$. Fail to reject the null hypothesis that both samples come from populations that have the same dollar values of building permits.

7. $\mu_R = 525$, $\sigma_R = 37.417$, $R = 437$. Test statistic: $z = -2.35$. Critical values: $z = \pm 1.96$. Reject the null hypothesis that the two samples come from populations with the same scores.

9. Reject the null hypothesis if $T_1 \leq 29$ or $T_1 \geq 61$. $T_1 = 38.5$. Fail to reject the null hypothesis. Conclude there is no significant difference in the median grade of the two pedagogies.

11. a.

Combination	Rank Sum for Treatment A
A B A B	4
A B B A	5
B A A B	5
B A B A	6
B B A A	7

b. AABB, $R = 3$; ABAB, $R = 4$; ABBA, $R = 5$; BAAB, $R = 5$; BABA, $R = 4$; BBAA, $R = 3$. Each value of R has a 1/3 probability.

c. If the distribution of A and B is uniform, then $P(A < B) = \frac{1}{2}$ or $P(B < A) = \frac{1}{2}$. This is the case for the middle 4 cases but not for the cases AABB and BBAA. In the case AABB, $P(B < A) = 0 < 0.1$ and in the case BBAA, $P(A < B) = 0 < 0.1$. In these two cases, it is possible to reject the null hypothesis at a 0.1 level of significance.

Section 12–5

1. Test statistic: $H = 7.240$. Critical value: $x^2 = 5.991$. Reject the null hypothesis that the three samples come from identical populations. There appears to be a difference in the number of House sittings per Parliament, for the three time periods.

3. Test statistic: $H = 5.717$. Critical value: $x^2 = 7.815$. Fail to reject the null hypothesis that the four samples come from identical populations. There is not sufficient evidence to support the claim that the prices of diamonds are different for different clarity levels. However, the data do not include diamonds of level 6 (the least clear). It may be useful to compare prices for diamonds of clarity 6 with the other diamonds.

5. Test statistic: $H = 2.075$. Critical value: $x^2 = 11.071$. Fail to reject the null hypothesis that the samples come from identical populations. No corrective action is necessary.

7. Test statistic: $H = 0.185$. Critical value: $x^2 = 5.991$. Fail to reject the null hypothesis that the three samples come from identical populations. There is not sufficient evidence to indicate that the samples of movie lengths come from different populations.

9. Test statistic: $FR = 12.25$. Critical value: $x^2 = 5.991$. Reject the null hypothesis and conclude there is a significant difference between at least 2 of the stores in median daily sales.

11. smallest $= 0$; largest $= 12.5$

13. The key to solving the problem is to use two key identities:

$$n_1 - n_2 = n \text{ and } R_1 + R_2 = \frac{n(n-1)}{2}.$$

Section 12–6

1. a. $r_s = 1$ and there appears to be a correlation between x and y.

b. $r_s = -1$ and there appears to be a correlation between x and y.

c. $r_s = 0$ and there does not appear to be a correlation between x and y.

3. $r_s = 0.855$. Critical values: $r_s = \pm 0.648$. Significant correlation. There appears to be a correlation between salary and stress.

5. $r_s = -0.067$. Critical values: $r_s = \pm 0.648$. No significant correlation. There does not appear to be a correlation between stress level and physical demand.

7. $r_s = 0.715$. Critical values: $r_s = \pm 0.591$. Significant correlation. There appears to be a correlation between the IQ scores of identical twins separated at birth.

9. $r_s = 0.363$. Critical values: $r_s = \pm 0.738$. No significant correlation. There does not appear to be a correlation between length and weight.

11. $r_s = 0.766$. Critical values: $r_s = \pm 0.507$. Significant correlation. There appears to be a correlation between the income levels for professions and the number of years of education they require. The results do not change if each income is multiplied by 1000.

13. $r_s = 0.786$. Critical values: $r_s = \pm 0.280$. Significant correlation. There appears to be a correlation between durations and intervals between eruptions.

15. $r_s = 0.736$. Critical values: $r_s = \pm 0.623$. Significant correlation. For persons who move from an initially sitting position to a supine position, there appears to be a correlation between body temperatures in those two states. Yes, but not if the estimates need to be exact.

17. $r_s = 0.143$. Critical values: $r_s = \pm 0.786$. No significant correlation. There does not appear to be a correlation between the temperature departures in the first and second springs of an El Niño. Neither of these variables can be used to predict the other.

19. a. ± 0.707
   b. ± 0.514
   c. ± 0.361
   d. ± 0.463
   e. ± 0.834

Section 12-7

1. $n_1 = 10, n_2 = 5, G = 3$, 5% cutoff values: 3, 12
3. $n_1 = 12, n_2 = 4, G = 4$, 5% cutoff values: 3, 10
5. $n_1 = 12, n_2 = 8, G = 10$, 5% cutoff values: 6, 16. Fail to reject the null hypothesis of randomness. A lack of randomness could result in large gains for the author and large losses to the casino. Fat chance.

7. $n_1 = 12, n_2 = 12, G = 12$, cutoff values (for $\alpha = 0.05$): 7, 19. Fail to reject the null hypothesis of randomness. There is not sufficient evidence to warrant rejecting the claim that the order is random.

9. $n_1 = 7, n_2 = 4, G = 7$, 5% cutoff values: 2, 10. Fail to reject the null hypothesis of randomness. It appears that we elect Liberals and Conservatives to become the party of the new prime minister in a sequence that is random.

11. $n_1 = 11, n_2 = 8, G = 8$, 5% cutoff values: 5, 15. Fail to reject the null hypothesis of randomness. There is not sufficient evidence to suggest that the sequence of monthly gains and losses is not random. Someone investing in a TSE 300–based fund could still make money—if the *magnitudes* of monthly gains, when they occur, are greater than the magnitudes of the losses.

13. $n_1 = 21, n_2 = 19, G = 7, \mu_G = 20.95, \sigma_G = 3.11346$. Test statistic: $z = -4.48$. Critical values: $z = \pm 1.96$. Reject the null hypothesis of randomness.

15. $n_1 = 49, n_2 = 51, G = 54, \mu_G = 50.98, \sigma_G = 4.9727$. Test statistic: $z = 0.61$. Critical values: $z = \pm 1.96$. Fail to reject the null hypothesis of randomness.

17. Minimum is 2, maximum is 4. Critical values of 1 and 6 can never be realized so that the null hypothesis of randomness can never be rejected.

Chapter 12 Review Exercises

1. Test statistic: $H = 4.037$. Critical value: $x^2 = 5.991$. Fail to reject the null hypothesis that the three populations are the same. There is not sufficient evidence to support the claim that the three populations are not all the same.

2. Test statistic: $T = 17$. Critical value: $T = 6$. Fail to reject the null hypothesis that both samples come from the same population distribution. But because the data are so variable, it would be imprudent to predict changes in car sales based on changes in truck sales.

3. $n_1 = 6, n_2 = 24, G = 12, \mu_G = 10.6, \sigma_G = 1.6873$. Test statistic: $z = 0.83$. Critical values: $z = \pm 1.96$. Fail to reject randomness.

4. Convert 2000 scores to ranks, and re-number 1999 ranks into a sequence beginning at 1. $r_s = -.800$. Critical values: $r_s = \pm 0.683$. Significant correlation. There is sufficient evidence to support the claim that 1999 Ranks and 2000 Scores are rank correlated.

5. Test statistic: $x = 4$. Critical value: $x = 0$. Fail to reject the claim that there is no difference in the number of search results before and after adding a third word to the search list. Simply adding more words to the search list does not appear to be an effective strategy to reduce the number of results. (*Hint:* Try prefixing the most important words in your search list with $+$.)

6. $\mu_R = 201.5, \sigma_R = 23.894, R = 163$. Test statistic: $z = -1.61$. Critical values: $z = \pm 1.96$. Fail to reject the null hypothesis that the drug has no effect on eye movements.

7. Test statistic: $H = 4.096$. Critical value: $x^2 = 7.815$. Fail to reject the null hypothesis that the four populations are identical. There is not sufficient evidence to support the claim that the samples come from populations that are not all the same.

8. a. Every answer "F" is followed by four answers of "T".
   b. $n_1 = 20 < n_2 = 5, G = 11$, cutoff values (for $\alpha = 0.05$): 5, 12. Fail to reject the null hypothesis of randomness.
   c. The pattern shows evidence of a sequence that is not random, but that evidence is not sufficient to warrant rejection of the null hypothesis of randomness.

9. $\mu_R = 144, \sigma_R = 16.25, R = 84$. Test statistic: $z = -3.69$. Critical values: $z = \pm 1.96$. Reject the null hypothesis of no difference between mail order and store prices.

10. The test statistic $x = 0$ is less than or equal to the critical value of 0. Reject the claim that the training has no effect on weight. There does appear to be an effect.

11. Test statistic: $T = 0$. Critical value: $T = 1$. Reject the null hypothesis that training has no effect on weight. There does appear to be an effect.

12. $r_s = 0.770$. Critical values: $r_s = \pm 0.648$. There appears to be a significant correlation between the two measures of health care.

13. Critical value $= 7.815$. Test statistic $= 9.6375$. Reject the null hypothesis and conclude there is a significant difference between at least 2 of the games in median ratings.

14. Critical value $= 28$. Test statistic $= 28$. Reject the null hypothesis and conclude the median amount of waste for the new lathe is significantly less than that of the existing lathe.

Chapter 12 Cumulative Review Exercises

1. a. $G = 21$, $n_1 = 20$, $n_2 = 10$. Critical values: 9, 20. Reject randomness.

   b. Test statistic: $z = -1.83$. Critical values: $z = \pm 1.96$. Fail to reject null hypothesis that $p = 0.5$. There is not sufficient evidence to support the claim that the proportion of women is different from 0.5.

   c. $0.165 < p < 0.502$

   d. There isn't sufficient evidence to support a claim of bias against either gender, but the sequence does not appear to be random. It's possible that the pollster is using selection methods that are unacceptable.

2. a. Test statistic: $x = 3$. Critical value: $x = 1$. Because the test statistic is not less than or equal to the critical value, fail to reject the claim that there is no difference between the weights before and after the program.

   b. Test statistic: $T = 6.5$. Critical value: $T = 6$. Because the test statistic is not less than or equal to the critical value, fail to reject the claim that there is no difference between the weights before and after the program.

   c. Test statistic: $t = 2.279$. Critical values: $t = \pm 2.262$. Reject the null hypothesis that there is no difference between the weights before and after the program. The program does appear to have an effect.

   d. Both nonparametric tests (the sign test and the Wilcoxon signed-ranks test) failed to detect a difference, whereas the t test resulted in the conclusion of a significant difference. The t test fully utilizes the magnitudes of

the differences, the sign test does not use the magnitudes at all, and the Wilcoxon signed-ranks test only uses the ordered rankings of the magnitudes. The two nonparametric tests will therefore be less sensitive in cases where the magnitudes have a potential impact on the test results.

3. a. 19176.6

   b. 11980

   c. 0

   d. 85550

   e. 21538.3 [*Hint*: If your calculator cannot handle the large x^2's, you can make the numbers manageable by converting the original data into units of 1000 (for example, convert 41810 to 41.810); then multiply the final answer for s times 1000.]

   f. The largest value of 85550: Its corresponding z score is greater than 3, and it is quite far from the next highest value. However, the data set is highly skewed, so this single high value does not necessarily suggest an input or data-collection error.

Note: Chapters 13E–15E are on the Pearson eText and CD only.

Section 13–2

1.

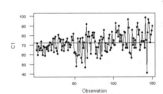

There is an upward trend indicating that the intervals between eruptions are increasing, with the result that tourists must wait longer. There is increasing variation, with the result that predicted times of eruptions are becoming less reliable.

3.

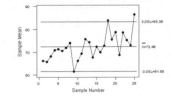

There are points lying beyond the upper control limit and there is an upward trend, so the process mean is out of statistical control. An out-of-control process mean indicates that we cannot make accurate predictions.

5.

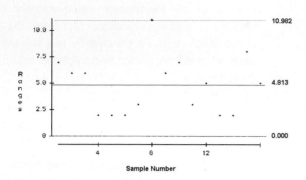

Sample Number

The process variation is out of statistical control because there is a point above the upper control limit.

7.

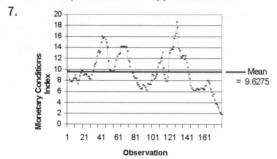

Yes, several patterns suggest that the process is not within statistical control. The last 39 observations are all below the mean, and apparently continue a downward trend that begins at about observation 130. There are also repeated sequences when 20 or more observations in a row are all one side or the other of the mean.

9.

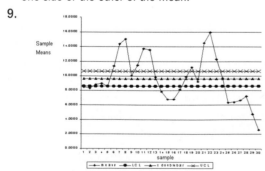

The process mean is far from statistical control. If the Bank of Canada has intended to bring monetary conditions under statistical control, it has not been successful.

11.

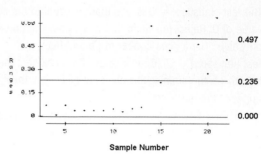

Sample Number

The process is out of control because there is a shift up, there are 8 consecutive points below the centre line, and there are points beyond the upper control limit.

13.

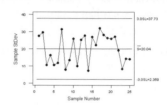

Using Table 13-1 values, $\bar{s} = 20.05$ and LCL = 2.366. The result and conclusions are very similar to those for the R chart.

Section 13–3

1. Process appears to be within statistical control.
3. Process appears to be out of statistical control because there is a pattern of an upward trend and there is a point that lies beyond the upper control limit.

5.

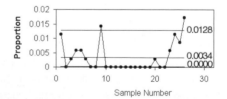

This process is out of statistical control because there are points above the upper control limit, and there are 10 consecutive points all lying below the centre line. Having samples with less than the mean proportion of defects is, in itself, a good result—but the chart shows that the situation is unstable. Causes for periods of instability should be identified.

7.

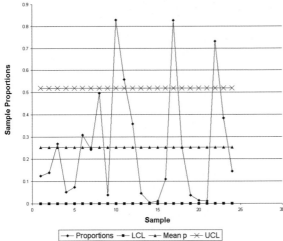

The process is out of statistical control because there are points beyond the upper control limit. Companies which rely on air transportation will be concerned about the apparent instability of that mode, with respect to the proportion of person-days lost due to work stoppages compared to alternative modes. This instability may result in air-based transportation service being unreliable.

9. $\bar{c} = 106/25 = 4.24$

Upper control limit $= 4.24 + 3\sqrt{(4.24)} = 10.4$

Lower control limit $= 4.24 - 3\sqrt{(4.24)} = -1.94$ which we round to 0.

Any 12-hour period with more than 10 patients waiting to be admitted is not in control. The only period that is not in control is the one with 15 patients.

11. a. Lower control limit $= 0.05 - 3\sqrt{(0.05(0.95)/100)}$
$= -0.0154$ which we round to 0. Upper control limit $= 0.05 + 3\sqrt{(0.05(0.95)/100)} = 0.1154$

b. Lower control limit $= 0.0123$. Upper control limit $= 0.0877$.

c. The chart with $n = 300$ requires more effort to collect the sample but is more sensitive to changes. With the upper limit of 0.0877, it would be more sensitive to a change of 5% to 10% compared to the chart with $n = 100$ which has an upper limit of 0.1154.

Chapter 13 Review Exercises

1.

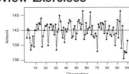

The process appears to be within statistical control.

2.

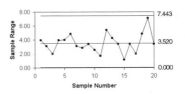

The process appears to be within statistical control.

3.

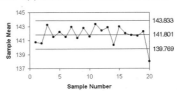

The process mean appears to be out of statistical control because there is a point lying beyond the lower control limit. The process should be corrected because too little epoxy cement hardener is being put into some containers.

4.

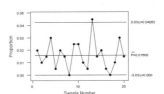

The process is out of statistical control because there is a point lying beyond the upper control limit.

5. $\bar{c} = 64/20 = 3.2$

Upper control limit $= 3.2 + 3\sqrt{(3.2)} = 8.6$

Lower control limit $= 3.2 - 3\sqrt{(3.2)} = -2.2$ which we round to 0.

Any day with more than 8 defective lipstick tubes is not in control. There is only one such day with 9 defective lipstick tubes. This is consistent with the results from question 4.

Chapter 13 Cumulative Review Exercises

1. a.

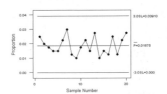

The process is in statistical control.

b. $0.0158 < p < 0.0217$

c. Test statistic: $z = 5.77$. Critical value: $z = 1.645$. Reject H_0: $p \le 0.01$. There is sufficient evidence to warrant rejection of the claim that the rate of defects is 1% or less.

2. a. 1/256

   b. 1/256

   c. 1/128

3.

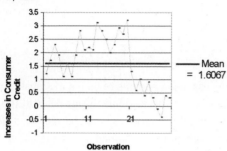

The process is out of statistical control because the run chart displays a shift up, followed by a shift down.

4. a. Yes. Test statistic: $r = 0.779$. Critical values: $r = \pm 0.361$ (approximate; assuming significance level $\alpha = 0.05$).

   b. $\hat{y} = 2.72 + 2.65x$

   c. $9.08 billion.

Section 14-2

1. 101.9

3. 89.2

5. 104.9

7. 86.1

9. 105.0

11. 86.6

13. 104.95

15. 86.35

17. Laspeyres = 109.1; Paasche = 109.4; Fisher = 109.25

19. 11.0%

Section 14-3

1. Seasonal variation

3. Irregular variation

5. Trend

7. Seasonal variation

9. Seasonal variation

11. Trend

13. Cyclical variation

15. Irregular variation

17. Trend

19. Irregular variation

21. Seasonal variation

23. Cyclical variation

25. Irregular variation

27. Trend

Section 14-4

1.

Quarter	Quarterly Index
1	0.9919
2	0.9969
3	1.0001
4	1.0111

3.

Year	Quarter	Deseasonalized Earnings
2004	1	710.90
	2	709.55
	3	708.57
	4	708.46
2005	1	735.74
	2	735.85
	3	740.27
	4	736.85
2006	1	755.75
	2	754.11
	3	754.46
	4	757.33
2007	1	785.19
	2	789.72
	3	788.22
	4	788.98
2008	1	815.19
	2	812.36
	3	809.06
	4	808.04

Quarter 1 of 2008 had the highest seasonally adjusted average weekly earnings.

		5.	7.	9.	11.	13.	15.	17.
Year	Quarter	Single Dwellings	Cottages	Double Dwellings	Row Housing	Apartments	Conversions	Total
2003	1	30,200	173	3,494	4,186	14,961	1,319	54,397
	2	29,455	205	3,003	5,105	14,628	1,192	53,893
	3	30,561	193	3,195	5,699	15,631	1,317	56,767
	4	31,740	220	2,962	5,457	16,064	1,780	57,644
2004	1	31,804	234	2,920	6,189	18,229	1,368	60,759
	2	32,621	234	3,079	5,184	19,267	1,397	61,976
	3	32,255	190	2,517	5,422	18,119	915	59,594
	4	32,568	255	3,340	6,357	15,865	1,268	58,860
2005	1	29,219	191	2,982	5,563	16,924	1,229	56,165
	2	30,221	174	2,779	6,383	18,553	1,029	59,152
	3	29,441	181	3,065	5,212	18,041	1,512	57,450
	4	32,290	155	2,700	6,007	23,418	1,079	66,049
2006	1	31,267	178	2,539	5,387	18,633	1,484	59,406
	2	28,551	168	2,623	5,289	18,127	1,225	56,003
	3	29,444	170	2,827	5,352	19,952	1,418	59,089
	4	29,728	152	3,032	5,661	19,753	1,001	59,409
2007	1	28,455	143	2,731	5,826	20,553	1,142	58,833
	2	28,724	139	3,030	5,566	20,465	1,172	58,922
	3	29,357	151	2,959	5,870	20,562	1,020	59,855
	4	29,342	132	2,624	6,032	20,812	970	60,209
2008	1	23,992	127	2,333	*5,751	*21,873	1,202	*55,236
	2	*25,363	126	*2,485	5,374	20,132	*1,728	54,851
	3	23,879	*162	2,435	5,347	18,867	1,562	52,041
	4	19,268	132	2,339	3,388	15,259	1,645	42,626

*Quarter of 2008 with the highest seasonally adjusted number of building permits.

	19.	21.
Month	Index for Single Dwellings	Index for Total Permits
Jan.	0.6178	0.6675
Feb.	0.6952	0.7066
Mar.	1.0819	1.0558
Apr.	1.2081	1.1544
May	1.3138	1.1964
June	1.2886	1.2283
July	1.1352	1.1177
Aug.	1.0849	1.042
Sept.	1.0489	1.0082
Oct.	1.0413	1.0509
Nov.	0.8542	0.943
Dec.	0.6302	0.8292

Section 14-5

1.

Year	Quarter	Predicted Mean Weekly Earnings
2009	1	791.43
	2	795.42
	3	797.97
	4	806.75
2010	1	806.25
	2	810.32
	3	812.92
	4	821.86

		3.	5.	7.
		Seasonally Adjusted Predictions		
Year	Quarter	Cottages	Row Housing	Conversions
2009	1	45	4,744	1,108
	2	183	5,885	1,467
	3	178	5,756	1,229
	4	75	5,138	1,469
2013	1	22	4,640	1,135
	2	89	5,756	1,502
	3	87	5,630	1,259
	4	37	5,026	1,504

9. $\hat{y} = 6856.8 + 850.5x$

11.

Month	Monthly Adjusted Forecasts
Jan.	5,063
Feb.	5,697
Mar.	8,867
Apr.	9,901
May	10,767
June	10,561
July	9,303
Aug.	8,891
Sept.	8,596
Oct.	8,534
Nov.	7,000
Dec.	5,165

13.

Month	Monthly Adjusted Forecasts
Jan.	5,332
Feb.	6,000
Mar.	9,337
Apr.	10,427
May	11,339
June	11,121
July	9,797
Aug.	9,363
Sept.	9,053
Oct.	8,987
Nov.	7,372
Dec.	5,439

MSE = 1,292,932

Section 14–6

		1.	3.
		Centred 4-Point Moving Averages	
Year	Quarter	Single Dwellings	Cottages
2003	1	none	none
	2	30,386	200
	3	30,706	206
	4	31,712	217
2004	1	32,173	216
	2	32,347	221
	3	31,831	217
	4	31,069	194
2005	1	30,303	191
	2	30,244	175
	3	30,653	174
	4	30,123	172
2006	1	30,123	168
	2	29,584	167
	3	29,023	164
	4	29,078	153
2007	1	29,054	146
	2	28,973	143
	3	28,082	141
	4	27,015	136
2008	1	25,523	140
	2	23,403	140
	3	none	none
	4	none	none

5. 52-point moving averages

7.

Total Permits	Alpha = 0.9
12,856	12,856
12,083	12,160
19,116	18,420
19,760	19,626
22,364	22,090
22,180	22,171
21,891	21,919
18,337	18,695
19,712	19,610
20,615	20,515
18,473	18,677
15,158	15,510
12,817	13,086
13,432	13,397
22,925	21,972
24,102	23,889

It is clear from the charts that $\alpha = 0.04$ leads to more smoothing.

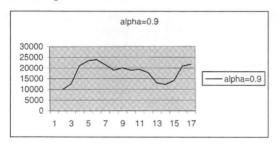

alpha=0.9

Chapter 14 Review Exercises

1. Seasonal variation
2. Cyclical variation
3. Irregular variation
4. Trend
5. Cyclical variation
6. Trend
7. Irregular variation
8. Seasonal variation
9.

Year	2001	2002	2003	2004	2005	2006	2007	2008
Price relative	100.7	100	105.3	114	132.2	138.8	144.7	163.9

10.

Year	2001	2002	2003	2004	2005	2006	2007
Price in 2002 $s	69	67.3	69.1	73.5	83.4	85.6	87.2

11. 43.7%
12. 26.4%
13. Laspeyres = 119.9; Paasche = 119.9; Fisher = 119.9. The general downward trend may be caused by other factors as well, such as health consciousness.

14. Quarterly indexes: 0.9857, 1.0008, 1.0025, 1.011

Year	Quarter	Deseasonalized Data
2005	1	8,769
	2	8,349
	3	8,210
	4	8,529
2006	1	8,957
	2	8,473
	3	8,313
	4	8,631
2007	1	8,949
	2	8,506
	3	8,366
	4	8,615
2008	1	8,990
	2	8,524
	3	8,466
	4	8,742

	16.	17.
Quarter	2008 Adjusted Forecast	2009 Adjusted Forecast
1	8,561	8,627
2	8,692	8,759
3	8,708	8,774
4	8,781	8,848
	MSE = 2,031.75	

18.

Raw Data	Centred 4-Point Moving Average
8,644	
8,355	8,463
8,231	8,509
8,622	8,541
8,829	8,566
8,480	8,592
8,334	8,590
8,726	8,599
8,821	8,612
8,513	8,608
8,387	8,618
8,709	8,622
8,862	8,648
8,531	8,680
8,488	
8,838	

19.

Month	Monthly Index
Jan.	1.0271
Feb.	1.0239
Mar.	1.02
Apr.	1.0179
May	0.9756
June	0.9661
July	0.9543
Aug.	0.9585
Sept.	1.0083
Oct.	1.0086
Nov.	1.0143
Dec.	1.0253

20.

Year	Month	Deseasonalized Data
2005	Jan.	8,442
	Feb.	8,412
	Mar.	8,478
	Apr.	8,474
	May	8,462
	June	8,473
	July	8,448
	Aug.	8,430
	Sept.	8,479
	Oct.	8,467
	Nov.	8,505
	Dec.	8,484
2006	Jan.	8,607
	Feb.	8,638
	Mar.	8,628
	Apr.	8,612
	May	8,577
	June	8,598
	July	8,529
	Aug.	8,564
	Sept.	8,584
	Oct.	8,618
	Nov.	8,584
	Dec.	8,563
2007	Jan.	8,611
	Feb.	8,634
	Mar.	8,607
	Apr.	8,623
	May	8,655
	June	8,611
	July	8,611
	Aug.	8,614
	Sept.	8,615
	Oct.	8,591
	Nov.	8,549
	Dec.	8,574
2008	Jan.	8,682
	Feb.	8,659
	Mar.	8,630
	Apr.	8,634
	May	8,648
	June	8,660
	July	8,755
	Aug.	8,734
	Sept.	8,665
	Oct.	8,666
	Nov.	8,704
	Dec.	8,722

	21.	**22.**
Month	2009 Adjusted Forecast	2008 Adjusted Estimate
Jan.	8,990	8,921
Feb.	8,961	8,893
Mar.	8,927	8,860
Apr.	8,908	8,841
May	8,538	8,473
June	8,456	8,391
July	8,352	8,289
Aug.	8,389	8,325
Sept.	8,825	8,758
Oct.	8,827	8,760
Nov.	8,877	8,810
Dec.	8,974	8,906

$$MSE = 1,479.95$$

23.

Raw Data	Centred 12-Point Moving Average
8,671	
8,613	
8,648	
8,625	
8,255	
8,186	8,463
8,062	8,477
8,080	8,496
8,549	8,509
8,540	8,521
8,627	8,530
8,699	8,540
8,841	8,547
8,844	8,558
8,801	8,566
8,766	8,579
8,368	8,586
8,307	8,592
8,139	8,593
8,209	8,592
8,655	8,591
8,692	8,592
8,707	8,598
8,780	8,599
8,845	8,605
8,840	8,609
8,779	8,612
8,777	8,610
8,444	8,607
8,319	8,608
8,218	8,614
8,257	8,616
8,687	8,618
8,665	8,619
8,671	8,618
8,791	8,622
8,918	8,634
8,866	8,643
8,803	8,648
8,788	8,654
8,437	8,667
8,367	8,680
8,355	
8,372	
8,737	
8,741	
8,829	
8,943	

Index

A

abuse of statistics
  bad samples, 10
  deliberate distortions, 11
  distorted percentages, 11
  guesstimates, 11
  loaded questions, 11–12
  misleading graphs, 12
  partial pictures, 11
  pictographs, 12
  precise numbers, 11
  self-selected samples, 10
  small samples, 10–11
acceptance regions, 366
acceptance sampling, 217
actual odds against, 123
actual odds in favour, 123
addition rule
  complementary events, 131–132
  compound events, 129
  formal, 129
  mutually exclusive events, 129–130
  notation, 129
adjusted coefficient of determination, 655–656
α (alpha) probability errors, 366, 369–371, 388–392
alternative hypotheses, 364–365
The American Statistician (Moser and Stevens), 474–475
analysis of variance (ANOVA)
  defined, 508
  F distribution, 509
  one-way. *See* one-way ANOVA
  overview, 508–509
  regression, 641
  table, 511–512
  two-way. *See* two-way ANOVA
arithmetic mean, 56–57
"at least one", probability of, 194–195
attribute data, 4, 5
average growth factor, 70

B

bad samples, 10
balanced design, 527

Bayes' Theorem, 146–148
β (beta) probability errors, 366, 369–371, 388–392
bimodal, 59
binomial experiments, 186
binomial probability distributions
  binomial probability formula, 188–189, 193–194
  computer statistics packages for, 190–191
  continuity corrections, 273–276
  defined, 186
  goodness-of-fit tests, 555–556
  hypergeometric distributions and, 215
  mean, 201–204
  negative, 200
  normal approximation, 270–276
  notation, 187
  probability of "at least one", 194–195
  rationale for formula, 193–194
  requirements, 186
  right questions, 191–193
  standard deviation, 201–204
  table for, 189–190
  variance, 201–204
binomial probability formula, 188–189, 193–194
bivariate data, 600
blinding, 16
blocks, 16, 525–527
Bonferroni test, 515–516
bootstrap method, 352
boxplots (box-and-whisker diagrams), 99–103, 105

C

capture-recapture method, 169, 458
categorical data, 5
causality, 606–607
census, 4
central limit theorem
  application, 262–264
  finite population correction factor, 265–266
  fuzzy, 264
  hypothesis testing and, 362–363
  notation, 261
  original population, 259–260
  sampling distribution of sample means, 258–261

central tendency, measures of
  average, 60
  comparison of, 63, 64
  defined, 55
  mean, 56–57, 60, 61–62
  median, 58, 60
  midrange, 59–60
  mode, 59, 60
  skewness, 63–65
  weighted mean, 63
centroid, 612
Chebyshev's theorem, 82
chi-square distributions, 339–343, 421–423, 548
class boundaries, 33–34
class midpoints, 33–34
class width, 33–34
classical approach to probabilities, 117, 118
classical method of testing hypotheses, 374–379
cluster sampling, 18, 20
coefficient of determination, 639–640
coefficient of variation, 87
combinations, 154
combinations rule, 156–158
complementary events, 121–122, 131–132
completely randomized design, 16, 525
composite sampling, 137
compound events, 129
conditional probabilities
  contingency tables, 136–139
  defined, 136
  overview, 135–136
  testing for independence, 140–141
confidence coefficient, 296
confidence interval for μ_y, 645
confidence interval limits, 299
confidence intervals
  defined, 296
  dependent samples, 445
  estimates, 295–297, 299–303
  hypothesis testing, 383–385, 423–424
  independent samples, 454–456
  one-sided, 338
  population proportions, 327, 330
  population variance, 342–343, 344–345

prediction. *See* prediction intervals
  round-off rule, 300, 327
  small independent samples/equal variances, 476–478
  small independent samples/unequal variances, 478–480
  Tukey's method, 516–517
  two proportions, 493–494
confounding, 16–17
Consumer Price Index, 91
contingency tables
  assumptions, 567
  conditional probabilities, 136–139
  critical values, 568–569
  expected frequencies, 568–569
  P-values, 571–572
  rationale for procedure, 570–571
  test of homogeneity, 572–575
  test of independence, 566–567
continuity corrections, 273–276
continuous (numerical) data, 5–6
continuous probability distributions
  central limit theorem. *See* central limit theorem
  density curve, 226
  exponential distribution, 280–283
  normal distributions. *See* normal distributions
  overview, 226–229
  standard normal distribution. *See* standard normal distributions
  uniform distribution, 227–229
continuous random variables, 175–176
control groups, 16
convenience sampling, 19
correlation
  assumptions, 601
  common errors, 606–607
  defined, 600
  formal hypothesis test, 607–612
  linear correlation coefficient *r*. *See* linear correlation coefficient *r*
  one-tailed tests, 610–611
  r for linear correlation, 609–610
  scatter diagrams (scatterplots), 601–603
  t for linear correlation, 608–609
Cost of Laughing Index, 91

counting
    approach, deciding on, 158–159
    combinations rule, 156–158
    factorial rule, 153–154
    fundamental counting rule, 151–153
    notation, 153
    permutations rule, 154–156
critical regions, 366
critical values
    chi-square distribution, 339–343
    contingency tables, 568–569
    defined, 297
    estimates, 297–298
    goodness-of-fit tests, 550–554
    hypothesis testing, 366, 369, 399
    Kruskal-Wallis test (H test), 710–712
    notation, 297
    rank correlation test, 724
    runs test, 737–739
    sign test procedure, 678–679
    two variances, 466–468
    Wilcoxon rank-sum test, 701–702
    Wilcoxon signed-ranks test, 690–691
cross product, 604
crosstabs. *See* contingency tables
cumulative relative frequencies, 37
cumulative relative frequency tables, 37

D
data
    bell-shaped, 81–82
    bivariate, 600
    centre, 32
        see also central tendency, measures of
        characteristics, 32, 106
    comparison, 49
    continuous, 5–6
    defined, 5
    description of, 48–49
    discrete, 5–6
    distribution, 32
        see also distributions
    exploration, 49
    exploration of, 100–103
    frequency tables, 33–38

graphs. *See* graphs
interval level of measurement, 7, 8
nominal level of measurement, 6, 8
numerical, 5–6
ordinal level of measurement, 6–8
outliers, 32, 98–99
qualitative, 5
quantitative, 5
randomness. *See* runs test
ratio level of measurement, 7, 8
reliability, 271–273
skewness, 63–65
symmetric, 202–203
time, 32
validity, 271–273
variation, 32
    see also variation, measures of
data collection technology, 230
deciles, 90, 93
decision theory, 181
degree of confidence, 296
degrees of freedom, 317
deliberate distortions, 11
denominator degrees of freedom, 466
density curves, 226–227
dependent events, 140–141, 145–146, 187
dependent samples (matched pairs)
    assumptions, 441
    confidence intervals, 445
    defined, 441
    hypothesis testing of, 442–445
    notation, 442
    Wilcoxon signed-ranks test, 691–693
dependent variables, 619
descriptive statistics, 32, 65–66, 93–94
destructive testing, 319
deviation, 73
discrete data, 5–6
discrete probability distributions
    binomial. *See* binomial probability distributions
    geometric distributions, 200
    hypergeometric distributions, 214–215
    overview, 174
    Poisson distributions, 207–211
    probability histograms, 177–178

random variables, 175–182
requirements, 177–178
discrete random variables, 175–176
distorted percentages, 11
distribution-free tests, 674
distributions
appropriate, choice of, 319, 399–402
bell-shaped, 100
binomial probability. *See* binomial probability distributions
chi-square, 339–342, 421–423
data, 32
discrete probability. *See* discrete probability distributions
exploratory data analysis. *See* exploratory data analysis
exponential, 280–283
geometric, 200
hypergeometric, 214–215
in hypothesis testing, 399–402
multinomial, 200–201
normal. *See* normal distributions
Poisson, 207–211
probability. *See* probability distributions
sampling, 258–261
skewed, 100
skewed vs. symmetric, 64
standard normal. *See* standard normal distributions
Student *t*. *See* Student *t* distributions
t distribution, 316–317
tails, 369
uniform, 100, 227–229
double-blind, 16
dummy variables, 654

E
efficiency rating, 675
empirical rule, 81, 202–203
estimate(s)
assumptions, 315–317
confidence intervals, 295–297, 299–303
critical values, 297–298
defined, 294
degree of confidence, 296

degrees of freedom, 317
estimator, 294
interval, 296
large samples, 294–308
margin of error, 298–301, 318–319
point, 295
population mean, 294–308, 315–322
population proportion, 326–332
population variance, 338–345
sample size, determination of, 304–308
small samples, 315–322
Student *t* distributions, 316–320
estimator, 294
events
complementary, 131–132
compound, 129
defined, 116
dependent, 140–141, 145–146, 187
independent, 140–141, 146
mutually exclusive, 129–130
rare event rule, 263, 360, 385
simple, 116
Excel
binomial probabilities, 195–196
boxplots, 101–103
coefficient of determination, 649
confidence intervals, 308, 322, 333–334, 345–346
contingency tables, 575–576
descriptive statistics, 65–66, 82–83, 93–94
finding values, 255
frequency table, 38
goodness-of-fit tests, 559–560
graphs, 49–50
hypergeometric distributions, 216
hypothesis testing, 385–386, 403, 415, 424–425
Kruskal-Wallis/Friedman tests, 716–718
linear correlation coefficient, 613
multinomial experiments, 560
multiple regression equation, 662
normality testing, 590–591
one-way ANOVA, 517–519
Poisson distributions, 211
position, measures of, 93–94

Excel (*cont.*)
    probabilities, 167
    probability in nonstandard distribution, 248
    probability when given z score, 239–240
    quartiles, calculation of, 93–94
    rank correlation, 728–729
    regression, 631–632, 649
    runs test, 740
    sign test, 684
    standard error of estimate, 649
    two dependent samples, 446
    two independent samples, 457–458, 482–483
    two population proportions, 495
    two variances, 448–469
    two-way ANOVA, 534
    usage guidelines, 23–24
    Wilcoxon rank-sum test, 704–705
    Wilcoxon signed-ranks test, 694–695
exclusive or, 129
expected frequencies
    contingency tables, 568–569
    goodness-of-fit tests, 550
    rationale for procedure, 570–571
expected values, 181–182
experimenter effects, 21
experiments, 16
    binomial, 186
    blinding, 16
    blocks, 16
    completely randomized design, 16
    confounding, 16–17
    control group, 16
    defined, 15
    design, 15–16
    double-blind, 16
    experimental units, 16
    Hawthorne effect, 21
    meta-analysis, 19
    nonsampling error, 19
    placebo effect, 16
    random selection, 16
    randomization, 17–20
    replication, 17
    rigorously controlled design, 16
    Rosenthal effect, 21
    sampling error, 19
    treatment group, 16
    variables, effects of, 16–17
explained deviation, 638–639
explained variation, 637–639
exploratory data analysis (EDA)
    boxplots, 99–100
    defined, 97
    outliers, 98–99
    tools, 100–103
exponential distribution, 280–283
extreme outliers, 105
extreme values. *See* outliers

F
F distribution, 465, 509
factor, 510
factorial design, 530–534
factorial rule, 153–154
factorial symbol, 153
false positive, 454
The Figure Finaglers (Reichard), 12
finite population correction factor, 265–266
Fisher, R.A., 553
five-number summary, 99
5% cutoff criterion, 734–736
formal addition rule, 129
formal hypothesis test, 363–367
fractiles, 90
frequency polygons, 43–44
frequency tables
    class boundaries, 33–34
    class midpoints, 33–34
    class width, 33–34
    construction, 34–35
    cumulative relative, 37
    defined, 33
    lower class limits, 33
    mean, 61–62
    relative, 36–37
    standard deviation, 78–79
    upper class limits, 33

Friedman test
    notation, 713
    test statistics, 713–714
    value of the test statistic, finding, 713
fundamental counting rule, 151–153

G

geometric distributions, 200
geometric mean, 70
goodness-of-fit tests
    assumptions, 549
    binomial distributions, 555–556
    critical values, 550–554
    defined, 549
    expected frequencies, 550
    normality testing, 585–588
    notation, 549
    P-values, 558
    Poisson distributions, 556–558
Gosset, William, 316
graphs
    box-and-whisker diagrams, 99
    boxplots, 99–103
    density curves, 226–227
    frequency polygon, 43–44
    histograms, 42–43
    misleading, 12
    ogive, 44
    Pareto charts, 46, 47
    pie charts, 46–47
    probability density function, 226–227
    probability histograms, 177–178
    relative frequency histograms, 43
    residual plots, 629–631
    scatter diagrams (scatterplots), 48, 601–603
    stem-and-leaf plots, 45–46
guesstimates, 11

H

H test, 709
harmonic mean, 69
Hawthorne effect, 21
highly unlikely results, 360
histograms
    defined, 42–43

normality testing, 583–585
probability, 177–178
relative frequency, 174
How to Lie with Statistics (Huff), 12
hypergeometric distributions, 214–215
hypotheses
    alternative, 364–365
    defined, 360
    null, 364–368
hypothesis testing
    accept/fail to reject, 368
    acceptance region, 366
    alternative hypotheses, 364–365
    assumptions, 398, 410, 420
    central limit theorem, 362–363
    chi-square distribution, 421–423
    classical method, 374–379
    confidence intervals, 383–385, 414, 423–424
    critical region, 366
    critical values, 366, 369, 399, 421
    decision making, 367
    distribution choice, 399–402
    F distribution, 465
    final conclusions, 367–368
    formal, 363–367
    goodness-of-fit. *See* goodness-of-fit tests
    highly unlikely results, 360
    independent and large samples, 452–456
    independent and small samples, 473–482
    informal example, 361–363
    large samples, 374–392
    left-tailed test, 369, 377, 381
    notation, 366
    null hypotheses, 364–368
    overview, 360–361
    P-value method, 379–383, 402–403, 414, 423
    population proportions, 410–415
    power of the test, 366, 387–392
    preliminary *F* test approach, 474–475
    rare event rule, 360
    right-tailed test, 369, 376, 380–381
    significance level, 366
    small samples, 398–403
    standard deviation, 420–423
    Student *t* distribution, 398–399

hypothesis testing (*cont.*)
  summary, 431
  test statistic, 366–367
  traditional method, 374–379, 411–413
  two population proportions, 488–495
  two-tailed test, 369, 378, 382, 423–424
  type I/type II errors, 365–366, 371
  variance, 420–423

I

inclusive or, 128–129
independent events, 140–141, 146
independent samples
  assumptions, 452, 474
  confidence intervals, 454–456, 476,
  478–480
  data sets, exploring, 452
  defined, 441
  equal variances, 475–478
  known variances, 474
  large, hypothesis testing, 452–456
  P-values, 454
  pooled estimate of σ^2, 475–476
  population variances known, 474
  preliminary *F* test approach, 474–475
  small, hypothesis testing, 473–482
  test statistics, 452–456, 475, 478
  unequal variances, 478–482
  Wilcoxon rank-sum test. *See* Wilcoxon
  rank-sum test
independent variables, 619
inferences
  overview, 440
  two dependent samples. *See* dependent samples
  two independent and large samples, 452–456
  two independent and small samples,
  473–482
  two proportions, 488–495
  two variances, 464–468
inferential statistics, 32, 294
    see also estimate(s); inferences
influential points, 625–626
interaction, 530–531
interval estimates, 296
interval level of measurement, 7, 8

K

Kruskal-Wallis test (*H* test)
  assumptions, 709–710
  critical values, 710–712
  defined, 709
  notation, 710
  value of the test statistic, finding, 710

L

large samples
  hypothesis testing, 374–392
  independent, 452–456
  population mean estimation, 294–308
law of large numbers, 117–119
lead margin of error, 494
least-squares property, 626–627
left-tailed test, 369, 377, 381
level of confidence, 296
Lilliefors test, 588–590
linear correlation coefficient *r*
  defined, 603
  interpretation, 605–606
  notation, 604
  properties, 606
  rounding, 603–605
loaded questions, 11–12
lower class limits, 33
lurking variables, 607

M

Mann-Whitney *U* test, 698
margin of error (E), 298–301, 318–319, 327, 329
marginal change, 625
matched pairs, 679–680
    see also dependent samples
mathematical expectation, 181
mathematical models, 628–629
maximum error of the estimate, 298
mean absolute deviation, 74
mean square (MS), 512–514
mean(s)
  binomial distributions, 201–204
  class size, 55
  defined, 56
  estimation, 294–308, 315–322

expected value, 181–182
explained deviation, 638
frequency table, 61–62
geometric, 70
harmonic, 69
multiple comparisons of, 515–517
notation, 57
population, 374–392, 398–403
probability distributions, 178–180
quadratic, 70
sample, 258–261, 295
sample size for estimating, 294–308, 315–322
square between, 514
square for error (MSE), 514
square for treatment, 514
square within, 514
total deviation, 637
trimmed, 71
unexplained deviation, 638
uniform distribution, 228–229
weighted, 63
measurement
    central tendency. *See* central tendency, measures of
    interval level, 8
    interval level of, 7
    nominal level, 8
    nominal level of, 6
    ordinal level of, 6–8
    position. *See* position, measures of
    ratio level, 7
    variation. *See* variation, measures of
median, 58, 60, 71, 682–683
meta-analysis, 19
midrange, 59–60
mild outliers, 105
Minitab 15
    binomial probabilities, 195–196
    boxplots, 101–103
    coefficient of determination, 649
    confidence intervals, 308–310, 322–323, 333, 345
    contingency tables, 576–577, 613
    descriptive statistics, 65–66, 83, 93–94
    finding values, 255
    frequency table, 38

graphs, 50
hypothesis testing, 386–387, 403–404, 415, 424
Kruskal-Wallis/Friedman tests, 716
multiple regression equation, 662
normality testing, 590–591
one-way ANOVA, 518–519
Poisson distributions, 211–212
position, measures of, 93–94
probability in nonstandard distribution, 248
probability when given z score, 239–241
rank correlation, 729
regression, 632, 649
runs test, 741
sign test, 684–685
standard error of estimate, 649
two dependent samples, 446–447
two independent samples, 483
two population proportions, 495
two variances, 469
two-way ANOVA, 534–535
usage guidelines, 23
Wilcoxon rank-sum test, 705
Wilcoxon signed-ranks test, 695
misleading graphs, 12
mode, 59, 60
model significance, 641–645
modified boxplots, 105
MS(between), 514
MSE (mean square for error), 514
MS(error), 514
MS(treatment), 514
MS(within), 514
multicollinearity, 657–658
multimodal, 59
multinomial distributions, 200–201
multiple coefficient of determination, 655–656
multiple comparison procedures, 515–517
multiple regression
    adjusted coefficient of determination, 655–656
    ANOVA P-value, 656
    best, finding, 658–662
    defined, 652
    multicollinearity, 657–658
    notation, 653
    t tests, 656–666

multiple regression equation, 652
multiplication rule, 145–146
mutually exclusive events, 129–130

N

negative binomial distributions, 194, 200
negatively skewed, 65
Nightingale, Florence, 46
nominal data, 680–681
nominal level of measurement, 6, 8
nonparametric statistics
    684-685. *See* Wilcoxon signed-ranks test
    advantages/disadvantages, 674–675
    overview, 674
    rank correlation. *See* rank correlation test
    ranks, 675–676
    runs test for randomness. *See* runs test
    sign test. *See* sign test procedures
    tests for multiple samples. *See* Friedman test; Kruskal-Wallis test
    Wilcoxon rank-sum test. *See* Wilcoxon rank-sum test
nonparametric tests, 674
nonsampling error, 19
normal distributions
    binomial distributions, approximation to, 270–276
    continuity corrections, 273–276
    defined, 230
    finding probabilities, 243–247
    finding values, 252–255
    nonstandard, 244–245
    standard. *See* standard normal distributions
    tests. *See* normality test procedures
normal probability plots, 583–585
normality test procedures
    formal, 585
    informal, 583–585
    Lilliefors test, 588–590
    x^2 goodness-of-fit tests, 585–588
notation
    addition rule, 129
    α (alpha), 366
    β (beta), 366
    binomial distribution, 187

central limit theorem, 261
counting, 153
critical values, 297
dependent samples, 442
factorial symbol, 153
Friedman test, 713
goodness-of-fit tests, 549
hypothesis testing, 366
hypothesis testing of population proportions, 410
Kruskal-Wallis test, 710
linear correlation coefficient r, 604
mean, 57
multiple regression equation, 653
percentiles, 91
population proportions, 326, 489
population variance, 343
probabilities, 116
rank correlation test, 723
regression equation, 619
runs test, 734
sign test, 677
standard deviation/variance, 77
two variances, 464
Wilcoxon rank-sum test, 700–701
Wilcoxon signed-ranks test, 690
z scores, 235
null hypotheses, 364–365
numerator degrees of freedom, 466
numerical data, 5–6

O

observational study, 15
odds, 123–124
ogives, 44
one-sided confidence intervals, 338
one-tailed tests, 610–611
one-way ANOVA
    assumptions, 510
    mean square (MS), 512–514
    multiple comparison procedures, 515–517
    rationale, 511
    sum of squares (SS), 512–514
    table, construction of, 511–512
    total sum of squares, 513–514
    variation between sample means, 513–514

ordinal level of measurement, 6–8
outliers
    data exploration with, 98–99, 100, 102
    defined, 32
    extreme, 105
    normality testing, 583–585
    regression equation, 625–626

P

P-values
    contingency tables, 571–572
    defined, 379
    goodness-of-fit tests, 558
    hypothesis testing, 379–383, 402–403, 414, 423
    independent samples, 454
    multiple regression, 656
paired samples, 441
parameter, 4
parametric tests, 674
Pareto charts, 46, 47
partial pictures, 11
payoff odds, 123
Pearson's index of skewness, 87
Pearson's product moment correlation
    coefficient, 603
percentiles, 90–93
permutations, 154
permutations rule, 154–156
pictographs, 12–13
pie charts, 46–47
placebo effect, 16, 455
point estimate, 327
Poisson distributions, 207–211, 556–558
pooled estimate of p_1 and p_2, 490
pooled estimate of σ^2, 475–476
population
    defined, 4
    mean, 294–308, 315–322, 374–392, 398–403
    median, 682–683
    original, 259–260
    vs. sample, 32
    standard deviation, 76, 420–423
population proportions
    assumptions, 328–329, 410, 489
    confidence intervals, 327, 330, 493–494

estimating, 326–332
hypothesis testing, 410–415
hypothesis testing about two, 490–495
margin of error, 327, 329
notation, 326, 410, 489
point estimate, 327
pooled estimate of p_1 and p_2, 490
rationale for procedures, 494–495
round-off rule, 327
sample size, 330, 332
population variances
    assumptions, 339
    chi-square distribution, 339–342
    confidence intervals, 342–343, 344–345
    defined, 77
    equal, 475–478
    equal vs. unequal, 474–475
    estimating, 338–345
    estimators of σ^2, 342–344
    independent, small samples and, 474–482
    notation, 343
    preliminary F test approach, 474–475
    round-off rule, 78, 343
    sample variance/size, 342–343
    unequal, 478–480
position, measures of
    deciles, 90, 93
    percentiles, 90–93
    quartiles, 90, 93
    z (standard) scores, 88–90
positively skewed, 65
power of the test, 366, 387–389, 390–392
practical significance, 377
precise numbers, 11
predicted values, 623–624
prediction intervals, 645–648
predictions, 622–624
predictor variable, 619
preliminary F test approach, 474–475
probabilities
    addition rule, 128–130
    "at least one", 194–195
    Bayes' Theorem, 146–148
    classical approach, 117
    classical approach to, 118

probabilities (*cont.*)
   combinations rule, 156–158
   complementary events, 121–122
   compound events, 129
   conditional, 135–141
   contingency tables, 136–139
   counting, 151–159
   dependent events, 140–141
   event, 116
   independent events, 140–141
   law of large numbers, 117–119
   multiplication rule, 145–146
   notation, 116
   odds, 123–124
   P-value method, 379–383
   permutations rule, 154–156
   rare event rule, 360, 385
   relative frequency approximation, 117
   relative frequency approximation of, 118
   rounding off, 122
   sample space, 116
   simple event, 116
   simulation, 118–119
   standardized tests, 131
   subjective, 117
   success, 187
   tree diagrams, 151–152
   values, 121
   z scores, 237–239
probability density function, 226–227
probability distributions
   binomial. *See* binomial probability distributions
   continuous. *See* continuous probability distributions
   defined, 176
   discrete. *See* discrete probability distributions
   mean, 178–180
   normal. *See* normal distributions
   requirements, 177–178
   round-off rule, 179–180
   standard deviation, 178–180
   variance, 178–180
probability histograms, 177–178
probability values. *See P*-values
proportionate sampling, 18
proportions, population. *See* population proportions
push polling, 326

Q
quadratic mean, 70
qualitative data, 5
qualitative variable, 654
quantiles, 90
quantitative data, 5
quartiles, 90, 93
queuing theory, 245

R
random samples, 17–18, 20, 119
random selection, 5, 16
     see also runs test
random variables
   continuous, 175–176
   defined, 175
   discrete, 175–176
   expected value, 181–182
   rationale for Formulas, 181
randomization
   cluster sampling, 18
   convenience sampling, 19
   random sample, 17–18
   simple random sample, 17–18
   stratified sampling, 18
   systematic sampling, 18
randomized block design, 525
range, 72–73
range rule of thumb, 79–80
rank correlation coefficient, 722
rank correlation test
   assumptions, 722–723
   critical values, 724
   defined, 722
   handling ties, 725–728
   notation, 723
   rank correlation coefficient, 722
ranks, 675–676
rare event rule, 263, 360, 385
ratio level of measurement, 7, 8
redundancy, 141
regression ANOVA, 641
regression equations
   assumptions, 619–620
   defined, 619
   dependent/predictor variable, 619

guidelines for use, 624–625

independent/response variable, 619

influential points, 625–626

least-squares property, 626–627

line, 619

marginal change, 625

mathematical models, other, 628–629

multiple, 652

    see also multiple regression notation, 619

outliers, 625–626

predictions, 622–624

residual plots, 629–631

residuals, 626–627

slope b_1, rounding, 620–622

y-Intercept b_0, rounding, 620–622

relative frequency

    approximation of probabilities, 117, 118

    histograms, 43, 174

    polygons, 43–44

    tables, 36–37

reliability, 271–273

replication, 17

residual plots, 629–631

residuals, 626–627

response variables, 619

right-tailed test, 369, 376, 380–381

rigorously controlled design,
    16, 525

root mean square (R.M.S.), 70

Rosenthal effect, 21

round-off rule, 330

    confidence intervals, 300

    probability distributions, 122, 179–180

    sample and population variance, 78, 343

rule of complementary events, 132

run, 733

runs test

    alternative application, 739–740

    assumptions, 733

    critical values, 737–739

    defined, 733

    5% cutoff criterion, 734–736

    fundamental principle, 734

    large sample cases, 736–737

    notation, 734

    rationale, 735–736

S

sample mean, 258–261, 295

sample size

    biased data, 361

    determination, 304–305

    determination of, 330–332

    estimating mean, 294–308, 315–322

    factors affecting, 305–308, 332

    hypothesis testing, 374–392, 398–403

    independence, 146, 187

    notation, 56

    population proportions, 326–332

    population variance, 338–345

    round-off rule, 330

    Student t distributions, 399

sample space, 116

sample statistic, 4–5

sample variance, 77

sample(s)

    bad, 10

    defined, 4

    dependent. *See* dependent samples

    independent. *See* independent samples

    inferences. *See* inferences

    large. *See* large samples

    multiple, tests for, 709–714

    paired, 441

    vs. population, 32

    random, 17–18, 20, 119

    self-selected, 10

    simple random, 17–18, 119

    small. *See* small samples

    standard deviation, 77–78

sampling

    acceptance, 217

    cluster, 18, 20

    composite, 137

    convenience, 18

    dependent events, 187

    error, 19, 263

    proportionate, 18

    stratified, 20

    systematic, 18, 20

    without replacement, 187

sampling distribution of sample means,
    258–261

scatterplots (scatter diagrams), 48, 601–603

Scheffé test, 515–516

self-selected samples, 10

sign test procedures

    assumptions, 677

    critical values, 678–679

    defined, 677

    matched pairs of sample data, 679–680

    nominal data, 680–681

    notation, 677

    rationale, 683–684

    single population median, 682–683

signal-to-noise ratio, 87

significance level, 366, 466

simple event, 116

simple random samples, 17–18, 119

simulation, 118–119

single-factor analysis of variance, 510

skewness, 63–65, 87

slope b_1, rounding, 620–622

small samples

    abuse of statistics, 10–11

    hypothesis testing, 398–403

    independent, 473–482

    population mean estimation, 315–322

The Small World Problem (Milgram), 57

Spearman's rank correlation coefficient, 723

Spearman's rank correlation test, 722

SS(between), 513

SS(error), 513

SS(factor), 513

SS(total), 513

SS(treatment), 513

SS(within), 513

standard deviation

    binomial distributions, 201–204

    Chebyshev's theorem, 82

    defined, 73

    empirical rule, 81

    finding with Formula 2-4, 74–76

    frequency table, 78–79

    hypothesis testing, 420–423

    interpretation, 79

    notation, 77

    population, 76

probability distributions, 178–180

    range rule of thumb, 79–80

    sample/population variance, 77–78

    uniform distribution, 228–229

standard error of the mean, 261

standard normal distributions

    defined, 231

    determining probabilities, 229–237

    finding z scores, 237–239

standard scores, 88–90

standardized tests, 131

STATDISK

    binomial probabilities, 197

    boxplots, 101–103

    coefficient of determination, 649

    confidence intervals, 309–310, 333, 345

    contingency tables, 576–577, 613

    descriptive statistics, 66, 83, 93–94

    finding values, 255

    frequency table, 38

    goodness-of-fit tests, 560

    graphs, 50–51

    hypothesis testing, 386–387, 404, 415, 424

    Kruskal-Wallis test, 715

    multiple regression equation, 663

    one-way ANOVA, 518–519

    Poisson distributions, 212

    position, measures of, 93–94

    probabilities, 168

    probability in nonstandard distribution, 248

    probability when given z score, 240–241

    quartiles, calculation of, 93–94

    rank correlation, 728

    regression, 633, 649

    runs test, 740

    sign test, 685

    standard error of estimate, 649

    two dependent samples, 447

    two independent samples, 458, 483

    two population proportions, 495

    two variances, 469

    two-way ANOVA, 534

    usage guidelines, 23–24

    Wilcoxon rank-sum test, 703

    Wilcoxon signed-ranks test, 693

statistic
    defined, 4
    sample, 4–5
statistical significance, 377
statistics
    abuses, 10–13
    defined, 4
    descriptive, 32, 93–94
    inferential, 32, 294
        see also estimate(s); inferences
    nonparametric. See nonparametric statistics
    test. See test statistics
    uses, 9
stem-and-leaf plots, 45–46
stepwise regression, 662
stratified sampling, 18, 20
Student *t* distributions
    population mean estimation, 316–320
    properties of, 399
subjective probabilities, 117
sum of squares (SS), 512–514
survey medium and results, 254
symmetric data, empirical rule for, 202–203
symmetric distribution, 64
systematic sampling, 18, 20

T

t distribution
    multiple regression, 656–666
    population mean estimation, 316–317
    test statistic for linear correlation, 608–609
Tainted Truth (Crossen), 399
test of homogeneity, 572–575
test of independence, 566–567
test of significance, 363–367
test procedures, normality. See normality test
    procedures
test statistics
    computation, 366–367
    Friedman test, 713–714
    goodness-of-fit tests, 550
    hypothesis testing/small samples, 398
    independent, large samples, 452–456
    independent, small samples, 475, 478
    Kruskal-Wallis test, 710

population mean claims, 362
proportions, 411
r for linear correlation, 609–610
rank correlation coefficient, 723–724
runs test, 737
sign test, 677
standard deviation/ variance, 420
t for linear correlation, 608–609
test of independence, 568
two dependent samples, 442
two population proportions, 490–492
Wilcoxon rank-sum test, 701
tests of normality. See normality test procedures
total deviation, 637–639
total sum of squares, 513–514
total variation, 513, 639
traditional method of testing hypotheses, 374–379,
    411–413
travelling salesman problem, 153–154
treatment, 510
treatment groups, 16
tree diagrams, 151–152
trimmed mean, 71
Tukey's method of simultaneous confidence
    intervals, 516–517
Turing test, 162
two-tailed tests, 369, 378, 382, 414,
    423–424
two-way ANOVA
    assumptions, 527–530
    factorial design, 530–534
    interaction, 530–531
    overview, 524–525
    procedure, 533
    randomized block design, 525–527
type I errors, 365–366, 371
type II errors, 365–366, 371

U

unbiased estimator, 77, 316
unexplained deviation, 638–639
unexplained variation, 637–639
uniform distribution, 227–229
unusual values. See outliers
upper class limits, 33

V

validity, 271–273
variables
  confounding, 16–17
  continuous random, 175–176
  dependent/predictor variable, 619
  discrete random, 175–176
  dummy, 654
  effects, 16–17
  independent/response variable, 619
  lurking, 607
  marginal change, 625
  predictor, 619
  qualitative, 654
  random. *See* random variables
  response, 619
variable significance, 657
variance(s)
  analysis of. *See* analysis of variance (ANOVA)
  assumptions, 464
  binomial distributions, 201–204
  comparison of two, 464–468
  critical values, 466–468
  data sets, exploring, 464
  defined, 77
  equal, 475–478
  equal vs. unequal, determination, 474–475
  F distribution, 465
  hypothesis testing, 420–423
  notation, 77, 464
  one-way analysis of. *See* one-way ANOVA
  population. *See* population variances
  preliminary *F* test approach, 474–475
  probability distributions, 178–180
  round-off rule, 78
  sample, 77
  within samples, 511
  between samples, 511
  single-factor analysis, 510
  test statistics, 464–466
  two-way analysis of. *See* two-way ANOVA
  unequal, 478–480
variation
  coefficient of determination, 639–640
  coefficient of variation, 87
  due to error, 511
  due to treatment, 511
  explained, 637–639
  measurement. *See* variation, measures of
  model significance, 641–645
  between sample means, 513–514
  standard error of estimate, 645–648
  total, 639
  unexplained, 637–639
variation, measures of
  overview, 71–72
  range, 72–73
  standard deviation. *See* standard deviation
The Visual Display of Quantitative Data (Tufte), 27

W

weighted mean, 63
Wilcoxon rank-sum test
  assumptions, 698–699
  defined, 698
  large sample procedure, 700–703
  notation, 700–701
  small sample procedure, 699–700
Wilcoxon signed-ranks test
  assumptions, 689
  critical values, 690–691
  defined, 688
  notation, 690
  one median, 689
  two dependent samples, 691–693
  value of the test statistic, finding, 689–690

X

x^2, goodness-of-fit tests, 585–588

Y

y-Intercept b_0, rounding, 620–622

Z

z scores
  defined, 88
  finding probability, 231–237
  finding z scores given probability, 237–239
  notation, 235
  position measurement, 88–90